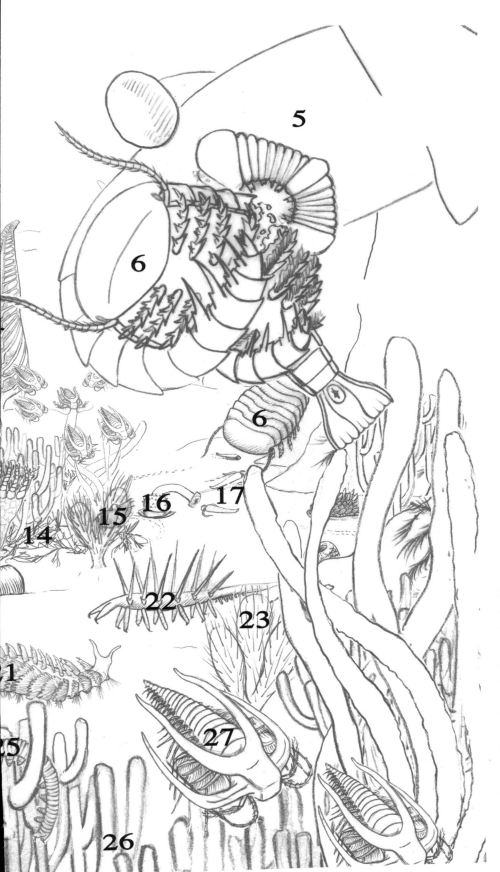

1. *Laggania cambria* - Unknown
2. *Leptomitus lineatus* - Demosponge
3. *Halichondrites elissa* - Demosponge
4. *Thaumaptilon walcotti* - Anthozoan
5. *Anomalocaris canadensis* - Unknown
6. *Sidneyia inexpectans* - Arthropod
7. *Olenoides serratus* - Trilobite
8. *Eldonia ludwigi* - Holothuroid
9. *Hazelia conferta* - Demosponge
10. *Canadaspis perfecta* - Crustacean
11. *Wiwaxia corrugata* - Annelid or Mollusc
12. *Echmatocrinus brachiatus* - Echinoderm
13. *Pikaia gracilens* - Chordate
14. *Marpolia spissa* - Cyanobacteria
15. *Nisusia burgessensis* - Brachiopod
16. *Ottoia proli=DEca* - Priapulid
17. *Haplophrentis carinatus* - Hyolith
18. *Opabinia regalis* - Unknown
19. *Eiffelia globosa* - Calcareous Sponge
20. *Burgessochaeta setigera* - Polychaete
21. *Canadia spinosa* - Annelid
22. *Hallucigenia sparsa* - Onychophora
23. *Pirania muricata* - Demosponge
24. *Margaretia dorus* - Green Algae
25. *Aysheaia pedunculata* - Onychophora
26. *Vauxia gracilenta* - Demosponge
27. *Marrella splendens* - Arthropod

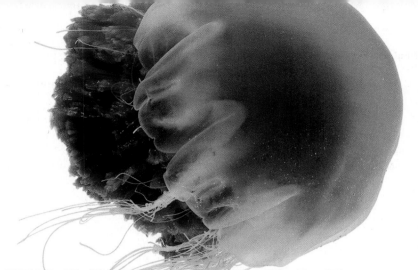

EVOLUTION and ECOLOGY of the ORGANISM

Michael R. Rose

*Department of Ecology and Evolutionary Biology,
University of California, Irvine*

Laurence D. Mueller

*Department of Ecology and Evolutionary Biology,
University of California, Irvine*

PEARSON
Prentice
Hall

Upper Saddle River, NJ 07458

Library of Congress Cataloging-in-Publication Data

Rose, Michael R.
   Evolution and ecology of the organism / Michael R. Rose, Laurence D. Mueller.—1st ed.
      p. cm.
   Includes index.
   ISBN 0-13-010404-3

   1. Evolution (Biology)   I. Mueller, Laurence D.   II. Title.
   QH366.2.R67 2006   576.8—dc22   2004028588

Publisher: Sheri L. Snavely
Executive Editor: Teresa Ryu Chung
Editor in Chief: John Challice
Project Manager: Karen Horton
Senior Media Editor: Patrick Shriner
Production Editor: Debra A. Wechsler
Executive Managing Editor: Kathleen Schiaparelli
Assistant Managing Editor: Beth Sweeten
Development Editor: Anne Reid
Editor in Chief, Development: Carol Trueheart
Marketing Managers: Andrew Gilfillan, Shari Meffert
Manufacturing Buyer: Alan Fischer
Director of Marketing, Science: Linda Taft MacKinnon
Director of Creative Services: Paul Belfanti
Manager, Composition: Allyson Graesser
Desktop Administration: Joanne Del Ben
Electronic Page Makeup: Clara Bartunek, Joanne Del Ben,
   Meg Montgomery
Creative Director: Carole Anson

Art Director: Jonathan Boylan
Interior Design: John Christiana
Managing Editor, Audio and Visual Assets: Patricia Burns
Art Development Editor: Jay McElroy
AV Production Manager: Ronda Whitson
AV Production Editor: Connie Long
Art Studio: Precision Graphics, Inc., Artworks
Copy Editor: Christianne Thillen
Director, Image Resource Center: Melinda Reo
Manager, Rights and Permissions: Zina Arabia
Interior Image Specialist: Beth Brenzel
Cover Image Specialist: Karen Sanatar
Image Permission Coordinator: Cathy Mazzucca
Photo Researcher: Sheila Norman
Editorial Assistant: Lisa Tarabokjia
Interior and Cover Design: Michael R. Rose and
   Laurence D. Mueller
Cover Illustration: Karina I. Helm

© 2006 Pearson Education, Inc.
Pearson Prentice Hall
Pearson Education, Inc.
Upper Saddle River, NJ 07458

Printed in the United States of America
10 9 8 7 6 5 4 3 2 1

ISBN 0-13-010404-3

Pearson Education LTD., *London*
Pearson Education Australia PTY, Limited, *Sydney*
Pearson Education *Singapore*, Pte. Ltd.
Pearson Education North Asia Ltd., *Hong Kong*
Pearson Education Canada, Ltd., *Toronto*
Pearson Educación de Mexico, S.A. de C.V.
Pearson Education—Japan, *Tokyo*
Pearson Education *Malaysia, Pte. Ltd.*

To the Mueller, Krieks, Parla, Horsey, Metal, and Rose Families,
Near and Far

MICHAEL R. ROSE

Michael Rose went to the University of Sussex in 1976 for doctoral studies with Brian Charlesworth on the fruit fly *Drosophila melanogaster*. There he began his work on the evolution of aging and created fruit flies with genetically postponed aging. In 1991, his book *Evolutionary Biology of Aging* appeared, offering a view of aging that was a complete departure from the views that had dominated the aging field since 1960. Rose received the Busse Research Prize from the World Congress of Gerontology. Among his other honors is a teaching award from the School of Biological Sciences at the University of California, Irvine. He has written popular articles for *Technology Review* and *Scientific American*, as well as a general-audience book, *Darwin's Spectre: Evolutionary Biology in the Modern World*. He has published a total of more than 200 scientific publications of all kinds. The 2004 book *Methuselah Flies* assembles selected articles from the last 25 years of work in the Rose laboratory. He is Professor of Ecology and Evolutionary Biology at the University of California, Irvine and the Director of the University of California Intercampus Research Program on Experimental Evolution.

LAURENCE D. MUELLER

Larry Mueller received his Ph.D. in 1979 from the University of California, Davis, where he studied under Francisco Ayala. Mueller then went on to do postdoctoral research in theoretical population genetics with Marcus Feldman at Stanford University. He was an Assistant and Associate Professor at the Washington State University before assuming his current position of Professor of Ecology and Evolutionary Biology at the University of California, Irvine. Mueller has published over 70 research papers in the fields of evolution, population genetics and population ecology. He is also the author of *Stability in Model Populations* with Amitabh Joshi. In his current research, Mueller uses experimental evolution to study problems like density-dependent natural selection and the evolution of late-life demographic patterns in *Drosophila*.

This book introduces biology students to the basic concepts of the spectrum of fields that we call Darwinian biology, a spectrum that includes population genetics, population ecology, community ecology, macroevolution, physiological ecology, systematics, and functional morphology. Charles Darwin first brought this type of science to fruition and all these fields owe their foundations to his pioneering work, directly or indirectly. Our primary goal in the book is to elicit the students' interest. Secondarily we want to prepare undergraduate students for more advanced specialist courses in Darwinian biology as they pursue their degrees. We have adopted the following means to achieve these ends:

- Evoking Darwinian theory using stepped-out equations and concrete graphics to foster quantitative intuition.
- Using examples illustratively rather than exhaustively, to support and sharpen the student's understanding.
- Using evocative text to give the student an appreciation for the drama of the science and the color of its material.
- Using a magazine format that allows text and graphics to combine synergistically, with close juxtaposition.
- Consistently emphasizing concepts over details and scientific reasoning over terminology.

We want to help students over the hump that keeps them from understanding state-of-the-art texts in evolution, ecology, physiology, and cognate fields. Our hope is that using this book will make students more interested in Darwinian science, and faculty more willing to teach it. For two decades biology curricula have been dominated by cell and molecular biology, with Darwinian biology relegated to passing mention in introductory courses or to small advanced courses taken by few students. This is ironic at a time when the abundance of sequence data from molecular biology pointedly confronts biology with its need for Darwinian theories and analytical tools. Yet that toolkit has mostly fallen into disuse, poorly understood even by many biology faculty. The time for the rediscovery of this other half of biology has arrived.

The general theme of the book is the interconnectedness of organism, environment, and evolution. In studying the book, students should develop an integrated understanding of the organism that is founded in evolution and ecology. Just as biochemistry and molecular biology provide the foundation for our understanding of the cell, we use evolutionary biology and ecology here to construct a foundation for understanding the organism.

With this in mind, *Evolution and Ecology of the Organism* (henceforth *EEO*) integrates the component Darwinian disciplines. Instead of three separate sections for Evolution, Ecology, and Organismal Biology, thematic interconnections have been developed that combine elements of all these areas throughout the text. In so doing, *EEO* follows the precedent of contemporary research, in which ecologists, for example, use molecular genetic data and phylogenetic analysis.

Level

One of the most important issues for any textbook is that of level. There are a number of excellent advanced textbooks in ecology, evolution, and organismal biology. They are not going to be displaced by *EEO*. Each of those books already approaches 1000 pages in length. In order to achieve a comparable scope and replace three of these advanced disciplinary books, *EEO* would then need to be 2200 pages long. As you can see, it is not such a weighty tome.

In addition, *EEO* does not compete with general biology texts, which introduce students to the diversity of life-forms, cells, and habitats. Such books are the factual starting point for biology degrees, and have only rudimentary introductions to the conceptual content and empirical results of evolutionary biology or ecology. *EEO* will be best comprehended and utilized by students who have already taken a good introductory biology course.

The content of *EEO* is primarily aimed at the biology major in the second or third year of the college or university curriculum. However, students who have taken AP biology can begin their university education with courses based on *EEO*. At the other end of the spectrum, graduate students, post-docs, and faculty from other areas of biology may find *EEO* a helpful source of basic information about Darwinian biology as well. But these qualifications aside, we would expect *EEO* most often to be assigned to second and third year biological science majors as part of their core curriculum. Such text adoptions might be for courses called "Ecology, Evolution, and Organism" or "Integrative Biology." Or it might be assigned in a sequence of courses called "Ecology," "Evolutionary Biology," etc. In the process of sending out the text for review and talking to our colleagues, we have heard of a number of variations on these themes.

A sensible question might be why don't these courses just use the traditional advanced texts in combination? For exceptionally well-prepared students, this may be workable. However, our experience is that the vast majority of biology majors become alienated from the study of ecology and evolution by the narrow specialization and quantitative detail of advanced texts in our disciplines. While advanced students may already understand the interconnections between population ecology, behavioral ecology, and population genetics, for example, intermediate-level students need an integrative framework within which to place the findings of these particular disciplines. We supply such an integration of Darwinian biology in a way that multiple separate texts cannot.

Format

EEO breaks with tradition with respect to format as well as content, to meet its twin goals of communication and integration. In *EEO*, the concepts are the focus. Because of this, these concepts are rendered primarily in visual form. Biology research is now image-driven, meaning that most of our discussions are based on images and photographs that convey our ideas. The use of dramatic colorful illustrations is commonplace in biology seminars. *EEO* is put together with graphics and photographs as the foreground elements and detailed text content as a background element. We try to make Darwinian biology lively and appealing.

Because *EEO* assembles the material so that at least half of it is conveyed graphically—and we consider that half to be essential content—we have designed the book so that all the one- or two-page spreads are art-centered and self-contained. These spreads are designed to convey a single message. That single message is embodied by a single declarative sentence, displayed prominently at the top of the page. In this way, information is focused into distinct nuggets, which we call Modules.

We are excited about this book. It has been a very important project for us and we have devoted long hours to its creation, looking forward to the day when students would sit down with the book and be enticed, we hope, by the fascinating world of Darwinian biology. We would appreciate any suggestions you might have about the book, how it works for you in your classroom, what additional content should be added, and so on. Please contact us at mrrose@uci.edu or ldmuelle@uci.edu and let us know what you think.

Acknowledgments

We are very grateful for the support and encouragement we have received from so many over the seven years we have been working on this book. The individuals who have directly contributed to this project are listed on the copyright page and in the acknowledgments list below. Still others contributed indirectly by supplying us with our formal and informal educations over the years, particularly our teachers, advisors, and colleagues, in many cases long before we actually started writing. Francisco J. Ayala, Brian Charlesworth, James F. Crow, Marcus W. Feldman, William D. Hamilton, Rudolf Harmsen, John Maynard Smith, and Sewall Wright have been our greatest direct influences. We know that we should have made greater use of the knowledge and the direction that we received from them, but time is running out on this edition of our book.

There are some individuals whom we would like to single out for particular thanks. Sheri Snavely, our Publisher at Prentice Hall, has stayed with us throughout this project, from her immediate enthusiasm when we first broached the project to her in Larry's office in 1997, to the last hectic days of New Year's 2005, as everything was racing toward a conclusion. Her belief in our dream saw us through the days when we wondered if we were ever going to accomplish the goals we had set for *EEO*.

Paul Corey stood behind Sheri and us, providing wisdom and sagacity in his very natural way. Teresa Ryu Chung was our principal Prentice Hall editor for years and years, and indomitably put up with our vacillations and digressions. Erin Mulligan then took over to help us prepare our final draft. In getting the book to press, Production Editor Debra Wechsler has been unbelievable in her care, diligence, and determination; the book is as close to our vision as it is thanks to her. Carol Trueheart, Editor in Chief of ESM Development was there for us through this process with helpful doses of sanity. Development Editor Annie Reid and copy editor Chris Thillen made our words smooth and eloquent when we were all too bound up in circumlocution or confusion. Sheila Norman and Jay McElroy found amazing photographs, both appropriate and dramatic. The design team was led by Carole Anson. Art Director Jonathan Boylan directed the design, making so much of the book beautiful, at least in our eyes. Patricia Burns and Connie Long from Artworks created the art program, key to making the book such a visual feast. Lisa Tarabokjia actually took our desperate phone messages, and Karen Horton scrambled to get us over the final finish line during the accuracy check.

We considered more than ten publishers for this project, and most of them would no doubt have been excellent, but we have found great fulfillment with Prentice Hall as a publishing partner. We have asked for a lot of help all the way through. What we have received has consistently exceeded our expectations, sometimes even our comprehension.

Our most important academic colleague on this project was George V. Lauder, who was originally slated to be an author. Unfortunately, he was called back to his natal stream at Harvard, which prevented him from continuing with us. He contributed the first draft of Chapter 2 and was heavily involved in planning the content and format of the book.

Our departmental colleagues in Ecology and Evolutionary Biology, UC Irvine, have played a wide range of supportive roles in this project. At one point, we saw the book virtually as an embodiment of our Department's collective knowledge. But then we recovered our sanity, having realized that you couldn't write a reasonable textbook with ten or more authors. Other colleagues at the University of California as a whole also inspired us to keep going. Darwinian biology is somewhat like a big scary Scottish clan, with all the intellectual fistfights, theoretical afflatus, and drunken color that you might expect. But it is our home, and we love it.

Michael Rose and Larry Mueller,
Irvine, California

Reviewers

The authors would like to thank the colleagues who reviewed the manuscript for this book at various stages of the project.

William G. Ambrose, *Bates College*

Mary Ashley, *University of Illinois, Chicago*

Stewart Berlocher, *University of Illinois-Urbana Champaign*

Robert Browne, *Wake Forest University*

Christina Burch, *University of North Carolina, Chapel Hill*

John A. Cigliano, *Cedar Crest College*

Sarah Cunningham, *University of California, Berkeley*

Mark D. Decker, *University of Minnesota*

Paul W. Ewald, *University of Louisville*

Susan Fahrbach, *University of Illinois-Urbana Champaign*

Scott Fay, *University of California, Berkeley*

Steve Frank, *University of California, Irvine*

Roberta J. Mason-Gamer, *University of Illinois, Chicago*

George Gilchrist, *College of William and Mary*

Katherine Goodrich, graduate student, *University of South Carolina*

Mark Hafner, *Louisiana State University*

Gregory R. Handrigan, *Dalhousie University*

Allan Larson, *Washington University*

Andrea Lloyd, *Middlebury College*

Robert Marquis, *University of Missouri, St. Louis*

Michael McDarby, *Fulton-Montgomery Community College*

Timothy A. Mousseau, *University of South Carolina*

Randolph Nesse, *University of Michigan*

Nicholas L. Rodenhouse, *Wellesley College*

Jay Rosenheim, *University of California, Davis*

Michael Ruse, *Florida State University*

Brody Sandel, *University of California, Berkeley*

Maria Servedio, *University of North Carolina, Chapel Hill*

Diane Wagner, *University of Alaska, Fairbanks*

BRIEF CONTENTS

CONTENTS

PART ONE INTRODUCTION TO DARWINIAN BIOLOGY   *0*

Chapter 1 Darwin, Ecology, and Evolution   *3*

PART FOUR ECOLOGY OF INTERACTING SPECIES *348*

Chapter 14 Parasitism and Mutualism *405*

Chapter 17 Conservation *507*

Chapter 18 Evolution and Ecology of Sex *531*

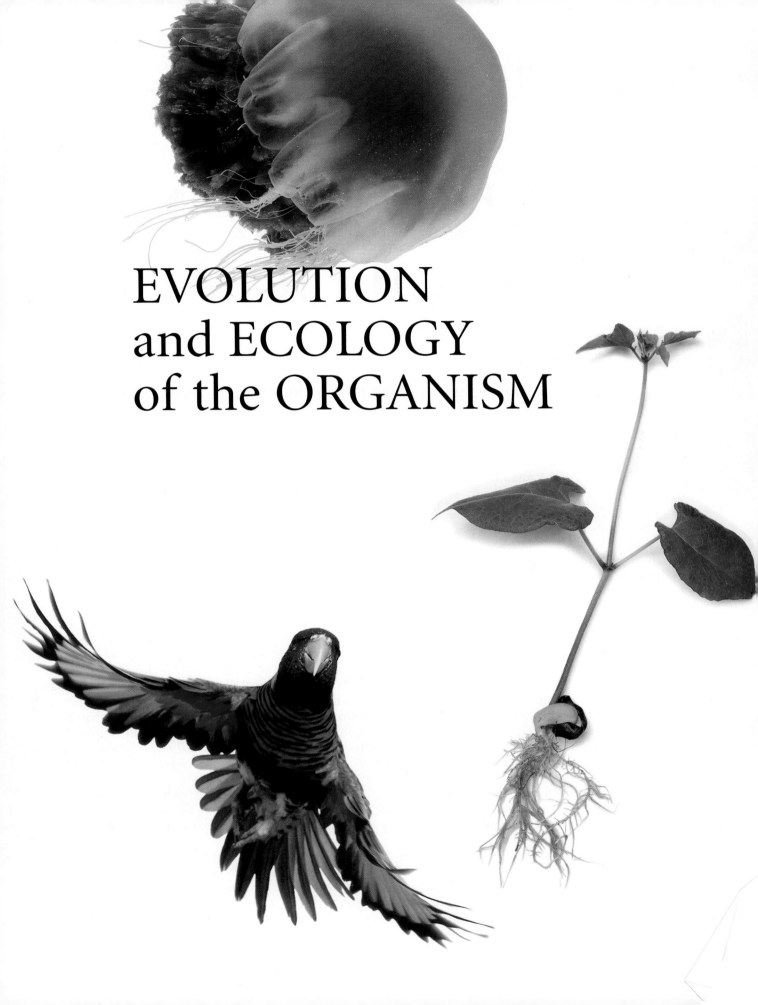

EVOLUTION
and ECOLOGY
of the ORGANISM

PART ONE

INTRODUCTION TO
DARWINIAN BIOLOGY

Have you ever looked through the two ends of a kaleidoscope? Depending on which end you look through, you get a very different impression of the contents of the device. But the actual contents do not change. Only your perspective changes. The different views of the inside of one toy are like the different views that two scientific disciplines can have of the world.

There is nothing unusual about two scientific fields studying the same things. The life sciences come in two parts, like the physical sciences. In the physical sciences, these two parts are known as physics and chemistry. They are usually taught by different departments in universities. They even have different Nobel Prizes. To speak of physicists and chemists as different kinds of scientists is normal in the sciences. They both deal with the nonliving world, but from different perspectives. Physicists talk about the theory of relativity, gravity, and the makeup of the sun. Chemists work on problems like the synthesis of compounds containing carbon, or the properties of plastics. There is some overlap in their interests, especially the physics of chemical bonds. But the two fields are thought of as separate.

The same division exists in the life sciences. Some biologists are primarily interested in topics like the makeup of the cell membrane, the replication of nucleic acids like DNA, and the chemistry of metabolism. Other biologists study mass extinctions of species, the ecology of competition, and the evolution of sex ratios. There is little overlap between these two areas—except for some topics in genetics, such as the evolution of DNA repair. These two areas of life science have no more to do with each other than physics and chemistry do,

perhaps less. But we have no names that signify the two as clearly as physics and chemistry do.

Biologists who study the biochemical foundations of cellular life are often referred to as *molecular biologists, cell biologists,* and *biochemists*. About 30 years ago, these were very different disciplines. Now it is hard to find dividing lines between them. Perhaps the most common name for these specialists is *molecular biologist*, which has the advantage of indicating that they are usually extremely interested in particular molecules, be they DNA, a protein, whatever. The name does not do justice to the great concern that these scientists have for the functioning of molecules within cells. But titles that might do this, like *cell-molecular biologist*, are too cumbersome. So molecular biologist tends to be the label of convenience.

The other type of biology is even harder to label. Thirty years ago, *population biology* was a common term. But since then, this end of biology has incorporated many theories and findings about organisms, cells, and molecules. This kind of biology is based on a general perspective that incorporates ideas about inheritance, selection, ecology, and evolution. It is a perspective that can be historically associated with one preeminent person: Charles Darwin (1809–1882). Darwin brought all of these elements together to form a new approach to biology. First published in his *Origin of Species* (1859), Darwin's approach to biology has not been lost since. To this day, many thousands of biologists are happy to acknowledge their intellectual debt to him. For that reason, here we will speak of this part of biology as *Darwinian biology*.

As the twentieth century gives way to the twenty-first, molecular biology is perhaps the most successful scientific discipline. Its most important competitor for prestige and money has been physics. But physics has been increasingly limited by the expense and difficulty of generating the extreme conditions in which smaller and smaller particles can be observed. Although molecular biology is expensive, its experiments cost thousands of dollars, not the millions or billions of dollars required by those of big-time physics. Thus molecular biologists have been able to acquire substantial funding from governments and corporations.

Molecular biology has also been very successful at attracting funding based on practical concerns. With the end of the Cold War, physicists were no longer as important militarily; powerful thermonuclear weapons became effectively obsolete. But molecular biologists can continue to play to concerns about cancer and other health problems, problems that they have promised to alleviate. Rapid progress in the development of anti-HIV drugs has perhaps been the most dramatic fulfillment of these promises. Medicine would have had grave difficulty even identifying, much less treating, HIV infection without modern molecular biology. Molecular biology is on the march, and it will contribute a great deal to the welfare of our species.

So what is the status of Darwinian biology, at this time in history? Darwinian biology has never truly dominated biology, not even during Darwin's lifetime. It may have been at a peak of influence from 1930 to 1970. By 1930, Darwinian biology was well-developed theoretically, and the subsequent 40 years were to see its ideas applied to most of the major problems of biology, a phenomenon known as "the modern synthesis." But since the 1960s, the technology and intellectual penetration of molecular biology have burgeoned dramatically. As one century turns into another, there is no question that molecular biology is the dominant force in the life sciences.

But Darwinian biology is showing signs of recovery. One of the most important of these is the breaking down of boundaries between specializations within Darwinian biology. Another sign of recovery is that organismal biology is enriching the study of ecology and evolution with important physiological information. A further symptom of health may be the reduced importance of heavily mathematical theory and an increased emphasis on experiments. At the level of research and graduate study, Darwinian biology is enjoying a renaissance.

This book introduces Darwinian biology to new students of biology. There are many advanced treatises of Darwinian biology, books that are read carefully by dozens of people worldwide. We have written some of these books ourselves. This book is not like them. This book is for the new visitor to Darwinian biology, not the settled inhabitant. It is focused on developing the ability to think in a Darwinian way. Many matters of biological fact were entirely unknown to Charles Darwin, including information crucial to Darwinian biology. But it is certain that he could have readily understood them, if he had traveled to our time, because he had a mental framework on which he could hang each particular fact. If you learn to think this way, as a Darwinian, you will have a mental framework that will always be of value when you are faced with a new piece of biology.

There are two basic ways to begin the study of any scientific field. One is historically, so that the actual development of ideas can be understood. One advantage of understanding science historically is that arguments and ideas that may seem utterly bizarre on their own will fall into place, into sequence. Chapter 1 is this kind of historical introduction to Darwinian biology. Another way to begin learning about a new field is to study its most elementary concepts and theories, in the abstract. That way, these concepts can be absorbed without any distractions. Chapter 2 takes this approach. Studying these two chapters, together, provides a more complete introduction to Darwinian biology than studying either of them separately.

Charles Darwin is the starting point.

Darwin, Ecology, and Evolution

The greatest watershed in the history of biology was not the invention of the electron microscope, or even DNA cloning. It was the career of Charles Darwin. Before Darwin, biology was an offshoot of theology, like most sciences before 1800. Darwin made biology part of natural science.

Who was this scientific revolutionary? As a man, Charles Darwin (1809–1882) was full of contradictions. Son of a very wealthy doctor, he abandoned the study of medicine because of squeamishness. Unlike his college chums, he had radical opinions on social issues, such as the abolition of slavery. He was almost painfully shy and never publicly defended his theory of evolution. Yet he was willing to publish books that were widely denounced from church pulpits and university lecterns. Perhaps the most intuitive biological theorist of all time, he was hopelessly confused about how inheritance operates. Reviled by some who were pillars of polite society, he is a controversial figure to this day. Yet his body lies in Westminster Abbey, where English royalty are entombed, not far from the resting place of Sir Isaac Newton (1642–1726), the principal creator of physics.

Darwin is the starting point for the fields of evolution, ecology, and organismal biology—but not because he was the first to put forward ideas or findings in these fields. He was not the first. He had forerunners from the Enlightenment, such as Jean-Baptiste Lamarck (1744–1829) and even his own grandfather, Erasmus Darwin (1731–1802). But Charles Darwin's thought departed radically from these earlier evolutionists. Instead of characterizing evolution in terms of mere patterns, he supplied materialistic explanations for evolution that transformed biology from natural theology to natural science. And he supplied two of the most important ideas that underlie biology: evolutionary descent with modification and the direction of evolution by natural selection. We will explain these later.

Here we introduce first Darwin and after that his thinking. We then show how Darwin's thought led to the type of biology that is the subject of this book. In some ways, it is good to have the foundations of the field so perfectly embodied in the career of one man, one humble Prometheus. ❖

DARWIN'S LIFE

1.1 There are many Darwin myths, perhaps because Darwin had an impact on the general culture

Charles Darwin is probably the most famous biologist of all time. Though the Greek philosopher **Aristotle** (384–322 B.C.) was essentially the founder of academic biology, most people, even most academics, do not think of him primarily as a biologist. Although the Austro-Hungarian monk Gregor Mendel (1822–1884) also made a huge contribution to the development of biology by discovering the foundations of genetics, Mendel is not a name that most people recognize. As for Darwin, his name appears on metal plates attached to cars, and people wear T-shirts with his image.

Broadly speaking, Darwin is a controversial figure among Christian and Muslim denominations. Some denominations accept his work, and some do not. The amazing thing is that so many religious authorities feel a need to state their opinion about Darwinism.

Starting in the late nineteenth century, **Darwinism** also had a great deal of impact on political and social discussions. Figures 1.1A and 1.1B show samples of political cartoons of Darwinism. Communists claimed Darwin as one of their inspirations, starting with his contemporary Karl Marx (1818–1883) and continuing through the 1980s, just before the collapse of Communism as the dominant ideology of Eastern Europe.

But other thinkers were influenced as well. *Social Darwinism* was an Anglo-American movement that argued for the elimination of the unfit from human society by starvation or neglect. Social Darwinism was not proposed by Darwin. It was essentially a variant of capitalist ideology, with a vague biological justification. This ideology had a great deal of influence on both sides of the Atlantic, blocking much valuable legislation for the protection of children and the care of the disabled.

The irony is that Charles Darwin himself was a liberal who hated slavery and was very concerned with the suffering of the unfortunate. He was no Communist revolutionary, but no Social Darwinist either.

The problem of Darwinism was that Darwin's ideas were much bigger than his public persona. They have also remained influential long after his death. As a result, his ideas have been appropriated by many different ideologies and personalities, and denounced by still more. Yet most of these commentaries were developed with very little understanding of Darwin's actual views, whether political or scientific. ❖

FIGURE 1.1A Nineteenth-Century Caricature of Darwinism

FIGURE 1.1B Victorian-Era Cartoon Satirizing Darwin

Darwin Myths

Many people think they know about Darwin. One version of his life is that he was a determined atheist who developed a perverse theory of life evolving in order to destroy Christianity. This myth supposes that Darwin was primarily a political radical who used biology to make mischief.

A competing myth is that while Darwin may have been a nasty radical in his youth and in his middle age, deep study of science left him convinced of the existence of a Creator. This myth is also em-

broidered with the story that Darwin took Communion or confessed on his deathbed, expressing regret about his theories. There is no documented evidence that such events ever took place.

Still, distortion and misappropriation are probably as much forms of flattery as imitation is. Such dubious accounts of Darwin's life reveal the fact that he has obsessed, impressed, and frustrated people for some time.

Amazingly, Charles Darwin was not the first evolutionist in the Darwin family. That distinction belonged to his paternal grandfather, **Erasmus Darwin** [shown in Figure 1.2A, part(i)]. Erasmus was a much-published physician who was interested in chemistry and biology as hobbies. He published what were then "popular science" books and had a wide influence. Mary Shelley cited him as an inspiration in the preface to her novel *Frankenstein*. This inspiration came from Erasmus's support for the view that species evolve into one another. However, Erasmus was not a well-trained biologist, and his biological ideas were generally sketchy or ill formed. His chemistry, by contrast, was first-rate.

But Charles never knew his grandfather, who died in 1802, seven years before Charles was born. Charles also hardly knew his mother, Sukey, who died when he was a child, probably of acute peritonitis. He was left to be brought up by older sisters. He had an older brother named Erasmus. Though Charles was spoiled as a young child, his physician father, Robert, sent him to a boarding school to enjoy the regimen of abuse and deprivation that was then the hallmark of an upper-class education. Father Robert amassed an ample fortune, which would pay for the education of Charles and his brother Erasmus, as well as all of their later adventures and labors. Neither had paid employment for a single day of their lives.

At the age of 16, Charles followed his brother to Edinburgh University, where he was supposed to study medicine. But cadavers disgusted Charles, and he turned his attention to biology. Leaving Edinburgh, he went to Cambridge University to study for service as a minister of the Church of England. At that time, he was absolutely conventional in his religious beliefs. His main interest was to live at a country parsonage where he could collect insects, especially beetles, one of his great passions as a young man.

Charles Darwin was mostly a gentleman student during his years in Cambridge. Hunting and horses were major concerns, and Charles never faced stringent exams or other challenges. But he did collect a great many beetles. He had a passion for collecting things, or shooting them, that would oddly prove central to his later life. Charles also "walked" with his professors a great deal, taking promenades beside them while they propounded their views on the academic questions of the day. From this experience, Charles Darwin apparently learned how to reason scientifically, his greatest single asset. Figure 1.2B shows Charles Darwin at various ages. ❖

(i)

(ii)

(iii)

(i)

(ii)

(iii)

FIGURE 1.2A Paintings of (i) Erasmus Darwin (grandfather of Charles), (ii) Robert Darwin (father of Charles), and (iii) the young Charles Darwin (with his sister Catherine).

FIGURE 1.2B Darwin at Different Ages As a young man (i), a middle-aged man (ii), and an old man (iii).

1.3 On the voyage of the *Beagle*, Darwin learned a lot of biology, but he did not discover evolution

Darwin did not become an Anglican minister, because he was offered a chance to go on a trip that was his fondest aspiration. This was the voyage of the *Beagle*. The **Beagle** was a British Royal Navy ship, commissioned to survey the biology and geology of South America. The ship is shown in Figure 1.3A. This commission reflected the global reach of the Royal Navy, which the British Empire used as its principal instrument of military domination. South America had broken free of Spain in the nineteenth century, and the British wanted to make sure that they knew enough about the geography of the continent so that the Royal Navy would face no unpleasant surprises as it sailed in South American waters.

The Captain of the *Beagle* was **Robert FitzRoy**. FitzRoy was truculent and obsessed with social rank. It was imperative for him that the ship's naturalist, the single most important person on board after the captain, be of the right social class. A middle-class, scholarly type would not do. Thus

FIGURE 1.3A HMS *Beagle*

Darwin, who was no professional biologist or geologist, got the job. He knew some biology. He knew how to collect scientific specimens. But most important for FitzRoy, Darwin was upper class.

It is a famous story that Darwin sailed around the world with the crew of the *Beagle*, which set sail late in 1831 and returned in 1836—a voyage of almost five years (Figure 1.3B). It is less generally known that Darwin collected a vast number of geological and biological specimens on this trip, enough to fill warehouses back in London. This feat was possible because the *Beagle* frequently met other Royal Navy ships, which transshipped Darwin's specimens back to London.

It is commonly thought that Darwin developed the theory of **evolution** on the rocky islands of the Galápagos Archipelago, right on the equator. This is not true. He developed his theory of evolution only after coming home to England. On first returning to London in 1836, Darwin was as much in the dark about the deep questions of biology as he had been at the time of his departure.

What Darwin did learn from his trip was the astonishing diversity of living forms, and their frequent occurrence as fossils located in rocks. Furthermore, some of these fossils were of forms that apparently no longer lived. Though Darwin remained an essentially **creationist** biologist during the voyage of the *Beagle*, the staggering heterogeneity among the plants and animals that he collected showed him the difficulty of understanding life. Darwin saw with his own eyes, as few biologists had up to that time, the range of living things on the planet Earth. It was an experience that was to color the rest of his life. ❖

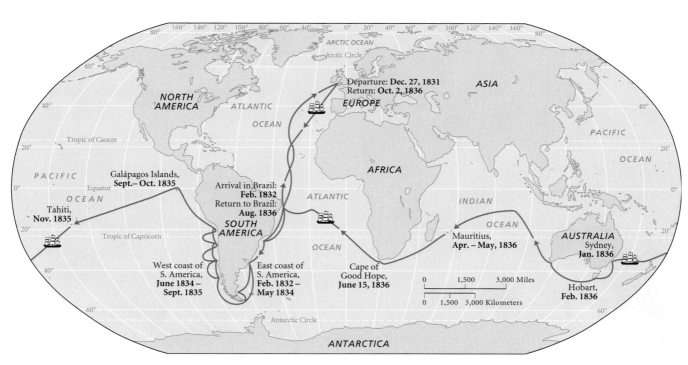

FIGURE 1.3B **The Voyage of HMS *Beagle***

1.4 Darwin intuited evolution from the differences between birds from the Galápagos Islands

The key moment in the history of biology came while Darwin was supervising the unpacking of his collections back in London. Darwin and the other members of the *Beagle*'s crew had collected a number of finches and mockingbirds from the Galápagos Islands, usually recording the specific island from which each bird came. Darwin had thought that these birds were so similar to each other that they were no more than different breeds within a common species of finch or mockingbird.

But the mockingbird specimens were examined more carefully by an ornithologist, John Gould. In March 1837, Gould sent Darwin a note in which he argued that the mockingbirds of the various Galápagos Islands were in fact distinct species. There was not enough overlap in their anatomy to indicate they were merely varieties of a common species. Moreover, the Galápagos finches and mockingbirds appeared to be very similar to species living on the mainland of South America, not far from the Galápagos. In principle, the islands could have been colonized by the mainland forms and then somehow changed, it seemed to Darwin. Darwin at first called this process of change from one form into another **transmutation**. Initially, Darwin's thinking about transmutation was little improved over the ideas of **Jean-Baptiste Lamarck** or grandfather Erasmus Darwin. But Charles's thinking progressed rapidly.

Figures 1.4A to 1.4D summarize Darwin's idea of transmutation applied to the birds of the Galápagos. Species originally only on the mainland colonize the Galápagos Islands, more accurately mapped in Figure 1.4E. This might have happened if a flock of birds were swept out to sea and then flew away from the mainland, perhaps because of an unusual storm.

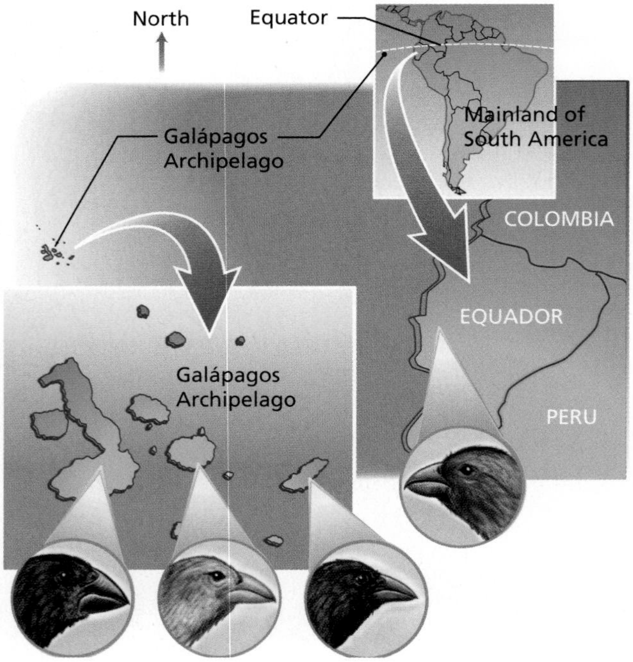

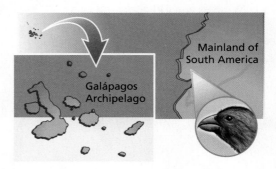

FIGURE 1.4B **Darwin's Evolutionary Scenario** The scenario starts with just one bird species found only on the mainland.

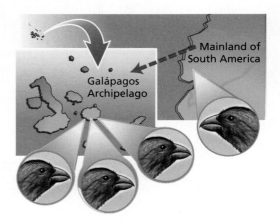

FIGURE 1.4C **Darwin's Evolutionary Scenario** Step two is the accidental migration of a flock from mainland to the Galápagos.

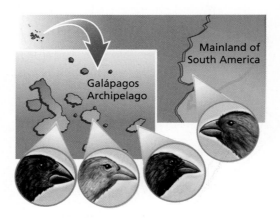

FIGURE 1.4D **Darwin's Evolutionary Scenario** Slow evolutionary diversification of birds on the Galápagos.

FIGURE 1.4A **Darwin's *Beagle* Collections** Similar bird species were found on the islands of the Galápagos and on the mainland of South America.

Once living on the different islands of the archipelago, populations on different islands may have evolved in response to the different habitats, following some law of evolutionary change that Darwin did not at first know. Long periods of divergence would then ultimately produce different species, as shown in Figure 1.4D.

Darwin's first great contribution to biological thought was his theory of evolutionary **descent with modification:** New species are produced from established species. Furthermore, new species do not remain the same as the old species. They become different. In this way, the Darwinian view of the history of life is both conservative and dynamic. It is conservative because new forms of life do not originate from nothing. Life comes from life—at least since the origin of life. Therefore, Darwinians are not surprised by the biochemical unity of life, its dependence on the same nucleic acids and proteins, because Darwinians regard all life as unified by a history of descent from common ancestors.

At the same time that the Darwinian view of life is conservative, it is also revolutionary compared to most traditional interpretations of biology. Theories of biology before Darwin usually assumed that species are fixed essences, unchanging and perfectly suited to their environments. Views like this may be found in the writings of **Plato**, Aristotle, and much

Christian theology. Darwin's simple idea that one type of bird may be a "transmutation" of another type of bird breaks from the assumption that species are fixed. Instead of species being created by a deity, or being present unchanged for all eternity, in the Darwinian view species are contingent creations of an historical process: evolution. Species do not have a special status in Darwin's theory. Their coming into being and their extinction may take more time than the creation or destruction of other things, such as individual organisms. But they still arise and disappear, slower than most familiar things, quicker than mountain ranges. The Darwinian view of life constrains life to evolve from life, except at its origin; but at the same time, it frees our view of life from static theories in which change is not possible. ❖

FIGURE 1.4E **The *Beagle's* Travels among the Galápagos Islands**

Darwin developed the concept of natural selection to explain the direction of evolutionary change

With his theory of evolution as descent with modification, Darwin solved many of the long-standing problems in biology: the underlying similarities of organisms, the superficial diversity of organisms, and the pattern of the fossil record, among other issues. But the mere plausibility of his evolutionary theory was not enough for Darwin. He wanted an observable process to explain how organisms evolve. Such a well-defined process would be a *mechanism* for the origin of species.

This need to go another step came from Darwin's training in geology, especially the writings of **Charles Lyell** (1797–1875), whose *Principles of Geology* was one of the few books that Darwin took with him on the voyage of the *Beagle*. Lyell argued that scientific explanations should use the cumulative effects of well-known processes. The alternative was to invoke catastrophes and miracles. This was not a merely hypothetical alternative in the nineteenth century. At that time, almost all scientists were devout Christians, if not religious ministers. It was natural to them to explain unusual fossil patterns in relation to Noah's Flood or some other biblical event. But Lyell explicitly rejected that kind of scientific explanation. For him, scientific explanations had to be based on processes that could be observed in the present, not on unknowable cataclysms from the distant past. Darwin adopted this stricture as his own, even though it made his intellectual struggles much harder.

Living a life of bachelor study in London in his late twenties, Darwin searched for more than a year for an explanation of evolution, without success.

Darwin found himself in some difficulty, because he did not know of a mechanism that could drive the evolutionary process—at least not at first. Living a life of bachelor study in London in his late twenties, Darwin searched for more than a year for an explanation of evolution, without success. Then one day in 1838 he read Malthus's essay on human population growth. **Thomas Robert Malthus** (1766–1834) was concerned that the European population would inevitably outgrow its ability to produce food, resulting in mass starvation, plague, and so on. Darwin read Malthus and thought instead of animals and plants living in nature. If they developed very large population sizes, then they too would be subject to very bad, very competitive conditions in which there would be too few resources for all to survive. This scenario, Darwin reasoned, would tend to favor those individuals who were better equipped to compete with their rivals, individuals with a greater ability to survive and prosper under bad conditions. These were the individuals most likely to reproduce. Darwin's reasoning to this point seems indubitable. A diagram of his logic is shown in Figure 1.5A.

The bold step came next. Don't offspring tend to resemble their parents? If it was primarily the more robust who survived and reproduced under difficult conditions, wouldn't their offspring tend to be better equipped for these conditions

How Darwin Used Malthusian Reasoning

Darwin was impressed with the antagonism between excessive reproduction and later mortality due to ecological processes. The later mortality opened up the possibility for natural selection to do its work.

Parents produce more offspring than can survive.

Ecological processes kill off many of the excess offspring in each generation.

FIGURE 1.5A From Malthus to Darwin—the Malthusian Problem of Limited Resources and Ecological Disaster

as well? Thus stressful environments should produce selective reproduction, or "natural selection," to use Darwin's term. Generally, **natural selection** can be defined as the net reproductive advantage of individuals with favored characteristics. Darwin immediately saw that natural selection might, under some conditions, produce sustained evolutionary change. Thus natural selection is a mechanism that might produce the evolutionary patterns of descent with modification that Darwin had already discerned. Natural selection could meet Lyell's stipulation about scientific explanation, because the process by which superior organisms prosper and inferior ones die was there to be seen in the everyday lives of plants and animals.

An example of natural selection from the Galápagos Islands is described in Figure 1.5B. In this example, drought conditions create selection in favor of those birds that can crack the seeds that remain available during drought. ❖

The Galápagos Islands are relatively barren and dry. In some years rainfall is rare. During these years, the plants produce fewer seeds. The seeds that are more common during drought are large, with thick husks. Only birds with large powerful bills can open and eat these seeds.

| Normal Conditions | Drought |

Large and small seeds available for large and small finches

Only large seeds, left over from earlier season

Under normal conditions, finches that are larger and smaller can survive on the Galápagos Islands. Under drought conditions, only larger finches, with bigger beaks, can survive, because they can open the large seeds that remain available from earlier seasons.

FIGURE 1.5B Selection on Birds Due to Drought on the Galápagos Islands

1.6 Despite publishing the *Origin of Species*, Darwin was a much-honored scientist during his life

Though Darwin published his other work on biology from 1838 to 1858, he was hesitant to publish his evolutionary theory. In the middle of the nineteenth century, most academic biologists were still creationists who believed either in the exact words of the Bible or something closely related to them. After Darwin had become a closet evolutionist, **Robert Chambers** (1802–1871) anonymously published the evolutionist book *Vestiges* in 1844. It contained a number of arguments in favor of evolutionary change, conceived crudely. Some of these arguments were remarkably weak. The book was greeted with a storm of protest, particularly from biologists. Darwin was mightily intimidated, because some of his own friends and colleagues joined in the savaging of the book. Instead of publishing his own evolutionary theories, he published on other topics, including a monumental study of barnacles. He received the Royal Medal of London's Royal Society for his work on barnacles and coral reefs.

So Darwin remained silent for twenty years. It is not clear that he would ever have published his ideas in his lifetime. He was a well-respected member of the Royal Society with considerable wealth. His will directed his colleagues to publish his draft essays on evolution in the event that he died. But posthumous publication was not the outcome of Darwin's story.

In 1858 Darwin received a letter from a young English naturalist, **Alfred Russel Wallace** (1823–1913), who was traveling in the South Seas, off southeast Asia. Wallace had independently discovered the theory of evolution by natural selection, and he wanted Darwin's comments and his help getting it published, if he thought it worthy. This event turned Darwin from fear of public hostility to an even greater fear of being scooped. But being an English gentleman, he could not tear up Wallace's letter and go ahead with publishing his own theory. Instead, it was arranged that Wallace's letter would be presented and published along with one of Darwin's unpublished essays. The table of contents of the journal involved is shown in Figure 1.6A. For the record, the two men were given codiscoverer status at the time.

But the initial publication of the theory of evolution by natural selection received little reaction from the scientific community. Some scientists were stunned. Others did not understand. Darwin set about writing a short book, which he

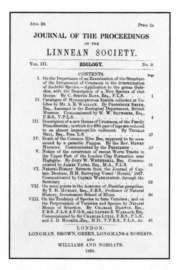

FIGURE 1.6A The Table of Contents of the Journal Which Published the Papers by Darwin and Wallace in 1858

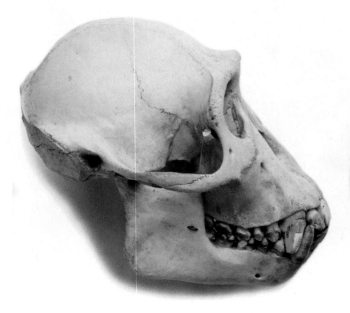

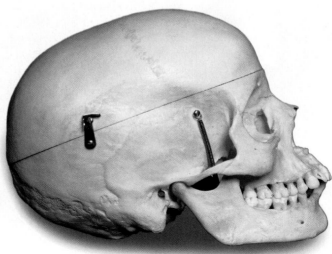

liked to call an "abstract," describing his theory. This book was *On the Origin of Species by Means of Natural Selection*, published by John Murray of London in 1859. Part of the title page is shown in Figure 1.6B.

The book caused a furor. It quickly acquired supporters and detractors. There were scientists for and against, clerics for and against, novelists for and against, journalists for and against. Few seemed to be neutral. It was a true intellectual sensation. Ever since, and more than any other single scientific theory, Darwin's theory of evolution has been in the public eye.

But Darwin went back to being a staid country scientist. He wrote books on inheritance, facial expressions, and earthworms. He did some elegant experiments in plant breeding. He very much wanted to figure out how heredity worked, but he never did. He refused to defend his theory of evolution before a public audience, though he wrote numerous letters in defense of it. Wallace and others—such as "Darwin's bulldog," T. H. Huxley (1825–1895)—played a more visible role than Darwin did.

When Darwin died in the spring of 1882, there was some controversy about how to honor him. His friends in the scientific establishment prevailed upon the authorities to have him buried in Westminster Abbey, near **Isaac Newton**. Darwin's burial slab there is large and dark. The only words are his name, Charles Robert Darwin, and the dates he lived—

February 12, 1809, to April 19, 1882. Never knighted, he has nonetheless been widely regarded as the second greatest English scientist, after Newton. ❖

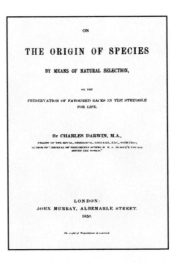

FIGURE 1.6B Title Page of the First Edition of Darwin's *Origin of Species*

DARWIN'S ECOLOGY AND EVOLUTION

1.7 Darwin used ecology to create evolutionary biology

The moment when Darwin made the connection between Malthus's ideas and evolution is one of the single most important episodes in the history of science. The consequences of that moment are still with us. But the intellectual core of Darwin's insight is worthy of a much closer look, because that core contains the essence of Darwinian biology, from molecular evolution to behavioral ecology.

The key to understanding this moment is the role of simple material processes in geology. Geologists deal with such striking things as mountains, rivers, and oceans. But they explain the existence of these objects by mundane, usually imperceptible, processes such as the deposition of sediments at the bottom of bodies of water, the slow buckling of layers of rock, and the erosion of land by wind or water.

When Darwin first had the idea of evolution, it was a basic intuition of change among species.

He thought that evolutionary change was the most appropriate way to explain the similarity of the birds of the Galápagos Islands. He even drew pictures of evolutionary trees in his notebooks. But he had no idea of *how* evolution occurred. His thinking was just pattern, without process to generate it.

As already mentioned, he was practically embarrassed about having a pattern without a process. His professional affiliation at that time was with geology. He was the Secretary of the Geological Society of London, in fact. And not just any kind of geologist. He was a self-conscious disciple of Charles Lyell, the leading gradualist of the day. Although Lyell's *Principles of Geology*, especially Volume II, raised the problem of the origin of species, it also conveyed the message that the only appropriate science was science based on material processes. Pattern alone would not do.

Darwin was a lonely man, wrestling with the problem of how evolution might work. He could not even tell his mentors, like Lyell, what he was doing. He needed to find evolutionary processes that were mere extensions of simple, everyday, material processes at work in the present.

This was where Malthus and his *Essay on the Principle of Population,* published in 1798, came in. In his essay, Malthus constructed a history-making argument about the problem of human **overpopulation**. Malthus was by no means the first to write about overpopulation. The philosopher **David Hume** (1711–1776) and the economist **Adam Smith** (1723–1790) had broached the subject of overpopulation earlier, but their writings were fairly superficial. Malthus took on the subject in detail. Malthus's main thesis was not that important to Darwin; Darwin was not trying to solve the problem of human overpopulation. The important thing was that Malthus supplied a semiquantitative ecological model. Malthus argued that the human population size was increasing geometrically while food production increased only linearly. This model was not correct, because food production has increased geometrically—not linearly—since the early 1800s. But his reasoning was like that of geologists in that Malthus wrote about everyday ecological processes: reproduction, death, disease, farming, and so forth.

All Darwin had to do was apply this ecological reasoning to the lives of animals and plants. Like humans, animals and plants have a great capacity to reproduce. But normally their population sizes do not explode. Therefore, there had to be checks to animal or plant population growth. These checks must take the form of deadly misfortunes. Nature must then select in some way. The better would be sorted out from the lesser. In realizing this, Darwin would start on a journey of discovery that would last him the rest of his life, as his understanding of evolution by natural selection progressively deepened.

Figure 1.7A shows graphically the parallel between Malthus's thinking about population growth and Darwin's basic ideas about natural selection. Darwin did not proceed by any kind of formal logic. Once Malthus had supplied him with the elements of ecological thinking, Darwin appropriated them wholesale to solve his problem of how evolution worked. Biology would never be the same again. ❖

Parents produce more offspring than can survive.

Ecological processes kill off many of the excess offspring in each generation.

Malthusian problem of limited resources

Parents produce more offspring than can survive.

Ecological processes kill off many of the excess offspring in each generation.

Surviving or "selected" adults are superior to adults that die before reproducing.

The offspring of the selected adults are themselves superior; this improvement will be inherited by future generations.

FIGURE 1.7A Darwin's Extrapolation from Ecology to Evolution

1.8 Malthus's essay led Darwin to apply ecology to other problems of evolutionary biology

As we have seen, Darwin got the idea of connecting ecology to evolution from reading T.R. Malthus. The essential Malthusian argument was that people reproduce faster than agriculture can be developed. In his words, "Population, when unchecked, increases in a geometrical ratio. Subsistence increases only in an arithmetical ratio." From this, Malthus concluded that the tendency for people to reproduce will necessarily be checked. It could be checked by famine, pestilence, war, and other disasters. Or it could be checked by social policies that prevent marriage and the birth of children out of wedlock. Malthus argued strongly for measures that would prevent poor people from marrying young and having children. He also advocated "moral restraint," which meant sexual restraint. The analysis of Malthus played an important role in nineteenth-century politics, often being used in arguments against supplying welfare.

But for Darwin, the ecological reasoning was more important. He was captivated by the idea that in nature, many more offspring are produced than can possibly survive. As shown in the box, "The Malthusian Moment," Darwin accepted Malthus's argument wholesale, where overpopulation was concerned.

Why was this essay such a revelation for Darwin? Darwin needed simple, observable machinery to drive evolution. He wanted a process like erosion to shape species, so that they would evolve materialistically. Malthus supplied the ecological machinery of reproduction, famine, and disease to get Darwin's evolution working, as Darwin himself explained in the quotation shown in the first box.

The Malthusian argument concerning overpopulation is an example of the effects of population growth on ecology, a major concern of Chapter 10. But Darwin used other ecological principles as well. For example, he considers the interaction between extreme environments and ecological competition in the passage quoted in the second box "Darwin Reasoning Ecologically," taken from the end of his chapter on the *struggle for existence,* his term for ecology. Here we see Darwin stretching beyond the ideas of Malthus. He was opening up the ideas of ecology, and generalizing them, to make evolution work. ❖

The Malthusian Moment

A struggle for existence inevitably follows from the high rate at which all organic beings tend to increase. Every being, which during its natural lifetime produces several eggs or seeds, must suffer destruction during some period of its life, and during some season or occasional year, otherwise, on the principle of geometrical increase, its numbers would quickly become so inordinately great that no country could support the product. Hence, as more individuals are produced than can possibly survive, there must in every case be a struggle for existence, either one individual with another of the same species, or with the individuals of distinct species, or with the physical conditions of life. It is the doctrine of Malthus applied with manifold force to the whole animal and vegetable kingdoms; for in this case there can be no artificial increase of food, and no prudential restraint from marriage. . . .

There is no exception to the rule that every organic being naturally increases at so high a rate, that, if not destroyed the earth would soon be covered by the progeny of a single pair. . . . the geometrical tendency to increase must be checked by destruction at some period of life. . . . Lighten any check, mitigate the destruction ever so little, and the number of the species will almost instantaneously increase to any amount. . . . The amount of food for each species of course gives the extreme limit to which each can increase; but very frequently it is not the obtaining food, but the serving as prey to other animals, which determines the average numbers of a species."

—Charles Darwin, 1859, *Origin of Species*
(Chapter III, "Struggle for Existence")

Darwin Reasoning Ecologically

Look at a plant in the midst of its range, why does it not double or quadruple its numbers? We know that it can perfectly well withstand a little more heat or cold, dampness or dryness, for elsewhere it ranges into slightly hotter or colder, damper or drier districts. In this case we can clearly see that if we wish in imagination to give the plant the power of increasing in number, we should have to give it some advantage over its competitors, or over the animals which prey on it. On the confines of its geographical range, a change of constitution with respect to climate would clearly be an advantage to our plant; but we have

reason to believe that only a few plants or animals range so far, that they are destroyed exclusively by the rigour of the climate. Not until we reach the extreme confines of life, in the Arctic regions or on the borders of an utter desert will competition cease. The land may be extremely cold or dry, yet there will be competition between some few species, or between the individuals of the same species, for the warmest or dampest spots.

—Charles Darwin, 1859, *Origin of Species*
(Chapter III, "Struggle for Existence")

It would have been ideal if Darwin had performed experimental studies of evolution. That way he could have seen how natural selection and inheritance work together, and he might have developed his theory accordingly. But in the years between Darwin's discovery of the theory of evolution by natural selection and the publication of the *Origin of Species*, no one had any notion of how to perform experimental evolution.

Instead, Darwin took advantage of the literature on breeding plants and animals. By Victorian times, enough was known of breeding to make this literature extremely useful for Darwin. The breeds of animal and varieties of plant could be treated as analogous to animal and plant species in Darwin's theory.

In the *Origin* and his other writings, Darwin based his key arguments on breeding. His most important point of emphasis was that a wide diversity of breeds and varieties could be derived from a single ancestral species. With respect to pigeon breeds, some people held that they were derived from several wild species. Darwin argued against this theory on several grounds. First, crosses of pigeon breeds were always fertile. If the original species were separate from each other, then they should not be able to interbreed. If that were so, then why should breeds derived from different wild species be able to interbreed? Second, the dozens of pigeon breeds were thought to derive from rock pigeons,

FIGURE 1.9A Breeds of the Common Pigeon, *Columba livia*

but there are only a handful of rock pigeon species. Third, the original wild species of rock pigeons do not resemble some of the domesticated breeds more than any other. That is, it is difficult to associate one particular wild species with particular domesticated breeds and another wild species with other domesticated breeds. Indeed, the general pattern is that all pigeon breeds seem to be related to *Columba livia*. (See the box, "Darwin and the Pigeons," for a quotation and Figure 1.9A for some examples of pigeon breeds.) Darwin's point in this argument is that breeding is capable of producing extensive variety, as embodied in living breeds.

Darwin also observed that domesticated breeds and varieties are not usually mere variants, established in a single step from a wild species. Rather, the diversity of breeds is to be explained by "man's power of accumulative selection; nature gives successive variations; man adds them up in certain directions useful to him" (*Origin,* chap. I). The point here is evolutionary **gradualism**. Like artificial selection, Darwin argued, evolution by natural selection cumulatively adds small modifications.

Finally, Darwin argued that artificial selection could powerfully modify breeds and varieies. "The great power of this principle of selection is not hypothetical" (*Origin,* chap. I). And since artificial selection can be demonstrably powerful, so can natural selection. ❖

Darwin and the Pigeons

Though Darwin accumulated a large collection of notes, letters, manuscripts, and published articles on animal breeds and plant varieties, he actually bred pigeons himself. He even joined two London pigeon clubs.

The following is an excerpt from Chapter I of the *Origin of Species*, which shows Darwin's obsession with pigeons:

> The diversity of the [pigeon] breeds is something astonishing. Compare the English carrier and the short-faced tumbler, and see

the wonderful difference in their beaks, entailing corresponding differences in their skulls. The carrier, more especially the male bird, is also remarkable from the wonderful development of the carunculated skin about the head, and this is accompanied by greatly elongated eyelids, very large external orifices to the nostrils, and a wide gape of mouth. The short-faced tumbler has a beak in outline almost like that of a finch; and the common tumbler has the singular and strictly inherited habit of flying at a great height in a compact flock, and tumbling in the air head over heels.

As a geologist, Darwin was well qualified to use information from the record of fossils. In this respect, as in so many others, the logic of his case for evolution was more or less devastating. Note, however, that Darwin did not regard the fossil record as a perfect record of evolution. As explained in the box "Darwin against the Fossil Record," he was a strong critic of the quality of fossils as documentation for evolution. His strategy, as a theorist, was to predict and describe the general features of the fossil record in relation to evolution.

Because Darwin's theory of evolution was based on local patterns of selection and inheritance, he predicted that fossils would not suggest global patterns of change. Instead, "species of different classes do not necessarily change together, or at the same rate, or in the same degree" (this and all subsequent quotations in this module are taken from *Origin,* Chap. X). Likewise, "Groups of species increase in numbers slowly, and endure for unequal periods of time." Because there is no coordinated driving force to evolution, when groups of species are successful, it is because of many individual instances of evolutionary good fortune. This makes evolutionary patterns slow to develop and haphazard in form.

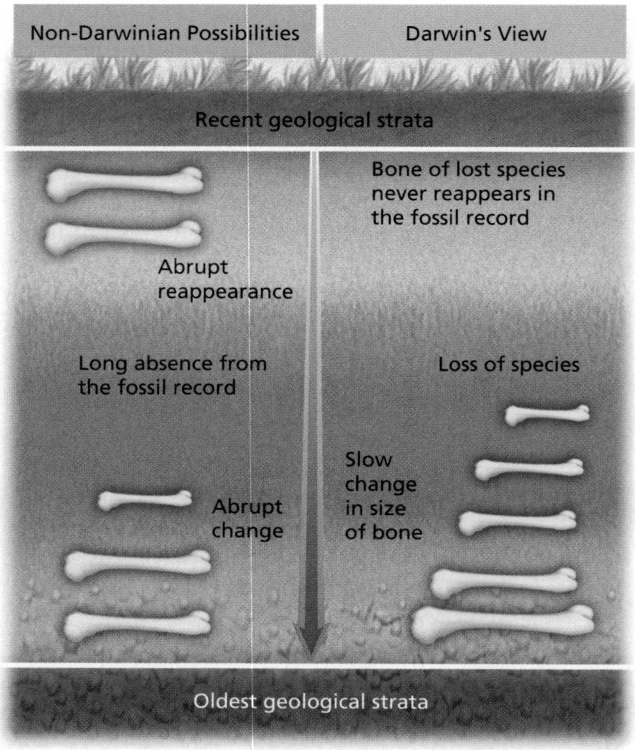

His strategy, as a theorist, was to predict and describe the general features of the fossil record in relation to evolution.

When a species disappears from the fossil record, it never reappears. Evolution is too historical to allow exact repetitions. Even though pterodactyls resemble large birds of prey, flying birds have never approached the size of pterodactyls, which were related to the terrestrial dinosaurs. Similarly, once an entire group has disappeared, like the dinosaurs, it does not reappear except in Steven Spielberg films.

Extinction is a slow process, in Darwin's opinion, especially in groups of species: "The utter extinction of a whole group of species may often be a very slow process, from the survival of a few descendants, lingering in protected and isolated situations." Again, because Darwin's theory of evolution lets each species grow or die off independently, in evolutionary terms, there is little likelihood that extinctions will be coordinated across large groups. (This, as it turns out, is not true every 50 or 100 million years. We take up these rare exceptions in Chapter 6.)

Darwin also identified a universal pattern in the succession of fossils: "The more ancient a form is, the more it generally differs from those now living." And "the organic remains of closely consecutive formations are more closely allied to each other, than are those of remote formations." That is, the fossil record is like a story that begins at one point and continues on to other points without ever doubling back on itself. Indeed, Darwin regarded such repetition as deadly for his theory. We know of no such examples of repeated evolution in long-separated times. There are no vertebrate fossils from 1 billion years ago.

Figure 1.10A shows Darwin's predicted evolutionary patterns and the patterns that he asserted do not happen. In most respects, Darwin's expectations are those of modern-day evolutionists too. ❖

FIGURE 1.10A Alternative Views of the Fossil Record, Illustrated by a Single Fossil Bone

Darwin Against the Fossil Record

Darwin regarded the fossil record as highly unreliable. Here are some of his reasons.

1. Only a small portion of the globe has been prospected for fossils.
2. Only some types of organisms fossilize well; examples of such organisms include vertebrates and shelled animals.
3. Far more species have lived than are preserved in all the museums of the world.
4. Because of the difficulty of fossilization, fossils will provide only a spotty record of evolution.
5. Species migrate from one location to another, preventing continuity of the local fossil record in most cases.

In most respects, Darwin's reasoning about evolution was impeccable. This probably reflected his background as a geologist, especially a Lyellian geologist. Lyell set high standards for scientific reasoning.

But there was a hole in Darwin's fabric of argumentation: inheritance. If there is variation for characteristics that determine survival or reproduction, then the organisms with the greatest tendency to survive and reproduce will have more reproduction. For this reason, Darwin continually emphasized **variation**. The word appears repeatedly in the chapter headings of the *Origin* and throughout the text of the book. Darwin also published a book devoted to the phenomenon of variation: *The Variation of Animals and Plants under Domestication*. In his words, "individual differences are highly important for us, as they afford materials for natural selection" (*Origin,* chap. II). But the transmission of these materials to the next generation is not explained in the *Origin*.

Everything Darwin writes about a struggle in nature leading to differences in the success of particular kinds of organisms would count for nothing if there was no way for differences to be inherited. Otherwise, the selection of particular types has no lasting impact on the population. Selection absolutely requires variation, as shown in Figure 1.11A.

The problem was that the pattern and nature of inheritance was an unsolved problem when Darwin wrote *Origin of Species* in the late 1850s. This lack of information led Darwin to speak vaguely about inheritance (see box). He also tried to sort out unrelated phenomena that, in the minds of biologists at that time, were connected with the problem of inheritance. For example, he had to argue against the view that only unimportant biological characters vary significantly. On the other hand, he found that some characters, such as the flowers of the genus *Rosa*, were extensively variable in a trivial fashion. This variation he considered of "no service or disservice to the species," and therefore unrelated to the action of natural selection.

But things became still murkier when Darwin turned to consider variation between individuals, between varieties, and between species. One of Darwin's characteristic points in the *Origin* is that species are little more than strongly differentiated varieties. He was fond of citing examples of varieties that had been regarded as species, and examples of species that had been regarded as varieties. ❖

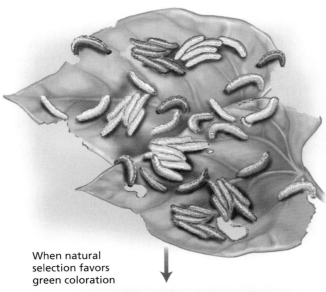

When natural selection favors green coloration

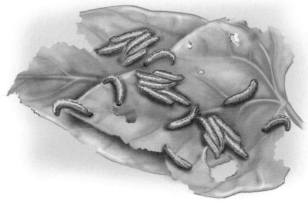

Variation makes evolutionary change possible, by supplying different organisms for selection to choose among, and so change populations.

FIGURE 1.11A Darwin's Concept of Variation as the Raw Material for Selection

Darwin's Thoughts on Variation and Inheritance in the *Origin of Species*

The effects of variability are modified by various degrees of inheritance of reversion. Variability is governed by many unknown laws, more especially by that of correlation of growth. Something may be attributed to the direct action of the conditions of life. ...Something must be attributed to use and disuse. The final result is thus rendered infinitely complex. (Chap. I)

I am convinced that the most experienced naturalist would be surprised at the number of the cases of variability, even in important parts of structure, which he could collect on good authority, as I have collected, during a course of years. (Chap. II)

1.12 Darwin tried to explain the mechanism of inheritance using his theory of pangenesis, but failed

Darwin struggled mightily to sort out the problem of inheritance. In the process, he accumulated as much, or more, information on the subject as anyone in the nineteenth century. Part of his problem may have been that he was willing to accept data that were not scientifically valid. This is not surprising, because nineteenth-century biology was grossly deficient as an experimental science. Most biologists just documented natural history; they did not do critical experiments.

The final result was that Darwin developed a hugely complex theory of inheritance to accommodate the varied and often shoddy information that he was given. He called this theory **pangenesis**. It is important to understand that pangenesis is not a theory of genetics. At the core of this theory are things that Darwin called **gemmules**. He thought that gemmules normally reside in all the tissues of life, shaping organs and processes (Figure 1.12A). In addition, he felt that the conditions of life in each organ or limb would shape the gemmules.

When reproduction occurs, according to Darwin's theory of pangenesis, gemmules migrate from the parts of the body to the gonads. Gametes are then made from mixtures of gemmules. Particular characters can be influenced by many gemmules. The quantitative balance of gemmules of a particular type determines the pattern of inheritance. For example, many gemmules specifying great height are needed to make offspring tall.

With all these gemmules moving about the body, it is reasonable to expect that gemmules should be found in the blood of mammals. Darwin and his cousin, Francis Galton, tested for gemmules in rabbits. They reasoned that rabbits with dark coloration should have gemmules for dark fur circulating in their bodies. So they gave white rabbits blood from dark rabbits and observed the offspring of these white rabbits (Figure 1.12B). If Darwin was right about the circulation of gemmules, the white rabbits that received blood from dark rabbits should have had offspring with some dark coloration, perhaps just a few flecks of dark. The dark color should have been there, but it was not. Darwin expressed the possibility that gemmules might circulate by some means other than blood, but it was a faint hope. His gemmules have never been found. Darwin never sorted out the mechanism of inheritance. ❖

Skin tans from exposure to the sun's radiation.

Gemmules carry signal from tanned skin to gonads.

Gametes produced by the gonads incorporate the migrating gemmules from the different parts of the body.

Gonads

●●● Gemmules for height
●● Gemmules for weight
●●● Gemmules for hair color
●●● Gemmules for eye color
●●● Gemmules for skin color

Gamete

Darwin also supposed that gemmules could vary in number, as part of his overall blending scheme of inheritance.

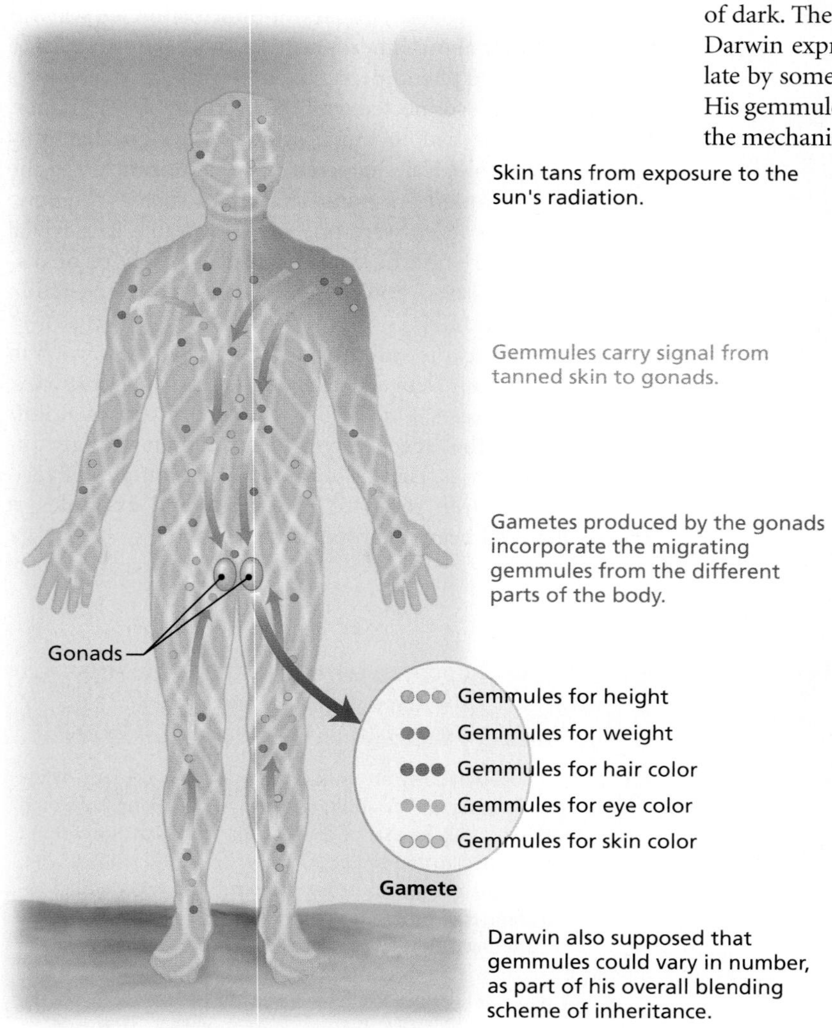

FIGURE 1.12A Darwin's Gemmules Moving About the Body

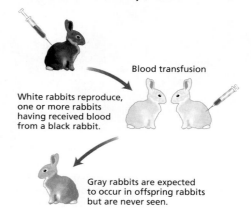

Blood from black rabbit is injected into white rabbit.

Blood transfusion

White rabbits reproduce, one or more rabbits having received blood from a black rabbit.

Gray rabbits are expected to occur in offspring rabbits but are never seen.

FIGURE 1.12B Blood transfusion experiments with rabbits did not reveal gemmules.

It is one of the great ironies in the history of science that in the same period when Darwin was struggling with the problem of inheritance, **Gregor Mendel** (1822–1884), an obscure monk in the Austro-Hungarian Empire, was solving it. The solution that Mendel found is now known as genetics. Mendel himself is described in the following box.

With the concepts of genetics, it would eventually prove possible to sort out the foundations of evolution. But it took a long time for that to happen. The first problem was that Mendel's genetic experiments were published in a journal that few scientists read. The second problem was that Mendel's main connection to the larger scientific community, a botanist called Karl Nägeli, was horrified by Mendel's experimental findings. He did little to communicate Mendel's findings. Indeed, he might have "spiked" them, trying to suppress genetics instead of spreading it.

Even though Darwin was still alive when Mendel's work appeared, he never read it or heard of it, so far as we know. Mendel himself died in middle age of an acute kidney infection, without ever receiving substantial scientific recognition. This is one of the most unfortunate failures of communication in the history of science.

But science has a way of finding the truth, even when it has been neglected or suppressed. Around 1900, several scientists independently rediscovered Mendel's work. They appreciated its immense significance immediately. However, there was still some confusion because Darwin had explicitly rejected theories like Mendel's, and Darwin's followers had continued in his footsteps. This gave rise to the entirely erroneous notion that genetics and Darwin's evolution were incompatible. As a result, evolutionists and geneticists fought with each other for a few decades.

Finally, evolutionary biologists and geneticists calmed down and saw that genetics and evolution worked perfectly well together. This reconciliation of Darwinism and **Mendelism**, together with systematics and other parts of biology, came to be known as the **modern synthesis**. ❖

The Strange Career of Gregor Mendel

The key person in the genetic revolution was the monk Gregor Mendel—one of the most obscure of all the great scientists, if not *the* most obscure. He died unnoticed in 1884. Born Johann Mendel in 1822, he became Gregor on taking orders in 1843. Mendel was rather bad at exams. Though he taught natural science at a high school, he never passed the examination for a teacher's license. Deep within the Austro-Hungarian Empire (in the town of Brno, now part of the Czech Republic), Mendel toiled away in the gardens of the monastery where he was first monk and then abbot.

An important part of Mendel's education was his study at the University of Vienna, where he learned an obscure branch of mathematics called *combinatorics*. Combinatorics is the mathematics of combining things by chance. For example, you do combinatoric experiments every time you shuffle a deck of cards and deal out four hands of five-card stud poker. Combinatorics lets you calculate the odds that you will get four aces and a king in your hand.

What was the significance of Mendel learning combinatorics? This knowledge made him better qualified than any other biologist then living to calculate the probability of particular outcomes from breeding experiments. Darwin, for example, who probably knew the literature of plant breeding better than Mendel, was unable to calculate mathematical odds with Mendel's precision.

THE BIRTH OF MODERN ECOLOGY

1.14 Predator-prey cycles and the origins of theoretical ecology

In the early part of the twentieth century, ecology remained a largely descriptive science. This would change as scientists from the physical and mathematical sciences, especially Alfred Lotka and Vito Volterra (Figure 1.14A), became interested in ecology. Lotka and Volterra brought with them an interest in developing general theoretical principles that would apply to many organisms. Perhaps because of their mathematical backgrounds, they were willing to ignore many complications that can characterize natural populations.

William Robin Thompson was an entomologist motivated by practical problems. Working in a biological laboratory in Paris, his goal was to control the corn borer insect using its naturally occurring parasites. Scientists in both the United States and France had independently noted that many host and parasite species appear to fluctuate in concert. Thompson took this to mean there should be some mathematical representation for that regularity. He explored these host-parasite relationships in his own work with simple equations. Many of Thompson's colleagues were apprehensive about developing theory that relied on parameters that had not been estimated for real populations. But Thompson's work influenced others, including Alfred Lotka.

Raymond Pearl, an established scientist in 1920, admired Alfred Lotka's work and invited him to give a series of lectures at Johns Hopkins University. These lectures then fostered a professional respect between Pearl and Lotka. Pearl would later help Lotka secure an unsalaried position at Johns Hopkins. Pearl also encouraged Lotka to develop a book of his theoretical work in ecology. This book, *Elements of Physical Biology*, was eventually published in 1925.

Lotka did not consider himself an ecologist, although his work was of great interest among ecologists. Lotka's grand vision was to found a field of physical biology that would accomplish the same goals as physical chemistry had for its discipline. Lotka's *Elements of Physical Biology* considers food webs, nutrient cycles, and the transfer of energy between organisms. However, Lotka is best known for a small part of his book that treats the dynamics of predator-prey systems. Lotka's grand scheme of establishing the field of physical biology was never realized.

About a year after Lotka's book came out, Vito Volterra published a nearly identical predator-prey model. Volterra was already an accomplished mathematician by 1925. His daughter was an ecologist who happened to be engaged to Umberto D'Ancona, a marine biologist. D'Ancona was responsible for getting Volterra interested in ecology after describing how he had noted a increase in predatory fish in the Adriatic Sea during World War I, when fishing had all but ceased. Volterra determined, as had Lotka, that interacting predator and prey species could give rise to oscillations in population size.

FIGURE 1.14A **Vito Volterra (1860–1940)** Held the chair of mathematical physics in Rome. Prior to his work on predator-prey models, Volterra was known for his work on elasticity and differential equations. He became interested in biological problems through his daughter's fiancé, who worked in ecology and marine biology.

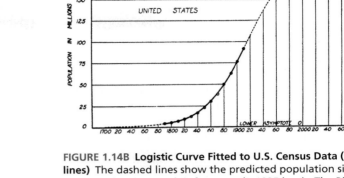

FIGURE 1.14B **Logistic Curve Fitted to U.S. Census Data (solid lines)** The dashed lines show the predicted population sizes from this fit. This graph is from Pearl's 1925 book, *The Biology of Population Growth*.

In the 1920s Raymond Pearl was driven to show that the logistic equation was an important law in ecology that described the growth of nearly all populations, including humans (Figure 1.14B). To that end, Pearl solicited Lotka's approval. Lotka found the logistic equation a useful starting point for describing population growth, but he did not give it the same status that Pearl did. Pearl argued for the utility of the logistic equation because it fit population growth curves from humans and fruit flies (see Figure 1.14B). These arguments were not universally accepted, even at that time. ❖

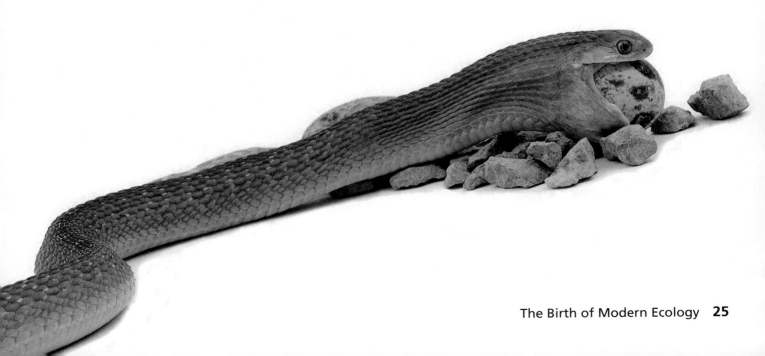

The Russian scientist, Vladimir Alpatov, had interests in the geographical distribution of invertebrates. In 1927 he secured a fellowship from the Rockefeller Foundation to study in the United States with Raymond Pearl for two years. During his stay, Alpatov worked on experimental research with fruit flies. In addition to his research accomplishments in the United States, Alpatov developed a very high respect for Pearl. Alpatov also encouraged his young student in the Soviet Union, Georgii Gause, to secure funding to work with Pearl in the United States. Gause was unsuccessful on the first attempt to secure a fellowship; the major reason given was Gause's young age—21. Pearl and Gause agreed that if Gause were to publish a book, it might elevate his scientific stature in the United States and thus help win a fellowship. Although Gause did not receive the fellowship, publication of his book *The Struggle for Existence* (from Darwin's original phrase) established Gause's scientific legacy.

Gause had initially started his experimental work with fruit flies that Alpatov had brought back with him from the United States. But Gause then switched to protozoans and yeast because they were easier to handle. In studying competition between species, Gause made small modifications in the logistic equation to account for the effects of a second species. By keeping two species of protozoans on the same resource, he was able to show that extinction of one species was the ultimate outcome, as predicted by the competition equations (Figure 1.15A). Gause was also able show that when competing paramecia were given two food sources, the levels of competition could be reduced and the species could coexist. It would have been easy to discount Gause's experimental results as overly simplistic. Gause, however, tried to tie his work to the larger picture of ecology and thus make his work appeal to a large audience.

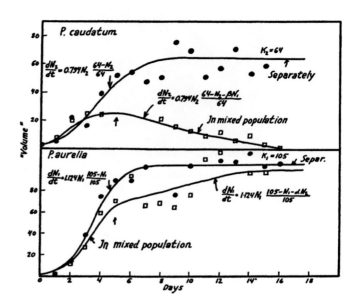

FIGURE 1.15A One of Gause's Original Figures from *The Struggle for Existence* This particular figure shows the change in the numbers of the two species of *Paramecium* when raised alone and when raised together. In this experiment *P. caudatum* is driven to extinction by *P. aurelia*.

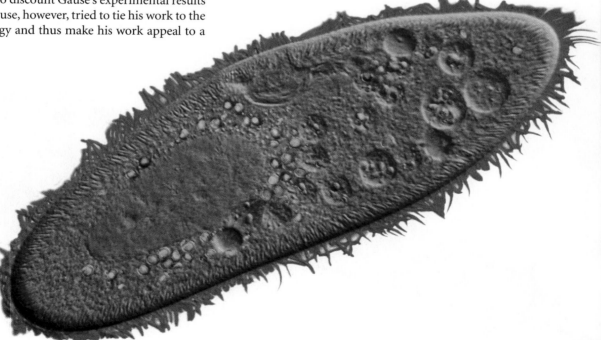

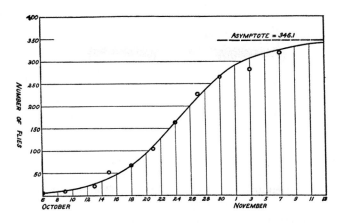

FIGURE 1.15B A Growth Curve from Pearl's Experimental Work, First Published in 1925 The y-axis shows the total numbers of *Drosophila* adults in a single culture over a period of about 1 month. The solid curve is a logistic equation fit to these data, which predict an equilibrium size of 346.

Gause is probably best known for the competitive exclusion principle, which suggests that no two species can occupy exactly the same niche, because one will be driven to extinction. Several others had in fact suggested this idea. In 1874 Karl Nägeli described the displacement of plant forms by competitors. Joseph Grinnell in 1917 suggested that "no two species regularly established in single fauna have precisely the same niche relationships." J.B.S. Haldane, the noted theoretical population geneticist, also stated the competitive exclusion principle in 1924. However, the clear experimental demonstration of this principle by Gause is almost certainly why he continues to get credit for this idea.

Experimental ecology also flourished under Raymond Pearl. Much of his experimental work on population growth and aging focused on fruit flies. That Pearl was working with fruit flies was in part acciden-tal. In 1919 he had actually started a mouse research colony. However, a fire in the laboratory destroyed the colony. Before beginning a new colony, Pearl talked to T. H. Morgan, who convinced him to work with *Drosophila*. Much of Pearl's experimental research with *Drosophila* was aimed at showing the utility of the logistic equation (Figure 1.15B). Pearl and his colleagues carried out an extensive program of experimental research with *Drosophila*. These studies included investigating the effects of nutrition and density on a variety of life-history traits. Pearl also looked at the effects of density on age-specific survival; this work established *Drosophila* as an important organism for experimental ecology. At the same time, *Drosophila* was also establishing itself as one of the most important model organisms for genetic studies. ❖

1.16 The controversy between density-dependent and density-independent population regulation

One of the greatest debates in ecology during the twentieth century concerned the role of density-dependent factors in population regulation. The debate lingered longer than necessary, partly because of the extreme positions taken by some of the protagonists. Curiously, the major protagonists in this debate were all from Australia. H.G. Andrewartha and L.C. Birch felt that environmental conditions were the major factors determining ultimate population numbers. In particular, they believed that favorable periods when populations might grow exponentially were short lived and to some extent unpredictable. The net result would be limited population increase from one season to the next.

Alexander Nicholson argued for the preeminence of competitive interactions between species. These interactions might be between members of the same species, or between the predators that feed on those animals. For Nicholson these competitive interactions, or governing reactions as he called them, determined the equilibrium that a population would reach.

Andrewartha and Birch were influenced by a series of studies on small insects called thrips (Figure 1.16A). These insects undergo large fluctuations in numbers in natural populations, but apparently they never completely deplete their food resources. Davidson and Andrewartha used a statistical regression model to predict these fluctuations in thrip numbers

(Figure 1.16B). This statistical model was based on environmental measurements of temperature and rainfall and appeared to be able to account for most of the variation in thrip population size. Davidson and Andrewartha took this result to mean that environmental factors alone were responsible for the regulation of population size, and therefore there was no need to invoke competition as a regulating factor. As statistical methodologies and sophistication improved in ecology, later studies of the same thrip data were able to demonstrate that future population sizes were in fact influenced by density.

Nicholson, on the other hand, had a long career devoted to the study of blowflies in carefully controlled laboratory settings. Under these conditions, the blowflies were allowed to increase in number until the levels of food supplied by the experimenters could no longer support continued growth of population size. Given these conditions, it is not surprising that the competition for food is crucial to understanding some of the complex dynamics of blowfly populations.

Today there can be no doubt that competition and density-dependent regulation are important in many natural populations. However, some populations are regulated by other factors, like predation, to a much greater extent than by competition. There is no single ecological explanation for the size of natural populations. ❖

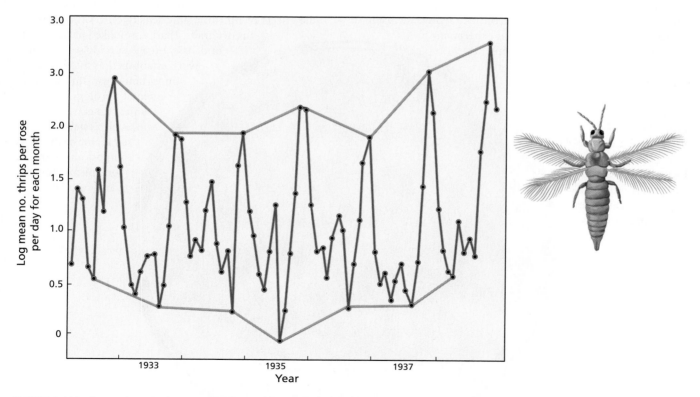

FIGURE 1.16A Fluctuations in the Numbers of Apple Blossom Thrips in Southern Australia Thrips are small insects, about 1 mm long, found in flowers of roses, fruit bushes, and other garden plants in Southern Australia. This figure, which is on a log scale, illustrates the large variation in numbers that are typical of this insect.

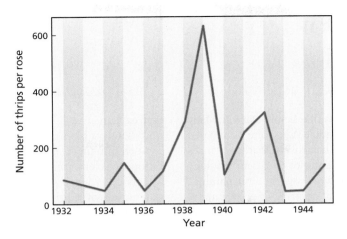

FIGURE 1.16B Population Density of Thrips over Many Years
Davidson and Andrewartha used a model that incorporated the current and previous years' temperatures and the current year's rainfall to predict the numbers of thrips. Based on the ability of these abiotic factors to accurately model the thrip populations, Davidson and Andrewartha concluded that competitive interactions were not very important factors in the regulation of thrip numbers.

DARWIN'S VIEW OF LIFE

1.17 Darwin's theories broke with prevailing biological doctrines, which were theological and vitalist in their foundations

It is hard to imagine what biology was like before Darwin. Many people know that the European Middle Ages were dominated by the authority of the Bible. Bitter controversies arose over access to it. The Bible stood at the center of Christian theology and defined the ultimate limits of knowledge for a millenium.

Less widely known is a medieval doctrine of great importance for the development of science in Europe. This doctrine held that God revealed Himself to people by two devices—the Bible and His Creation. Thus God could also be understood by studying His works in the world, in addition to studying His Word. In a sense, this view made European science fundamentally theological before Darwin.

Indeed, there was a long tradition of theologically informed European science. Isaac Newton was an intensely mystical Christian who hated atheism with a passion. The Polish astronomer Nicolas Copernicus (1473–1543) advocated a sun-centered solar system within a religious, if not mystical, framework. Darwin's science professors at Cambridge University were often clergymen, many of whom saw their study of science as a devout occupation. When Darwin was a young man, his career fantasy was to preach to a small country congregation while collecting insects during his spare time. Science and Christianity were intertwined well into the nineteenth century.

At the time of the Renaissance, about A.D. 1400–1600, European thought acquired numerous ideas from the classical world of 1000 B.C. to A.D. 400. It is commonly believed that classical thinking led to a great improvement in European thinking, but this view is not altogether correct. A particular problem for biology was that classical theories were based on spiritual forces investing simple biological processes, and those forces were used to explain everything from growth to aging to physical processes. If there were distinct vital spirits that explained each biological process, how could a scientist make sense of biology?

Such thinking was so difficult to reconcile with traditional European science that Darwin and other nineteenth-century leaders of Western science recoiled from it. They preferred watered-down Christian science to the ancient scientific thinking, which they found repugnantly mystical. The rejection of such classical ideas was part of a general scientific movement that was started primarily by the physicist and astronomer Galileo. This movement is now known to us as **materialism**. In science, materialism is the doctrine that observations are to be explained in relation to the action of simple, observable processes—not spiritual forces such as angels or demons. The triumph of materialism played a large part in the acceptance of Darwin's ideas, because he did not invoke gods, spirits, or other magical agents in living processes.

Geology supplied one of the important proving grounds for materialism. There were two main doctrines in geology to explain the formation of mountain ranges, lakes, and other features of the Earth. The **catastrophists** believed that long past geological upheavals, such as those arising from Noah's Flood, could explain the creation of geological features. The **uniformitarians**, or **gradualists**, argued instead that the slow, cumulative action of everyday processes like sedimentation and erosion was a sufficient explanation of geology. In particular, uniformitarians assumed that the geological processes at work in the present were the same as those at work in the past. The tension between these two main points of view hinged on materialism. The uniformitarians were committed to a materialistic view of the world. The catastrophists were closer to a biblical view, in which God could arbitrarily achieve anything.

Darwin began his scientific career as a geologist siding with Charles Lyell, the great gradualist geologist (see box). One way to describe Darwin's career is to say that he extended Lyell's gradualist beliefs to the realm of biology. ❖

Sir Charles Lyell, 1797–1875, the Founder of Modern Geology

FIGURE 1.17A
Sir Charles Lyell

Trained as a lawyer, Charles Lyell (1797–1875; shown in Figure 1.17A) was a gentleman English scientist who found teaching at King's College, London, too much of a distraction from research. He was central to the success of England's Geological Society, the world's first. Starting in 1830, his *Principles of Geology* created the foundations of modern geology. Later in life, he was a key scientific colleague of Charles Darwin's, though he mostly sat on the fence where the doctrines of the *Origin of Species* were concerned.

There have been many theories of evolution other than Charles Darwin's. But some of these other theories of evolution have not been materialistic. The significance of Darwin's materialism requires some explanation.

First, let's consider the role of materialism in scientific theories generally. Some elaboration of the general concept of materialism is provided in the box "Materialism." As noted earlier, there is a strong connection between materialism and ideas like gradualism. On the other hand, catastrophism and related doctrines of dramatic coordinated change fit well with theistic and other supernatural doctrines. After all, an all-powerful supernatural being should be able to change trivial details like the nature or number of the species living on a small planet orbiting a mediocre star. How dramatic changes in life-forms could be produced by material causes alone was not clear in Darwin's lifetime. (We now know how this might happen, as discussed in Chapter 6 on speciation and extinction.) So catastrophism and materialism did not tend to combine well in theories of biological evolution until recent times. Materialistic doctrines in science instead have the same general properties as Lyell's gradualism: Ordinary things happen in ordinary ways, step by step, without large discontinuities.

Second, there are the nonmaterialistic theories of evolution. Many theories of evolution are based on vague or mystical assumptions. Some of these theories assume cosmic progress, a tendency for the universe to improve. Much of ancient Greek thought made this assumption, reflecting the then prevalent Greek assumption that the cosmos as a whole was benign. Even in the twentieth century, intellectuals often assumed some kind of mystical direction to evolution, a way of thinking that ranged from the writings of the German Oswald Spengler (1880–1936) to those of the French philosopher Henri Bergson (1859–1941). Jean-Baptiste Lamarck was the preeminent biologist who assumed that there was a general tendency to upward progress in response to environmental challenge.

Third, Darwin and his closest colleagues reacted to the nonmaterialistic elements in **Lamarckism**—Lamarckian thinking—with revulsion, almost disgust. Lamarck's scheme is diagrammed in Figure 1.18A. (We discuss Lamarck's ideas further in subsequent modules.) Darwin's intellectual forebears were men like Adam Smith and Sir Charles Lyell, who argued in concrete terms, who favored mechanical examples and analogies. People like Smith and Lyell were materialists who disliked cosmic ideas and speculations. Darwin's work on evolution was resolutely driven by a need to conform to the same intellectual standards as Smith and Lyell did, and Darwin utterly forswore any kind of overarching direction to evolution. There is nothing mystical about Darwin's thinking; everywhere he found such a hole in his reasoning, he worked very hard to patch it with a materialistic argument or explanation. Darwin's goal was to propose a materialistic theory of evolution, or he wasn't going to put forward a theory of evolution at all. ❖

Hot weather requires thin fur, light color.

So the rabbit grows thinner fur, with a lighter color, in the same generation.

Rainy weather favors thicker fur to keep water away from skin.

So the rabbit grows thicker fur, in the same generation.

Some physiological processes are like this—human tanning is one example, and the production of digestive enzymes by most animals is another. But many responses to the environment are not so immediate. Genetic evolution is required instead.

FIGURE 1.18A Lamarck's Theory of Evolution Jean-Baptiste Lamarck proposed a theory of change among living things some time before Darwin, in the eighteenth century. His theory of evolution also starts from the effects of the environment. But Lamarck supposed that organisms respond directly to the effects of the environment. There is no natural selection in Lamarck's scheme, only direct physiological response to the environment.

Materialism

Put simply, materialism is the doctrine that everything is made of matter. But the matter-energy interchangeability revealed by modern physics undermines any such simple theory of materialism. Matter just isn't what it used to be. A more subtle view is that matter is whatever stuff science can study, and therefore energy is also materialistic, even though it isn't matter.

A secondary meaning of materialism is the de-emphasis of mental and other nonmaterial states in causal explanations. To a materialist, you didn't eat the apple because "you felt hungry." Instead, you ate the apple because nerves in your stomach signaled to your brain that your stomach was empty, and thereafter your behavior was modified by the actions of further cerebral neurons. Materialistic explanation has as its hallmark the use of simple physical processes to explain more complex events.

1.19 Natural selection supplies direction to Darwin's evolution, but is not itself directed

In Darwin's theory, natural selection supplies a direction to each process of evolution. Darwin ruled out cosmic directing forces like God, or urges to perfection. Indeed, a shorthand way of understanding natural selection is that it is the ultimate supervisor of the living world, scrutinizing and guiding life on Earth. Much of Darwin's language implies as much.

The unusual thing about Darwin's theory of evolution by natural selection was that the natural selection that was supposed to drive the process was not a unified force, acting in parallel on all living things. Instead, natural selection acts within each species, separately, as sketched in Figure 1.19A. Darwin's entire theory rests on the evolutionary fates of individual species being separate from each other. This separation does not mean that Darwin saw no ecological interconnection between species. He certainly saw species as ecologically interconnected. But the evolutionary response of each species was, for Darwin, inherently its own. It did not share its selectively driven response to ecological circumstances with other species.

Furthermore, Darwin saw the individual members of each species as struggling with the other members of the species—for light (in the case of plants), for food (in the case of animals), and for mating opportunities (in the case of sexually reproducing organisms). Even within species, Darwinian evolution depends on the individual fates of organisms; see Figure 1.19B, in which the selective fates of the two birds are entirely separate. In this view, Darwin showed clear signs of his intellectual parentage in the laissez-faire capitalism of the eighteenth and nineteenth centuries. Darwin reasoned much like Adam Smith, the father of economics, with natural selection playing the role of Smith's economic competition in Darwin's discussion of biological evolution. Thus Darwin's theory of evolution was atomized. Species were distinct evolutionary units, with individual fates, and the organisms making up species also underwent selection as groups of individuals.

Instead of some overarching force for evolutionary progress, Darwin supplied a natural selection acting separately within each species. And Darwin did not suppose that this action would be globally coordinated. Instead, each species was its own story—a story ecologically interacting with the stories of other species, but still its own story. The "direction" of Darwinian evolution is a meaningful concept only when applied to single species. Evolution as a whole has no direction, only the ensemble of directions unique to each species. And even the direction of natural selection within a species is changeable, rather than a fixed thing. For this reason, it is incorrect to think in terms of evolution on a grand scale. In fact, there are millions of distinct evolutionary processes, at least one for each species.

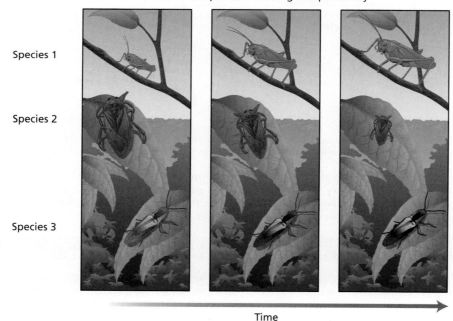

Three competitors evolving independently

Species 1

Species 2

Species 3

Time

FIGURE 1.19A Darwin supposed that evolution occurred independently in separate species. We can think of each species as a discrete unit, possibly having effects on other species but remaining a separate entity during its evolution. An important scientific bonus of this is that we can study evolution one species at a time.

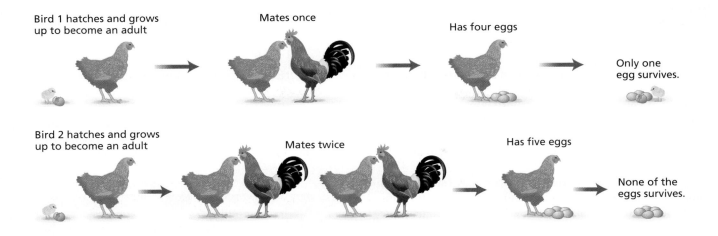

Bird 1 hatches and grows up to become an adult → Mates once → Has four eggs → Only one egg survives.

Bird 2 hatches and grows up to become an adult → Mates twice → Has five eggs → None of the eggs survives.

FIGURE 1.19B Darwin also supposed that natural selection acted separately on the individuals within populations; the process of differential survival and reproduction acts on individual organisms, killing them or their offspring, letting them mate independently.

In short, there cannot be "A Direction" to evolution, due to the splintering of the evolutionary process. Natural selection supplies "directions" to the evolutionary process, but these are always specific, not general. This feature of Darwin's theory of evolution by natural selection allied it firmly with modern, materialistic science, leaving behind the theologically based creationism of Darwin's own teachers. ❖

Because evolution by natural selection is not driven by a cosmic ordering principle, it has disorderly patterns. There is no Grand Conductor in Darwin's theory. Therefore, Darwin's view was that the extinction and origin of new species does not occur according to some overall imperative. This can be hard to understand, because the human mind naturally seeks coherent, simple stories.

Darwin *does* offer stories. The Galápagos finches are a story of migration, in which a specific finch population migrated from continental South America to a challenging new habitat. There the descendants of the original finch migrants overcame considerable difficulties, diversified into new species, and became one of the important life-forms of an important island group. But it could have been any of a variety of continental bird species that migrated and diversified on the Galápagos Archipelago. In fact, some other bird species did so. Mockingbirds (genus *Mimus*) are another group that migrated to the Galápagos, as shown in Figure 1.20A. They diversified too. There was no great historical necessity to the migration of a finch population; it was an historical accident.

The messiness of ecology led Darwin to another important conclusion: Evolution by natural selection does not produce perfection. "Natural selection will not produce absolute perfection, nor do we always meet, as far as we can judge, with this high standard under nature" (*Origin*, Chap. VI). Drawing on his extensive knowledge of biogeography, Darwin brings up the example of the flora and fauna of New Zealand. Darwin argued that these species were biologically successful

FIGURE 1.20A **A Galápagos Mockingbird, *Mimus parvulus***

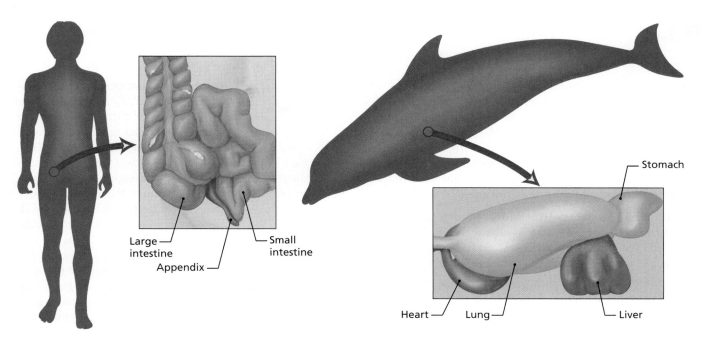

FIGURE 1.20B Two Examples of Imperfection among Living Things (i) The vermiform appendix in humans, which has no known function and can cause death when it ruptures; (ii) lungs in cetaceans, air-breathing aquatic mammals that are thereby forced to come to the surface to breathe, which can result in drowning.

among the other species of New Zealand, but still inferior to the invading species introduced by Europeans, which have been more successful than the endemic species in New Zealand. Darwin also considered the production of pollen by conifers, in which a few new seedlings result from the production of millions of pollen grains—highly inefficient. But in considering these examples, Darwin was complacent. He did not expect perfection. Indeed, he seems to have relished examples of imperfection, because they suited his view of life as a product of ecologically driven evolution, as opposed to a divine creation. If everything had been perfect in all living things, then a divine creation might still have been a credible theory compared with material evolution. Since Darwin's time, we have learned vastly more about the imperfections of life, even though living things still are remarkably efficient

contrivances much of the time. Figure 1.20B shows two examples of imperfection among mammalian species.

Although natural selection is a process that can theoretically change the morphology and behavior of a species with great speed, Darwin expected that it usually would not work quickly. Because it is not a directed process, because it has no inertia or intention, natural selection will act mindlessly and inconsistently. Only over a very long period are we likely to see natural selection produce a sustained change in a species. Evolution by natural selection is neither efficient nor swift. Instead, it is usually inconsistent and unfocused, because it lacks the directed properties that make human actions deft and efficient. Natural selection is not to be thought of as a careful farmer of life. It is not that efficient, at least not in the short run. ❖

1.21 Darwin argued that all order in the history of life was a result of evolution by natural selection

Even though Darwin did not view natural selection as all-powerful or consistent in direction, he still regarded evolution by natural selection as the sole and sufficient source of order in the living world. In particular, he held fast to some strikingly bold interpretations of life, all of them derived from his theory of evolution by natural selection. We list some of these here. They are all taken from the last chapter of the *Origin*, "Recapitulation and Conclusion."

1. *Complex organs* Darwin regarded complex organs—"organs of extreme perfection"—such as the human eye, as a challenge for his theory. But he held fast to his interpretation of them as products of gradual evolution:

> Nothing at first can appear more difficult to believe than that the more complex organs and instincts have been perfected, not by means superior to, though analogous with, human reasons, but by the accumulation of innumerable slight variations, each good for the individual possessor. Nevertheless, this difficulty, though appearing to our imagination insuperably great, can not be considered real if we admit the following propositions, namely, that all parts of the organisation and instincts offer, at least individual differences—that there is a struggle for existence leading to the preservation of profitable deviations of structure or instinct—and, lastly, that graduations in the state of perfection of each organ may have existed, each good of its kind.

This argument is illustrated by the example of the evolution of the vertebrate eye in Figure 1.21A.

2. *Geographic distribution* Another problem for Darwin's non-creationist theory was how similar life-forms came to be found at widely different locations. This geographic distribution was not a problem for creationism, because it could easily be supposed that the Creator would want some mammals here, and some insects there, while creating the flora and fauna of the different continents. Again, despite this problem, Darwin was undeterred:

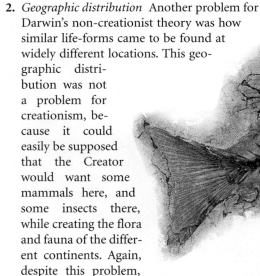

Turning to geographical distribution, the difficulties . . . are serious enough. All the individuals of the same species, and all the species of the same genus, or even higher group, are descended from common parents; and therefore, in however distant and isolated parts of the world they may now be found, they must in the course of successive generations have travelled from some one point to all the others. We are often wholly unable even to conjecture how this could have been effected. Yet, as we have reason to believe that some species have retained the same specific form for very long periods of time, immensely long as measured by years, too much stress ought not to be laid on the occasional wide diffusion of the same species; for during very long periods there will always have been a good chance for wide migration by many means. A broken or interrupted range may often be accounted for by the extinction of the species in the intermediate regions.

Figure 1.21B illustrates this comment of Darwin's.

3. *Absence of intermediate fossil forms* Although Darwin was a geologist, one of the biggest problems facing his theory was the frequent absence of intermediate forms among known fossils. That is, the fossil record did not seem to follow the gradual pattern of change from one species to another assumed by Darwin. Here Darwin's geological expertise stood him in good stead:

> Although geological research has undoubtedly revealed the former existence of many links, bringing numerous forms of life much closer together, it does not yield the infinitely many fine gradations between past and present species required on the theory . . . I can answer these questions and objections only on the supposition that the geological record is far more imperfect than most geologists believe. . . . Only a small portion of the world has been geologically explored. Only organic beings of certain classes can be preserved in a fossil condition, at least in any great number. Many species when once formed never undergo any further change but become extinct without leaving modified descendants.

Nervous tissue is photosensitive, so a transparent patch of skin will allow a simple chordate to detect shadows.

Crude focusing will allow a simple eye to detect rough shapes, an improvement over mere photosensitivity.

A good lens and well-defined retina would allow vision of the entire environment, selection favoring improved acuity.

Though the vertebrate eye is amazingly good, less efficient forms of photosensitivity would also be favored by natural selection. The inherent photosensitivity of neural tissue allows the initial evolution of crude photosensitivity.

FIGURE 1.21A The eye illustrates the evolution of "organs of extreme perfection."

Initial distribution

Final distribution

FIGURE 1.21B Species that are related to each other but widely dispersed can be explained by the loss of intermediate populations.

1.22 The Darwinian universe and the organisms within it undergo materially important change

Corresponding to the Platonic notion of merely superficial variation is the idea of merely superficial change. If there is indeed nothing new under the sun, then understanding life on Earth is only a question of cataloging all the unchanging species, together with all their essential characteristics. With this model for science, apparent change in the universe is only the shuffling of unchanging objects in an essentially unchanging world, as shown in Figure 1.22A.

The Darwinian universe is radically different. Though there are constraints on the kind and degree of change allowed in Darwinian evolution, the principle of material change is fundamental. In the Darwinian universe, species are allowed

Though there are constraints on the kind and degree of change allowed in Darwinian evolution, the principle of material change is fundamental.

to evolve. Small dog-sized herbivores can evolve into modern horses that are much larger and faster than their distant ancestors. Shrewlike mammals of 70 million years ago have descendants that are enormous whales, fleet bats, and humans that play chess. Entire groups of organisms, like dinosaurs, evolved and flourished over more than 100 million years, only to be wiped out in a few million years of cataclysm. This principle of material change is sketched in Figure 1.22B.

But it is not only the gross anatomy of living things that changes in the Darwinian universe. Fine morphology and molecular biology evolve too. A gill arch in the fishes of several hundred million years ago has evolved into the middle

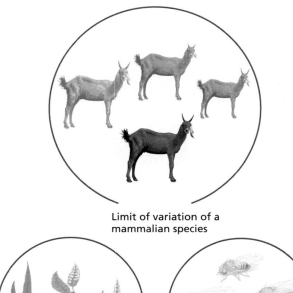

Limit of variation of a mammalian species

Limit of variation of a plant species

Limit of variation of an insect species

Members of biological species vary, but without any essential changes—merely superficial changes, such as incidental variation.

FIGURE 1.22A Western Idea of the Universe before Darwin Biological species never really change.

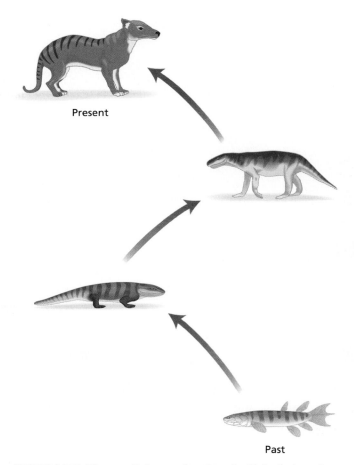

Present

Past

FIGURE 1.22B Western Universe after Darwin Biological species undergo important changes through evolution over long periods of time.

ear of their mammalian descendants. Proteins that serve one function in bacteria evolve new functions in plants. Proteins such as globins are duplicated, descendant proteins taking on new functions in a more complex respiratory molecule, hemoglobin.

The depth, panoply, and scale of Darwinian change present a devastating rebuttal to the idea that there is nothing new under the sun. The history of life has been full of change, reversals, parallelisms, and upheavals. Even our own biological history is a hallmark of evolutionary change. Few gross morphological changes have occurred as rapidly as the expansion of the human brain over the past 2 million years. Darwinism shows us that the spectrum of life is not a catalog but a giant novel, full of incident and surprise. ❖

1.23 Despite great need, it took a long time for biology to be transformed by the Darwinian revolution

Before Darwin, the biological sciences had an abundance of minor theories and unreliable data. On the theoretical side, some Greeks of classical times used ideas about internal heat, drying, and coldness to explain growth, maturation, disease, and aging. In the late middle ages, the intellectual followers of Aristotle used elaborate theories of vital forces to explain all living processes, these theories being couched in terms of various types of "will" to be found in nature, in both animate and inanimate things. Gravity, for example, was explained as a will existing in many objects that made them want to return to the Earth. The quality of biological thought just before Darwin was not radically improved over that of Plato or Aristotle, even though the quantity of biological information was much greater.

Nor was the idea of an experimental method important to Darwin's breakthrough. Darwin did perform some experiments, particularly in plant breeding, but these came long after his theory of evolution was well developed. The young Darwin, the Darwin who developed evolutionary theory, was no more of an experimentalist than Aristotle was, perhaps less.

These points raise the question, why did biology have to wait so long for Charles Darwin to revolutionize the field? There are at least three possible reasons that the theory of evolution by natural selection took so long to arise.

The first is that the basic ideas of species and adaptation needed to be developed. These ideas are present in classical Greek writings, but the work of European biologists after the Renaissance clarified them. In particular, the idea of an organism's suitedness to its natural habitat was central to European biology before Darwin.

A second possible reason the world had to wait for Darwin is that many of the ideas in his theory of evolution by natural selection were taken from eighteenth-century economics. Darwin transposed these ideas, both wittingly and unwittingly, to biology.

A third possible reason for the long wait may simply be luck. Darwin had several predecessors whose ideas were very close to his. For example, in 1831 Patrick Matthew published an obscure book on naval arboriculture that had an appendix on evolution by natural selection, but it was completely ignored. And there were others like him. Perhaps Darwin was just the first of a group of independent discoverers to develop evolutionary ideas enough to attract general attention. The very similar ideas of A. R. Wallace support this interpretation.

Whatever the source of Darwin's precedence, the Darwinian view of life transformed biology. Before Darwin, God was the most important explanatory principle of biology. Living things were manifestations of God's Book of Creation, testimonies to His Grace and Perfection.

After Darwin, biology was freed from theological components—or rather, it had the potential to be free of such components. To this day, some biologists are determined to interpret life based on a deity. However, these individuals are decreasing in number among the ranks of biologists. The burgeoning of Darwinian thinking is arrested only when powerful political regimes seek to impose particular religious or ideological views. Then, Darwinists are killed for their beliefs. This happened to population geneticists under Stalin's regime, even though Darwin was a hero in the Soviet Union. At that time, Stalin was interested in Lamarckian research, making him hostile to the genetically-based form of Darwinism that was established by the modern synthesis of genetics and evolutionary theory after 1930. But in the developed countries of Europe, North America, Japan, and Australasia opposition to Darwinism is less extreme. Australia even has a town named Darwin. To get some flavor of the worldwide view of Darwinism today, see the accompanying box. ❖

The world celebrated the centennial of Darwin's death in 1982, and the celebrations were various and surprising. Within the English-speaking world, Darwin's intellectual home, Darwin was feted as a great scientist by university professors and scientists in decorous symposia that were later published in thick hardcover books. This is the way American evolutionary biologists expected Darwin's centennial to proceed around the world.

But it didn't. Behind the then-Iron Curtain, in the Soviet Bloc countries, Darwin was celebrated as an official hero of communism. Because Karl Marx regarded Darwin's theories as the starting point for his "dialectical materialism," Darwin's vast scientific standing was exploited by the Soviets to bolster their ideological system only a few years before its collapse.

Perhaps even more amazing for biologists were the events in Italy and Spain. In Florence, for example, a conference on the Darwinian Heritage was organized by nonscientists. The Gramsci Institute, a political organization, organized civic celebrations in a variety of Italian cities. In Genoa alone, organizations ranging from the provincial government to an association of shipyard workers produced three lecture series, three roundtable discussions, six pamphlets, and five television broadcasts. The national weekly magazine *L'espresso* published a special supplement on the centennial in April 1982. In Spain, Barcelona held many events open to the public—a museum exhibit, a lecture series, a film series, and an exhibit of books. Comparable events occurred in other European countries.

Darwin has become a cultural hero, especially among those who reject priests and monarchs. Antiauthoritarians in many countries use Darwin as a buttress for their rejection of traditional Western religious and philosophical thought.

As a final note, Figure 1.23A shows some Darwin memorabilia, his scientific tools in particular.

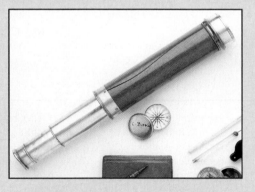

FIGURE 1.23A Scientific Instruments and Other Darwin Memorabilia

SUMMARY

1. Evolutionary biology, ecology, and organismal biology were founded largely by one man, Charles Darwin. Darwin came from the English landed gentry, and his father intended that Charles would study medicine. But Charles ended up as the naturalist aboard a ship that circumnavigated the world. On his return, Darwin stumbled on the idea of evolutionary change. Later he developed the concept of natural selection as a mechanistic explanation of evolution. These two concepts were Darwin's focus, and his legacy to all biologists.

2. The theory of natural selection is founded on ecological ideas that Darwin got from Thomas Robert Malthus, especially the tendency of all organisms to overpopulate. Darwin reasoned that there must always be a check to overpopulation, and this check will sometimes cause differential survival of those best suited to the environment, especially the hardy. This differential survival will change the composition of the parents of the next generation. Because offspring tend to resemble their parents, the offspring of the next generation will change, too. Thus evolutionary change can occur as a result of a natural selective process.

3. Darwin's theory of evolution by natural selection had a number of important features that would shape the future of biology. It was a materialistic theory that did not require the guiding hand of a deity or other higher intelligence. It was a gradualist theory. Abrupt change was not predicted. It was not a globally directed theory; individual species were expected to evolve independently, even when these species interacted with each other. Darwin's evolution was not a single process, affecting all species in a coordinated manner. At its root, it was an ecological theory, and the success of Darwinism made it possible for modern ecology to develop in the twentieth century.

4. At the core of Darwinism is the idea of variation. In Darwin's scheme, living things vary in ways that are materially important for survival and reproduction. This variation then supplies the raw material for the process of natural selection. Such true variation powers a process of true change, shaped by natural selection. In emphasizing variation and change, Darwinism broke with many elements of classical Greek thought.

REVIEW QUESTIONS

1. Why did Charles Darwin abandon the study of medicine?

2. What was Darwin's role on the *Beagle*?

3. Where on the Galápagos Islands did Darwin discover the theory of evolution by natural selection?

4. Did anyone else think of the theory of evolution besides Darwin?

5. Was *Origin of Species* a well-known book in Darwin's lifetime?

6. What views did Darwin share with the geologist Charles Lyell?

7. In Darwin's view of life, what is the general trend of evolution?

8. Why is ecology important for the theory of natural selection?

9. How did Darwin explain the evolution of complex structures like the vertebrate eye?

10. Why were materialism and gradualism important to the development of Darwin's theory of evolution by natural selection?

KEY TERMS

Aristotle
Beagle
catastrophism
Chambers, Robert
creationism
Darwin, Charles
Darwin, Erasmus
Darwinism

evolution
Fitzroy, Robert
Galápagos Islands
gemmule
gradualism
Hume, David
Lamarck, Jean-Baptiste
Lamarckism

Lyell, Charles
Malthus, Thomas Robert
materialism
Mendel, Gregor
Mendelism
natural selection
Newton, Isaac
overpopulation

pangenesis
Plato
Smith, Adam
transmutation
uniformitarian
variation
Wallace, Alfred Russel

FURTHER READINGS

Browne, Janet. 1995. *Charles Darwin: Voyaging*. New York: Knopf.

Browne, Janet. 2002. *Charles Darwin: The Power of Place*. New York: Knopf.

Clark, Ronald W. 1984. *The Survival of Charles Darwin: A Biography of a Man and an Idea*. New York: Random House.

Darwin, Charles R. 1859. *On the Origin of Species by Means of Natural Selection, or The Preservation of Favoured Races in the Struggle for Life*. London: John Murray.

Desmond, Adrian, and James Moore. 1991. *Darwin: The Life of a Tormented Evolutionist*. New York: Warner.

Himmelfarb, Gertrude. 1959. *Darwin and the Darwinian Revolution*. New York: Norton.

Malthus, Thomas Robert. 1798. *An Essay on the Principle of Population*. Repr., New York: Norton, 1976.

Rose, Michael R. 1998. *Darwin's Spectre: Evolutionary Biology in the Modern World*. Princeton, NJ: Princeton University Press.

Wassersug, Richard J., and Michael R. Rose. 1984. "A Reader's Guide and Retrospective to the 1982 Darwin Centennial." *Quarterly Review of Biology* 59:417.

Weiner, Jonathan. 1994. *The Beak of the Finch: A Story of Evolution in Our Time*. New York: Knopf.

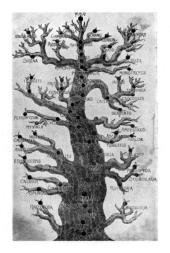

The evolutionary tree unifies biology.

2

Evolutionary Trees in the Ecological Garden

The history of life is magnificent in its depth and in its variety. There have been living things on this planet for almost 4 billion years. In the last billion years, multicellular plant and animal life has evolved. The historical details of that evolution were one of the main interests of biologists over the last century.

Here we present the ideas that biologists use to make sense of biological history. At the core of this group of ideas is the concept of the evolutionary tree. We introduce that concept in depth. Then we connect the evolutionary tree concept to some of the most important historical topics in biology: fossils, biogeography, and development. Finally, we show how to apply the tree concept to one of the

venerable tools of evolutionary biology, the comparative method. The evolutionary tree is one of the most unifying explanatory concepts in biology.

Determining a tree for a group of organisms may be the first step in understanding the evolution of the group. But Darwin's tree concept is also one of the best devices to understand some of the most important findings of biology.

Evolutionary trees do not grow in isolation. Their growth depends in part on the ecological conditions facing the evolving species. We refer to these conditions as the "ecological garden." As in Darwin's original theory of natural selection, which concerned the fate of individual organisms, the fate of species depends on their ecological circumstances. ❖

THE TREE CONCEPT

2.1 The history of life could have followed a variety of patterns, including an absence of evolution

The history of life did not have to grow like a tree. Other patterns might have arisen in the history of living things. We can call these possibilities "systems of life," equivocating as to whether they have to be evolutionary. One system of life, for example, would be reincarnation, in which particular organisms, or even entire species, reappear at intervals. This circular system of life has been common in the thought of south Asian cultures. Before the advent of Christianity in Western civilization, there were various mythological systems of life in western Eurasia and northern Africa. Many specific systems of life were also developed by the aboriginal populations of the Americas and Africa, some very elaborate. But in Western thought, there have been four main systems of life: the traditional Judeo-Christian or Biblical system, the Lamarckian system, the pre-Darwinian system of Lyell, and the Darwinian system. We will look at each of these in turn, beginning with a brief consideration of the first two.

The Biblical or Creationist System of Life Over the last thousand years, most Western scholars accepted the Biblical scheme of creation as the definitive system of life. This system of life is based on the first book of the Bible, Genesis. Genesis describes the creation of life by an omnipotent deity, after that same deity has created the world in which these living things are to live. There is no mention of extinction in this system of life, except possibly at the time of the great flood of Noah, when it is conceivable that terrestrial forms of life might have drowned. On the other hand, Noah's task was to preserve all forms of life that might otherwise have drowned in the great flood, which suggests that they should not have gone extinct.

To a first approximation, then, the Biblical system of life is one in which the original creations of God are supposed to survive indefinitely on Earth. This **creationist** system is shown in Figure 2.1A , in which the bottom of the figure represents the moment of creation, and the top of the figure represents

The present: The surface layer

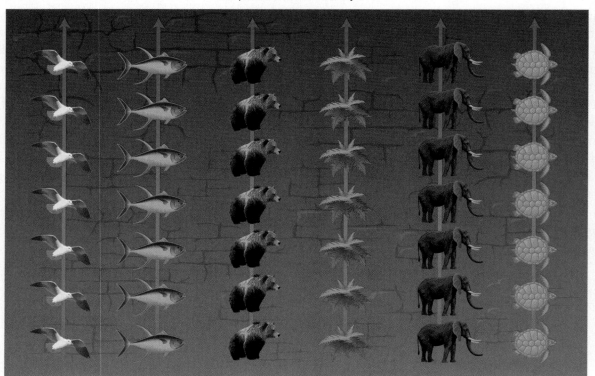

The origin of life: The Creation

Time

FIGURE 2.1A The Creationist System of Life

the present. After the initial creation, no new life-forms arise, and none of the original forms go extinct.

The scientific problems facing this scheme are immense. Much fossil evidence seems to contradict it. Why are there no fossil vertebrates from more than 600 million years ago ? Why are there so many more multicellular fossils in rocks less than 500 million years old, compared with the billion years before that? What happened to the dinosaurs? What happened to the trilobites? Mammoths? Saber-toothed tigers? Extremely few professional biologists still accept the Biblical system of life.

The Lamarckian System of Life

As European scientists found more and more fossils of life-forms that they had never seen before, crude schemes of evolution were proposed in the eighteenth century. The first evolutionary scheme to have much influence was that of Jean-Baptiste Lamarck, introduced in Chapter 1.

Lamarck was not entirely consistent. For one thing, he had several proto-Darwinian ideas that were not formally incorporated in his thinking. But the essential model for the Lamarckian system of evolution can be roughly summarized. It is shown in Figure 2.1B.

The basic Lamarckian concept was "linear" progressive evolution. Lamarck thought that life was frequently produced by spontaneous generation from inanimate precursors. Once a particular lineage came to life, Lamarck thought that it then tended to evolve toward more and more complex forms. New species were not supposed to be created by the splitting of lineages. Interestingly, Lamarck also did not suppose that species went extinct. They were supposed to evolve into new life-forms instead.

With these assumptions, the diversity of species present at any one time was explained by the particular ensemble of lineages evolving in parallel. The features of any one species were explained by Lamarck in terms of its progress toward complexity and its immediate adaptation to the environment in which it lived, as we saw in Chapter 1. This process of adaptation was based on use and disuse, not selection. ❖

The present: The surface layer

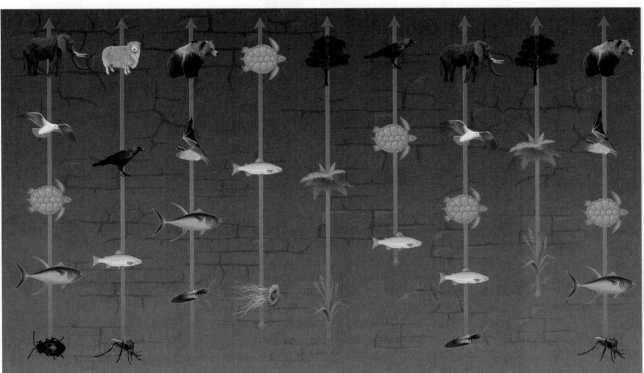

Many origins of life: ◆
No extinctions

Time

FIGURE 2.1B The Lamarckian System of Life

By the nineteenth century, biologists were aware of enough of the fossil history of life to find the Biblical scheme untenable. At the same time, the beneficial contrivances of living things, from their physiology to their anatomy, left many scientists convinced that some type of beneficent creation was required to explain life.

Most students of the fossil record, such as the geologist Charles Lyell, accepted the principle that extinction had occurred. Yet, if extinction occurred, then new species must originate, because there was little evidence that the Earth was progressively losing species. Therefore a consensus developed that the Creator allowed some species to go extinct, with new species being created to take their place, in some way. This Lyellian system of life is shown in Figure 2.2A. This system of life, however, was not an evolutionary scheme. It was not necessarily supposed that species change, once they have been created. An important feature of this system of life was that it faced no difficulties in explaining the premise that species went extinct, because falling numbers of organisms can arise from everyday ecological processes, like exhaustion of food supplies. So the extinction process was not a mystery. It fits the pattern of good gradualist science. It was the origin of new species that was the abiding mystery.

The mystery concerning the origin of new species was Darwin's particular interest. It led him to the fourth major Western system of life. As we saw in Chapter 1, the Darwinian system is a simple materialistic scheme by which life evolves from life without either a cosmic drive to perfection or an intervening deity. ❖

The present: The surface layer

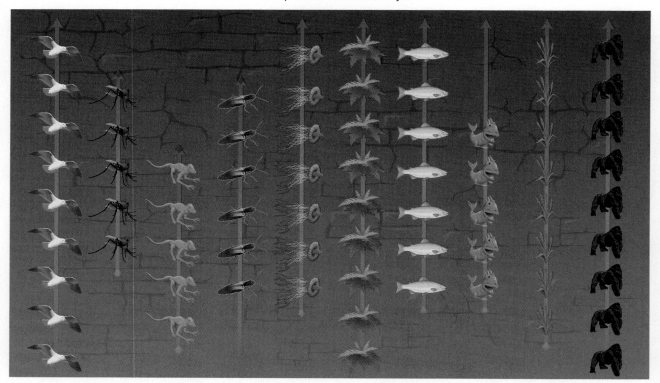

Many orgins of life: ◆
Many extinctions: ◆

Time

FIGURE 2.2A The Lyellian System of Life

The only figure in the *Origin of Species* was an evolutionary tree

We now know that Darwin's first step toward developing his system of life was the concept of an **evolutionary tree**, a representation of the history of life as a treelike pattern of diversification. Despite this, he began the *Origin* with the problem of natural selection. This beginning probably reflected the influence of Lyell, who argued for the preeminence of slow, cumulative change as the foundation of natural processes, geological and biological.

Nonetheless, the centrality of the evolutionary tree is revealed by the fact that the only figure in the *Origin* is an abstract evolutionary tree in Chapter IV, "Natural Selection." This figure is reproduced here as Figure 2.3A. Darwin's explanation of how evolution worked is somewhat disingenuous. He consistently commingles the ideas of natural selection and evolution. Yet he knew very well that evolution is a pattern of change that results from underlying processes, with natural selection one of those processes.

For now, we are going to focus specifically on the pattern of evolution, separately from its mechanism. The bare hypothesis of evolution is enough of a revolution in our understanding of the history of life. We will take up natural selection in detail in Chapter 4, although some general features of its role in evolution are discussed in this chapter.

A sample of Darwin's own explanation of his figure is given in the accompanying box. An important feature of the figure is the species lineages that go extinct. Darwin used no particular notation to indicate extinction, just the termination of a branch of the evolutionary tree. When a new lineage originates, Darwin gave it its own letter of the alphabet; but he indicated successive forms within evolutionary lineages by a superscript only. Note that Darwin did not suppose that all lineages have to diverge into multiple descendant lineages. His "F" lineage never undergoes speciation, but survives over many thousands of generations.

In the following modules, we will unpack the evolutionary tree concept in its modern form and then apply it to some of the central problems of biology. ❖

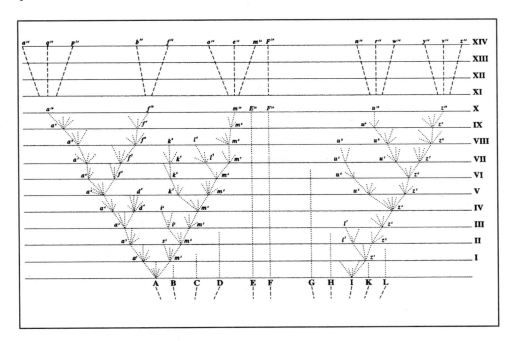

FIGURE 2.3A Darwin's Evolutionary Tree from *Origin of Species*

Darwin's Introduction of the Evolutionary Tree Concept

The accompanying diagram will aid us in understanding this rather perplexing subject. Let A to L represent the species of a genus large in its own country; these species are supposed to resemble each other in unequal degrees, as is so generally the case in nature, and as is represented in the diagram by the letters standing at unequal distances.... Let (A) be a common, widely-diffused, and varying species.... The little fan of diverging dotted lines of unequal lengths proceeding from (A), may repre-sent its varying offspring.... After a thousand generations, species (A) is supposed to have produced two fairly well-marked varieties, namely a^1 and m^1.... the diagram illustrates the steps by which the small differences distinguishing varieties are increased into the larger differences distinguishing species.

—Charles Darwin, 1859, *Origin of Species*
(Chapter IV, "Natural Selection")

2.4 Modern evolutionary trees represent species as growing, splitting, and truncated branches

Modern-day evolutionary biologists use much of Darwin's original reasoning, but they do so following somewhat different conventions and meanings. From this point on, we will focus on modern usage, rather than Darwin's.

Figure 2.4A presents some of the typical graphical conventions built into evolutionary trees. Time flows from the past, at the bottom of a tree, toward the more recent past—even the present, which is located higher up on the diagram. Branches that grow from the bottom of the tree to the top without splitting or truncating, as shown in Figure 2.4B, represent the evolution of individual species. These species may

change morphologically, because evolutionary change does not require the evolution of new species. Or they may remain unchanged. Such unchanging species are sometimes called "fossil species."

When new species evolve from existing or *extant* species, and the extant species survive, the evolutionary event is called **speciation**. Speciation is graphically represented as a fork in a branch, where two or more branches grow out of one branch.

When species go extinct, their branch is truncated. Proceeding from the bottom of Figure 2.4A to the top, the branches representing species that go extinct do not grow to

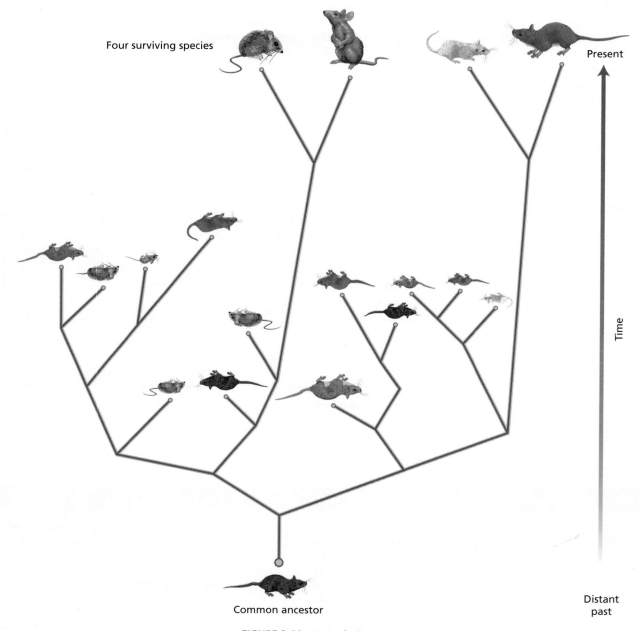

Four surviving species

Present

Time

Common ancestor

Distant past

FIGURE 2.4A An Evolutionary Tree

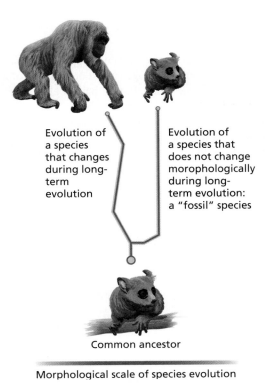

Evolution of a species that changes during long-term evolution

Evolution of a species that does not change morophologically during long-term evolution: a "fossil" species

Common ancestor

Morphological scale of species evolution

FIGURE 2.4B Evolution of Long-Lasting Species

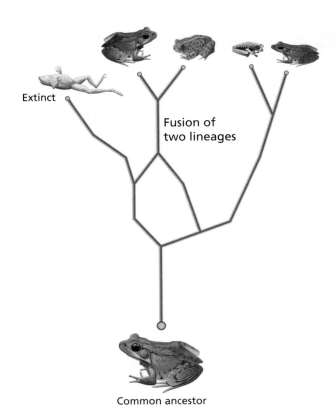

Extinct

Fusion of two lineages

Common ancestor

FIGURE 2.4C Evolutionary Anastomosis, Also Known as Reticulate Evolution, or Branch Fusion

the top of the figure. Extinction "prunes" branches from the evolutionary tree.

Several points should be borne in mind about evolutionary trees. The most important is that trees represent normal ecological and evolutionary process. Extinction still happens

due to failures of survival or reproduction among individual members of a species. Speciation depends on the reproductive biology of individual organisms. We will discuss speciation in Chapter 6. There is no general "forking principle" that explains speciation.

In both real trees and evolutionary trees, two branches can grow together to make interspecies hybrids or symbiotic fusions called **anastomoses** or reticulate evolution. This process is shown in Figure 2.4C. In nature this process is rare among animals, more common among plants and microbes. It is now generally agreed that the eukaryotic cell evolved as a result of repeated symbiotic fusions of simple microbes, an evolutionary event of great importance. But a reasonable bet is to assume that anastomosis does not occur unless there is evidence in its favor. With this qualification, much of your intuition about the ways in which trees grow can be transferred from maples to evolution. ❖

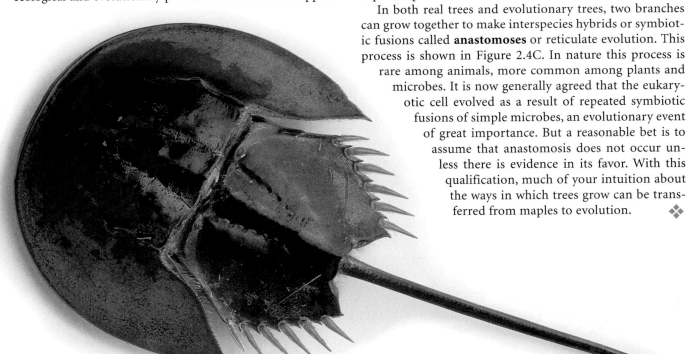

Evolution is often best understood using tree diagrams. How then do we get such diagrams? How do we infer the best evolutionary tree for horses? For insects? For flowering plants?

Often the best way to estimate which tree is best is to use the principle of **maximum parsimony**. This principle is the preference for a simple explanation over a complex one. The maximum parsimony principle is much like the ideas that

motivate the criminal justice system. Suppose you have two suspects. One of them fits an eyewitness description of a man fleeing from the crime scene, but he has an alibi. The other suspect does not fit this description, but is known to have quarreled with the victim, has the victim's blood on his clothes, and has no alibi for the time when the victim died. Most people would guess that the second suspect committed

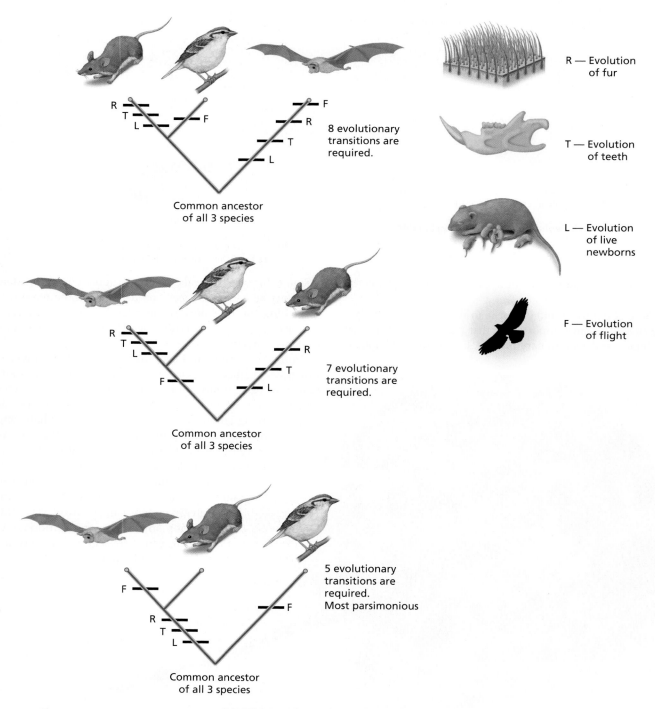

FIGURE 2.5A Alternative Evolutionary Trees

the crime, because there is only one piece of information that doesn't fit him, whereas the first suspect fits the crime profile in only one respect. Otherwise, a complex explanation would have to be provided to explain away the three points of evidence that implicate the second suspect. Such complex explanations are sometimes found in trials. For example, it could be discovered that the first suspect managed to frame the second suspect. In the absence of such exculpatory explanations for incriminating evidence, it is more parsimonious to indict the second suspect.

We can apply this same type of reasoning to evolutionary relationships. Let's say that you are from Mars, studying the animals of Earth. You have used your spaceship's tractor beam to catch three small animals: a mouse, a bat, and a sparrow. Being a Darwinian extraterrestrial, you wonder which of these three species are more closely related. There are three possibilities: mouse and sparrow closely related with bat distantly related; bat and sparrow closely related with mouse distantly related; or mouse and bat closely related with sparrow distantly related. These three possibilities are shown as three different evolutionary trees in Figure 2.5A.

How to decide? Consider a set of five characters , as listed in the table of Figure 2.5B: fur (vs. feathers); live offspring (vs. eggs); flight (vs. walking); warm-blooded (vs. cold-blooded); and teeth (vs. no teeth). All three animals are warm-blooded, so that character gives us no information, and we can ignore it. Obviously, bats and sparrows share the dramatic character of flight. But for three characters (teeth, fur, and live offspring),

only bats and mice have the same attributes.

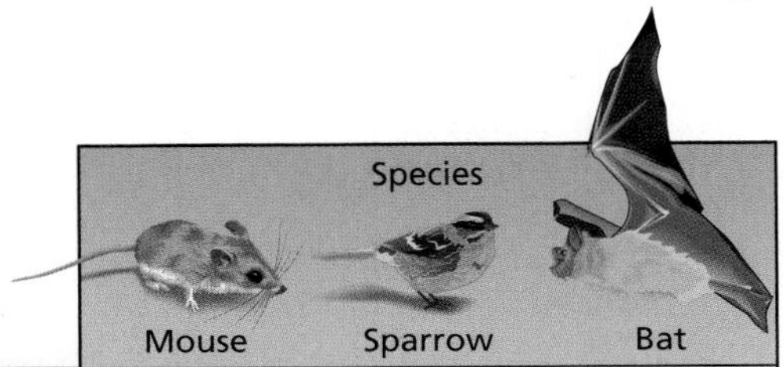

Look again at the evolutionary trees in Figure 2.5A. Note that they include bars representing evolutionary events—such as the evolution of fur—that would be required for each tree. These evolutionary events are required to explain the patterns of each tree. If these evolutionary events did not occur, we would have anomalies in our evolutionary trees, such as sparrows with teeth and fur. The first evolutionary tree, at the top of the diagram, requires eight separate evolutionary events to generate the differences between bat, sparrow, and mouse. The next tree down requires seven evolutionary events. The bottom tree requires only five evolutionary events to account for the observed distribution of characters. That is the most parsimonious tree, so most evolutionary biologists would accept the evolutionary tree in which bats and mice are most closely related.

This example shows us how to determine the most parsimonious tree. (1) We create a matrix that associates particular character states with particular species. (2) We draw all possible evolutionary trees. (3) We place on each of these trees the evolutionary events required to generate the observed matrix of species characteristics. (4) We score the parsimony of an evolutionary tree as the number of distinct evolutionary transitions that have to be hypothesized to generate the observed pattern of differentiation among species (Figure 2.5A gives the number of hypothesized evolutionary transitions required for each tree in our example). (5) We compare the number of evolutionary transitions associated with each tree to discover which tree has the fewest transitions. That tree is the most parsimonious.

It is at least conceivable, however, that the most parsimonious tree is not the correct one. One problem is that an inappropriate collection of characters may have been chosen. For example, the extraterrestrial biologist might have chosen only those characters involved in the evolution of flight, in which case parallel selection for efficient flight in both bat and sparrow might have led to their parsimonious assignment to a common branch of evolution, separate from that of the mouse. In short, trees constructed using maximum parsimony are often better trees, though they may still be incorrect in some of the inferred branches. Adding more species to an evolutionary tree, and more characters, should improve the reliability of a tree.

	Species		
Character		Mouse	Sparrow	Bat
Teeth	yes	no	yes
Flight	no	yes	yes
Warm-blooded	yes	yes	yes
Fur	yes	no	yes
Live offspring	yes	no	yes

FIGURE 2.5B **Character by Species Table for Classification**

SOME IMPORTANT TREES

2.6 The origin of life and the three domains

The origin of life seems like an inherently impossible topic for a scientific inquiry. After all, no biologists were around at the time. But it is precisely under such difficult conditions that tree making can reveal evolutionary history. In this case, the evolutionary history lies at the foundations of life itself.

For a long time, biologists divided life into two main groups: prokaryotes and eukaryotes. **Prokaryotes** are single-celled microbes that lack nuclei and other cell organelles. **Eukaryotes** are both single-celled and multicellular, with nuclei and other organelles. Crudely speaking, prokaryotes were seen as simple and eukaryotes were seen as complex. Naturally, the origin of life was posed as a problem in the early evolution of prokaryotes, with eukaryote evolution seen as the next major step in the evolution of life.

Figure 2.6A provides the crude timeline for the evolution of life on Earth. A striking feature of this timeline is the small amount of time between the origin of Earth and the first evolution of life. Another striking feature of the timeline is the late evolution of animals and plants. Most of the time, the evolution of life has been microbial.

A major area of biological research has concerned how life first evolved. One of the earliest findings was that amino acids and other components of cells arise randomly when water, hydrogen, methane, and ammonia are mixed together and subjected to electrical sparks. This result showed that the carbon molecules that are basic to life probably arose from chemical accidents alone.

Interest has shifted to how simple proteins and nucleic acids formed the first cells. One idea is that ribonucleic acid (RNA) may have been the first molecule of heredity, because it is simpler than deoxyribonucleic acid (DNA) and it can function as a catalyst for chemical reactions, like proteins do. Another proposal is that natural selection could have acted on cell components before full cellular integration, because durable or self-replicating cell constituents would tend to last longer. Finally, it has been suggested that complex cells ultimately evolved because of cooperation among components that evolved separately first.

One of the bigger surprises in the study of early evolution has been the realization that life is divided into three major branches, or **domains**. Two of these were already well known. The **Bacteria** are the familiar prokaryotes—including *Escherichia coli*, which inhabits our guts. The **Eukarya** are the well-known eukaryotes—from protozoa, to fungi, to plants, to animals. The third major domain is made up of the **Archaea**, formerly the archaebacteria. The Archaea have a mix of attributes. Like Bacteria, they have no nuclei or organelles. But they differ from Bacteria in that they have histone proteins associated with their DNA and introns in some genes, like Eukarya. The Archaea have hydrocarbons that are not found in Bacteria or Eukarya, and, in some species, the ability to survive at very high temperatures. An evolutionary tree for the three domains

is shown in Figure 2.6B. Due to the great diversity of each domain, modern tree builders like to identify multiple **kingdoms** within each taxonomic domain. ❖

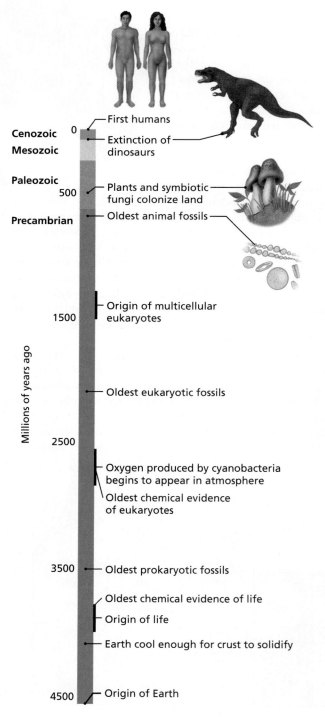

FIGURE 2.6A Timeline of the History of Life

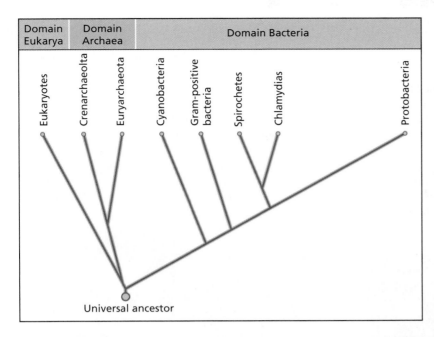

FIGURE 2.6B **Major Branches of the Tree of Life**

The most important example of anastomosis in the evolution of life was the origin of the eukaryotes. For some time it had been noticed that mitochondria, the power plant of the eukaryotic cell, and chloroplasts, the photosynthesizing organelle of plants, were similar to bacterial cells. Figure 2.7A shows pictures of mitochondria and chloroplasts alongside free-living cyanobacteria. Mitochondria and chloroplasts have closed circles of DNA, like bacteria, and they are usually small, like bacteria. These features suggested to some biologists, like Lynn Margulis, that the eukaryotic cell might have evolved from the symbiotic combination of bacterial cells. This was called the **endosymbiont** theory of eukaryotic evolution.

But most biologists were not convinced by the cytology of eukaryotic cells, that is their crude features under a microscope. Fortunately, tree construction would persuade virtually everyone that the endosymbiont theory is correct.

There are two basic possibilities for the position of organelles in a phylogenetic tree. The first is that eukaryotes evolved their organelles from within. All the genes in organelles should then be closely related to the genes found in the nucleus. In this scenario, the eukaryotes evolved using their own resources, with no major contribution from bacteria.

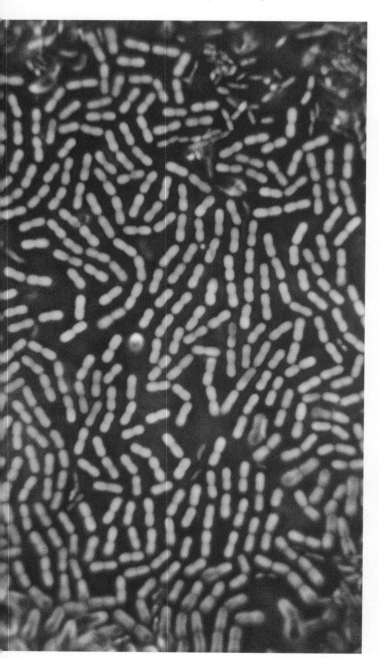

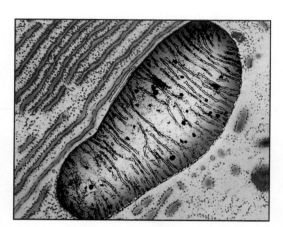

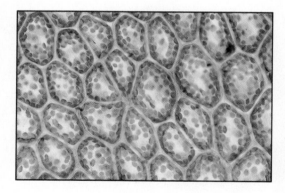

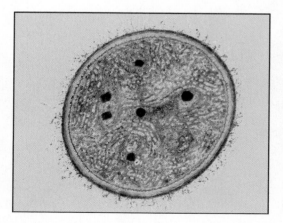

FIGURE 2.7A Mitochondria, Chloroplasts, and Cyanobacteria Related to Chloroplasts, Proceeding from Top to Bottom

The second basic possibility is that eukaryotic organelles were evolutionarily derived from bacterial cells that had originally evolved on their own. If this is correct, then most genes in organelles should be closely related to the genes found in bacteria, not the nuclei of eukaryotes.

Based on tree structure, the first hypothesis would place eukaryotic nuclei and organelle genes on the same branch of the tree of life. The second hypothesis would place eukaryotic nuclei on one main branch, with prokaryote and organelle genomes together on the other branches of the evolutionary tree.

The actual tree is shown in Figure 2.7B. As the figure reveals, eukaryotic nuclei do not share recent ancestors with the genes of their organelles. Instead, those organelles are closely related to bacterial genomes. This finding dramatically supports the endosymobiont theory for the origin of eukaryotes.

A second point that emerges from this result is that the origin of eukaryotes depended critically on symbiosis. This fact suggests that, in general, symbiosis may be as important as competition and predation in the evolution of life. ❖

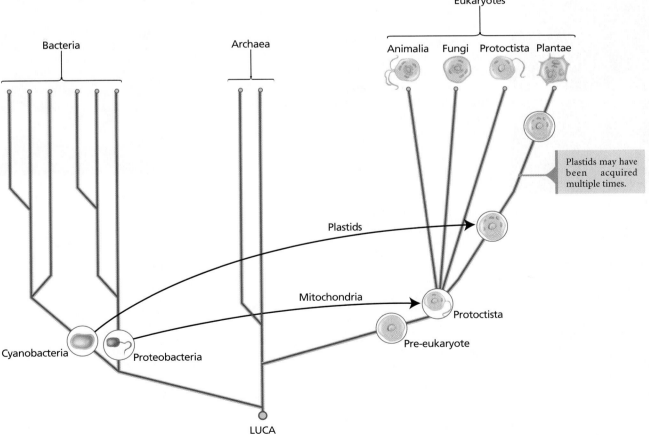

FIGURE 2.7B **Evolutionary Tree of Eukaryotic Nuclei, Bacteria, and Eukaryotic Organelles**

There are many things to learn about prokaryotes. Most important is the considerable diversity among prokaryotic groups. The most profound feature of this diversity is the evolutionary split between Bacteria and Archaea, already discussed. But there is also considerable diversity within these two domains.

Figure 2.8A shows the phylogeny of the Archaea, together with photos of some particular species from this group. The range of environments inhabited by Archaea is their most impressive feature. The **methanogens** live in anaerobic environments, like swamps and animal guts, where they use carbon dioxide and hydrogen to produce energy, with methane as a by-product. They are thereby responsible for the characteristic fragrance that bubbles out of such habitats. **Halophilic** species live in environments with a lot of salt, such as salt lakes. **Thermophiles** live in or near hot springs, on land as well as in oceans. Other Archaea live in less extreme environments.

Figure 2.8B shows the phylogeny of the Bacteria. These are the most abundant prokaryotes. The main groups of Bacteria have been given kingdom status. Each of the bacterial kingdoms is probably more ancient than either the plant or animal kingdoms. Bacteria take on many ecological roles, but they are best known to us as pathogens. *Bdellovibrio* causes cholera; *Helicobacter pylori* infects stomachs, where it causes ulcers; and *Chlamydia trachomatis* causes both venereal and eye diseases. Other bacteria are symbiotic, the best known being the *Rhizobium* bacteria, which take nitrogen from the atmosphere and supply it to the plants in which they live.

Still other bacteria are free living. Some of the extremely small mycoplasma species live in soil. Other free-living bacteria form multicellular fruiting bodies that produce spores. Cyanobacteria are able to photosynthesize. Their relatives evolved into chloroplasts.

An important feature of prokaryotic evolution is that horizontal gene transfer between species is common. Prokaryotes have several forms of parasexuality: **conjugation**, **transduction**, and **transformation**. During conjugation, one cell receives a plasmid from another cell. During transduction, DNA is transmitted between cells by a virus. Transformation occurs when linear molecules of DNA are excreted by one cell and ingested by another cell, where they are recombined with the genome of the recipient cell. All of these forms of sex are considerably less particular than eukaryotic sex. (They are discussed further in Chapter 18.) Useful genes may be passed from one species to another during any of these three parasexual processes. The best-known example of this process is the spread of antibiotic resistance genes among bacteria, a major problem for public health discussed further in Chapter 22. However, the long-term importance of horizontal gene transfer for bacterial evolution is hard to evaluate. Unlike the endosymbiotic origin of eukaryotes, which is now fairly obvious, the exchange of only one or a few genes during bacterial evolution is far harder to detect. But this difficulty of detection does not mean that gene transfer has been unimportant in bacterial evolution. ❖

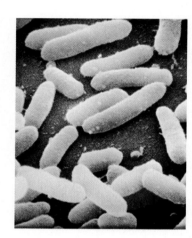

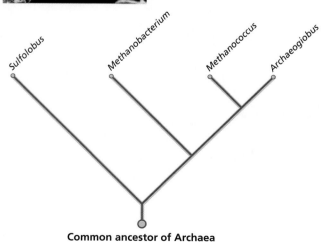

FIGURE 2.8A Evolutionary Tree of the Archaea

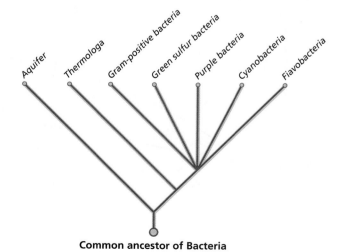

FIGURE 2.8B Evolutionary Tree of the Bacteria

The traditional classification of eukaryotes was a tree based on a protistan base primarily made up of single-celled organisms, with fungi, plants, and animals radiating out from those species. This image is shown in Figure 2.9A. It is almost entirely incorrect.

A better view of evolution in Eukarya is supplied by Figure 2.9B. Based on long-standing evolutionary differentiation, many protist groups are worthy of the designation kingdom. This viewpoint is not too surprising, because there was a long evolutionary period of protistan evolution before animals, fungi, or plants evolved.

The diversity of protistan forms needs to be appreciated. Diplomonads and parabasalids lack mitochondria. However, this may be a secondary loss during the evolution of these species, rather than a failure to acquire mitochondria during evolution. Protistans also vary considerably with respect to their locomotion. Some move their bodies in waves, others have long flagella, and still others have cilia. There are still more variations on these themes. The ciliates are a major group of protists that have cilia. The reproductive systems of protists vary too. Some are asexual (e.g., some *Ameoba* species), some have sexual processes where entire cells fuse to form zygotes (e.g., *Chlamydomonas*), and some ciliate species have a complex sexual process involving different types of nuclei within each cell.

Some of the first land plants to evolve were the bryophytes, which include the liverworts, hornworts, and mosses. These plants lack a complex vascular system and do not produce seeds. Vascular systems evolved later among the ferns and horsetails, among other groups. Seed production evolved among the ancestors of the gymnosperms (e.g., conifers) and angiosperms, the flowering plants. Most plant evolution has occurred during the last 600 million years.

Animal evolution has occurred primarily during the last billion years or so. Although animal evolution is commonly taught in relation to the evolution of the main vertebrate groups (jawless fish, cartilaginous fish, bony fish, amphibians, reptiles, birds, and mammals), vertebrates are actually a

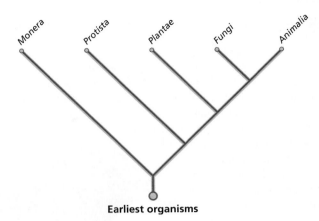

FIGURE 2.9A Historical Division of Eukaryotes into Protozoa, Fungi, Plants, and Animals

minor group of animals, with few species and few individual organisms. The most abundant animal group is the arthropods, made up of insects, spiders, crustaceans, and so on. There are far more arthropod species than vertebrate species.

Finally, we come to the fungi, once grouped with plants and now recognized as an independent group of organisms. Fungi are primarily known for their major role in decomposition. This sounds like a relatively passive ecological role. But some fungi actively catch animals for food, while still other species have flagella and can locomote. A number of fungi attack plants, including agricultural crops. In humans, fungal infections cause irritating disorders like athlete's foot as well as persistent lung infections that can kill.

The spectrum of eukaryotic life is vast. Students of biology have to be careful not to let their biases and practical interests distract from the important patterns of evolution. However interesting we may find dinosaurs, they were a tiny part of the tree of life. There have been far more parasitic wasp species than dinosaur species in the history of life, and those parasitic wasps have probably been more important in the ecology of the planet.

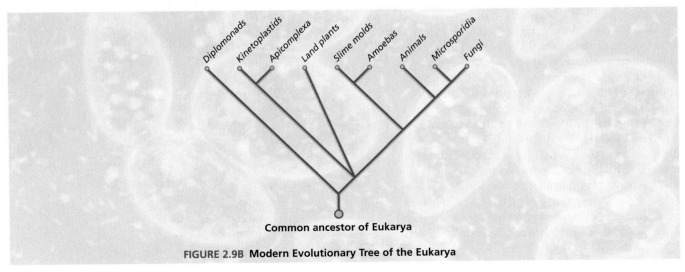

FIGURE 2.9B Modern Evolutionary Tree of the Eukarya

USING TREES TO STUDY EVOLUTION

2.10 The classification of species can be explained elegantly with Darwin's evolutionary tree concept

The classification of species has roots in classical Greek thinking, especially the theory of ideal types that we know best from Plato. Originally classification was an entirely non-evolutionary tool for grouping living things. But it provides us with some of the best evidence in favor of evolution.

It is an obvious fact about life that organisms can be grouped together. A large number of birds weigh several pounds, have entirely black feathers, and call with a distinctive sound, which can be rendered roughly as "caw! caw!" In everyday English, these birds are called crows (Figure 2.10A). We will see hundreds of them in our lives. There are a number of species of crow, from carrion crows to ravens to rooks. But they all have easily recognizable plumage, shape, and call.

To a Platonic thinker, these birds all share an inner "crowness." For this reason, all these birds are grouped together in the genus *Corvus*. Different species, such as *Corvus graculous*, are recognized for their differences from other crows. Crows are also examples of birds, and for that reason they are placed in the taxonomic class called Aves, which brings together all animals with two walking feet, wings (usable for flight or not), and feathers.

There are several notable things about conventional **biological classification**: (1) It groups very similar organisms into a species. (2) It goes on to cluster species together in larger and larger aggregations: genus, family, order, class, phylum, and **kingdom**. As shown in Figure 2.10B, this clustering gives a kind of Chinese box structure to biological classification. (3) Some of the affiliations of organisms become fairly subtle. For example, humans, crocodiles, and the wormlike *Amphioxus* (Figure 2.10C) are all grouped as members of the Phylum Chordata. This grouping is based on a structure called a *notochord*, a proto-backbone that appears in the development of chordates. This obscure structure is not as obvious as the milk of a mammal or the feathers of a bird. (4) This classification approach is not based on evolutionary theory. Indeed, there is no deep theory to traditional biological classification. It is based entirely on mere similarity.

The incredible thing about biological classification is that it works so often. Organisms do cluster fairly well, with further clustering within larger clusters. There are no bats with feathers. There are no insects with calcified bones. Living things obey unseen rules that make classification work. Why?

The answer is evolution, the tree of life. As shown in Figure 2.10D, we can make sense of the clustering of organisms in relation to their descent from **common ancestors**. In the figure, each genus of the imaginary Family Treeidae shares a common ancestor. Common features of the members of the Genus Luckius that are not shared with other members of the Family Treeidae can be explained evolutionarily as deriving from the features of the common ancestor of

FIGURE 2.10A A Crow

that particular genus. Thus a Darwinian would explain unique characters that are shared among groups of organisms as a result of derivation from a common ancestor that possessed those characters. The descendants of such common ancestors then evolve into multiple descendant species that differentiate from each other in many respects, yet share features that were retained from their common ancestor.

In other words, biological classification is a reflection of evolutionary history. Because evolutionary history follows Darwinian rules, some of the critical assumptions of biological classification are guaranteed. Species are not members of more than one genus, family, order, class, or phylum. And multiple characters can be used to identify organisms as belonging to species, genera, families, and so on.

The explanation of biological classification is an application of the evolutionary tree concept that supports the foundations of the science of biology. Before the tree concept came along, biological classification was based on convenient similarities. Now biological classification actually makes sense. ❖

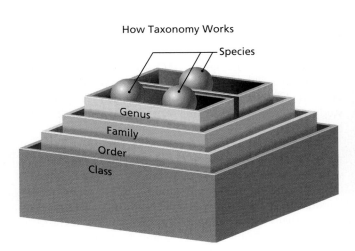

FIGURE 2.10B **Taxonomy Works like Chinese Boxes**

FIGURE 2.10C *Amphioxus* **in Its Natural habitat**

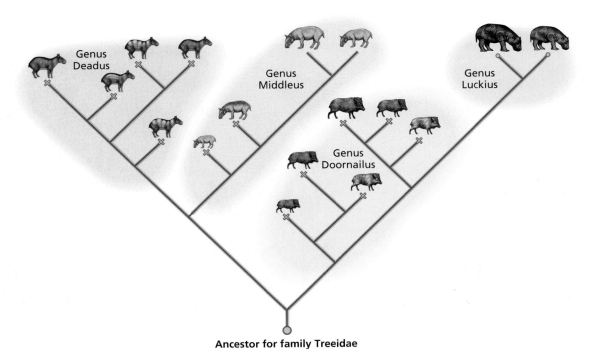

Ancestor for family Treeidae

FIGURE 2.10D **Trees and Taxonomy**

2.11 Fossil differentiation often follows tree patterns

Fossil specimens are used to construct evolutionary trees, when they are available. The hard structures of many animals can become fossils. That is, sedimentary and other slow geological processes convert shells and bones into rocks that retain the original shape of these hard structures. In some rare cases, geological events allow the fossilization of softer body parts. This occurred at the Burgess Shale site in the Canadian Rockies (Figure 2.11A). This one fossil deposit has told us more about the early evolution of animals than any other. Some illustrations of these animals are shown on the book cover. But for most animals, only thick shells, like those of mollusks, or bones, like those of vertebrates, fossilize.

In taxonomic groups with hard structures, we can follow the evolution of these hard structures using reasoning based on the tree concept. With abundant fossil specimens, it is often possible to reconstruct in detail the evolutionary sequences linking one species with another, especially when the fossils can be accurately dated. Entirely novel structures may evolve, allowing us to identify either the branching off of new species or profound evolutionary change within an unbranching part of the evolutionary tree.

Evolutionary biologists sometimes physically array fossil specimens into tree patterns. Consider Figure 2.11B. The fossils placed at the bottom of the evolutionary tree come from the oldest, usually deepest, fossil strata or layers. Species that evolved somewhat later come from the middle layers of geological deposits. The species that have evolved most recently will either come from the fossils nearest the surface of the Earth, or they will be represented by living organisms.

Fossils can reveal specific features of the evolutionary process. Instead of speculating about the properties of common ancestors based on present-day forms of life, evolutionary biologists may be able to examine fossils from deep strata that show hypothesized features. Such fossils provide a glimpse of the evolutionary process at a point in time when the common ancestor might have lived.

Fossils can also reveal the speed with which hard structures evolve. One of the more spectacular examples of rapid evolution is the expansion of the human cranium over the last 2 million years. The human skull is an excellent structure for fossil studies, because it fossilizes well. On the other hand, humans have only recently become abundant. For most of our evolution, humans were relatively rare animals—compared to pigs, for example—thus impeding the search for our fossils. Figure 2.11C gives a human evolutionary tree. (We will consider this example of evolution in some detail in Chapter 21.) Paleontologists, the scientists who study fossils, have many interesting stories to tell about human evolution.

Dinosaurs and humans are the most famous concerns of **paleontology**, but there are many other taxonomic groups with which we have made considerable scientific progress. One of the most-studied groups is the hard-shelled Foraminifera. These small oceanic creatures make extremely durable shells that are among the most common microfossils in the world. Ironically, the seemingly obscure study of "foram" fossils is valuable in prospecting for oil, one of the most remunerative businesses of modern times. ❖

FIGURE 2.11A **The Animals of the Burgess Shale, a Traditional View**

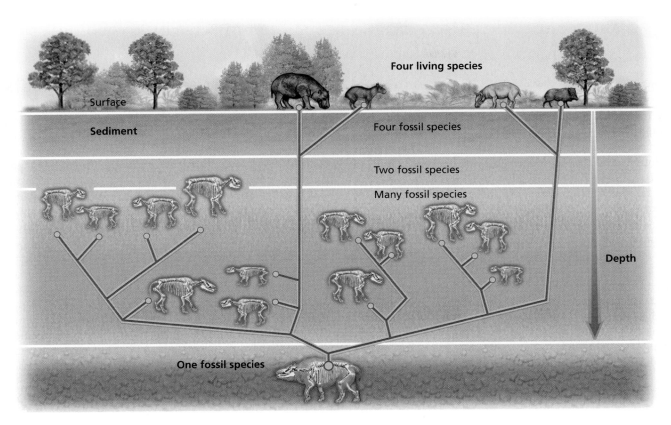

FIGURE 2.11B Trees and Fossils

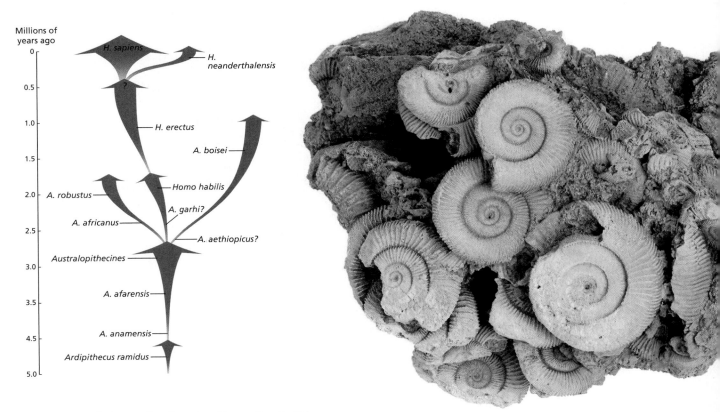

FIGURE 2.11C **Evolutionary Tree Based on Human Fossils** This tree is revised frequently by paleoanthropologists.

2.12 Biogeographic patterns can be explained in relation to geology, migration, and evolutionary history

All evolution is local. New species evolve from already existing species at particular times and places. This small point has important implications for our understanding of the geography of life.

In some cases, the biology of evolving species obscures the evolutionary basis of **biogeography**, the distribution of species over the Earth. Birds that have evolved in one locale may be able to fly over most of the Earth's surface. Albatross (species of the genus *Diomedea*), for example, are very strong long-distance fliers, able to travel over thousands of miles of ocean, so their present-day distribution on Earth does not reveal much about where they evolved. Likewise, some moths and butterflies can fly thousands of miles during their migrations. The monarch butterfly (*Danaus plexippus*), for example, seasonally migrates between Mexico, the midwestern and northeastern areas of North America, and the southern states of the United States. Oceanic species, like some of the whales, swim between the poles, as do many bony fishes. Even plants may have pollen that can be carried on winds over hundreds of

miles. The ubiquity of the dandelion (*Taraxacum officinalle*) is just one example of the effects of aerial dispersal. Even if evolution is local, dispersal may obscure that fact. (The biology and impact of dispersal is considered further in Chapter 11.)

But some species reveal their evolutionary history in their distribution—especially their distribution before humans started deliberately and accidentally dispersing species from continent to continent. The classic example is the abundance of marsupial mammals in Australia. Elsewhere, marsupials are extremely rare, the most notable group outside of Australia being the opossums of America (*Didelphis marsupialis*). Before humans arrived in Australia some 30,000 to 60,000 years ago, the only significant group of placental mammals on that continent were the bats (Order Chiroptera), which could have reached Australia readily by flight.

How can we explain the biogeography of this distribution? The standard evolutionary explanation involves the isolation of Australia due to continental drift. Marsupial mammals evolved before placental mammals and spread over the surface

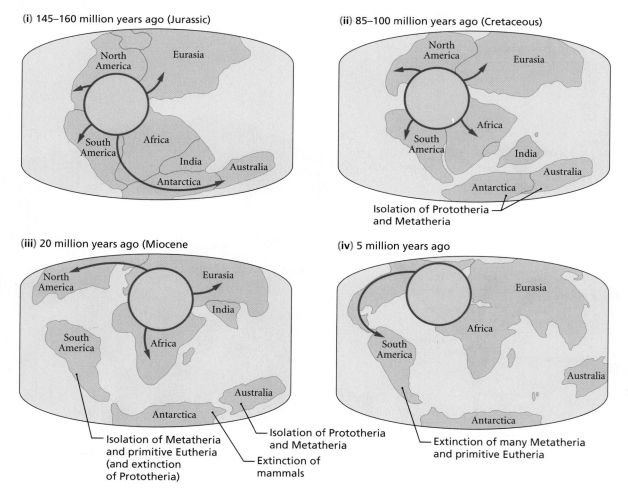

FIGURE 2.12A The Biogeography of Mammalian Evolution   The disk shows the main line of mammalian evolution that eventually produced the Eutheria, the placental mammals.

of the world when the continents had not drifted too far apart. Placental mammals evolved later, but the terrestrial mammal species could not reach Australia due to the absence of land bridges—such as those that have connected all the other continents—in the last 50 million years. The isolation of Australia protected the marsupials there from ecological elimination by the placental mammals. This biogeographical hypothesis is illustrated in Figure 2.12A.

The ultimate corroboration of this model is supplied by the animals of New Zealand. Before humans arrived around 800 years ago, the fossil record of New Zealand shows no evidence of colonization by any type of terrestrial mammal. Instead, the birds that reached New Zealand apparently evolved to take on ecological roles similar to those of terrestrial mammals. One example of this is nighttime ground foraging by kiwis (genus *Apteryx*) instead of rodents. A number of New Zealand bird species are completely flightless (see Figure 2.12B). Some of these birds were small and timorous, like kiwis. Others were quite large, like the moa, a group of bird species driven to extinction by the first human colonists of New Zealand, the Maori. Some moas weighed several hundred pounds, remarkable for a bird. The ecology and evolution of the moa must have hinged on the complete absence of terrestrial mammals in New Zealand. The extinction of the moa after human colonization and the absence of birds like them elsewhere in the world suggests that moas could not have survived with dogs, cattle, and monkeys as competitors. Ostriches, for example, are not as large as the big moa species and probably much better at running.

Another way to look at biogeography is to realize that it shows the imperfection of evolution by natural selection. By accidents both constraining and fortuitous, new species evolve from other local species in response to local ecological conditions. Evolution works with the materials at hand, like the birds of New Zealand, not with the species best suited to the ecological circumstances—presumably, placental mammals in the case of New Zealand. If evolution were perfect, it is arguable that the kiwis should have been rodents. But we don't expect evolution to be perfect. ❖

FIGURE 2.12B **Flightless Birds of New Zealand**

2.13 Developmental patterns can be explained using evolutionary trees

Figure 2.13A, below, shows one of the famous features of life. The embryos of many groups of organisms often show considerable similarity when they are very young. The example shown is a drawing of vertebrate embryonic development. (Illustrations of this kind are not regarded as entirely reliable by biologists, but they were important in convincing biologists of the conservation of patterns of development.) This is one of the most surprising facts in biology, if you are not an evolutionary biologist.

Darwin offered an explanation for this pattern in the *Origin*. His argument distinguished two cases. In the first case, the embryo lies within a protective structure during development. An eggshell is an example of such a protective structure, and so is the uterus of a placental mammal. Another example is a seed. In all these cases, the developing embryo will not be shaped by natural selection for a particular type of locomotion, feeding, or growth. Therefore, there will be no immediate selection for the embryo to resemble the adult. The embryonic stage is not shaped by the selection imposed on the free-living, unprotected life stages. Therefore, natural selection is likely to leave the embryo relatively unchanged from the features in the common ancestor of the taxonomic group to which the embryo's species belongs.

One way to look at the embryos of vertebrates as diverse as fish, reptiles, birds, and mammals is that all these embryos develop in protected environments, whether inside an egg or inside a uterus. These embryos live in a manner that is not as different from that of the ancestral wormlike aquatic chordate as the ways of life of many vertebrates today, which may fly, run on legs, and so forth. Thus Darwin essentially argued that many embryos reflect ancestral features largely unchanged by natural selection.

The key test of Darwin's interpretation, as he realized himself, lay in the second case that he distinguished—the embryology of species in which larvae are free living. Free-living larvae may be subject to selection for feeding, avoiding predation or herbivory (if they are plants), and so on. Under these conditions, Darwin predicted that natural selection would tend to obscure an organism's evolutionary ancestry.

A good example of this principle is the larval stages of butterflies, moths, flies, and other insects. Most larval development in these species takes place during a free-living stage in which food must be found and eaten, predators avoided, and so on. Caterpillars, for example, take on a wide range of sizes, coloration, "hairiness," and so on. Instead of

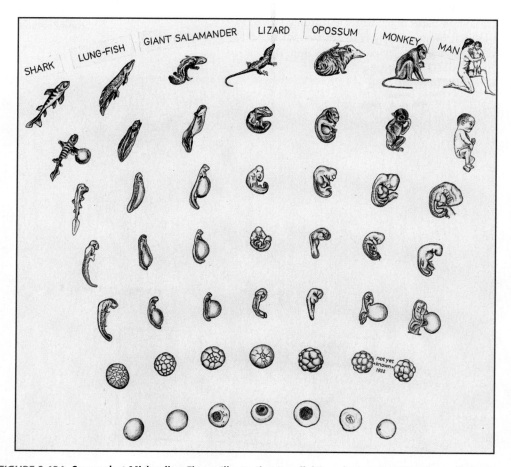

FIGURE 2.13A Somewhat Misleading Figure Illustrating Parallel Development in Vertebrate Embryos

FIGURE 2.13B **Trees and Development** Evolutionary history leaves traces in development.

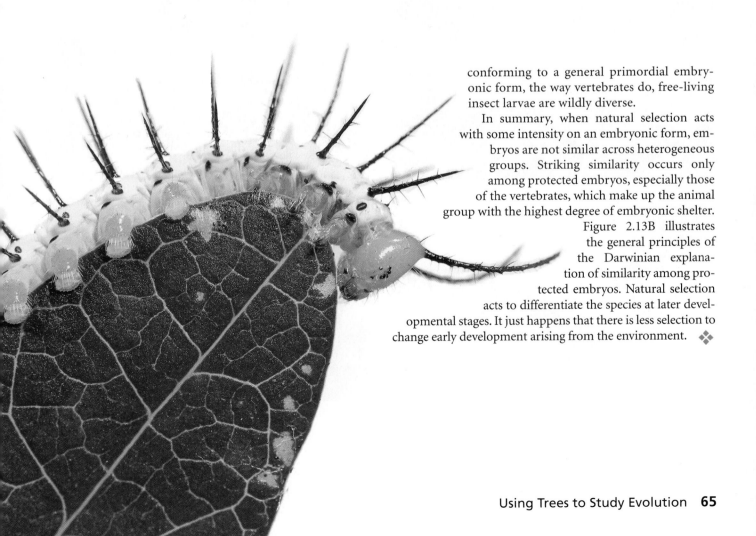

conforming to a general primordial embryonic form, the way vertebrates do, free-living insect larvae are wildly diverse.

In summary, when natural selection acts with some intensity on an embryonic form, embryos are not similar across heterogeneous groups. Striking similarity occurs only among protected embryos, especially those of the vertebrates, which make up the animal group with the highest degree of embryonic shelter. Figure 2.13B illustrates the general principles of the Darwinian explanation of similarity among protected embryos. Natural selection acts to differentiate the species at later developmental stages. It just happens that there is less selection to change early development arising from the environment. ❖

THE COMPARATIVE METHOD

2.14 The comparative method uses the pattern of adaptation among species and their environments to infer the evolutionary causes of particular adaptations

The **comparative method** is one of the venerable parts of evolutionary biology. Indeed, it is such an obvious biological idea that ancient Greek scientists, such as Aristotle, often used it, despite the lack of scientific foundations for it in Greek science. Closely allied to the comparative method is the idea of **adaptation**. Adaptations are the products of natural selection, while adaptation is the response to the process of natural selection. Normally these two meanings of the term *adaptation* are quite close, but sometimes they are different.

Nothing is more natural than explaining the features of organisms as adaptations produced by natural selection. The wings of flying animals are adaptations that permit efficient flight. The roots of plants are adaptations that extract moisture from the ground. The sticky surfaces of pathogens enable these organisms to bind to the tissues of their hosts. There is really no limit to our ability to invent stories of this kind, whether they are right or wrong.

The problem is that it is not obvious how to separate correct Darwinian explanations from incorrect ones. For example, it might be supposed that six legs are adaptations to insect life, which explains why almost all adult insects have six legs instead of two, four, or ten. Then the eight legs of spiders could be explained as particularly well suited to the spider way of life. But are these really valid explanations? Can every feature of life be explained well by the comparative method?

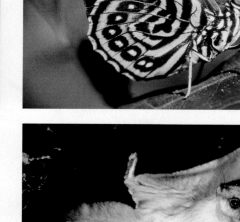

FIGURE 2.14A Wings and Related Structures Used in Flight and Gliding

Take coughing as an example. It is easy to speculate that coughing is an adaptation produced by natural selection on humans that enables us to clear our throats and lungs of debris when food has entered our respiratory tract. We cough to get the food out. But we cough even more when we have viral and other respiratory tract infections; and some of this coughing is "dry hacking," which produces very little fluid or mucus. Some biologists have suggested that behavior such as coughing may be produced by microbes to help them spread to new hosts. Or it could be that we are attempting to remove microbes from our respiratory tract, especially our lungs, to forestall suffocation. It is not our concern, right now, to decide between these explanations for coughing. Both could be right. (We discuss this type of problem further in Chapter 22.) But they illustrate the problem that there may be more than one way to use the concept of natural selection to generate a reasonable explanation for an organismal feature.

The tool that biologists have tried to use in resolving this problem is the comparative method. With the comparative method, as it was traditionally employed, biologists sought to associate particular environments with both natural selection and adaptations. To use a fairly obvious example, a biologist might note that animals that fly typically have morphological structures that produce aerodynamic lift forces. This is true of bird and bat wings. It is also true of insect wings, flying squirrel fur, and the highly modified fins of flying fish. Even some plant seeds have structures that supply aerodynamic lift. All of the structures shown in Figure 2.14A are associated with flight.

To finish off a traditional comparative analysis, a biologist might then point to moles, flightless insects, plants that produce immobile seeds, and other nonflying species, noting that these species do not have structures like wings. This observation would be offered as comparative evidence that wings are indeed adaptations for flight in species whose way of life depends on such locomotion.

But what about the exceptions? Some birds, like ostriches and penguins, have wings but do not fly (see box). Perhaps wings are used for things other than flight. For example, many birds and insects use wings as part of their courtship, the right motion being required to arouse the opposite sex. Some male insects even use their wings to produce a humming sound during mating. In these cases, is the wing still "for" flight? ❖

Why Does the Ostrich Still Have Wings?

One solution to the problem of why ostriches still have wings is to suppose that the development of birds tends to lead to the growth of wings. Originally, wings might have evolved as an adaptation for flight. Or they might already have existed for other reasons, such as selection for body covering, and then later were selected for use in flight. Either way, we might suppose that the common ancestor of birds evolved wings that were efficient for flight, so that almost all members of the bird, or avian, group used wings for flight. Now suppose that some birds became so large that they couldn't fly anymore. Suppose that they lived in dense jungles where flight was not efficient. They might still keep their wings, possibly much smaller than before. In other words, wings might be vestiges of earlier evolution.

There are three ways in which evolutionary history can account for wings in flightless animals:

1. The wings could be used for something other than flight. Courtship has already been mentioned as an alternative use for wings, a use that would be important for reproduction. This use might have been established before the loss of flight.

2. Developmental changes that would eliminate wings might have detrimental secondary effects, such as bad leg development. In other words, wings aren't useful anymore, but getting rid of them creates other problems .

3. Having wings is just irrelevant to natural selection , so they are kept for no functional reason. This type of evolution is discussed more in Chapter 3.

Whichever pattern explains the wings of the ostrich, note that all of these hypotheses hinge on the interaction between history and selection in the comparative biology of adaptations. History can provide a resolution. Our next interest is how historical information can be efficiently and objectively incorporated in comparisons of species.

2.15 Evolutionary trees can be used to test hypotheses of adaptation objectively

The comparative method can be used objectively to test hypotheses of adaptation. The key is to employ evolutionary trees. Good evolutionary trees define the biological history of a taxonomic group. From these trees we can easily see how history makes the adaptations of different species depend on the derivation of these species from common ancestors.

Consider Figure 2.15A. If we have four species of a particular type, say four primitive birds, they could all have evolved wings for flight independently. This first possibility is shown in the tree to the left in Figure 2.15A. Or, the four bird species could all have descended from an ancestral bird species that had already evolved flight. This second possibility is shown in the tree on the right in Figure 2.15A. In the first case, winged flight evolves four times. In the second case, winged flight evolves only once. Evidently the case where flight evolves once is more parsimonious. But that is not our concern here.

Evolution of flight

Common ancestor
Parallel evolution

Common ancestor
Ancestral evolution

FIGURE 2.15A **Trees and Adaptation**

When these evolutionary tree patterns are used comparatively, the correct choice of tree is very important. For example, if the ancestral species changed its ecology from ground dwelling to tree dwelling, then its evolution of wings for flight might be explained by living in trees. On the other hand, if the ancestral species moved into trees but did *not* evolve flight, with winged flight evolving in the four descendant species, then it would be less reasonable to suppose that winged flight evolved because of tree dwelling.

The ideal pattern for comparative inferences is that of Figure 2.15B. In Figure 2.15B, an adaptation (the purple bar) evolves multiple times. And in most of the cases when it evolves, it follows a well-defined ecological change (the turquoise bar).

In an evolutionary tree like that of Figure 2.15B, we have several types of important information:

- *Evolutionary relationships* We have the actual evolutionary relationships of the species. That is, we know the correct evolutionary tree.
- *Ecological events* We also know the major ecological events in the history of these species.
- *Timing of adaptations* And finally we know when the species involved actually evolved their particular adaptations.

Under conditions like these, we can make legitimate comparative inferences. We can even get an objective feeling for the validity of our inference, from the number of times an evolutionary transition occurs independently. The more times the same evolutionary transition occurs in response to a particular ecological transition, the more confidence we have in our conclusions. ❖

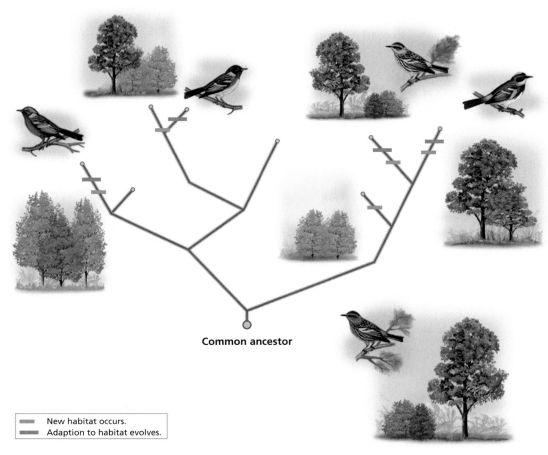

| | New habitat occurs. |
| | Adaption to habitat evolves. |

FIGURE 2.15B Repeated Adaptation to a New Habitat

Homology: When similar features among related species are inherited from their common ancestor

One of the most important patterns in evolutionary history is **homology.** Homology occurs when a structure present in an ancestral species is retained in descendant species, possibly with considerable evolutionary modification.

The qualitative pattern of homologous evolution is shown in Figure 2.16A. The ancestral species has a rib bone, call it A. The particular form of the rib bone in the ancestor is A1. In the species that descend from the common ancestor, A has been modified, taking on specific forms A2, A3, and A4. In one species, the structure has been completely lost. But this loss does not affect the homology involving the same structure in the other species. A homology may be involved in the evolution of a group of related species even if the structure is entirely lost in a few cases, although the species that lost the structure are not described as possessing the homology.

Figure 2.16B shows a well-known case of homology, the vertebrate forelimb. We humans experience this homology by having arms. Other vertebrates have fins and wings, but these structures are still homologous to our arms. There are many kinds of evidence for homology in a case like this. The best kind of evidence is illustrated by vertebrate fossils. Vertebrate bones fossilize well, so there is a good fossil record for the last 500 million years of vertebrate evolution. It is not difficult to follow the evolutionary expansion and shrinkage of particular bones, including the bones of the forelimb, in each of the vertebrate lineages. This information has left biologists confi-

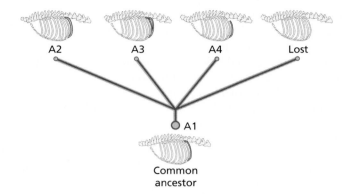

FIGURE 2.16A Evolution of Homologous Structures in the Descendants of a Common Ancestor

dent that the different vertebrate forelimbs all derive evolutionarily from the forelimbs of ancestral species.

Another type of evidence for homology comes from the idiosyncratic details of evolutionary history. These are the telltale clues that biologists use when they don't know as much about fossil history as we do with vertebrates. Homologies among organs that do not readily fossilize are usually inferred from the relative positions of structures. The vertebrate heart, for example, varies widely in its gross structure (Figure 2.16C). Some contemporary vertebrate hearts

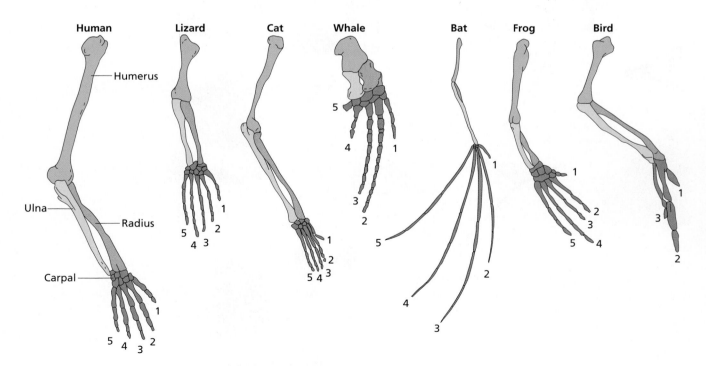

FIGURE 2.16B The skeletal structures of forelimbs in various vertebrate animals show the homologies among bones.

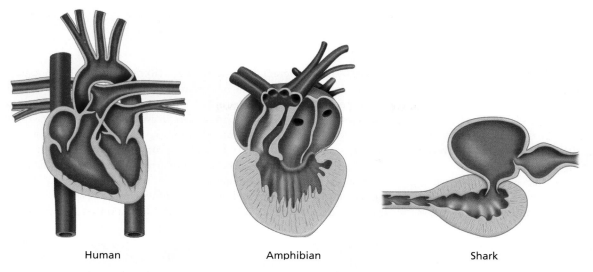

FIGURE 2.16C **The Hearts of Different Vertebrate Species: Human, Amphibian, and Shark**

have only one ventricle, the large pumping structure, while others have two ventricles. From the similar connections of ventricles to the heart's auricles and the arteries that lead from the heart, it is inferred that all vertebrate ventricles are homologous structures. Similar inferences are made about the components of the vertebrate brain—the medulla oblongata, the cerebellum, and the cortex—although fossil skulls sometimes give additional information. To revert to the forelimb example, vertebrate forelimbs usually have three main bones, called humerus, radius, and ulna. One humerus bone supports the first, proximal, part of the forelimb; but two bones, the radius and the ulna, support the second. This structural pattern serves as a clue that enables us to find the structural parallels between the vertebrate bones. Other cases are harder, but this general principle can be extended even as far as structural patterns of enzyme evolution.

An interesting variation on the homology concept is **serial homology.** Serial homology arises when a species has repeated structural features. When a common ancestral species has relatively simple repetitions of the same structural elements, it may be possible to detect the evolutionary divergence of individual segments in the descendant species.

The best-known example of serial homology is the structural evolution of the arthropod body plan, shown in Figure 2.16D. It is generally agreed that the ancestral arthropods had repeated segments, each bearing similar jointed limbs, like a centipede does today. In the descendant arthropod species, these segments diverged evolutionarily, producing antennae, jaws, other feeding structures, claws, walking legs, wings, swimmerets, and so on. These structures are homologous to each other because they are modifications of a common structure that was repeated serially in the ancestral species. The next time you eat a whole lobster or crayfish, you might look at its limbs, noticing how they vary down the length of the body. ❖

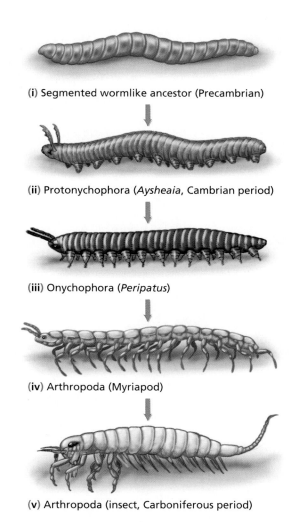

(i) Segmented wormlike ancestor (Precambrian)

(ii) Protonychophora (*Aysheaia*, Cambrian period)

(iii) Onychophora (*Peripatus*)

(iv) Arthropoda (Myriapod)

(v) Arthropoda (insect, Carboniferous period)

FIGURE 2.16D **Homologous Segmentation of Various Arthropod Species**

The concept of **homoplasy** is the opposite of homology. With homology, derivation from a common ancestor is responsible for shared characteristics in a group of species. With homoplasy, similarity is produced in multiple evolutionary lineages by the action of natural selection specifically favoring the similar features. **Convergent evolution** is one case of homoplasy.

Bear in mind that evolution acts independently on each species, though of course they may have ecological interactions with each other. Nonetheless, homoplasy is fundamentally different from homology. Homoplasy can be thought of as a product of natural selection. In the species of Figure 2.17A, there are a number of homoplasious patterns of evolution. The hallmark of homoplasy in evolutionary trees is the evolution of common features among species that have already evolved for some time *without* those common features. In other words, homoplasy produces similarity among species when homology cannot explain it.

Homoplasy supplies us with many of our most revealing examples of natural selection in action. An important historical

Marsupial mammals

Plantigale

Marsupial mole

Sugar glider

Wombat

Tasmanain devil

Kangaroo

Eutherian mammals

Mole

Deer mouse

Flying squirrel

Woodchuck

Wolverine

Patagonian cavy

FIGURE 2.17A Homoplasious Adaptation in Eutherian and Marsupial Mammals

example is the evolution of industrial **melanism**, the dark coloration of many European butterflies and moths in the early twentieth century. A notable feature of this evolutionary event was that many species of butterflies and moths evolved darker coloration in parallel (Figure 2.17B). In the early nineteenth century, darker forms were little known in England. By the early twentieth century, darker forms were extremely common among many species found in the industrialized parts of England, where the surrounding forest had lost lichen from its bark.

Much of the habitat of the butterflies and moths became dark and sooty. For these insects to remain camouflaged, they had to evolve dark coloration. Many of them did so. We will discuss this case in detail in Chapter 4. But the homoplasious evolution of darker coloration in many butterfly and moth species that had been light-colored just two centuries ago told evolutionary biologists that natural selection, and not common ancestry, must have been involved. Homoplasy is very important in the comparative analysis of evolution by natural selection. ❖

FIGURE 2.17B **Dark Melanic Moths and Speckled Lighter Moths from Europe**

SUMMARY

1. The evolutionary tree is one of the most important concepts in biology. With the evolutionary tree, we can conveniently represent evolutionary histories using diagrams that have many intuitive features. Evolutionary trees grow outward only, never doubling back. Just like real trees, they only rarely involve branches growing together. Extinctions are easily represented as truncated branches. Speciation is shown as forks, usually with two descendant species continuing onward, and thus upward.

2. Evolutionary trees are often constructed using the rule of maximum parsimony. Maximum parsimony is a rule that tells evolutionary biologists to prefer evolutionary histories with the fewest number of independent evolutionary events. Using parsimony, for example, a biologist does not suppose that fur independently evolved in each mammalian species. Instead, she assumes that the common ancestor of the mammalian species evolved fur, and it has since been retained by all the mammalian species that descend from their common ancestor.

3. The success of biological classification can be explained in relation to evolutionary history. Groups of organisms that share numerous attributes probably share common ancestors. Larger groups of organisms that share attributes are likely to have a relatively ancient common ancestor, compared to smaller groups that are similar to each other. Small, similar groups of species are likely to have had a recent common ancestor.

4. The geographical distribution of species indicates that species arise by local evolutionary processes, not by a global creation. Large, flying species are distributed throughout the world, because they can fly to distant areas once they have evolved in one region. But terrestrial animals have localized distributions that can be explained best in relation to local evolution.

5. In some groups, developmental patterns are conserved in protected larval or embryonic stages due to an absence of selection on them.

6. The comparative method contrasts species in order to understand adaptation. Recent applications of the comparative method have benefited from the use of evolutionary trees. Homology occurs when a structure evolves from a common ancestral feature, changing somewhat from one descendant species to another. Homoplasy occurs when similar characters evolve independently in separate species, homology not being involved.

REVIEW QUESTIONS

1. Are the fins of dolphins and sharks examples of homology or homoplasy?

2. Which evolved first, bacteria or fish?

3. Was the origin of eukaryotes an example of evolutionary tree branches terminating or fusing?

4. The terrestrial fauna of Australia illustrate which principles of biogeography?

5. Why are bats unusual among the mammals of Australia?

6. Why is Darwin's theory of evolution compatible with the system of biological classification?

7. Why does the principle of maximum parsimony help us to choose between evolutionary trees?

8. Extinction corresponds to what type of change in the evolutionary tree?

9. Does adaptation foster homology or homoplasy?

KEY TERMS

adaptation	conjugation	fossil record	paleontology
anastomosis	convergent evolution	halophile	prokaryote
Archaea	creationism	homology	serial homology
Bacteria	domain	homoplasy	speciation
biogeography	endosymbiosis	kingdom	thermophile
biological classification	Eukarya	maximum parsimony	transduction
common ancestor	eukaryote	melanism	transformation
comparative method	evolutionary tree	methanogen	

FURTHER READINGS

Darwin, C. 1859. *On the Origin of Species by Means of Natural Selection.* London: John Murray.

Ford, E. B. 1971. *Ecological Genetics,* 3rd ed. London: Chapman and Hall.

Harvey, P. H., and M. D. Pagel. 1991. *The Comparative Method in Evolutionary Biology.* Oxford, UK: Oxford University Press.

Larson, A., and J. B. Losos. 1996. "Phylogenetic Systematics of Adaptation." In *Adaptation*, edited by M. R. Rose and G. V. Lauder, 187–220. San Diego: Academic Press.

Novacek, M. J. 1996. "Paleontological Data and the Study of Adaptation." In *Adaptation*, edited by M. R. Rose and G. V. Lauder, 311–59. San Diego: Academic Press.

Romer, A. S. 1970. *The Vertebrate Body,* 4th ed. Philadelphia: Saunders.

Schluter, D. 2000. *The Ecology of Adaptive Radiation.* Oxford, UK: Oxford University Press.

PART TWO

MACHINERY OF EVOLUTION

The great spectacle of biological evolution is one of the most inspiring things a young biologist can be exposed to: the colors of birds, flowers, and insect wings; the vast scale, from virus particles to blue whales; the intricate shapes of shells, bones, and fossils. The diversity of life has a majesty like that of a great symphony, detailed and overpowering. This diversity is the main topic of most introductory biology courses. However, the beauty of life evolving on this small blue planet tends to inspire wonder more than rational thought.

Yet at its core, evolution runs on machinery that is utterly repetitive, implacable, hardly beautiful at all. The workings of evolution resemble the ledgers of a Victorian firm owned by Ebenezer Scrooge more than they do the sheet music of Wolfgang Mozart. Evolution works by scrutinizing populations with a jaundiced eye. The feeble and the barren are discarded with an uncaring brutality. Like a merciless robot, evolution produces life but does not care for it. Individuals die, species go extinct; yet the evolutionary process always goes on.

The biggest problem with understanding how evolution works is that it is invisible. You can see organisms right in front of you. With a microscope you can see cells, and you can pick up a white, stringy hunk of DNA that you can see with the naked eye. The processes of ecology are visible in the birth, death, feeding, and decomposition of plants and animals. But evolution is the ghost at the banquet of life. Its machinery is hidden.

There is nothing unusual about scientists looking for the hidden machinery underlying a process. For a century, first atomic physicists, then nuclear physicists, and now particle physicists have been looking for smaller and smaller particles to explain matter. They seek the hidden machinery of matter and energy. In the same way, biologists have sought the hidden machinery of life's evolution.

At the core of this machinery is a genetic engine, a whirling thing that contains the information used to produce each generation, the record keeper for the entire process of evolution, containing the fuel on which evolution works. Genetic transmission works with greater precision and reliability than any other strictly biological process. Like the atoms of physics and chemistry, genes are building blocks; without them, life could not exist. To have a solid understanding of evolution, you have to know basic genetics. This we supply in Chapter 3.

But genetics does not determine the direction in which the evolutionary leviathan advances. The bit of machinery that guides evolution, its steering wheel and its accelerator, is natural selection. Natural selection is the discriminator within the evolutionary machine. It is natural selection that chooses among the individual organisms in a population, determining who dies, who has a chance to reproduce, how much they reproduce, and finally the duration of their lives. This selection process ultimately adapts populations to the environments that they inhabit.

Evolution was first "seen" around the year 1800 by people like Jean-Baptiste Lamarck and Erasmus Darwin, who were introduced in Chapter 1. But evolution was not properly understood and analyzed until Charles Darwin discovered natural selection as part of the machinery of evolution. All students of biology need an understanding of natural selection. We supply you with that understanding in Chapter 4.

The fundament of the machinery of evolution lies at the molecular level. Life is an interplay of molecules—a process built up from a swamp of physics and chemistry, but not just physics and chemistry. Molecular evolution supplies the foundations for all processes of evolution, because nothing happens in evolution unless the molecular composition of organisms changes. But there is a great inscrutability to molecular evolution; because it can proceed with or without natural selection, it is very difficult to determine whether selection has acted at the molecular level. The ultimate machinery of evolution is DNA, the molecule of heredity. At this deep level of the evolutionary machinery, we can see how evolution plays out—much like we can learn how a computer works by examining the lines of code that make up its programs. Evolution at the molecular level is the topic of Chapter 5.

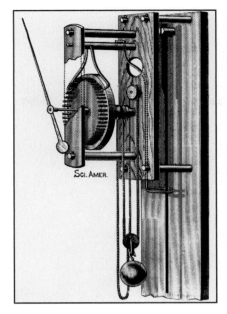

Sci. Amer.

The ultimate regulators of the evolutionary machinery are like two conveyor belts. One belt brings species into the machinery of evolution; the other belt takes species out of the machinery. Speciation adds species. Extinction eliminates species. For a long time, evolutionists thought of speciation as mysterious and extinction as straightforward. Now we know that speciation is not as inexplicable as we thought, while extinction appears to be considerably more complex than we imagined. The creation and destruction of species are the most dramatic actions of the evolutionary machinery. In Chapter 6 we introduce the machinery of speciation and extinction.

Genetics, natural selection, molecular evolution, speciation, and extinction all play a role in the evolutionary machinery—sometimes in isolation, but often not. In considering speciation, for example, issues of genetics, selection, and molecular evolution all play an important role. We will separate these processes from each other while introducing them; but in nature they normally operate together, like a giant factory with many machines grinding, and spinning, and stamping.

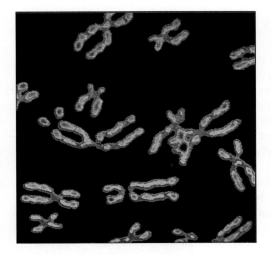

The genetic engine supplies information for development and variation for selection.

3

The Genetic Engine

The evolutionary machinery has genetics as its core. Genetics in turn has two facets. One is the role of genetics as the keeper of the library of information for making each organism, its complete *genome.* Now that the human genome has been completely sequenced, anyone can marvel at the fact that our biological coherence depends on some billions of nucleotide pairs of DNA. Without genetics, life would have to be renewed by some external force in every generation. With the information in the genome, life has momentum, continuing on from parent to offspring. Thanks to this genetic transfer of genomic information from one generation to the next, each offspring is the cumulative product of millions of years of evolution.

Almost as important as the genetic transmission of the genome is the role of genetics in providing *variation*—heterogeneity in the biological charac-ters of the individuals who make up the population. Some of this variation has nothing to do with heredity, and is called *environmental,* though this term includes all nongenetic sources of variation. Environmental variation can be very important. Human learning, for example, is environmental variation. But there is also genetic variation. Such genetic variation is essential to the process of evolution, because it provides the raw material for natural selection. We have already seen how Darwin emphasized the importance of heredity.

Thus genetics faithfully transmits the information built up by evolution over long periods, and it supplies the fuel for evolution—genetic variation. These two roles of genetics are key to the machinery of evolution, and so they are integral to this part of the book. To understand the role of genetics in evolution is a fair start to understanding the evolutionary process as a whole. ❖

HOW GENETICS WORKS

3.1 Genetics is central to modern biology

The field of genetics was probably the greatest achievement of twentieth-century biology. It is doubtful that modern biology could have been created without genetics. Too many basic questions about life had no good answers before the creation of genetics.

As we have seen, the person who started the genetic revolution was the monk **Gregor Mendel** (1822-1884), one of the most obscure of all the great scientists (Figure 3.1A). It is not clear whether Mendel's genetic theory of inheritance preceded his plant-breeding experiments, or whether he formed his ideas from his data. It is certain that by the time he presented his first papers on genetics, he had both a remarkably clear model as well as garden pea data that beautifully illustrated his theory. These papers were presented locally, in Brno, and circulated to some of the leading botanists of Mendel's day. Darwin had an unread copy of one of Mendel's papers. Yet no leading scientist appreciated the implications of Mendel's work during his lifetime.

All that changed in 1900, when several botanists independently rediscovered Mendel's work. Very quickly these scientists and their colleagues abandoned the old models of heredity and formed a new discipline within biology—genetics. Basic discoveries of genetics have since flowed rapidly. A partial list would include the discoveries listed in Table 3.1A.

FIGURE 3.1A **Gregor Mendel**

Even this list does very poor service to more than a century of brilliant work, in which many scientists played major roles. The point is that we can view the twentieth century as a century in which the development of biology has been dominated by the unfolding of the research of geneticists and their allies.

A warning needs to be provided here: It is easy to think about genetics strictly as the triumphs of molecular genetics, starting with the double-helix DNA model of Watson and Crick. Scientific progress had been rampant, however, for the 53 years preceding publication of the Watson-Crick DNA model in the journal *Nature*. In particular, the foundations of

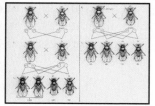

FIGURE 3.1B **Thomas Hunt Morgan**   Morgan was the founder of fruit fly genetics. This is one of his drawings.

genetics have important elements of evolutionary reasoning, beginning in 1908 with the derivation of the Hardy-Weinberg equilibrium, a key concept in population genetics. Molecular genetics would be difficult to sort out without the use of important principles from population genetics.

In this chapter, we will introduce the essential model of Mendelian genetics and then present some of the most basic ideas of population genetics and quantitative genetics. ❖

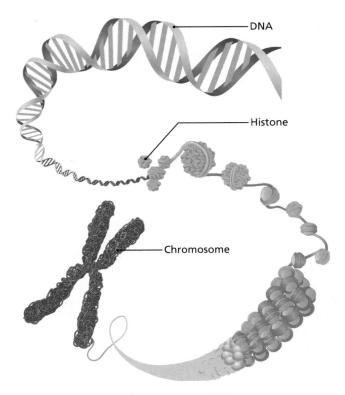

FIGURE 3.1C **The Structural Features of DNA**

TABLE 3.1A Abbreviated List of Twentieth-Century Genetic Discoveries

- Mendel's genetic theory of inheritance is rediscovered (1900).

- Genetic linkage between characters is discovered (1911).

- Fisher develops quantitative genetics for continuous characters (1918).

- Muller demonstrates the physical basis of mutation (1928).

- Morgan establishes the chromosomal basis of genetics (1931; *Figure 3.1B*).

- Avery, MacLeod, and McCarty determine that DNA is the hereditary molecule (1944).

- Watson and Crick propose that the DNA double helix codes for genes (1953; *Figure 3.1C*).

- DNA replication is worked out (1958).

- The genetic code is determined (1966).

- Recombinant DNA allows rapid cloning of DNA (1972).

- First rapid gene sequencing is performed (1975).

- Organisms are genetically engineered by the insertion of transposable DNA (1982).

- Polymerase chain reaction (PCR) is used to clone small amounts of DNA (1985).

- The entire human genome is sequenced (2000).

Reproduction may transmit one or two copies of the hereditary information to the next generation

However the specifics of inheritance work, species vary greatly in the amount of hereditary information that is copied from parent to offspring. There are considerable differences in the size of the genome that each species carries, from the small genomes of viruses to the enormous genomes of some cereal plants and some salamanders. This variation in genome size is examined further in Chapter 5. The rest of the variation in DNA content is in the number of copies of the basic genome that each cell has.

A bit of terminology is important here. An organism or cell that has only one copy of the basic genome is called **haploid** (Figure 3.2A). Most organisms that we are used to seeing are not haploid. However, almost all gametes are haploid, so you may already have some familiarity with haploid cells. Like most animals, humans are **diploid**, possessing two copies of the basic genome (Figure 3.2B). (We are neglecting hereditary differences between male and female, which we will consider later in this chapter.)

Organisms that have more than two copies of the genome are called **polyploid** (Figure 3.2C). Most of these have higher multiples of two copies, with four and eight copies being common. Some species that consist only of females are triploid, with three haploid genomes. This condition is known in some fish and some lizards. We discuss it further in Chapters 6 and 18. Bacteria are sometimes considered haploid, but it is also common for them to have a variable number of additional copies of their basic genome. This variable ploidy in bacteria is called **meroploidy**.

Polyploid

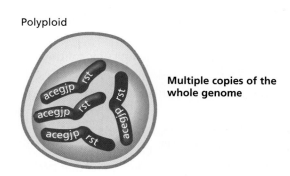

Multiple copies of the whole genome

FIGURE 3.2C Polyploid Genome

Haploid

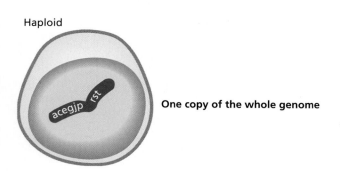

One copy of the whole genome

FIGURE 3.2A Haploid Genome

Diploid

Two copies of the whole genome

FIGURE 3.2B Diploid Genome

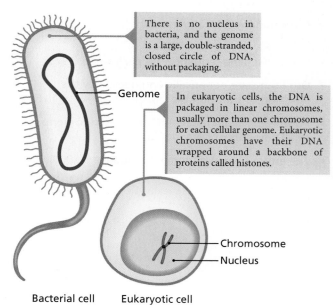

There is no nucleus in bacteria, and the genome is a large, double-stranded, closed circle of DNA, without packaging.

Genome

In eukaryotic cells, the DNA is packaged in linear chromosomes, usually more than one chromosome for each cellular genome. Eukaryotic chromosomes have their DNA wrapped around a backbone of proteins called histones.

Chromosome
Nucleus

Bacterial cell   Eukaryotic cell

FIGURE 3.2D Genome organization is very different in bacterial cells as compared to eukaryotic cells.

Even though we consider ourselves diploid organisms, let's bear in mind that many organisms, humans included, regularly alternate between haploid and diploid genomes. In plants, the haploid stage of the life cycle can produce a substantial plant. Both mosses and ferns have haploid plants. Angiosperms, on the other hand, have vestigial haploid forms contained within flowers. Usually, haploidy is associated with the production of gametes, anzd fertilization then creates a diploid organism as a result of the union of two haploid cells (the gametes).

Some organisms have specific tissues with odd *ploidy* (number of genome copies). Plant **endosperm**, a tissue involved in fertilization, is triploid. The larvae of fruit flies have salivary gland cells that contain hundreds of copies of the genome—a condition called **polyteny**. Ploidy is thus a complicated business. The tissues and life-cycle stages of animals and plants can vary significantly in number of genome copies. ❖

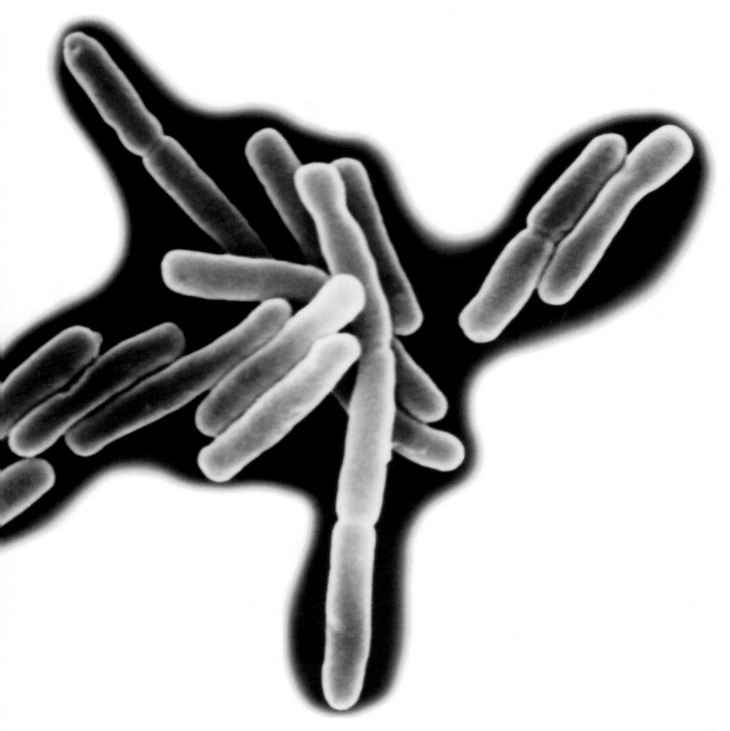

3.3   Sexual reproduction recombines chromosomes containing many discrete loci

In the vast majority of organisms, Mendelian genetics is the system of inheritance. Numerous bacteria do not have Mendelian genetics, as well as some asexual organisms that retain only vestiges of it. But for most organisms of interest, Mendelian genetics defines inheritance. Alternation of ploidy is a basic feature of the Mendelian system. Another basic feature is the **chromosome**. Think of chromosomes as strings of genetic **loci** that can be defined as locations for particular genes. Each locus contains **alleles**, which are the specific versions of genes, defined by DNA sequences.

During the processes of fertilization, development, and meiosis, each chromosome in the cell will be present in one, two, and four copies. This is indicated in the genetic cycle diagram of Figure 3.3A by n, $2n$, and $4n$, where n is the number of chromosomes in the haploid genome. Combinations of genes are scrambled, or randomized, at three points during this cycle. In this way, Mendelian genetics generates variation, as we will now describe.

First, start with the pool of **gametes** that precedes fertilization in Figure 3.3A. It is a basic tenet of Mendelian genetics,

and a demonstrable empirical fact, that during fertilization gametes tend to combine at random with respect to their genetic makeup. (There are some exceptions, but they are minor.) Thus the **zygote** that is formed during fertilization is a random combination of two gametes among many. This process is analogous to a card game in which the "hand" has only two cards, but the deck contains thousands or millions of different cards, the cards being gametes and the hand being the zygote. This random combination of gametes is the first point at which Mendel's genetic combinatorics can generate large amounts of variation, following orderly rules of probability.

The second point in the cycle at which randomizing factors play a key role is the recombination of chromosomes during **meiosis** (Figure 3.3B). Meiosis works with recombination as follows. At the start of meiosis, each of the two chromosomes of the diploid cell produces an additional copy of itself. There are thus four copies of the chromosome present in the cell. Chromosomes have a tendency to break, but once they have broken into pieces, they usually find matching broken ends

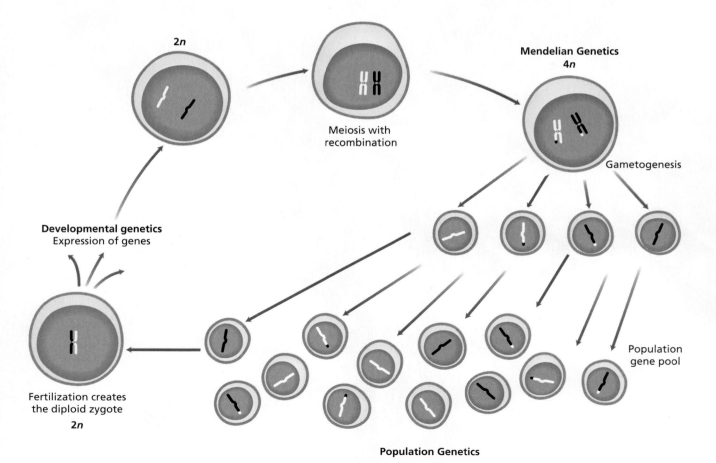

FIGURE 3.3A The Genetic Cycle Different fields emphasize different phases of this cycle: Mendelian genetics, developmental genetics, and population genetics. The black and white structures are the chromosomes; pieces of black or white joined to chromosomes of the opposite color indicate recombination.

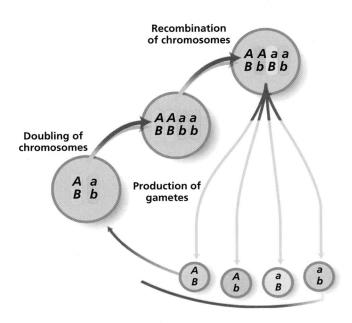

Recombination
of chromosomes

A A a a
B b B b

Doubling of
chromosomes

A A a a
B B b b

A a
B b

Production of
gametes

A
B

A
b

a
B

a
b

FIGURE 3.3B This close-up of Mendelian genetics shows the recombination and segregation of specific genetic alleles from two loci, both located on a single chromosome. Here the chromosomes are represented by colored oblong shapes with letters on them.

(which are effectively "sticky") and reconstitute themselves. If these broken ends are all from the same chromosome, no genetic recombination occurs. But different chromosomes of the same type can break and rejoin, forming new composite chromosomes that never existed before. In Figure 3.3B we show this process for two genes, each with two alleles. At the *A* locus we show an *A* allele and an *a* allele. Similarly, at the *B* locus the two alleles are *B* and *b*. A chromosome with the *A* and *B* alleles side by side can recombine so that the *A* and *b* alleles are put together on the same chromosome.

Third, with **independent assortment** of chromosomes, different gametes can be produced from the same sets of chromosomes. Recombined chromosomes join in varied combinations to create gametes with combinations of chromosomes that never existed before. (This combination of multiple recombined chromosomes is not shown in Figure 3.3A or Figure 3.3B.) And this process is superimposed on any process of chromosomal recombination that may have occurred. Thus at three points, Mendelian genetics scrambles the alleles of the strings of genetic loci that make up entire genomes. ❖

GENES IN POPULATIONS

3.4 Genes specify phenotypes, then the phenotypes are selected, which changes gene frequencies

For evolution, Mendelian processes define the basic foundations for heredity and variation. From these foundations, the process of evolution creates adaptations, species, new ways of life. Our interest now lies in how the evolutionary machinery uses genes.

First, we need to know the frequencies of alleles and gene combinations. These frequencies reveal the genetic substratum to the evolutionary process. Second, we need to know how genes determine the characteristics of organisms, a process called **gene expression**. Knowing gene expression gives us information about the consequences of having particular genes, in particular combinations. Third, we specifically need information about the consequences of particular genes for survival and fertility, because these consequences generate natural selection, the directional motor of the evolutionary machinery.

Some useful terminology helps to define the operation of evolutionary machinery. We use the term **genotype** to define that part of the genetic makeup of the organism in which we are interested—for instance, a haploid genotype or a diploid genotype. Sometimes the genotype of interest may be a large part of the genome. Sometimes it may be just one Mendelian locus.

From the genotype, in each particular environment, an organism develops. The features of interest for a particular organism are called the **phenotype**. Sometimes we say phenotypes when we mean several different attributes of the organism, such as height or color. But there is really only one phenotype in total, just as there is only one complete genome.

Sometimes the evolution of genes is divorced from the phenotype. It is possible that variation does not matter for natural selection. For example, exactly how your fingerprints curve over your fingertips may not make any difference to your survival or your reproductive success. But this absence of effect is a very important point for the evolution of the genes that affect fingerprints, in that their evolution will then be uncoupled from natural selection. We will consider this situation later in this chapter, and elsewhere.

In studying genes in populations, we are attempting to learn the genotypic components of the evolutionary machinery. Genotypes can determine phenotypes that have consequences for survival or reproduction. Such consequences generate natural selection. Selection then changes allele frequencies, an essential feature of evolution. Allele frequencies define selection, and selection in turn changes these allele frequencies. The process feeds back on itself. (The process is not, however, circular—because selection does not determine earlier allele frequencies.) The phenotypes (P) of one generation undergo selection and other processes, which determine the genotypes (G) of the next generation, which determine the phenotypes of the following generation, and so on.

$$G_1 \rightarrow P_1 \rightarrow G_2 \rightarrow P_2 \rightarrow G_3 \rightarrow P_3 \rightarrow$$ ❖

The Population Concept

To monitor the evolutionary process, allele and genotype frequencies are some of the most basic variables that we need to follow. These variables are the particular interest of *population genetics*. Population genetics is based on the Mendelian model for inheritance coupled with the concept of population. This population concept therefore requires some closer attention.

The key concept that allows us to move from genetics to evolution is **population**. Loosely speaking, a population is just members of a sexual species living within easy traveling distance of each other. "Easy traveling distance" strictly relates to the biology of a species. For an albatross, that may be thousands of miles over open ocean, almost regardless of the weather. For a small worm, a lifetime's easy traveling distance may be a few dozen yards. The issue is whether or not organisms can mate with each other. If mating is a reasonable possibility, then two organisms are members of the same population. If mating is unlikely to occur, then they are not. The mating pattern defines the scope of the population in *space*, as shown in Figure 3.4A.

When organisms belong to the same sexually reproducing population, then the fates of their genes are intertwined. They may have ancestors in common. They may later have descendants in common. This is because the Mendelian cycle of fertilization, recombination, and gamete production ensures that genes are shuffled among organisms (Module 3.3). Indeed, whereas organisms may seem to constitute the population, in another sense the animals and plants that we see with our eyes are only the fronts, or masks, for the genes that define these organisms and determine their fates. The continuous transmission of genes from one generation to another defines the scope of the population in *time*, as shown in Figure 3.4B.

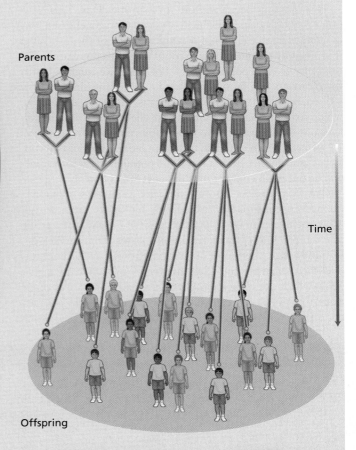

FIGURE 3.4A Populations in Space The animal species is indicated in the ellipse that qualitatively illustrates its approximate range in space.

FIGURE 3.4B Populations through Time, from One Generation to the Next. Parents are joined by *v*'s. Lines connect offspring to parents. Note that a man and a woman have other partners with whom they have children. Some other adults don't have any children.

3.5 The evolutionary state of a population is defined by its genotype frequencies

The genes at a locus come in different flavors or alternative forms, which we have referred to as *alleles*. These alleles have different DNA sequences. The different sequences sometimes specify the production of different amino acid sequences in proteins. But sometimes they do not. Instead, they may lead to increases or decreases in the total amount of a particular protein. Or the variant DNA sequence may change when the protein is made, perhaps in response to a temperature change. Another possibility is that the alleles code for RNA that does not make protein. For example, the RNA in ribosomes is used to synthesize proteins, but does not itself code for protein. So allele differences may have quite heterogeneous effects on the molecular and cell biology of organisms.

Mendel and the geneticists who came before the double-helix model of DNA knew none of these molecular details. They were able to detect discrete genetic variants in their breeding experiments. They then treated those variants as alleles, with no knowledge of the molecular biology of the gene.

Genetics thus began without mechanistic detail. This viewpoint probably helped the birth of population genetics. So long as genes were not known biochemically, one of the few things that a scientist could do was to count them in a population. For example, given a population of 124 mice, we might identify three carriers of a fairly rare allele for albinism using the gene sequences present at the locus. Because mice are diploid, we would have to sequence a total of 248 alleles, and 3 albinism alleles out of 248 would give us an albinism allele frequency of 1.2 percent. From data like this, population geneticists reconstruct and predict the evolution of populations. This activity will be our main concern over the next few chapters.

But first we need to be quite sure about how allele frequencies are estimated. In genetics, biologists often do not assay the alleles of an organism directly. Instead, they group phenotypes into distinct genotypic classes. That is, they determine what phenotype is associated with each genotype, and then calculate the allele frequency as shown in Figure 3.5A, which uses the example of a small population of 34 mice. This frequency calculation process is somewhat like accounting. It is the foundation of population genetics, which

in turn is the key to understanding the machinery of evolution, since evolution depends on changes in allele frequencies, as we will describe in detail.

These calculations can also be applied to real genetic data as in the following box.

Mice homozygous for allele *a*: 9

Mice heterozygous for allele *a* and *A*: 16

Mice homozygous for allele *A*: 9

The *aa* homozygotes carry 9 × 2 = 18 *a* alleles.

The *Aa* heterozygotes carry 16 × 1 = 16 *a* alleles.

The total number of *a* alleles = 18 + 16 = 34 alleles.

The *AA* homozygotes carry 9 × 2 = 18 *A* alleles.

The *Aa* heterozygotes carry 16 × 1 = 16 *A* alleles.

The total number of *A* alleles = 18 + 16 = 34 alleles.

There are a total of 68 allele copies at this locus.

The frequency of *A* = 0.5. The frequency of *a* = 0.5.

FIGURE 3.5A A Small Mouse Population with Two Alleles, *a* and *A*

Population Genetics of a Human Blood Type

To gather this information on human **blood groups**, the blood from 730 individuals was tested and classified as either type 'M', 'MN', or 'N'. Each of these types corresponds to a single genotype, permitting us to directly estimate allele frequencies as shown in the table. Note that the M blood group corresponds to the homozygous $L^M L^M$ genotype, while the N blood group corresponds to the $L^N L^N$ blood group. Since the allele frequency (L^M) + allele frequency$(L^N) = 1$, we can use the following results to estimate the frequency of L^N as $1 - 0.18 = 0.82$.

Allele Frequencies of a Human Blood Group

Blood Group	Genotype	Number	Frequency
M	$L^M L^M$	22	0.030
MN	$L^M L^N$	216	0.296
N	$L^N L^N$	492	0.674
Total		730	1.00

Frequency of allele L^M = (2 x 22 + 216)/1460 = 0.18

Data from a study of Australian aborigines; published in Ayala and Kiger (1980, p. 603).

Early in population genetics, the question came up as to whether gene frequencies would have some inherent tendency to change, perhaps as a result of the blind operation of the Mendelian machinery during sexual reproduction. By 1908, G.H. Hardy and W. Weinberg had independently worked out that there was no such tendency for allele frequencies to change, providing there were no perturbing outside forces, like selection or migration. This idea is known as the **Hardy-Weinberg equilibrium,** and we can demonstrate it mathematically.

Understanding this demonstration first requires an understanding of two basic rules of probability:

1. If two *independent* events together cause a third event to occur, then the probability of the third event is the *product* of the probabilities of these two events.

    For example, if a car crash requires you to look away from oncoming traffic *and* it requires you to lose control of your vehicle temporarily, then the probability of the crash is the *product* of these two unlikely (we hope) events. This is the **multiplication rule**.

2. When *either* of two events suffices to cause an outcome, then the probability of this outcome is the *sum* of the probabilities of these individual events.

    Thus if you can crash on an icy mountain highway by *either* ignoring a curve *or* by losing traction on a patch of ice, then the probability of such a crash is the *sum* of these two events. This is the **addition rule**.

To demonstrate the Hardy-Weinberg equilibrium, we need to calculate the consequences of Mendelian genetics for allele frequencies. Thus we need to go through a complete cycle of reproduction, from the parents of the first generation through their production of gametes, fertilization, and the creation of the next generation of adults. This calculation is shown in Figure 3.6A.

It is crucial to keep track of the different combinations of alleles in each mating, their probabilities of occurrence, and the genotypes of the progeny that they produce. If we follow all these variables carefully, then we can see that the frequencies of the two alleles (A and a, given by p and q respectively) do not change from one generation to the next. This result also holds true if there are more than two alleles at a locus, but the calculations get much more complicated. In all these calculations, it is an absolute requirement that the population size be large enough so that we can calculate probabilities exactly. When this is not true, additional evolutionary processes arise, as described later in this chapter. Likewise, there must be no biases coming from selection or differences in reproductive success. These are discussed in Chapter 4. From these stipulations, you will realize that the Hardy-Weinberg equilibrium is the simplest case considered by population genetics.

The attractive thing about the Hardy-Weinberg equilibrium is that it allows biologists to calculate the frequencies of the genotypes at a genetic locus from the frequencies of the alleles that make up those genotypes. If you keep track of order, there are four diploid genotypes at a locus: AA, Aa, aA, and aa. At Hardy-Weinberg equilibrium, the frequencies of these genotypes are given by the product of the frequencies of their alleles: p^2, pq, qp, and q^2, respectively. At one locus, we do not normally keep track of order, so the frequency of the Aa genotype is given as $2pq$, the sum of pq and qp. This enables us to understand evolution in terms of allele frequencies, instead of genotype frequencies.   ❖

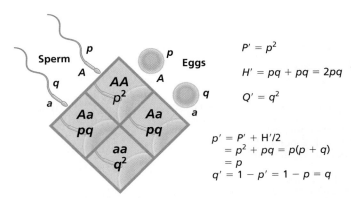

Start with three possible genotypes: AA, Aa, and aa. Say these genotypes have frequency: P, H, and Q.

The frequency of A is $P + H/2 = p$
The frequency of a is $Q + H/2 = q$

Therefore, the frequency of A gametes is p and the frequency of a gametes is q.

With random mating, gametes combine by the rules of independent assortment like cards and coins.

$$P' = p^2$$
$$H' = pq + pq = 2pq$$
$$Q' = q^2$$

$$p' = P' + H'/2$$
$$= p^2 + pq = p(p + q)$$
$$= p$$
$$q' = 1 - p' = 1 - p = q$$

We represent frequencies of alleles and genotypes in the next generation by a prime , so p' indicates the frequency of allele A in the next generation.

FIGURE 3.6A Hardy-Weinberg Law Gene frequencies do not change because of genetic segregation alone—nor do genotype frequencies at each locus, once Hardy-Weinberg equilibrium has been achieved.

When the alleles of different loci are combined randomly, they are in linkage equilibrium

It would be nice if the Hardy-Weinberg equilibrium applied to genotypes that are combinations of loci. But it doesn't. Say we are interested in two loci: *A* and *B*. Suppose there are only two alleles at each of these loci: *A-a* and *B-b*. Let's also suppose that these alleles have frequencies *p-q* and *r-s*, respectively. Then we might expect that the frequency of the *AABB* genotype would be p^2r^2, by analogy to the frequency of *AA* being p^2. However, there is no principle like the Hardy-Weinberg law that always applies to calculating the frequency of genotypes across more than one locus.

Genotype frequencies can differ substantially from expectations based on random combination of alleles over loci. However, as we will show later in the chapter, there is a tendency for the frequencies of gametes to evolve so that they *do* come to follow such simple expectations. But this tendency is not as quickly or as reliably expressed as the Hardy-Weinberg equilibrium is.

However, we can calculate what genotype frequencies would be like *if* alleles combined at random when two or more loci are involved. This hypothetical situation is called **linkage equilibrium** in evolutionary biology. Deviation from this ideal of random combination is called **linkage disequilibrium**.

In the display in Figure 3.7A, we calculate a simple example of genotype frequencies when there is linkage equilibrium. Note that the four combinations of the same gamete type (*AB/AB*, *ab/ab*, *Ab/Ab*, *aB/aB*) have half the frequency of the

six combinations of different gametes (*AB/ab*, *Ab/ab*, *aB/ab*, *AB/Ab*, *AB/aB*, *Ab/aB*). This is just the working out of the laws of probability, not some peculiar biology. The following box explains this in greater detail. ❖

FIGURE 3.7B Can You Do This?

There are many loci in a genome, and many possible alleles at each locus.

These alleles need not be combined randomly, like cards in a shuffled deck. When combined randomly, the alleles are in *linkage* equilibrium. When not combined randomly, the alleles are in *linkage* disequilibrium.

The simplest case involves two loci, each with just two alleles. If each allele has a frequency of 0.5, there are four possible chromosomes, each with frequency 0.25.

When chromosomes are combined randomly, the fully homozygous combinations occur half as often as the heterozygous combinations.

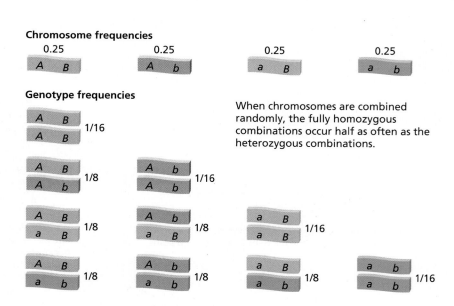

FIGURE 3.7A Linkage Equilibrium or Gamete Phase Equilibrium

Linkage Equilibrium and Disequilibrium

Linkage equilibrium is like Hardy-Weinberg equilibrium. Recall that at Hardy-Weinberg equilibrium, the frequencies of genotypes are given by the product of the frequencies of their alleles: p^2, $2pq$, and q^2.

At linkage equilibrium, the frequencies of the four possible genotypes are as follows, when the frequencies of alleles B and b at the second locus are r and s, respectively:

Gamete	AB	Ab	aB	ab
Linkage equilibrium frequency	pr	ps	qr	qs
Actual gamete frequency	P_{AB}	P_{Ab}	P_{aB}	P_{ab}

As for Hardy-Weinberg equilibrium, there is linkage disequilibrium when P_{AB} is not equal to the product pr, and so on for the other gamete frequencies. For this reason, linkage equilibrium is sometimes called "product" equilibrium. Note the parallel between such products in genotype frequencies and the products calculated using the multiplication rule for coin tossing, poker, and so forth. Both reflect independent probabilities, when there is linkage equilibrium or when coins are tossed fairly and cards are dealt by the rules.

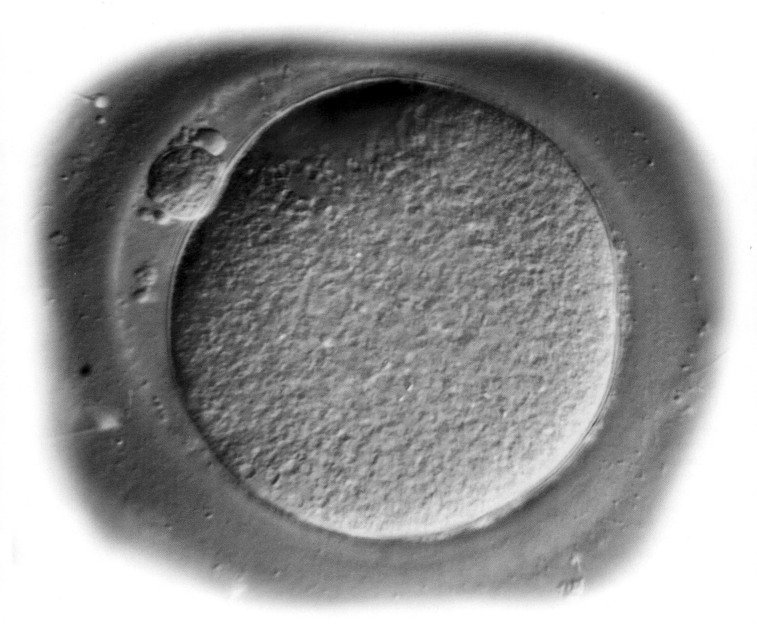

QUANTITATIVE CHARACTERS

Quantitative characters have to be studied statistically

Most Mendelian genetics focuses on qualitatively distinct characters, like eye color or flower color. More recently the genetic character of choice is the DNA sequence. The DNA sequence is indeed the most fundamental and accurately recorded character of all. It is the genotype itself.

But to study evolution often requires that we study messy characters like fertility, size, and resistance to lethal stress. These characters cannot usually be scored qualitatively. Instead, they need to be evaluated quantitatively. We need special tools to make sense of them.

Those tools come from statistics. Indeed, much of statistics had its historical origins in the work of early evolutionists, like **Francis Galton**, **Karl Pearson**, and **R. A. Fisher** (Figure 3.8A). These evolutionists wanted to understand such characters as intelligence and size, and there were no statistical tools already available. So they created them. Some of the terminology of statistics reflects its origin in the words of evolutionists. One example is the word *regression*, which originally referred to a pattern of inheritance and now refers to a statistical method for calculating the best line to plot through graphical data. This combination of evolutionary interests and statistical tools gave birth to the field of quantitative genetics.

(i)

(ii)

(iii)

FIGURE 3.8A **(i) Francis Galton, (ii) Karl Pearson, and (iii) R. A. Fisher**

Before introducing quantitative genetics, we will explain some elementary statistical ideas. These ideas are the essential tools for understanding patterns of gene expression in populations. (For more detail see the Appendices, following chapter 22.)

What is a **quantitative character**, when compared to a *qualitative* (or "Mendelian") character? It is a character that has no clear categories. It is fairly easy to say that a color is red or blue. It is much harder to say whether a mouse is large or small.

Therefore, we measure quantitative characters. We might have 122 weight records for a laboratory population of 124 mice. (You will lose one or two mice during weighing.) These 122 numbers are our record of the quantitative character body weight in these mice. They are the raw data that we want to make sense of. This is the starting point for further research on quantitative characters.

An important point about the study of quantitative characters is that the conditions of measurement are often more important than they are for typical Mendelian characters. Mendelian characters are not normally changed by minor differences in laboratory handling. The eye color of a laboratory animal will not normally change depending on whether the animal is hot or cold, recently fed or not, and so on. But such factors might be important for a quantitative character. Body weight, for example, will be affected by recent feedings. Quantitative characters are not only harder to measure, but those measurements themselves are not always reliable, unless efforts are made to control and standardize handling.

Once we have a collection of numbers that are measurements of a quantitative character in a group of organisms, what is the next step? In quantitative genetics, we normally calculate two important pieces of statistical information.

First, we usually calculate a **mean** for these measurements. Scientists use different kinds of mean. One is the **median,** which is the value at which half the measurements are above and half are below. For example, we might determine the median height among a group of corn plants. Most of the time, geneticists employ the **arithmetic mean**, which is the sum of all observations divided by their number. For example, the 122 weighed mice will have an arithmetic mean weight.

Means provide crude summaries of the features shared by groups of organisms. For example, the arithmetic mean is used to give a sense of "where" a population is located. If one group of mice has mean weight of 10 grams and another group has a mean weight of 9 grams, we usually say that the first group is heavier than the second. We say this even though the second group probably has members that weigh more than 10 grams.

The second important piece of information that we normally collect concerning a quantitative character is its

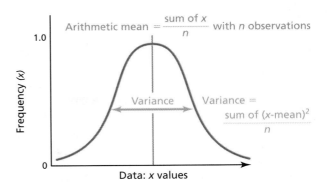

FIGURE 3.8B The variance of a random variable may be estimated as shown.

variance. Formally, a **variance** (see Figure 3.8B) is the sum of squared deviations of the observations from the arithmetic mean, divided by the total number of observations. A *deviation* is the difference between an observation and the mean. (These deviations are squared because the sum of all deviations by itself is zero.)

Just as a mean gives us a sense of "where" a population is with respect to a character, its variance gives us a sense of how dispersed the population is about that location. For example, a group of mice that has a mean weight of 20 grams might have a variance of 1.2, while another group of mice might have a mean weight of 19.5 grams but a variance of 3.8. The second group can be considered more variable.

Using these concepts of mean and variance, quantitative genetics studies quantitative characters. Figure 3.8B summarizes these concepts and gives their formulas. The Appendices give additional information about statistics. ❖

When environmental (*E*) and genetic (*G*) influences on a phenotype (*P*) are independent, $V_P = V_G + V_E$

Genetic research started with simple characters that could be scored qualitatively. As we have noted, color was a favorite character—the color of eyes and other parts. But geneticists also studied other discrete differences, like wrinkling of peas, loss of hair, dwarfism, and so on. In many cases, these distinctive phenotypic differences were based on differences in the alleles present at single genetic loci. Because Mendel himself performed this kind of research, it is called *Mendelian genetics*.

But most characters of interest to biologists, especially evolutionary biologists, do not vary in this discrete Mendelian fashion, as we have already mentioned. Instead, they vary *continuously*. This means that it is hard to pick out distinct groups. For example, we talk about people as short or tall. But there are no such distinct groups, leaving aside dwarfism. Almost everyone falls within a smeared distribution, in which every height is represented from the very short to the very tall. And height is just one example of a continuously varying character. Others include weight, endurance, hand strength, running speed, resistance to disease, fertility, longevity, and so on.

How can we understand the genetics of such continuously varying characters? The first step is to realize that most characters are determined by both genetic and environmental factors. For much of the twentieth century, a controversy raged over the "nature vs. nurture" issue, particularly with respect to child rearing. In other words, does nature (genes) determine an organism's characteristics, or does nurture (environment)? Scientifically, this controversy is now dead, because almost all biologists qualified to address this issue agree that the answer is that *both* genes and environment are important, not one or the other (Figure 3.9A). This conclusion is embodied in a simple equation, where *P* refers to the phenotype, *G* refers to the genotype, and *E* refers to all other influences, from the physical environment to disease to development:

$$P = G + E$$

This equation summarizes the theoretical starting point for quantitative genetics. It is not, however, always true. As presented in Figure 3.9B, when there is an interaction between genetic and environmental effects, the equation fails. However, it does not fail in such a way that *G* or *E* alone determine *P*. The importance of both components remains.

A very simple statistical law is useful for understanding the action of genes and environments. When we have a variable *A* that is determined by an equation like $A = B + C$, and *B* and *C* are independent, then the variance of *A* is equal to the sum of the variances of *B* and *C*. That is, variances accumulate as additional causal factors are added in. Recall that variances (*V*) are the averages of the squared deviations from the mean. That is, the more heterogeneity there is for a character, the greater its variance. In the genetic situation, then,

$$V_P = V_G + V_E$$

This equation means that variation in a character like height has both genetic and environmental sources. The **phenotypic variance** (*V_P*) is equal to the sum of the genetic variance (*V_G*) and the **environmental variance** (*V_E*). ❖

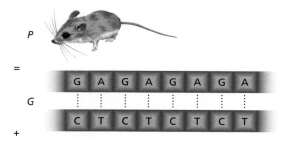

FIGURE 3.9A Phenotype = Genotype + Environment

FIGURE 3.9B When Genes and Environment Don't Add Up If the populations of a species evolve in both desert and mesic (not dry) habitat, then plants from these two habitats may respond to a lack of water very differently. In this example, the mesic plants grow very poorly when denied additional water, but desert plants are not affected very much.

3.10 The genes that make up a genotype may determine the phenotype additively or nonadditively

One concern of biologists is how genes determine characters. In developmental genetics, this topic would be called the problem of gene expression. In evolutionary genetics, this topic is the mapping from genotype to phenotype that is half of the basic cycle of the evolutionary process. The topic is also studied in quantitative genetics.

A central difficulty of genetics is that the same allele does not always have the same effect on the phenotype. The effects of an allele may be modulated by other alleles in the genotype. The simplest form of this modulation is called **dominance** (see Figure 3.10A). In genetic dominance, the expression of one allele largely or completely dominates the expression of another allele at the same locus. For example, diploid loci that code for pigmentation usually have dominance of nonwhite alleles over white or albino alleles. This is true in animals as different as mammals and insects. In these cases, having only one copy of an allele for normal pigment is enough to produce an almost normal coloration. This type of genetic dominance was much studied by early geneticists, and is still of interest to geneticists today.

Of greater medical interest is the fact that many human *genetic diseases* also exhibit dominance. This dominance comes in two forms. In one form, alleles causing genetic diseases may do so when they are present in just one copy. This is true of Huntington's disease, for example. Chapter 4 consider these diseases in more detail.

Another form of dominance in human genetic disease is **recessiveness** of the disease-causing allele, so that it takes two copies to cause the disease. Individuals who have only one copy of the disease allele are disease-free "carriers." Recessiveness is a very common pattern, illustrated by such genetic diseases as cystic fibrosis and Tay-Sachs disease. We will discuss these diseases further in Chapter 4, also. Genetic dominance is not just interesting; it can also be deadly.

Alleles that are not at the same genetic locus can also interact. This interaction is called **epistasis**. With epistasis, alleles at loci located far away in the genome, perhaps on another chromosome, can alter patterns of gene expression at another locus. This type of genetic interaction is not as easy to characterize as dominance. The simple thing about dominance is that it involves just two alleles, and they are "across from" each other genetically on matching (homologous) chromosomes. With epistasis, the important allele could be located millions of nucleotides down the chromosome, and there is no simple way to find it. Experiments with laboratory and agricultural organisms have also shown that epistasis can be important in determining such important characters as early survival, growth, and reproduction, as shown in the accompanying box. ❖

This experiment measured the viability levels of fruit fly (*Drosophila*) larvae with different genotypes at two loci: alcohol dehydrogenase and alpha-glycerophosphate dehydrogenase. S and F refer to the mobility of the proteins made by their corresponding alleles, S meaning slow, F meaning fast. This particular experiment was performed with alcohol absent. The values shown are average viabilities for the genotypes at the corresponding genotypes that align with the row and column of the viability's position.

Note that in some gene combinations there is superiority of the heterozygote. In other combinations, the heterozygote is not superior. This is epistasis, because the effects of alleles at each of the two loci are modified by the genotype at the other locus. This is an example of "nonadditive" inheritance.

(This experiment by Cavener and Clegg is presented in more detail in many genetics textbooks, including those in the list of readings at the end of the chapter.)

| | | Alcohol Dehydrogenase Genotype | | |
		SS	SF	FF
	SS	0.99	1.06	0.86
Alpha-glycerophosphate	SF	1.08	1.00	0.94
dehydrogenase genotype	FF	0.77	1.16	0.75

Expression of genes

Additive

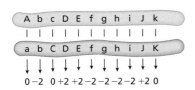

Example *AA* = +2, *Aa* = 0, *aa* = −2

Total genotypic value equals −4. We assume, as an illustration, that alleles represented by capital letters add +1 to the phenotype, while lower case alleles decrease the phenotype by −1.

Full dominance

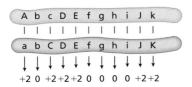

Total genotypic value is +24. We assume, as an illustration of dominance, that alleles represented by capital letters add +2 to the phenotype, whether there are one of two copies of the allele. The lowercase alleles have no effect.

FIGURE 3.10A Additive versus Fully Dominant Inheritance

The resemblance of relatives is determined by the ratio of the additive genetic variance to the phenotypic variance

What can we predict with variances? What can we learn from them? In fact, variances can be highly revealing. Suppose you wanted to know how much relatives are likely to resemble each other. Consider the simplest genetic situation: no interactions between different alleles, no interaction between genes and environment, and linkage equilibrium. This is the simplest, essentially ideal, case for quantitative genetics. Under these conditions, the genetic var-iance is reduced to the **additive genetic variance,** which is simply the genetic variance when there is no dominance or epistasis. We can define the **heritability** (h^2) of a character as the ratio of the additive genetic variance (V_A) to the phenotypic variance (V_P) that we have already seen:

$$h^2 = V_A/V_P$$

Heritability indicates the relative importance of inheritance in determining quantitative characters. When heritability is zero, inheritance has no importance. When heritability is 1.0, inheritance has overwhelming importance. The surprising thing about heritability is that it has a simple quantitative relationship to the **resemblance between relatives**. Resemblance between relatives means the similarity of biological characters between parent and child, brother and brother, grandparent and grandchild. This is a quantitative similarity. We can represent it graphically by plotting, for example, the average values of parental characters against the values of offspring characters, as shown in Figure 3.11A. In this type of graph, each point represents the quantitative characters that come out of an entire family. The collection of plotted points gives the data for the collection of families that have been studied. This kind of plot is just like any other in science. If we have two variables that are closely and positively related, we expect the points to fall near a rising line.

In Figure 3.11A, we show hypothetical data for adult weight in Old English Sheepdogs, a breed that you may have seen in children's movies, if not in real life. Usually the pattern of data like this is characterized by linear regression of the y-dimension data on the x-dimension data. (Linear regression, or least-squares linear regression, gives the straight line that comes closest to fitting the scatter of data plotted in two dimensions; see the Appendices.) In this example, the data are plotted with the average weight of the parents on the x-axis and the average weight of the offspring on the y-axis. A linear regression of average offspring weight on the average weight of their parents gives the best straight line for the fit of offspring to parent data. The slope of this straight line measures the strength of the relationship between parent and offspring weight when they are measured at the same age. Genetic theory shows that the slope of this regression is equal to the heritability. Thus characters with higher heritability, like body weight, should have larger slopes relating offspring phenotypes to parental phenotypes. Characters with lower heritabilities, such as most behavioral characters, should have shallower slopes relating offspring to parent, which would mean that genetics are relatively less important in determining such characters. Table 3.11A gives some examples.

It is also interesting that when the data are re-plotted using the weight of just one parent, instead of two, the expected slope falls to half the heritability. This makes sense, intuitively. If we know the phenotype of half the parents, we have half the information needed to predict the phenotype of the offspring. The importance of heritability in the regression scales with the amount of genetic information available.

Size characters tend to have higher heritabilities than do characters closely related to fertility. Analysis of many more characters than those shown here confirms this pattern. This may indicate that natural selection has used up genetic variability for characters like fertility, for the reasons discussed in Chapter 4. ❖

TABLE 3.11A Some Examples of Heritabilities

Morphology

Human height: 0.65	Adult weight in cattle: 0.65
Pig growth rate: 0.40	Adult weight in poultry: 0.55

Fertility

Pig litter size: 0.05	Egg production in poultry: 0.10
Mouse litter size: 0.20	Fruit-fly egg production: 0.20

From D. S. Falconer and T.F.C. Mackay, *Introduction to Quantitative Genetics*, 4th ed. (Harlow, Essex: Longman, 1996).

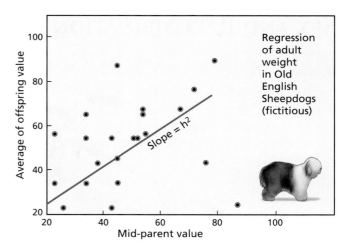

The slope of the line predicting an individual's phenotype from the phenotypes of its relatives has the following values:

h^2 when the corresponding phenotypes of both parents are used to make the prediction

$h^2/2$ when the phenotype of only one parent is used to make the prediction

For characteristics like size and weight, heritability is typically 0.6 – 0.8. For such characters as fertility or running speed, heritability is typically 0.2 – 0.6.

FIGURE 3.11A How Much Will Organisms Resemble Their Relatives? This depends on "heritability" or h^2. When inheritance is additive, $h^2 = V_G/V_P$.

Sex and Recombination

From an evolutionary standpoint, the important thing about genetic processes is that they shuffle genes, just like cards are shuffled in card games. This shuffling process does not create any new genes, and it is not usually biased. Like the dealer in a card game, the genetic machinery makes new combinations of alleles—like new combinations of cards in "hands"—and thus new genotypes.

From card games, you may know that even with the same 52 cards in a deck, the number of different hands that you might be dealt is extremely large, if there are a lot of cards per hand. Blackjack, which has relatively few cards in play at any one time, may be at the simpler end of the spectrum. But in bridge, where each player gets 13 cards, the likelihood that you will get two hands with exactly the same combination of 13 cards in any one game is vanishingly remote.

In genetics, the number of genes per genotype is far higher than the number of cards in the hands of any card game. Humans have about 20-25 thousand genes that code for important functions, and each of these genes may have many different alleles. If you calculate the possible genotypes, the total number that might occur is very large. By some estimates, there are more possible human genotypes than there are atoms in the universe.

In this sense, the genetics of populations are very big card games indeed, and we have little prospect of winning by "counting cards." In other words, it is unlikely that we can understand life by directly calculating all of its genetic possibilities. Instead, we try to understand how the evolutionary machinery works in some average, or typical, sense. What does the evolutionary machinery normally do?

As mentioned, a notable feature of genetics is that it usually operates without any bias to its shuffling. Because of this, it tends to randomize genotype frequencies, within loci and between loci. Neither of these statements is absolute, however. Genetic processes are sometimes biased, as we will discuss in Chapter 5. And the tendency to randomize genotype frequencies is not absolute. Indeed, sustained inbreeding tends to produce odd combinations of genes, though sustained inbreeding is not the rule in the natural world. The idea of genetics as randomizing is no absolute law of science, but it is a general rule of thumb.

For example, consider a piece of the genetic machinery that should already be familiar. We have seen in Module 3.6 what one generation of random mating can do to genotype frequencies at a single locus. One generation of random mating leads immediately to the Hardy-Weinberg equilibrium. At this equilibrium, we can calculate the frequency of genotypes from the frequencies of the alleles at the locus. If the frequency of allele A is p, then the probability of A occurring twice in a genotype (AA) is p^2. This is just like the probability of getting two heads in a row when tossing a coin: $\frac{1}{2} \times \frac{1}{2} = \frac{1}{4}$. And so on. The Hardy-Weinberg equilibrium is a kind of "even" or "smoothed-out" state, in which probabilities are well behaved. Alleles are randomly associated with each other; this we have already seen.

Figure 3.12A shows two different populations in which the frequencies of alleles A and B are both 0.5, at their two respective loci. In part (i), the improbable case, all the uppercase alleles are in one genotype and all the lowercase alleles are in the other genotype. In part (ii), the probable case, the genotype frequencies reflect the random combination of alleles into genotypes, with no unusual biases or associations. Intuitively, we expect the evolutionary machinery to undermine the improbable pattern of genotype frequencies, and so produce something like the random combinations of alleles of this example. Next we consider exactly how this happens. ❖

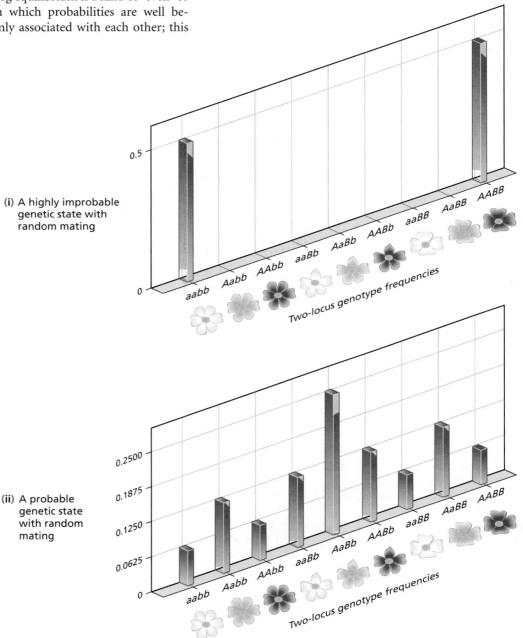

(i) A highly improbable genetic state with random mating

(ii) A probable genetic state with random mating

FIGURE 3.12A Two Genetic States, One Improbable and One Probable

3.13 With random mating, sex-chromosome genes that start out at different frequencies move toward the same frequency

One of the simplest cases of genetics randomizing genotypes occurs when there are two sexes. In principle, though rarely in practice, the two sexes can have very different frequencies for alleles located on the **X chromosome**. Let's take a hypothetical example. Imagine a group of female space-travelers whose spaceship crashes on an all-male prison planet. If the females came from a different solar system, which had been colonized long before, then they might have X-chromosome alleles that are totally different from those of the men in the prison. All the women would have one allele, and all the men would have another allele. The interesting question then is, what will happen to the X-chromosome allele frequencies of the two sexes if they start mating with each other and thereby create their own autonomous population on the prison planet?

A key factor is that, after the founding generation, the XY males of each generation will have the X-chromosome allele frequency of the XX females of the previous generation. This occurs because males receive their **Y chromosome** from their father and their X chromosome from their mother. You can see this in Figure 3.13A, which shows the outcomes of all possible matings when there is one X-chromosome locus having just two variant alleles. When mating is random, the male X chromosomes will be a random sample of the X chromosomes of the females of the preceding generation.

But the female case is different. Daughters do not get a Y chromosome from their fathers. Their XX genotype comes from a paternal X chromosome paired with a maternal X chromosome, exactly one from each parent.

With random mating, this pair of X chromosomes is a random combination of X chromosomes from both males and females, the two sexes equally represented. Therefore, for daughters, the frequencies of alleles on the X are averages of

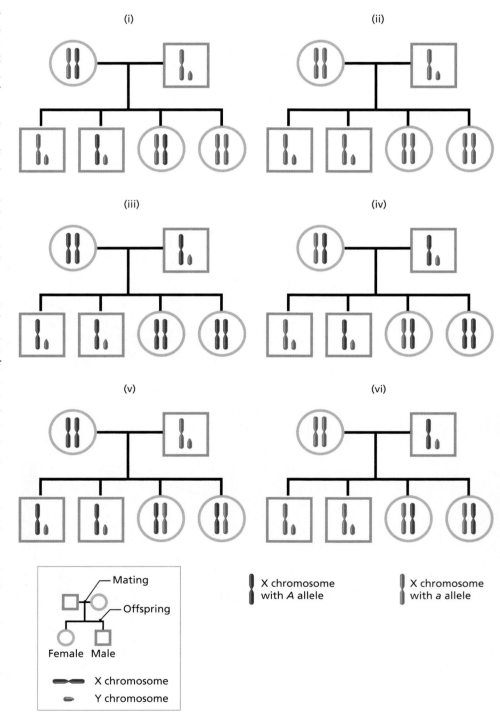

FIGURE 3.13A **All Possible Matings When There Is Polymorphism at a Sex-Linked X Chromosome Locus**

allele frequencies from both sexes. This averaging tends to reduce the difference between male and female allele frequencies for genes on the X chromosomes, because the average of two different numbers is equidistant between them. (The average of 2 and 8, for example, is 5.) This averaging does not immediately eliminate the difference in allele frequency between the sexes, however. Instead, as Figure 3.13B shows for allele *a*, the difference between them progressively falls, as allele frequencies bounce back and forth between males and females. Eventually, there would be no difference between male and female X-chromosome allele frequencies in the new population on the prison planet. ❖

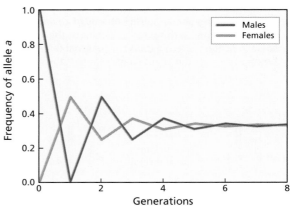

FIGURE 3.13B **Allele Frequency Change at a Sex-Linked Locus**

Sex and Recombination **103**

A simple analogy for understanding genetics is that genes are organized like sentences in books, where chromosomes are the books. There are 23 pairs of human chromosomes, together containing about 20-25 thousand genes. This means that each chromosome has an average of about 1,000 genes. These genes are not strung along, one after another, with no gaps. At many places along the length of a human chromosome, there are large gaps where the DNA does not code for useful RNA or protein. Because of these large gaps, pairs of chromosomes sometimes recombine their genes without much effect on gene-coding regions.

Let's continue with the book analogy to see how this happens. Imagine two novels *of the same length* being printed at the same time by a book factory. If a printer mixes up the printing plates, then instead of producing copies of *Death's Dishonor* and *Summer Swoon*, they would print some books in which the text comes from *Death's Dishonor* for the first 128 pages, and then *Summer Swoon* for the last 211 pages, as well as the reciprocal switch of *Summer Swoon* for the first 128 pages, followed by *Death's Dishonor* for the last 211 pages. The printer's recombination occurred between pages 128 and 129. At that point, the two stories *cross over*. In the same way, chromosomes of the same type (homologous) can be recombined by breaking in two between genes and then rejoining by crossing over between chromosomes.

Chromosome **recombination** has varied results. First, genetic recombination may make no difference, because pairs of chromosomes may be carrying the same alleles at a genetic locus, as shown in Figure 3.14B. In our two novels, this case is

analogous to recombination between two books that begin with exactly the same words, at least up to the point of recombination. If the printer's mix-up takes place before the two stories differ, then it makes no difference.

But when physical recombination occurs and there are different alleles on the recombining chromosomes, the genetic system acts to shuffle the alleles. This shuffling tends to break down unique or unusual genetic combinations, rendering them only as common as they would be if the alleles had combined at random, irrespective of their chromosomal location.

For example, as shown in Figure 3.14A, A and B alleles could be associated with each other, and a and b likewise. These alleles would show **coupled phase** association with each other, according to uppercase or lowercase. Similarly, associations between A and b or a and B, which are called **repulsion phase** associations, could occur. Such coupled phase associations and repulsion phase associations can contribute to **linkage disequilibrium**. Linkage disequilibrium is measured as the difference between actual genotype frequencies and the frequencies expected from random combination of alleles, as described earlier. Either type of association tends to be broken down by recombination, as shown in Figures 3.14C and 3.14D, reducing linkage disequilibrium.

Figure 3.14E shows this process of randomization among genes undergoing recombination at different rates ($r = 50\%$, 10%, and 1%). The more recombination, the faster the association between alleles disappears. But in all these cases, recombination does eventually destroy nonrandom association, making linkage disequilibrium fall to zero. ❖

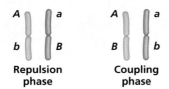

Repulsion phase   **Coupling phase**

FIGURE 3.14A Coupling and Repulsion Phases of Linkage Here and in the other figures on this page, the color of the chromosome indicates maternal (pink) or paternal (blue) origin of the chromosome, not the allelic composition of the chromosome.

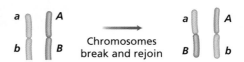

Chromosomes break and rejoin

FIGURE 3.14C Recombination matters in coupling phase, when it produces repulsion-phase gametes.

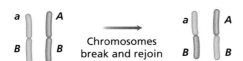

Chromosomes break and rejoin

FIGURE 3.14B Physical recombination of chromosomes may not change genetic makeup.

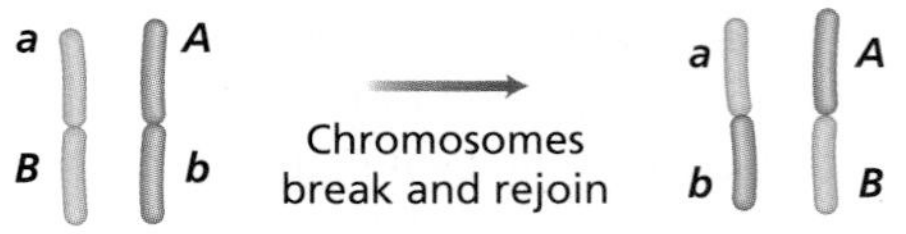

Chromosomes break and rejoin

FIGURE 3.14D Recombination matters in repulsion phase, when it produces coupling-phase gametes.

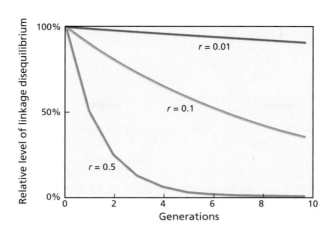

FIGURE 3.14E Decay in Linkage Disequilibrium through Time

INBREEDING

3.15 Inbreeding is a bad thing in normally outbreeding natural populations

Inbreeding occurs when related individuals mate. This process is very important for the genetics of families and for the genetics of populations. Inbreeding is not an all-or-nothing event. Instead, the degree of inbreeding is quantitative. The pedigree in Figure 3.15A diagrams a case of brother-sister mating, from which three daughters were produced. The pedigree in Figure 3.15B shows a first-cousin mating that produces two daughters and a son. Both of these pedigrees are examples of inbreeding. The critical point in the definition of inbreeding is whether individuals share alleles that they inherited from a common ancestor. Only if there is biological descent from at least one common ancestor can there be inbreeding. (A stepfather's rape of his stepdaughter is not usually biological inbreeding, even when the law considers it incest.) Having a great-grandfather in common, as may be true of second cousins, is inbreeding even if it is not considered incest. From here on, our discussion will consider inbreeding only, whatever the social or legal conventions concerning incestuous mating.

Within biology, degrees of inbreeding are distinguished according to the probability that two mating individuals have an allele in common from an ancestor. This is a probability that can be calculated, as we will discuss shortly. For now, the important point is that inbreeding varies quantitatively.

Inbreeding has many important effects on the evolutionary process—reducing **heterozygosity**, decreasing genetic variance, and so on. In medicine, however, inbreeding is better known from its effects on health. Many human genetic diseases are caused by recessive alleles. In such cases, the diseases occur only when the patient has two copies of a defective allele, resulting in a failure to produce the normal protein coded for by the genetic locus. Such disorders are called **recessive genetic diseases**. Some of the most common and devastating human genetic diseases, like cystic fibrosis and Tay-Sachs (see Chapter 4), are caused by a lack of normal alleles at a single locus.

Inbreeding greatly increases the frequency of genetic diseases that arise from **homozygosity** (having two copies of the same allele). Natural selection normally keeps disease genes at very low frequencies, so only a few individuals will have even one copy of a gene for a recessive genetic disease. But because there are thousands of loci that can cause recessive genetic disease, even though disease alleles are rare at each locus, each of us carries one copy of a few alleles for recessive genetic diseases among all our loci. Fortunately, these alleles are heterozygous,

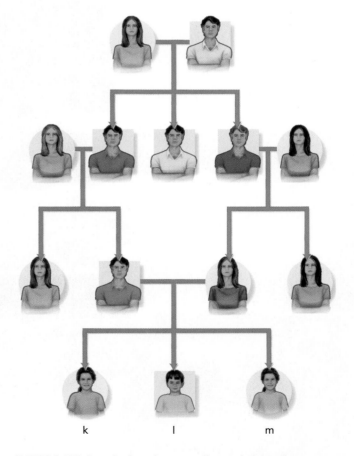

FIGURE 3.15B Females k and m, as well as male l, are the offspring of a first-cousin marriage. As a result, they will be moderately inbred.

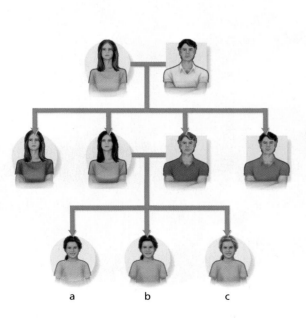

FIGURE 3.15A Females a, b, and c are the inbred offspring of full siblings. On average, they will be very inbred.

so they cause us few medical problems. Statistically, we are unlikely to mate with an individual carrying the same disease alleles, if we are not related to each other. For this reason, most of our offspring are normal regarding genetic disease.

Things are different for inbred individuals. Parents that share a common ancestor are much more likely to have inherited the same genetic disease genes, albeit in heterozygous form. When such related individuals have children, their children are at a much greater risk of being homozygous for one or more alleles conferring a genetic disease. These inbred offspring are more likely to suffer medical complaints and more likely to die young. Such victims of inbreeding also tend to suffer a range of morphological oddities, such as extra toes or fingers, along with intellectual retardation. ❖

How Much Inbreeding Is There in Nature?

Inbreeding is common on farms and in scientific laboratories. But how common is it when humans don't interfere with animal or plant breeding?

It is fairly well established that species with very limited dispersal are often inbred in the sense of having very small populations. Small cave invertebrates, mostly-selfing plants like peas, plants with very local pollination, and selfing worms like nematodes all may have high levels of inbreeding. With such inbreeding comes high levels of homozygosity and accidental differentiation of local subpopulations.

On the other hand, outbreeding is extremely common. Animals and plants that seem like they should be inbred often are not. The humble fruit fly *Drosophila melanogaster*, which doesn't seem like much of a flyer and can't walk very fast, apparently mates over a large area, so that it exchanges genes over hundreds of square miles.

Pollen can spread surprising distances, genetically uniting plants over wide areas. This is to say nothing of organisms that obviously disperse and mate over large areas, like most birds and many mammals. We currently do not know enough about how common inbreeding is relative to outbreeding. This is an important question, because the human impact on the environment accumulates unchecked. We do not know if we are causing the extinction of distinct populations or merely reducing the abundance of a species that disperses widely. If species mate widely, then their homozygosity will be much less than it would be if the species are broken up into local breeding populations. (This is considered further in Module 3.20.) Less homozygosity, in turn, should reduce inbreeding depression. With less inbreeding depression, endangered species should be less likely to go extinct.

3.16 The degree of inbreeding can be calculated from the probability that parents share alleles from a common ancestor

Because inbreeding is one of the most important processes of population genetics, it is convenient that biologists can calculate the degree of inbreeding fairly easily. These calculations revolve around the probability that related individuals will have offspring that are homozygous. The most important thing in these calculations is to keep track of who is mating with whom. Here we will assume that we know the exact truth about every individual's parentage.

Figure 3.16A shows what happens when two half-siblings mate. Half-siblings have only one parent in common—in this case, their mother. We can consider one locus at a time, because we will be calculating average probabilities for the genetics of this locus. Many loci may be made homozygous by inbreeding, but the average frequency of homozygosity will be close to the results of the one-locus calculation.

We need to follow the alleles that each half-sib inherits from its mother, so we label the mother's alleles a_1 and a_2 in Figure 3.16A. We can ignore the genotypes of the fathers when we can assume that they are not related to each other or to the mother. For this reason, the alleles that come from the fathers are represented by dashes.

Mendelian genetics is like a card game, as we remarked earlier in the chapter. When the mother has a son with one father, the son has a probability of $\frac{1}{2}$ of getting allele a_1 from his mother and a probability of $\frac{1}{2}$ of getting allele a_2 from her. The same thing is true of the daughter by the other father.

Under these conditions, there are two scenarios by which the half-sibs can have an allele in common: Each receives a copy of a_1 or each receives a copy of a_2. The probability of the first scenario is $\frac{1}{4}$, because the chance of each of them receiving an a_1 is $\frac{1}{2}$, and these two genetic transmission events are independent of each other. Therefore, we multiply their probabilities together to get the probability that both occur: $\frac{1}{2} \times \frac{1}{2} = \frac{1}{4}$. Likewise, the probability of the second scenario, in which the two half-sibs have the a_2 allele in common, is also $\frac{1}{4}$. We then sum these two probabilities to get a probability of $\frac{1}{2}$ that the half-sibs have a maternal allele in common at this locus, whether a_1 or a_2.

Mother and two unrelated fathers

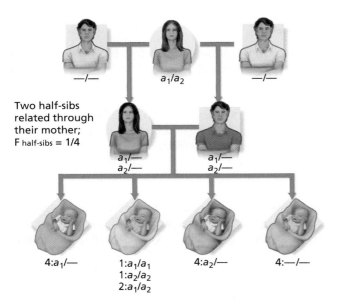

Two half-sibs related through their mother; $F_{half-sibs} = 1/4$

Offspring of the two half-sibs; 16 possible outcomes of which one is homozygosity for allele a_1 and one homozygosity for a_2

One-sixteenth of the offspring: a_1/a_1.
One-sixteenth of the offspring: a_2/a_2.

One-eighth of the offspring are homozygous because of inbreeding alone.

FIGURE 3.16A Genetic effects of Incest

Probabilities of genetic outcomes

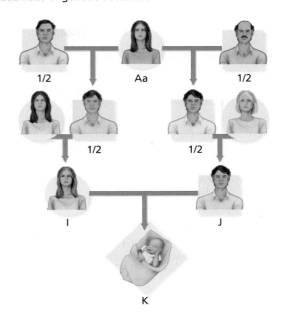

The inbreeding coefficient can be estimated by multiplying 1/2 by 1/2 by 1/2 by 1/2 to get 1/16. The 1/2 comes from Mendelian segregation.

FIGURE 3.16B A Simple Rule for Estimating the Inbreeding Coefficient, F.

In the case that the half-sibs have an allele in common, there is a probability of $\frac{1}{4}$ that a child of theirs would be homozygous for that locus. (From the rules of Mendelian segregation, two heterozygotes have a chance of $\frac{1}{4}$ of producing an offspring that is homozygous for a particular allele when there are only two alleles.) Multiplying this probability by the chance that the half-sibs do have an allele in common, 1/2, we find that the probability of homozygosity arising from half-sib inbreeding is 1/8.

If we choose an allele at random from one of two half-sibs just described, there is a 50 percent chance that this allele is a maternal allele. As we have shown earlier, the chance that the two half-sibs share a maternal allele in common is also 50 percent. Thus the chance that a randomly chosen allele in two half-sibs is identical by common ancestry is $\frac{1}{2} \times \frac{1}{2} = \frac{1}{4}$. This probability is sometimes called the **inbreeding coefficient** or *F* value. In large populations that are mating at random, the average inbreeding coefficient will be close to zero. In organisms that have been inbreeding for a number of generations, *F* may approach 1, its maximum value. In a sense, the inbreeding coefficient is a measure of deviation from random mating.

Inbreeding coefficients can be relatively easy to calculate. You take the two individuals whose inbreeding coefficient you wish to calculate, and you follow the pedigree connecting them. For each connection in the pedigree, you multiply by $\frac{1}{2}$. Thus two individuals related by just one common parent—such as the half-sibs we have just discussed—have an *F* value of $\frac{1}{4}$ (Figure 3.16A). When they have just one common grandparent, they would have an *F* value of 1/16 (Figure 3.16B). With multiple parents and grandparents in common, these calculations have to be repeated for each independent pathway through the pedigree. This can get complicated. However, most types of inbreeding involve the same patterns. ❖

3.17 When relatives mate frequently, homozygotes increase in frequency while heterozygotes decrease in frequency

Let's consider cases in which inbreeding is sustained from generation to generation. In these situations, groups of inbreeding organisms are called **inbred lines**.

Although human inbreeding causes genetic disease, there are some good reasons for inbreeding in agriculture. We regularly eat parts of inbred organisms, both plant and animal. When inbreeding is sustained, the genetic polymorphism of randomly mating, or **outbred**, populations is lost (Figure 3.17A). Inbred lines tend to *fix* single alleles at each locus. (The term **fixation** in genetics refers to 100 percent frequency.) Which allele will be fixed is not usually predictable. But if several different inbred lines are produced, plant or animal breeders can choose the line with the best attributes for their purposes. The advantage of working with inbred lines is that inbred lines remain "true to breed." Their lack of genetic variation ensures that deviant offspring will be rare.

Figure 3.17B presents an example of how inbreeding can reduce the heterozygote frequency. An outbred population of plants has six genotypes at locus *A*. All these genotypes are represented in the original population. In three inbred lines derived from the outbred population, only two genotypes are represented, one genotype by two different inbred lines. Both of these genotypes specify short plants, though this is only coincidental. In this way, by accident, inbreeding has produced a directional change in plant height.

Some plants can **self-fertilize**; that is, they can be fertilized by pollen from their own flowers. **Selfing** is a form of inbreeding. To breed with another plant, plants are often dependent on **pollinators** like bees and birds, which carry pollen from one plant to another. When both selfing and external pollination are possible, all that is required for inbreeding is the loss of pollinators. When all the pollinators are lost, reproduction must occur by selfing, if it is to occur at all. Self-fertilization is not as common in animals, but it does occur in some snails and worms. In some species of nematodes—very simple worms—reproduction is normally by self-fertilization. This makes inbreeding, along with homozygosity, very common in these species; and *F* (the inbreeding coefficient) approaches 1. ❖

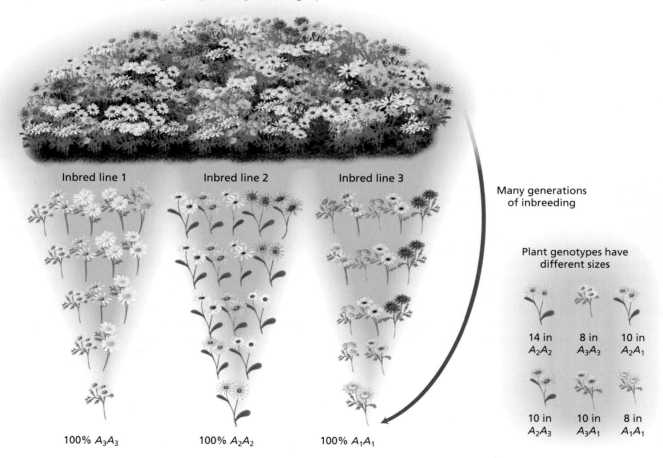

Alleles A_1, A_2, and A_3 at Hardy-Weinberg equilibrium

Inbred line 1   Inbred line 2   Inbred line 3

Many generations of inbreeding

Plant genotypes have different sizes

14 in A_2A_2 8 in A_3A_3 10 in A_2A_1

10 in A_2A_3 10 in A_3A_1 8 in A_1A_1

100% A_3A_3 100% A_2A_2 100% A_1A_1

FIGURE 3.17B Differentiation of Inbred Lines from Each Other Again we suppose that several inbred lines are derived from an outbred population. When different genotypes are associated with different phenotypes, such as height, then inbreeding will produce inbred groups with consistent differences in such characters.

Outbreeding population

FLOWER SALE

A_3A_7

A_3A_4

A_7A_0

A_5A_3
A_2A_3

A_5A_5 A_5A_6 A_1A_3 A_8A_5

38 genotypes

First year: Planting the plants from the green house.

Inbred derivative

Second year: Both families have only three plants survive.

Third year: Both families lose one plant from their garden.

Fourth year: Both families lose another plant leaving each family with just one.

A_3A_3 A_4A_4

FIGURE 3.17A Creating Inbred Lines Imagine a plant that has one generation per year. If we start with a greenhouse population of 38, two families might buy plants for their gardens. The Gardenias might buy 5 plants, while the Mudges buy 3. If these plants mate only within their home gardens, and some plants are lost each year because of bad gardening, then the two families may end up with genetically distinct plants in their gardens.

Inbreeding   **111**

Inbreeding tends to reduce the variance of quantitative characters within inbred lines

It is fairly easy to understand that inbreeding will tend to fix particular alleles at individual loci, increasing the level of genetic homozygosity in a population. It may be more difficult to understand that similar effects occur where the inheritance of quantitative characters is concerned; but they do.

Recall that the variation of a quantitative character has two basic components: genetic variation and environmental variation. For now, we will assume that changes in the mating system will not affect the environmental component of variation. It is the effect on genetic variation that is the basis for predicting that inbreeding will tend to reduce the variance of quantitative characters.

To understand the effect of inbreeding, it is important to distinguish its effect on a group of inbred lines from its effect on a single line. An entire collection of different inbred lines may possess a great deal of genetic variation. Just think of the enormous variation among all the different breeds of dog (see Figure 3.18A). Domestic dogs are descendants of one, or a few, populations of dog that humans domesticated 50 to 150 thousand years ago, when these dogs were essentially wolves. Yet an individual breed of dog does not contain all this variation. The effect of inbreeding on the individual breed is very different from its effect on an ensemble of inbred lines.

Let us focus on a single inbred line. As inbreeding proceeds, each locus has an increased likelihood of becoming homozygous, as we saw in the preceding module. Over all loci, many will become homozygous as a result of inbreeding. When loci that affect quantitative characters become homozygous, there will be less quantitative genetic variance (V_G) affecting those quantitative characters. This occurs because quantitative genetic variance requires genetic polymorphism. As Figure 3.18B shows, sustained inbreeding reduces the genetic component of the phenotypic variance (V_P) during sustained inbreeding, making the phenotypic variance approach the value of the environmental variance (V_E). This reduction in variance increases the predictability of the characters of the inbred line.

If these inbred lines are horses, dogs, cows, or tomatoes, this increased predictability may increase the value of the inbred line. Indeed, dog shows and similar competitions for agricultural animals often focus on specific standards that have been established for breeds. Deviations from those standards result in lost points during competition. Such deviations can be prevented best by maintaining the purity of the breed, avoiding any crosses with animals from other breeds or crosses with mongrels. Long-standing human practices have thereby fostered the continued inbreeding of dogs as well as other animals and plants. ❖

FIGURE 3.18A Three Inbred Dog Breeds Proceeding from top to bottom: American beagle, Belgian sheepdog, Pekingese.

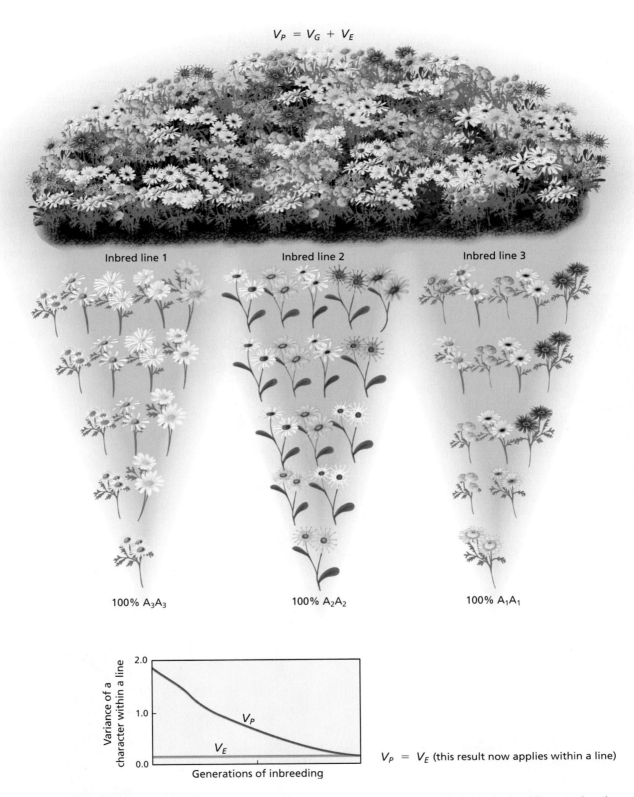

$$V_P = V_G + V_E$$

Inbred line 1

Inbred line 2

Inbred line 3

100% A_3A_3

100% A_2A_2

100% A_1A_1

$V_P = V_E$ (this result now applies within a line)

FIGURE 3.18B Uniformity within Inbred Lines Inbreeding bleeds the genetic variation out of individual inbred lines, so that the phenotypic variance (V_P) eventually equals the environmental variance (V_E). This is how breeders create lines that breed pure; examples of such inbred lines range from dog breeds to flower varieties.

Inbreeding tends to reduce the average value
of beneficial characters

Inbreeding does more than reduce genetic variances. It also reduces the value of characters that are related to fitness or function, when populations begin as outbred. This reduction is known as **inbreeding depression**.

Figure 3.19A shows the impact of inbreeding on the dog pelvis. In highly inbred large breeds, like German shepherds and Saint Bernards, the joint connecting the femur to the pelvis is undermined. The head of the femur tends to become detached—a condition known as *hip dysplasia*. High levels of inbreeding in dogs also produce infertility, inappropriate aggression, blindness, and so on, as Figure 3.19B shows. The phenomenon is well known in a variety of agricultural animals and plants, and it can be produced in laboratory breeding experiments at will.

Inbreeding depression has an opposite. When unrelated inbred lines are crossed, their hybrids usually show considerable superiority. This is called **hybrid vigor**. For example,

mongrels usually have fewer health problems than purebred dogs do. Most grains used in agriculture are hybrids of inbred lines. These grains produce higher yields, and they are more resistant to disease. Inbreeding depression and hybrid vigor are thus fundamental for agricultural breeding programs.

The puzzle is why inbreeding depression should be so common. One clue comes from organisms that normally inbreed in nature, such as self-fertilizing worms. These species do *not* show inbreeding depression or hybrid vigor. Inbred lines of such organisms show no decline in function or fertility, nor do crosses of their inbred lines always give hybrid vigor. Therefore, mere homozygosity does not cause inbreeding depression.

Wild populations of outbreeding organisms are fairly heterozygous. Inbred lines, on the other hand, are highly homozygous. Therefore, relative to inbreeding, outbreeding will tend to select for alleles that produce higher fitness when heterozygous. Long-standing inbreeding, by contrast, will select strictly on an

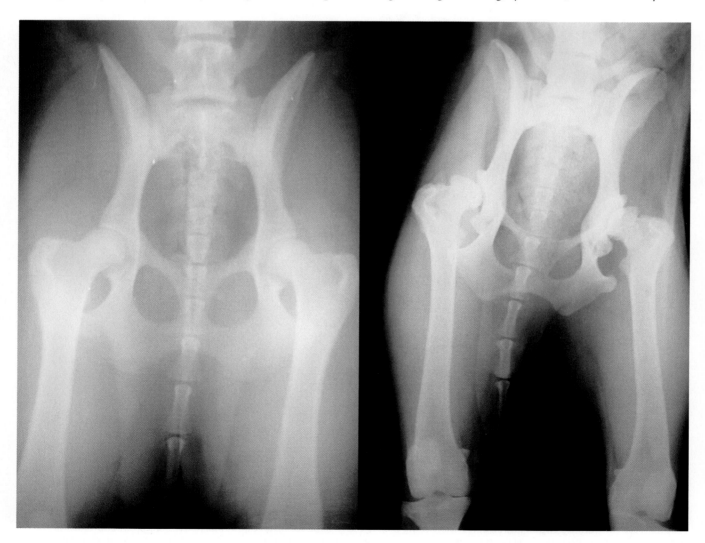

FIGURE 3.19A Hip Dysplasia in a Dog The x-ray of a normal dog is shown on the left. The dog with hip dysplasia is shown on the right.

allele's effect when homozygous, because that is the normal genetic situation with sustained inbreeding.

Consider a rare allele that is fairly benign when present in heterozygous form, but a disaster when there are two copies of the allele in the genome. Some human **genetic diseases** are thought to fit this pattern, one example being cystic fibrosis. Many people carry one copy of the gene for this genetic disease, but they show few, if any, bad effects. They are only carriers. Those with two copies of the gene suffer a debilitating and life-shortening disease, a disease that renders males almost totally infertile. And for evolution, infertility is a major problem. (Cystic fibrosis, and other examples of genetic disease, are discussed further in Chapter 4.) Inbreeding among carriers of the genetic disease cystic fibrosis would produce many more afflicted individuals than normally occur in the general population. But if humans were always inbred, this situation would not occur—because selection would then virtually eliminate the gene for cystic fibrosis. ❖

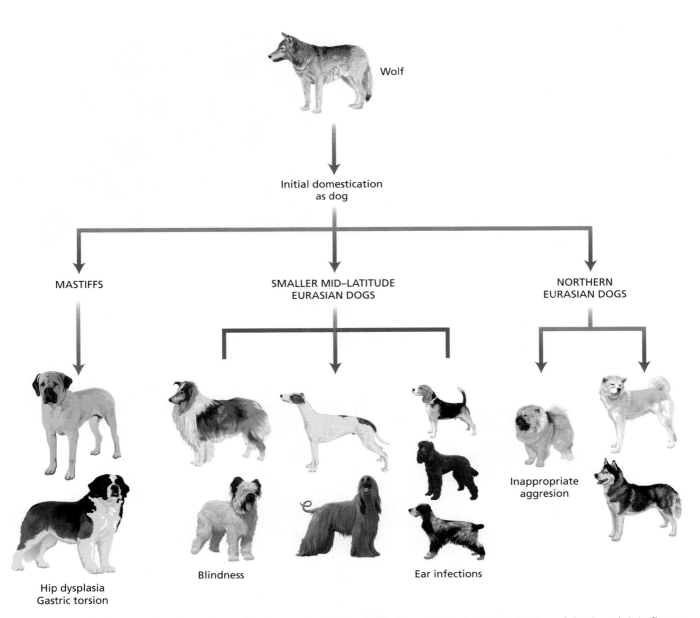

FIGURE 3.19B The History of Dog Breeding, with Annotations for Health Problems in Purebreds The history of dog breeds is in flux as of this writing, but that does not affect the health problems of inbred breeds. Mongrels are usually free of the health problems of purebred dogs.

Inbreeding can be more subtle than the mating of relatives or the systematic inbreeding of agricultural animals. It can arise from the geographical distribution of a species. That is, who mates with whom can be made nonrandom by the mere accident of location.

Figure 3.20A illustrates this principle of **subdivision of populations**. In our hypothetical example, plants are living on the side of a large mountain. Soil suitable for the growth of this plant is present only at a few locations—one pasture high on the mountain, and one on each of the east and west mountain sides. If these plants have pollinators that do not disperse very far—such as small beetles—then the three plant populations may be largely isolated from each other. When this occurs, chance alone will cause genetic differentiation of these small populations, as we will discuss

in the next module. There are four possible patterns of flowering—no flowers, pink flowers, yellow flowers, and blue flowers—determined genetically in part. The three populations differ in the genes that affect flowering, such that each population is a bit differentiated from the other. This differentiation gives rise to more variation in the species as a whole than would be expected from study of just one of the isolated populations. Conversely, there is less variation within each of these isolated populations than there is at the level of the species as a whole.

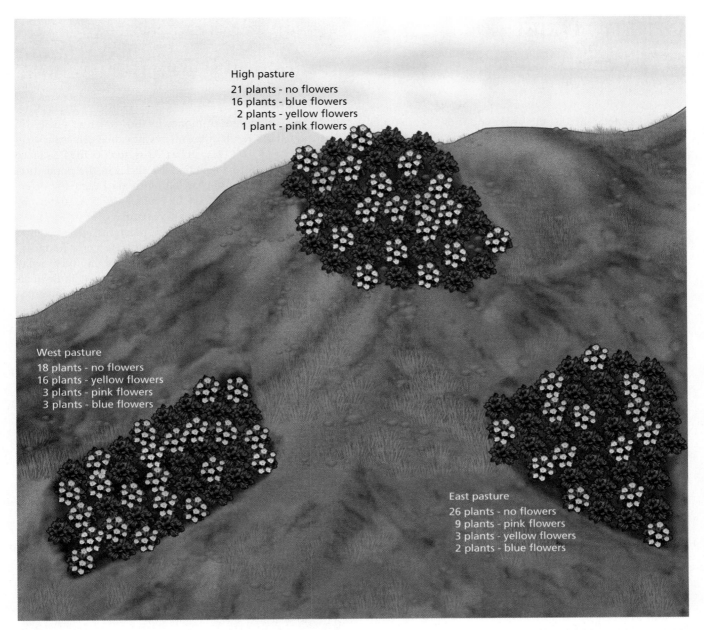

High pasture
21 plants - no flowers
16 plants - blue flowers
2 plants - yellow flowers
1 plant - pink flowers

West pasture
18 plants - no flowers
16 plants - yellow flowers
3 plants - pink flowers
3 plants - blue flowers

East pasture
26 plants - no flowers
9 plants - pink flowers
3 plants - yellow flowers
2 plants - blue flowers

FIGURE 3.20A Some populations in nature often contain more of one genetic variant than do other populations of the species; genetic variation is in this case described as subdivided.

Species with very large populations that frequently intermate do not have this type of population structure. Seagulls and pine trees, for example, usually mate widely, because both marine birds and conifer pollen can travel great distances, breaking down local subdivision and differentiation of populations.

The impact of population subdivision is known as the **Wahlund effect**. Figure 3.20B gives a calculation showing the effect on heterozygosity of variance in allele frequency between subpopulations. The more variance in allele frequency, the greater the depression in the frequency of heterozygotes in the population as a whole. Conversely, the isolation of subpopulations increases the frequency of homozygous individuals. ❖

$$H = 2pq - 2V_q$$

The frequency of heterozygotes

The normal frequency of heterozygotes, with no population subdivision

Twice the variance in the allele frequency over the subpopulations

FIGURE 3.20B Wahlund Effect The more subpopulations vary in allele frequency, the fewer heterozygotes will be present in the population as a whole.

GENETIC DRIFT

3.21 Genetics is like card games, and genetic drift is like a trip to Las Vegas

Think of **genetic drift** as the working out of chance on the next level up from inheritance in individual families. Mendelian genetics considers the inheritance of particular characters with known parents. It is very much like a game of chance, such as coin tossing or cards (Figure 3.21A). If we mate two heterozygotes, *Aa* and *Aa*, what are the chances that we will get an *AA* genotype with just one child? We know that the chances are $\frac{1}{4}$, because the chance that the gamete from one parent is *A* is $\frac{1}{2}$, and the same for the other parent. With independent production of gametes by the two parents, the probability of this happening twice is just the product $\frac{1}{2} \times \frac{1}{2}$, which equals $\frac{1}{4}$. This problem is just the same as the chance of getting two heads in a row when we toss a coin.

> *Just as the amount of money in your wallet or purse will rise and fall as a result of the many individual games that you play in Las Vegas, so does the frequency of individual alleles in the population rise or fall.*

But at the next level up, in the genetics of whole populations, the situation is different. In the genetics of populations, the effects of the individual genetic processes combine. And there are as many of these processes as there are individuals producing gametes from which the next generation is made. Population genetics without selection is like going to Las Vegas for a weekend's gambling. (Figure 3.21B). You will be playing many rounds of poker or blackjack, putting quarters in slot machines, each play like the production of gametes by a single mated couple. The financial outcome for you, whether you will be richer or poorer, is a higher-level chance process. In this sense, we can describe genetic drift as a population's random production of gametes and zygotes to create the next generation, a higher-level process laid on top of the lower-level process of genetics itself.

Just as the amount of money in your wallet or purse will rise and fall as a result of the many individual games that you play in Las Vegas, so does the frequency of individual alleles in the population rise or fall. Both are essentially determined by "luck," which is to say by nothing in particular.

In small populations, allele frequencies change in discrete steps. To understand why, consider a very small population of just two individuals (Figure 3.21C). If we consider an autosomal locus with two alleles, *A* and *a*, there are five possible values for the frequency of the *A* allele: $0, \frac{1}{4}, \frac{1}{2}, \frac{3}{4}, 1$. Two configurations will produce an allele frequency of $\frac{1}{2}$: The population may be composed of two heterozygotes, or it may have one *AA* homozygote and one *aa* homozygote. When these two individuals reproduce to create the next generation of two individuals, the frequency of the *A* allele may change to any

FIGURE 3.21B Las Vegas around the End of the Twentieth Century

FIGURE 3.21A A Game of Cards in the Eighteenth Century

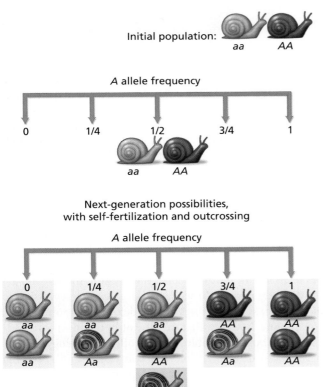

one of these five values. Those of you paying close attention may not understand how a sexually reproducing species with one *AA* individual and one *aa* individual could produce two *AA* offspring, for instance. The answer is that our simple model permits self-fertilization, although it can be easily modified to account for sexual reproduction without self-fertilization.

The important point is that with random genetic drift, as long as both alleles are present, allele frequencies can change from their current values to any of the other possible values. As it turns out, the probabilities of changing to any one of these other values are not all equal, and they depend on the population size. In a population of size two at an allele frequency of 0.5, the chance of one allele frequency going to zero in the next generation is 1/16. However, in a population of 100 at the same allele frequency, the chance of going to zero in one generation is about 10^{-61}. ❖

Initial population:
aa   AA

A allele frequency

0   1/4   1/2   3/4   1

aa   AA

Next-generation possibilities,
with self-fertilization and outcrossing

A allele frequency

0   1/4   1/2   3/4   1

aa   aa   aa   AA   AA
aa   Aa   AA   Aa   AA

Aa
Aa

FIGURE 3.21C Genetic Drift with a Population Size of Two Six different population types can arise, including one that is the same as the initial population type: *aa, AA*.

3.22 Populations can undergo evolutionary change from genetic drift alone

Let's take a closer look at genetic drift. In any one generation, some alleles will be lost due to accidents of Mendelian segregation during gamete production. A parent with both *a* and *A* alleles, a heterozygote, may have offspring that receive only the *a* allele. On average, this particular accident will tend to be canceled out by the opposite happening to another heterozygous parent, which has offspring receiving only *A* alleles.

Think of a population as a column of alleles, as shown in the first column in Figure 3.22A. In any one generation, there might be *N* of the *A* allele and *n* of the *a* allele. (The total number of individuals in a diploid population will be one-half of *N* + *n*.) In this case, we have 10 *A* alleles and six *a* alleles, and eight individuals. Figure 3.22B shows in detail what happens as the frequency of the *A* alleles increases and decreases, the total number of alleles holding constant, over two generations. (Notice that both figures start with 10 *A* alleles and six *a* alleles.)

Figure 3.22A provides a way of visualizing changes in allele frequency over several generations. Note that we group all of the *A* alleles with each other, and all of the *a* alleles with each

other. This grouping enables us to represent the process of accidental sampling as an expansion or contraction of the *A* and *a* parts of the column.

Figure 3.22A follows the genetic drift of the population as a whole for generations 1 to 7. Various chance processes affect allele frequencies. Some heterozygotes produce more of allele *A*, and some

produce more *a* alleles. These are accidents of genetic segregation. Other chance events occur as well. For example, different families have different levels of fertility. Some offspring do not survive to adulthood. All these chance events make allele frequencies fluctuate.

Intuitively, we can see that there will be more fluctuation in gene frequencies when populations are smaller. With many individuals producing gametes and many families, accidental biases in favor of one allele over the other will average out. For this reason, we expect genetic drift to be more rapid with inbreeding, and slow when population size is very large. Again, this is somewhat like what happens on a trip to Las Vegas. If you start with a large amount of money, you are not as likely end up broke. Your large cash reserve should allow you to survive strings of bad luck without going broke. (Please note, however, that we are not in the business of advising you how to gamble. We just offer an analogy.) Think of genetic drift in the same way you would think of being "ahead" or "behind" in a gambling situation, and you have its essential features. ❖

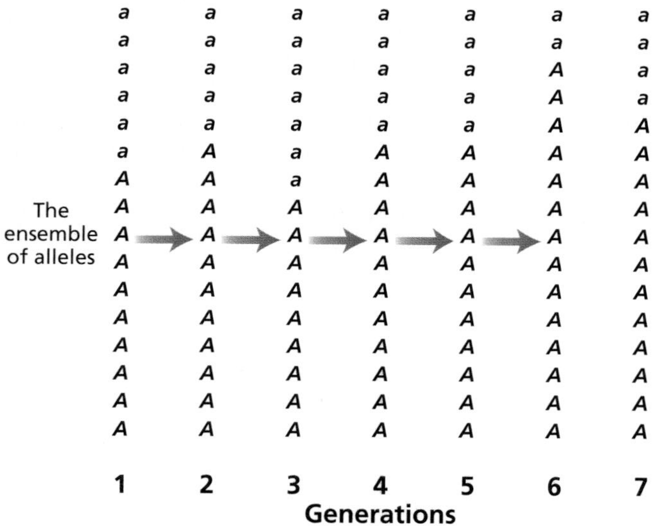

	a	a	a	a	a	a	a
	a	a	a	a	a	a	a
	a	a	a	a	a	A	a
	a	a	a	a	a	A	a
	a	a	a	a	a	A	A
	a	A	a	A	A	A	A
	A	A	a	A	A	A	A
The	A	A	A	A	A	A	A
ensemble	A	A	A	A	A	A	A
of alleles	A	A	A	A	A	A	A
	A	A	A	A	A	A	A
	A	A	A	A	A	A	A
	A	A	A	A	A	A	A
	A	A	A	A	A	A	A
	A	A	A	A	A	A	A
	A	A	A	A	A	A	A
	1	**2**	**3**	**4**	**5**	**6**	**7**

Generations

FIGURE 3.22A Genetic Drift in a Population of Eight Organisms with One Diploid Locus and Two Alleles, *a* and *A*

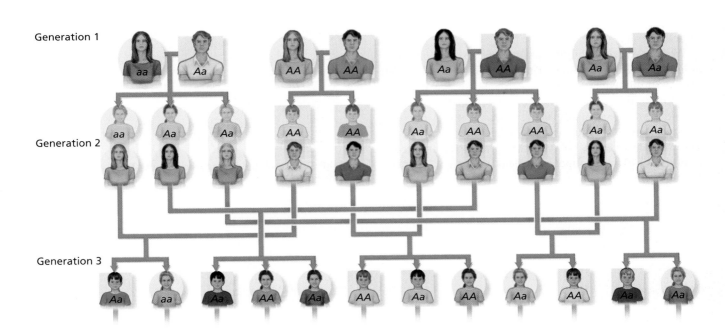

Generation 1

aa   Aa   AA   AA   Aa   AA   Aa   Aa

Generation 2

aa   Aa   Aa   AA   AA   Aa   AA   AA   Aa   Aa

Generation 3

Aa   aa   Aa   AA   Aa   AA   Aa   AA   Aa   AA   Aa   Aa

FIGURE 3.22B **An Example of How Accidents of Genetics and Reproduction Can Produce Genetic Drift**

Because genetic drift makes alleles bounce up and down in frequency, accidental loss of alleles is possible. A key factor is population size. Large populations—for practical purposes, those of more than 1000 breeding individuals—make the loss of alleles unlikely, for two reasons. First, genetic drift is very "slow" in such large populations, as mentioned earlier. Chance accidents will average out, so allele frequencies will not fluctuate much. Second, large populations receive more genetic mutations per generation, because they have large numbers of individuals having new offspring, which could carry new mutations. With frequent mutation, lost alleles will be regenerated by the mutational process alone. Thus genetic drift is expected to lead to rapid loss or fixation of alleles only when populations are relatively small—for practical purposes, fewer than 100 individuals.

When populations are small, genetic drift can cause the frequencies of alleles to wander erratically. Thus it is not surprising that some alleles might accidentally fall to very low frequencies, or even disappear altogether from the population. We can think of this process as analogous to the path of a drunk wandering down an unfenced pier in the dark. The drunk is highly likely to fall off one side of the pier or the other. In our analogy, when the drunk falls off one side, one allele is lost (for example, A), and when the drunk falls off the other side, the other allele is lost (for example, a). When one of two alleles is lost, the other becomes fixed, as shown in Figure 3.23A, where the loss of a would be equivalent to the fixation of A. (If there are more than two alleles, the geometry of the pier has to be more complicated, but the basic evolutionary event is still analogous to falling off an edge.) In very large populations, mutation acts as fencing along the sides of the pier; but small populations have too few mutations and so lack fencing.

What is the probability that an allele fixes? More common alleles fix more often. The probability that an allele fixes is simply equal to its initial frequency. For example, new mutations necessarily have a frequency of $1/2N$ in diploid populations of size N, because at first they are present in just one copy out of all $2N$ alleles at a locus. Therefore, they have a chance of fixing of only $1/2N$, compared to all other alleles in the population. This is a kind of "fairness"

because, without selection, every allele in the population is equal. Which allele "wins" and becomes fixed by drift is then an accident, like picking a card from a shuffled deck of $2N$ cards. Because there are a total of $2N$ alleles in a population, they each have a chance of $1/2N$ of fixing, just like one card out of $2N$ cards has a chance of $1/2N$ of being picked at random. Figure 3.23B shows some possible gene-frequency trajectories with genetic drift. ❖

FIGURE 3.23A Genetic Drift Leads to Loss or Fixation of an Allele This process can be compared to the path of a drunk wandering down a pier at night. The probability that allele A will be fixed = p_i, its initial frequency. Initially common alleles are more likely to be fixed accidentally.

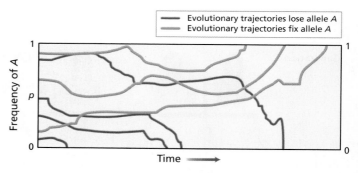

FIGURE 3.23B Multiple Examples of Genetic Drift through Time

SUMMARY

1. Genetics is perhaps the greatest achievement of twentieth-century biology. At its most fundamental level, the genetic engine transfers hereditary information from one generation to the next. That hereditary information is packaged into chromosomes, either prokaryotic or eukaryotic in configuration. Some gametes, cells, or organisms have one complete set of chromosomes (a haploid genome), while others have two sets (a diploid genome) or more.

2. The genetics of populations are determined by the frequencies of genotypes, and thus the frequencies of genes. With random mating for one generation, and no selection, the frequencies of the genotypes at a single locus can be calculated from the product of gene frequencies. The genotype frequencies are stable at this Hardy-Weinberg equilibrium. When the frequencies of gametes are the product of the frequencies of the alleles at multiple loci, there is linkage equilibrium.

3. Quantitative characters are affected by multiple genetic loci and multiple environmental influences. Sometimes these effects are additive; sometimes they are not. When inheritance is additive, the resemblance between relatives is determined by the genetic variance.

4. With random mating, the frequencies of genes on sex chromosomes evolve toward equality. In the absence of selection, random mating and genetic recombination lead to the evolution of linkage equilibrium.

5. Inbreeding occurs when relatives mate. If inbreeding occurs more than expected by chance, the level of homozygosity rises in the population, while the level of heterozygosity falls. Inbreeding leads to increased genetic disease. It also depresses functional characters, such as survival and fertility. Inbreeding increases when populations are subdivided.

6. Genetic drift results from the combination of individual genetic accidents. Genetic drift leads to the accidental fixation or loss of particular alleles by chance alone.

REVIEW QUESTIONS

1. The person who discovered genetics was …?
2. Gametes are produced using what cellular process?
3. Hardy-Weinberg equilibrium arises after how many generations, with random mating and no selection?
4. Quantitative characters are influenced by which two major factors?
5. Evolution proceeds toward linkage equilibrium because of what process?
6. What are the effects of inbreeding on the genotypic composition of a population?
7. What does genetic drift do to gene frequencies?
8. Heritability measures what biological tendency?
9. Why do mutts live longer than purebred dogs?

KEY TERMS

addition rule
allele
arithmetic mean
blood group
chromosome
coupled phase
dominance
diploid
endosperm
epistastis
Fisher, R. A.
fixation
Galton, Francis
gamete
gametogenesis
gene expression

genetic disease
genetic drift
genotype
haploid
Hardy-Weinberg equilibrium
heritability
heterozygosity
homozygosity
hybrid vigor
inbred line
inbreeding
inbreeding coefficient, F
inbreeding depression
independent assortment
linkage disequilibrium
linkage equilibrium

loci
mean
median
meiosis
Mendel, Gregor
meroploidy
Morgan, Thomas Hunt
multiplication rule
outbred
Pearson, Karl
phenotype
pollinator
polyploid
polyteny
population
quantitative character

recessive
recessive genetic disease
recombination
repulsion phase
resemblance between relatives
self-fertilization, selfing
subdivision of populations
variance
variance, environmental
variance, genetic, additive genetic
variance, phenotypic
Wahlund effect
X chromosome
Y chromosome
zygote

FURTHER READINGS

Ayala, Francisco J., and John A. Kiger. 1980. *Modern Genetics*. Menlo Park, Calif.: Benjamin Cummings.

Falconer, D. S., and T.F.C. Mackay. 1996. *Introduction to Quantitative Genetics*, 4th ed. Harlow, Essex: Longman.

Hartl, Daniel L. 2000. *A Primer of Population Genetics*, 3rd ed. Sunderland, MA: Sinauer.

Hartl, Daniel L., and Elizabeth W. Jones. 1999. *Essential Genetics*, 2nd ed. Sudbury, MA: Jones and Bartlett.

Hedrick, Philip W. 2000. *Genetics of Populations*, 2nd ed. Boston: Jones and Bartlett.

Whitehouse, H.L.K. 1969. *Towards an Understanding of the Mechanism of Heredity*, 2nd ed. London: Edward Arnold.

Nature Red in Tooth and Claw

4

Natural Selection

Charles Darwin had an incomplete understanding of natural selection in the wild. He never found any good examples of the process. His touchstone was instead artificial selection—the type of selection practiced on farms and ranches.

Today we understand the machinery of natural selection quite well, and we have found excellent examples of natural selection in the wild environment. We return to Darwin's original theory of natural selection at the beginning of this chapter. In Darwin's time, the best examples of selection were those of animal and plant breeders, though the type of selection that they practiced was contrived. Much the same is true today, so we will use artificial selection to make the process of selection absolutely transparent. You can think of natural se-

lection as the unforced version of the general process of selection.

Natural selection has two faces, one pointed toward the phenotypes of organisms, one facing their genes. We will consider various types of natural selection, divided according to their effect on the phenotype or their effect on the genetic locus.

Observers of natural selection have studied it in two settings: the laboratory and the wild. Some scientists have strong preferences for one or the other. But it is probably more reasonable to admit that the study of natural selection under both controlled laboratory conditions and actual conditions in nature have jointly helped us to understand natural selection. Indeed, we have too few well-understood examples to neglect either arena of study, the laboratory or the wild. We will look at both. ❖

DARWIN AND NATURAL SELECTION

4.1 Darwin did not expect to observe natural selection

Charles Darwin's *Origin of Species* uses **natural selection** to explain many of the features of organisms. These features are now called adaptations. Adaptations were well known before Darwin ever wrote. Indeed, he learned a great about them from William Paley's theological works. For theologians and many biologists before Darwin, adaptations were instances of God's beneficence, demonstrating a provident creation. Darwin explained adaptation using natural selection, a material process. In Chapter 1 we consider the cultural impact of this change in explanations. Here we consider the scientific issues that arise from Darwin's innovation.

The problem Darwin faced was that he had no direct examples of natural selection. Indeed, he did not expect to have such examples, because he assumed that natural selection would take many generations to act noticeably, with each generation's selection

> *For theologians and many biologists before Darwin, adaptations were instances of God's beneficence, demonstrating a provident creation.*

causing undetectable amounts of change. Darwin expected such gradual change because his notion of causation was gradualist, deriving from the geological doctrines of the geologist **Charles Lyell**, as expounded in Lyell's treatise *Principles of Geology.*

Lyell had argued that large-scale geological change was produced by the slow cumulative action of everyday geological processes such as subsidence, erosion, and sedimentation. These processes are hard to detect over a short period of time, but they could nonetheless eventually produce mountains, valleys, and other major geological formations. Likewise, Darwin expected that very small changes in the composition of populations would be wrought by natural selection in each generation. But these changes could nevertheless finally produce animals and plants with very different morphologies and physiological functions.

Thus, Darwin did not expect to "see" natural selection himself. Nor did he expect any other biologist to be able to detect natural selection directly. Therefore, he had an enormous problem in arguing for the importance of natural selection. He could argue that a great many facts of biology could be explained using the two principles of evolution and natural selection. These principles were plausible because of their great explanatory value. But such *explanatory plausibility* is rarely enough to establish a scientific theory.

A major requirement that is usually added to explanatory appeal is *mechanistic plausibility:* Were there demonstrable processes that could generate the process assumed by the theory? In geology, for example, the plausibility of processes like sedimentation and erosion can be established by setting up a laboratory apparatus in which such processes are measurable under controlled conditions. Likewise, experimental science was begun by Galileo using simple experiments with rolling balls on inclines, among his other ingeniously simple demonstrations. Such demonstrations of mechanism are indispensable for experimental science. For Darwin to successfully convince other scientists of evolution by natural selection, he had to produce a similar demonstration of mechanism.

The concrete demonstration of natural selection that Darwin used was artificial selection. In **artificial selection**, the breeder plays the role of nature, choosing the attributes that will determine the survival or reproductive success of the stock being bred. Most artificial selection is performed with agricultural species, such as grains, potatoes, tomatoes, chickens, cattle, and pigs. But it is also important in the breeding of less obviously useful domesticated species, such as dogs, pigeons, and sometimes even cats.

No one could dispute that breeders practiced selection. And none could dispute that it was often very successful. The increases in value in breeds of livestock were well established in the nineteenth century, and Darwin pointed them out in Chapter I of the *Origin of Species*. It seems that artificial selection readily supplied the support that Darwin needed for the mechanistic cogency of this theory of natural selection.

However, Darwin encountered some problems with using artificial selection to support his theory of evolution. First, much of the selection that breeders had been practicing was not intentional. For example, breeders often selected for increased docility in their handling of domesticated animals, even when they did not intend to do so. Overly violent livestock, particularly some males, would be destroyed or castrated because they were too much of a nuisance. By eliminating these animals, breeders unconsciously selected for docility, a well-known hallmark of domesticated breeds compared with their wild cousins.

Another problem with breeding by humans is that it has often involved inbreeding of the domesticated species. The negative effect of inbreeding is well known in dog breeds, as we noted in the preceding chapter. But it is also common in varieties of cultivated plants, such as roses. Inbreeding makes plant varieties highly susceptible to infection and other problems. The Irish potato famine was caused by the susceptibility to blight of the variety of potato that was ubiquitous in Ireland (Figure 4.1A). Thus breeding does not necessarily guarantee the best qualities. This fact undermines the analogy to natural selection, which is supposed to improve each species, according to Darwin.

These problems, although of great scientific interest, did not negate the basic point that Darwin wished to draw from the practice of artificial selection. This point was that the directly observable action of selection, as practiced by breeders, could produce material improvement in their stocks and varieties. Darwin's natural selection could then be reduced to the mechanistic processes of artificial selection, except that nature was to supply the careful scrutiny that the human breeder supplied in artificial selection. ❖

SEARCHING FOR POTATOES IN A STUBBLE FIELD.

FIGURE 4.1A The Irish Potato Famine

Darwin's original theory of natural selection made nature, through the struggle for existence, the breeder or selector

Reduced to its essence, Darwin's natural selection is just artificial selection, with Nature—almost as a personified agent—replacing the human breeder. Darwin clearly saw artificial and natural selection as two forms of the same thing, as revealed by the second quotation in Figure 4.2A. Notice the phrase "variations useful to man" before the suggestion that there might be "variations useful in some way to each being in the great and complex battle of life."

But the important thing is Darwin's argument, in the first quotation, concerning why nature would act as a selector. The first point is that "more individuals are produced than can possibly survive." In other words, there is a potential reproductive excess. Hence, point two—there must be ecological factors that hold the size of populations in check. These factors define a "struggle for existence," involving competition, predation, and an inimical environment. Life in a state of nature is nasty, brutish, and short, making ecology a stern breeder.

The raw material that natural selection acts on is not, however, always ideal. In the quotation given in Figure 4.2B, Darwin uses a metaphor of stones falling from the side of a cliff to characterize what we would today call hereditary variations. These stones—from which natural selection must build—are not hewn to any functional purpose. They arise by accidents of physical, especially geological, processes. Likewise, the variants that breeders or nature use in the course of selection arise accidentally. Yet the power of natural and artificial selection is such that they can nonetheless use accidental genetic variation to mold attributes that are beneficial, either for the fitness of the organism or for human purposes, respectively. ❖

Darwin begins with the "struggle for existence," which we would now call ecology:

"as more individuals are produced than can possibly survive, there must in every case be a struggle for existence, either one individual with another of the same species, or with the individuals of distinct species, or with the physical conditions of life. It is the doctrine of Malthus applied with manifold force to the whole animal and vegetable kingdoms."

(*Origin of Species,* Chapter III)

This struggle for existence sets the stage for the action of natural selection:

"Can it, then, be thought improbable, seeing that variations useful to man have undoubtedly occurred that other variation useful in some way to each being in the great and complex battle of life, should occur in the course of many successive generations. If such do occur, can we doubt (remembering that many more individuals are born than can possibly survive) that individuals having any advantage, however slight, over others, would have the best chance of surviving and of procreating their kind? On the other hand, we may feel sure that any variation in the least degree injurious would be rigidly destroyed. This preservation of favorable individual differences and variations, and the destruction of those which are injurious, I have called Natural Selection."

(*Origin of Species,* Chapter IV)

FIGURE 4.2A Darwin's Version of Natural Selection

Darwin never knew about genetics. He usually referred to hereditary variants as "variations" or "variability." But he had a good intuition for the way natural selection uses hereditary variation during evolution:

"I have spoken of selection as the paramount power, yet its action absolutely depends on what we in our ignorance call spontaneous or accidental variability. Let an architect be compelled to build an edifice with uncut stones, fallen from a precipice. The shape of each fragment may be called accidental; yet the shape of each has been determined by the force of gravity, the nature of the rock, and the slope of the precipice—events and circumstances, all of which depend on natural laws; but there is no relation between these laws and the purpose for which each fragment is used by the builder. In the same manner the variations of each creature are determined by fixed and immutable laws; but these bear no relation to the living structure which is slowly built up through the power of selection, whether this be natural or artificial selection."

(*The Variation of Animals and Plants Under Domestication,* p. 236 of the 1896 edition)

FIGURE 4.2B How Darwin's Natural Selection Uses Variation

Natural selection is often difficult to detect in nature, although we will examine examples where it is detectable in Modules 4.22–25. It is even difficult to understand how natural selection might be working in nature. Therefore, a close look at artificial selection is a better starting place for understanding selection.

With artificial selection, we know explicitly the character(s) that are undergoing selection. For example, in Figure 4.3A we are selecting for body weight in mice. Once we identify the selected group (individuals that display the character being selected, in this case greater body weight), the rest of the population can be used for some other purpose.

With artificial selection, we can determine the quantitative magnitude of selection: It is the deviation of the selected group from the population as a whole, known as the **selection differential (S)**. In Figure 4.3A,

the selection differential is 6 grams. The individuals in the selected group are mated to each other, and their offspring are reared. The average difference between the offspring of the selected group and the rest of the breeding population is the **response to selection (R)**. In Figure 4.3A, the response to selection is 2 grams.

The only feature of this process that is not mere bookkeeping is that the offspring resemble the selected parents. This, of course, depends on heredity. To predict the response to artificial selection in quantitative terms, the only genetic parameter that we need is the **heritability** (h^2), defined in Chapter 3. The product of heritability and the selection differential gives the predicted response to selection. This formula is displayed in the accompanying box. Because heritability gives the predicted resemblance of offspring as a function of the average character value of the parents, this formula makes intuitive sense. ❖

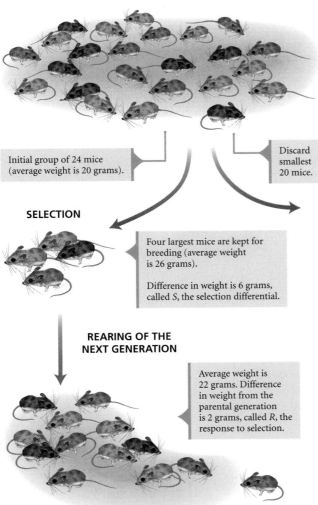

Initial group of 24 mice (average weight is 20 grams).

Discard smallest 20 mice.

SELECTION

Four largest mice are kept for breeding (average weight is 26 grams).

Difference in weight is 6 grams, called S, the selection differential.

REARING OF THE NEXT GENERATION

Average weight is 22 grams. Difference in weight from the parental generation is 2 grams, called R, the response to selection.

FIGURE 4.3A Artificial Selection on Mouse Body Weight

Quantitative Genetics Predicts the Response to Artificial Selection

The response to artificial selection can be predicted from the heritability of a character (h^2):

$$R = h^2 S$$

In words, this equation says

(response to selection) is equal to (heritability) times (selection differential).

Multiple generations of artificial selection can change a character substantially

As Darwin suspected would be true of natural selection, artificial selection over multiple generations leads to progressively greater deviation of selected characters. The gains from selection in just one generation of artificial selection are usually measurable, but they are not often very great. When artificial selection is applied generation after generation, we gain a quantitative picture of the power of selection in general—including natural selection, when it is strong and sustained.

There are two basic ways to keep track of the response to artificial selection. First, in Figure 4.4A, we see a common type of graphical plotting of selection data. The average phenotype of the selected line is shown for each generation. The qualitative expectation is that more generations of selection will give a greater response to selection. However, there is no simple way to predict what the selection response will be, generation by generation.

To make a quantitative prediction over multiple generations of selection, we must plot (on the y-axis) the **cumulative selection response** (ΣR), measured as deviations of the offspring of selected organisms from the average of the total population in each generation, summed over the generations of selection that have occurred. The x-axis must show the **cumulative selection differential** (ΣS), measured as the sum of the deviations of the selected parents from the average of the total population in each generation. The cumulative selection differential accumulates over all generations, just as the mileage of your car accumulates over all the days you drive it. Figure 4.4B shows this kind of plot for selection on mouse body weight.

The predicted response to selection is then given by the sum of selection differentials times the heritability, a quantity that is the analog of the one-generation calculation. This is shown in the accompanying box, "Estimating Heritability from the Response to Artificial Selection." ❖

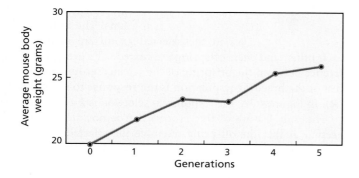

FIGURE 4.4A Response to Selection on Mouse Body Weight Response to selection per generation.

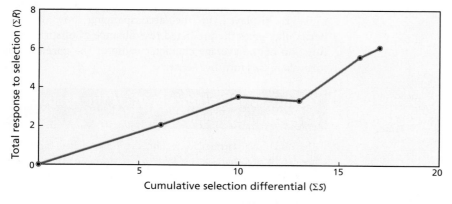

FIGURE 4.4B Response to Selection on Mouse Body Weight Cumulative response to selection versus cumulative selection differential. (Both variables are plotted in grams of increased weight relative to the weight of the initial population of mice before selection.)

Estimating Heritability from the Response to Artificial Selection

We can reverse the equation for the heritability of a quantitative character. If ΣR is the total cumulative response to artificial selection and ΣS is the total of the selection differentials added together

$$h^2 = \Sigma R / \Sigma S$$

THE CYCLE OF NATURAL SELECTION

4.5 Natural selection will sometimes have more impact than artificial selection, sometimes less

With the example of artificial selection to guide us, we can think about natural selection with greater focus. Think of the life cycle of a mouse population as if all the young mice grow up together, become older mice, undergo selection together, reproduce, and then die. This is not usually true of the mice of North America, but it is true of the "marsupial mice" of Australia discussed further in Chapter 7. In any event, the simpler pattern is easier to visualize.

Natural selection within the life cycle is like artificial selection in many respects. As in the artificial selection example, the mice will be selected for particular attributes, possibly size. Larger mice might survive cold temperatures better, for example. Larger mice would then be more likely to survive to reproduce, and the offspring of the next generation should grow up to be larger, following the cycle shown in Figure 4.5A.

But there are also differences between natural and artificial selection. The most important of these is that natural selection normally acts on many organisms, perhaps millions or billions in a particular locale. Populations of insects and grasses and bacteria, among other organisms, can reach very high numbers—well into the trillions. In contrast, breeders practicing artificial selection rarely work with more than thousands of organisms. Often they are limited to a few hundred organisms, especially a small number of males.

Because breeders work with small populations, artificial selection is often slowed down, or stopped, by worsening inbreeding depression. Especially with large mammals, such as cattle and sheep, inbreeding is likely to be a problem. Dogs, as we have seen in Chapter 3, are highly inbred, causing numerous veterinary problems.

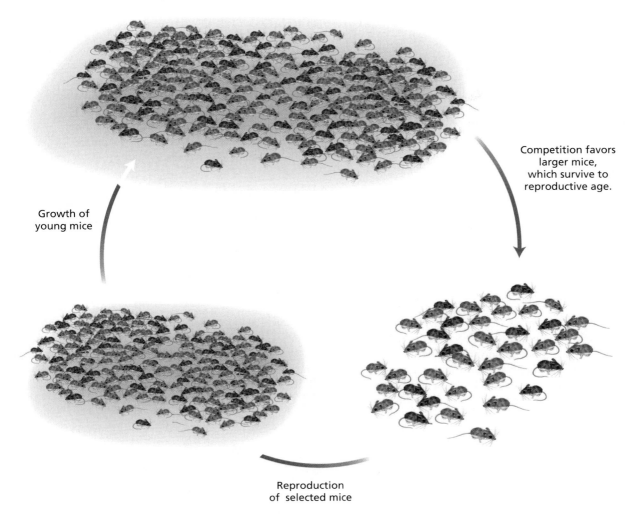

Growth of
young mice

Competition favors
larger mice,
which survive to
reproductive age.

Reproduction
of selected mice

FIGURE 4.5A The Cycle of Natural Selection

Related to the problem of inbreeding depression, but distinct from it, is the problem of exhaustion of genetic variation. When artificial selection is sustained for a long period of time, heritability may fall due to a loss of genetic variation. Under these conditions, the progress of selection stops. There will not be enough heritability for the population to respond to selection.

Because natural selection is often free of problems of small population size, it sometimes continues to act over thousands or millions of years. In this respect, natural selection should ultimately prove more powerful than artificial selection.

A problem impeding the action of natural selection is that it is unlikely to be consistent in its action. For example, suppose that Australian mice are selected for larger body size because of intermittent frosts. But Australia is generally a fairly warm place, so some years may be largely free of frost. In such years, there will be no selection for increased size resulting from frost. Natural selection for larger body size will be on vacation in those years. In these circumstances, environmental fluctuation dilutes natural selection.

An additional complication arises from the multitudinous sources of mortality that organisms face in nature. It may

happen that a large mouse made it through the winter frosts because of its large size, but it was eaten by a snake in the spring. Or perhaps it was picked off by a hawk. Or it could have succumbed to a gastrointestinal infection. Even in an environment that is constant on the whole, many factors are contributing to variation in survival or fertility. The action of this range of factors in the "struggle for existence" will dilute selection arising from any one of them. We consider this issue further in Module 4.9.

For all these reasons, natural selection is usually not as focused as artificial selection; and this relative lack of focus will leave it weaker. The rate of progress in response to natural selection should therefore be considerably slower than that normally achieved by artificial selection. However, as we will show in Module 4.22–25, even natural selection can sometimes act with great speed and power.

In the meantime, we will break down some of the elements in the cycle of natural selection, particularly with a view to defining the factors that establish natural selection and delineating the consequences of natural selection at different points in its cycle. ❖

4.6 Natural selection requires genetic variation for characters related to fitness

For the cycle of natural selection to start, a basic requirement is genetic variation that affects characters related to fitness. It is not enough just to have genetic variation. DNA sequences can differ in their sequence of nucleotides, but this difference may have no effect on the phenotype of the organism. Even a genetic difference in phenotype may have no selective importance. The specific patterns of human fingerprints, for example, probably have no importance for natural selection. Finally, **phenotypic variation** for fitness-related characters will not necessarily have any importance for natural selection if that variation is due to the environment rather than genes; in other words, if $V_P = V_E$.

For example, Figure 4.6A contrasts the consequences for selection on body weight of the presence or absence of genetic variation for the character. In part (i), the variability for body weight in mice is entirely environmental. Different levels of nutrition might have produced this variation in body size, for example. There is no genetic variation, and there is no genetic difference in the mice before and after selection. In part (ii), there is genetic variation for body weight. In this case, the selected mice are genetically different from the original group of mice. One of the big questions about natural selection is how often fitness-related characters have significant amounts of genetic variation (V_G), particularly selectable genetic variation (V_A). It turns out that it is surprisingly common for components of fitness—such as female fertility, male mating success, development time, and longevity—to have significant amounts of selectable genetic variation. Characters that are related to fitness but are not components of it—such as stress resistance and endurance—show even more selectable genetic variation. The box, "Is there Genetic Variation for Characters Related to Fitness?" discusses the evidence for this genetic variation. ❖

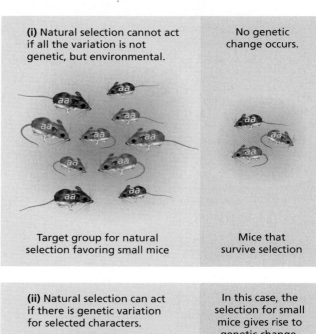

(i) Natural selection cannot act if all the variation is not genetic, but environmental.

No genetic change occurs.

Target group for natural selection favoring small mice

Mice that survive selection

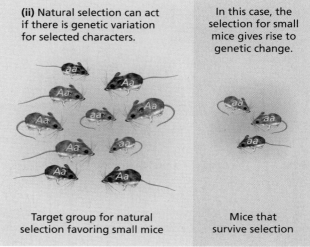

(ii) Natural selection can act if there is genetic variation for selected characters.

In this case, the selection for small mice gives rise to genetic change.

Target group for natural selection favoring small mice

Mice that survive selection

FIGURE 4.6A Natural selection requires genetic variation.

Is There Genetic Variation for Characters Related to Fitness?

Recall our discussion of the inheritance of quantitative characters in Chapter 3. We pointed out that there was more genetic variation for morphological characters than there was for fitness-related characters.

But is there enough genetic variation for selection to act on characters related to fitness?

The answer is yes. There are two lines of evidence. The first is that studies of genetic variation themselves reveal significant heritabilities for such characters as viability and fertility. The second line of evidence is that it is possible to apply artificial selection on fitness-related characters, like fecundity or longevity, and obtain a detectable response.

If there is selectable genetic variation on which natural selection can act, it will act to change the components of fitness. Figure 4.7A presents an example involving survival probabilities related to different sizes of mice.

In the absence of genetic variation, differential mortalities and fitnesses may also produce a temporary change in fitness characters. But note that the assertion here is that *if* there is selectable variation, there will be change in fitness (not *only if* there is such selectable variation). Naturally selected organisms will have detectable superiority in their fitness, and these organisms can be discriminated *if* there is selectable genetic variation (but not *only if*).

An intriguing question about natural selection is how precisely it discriminates between individuals of different phenotypes. When Darwin wrote about natural selection, he used phrases like "careful scrutiny" and "rigidly destroyed" to convey his sense of Nature as a personified breeder of infinite power and patience. There is now some controversy about this idea among biologists, as highlighted in the box, "How Powerful Is Natural Selection?"

Some biologists tend to follow Darwin, assuming that natural selection is sensitive to very slight differences between phenotypes. Others emphasize the chanciness of natural selection. In particular, a common theme in modern evolutionary biology involves the limitations to the power of natural selection. We will take up this theme repeatedly. ❖

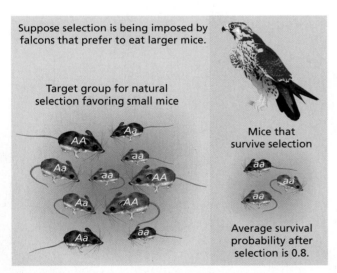

The intial group of mice has a range of survival probabilities that go with their different sizes as follows:

AA — average size 30 grams; average survival probability is 0.0 due to falcons and other risks

Aa — average size 24 grams; average survival probability is 0.1 due to falcons and other risks

aa — average size 18 grams; average survival probability is 0.8 due to falcons and other risks

The initial survivor probability of these mice is 0.28.

FIGURE 4.7A Natural selection changes probabilities of survival or reproduction.

How Powerful Is Natural Selection?

This question can be rephrased as, "How much of a difference in probabilities of survival or their reproductive output is there likely to be between individuals having different phenotypic characteristics?"

The best answers to this question come from studies of natural selection in the wild, which we will discuss later in this chapter. But recent research with birds and with human medical disorders does indicate large differences in survival rates between birds with different beak sizes and between humans with genetically different enzymes.

To complete a cycle of natural selection, the average fitness of the offspring must be changed, as shown in Figure 4.8A. This will not happen if the selected parents are not genetically different from the rest of the population. But there is still the question of whether or not genetically different parents will have genetically, and thus phenotypically, different offspring.

The question is essentially answered by quantitative genetics, as described in Chapter 3. The key concept is the heritability (h^2). Given a distinct group of selected individuals, with artificial or natural selection, the offspring will be different according to the equation for the response to selection: $R = h^2S$. (Recall what this equation says, in words: The response to selection is equal to the heritability times the selection differential.) In this respect, there is no difference between artificial and natural selection. In both cases, quantitative genetics predicts a response to selection under the same conditions of (1) a significant selection differential (S) and (2) significant heritability (h^2).

Once again, the quantitative values of selection differential and heritability are the key issues for natural selection. If natural selection can carve out a selection differential for a character, and there is heritability for that character, then there will be a response to natural selection. The accompanying box summarizes this idea. ❖

How Much Will the Offspring of Selected Parents Resemble the Parents?

One of the most common points of confusion about natural selection is the extent to which the selected offspring will resemble their parents.

It turns out that we have already solved this problem. The heritability of the selected character(s) tells us the extent to which the offspring will deviate from the mean. Using the heritability concept explained in Chapter 3 and the artificial selection arithmetic of Module 4.3, we can expect the response to natural selection to approximate the heritability times the difference between the selected group and the entire population, before selection.

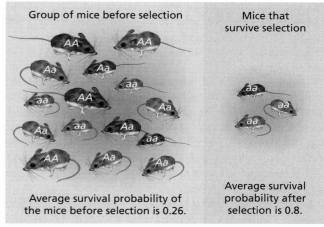

Group of mice before selection

Mice that survive selection

Average survival probability of the mice before selection is 0.26.

Average survival probability after selection is 0.8.

The frequency of *A* and *a* alleles is 0.5, which will not change much in the absence of selection. So their offspring would have had average survival probabilities of 0.26 as well.

Selection has fixed the *a* allele, so all the offspring of the selected parents will have an average survival probability of 0.8, much greater than 0.25.

FIGURE 4.8A Selected parents have offspring that are different from the offspring that the population would have had without selection.

PHENOTYPIC PATTERNS OF NATURAL SELECTION

4.9 Natural selection acts powerfully on just a few characters at a time

Although the cycle of natural selection is an integrated process, it has two very different faces. One is pointed at the gene, and the ways in which different patterns of inheritance determine the outcome of selection. We will take up this topic later in this chapter. The other face is pointed at the phenotype of the organism—its external characteristics (Figure 4.9A). This phenotypic selection is the focus of the present portion of Chapter 4.

It is a common error to suppose that natural selection precisely targets the ideal phenotype for each species, and then "rigidly destroys" any deviations from this ideal. This strict discrimination among organisms is not usually the case. In reality, there is a great deal of sheer chance in life and death. Organisms that natural selection favors might die for completely accidental reasons. They might be trampled by large animals. They might die of desiccation in a freak drought. For these reasons, it is important to understand that natural selection is *not* some kind of perfect winnowing process in which only the best-adapted reproduce and all others are eliminated.

When natural selection acts effectively, it is likely to do only a few things with any intensity. Only strong selection can dominate over accidents in shaping a population. When selection is weak or variable in direction, it cannot act effectively. An additional factor limits the capacity of natural selection to reshape a population—the number of reproductive deaths the population can sustain and still maintain itself. If multiple selection processes act on several characters at the same time, then too few organisms may survive to reproduce. For example, as Figure 4.9B shows, a shrub might be selected for (1) resistance to drought, (2) growth under conditions of poor nutrients, and (3) mechanical damage from large animals trampling the plant. Suppose that the first factor kills 80 percent of the shrubs, the second kills 70 percent of the remainder, and the

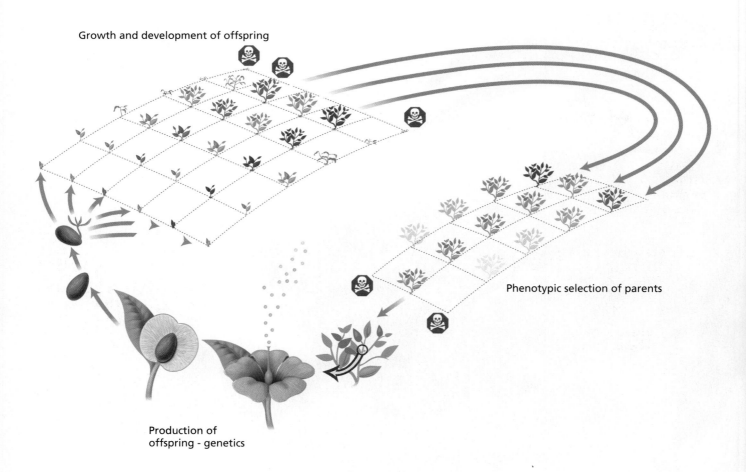

FIGURE 4.9A **Phenotypic and Genotypic Facets of Natural Selection**

third kills 40 percent of the shrubs surviving the first two selective factors. With these assumptions, only 3.6 percent of the population survives selection. For the population to maintain itself, each shrub would have to produce almost 30 offspring that survive to adulthood. If these plants reproduce less than that, the population will die out.

On the other hand, the population might be able to maintain itself if it is subject to just one of these selective pressures. Then the population size would be reduced to 20–60 percent of its level without selection. Darwin's "reproductive excess," a concept he took over from Malthus, might be enough to sustain the population during selection.

A further limitation on natural selection, as opposed to artificial selection, is that the focus of selection is likely to change from time to time, as shown in Figure 4.9C. Some of this change will occur within the life of a single organism; some of it will change only from generation to generation. In the latter case, the entire direction of natural selection may change from generation to generation. Both of these possibilities will weaken the power of selective mechanisms.

In the following modules, we consider three different phenotypic patterns of natural selection: directional selection, stabilizing selection, and disruptive selection. ❖

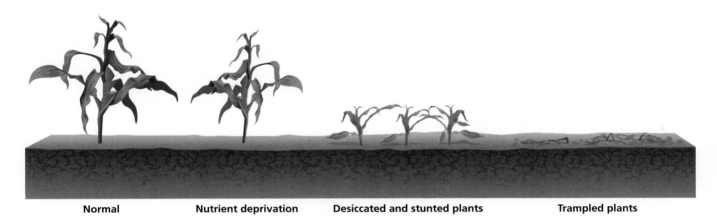

Normal   **Nutrient deprivation**   **Desiccated and stunted plants**   **Trampled plants**

FIGURE 4.9B **Multiple Processes of Natural Selection**

Hot weather selects
for thin fur, light color.

Overcast weather favors
darker fur for camouflage.

Rainy weather favors thicker
fur to keep water away from skin.

FIGURE 4.9C **Variable Natural Selection**

Directional selection is the type of selection that the phrase "natural selection" calls to mind. In directional selection, only the biggest, the fastest, or the smallest are able to survive and reproduce. **Directional selection** favors a particular phenotypic extreme, which most of the population does not attain. Selection for the phenotypic extreme results in strong selection against most members of the population. The result is the progressive movement of the population toward the extreme favored by natural selection.

Figure 4.10A diagrams a simple form of directional selection—the distribution of a particular character, represented by values on the horizontal axis. The peak shows the most common character values. The mean of the population will be near this peak. In this case, selection strongly favors individuals that have a high value of the character. Indeed, there is an absolute threshold, indicated by the vertical bar, for successful reproduction. Individuals with phenotypic values below this bar do not reproduce at all. This type of stringent selection is expected to increase the average value of the selected character, assuming that the character has significant heritability.

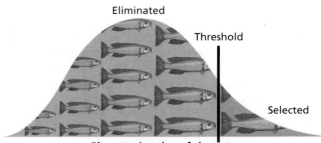

Phenotypic value of character

- With directional selection, the individuals with the most extreme phenotypes survive the selection process.

- The selected phenotypes do not necessarily belong to those individuals that will have correspondingly extreme offspring.

- Phenotypic selection is only about phenotypes, not underlying genotypes, or anything else.

FIGURE 4.10A Directional Selection

How often does natural selection take the form of directional selection? This is an open question for most phenotypic characters. Will natural selection always favor the larger organisms, the organisms that best resist cold, and so on? Isn't natural selection more likely to favor a compromise between high and low values for most characters? (This possibility is considered next.)

But there is one character for which natural selection will be consistently directional. That character is **fitness**. Natural selection always favors phenotypes that have higher Darwinian fitness. Therefore, there is at least one character for which directional selection will always be the pattern of selection. (Later in the chapter, we give additional examples of directional selection in the wild.) ❖

If directional selection is the type of selection that we can readily associate with Darwin's concept of progressive natural selection, then *stabilizing selection* can be associated with **Aristotle's** original model of selection. Aristotle was interested in the stability of species, in why they retained their typical anatomy and physiology. His explanation, more than two thousand years ago, was that deviant individuals would be less successful in life. They would be less likely to survive or to reproduce. Thus selection would act to stabilize the species, eliminating "monsters." In modern quantitative terms, this sort of stabilizing selection can be represented as in Figure 4.11A. **Stabilizing selection** eliminates individuals at the extremes of the distribution of a quantitative character, favoring those with intermediate phenotypes.

Aristotle was interested in the stability of species, in why they retained their typical anatomy and physiology.

Note that stabilizing selection is unlikely to change the average value of a character much. Instead, its main effect is on the variance of the population. In each generation, the reproducing parents will have a *lower* variance for the selected phenotype, compared to the variance that might have existed in the absence of stabilizing selection. This type of **phenotypic selection** can be thought of as conservative. As Aristotle supposed, it is likely to help conserve species attributes.

Several examples of stabilizing selection are known to biologists. One for which we have excellent data is stabilizing selection on the weight of human newborns, shown in Figure 4.11B. However, there is evidence of stabilizing selection for many other characters—particularly morphological characters, such as total body weight and the size of body parts in both animals and plants—from bone lengths to gall size. ❖

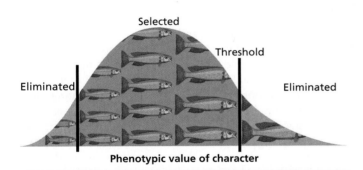

- With stabilizing selection, the individuals that have the most extreme phenotypes are eliminated by natural selection.

- The immediate phenotypic effect of stabilizing selection is to reduce the **variance** of the selected group compared to the population's distribution before selection.

- This type of selection may not change the mean of the phenotypic distribution very much.

FIGURE 4.11A Stabilizing Selection

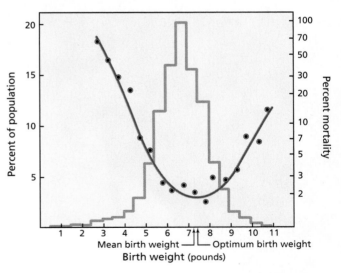

FIGURE 4.11B Mortality and Birth Weight The average human birth weight is about 7 pounds. The mortality rate (red line) is much higher among very small and very large babies than among babies of average size. The best birth weight is close to the population average. Data from Cavalli-Sforza and Bodmer (1971) and references therein.

4.12 Disruptive selection favors organisms that have character values at both extremes of the phenotypic distribution

Natural selection can do things that are very difficult to conceive intuitively, at least at first. Logically, if natural selection can select *against* organisms at either end of a distribution of phenotypes, then it is conceivable that natural selection might select *for* organisms at the extremes of the distribution. This pattern would be the opposite of stabilizing selection, with its conservative pattern. For this reason, selection against the middle of a distribution is called **disruptive selection**. Figure 4.12A shows this pattern.

How could disruptive selection ever arise? One general context in which it might arise would be if a predator or a herbivore preferentially fed on the most common type of food. Consider plant evolution. Assume that the plants that produce medium-sized seeds are the most abundant, and therefore their seed is preferred by seed-eating birds. Under these conditions, plants that gave smaller or larger seeds might have a selective advantage. Figure 4.12B gives an actual example of disruptive selection involving bill size in a seed-eating bird. However, examples of disruptive selection are not common.

Like stabilizing selection, disruptive selection changes the variance of the population. The difference is that disruptive selection is likely to *increase* the variance. In addition, disruptive selection can change the phenotypic distribution, favoring two-peaked, or **bimodal**, phenotypic distributions. The example in Figure 4.12B is a case in which such bimodality has evolved. ❖

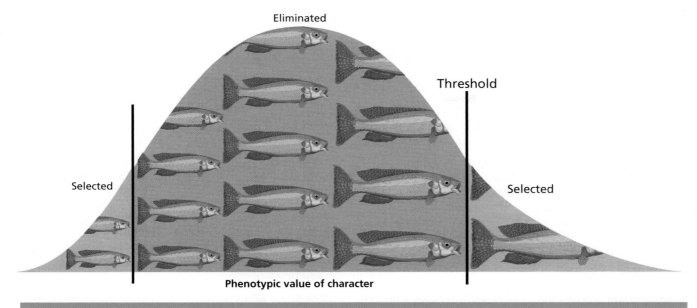

Eliminated

Threshold

Selected

Selected

Phenotypic value of character

- Disruptive selection is the precise opposite of stabilizing selection: Selection eliminates the individuals from the middle of the phenotypic distribution, keeping only the individuals from the extremes.

- A key feature of this type of selection is that it produces a bimodal distribution of selected individuals, unlike the unimodal distributions produced by directional and stabilizing selection.

FIGURE 4.12A Disruptive Selection

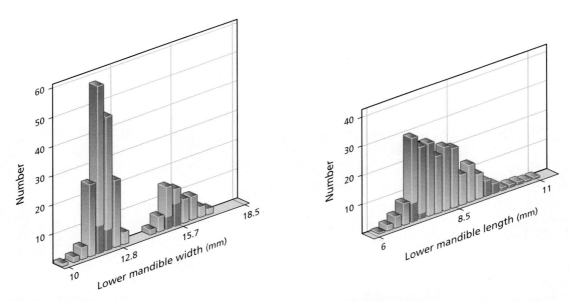

FIGURE 4.12B Disruptive Selection on Bill Size in the Black-Bellied Seedcracker (*Pyrenestes ostrinus*) The red portion of each bar represents juveniles that did not survive to adulthood; the green portion represents juveniles that did survive. The survivors were those individuals with bills that were either relatively large or relatively small. Data from Bates Smith (1993).

GENETIC MECHANISMS OF NATURAL SELECTION

4.13 Genetics complicates the action of natural selection

The basic principles of selection are the same in all living things. But the specific effects of selection are much more complicated in organisms that have sex, because the genetic engine that sex introduces into the evolutionary machinery makes inheritance tricky. For this reason, let's first look at organisms that do without the complications of sex.

Without sex, mothers have daughters that are genetically identical to themselves. The daughters are **clones** of the mother. The main complication to this *clonal* pattern of reproduction is **mutation**, which occurs when specific DNA sequences that determine the phenotype are chemically changed or miscopied, producing a daughter with a different gene, or genes. For most genes, mutation occurs at a rate of once every 10^5 to 10^8 acts of reproduction. On the other hand, there are at least a thousand genes in even the simplest organisms, so that the rate of mutation over all the genes of an organism might be 10 to 10^{-4} mutations for each act of reproduction. This can create many mutations in microbial populations that might contain billions or trillions of individuals, which is good from an evolutionary point of view because natural selection needs varied genotypes that confer varied biological abilities to survive and reproduce.

Put another way, natural selection exploits genetic variance to increase fitness. (More on this later.) In **clonal selection**, mutation supplies that genetic variance. The mutations that are big genetic improvements will be found and fixed by natural selection.

Purifying selection may be the easiest form of natural selection to understand, perhaps because it is like correcting bad grammar, discarding defective products, or buying a used car.

FIGURE 4.13A Aristotle, the Founder of Biology

We sort, evaluate, and eliminate. The idea that a similar process goes on in nature appeals to the human mind.

The idea of natural selection is much older than Darwin. Aristotle (Figure 4.13A) used the idea of selection to explain the preservation of the typical form of each species. He argued that when highly deviant offspring are born, they will be defective in survival and reproduction. Such deficiencies will then prevent the appearance of monstrosities in subsequent generations. (Some fanciful monsters are shown in Figure 4.13B.) In this way, Aristotle proposed, species are kept separate. Notably, this *purifying* type of selection will also keep the members of the species functional, adapted to their particular way of life.

The main targets of purifying selection are mutations,, especially mutations that decrease fitness. Many mutations generate such deleterious effects. Some of these may be small or specific, such as losses of metabolic pathways that break down toxins or that extract energy from specific sugars. Other mutations may have pervasive and disastrous effects, creating anatomical monsters, sterilizing, or killing. Large or small in their effects, all these mutations are targets of purifying selection.

Natural selection may not consistently favor a single allele. This occurs, for example, in cases of **heterozygote superiority**, when heterozygotes, which have two different alleles, have the highest fitnesses in diploid sexual populations. Because of the sexual process, a population that starts out consisting entirely of heterozygotes will generate 50 percent homozygotes in the next generation. Selection will not be able to eliminate this half of the population immediately, unless all the homozygotes have zero fitness. Even when this unusual situation applies, the population will keep both alleles at frequencies of 0.5. Selection cannot eliminate genetic variation with heterozygote superiority, because the genotype with the greatest fitness itself is a repository of genetic variation. (Less extreme cases of heterozygote superiority are discussed later in the chapter.) But natural selection still acts to maintain genetic variation, by a pattern called **balancing selection**.

Balancing selection is important because there is abundant genetic variation for virtually every character that is related to fitness in outbred sexual populations: viability, fertility, running speed, seed production, height, weight, and so on. Why is there so much genetic variation for fitness-related characters? Purifying selection is expected to purge populations of genetic variation; but balancing selection is not. Balancing selection has therefore been proposed as a possible explanation for genetic variation in important characters. We give an important example of balancing selection later in this chapter. ❖

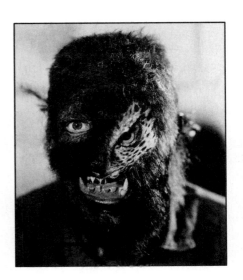

FIGURE 4.13B A Gallery of Humanoid Monsters

Selection in asexual populations increases mean fitness until the genetic variance in fitness is used up

If every organism is the same, selection cannot work. Instead, when there is variation between organisms, selection takes the different groups in a population and makes them compete with each other. The key to this competition is differences in *net reproduction*, or **fitness**. **Net reproduction** is the product of total reproduction times the viability (survival probability) of offspring. For example, if a flatworm produces 12 offspring, but only 1/6 of these survive to become adult flatworms, then net reproduction is 2. Most microbes reproduce by splitting in two, so their total reproduction is always 2. In this case, net reproduction then varies only because viability varies.

Figure 4.14A illustrates an asexual population of microbes in which different groups of organisms have different viabilites, or survival probabilities. In this example, all reproduc-tion is by splitting in two, or **fission**. Therefore, total reproduction is always two. In the illustration, the pink cells have a survival probability of about $1/4$; that is, one in four of these cells will survive to reproduce. The green cells have a survival probability of $1/2$, and the blue cells have a survival probability of 1. Their corresponding fitnesses can then be obtained by multiplying these survival probabilities by the total reproductive output of two. This gives fitnesses of $1/2$ for pink cells, 1 for green cells, and 2 for blue cells.

What happens in this population over time? From the fitness numbers just given, we intuitively expect the pink cells to be eliminated from the population. Indeed, by generation 10, all the pink cells are gone (see Figure 4.14A). During these 10 generations, the average fitness of the population steadily increases. After 100 generations, the mean fitness becomes very close to 2.00, the same as the fitness of the blue cells. This is natural because, by generation 100, the population is made up almost entirely of blue cells.

What about variance? The variance for fitness is high in the population for the first few generations. But by generation 10, it has already fallen a lot. And by generation 100, there is hardly any variance left. Again, this is natural, because by generation 100 almost the entire population is blue cells, which all have the same fitness. Selection has fed on the variance for fitness, and the consequence is a high mean fitness.

Notice that the mean fitness and the variance of the population changed, but *not* because of any change in the fitnesses of pink, blue, and green cells. Those stayed the same. The population changed in *composition*, but the different types of cells did not change. Selection works with what it is given. It is an editor, not a writer. ❖

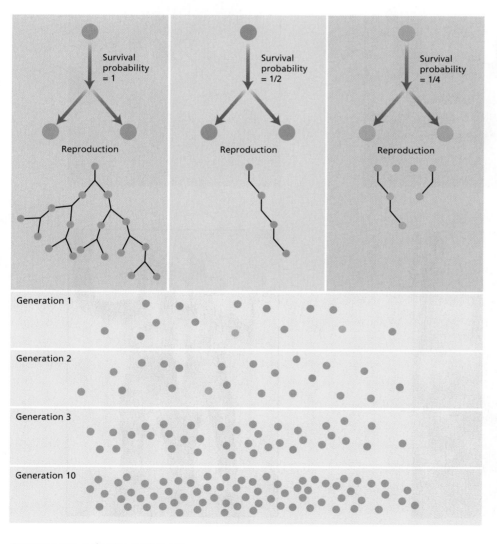

FIGURE 4.14A Selection without Sex

The role of selection is that of a filter or sieve. Some genotypes don't make it through the filter as often; they are selected against. These differences are summarized by numbers giving the chance that different genotypes will make it to adulthood; that is, fitnesses. In normal notation, we write that W_{ij} is the fitness of genotype ij, where i and j are the two alleles at the diploid locus of interest. Typical values for fitnesses (W's) range from 0 for lethal genetic diseases, to values over 1.0 for genotypes of superior fitness.

In Figure 4.15A, the bookkeeping of genetics and fitness is laid out for cases where there is a consistent relationship between genotype and fitness. Fitness (W) is graphed for three cases: where allele a is dominant (part i); where the heterozygote Aa produces characteristics that are intermediate between those produced by aa and AA (part ii); and where the allele A is dominant (part iii). Note that in all three cases, the A allele is favored by natural selection, making the AA genotype as good as any other genotype, or better.

Let us start with the basic genetic model of Chapter 3: one diploid locus with two alleles. We suppose a very large population, discrete generations, and random mating. These are the model assumptions of the Hardy-Weinberg Law, so we can assume that gene frequencies do not change unless there is selection. The consequences of directional selection are shown in Figure 4.15B, which graphs the change in the frequency of a favored allele A in a population over many generations. In overview, selection makes the favored A allele sweep through a population, eliminating the a allele—which has less fitness. In the example shown, the more copies of the A allele, the more fit the genotype. In other words, AA is more fit than the heterozygote Aa, which is more fit than aa. And with greater fitness comes the attainment of A allele frequencies near 1.0, a state called **fixation**. ❖

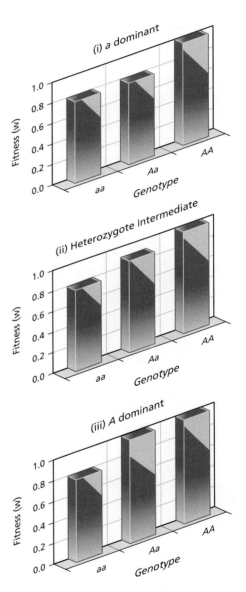

FIGURE 4.15A **Directional Selection with Sex**

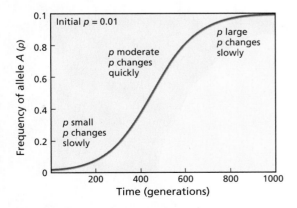

FIGURE 4.15B **The Dynamics of Directional Selection:**
$W_{AA} > W_{Aa}$ and $W_{Aa} > W_{aa}$.

Because selection is an editor and not a writer, it works with what it is given—with the types of organisms that make up each population. Mutation is the initial creator in the evolutionary process. It produces all the "first drafts" of life. The only problem is that mutation occurs at random with respect to the direction of selection, so mutations often reduce fitness. This is a pattern called **deleterious mutation**. It is like noise introduced into the evolutionary process. Natural selection then makes messages from this noise by editing out deleterious mutations.

But deleterious mutations keep on occurring. So, to change metaphors, natural selection is always pushing the boulder of mean fitness toward the top of an evolutionary hill, but never quite getting there. Mutation is like the "king of the hill," always preventing natural selection from making it all the way to the top. This "top" has a concrete meaning in asexual populations. It is the fitness of the type with the highest fitness. Natural selection tends

Mutation is like the "king of the hill," always preventing natural selection from making it all the way to the top.

to increase the frequency of this type toward 100 percent, or fixation—as we just saw. Recurring deleterious mutation frustrates that tendency.

In sexual populations, individuals with the best alleles across all loci are produced only transiently, and then rarely, as the sexual process shuffles the best alleles among individual genotypes. In this sense, it is unlikely that natural selection will fix the best possible genotype in a sexual population. With sex, deleterious mutations further degrade the fitnesses of members of the population.

We will look at the effects of mutations in sexually reproducing populations further in Chapter 18. For now, note that many mutations are highly deleterious. They are the source of some common human genetic diseases, as described in the box, "A Catalog of Misery."

Few things make having children more terrifying than reading about genetic diseases. But

Aristotle's cool reasoning has to be faced. In every reproductive act, there is the chance that our offspring are going to be targets of purifying selection.

Pulling back from the clinical horror of some genetic afflictions, we can see certain patterns from these examples of human genetic diseases. The first is that, even if the predisposing gene is not very rare, the syndromes themselves are rare. Cystic fibrosis is the most common lethal genetic disease in the United States, where it occurs in fewer than one in every 2000 live births. Even in modern medical surroundings, this is a small part of the full spectrum of disease. Perhaps part of our unease with genetic diseases is the sense of inevitability about them—making them unlike contagious diseases. Another part may be that they primarily afflict, and kill, children.

But our small list of genetic diseases is large enough to reveal the error in both of these conclusions. PKU is a genetic disease that is almost entirely treatable. There is nothing inevitable about the pathologies of genetic diseases. Huntington's disease makes a different point: Genetic diseases may strike older individuals exclusively. Genetic diseases do not stalk the sleeping babe alone. Indeed, many of the disorders of the elderly may be genetic in origin.

But the unavoidable conclusion about genetic diseases is that they can indeed turn life monstrous, frustrated, and barely—if at all—sustainable. And with survival, fertility may go, too. For some disorders, like Tay-Sachs disease, impaired fertility is medically irrelevant because victims usually die so young. But in the case of cystic fibrosis, fertility is now an overt concern of patients who survive into adulthood. On the other hand, there is the societal question regarding the fertility of individuals with some disorders, like Huntington's disease, who appear to be able to produce many offspring like themselves. Their afflicted offspring may impose huge medical costs on posterity. For all these reasons, and more besides, the problems of deleterious mutations are of great concern. ❖

A Catalog of Misery: Some Common Human Genetic Diseases

AT syndrome, or ataxia-telangiectasia syndrome, causes progressive loss of coordination, decreased resistance to infection, increased risk of cancer, and what is called the acceleration of aging, among other pathologies. Intelligence is normal. The full syndrome is caused by two defective mutations at the *ATM* gene, though a single copy may increase the risk of cancer. The frequency of the mutation is estimated as 1–3 percent. There is no specific medical treatment that relieves those afflicted with AT syndrome.

Cystic fibrosis is the single most common lethal genetic disease in the United States. Cystic fibrosis causes the secretion of thick mucus, disrupting the functions of pancreas, liver, intestine, and especially lung. Life expectancy without modern medical care is less than 10 years. With modern medical care, life expectancy is now about 30 years. Only 2 percent of males are fertile, while females have a less severe reduction in fertility. Intelligence is normal.

Cystic fibrosis is caused by two copies of a mutation of the *CFTR* gene. A single copy has no bad effects. Cystic fibrosis mutations are carried harmlessly in single copies by about 12 million Americans; one in 2300 children is born with the condition each year, having received copies of the mutation from both parents. The frequency of the gene in the U.S. population as a whole is about 2 percent. There is no known cure, though much research has been targeted at curing this particular genetic syndrome.

Huntington's disease, also known as Huntington's chorea, results in progressive deterioration of the central nervous system. This deterioration usually begins at from 30 to 50 years of age. From the first symptoms of neurological impairment, Huntington's disease takes 15–20 years, or longer, to kill its victims. During this period, the loss of coordination, mental function, and self-control becomes catastrophic.

The disease is caused by a single faulty copy of the *HD* gene. Because disease onset occurs later in life and progresses fairly slowly, fertility is roughly equal to that of normal individuals; life span is not dramatically curtailed. The frequency of the *HD* mutation varies considerably between populations, over the range of 0.0001–0.01 percent. There is no medical treatment to prevent the start of the disease or to halt the progression of the disease; there is only treatment of the symptoms.

Phenylketonuria (PKU) results from the lack of the enzyme phenylalanine hydroxylase, which is responsible for the conversion of the amino acid phenylalanine to another amino acid, tyrosine. Interruption of this biochemical conversion results in the accumulation of phenylalanine to toxic levels, in turn causing neurotoxicity and eventually severe mental retardation.

PKU is caused by possession of two mutant copies of the gene coding for the phenylalanine hydroxylase enzyme. Carriers of single copies do not get the disease. The frequency of the PKU gene is about 1 percent in the United States. PKU can be treated very successfully by eliminating phenylalanine from the diet, starting with newborn infants. Screening of newborns for PKU is routine in the United States and other countries.

Tay-Sachs disease causes the accumulation in the brain of a fatty substance known as ganglioside GM2. This accumulation results in juvenile blindness, deafness, paralysis, and severe mental retardation. Death usually occurs before the age of 5. Tay-Sachs disease is caused by the lack of the enzyme hexosaminidase A, which helps degrade gangliosides.

Having two copies of mutations deficient in the production of this enzyme result in the disease. In Ashkenazi Jews and some French Canadians, the frequency of these mutations reaches 3–4 percent. In other groups, the frequency of Tay-Sachs disease mutations is much lower. There is no treatment available for infants suffering from Tay-Sachs disease.

4.17 When heterozygotes are superior, selection maintains genetic variation

Selection and genetics are both blind mechanisms, without foresight. These mechanisms reveal their blindness clearly when there is heterozygote superiority in sexual populations. In such cases, the homozygotes are inferior; this is shown in Figure 4.17A. It would be best if selection and genetics could work together to guarantee the fixation of the heterozygote. But this does not happen. Mendelian genetics ensure that heterozygotes produce abundant homozygotes, all with lower fitness. Yet the superiority of the heterozygote ensures that these homozygotes will be selected against. In short, selection pushes for the fixation of the heterozygote, while genetics regenerates the disfavored homozygotes every generation. Selection strives for what can never be, while genetics heedlessly produces the homozygotes that selection will penalize. Out of this stalemate, genetic variation is maintained.

The eventual outcome is stable, an equilibrium with one redeeming feature. At the equilibrium, when selection does not change gene frequencies any more, the average fitness is at its maximum value relative to the value of average fitness at all other gene frequencies, assuming Hardy-Weinberg proportions among the genotypes. In other words, even though superior heterozygotes frustrate the best possible outcome, selection still works to produce the highest average fitness, among the genetic states that the sexual population can attain, as shown in Figure 4.17B.

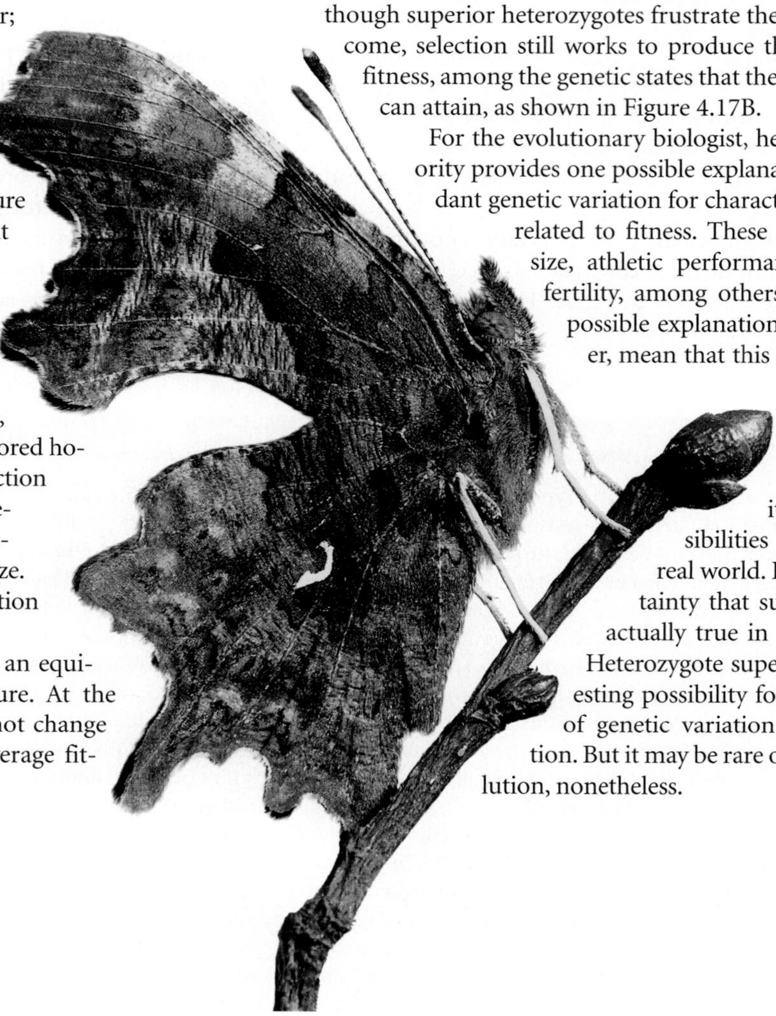

For the evolutionary biologist, heterozygote superiority provides one possible explanation for the abundant genetic variation for characters that are closely related to fitness. These characters include size, athletic performance, survival, and fertility, among others. This status as a possible explanation does not, however, mean that this explanation *has* to be true. (A very important feature of theoretical science is that it offers varied possibilities for explaining the real world. But there is no certainty that such a possibility is actually true in a particular case.) Heterozygote superiority is an interesting possibility for the maintenance of genetic variation by natural selection. But it may be rare or common in evolution, nonetheless.

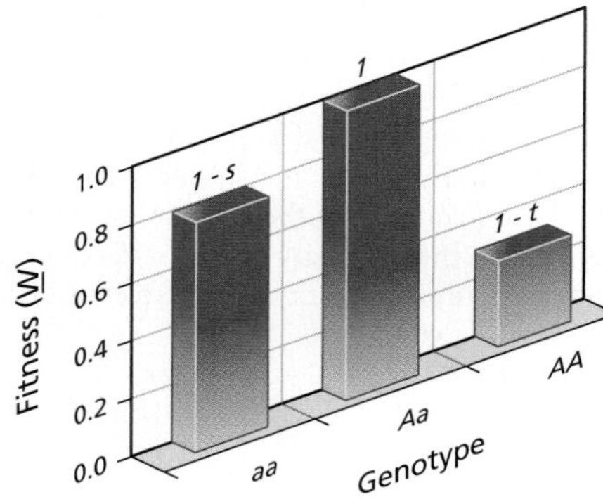

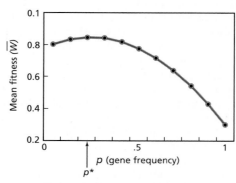

FIGURE 4.17A Fitnesses with Heterozygote Superiority

The evolutionary equilibrium is at p^*. At this gene frequency, mean fitness is at a maximum.

FIGURE 4.17B The Evolution of Mean Fitness with Heterosis

Another way to maintain genetic variability is when selection favors rare genotypes. This is a form of **frequency-dependent selection**. Selection gives the rare a "boost up" in a kind of Darwinian affirmative action. But how might selection do this?

Perhaps the most easily understood kind of favoritism of the rare occurs when predators learn to seek a particular type of prey because it is common. In the example shown in Figure 4.18A, the bird is seeking caterpillars crawling on the forest floor. The bird faces a problem: There are two kinds of caterpillars, green and brown. Both are equally camouflaged, because there are leaves on the ground, which are green, and the ground is otherwise brown dirt. But when there are many green caterpillars, it is easier for the bird to look at moving green objects and find food, compared to looking for the rare brown caterpillars. Eating the green caterpillars rewards the bird, and it associates the green color with the pleasure of feeding. So the bird develops a search image for green caterpillars, and eats a great many of them. But with time there are fewer and fewer green caterpillars. The brown caterpillars

have been left alone, and they are now common. Birds will then be better off searching for brown caterpillars. So they switch to the brown caterpillars as prey, causing the numbers of the brown to drop. The caterpillar color variants are being selected for when they are rare, and selected against when they are common. This pattern of selection should maintain genetic variability for coloration.

Figure 4.18B shows a well-studied example of frequency-dependent selection—right- and left-handed scale-eating fish. The data show that the population tends to return to balanced frequencies of the two types of fish whenever the population deviates too far from equal proportions of the two types.

This idea can be extended to other types of selection. Common vertebrate pathogens face defeat by the responsive vertebrate immune system, which develops host defenses specific to common pathogens. New mutant forms of the pathogen then have an advantage because they do not have the same molecular cues for the vertebrate immune system. Several pathogens evolve this way, including HIV, the cause of AIDS. HIV continually generates new genetic variants that elude the human immune system.

Together with heterozygote superiority, frequency-dependent selection can explain the maintenance of genetic variation. But like heterozygote superiority, it is not known how common frequency-dependent selection is in nature. For now, these two genetic mechanisms of natural selection remain of interest as possibilities for evolution. ❖

FIGURE 4.18A Most predators seek new prey like the prey they have already eaten, which leads to frequency-dependent selection.

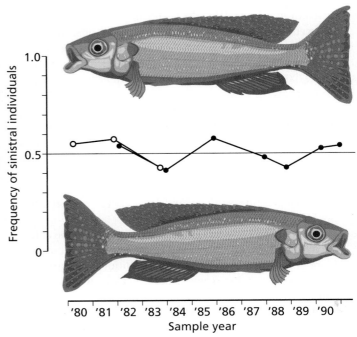

FIGURE 4.18B Frequency-Dependent Selection in Scale-Eating Fish That Scrape the Sides of Other Fish

NATURAL SELECTION IN THE LABORATORY

4.19 Natural selection in the laboratory offers a view of what is possible in evolution

Earlier in this chapter, we discussed the use of artificial selection in breeding and in scientific experiments. In artificial selection, the experimenter or breeder chooses the target of selection. The entire breeding process is controlled, so that the main determinant of the outcome of the selection procedure is the amount and nature of the genetic variability present in the selected population.

Natural selection in the laboratory is a different kind of procedure. Instead of carefully controlling all aspects of selection, the experimenter sets up a particular kind of environment that may force the evolution of the experimental population. For example, a microbial population may be placed in a hot incubator for 1000 generations, to see if the microbe will evolve in response to this selection regime. In an experiment like this, the experimenter does not control anything more than a general feature of the environment. But this environment then makes natural selection act, picking among the cells with greater fitness under hot conditions. Any genetic tendency to such greater fitness leads to an increase in the frequency of cells that are better adapted to heat.

Natural selection observed in the wild clearly reveals how natural selection actually works. Artificial selection observed in the laboratory gives biologists a close look at how selection can operate, even though it provides no guarantee that natural selection will ever act in exactly this manner. So why perform natural selection in the lab?

This question is a particular example of a very general question in science. Why do scientists generally perform experiments under carefully controlled conditions, with good instruments, stable laboratory conditions, and repeated observations? They do so because it is very difficult to gather accurate data in natural environments. Should chemists do their experiments in rainstorms, outdoors? (See Figure 4.19A.) Should molecular biologists do biochemical experiments on banana splits in restaurants? Perhaps, if those particular circumstances are interesting to them, they should. But if scientists aren't interested in such scenarios, then it seems counterproductive to insist on their performing experiments under those conditions. Laboratories provide better conditions. Temperature, sunlight, and other environmental factors can be controlled.

Likewise, evolutionary biologists will normally be able to do natural selection experiments better in the laboratory. More data will be gathered under conditions that are better controlled and better known. But what can we learn from natural selection in the laboratory?

Biologists have many questions about evolution. How fast can natural selection change characters? Can it be reversed easily? If the question relates to what natural selection or evolution actually does in real populations, then natural selection in the lab is not useful. The lab will not tell us what actually happens in nature.

But observing selection in nature is often difficult, because so much environmental change is going on at the same time as selection is working. Usually the best way to see the *potential* power of selection is to study it in the laboratory, where its effects can be seen clearly against a background of environmental stability. In addition, biologists can replicate the same evolutionary process over and over. This helps us to answer major questions of principle about evolution.

We often want to know whether or not some selective process can occur at all. Can natural selection ever change X? Or our hypotheses may concern the consequences of a particular type of selection. If we impose environment Y on population Z, will that population evolve in response? Sometimes an evolutionary theory might even make a specific prediction about the response to a particular type of selection. For example, we might predict that bacteria kept in hot conditions would evolve greater Darwinian fitness at high temperatures after some hundreds of generations in the

FIGURE 4.19A A Chemist in a Rainstorm

heat. One way to think of this situation is that the laboratory can be used to discover what is *possible* for evolution by natural selection. What is *actually* occurring is better studied under natural conditions.

Of course these stipulations and limitations are quite general to the study of evolution, ecology, and organismal biology. All these fields have problems with interpreting experiments in the laboratory. However, these problems are sometimes little more than confusion about the difference between the possible and the actual. In the laboratory, we never learn anything beyond the merely possible, whether we are biologists, chemists, or physicists. ❖

Bacterial evolution in the laboratory shows that the response to selection is very powerful at first, but tends to slow down

The power of selection in the absence of sex has been studied systematically in the laboratory of **Richard E. Lenski,** a leading evolutionary microbiologist. The classic bacterial evolution experiment from this group is still one of the best examples of the action of selection without the complications of Mendelian genetics. Perhaps the single most important reason this experiment is a classic is that it used massive replication and numerous generations. The basic data involve 12 independent populations, evolving for 10,000 generations under identical conditions. (In human terms, that many generations would be about 250,000 years—a very long time indeed.) Not only were there many populations and many, many generations, but the size of each population was also large—fluctuating between 500,000 and 50 million cells. If we assume that the average population size per generation was 7–10 million, then the 10,000 generations of evolution involved about 1 trillion cells.

Figure 4.20A shows Lenski's experimental procedure, including the system of replication, the method of population cultivation, and the way that fitness was estimated. The basic point, however, is that these bacteria were given a novel environment to adapt to in the laboratory: growth medium with basic nutrients and a small amount of glucose for metabolic fuel. Over 10,000 generations, the descendants of the original bacteria became much better at competing with their ancestors under these conditions, as

shown in Figure 4.20B. This result could be summarized as "fitness increased," or "the bacteria adapted to the laboratory conditions imposed." Either way, a lot of genetic change took place, especially in the first 2000 generations. But the rate of improvement fell, becoming much slower in later generations, as the graph of Lenski's results shows. Note how the curve is steep at first, as the populations adapt relatively quickly to the novel environment, but then flattens out after the first 2000 generations. This slowing is also expected: Initial adaptation should be faster, because the populations begin some distance from efficient exploitation of their environment, making selection stronger. ❖

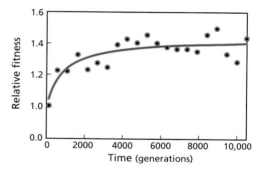

FIGURE 4.20B **The Data for Fitness over 10,000 Generations of Microbial Laboratory Evolution**

(i) Experimental Design

Samples were frozen
for later use.

E. coli clone

12 derivatives

For 2000
generations,
fitness was checked
every 100
generations.

12 independent lines
evolved for 10,000 generations

For the next 8000
generations,
fitness was checked
every 500
generations.

(ii) Experimental Methods: Population Culture

(iii) Experimental Methods: Fitness Measurement

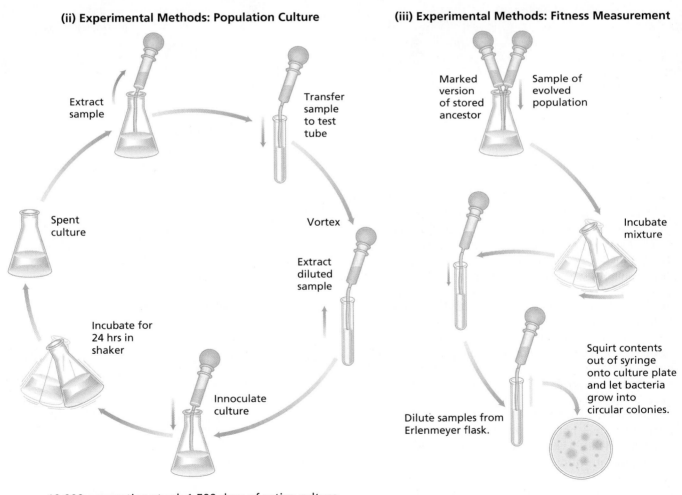

Extract
sample

Transfer
sample
to test
tube

Spent
culture

Vortex

Extract
diluted
sample

Incubate for
24 hrs in
shaker

Innoculate
culture

Marked
version
of stored
ancestor

Sample of
evolved
population

Incubate
mixture

Squirt contents
out of syringe
onto culture plate
and let bacteria
grow into
circular colonies.

Dilute samples from
Erlenmeyer flask.

10,000 generations took 1,500 days of active culture

FIGURE 4.20A The Laboratory Evolution of Fitness in a Microorganism

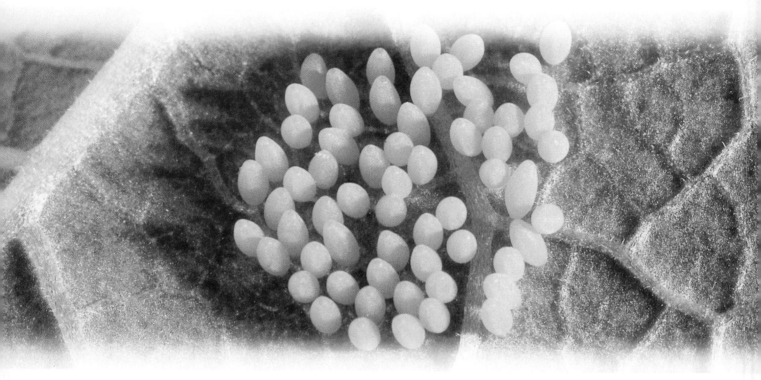

It is easy to show that sexual populations can respond quickly to directional selection. As in experiments with asexual bacteria, it is usually most convenient to use laboratory organisms. Like Mendelian geneticists, evolutionary geneticists like to work with the fruit fly *Drosophila melanogaster*. Many experiments have shown that directional selection on fruit flies can produce dramatic and sustainable genetic change. In the present example, Chippindale et al selected on fly development.

The focus of selection in our example is speed of development in the fly. The term *development* refers to the complete progression, from development in the insect's egg, to larval growth and development, to the pupal transition from larva to adult, to the initial maturation and copulation of the adult. In human terms, this is the progression from fetus to first-born. This developmental process takes the fly through a complete life cycle. Normally, the entire process takes 11–12 days in fruit flies, with standard temperatures of about 25°C and lots of good fly food. The question was, could development be speeded up using selection?

The flies used for selection came from a group of five populations that normally develop at a leisurely pace, labeled the CB flies. These flies usually take at least four weeks to complete an entire generation. B flies were the ancestors to the CBs and were used as controls for environmental fluctuations. Such controls are very important with small invertebrate animals, which can be affected by subtle features of the lab environment. The ACB flies were derived from the CB flies, and then subjected to selection for faster development, as shown in part (i) of Figure 4.21A.

(i) Selection Design

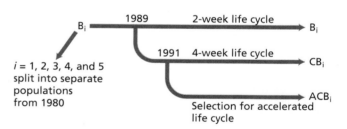

(ii) Selection Method

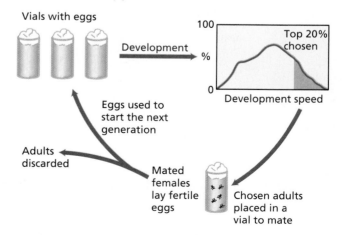

FIGURE 4.21A Selection for Rapid Development in *Drosophila*

There were five distinct sets of matched B, CB, and ACB populations, for a total of 15 populations. This quantity was important, because each population is a distinct, unique entity in evolution, like a corporation in a modern economy or a football team. To study evolution scientifically, we have to study multiple populations, not just one population. Only populations evolve, not individuals.

The selection method imposed on the ACB flies involved choosing the fastest 20 percent of flies completing maturation to the adult stage, as shown in part (ii) of Figure 4.21A. These flies then had to get mated quickly, so that the fast-developing females could lay fertilized eggs. This selection procedure was followed for 125 generations, about three years. As shown in Figure 4.21B, development time was dramatically reduced in the selected ACB lines. Directional selection led to the evolution of "faster" flies—flies that got to their nuptials sooner. ❖

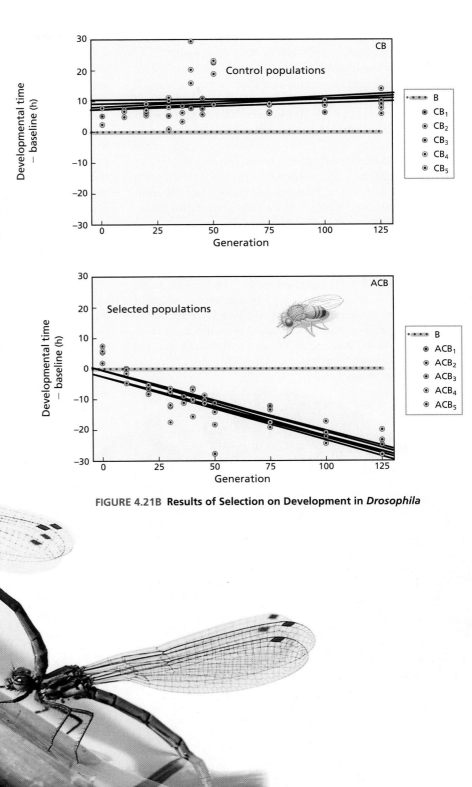

FIGURE 4.21B **Results of Selection on Development in *Drosophila***

NATURAL SELECTION IN THE WILD

4.22 The evolution of antibiotics illustrates the basic principles of natural selection in the wild

To study what natural selection actually does, scientists must examine how populations evolve in nature. We will be considering several laudable studies that demonstrate how much can be learned by studying how natural selection operates in the wild. But many of the basic ideas about how natural selection works are illustrated by the evolution of **antibiotic resistance**. When antibiotics were first used medically, at the time of World War II, they were very successful in clearing up bacterial infections. Very few bacteria were able to resist being killed by antibiotics. And there was not enough genetic variation in bacterial populations to mount a successful evolutionary defense against the antibiotics.

But within a few years, antibiotic resistance started to appear. The use of antibiotics had imposed selection for genetic variants of normal bacteria, variants that had the capacity to resist being killed by the antibiotics. For this reason, in the treatment of some infections, a particular antibiotic might fail. And over the decades since World War II, resistance to antibiotics such as penicillin (obtained from the mold *Penicillium*, Figure 4.22A), one of the first introduced, became common. Medicine had created a new selective environment against

which bacteria were at first helpless. But with time, natural selection exploited initially rare genetic variants to produce increasingly resistant strains of bacteria. Today these bacteria pose a considerable threat to the medical battle against bacterial diseases, from gonorrhea to staphylococcus infections.

But this tale gets more complicated with respect to the bacteria as organisms evolving in "nature." (Their habitat is of course our bodies.) The pattern of medical use of antibiotics is one of the basic selective factors for the bacteria. Sensitive bacteria undergoing antibiotic attack die off in large numbers, at first. However, some may survive the first 24 hours of treatment, perhaps because the tissues in which they are located receive less of the antibiotic. These bacteria may be partially resistant, as well. If antibiotics were then withdrawn from the patient, relatively more bacteria that were partly resistant would have survived. If their descendants remain within the body, they could produce a later bout of infection; but these descendants might be resistant to further antibiotic treatment.

For this reason, doctors tell their patients to finish the complete course of antibiotic treatment—for 7, 10, or 14 days. This type of prescription is very different from symptomatic medication, from aspirin to Demerol, for which doctors often specifically warn against continued medication because of potential side effects or addiction. The medical doctors are trying not only to kill off their targeted pathogen but also to reduce the chance that some bacteria will evolve increased resistance because of the antibiotic medication.

Despite the great care that Western doctors have taken to prevent the spread of antibiotic resistance among bacteria, it has in fact spread—and spread widely. Doctors try to prescribe different antibiotics when their first prescriptions do not work, but bacteria are now often resistant to multiple antibiotics. Bacterial evolution by natural selection has modern medicine on the run. We discuss this problem in more detail in Chapter 22.

But this fight has not been entirely fair. We have learned that bacteria are not as asexual as we had supposed. It turns out that bacteria exchange DNA with each other, particularly using plasmids. **Plasmids** are large circles of DNA, somewhat like the large circle of DNA that is each bacterium's genome. Most bacteria can live without their plasmids. In fact, plasmids may

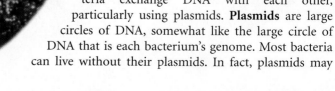

FIGURE 4.22A *Penicillium notatum*, a Source of Antibiotics

often be parasites within the bacterial cell. Plasmids are transmitted from one bacterium to another by a bridge called the *pilus*, a microscopic analog of a penis. The pilus injects copies of plasmids from one cell into another, by a process called *conjugation* (Figure 4.22B; see also Chapter 18). The donating bacterium usually keeps copies of the plasmid, so plasmids can rapidly accumulate in bacterial populations. We now know that some plasmids carry genes for resisting antibiotics, so that the use of antibiotics must have selected for bacteria carrying such plasmids. This is a more complicated story than that of simple selection on bacteria, but it reveals a profound truth about organisms without genetic organization of reproduction: They may be somewhat sexy anyway. A clone may not be a clone, after all. This is taken up further in Chapter 18. ❖

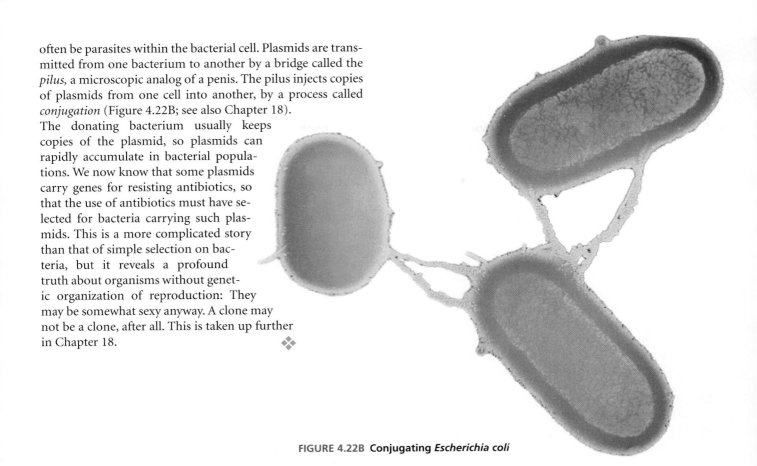

FIGURE 4.22B Conjugating *Escherichia coli*

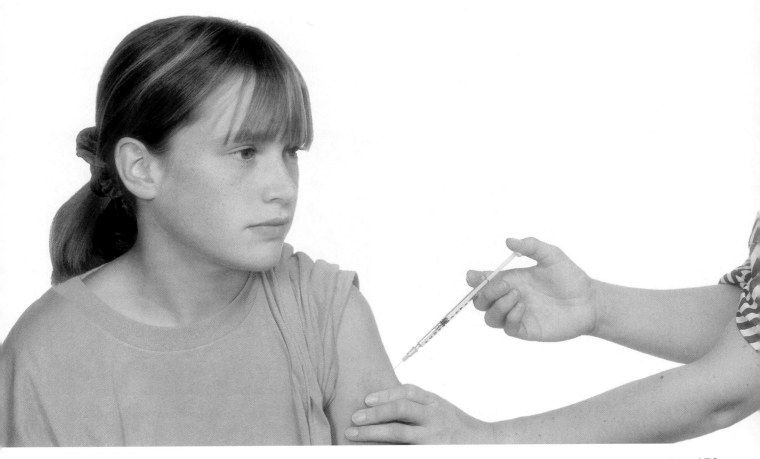

In the nineteenth century, while Charles Darwin lived in the leafy surroundings of southeast England, the "satanic mills" of northern England's Industrial Revolution were creating one of the most important examples of directional selection in the history of evolutionary biology. This example was **industrial melanism**. The term refers to the darkening of butterflies and moths in the industrial regions of Western Europe during the period when coal was the main fuel. *Melanin* is a darkening pigment. This phenomenon of industrial melanism was discovered in the collections of English amateur lepidopterists (those who study the butterflies and moths of the insect order Lepidoptera). These were mostly middle-class naturalists with the time and athletic ability to collect flying insects using nets. Sometimes their collections were immense, including numerous species gathered over several decades.

In several of the affected moth and butterfly species, the melanism pattern was striking. Beginning sometime after the introduction of extensive coal burning in the mid-nineteenth century, butterflies and moths of these species started to exhibit more and more dark *morphs*, or forms, with a lot of pigment. Over the course of the late nineteenth century, these morphs became more common, until they were in the majority of some species in the period from 1920 to 1950. This change is illustrated in Figure 4.23A. Much later, with the introduction of environmental laws and reduced production of coal soot, the frequency of dark morphs declined.

How can these evolutionary changes be explained? The record of butterfly and moth collections in England is good enough that it is unlikely to be a product of accident or fashions in butterfly collecting. In some species, dark morphs were not in any of the collections from early in the nineteenth century. It was proposed that coal soot might have changed the physiology of some butterflies and moths, making them darker. But laboratory rearing of the offspring of dark morphs under controlled conditions still gave dark morph progeny.

H.B.D. Kettlewell, a dedicated evolutionary lepidopterist working in the middle of the twentieth century, spent years lying in the grass of damp English meadows watching butterflies and moths. What he found was a beautiful example of natural selection at work. In regions with little coal burning, tree trunks were usually mottled with light-colored lichen. Moths and butterflies rested on these tree trunks, where the light-colored morphs would blend in with their surroundings (Figure 4.23B). The dark morphs were much more visible to Kettlewell. They were also more visible to birds, who picked them off the trees in greater numbers than the light morphs. But with coal burning, soot covered the tree trunks and most lichen died off. Under these conditions, the dark morphs were camouflaged, while the light morphs were picked off by birds in greater numbers. We have since learned that some of Kettlewell's experiments were artificially staged; moths were glued onto tree trunks, among other contrivances. But there is no evidence that his essential conclusions are not valid.

Natural selection was selecting moths and butterflies with appropriate protective coloration, and birds were playing the role of selective agent. The pigment change in moth and butterfly coloration is one of most straightforward examples of directional selection now known to evolutionary biologists. If Darwin had lived near Manchester, instead of south of London, he might have seen this vindication of natural selection with his own eyes. ❖

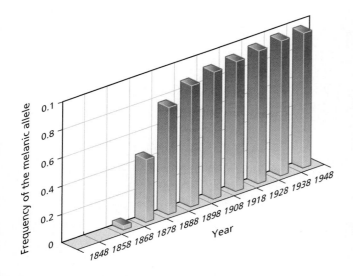

FIGURE 4.23A Estimated Pattern of Substitution of the Gene for Melanism in a Moth Data from Ridley (2004, p. 110).

FIGURE 4.23B Melanic and Speckled Moths on an Unpolluted Tree Trunk

Darwin's visit to the Galápagos Islands in 1835 uncovered several interesting organisms for biological research, such as the very long-lived Galápagos tortoise. But the group that scientists have studied most is the group of finches named after Darwin—Darwin's finches, of the genus *Geospiza*. **Peter Grant** of Princeton University has for some time led a group of biologists who have studied these birds, banding each bird when young and carefully keeping track of deaths.

In the late 1970s a major drought struck the Daphne Major Island of the Galápagos archipelago. Many of the plants on the island produced few or no seeds. In the species *Geospiza fortis*, a ground finch that depends on seeds for food, the population size fell precipitately—from about 1400 to a few hundred—over just two years, as shown in Figure 4.24A. At the same time, the sex ratio shifted from 1 male:1 female to 6 males:1 female. Some details of this ecological disaster are shown in Figure 4.24B. This is the kind of large-scale misfortune that the demographer Malthus had foreseen, though its causes in this case were meteorological and not crowding. In a Darwinian view of life, we would expect intense natural selection to occur under these conditions.

What Grant's group observed was a dramatic increase in the average body size and the average beak size of the ground finch population. This change occurred because small seeds were rare during the drought. However, even during the drought, large seeds with thick husks were still available. Only the large birds with large beaks could successfully crack open the large seeds and eat their contents. Because smaller birds with smaller beaks had few seeds that they could eat successfully, they either starved or died of exposure, from lack of the caloric reserves to survive lower nighttime temperatures. (See Figure 4.24C for photos of some of the finches.) From this differential pattern of death, the finch population rapidly changed in response, presenting a clear-cut case of natural selection. ❖

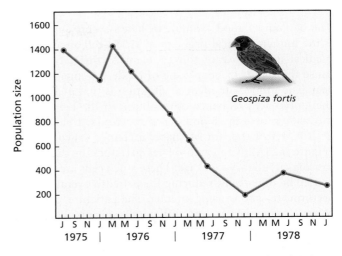

FIGURE 4.24A A Darwin's Finch Species (*Geospiza fortis*) Dying off During a Drought

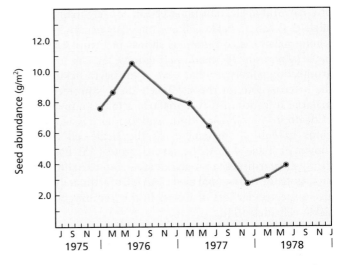

FIGURE 4.24B Falling Seed Abundance during a Drought on the Galápagos Islands

FIGURE 4.24C Various Species of Darwin's Finches

The best-understood example of heterozygote superiority is the human medical problem of **sickle-cell anemia.** The medical importance of this problem is considerable, because thousands of people die of sickle-cell anemia every year. The immediate medical situation is that a change in the amino acid sequence of a subunit of the hemoglobin molecule causes the blood's erythrocytes (red blood cells, or RBCs) to bend out of shape, a change called sickling (Figure 4.25A). The erythrocytes get stuck in capillaries, causing circulatory blockage. These stuck cells are then degraded by other cells, reducing the patient's overall level of erythrocytes and causing anemia. The patient experiences chronic pain, anemia, and difficulty respiring, with death usually coming before reproduction.

This particular case handily, though tragically, illustrates the anomalies of balancing selection with heterozygote superiority. Unlike the genetic syndromes of purifying selection, in which natural selection acts to reduce the frequency of a harmful allele, natural selection actively maintains the allele that causes sickle-cell anemia. This occurs in regions where malaria is common, as shown in Figure 4.25B. The malarial parasite, *Plasmodium* (Figure 4.25C), is not as good at infecting genotypes that have erythrocyte sickling. Even the heterozygote for the sickle-cell allele is protected from malarial infection. Because **malaria** remains a major cause of death in Africa, Asia Minor, and southern Asia, these regions have high frequencies of the sickle-cell allele. As shown in Table 4.25A, the heterozygote (*AS*) has significantly greater fitness in malarial areas compared to the homozygote for the normal allele (*AA*). Unfortunately, there is also a spectacular loss of fitness in the homozygote for the sickle-cell allele (*SS*).

If we didn't know that the heterozygote is of higher fitness in malarial areas, we might think that sickle-cell anemia is a genetic

TABLE 4.25A	Human Polymorphism for Hemoglobin		
S—sickle-cell hemoglobin			
A—normal hemoglobin			
Genotypes	*AA*	*AS*	*SS*
Initial genotype frequencies	0.77	0.21	0.02
Death due to sickling	0	+	+++
Death due to malaria	+++	+	+
Fitness in malarial areas	0.88	1.0	0.14

disease that is similar to Tay-Sachs disease. But the tip-off is the greater frequency of the disorder. Sickle-cell anemia affects about 2 percent of the population in malarial regions. Cystic fibrosis, the most common genetic disease, attacks only about 0.04 percent of the U.S. population—a population that is relatively more afflicted with this disorder compared with other countries. So sickle-cell anemia is two orders of magnitude more common than the most common genetic disease. This tells us that selection must have played a role in establishing sickle-cell anemia in malarial populations. This is a case where selection actively fosters human misery. ❖

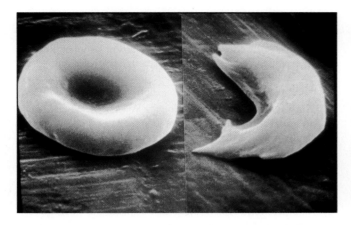

FIGURE 4.25A Sickled and Normal Red Blood Cells (RBCs)

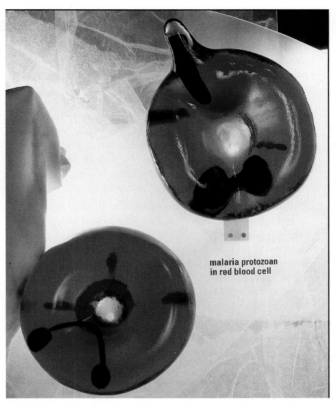

malaria protozoan
in red blood cell

FIGURE 4.25C *Plasmodium*, the Pathogen that Causes Malaria

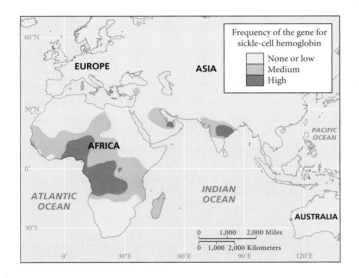

Frequency of the gene for
sickle-cell hemoglobin

- None or low
- Medium
- High

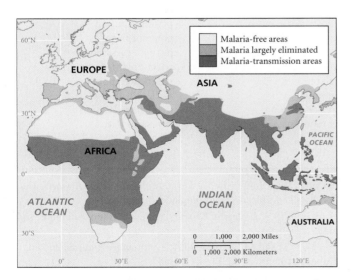

Malaria-free areas
Malaria largely eliminated
Malaria-transmission areas

FIGURE 4.25B The Geography of Sickle-Cell Anemia and Malaria
in Africa

Natural Selection in the Wild   **163**

SUMMARY

1. Darwin's views concerning natural selection were thoughtful, though not always correct. Darwin did not expect to observe natural selection, because he thought that its action would be too protracted. Although this is often the case, we now know that it is not always true. But this lack of faith in the speed of natural selection left Darwin short of examples of the action of selection in nature. He compensated by heavily using the literature on artificial selection. Perhaps as a result, Darwin cast nature as a kind of breeder or selector. He emphasized that many generations of breeding can produce breeds of animal and varieties of plants that are strikingly different.

2. The way selection works is transparent in artificial selection. This helps us to understand natural selection, which parallels artificial selection. Both require the presence of genetic variation for the selected character. But there are some differences between artificial and natural selection. Natural selection can act on more organisms than artificial selection can. Natural selection usually fluctuates more in direction. Which type of selection will act with the greater power is uncertain.

3. Selection operates according to phenotype in different ways. Sometimes selection favors a directional change toward one extreme or another. At other times selection favors intermediate phenotypes, by a pattern called stabilizing selection. Selection can alternatively penalize intermediate phenotypes, by a pattern known as disruptive selection.

4. The genetics of natural selection are also varied. When there is no sexual reproduction in a population, natural selection efficiently increases the frequency of the clone with highest fitness, using up the genetic variance of fitness. Deleterious mutations lead to purifying selection and to reduction in genetic variation. Heterozygote superiority and frequency-dependent selection can maintain genetic variation.

5. Natural selection can be studied in the laboratory with good environmental control and replication. This research shows us what is possible in evolution, not what actually occurs. Bacterial evolution of fitness suggests that fitness evolution may decelerate through time. The life-history characters of fruit flies evolve readily in laboratory experiments.

6. Natural selection in the wild is illustrated by a few important examples: antibiotic resistance in bacteria, industrial melanism, the beaks of Darwin's finches, and sickle-cell anemia in humans.

REVIEW QUESTIONS

1. Why did Darwin discuss artificial selection in the *Origin*?

2. Does the speed of natural selection always fit Darwin's expectations?

3. What is a phenotype?

4. Why does heterozygote superiority lead to genetic variation?

5. Industrial melanism is a case where selection focused on what type of adaptation?

6. Why might natural selection favor an intermediate phenotype?

7. If natural selection is steady, always applying the same pressure, what would the pattern of evolution look like?

8. Are natural environments always moderate in their selection pressures?

9. The evolution of sickle-cell anemia in humans is an example of what kind of selection?

KEY TERMS

achondroplasia	cumulative selection response	heritability	phenotypic selection
adaptation	cystic fibrosis	heterozygote superiority	phenotype variation
Aristotle	deleterious mutation	Huntington's disease	phenylketonuria (PKU)
artificial selection	directional selection	industrial melanism	plasmid
AT syndrome	disruptive selection	Kettlewell, H.B.D.	polymorphism
balancing selection	dwarfism	Lenski, Richard E.	purifying selection
bimodal	fission	Lyell, Charles	selection differential (S)
Chippindale, Adam K.	fitness	malaria	selection response (R)
clonal selection	fixation	mutation	sickle-cell anemia
clone	frequency-dependent selection	natural selection	stabilizing selection
cumulative selection differential	Grant, Peter	net reproduction	Tay-Sachs disease

FURTHER READINGS

Bates Smith, T. 1993. "Disruptive Selection and the Genetic Basis of Bill Size: Polymorphism in the African Finch *Pyrestes*." *Nature* 363: 618–620.

Cavalli-Sforza, L. L., and W. F. Bodmer. 1971. *The Genetics of Human Populations*. San Francisco: W.H. Freeman.

Darwin, C. 1859. *On the Origin of Species by Means of Natural Selection*. London: John Murray.

Darwin, C. 1896. *The Variation of Animals and Plants under Domestication*, facsimile edition. New York: Appleton.

Falconer, D. S., and T.F.C. Mackay. *Introduction to Quantitative Genetics*, 4th ed. Harlow, Essex, England: Longman.

Ford, E. B. 1971. *Ecological Genetics*, 3rd ed. London: Chapman and Hall.

Freeman, S., and J. C. Herron 2001. *Evolutionary Analysis*, 2nd ed. Upper Saddle River, NJ: Prentice-Hall.

Ridley, M. 2004. *Evolution*, 3rd. ed. Cambridge, MA: Blackwell.

Rose, M. R., and G. V. Lauder, eds. 1996. *Adaptation*. San Diego, CA: Academic Press.

Understanding life based on molecular sequences

5

Molecular Evolution

In the year 2000, then-President Clinton and Prime Minister Tony Blair jointly announced that the sequence of the human genome had largely been determined. There were a few missing parts to the sequence, but the string of nucleotides making up all 23 human chromosomes was largely known by the time of the announcement. Many journalists wrote this story up as if it was an unheralded thunderclap, a dramatic new turn in mankind's knowledge of itself.

But biologists had been working on genomic projects for years, often under the heading of "molecular evolution." The project of understanding life based on molecular sequences was conceived before 1950, and it received its greatest success with the development of the double-helix model for DNA by James Watson and Francis Crick in 1953. Rather than marking a beginning, the sequencing of the human genome was more the completion of a vision first glimpsed half a century earlier. No-

tably, Watson himself was the first to have guided the human genome sequencing project, before political enemies forced his resignation.

To understand the significance of the "genomic era," you need to understand molecular evolution. You need to learn what DNA sequences can, and cannot, tell you about life. You need some background concerning the information in genomic DNA—what is "junk," perhaps, and what is revealing.

We will cover all these topics in this chapter. First we will survey the overall structure of the genome, what it is made of, and which processes have contributed to its evolution. Then we show that much of gene evolution and genome evolution has been neutral, unimportant with respect to natural selection. In the final section, we deal with the role of selection in molecular evolution. By that point, you should begin to understand the true importance of sequencing the entire human genome. ❖

GENES AND GENOMES

5.1   The genome is not a huge library of information

DNA was established as the material of heredity in the 1940s and 1950s. The collection of all the DNA in the cell is called the **genome**. Until the 1970s, the common view of the genome was that it was a vast library (Figure 5.1A) of genetic information encoded by base pairs of DNA. Most multicellular animals and plants have billions of pairs of DNA nucleotides in each cell, enough for millions of genes. Therefore, biologists thought that there must be large numbers of genes, more than enough to specify physiological functions in great detail.

We now know that almost nothing about this view of the genome is correct. There is indeed a vast amount of DNA in many genomes, but most of that DNA does not code for amino acid sequences. This does not necessarily mean that the noncoding DNA is nonfunctional. Some of it is involved in the control of **genetic transcription**, the copying of the DNA sequence from the chromosome to *messenger RNA*. But any such information is secondary to protein encoding. Figures 5.1B and 5.1C show the contrast between the old and new views of genome structure.

The number of protein-coding genes is several orders of magnitude *smaller* than we used to think. Instead of millions of genes per genome, we know now that the number of genes ranges from a few thousand among bacteria to about 40,000 or less in vertebrates. Commonly studied genomes, like those of *Drosophila* or the nematode *Caenorhabditis* (Figure 5.1D), have 10,000 to 20,000 genetic loci. This is comparable to the number of parts in a modern car or airplane. In regard to gene numbers, genomes are extremely compact.

Methods for rapidly sequencing DNA were discovered in the 1970s. This finding led to a vast expansion of gene-sequencing activities. The first important result of this burst of sequencing activity was the discovery of DNA sequences within genes that did not code for amino acid sequences. Instead, as shown in Figure 5.1E, the noncoding segments of DNA are transcribed into **messenger RNA (mRNA)** and

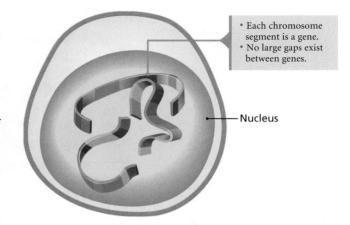

FIGURE 5.1B **Old View of the Genome, Still Accurate for Many Microbes**

* Each chromosome segment is a gene.
* No large gaps exist between genes.

Nucleus

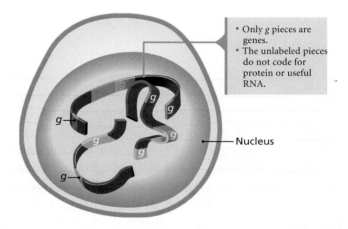

FIGURE 5.1C **New View of the Genome, Correct for Most Organisms, Especially Animals and Plants**

* Only *g* pieces are genes.
* The unlabeled pieces do not code for protein or useful RNA.

Nucleus

FIGURE 5.1A **A Library of Books**

FIGURE 5.1D **The Nematode *Caenorhabditis elegans***

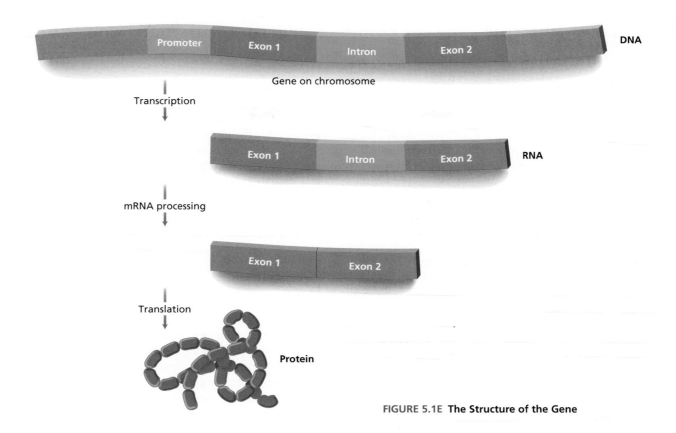

FIGURE 5.1E **The Structure of the Gene**

then excised from messenger RNA before it is used to guide **translation,** the assembly of the amino acid sequences of proteins from the coded instructions of the messenger RNA. The noncoding DNA sequences within genes are called **introns**. The coding DNA sequences are called **exons**.

Among the DNA found within introns and other noncoding regions were **transposable elements**—DNA sequences that move around within genomes. This movement does not follow any simple genetic rules (Figure 5.1F). Transposable elements are even thought to move from species to species.

Both introns and transposable elements were major anomalies for the old view of the genome as a well-organized library of functional information. They suggested that the "texture" of the genome was like Swiss cheese—full of holes and lacking in structural rigidity. ❖

(i) Conservative transposition   (ii) Duplicative transposition

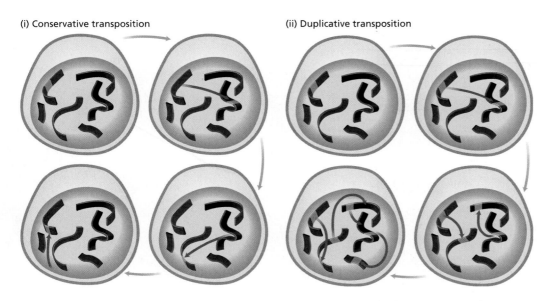

FIGURE 5.1F **Two Types of Transposition**   With conservative transposition (i), the numbers of transposable elements does not increase. Duplicative transposition (ii) permits copy number increases.

5.2 The eukaryotic gene is a complex structure with many nucleotides that do not code for amino acids

Although introns are common features of eukaryotic gene structure, they have a number of features that are unlike coding DNA. Introns are not universal. Most of the genes of bacteria entirely lack introns. There is nothing about genetic function that appears to require introns. In itself, this tells us that introns are not functional parts of genes *in the same way* that coding sequences are.

The DNA flanking the exons of eukaryotic genes plays a functional role. Both the DNA that precedes the transcribed portion of genes and the DNA that follows influence the functioning of proteins. In short, this DNA has molecular-genetic regulatory functions. The DNA that comes just before genes plays an important role in determining the situations in which transcription occurs, such as starvation, development, aging, and so on. The most important regulators of transcription are DNA sequences called **promoters.** The DNA that follows exons also plays a role in the stability of the RNA transcript before it is processed to remove introns and then used for translation. The additional regulatory DNA sequences extend what can be considered the gene beyond the exons that code for amino acids.

Eukaryotic genes—comprising exons, introns, and flanking regulatory sequences—can be very long, containing thousands of nucleotides. In addition, related genes may be clustered together. Such gene clusters can have complex interactions, and the DNA flanking a gene cluster can have regulatory functions.

Introns are highly variable in their location. As shown in Figure 5.2A, the introns of the actin gene family are variable in their site. Organisms like *Saccharomyces pombe,* which is a close relative of brewer's yeast that reproduces by dividing in two, entirely lack introns in their actin genes.

Rates of intron evolution are very different from rates of exon evolution. Exons evolve at a rate of about one substitution per billion years per nucleotide. Introns evolve at a rate about ten times greater than that. This difference in rates of evolution suggests, to a first approximation, that the evolution of intron sequences proceeds at a rate determined either by genetic drift or by some rapid form of natural selection. We will consider this issue in more detail later in this chapter.

The origin of introns has been a source of argument. One theory is that introns are the residue of genetic reorganizations that brought together small exons. Furthermore, it has been proposed that these small exons might represent distinct functional elements, possibly ancient "proto-proteins." These small proto-proteins

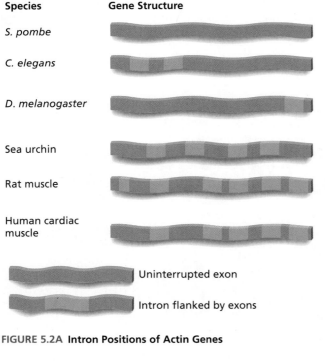

| Species | Gene Structure |

S. pombe

C. elegans

D. melanogaster

Sea urchin

Rat muscle

Human cardiac muscle

Uninterrupted exon

Intron flanked by exons

FIGURE 5.2A Intron Positions of Actin Genes

may have functioned together with the amino acids encoded by other proto-proteins. Exons could be relics of these cooperating proto-proteins, assembled together later by recombination and selection. This is the *domain* model of intron evolution. This theory proposes that genes code for proteins that are combinations of simpler bits and pieces that come from ancient proteins.

The idea is somewhat as if your car were made of parts that came from different vehicles: the engine from one vehicle, the passenger compartment from a second, and the trunk from still a third. The construction of the car would then involve the assembly of these different elements.

In the case of genes, presumably they evolved when different exons were brought together within a larger gene structure, with the introns representing the relics of the DNA that used to separate the ancient genes long ago.

One of the biggest problems facing this theory is that fairly simple organisms, like bacteria, have very few introns. If ancient organisms were like bacteria in their gene structure, then this type of theory must be wrong. On the other hand, there is the possibility that the genomes of today's bacteria are evolutionary products of selection for removal of introns. ❖

Transposable elements are mobile genes that make copies of themselves to move about the genome

For decades, geneticists thought that the genes of a species occupied fixed positions on chromosomes, much as ancient astronomers thought that stars were stationary celestial bodies. Geneticists knew that genes changed positions as evolution changed one species into another, but this was regarded almost as a cataclysmic event, unlike normal genetics. One of the best pieces of evidence in favor of such a conservative view of genome structure was that the order of genes along pieces of chromosomes tended to be the same within groups of closely related species, such as species of the fruit-fly genus *Drosophila* or the great ape species— orangutans, gorillas, chimpanzees, and humans.

One of the first anomalies for this view of genes as static compartments of information came from studies of North American maize ("Indian corn") by Barbara McClintock, a pioneering American plant geneticist and Nobel Laureate. The kernels of native American-cultivated maize show extensive **variegation** in color pattern. Each kernel is genetically dis-

tinct. Some kernels are yellow, but others, on the same ear of corn, might be dark brown (Figure 5.3A). Further, this pattern was not predictable from one ear of corn to another—unlike normally inherited color patterns in most plants and animals. McClintock proposed that there were *controlling elements* that caused the variegation in the kernel color, elements that we now know are transposable elements: DNA sequences that move about the genome, making new copies of themselves and inserting themselves in sites that they did not previously occupy. In this case, these tranposable elements are inserting into pigment genes, disrupting their function.

Transposable element insertions are known to be selected against when they occur within exons. For example, the *white* locus of *Drosophila* is known to have suffered repeated exon insertion by transposable elements, all deleterious because they impair vision. (A white-eyed fly is shown in Figure 5.3B.) On the other hand, transposable elements that insert in introns appear to have fewer deleterious effects.

FIGURE 5.3A Corn, Showing the Effects of Transposition on Kernel Color

FIGURE 5.3B **White-Eyed Male Fruit Fly** This type of mutant can be produced by the insertion of a transposable element in the eye color gene.

Typical transposable elements code for **transposase**, a protein that allows the element to make new copies of itself and insert them in the genome at various locations. Sometimes transposable elements code for additional proteins that are also indispensable for their life cycle. Still other transposable elements code for proteins that are unrelated to the replication of the transposable element, such as proteins that help cells resist antibiotics.

Some transposable elements cannot produce transposase. Their transposition depends on the presence of transposable elements that still produce transposase. Because transposition causes frequent mutations, it is common to find that groups of transposable elements include passively transposing mutants that cannot transpose on their own, along with transposable elements that remain intact, as Figure 5.3C shows.

There are distinct transposable element life cycles, two of which are displayed in Figure 5.3D. **DNA-based transposition** is shown in part (i) of the figure. Chromosomal DNA is copied by DNA replication to form extrachromosomal DNA. Some of the extrachromosomal transposable element DNA is then inserted at a new site in the genome.

A second class of transposable elements is made up of the **retrotransposons,** shown in part (ii) of Figure 5.3D. These elements, also known as **retroposons**, reside in the genome as DNA. But their replication requires transcription and the formation of an RNA intermediate, as part (ii) of Figure 5.3D shows. This RNA intermediate is then used to guide the synthesis of the corresponding DNA sequence, using the protein **reverse transcriptase.** Reverse transcriptase may be incorporated in the retrotransposon undergoing reverse transcription, or it may come from another transposable element. The DNA produced with the help of the reverse transcriptase is then incorporated in the host genome. Some of these elements consist of little more than a promoter sequence for transcription and flanking sequence information for incorporation of the reverse transcribed DNA back into the genome. An example of this type of element is *Alu I*, which is present in humans in hundreds of thousands of copies in each of our nucleated cells. ❖

(i) DNA-based transposon life cycles

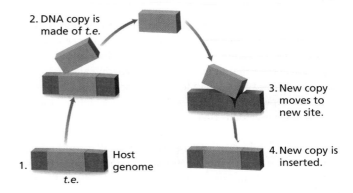

Intact transposable element

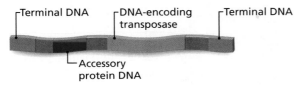

Passively transposing elements

Inert relics of transposable elements

FIGURE 5.3C **Polymorphism within a Single Class of Transposable Elements.** Transposition requires transposase and both terminal DNA sequences. Elements lacking their own transposase may get it from another transposable element.

(ii) RNA-based retroposon life cycles

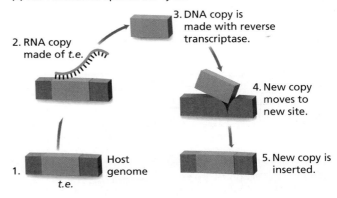

FIGURE 5.3D **Life Cycles of Transposable Elements (*t.e.*)**

Because of its double-helix structure, DNA has an inherent tendency to recombine. However, this tendency to recombine is physically limited by sequence similarity. Dissimilar DNA sequences are unlikely to recombine with each other, because their helices will not pair well. But similar DNA sequences tend to pair. This produces normal recombination during meiosis, in which homologous alleles are swapped between chromosomes due to pairing of helices from different chromosomes.

But the situation is different when DNA sequences are arrayed *in tandem*, with repeated sequences following one after another along a chromosome, as shown in Figure 5.4A. Many such **tandem arrays** of DNA sequences are known to occur in nature. These may be simple sequence repeats, or they may be repeats of sequences that code for proteins or ribosomal RNA, as shown in Figure 5.4A.

The excitement comes from the things that DNA can do when sequences occur in tandem arrays. Figure 5.4B shows that DNA sequences in tandem arrays can be misaligned during meiosis, before recombination occurs. With this misalignment and recombination, which is called **unequal recombination** or **unequal crossing over**, one of the two products of the recombination event will contain more copies of the repeated sequence, while the other recombination product will have fewer copies of the repeated sequence.

At this point there are two main possibilities:

1. The repeated sequence is of no functional significance. Then the number of copies of a repeated sequence may rise or fall with accidents of unequal recombination. This is a drift process like that of genetic drift, involving alternative alleles at a normal Mendelian locus, as described in Chapter 3.

2. The repeated sequence is a functional genetic sequence, such as a genetic locus that codes for a protein, or a sequence that is used to transcribe the RNA components of a ribosome. Then natural selection will oppose unequal recombination that leads to a lack of the functional sequence. It may also select against high numbers of the sequence, perhaps because too much production of a protein results. This will give rise to *stabilizing selection*, described in Chapter 4, in which an intermediate copy number would be the state favored by selection.

Simple nucleotide repeats
ATATATATATATATATATATATATATATAT

Longer nucleotide repeats
ATGCCCATGCCCATGCCCATGCCCATGCCC

Single-gene repeats

Multiple-gene repeats

FIGURE 5.4A Tandem Array Structures

(i) Misaligned tandem arrays

(ii) Recombination of misaligned tandem arrays

(iii) Unequal sizes of resulting recombinant tandem arrays

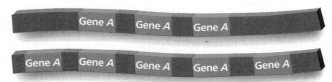

FIGURE 5.4B Unequal Crossing Over in Tandem Arrays

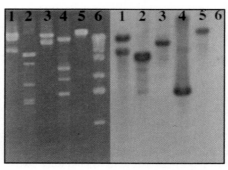

Tandemly repeated DNA sequences have many applications in genetics. They are used as highly variable markers in molecular ecology, particularly to determine who is mating with whom in natural populations. They are also used in *genetic fingerprinting*, to distinguish between suspects in criminal cases, particularly rape, where the perpetrator almost always leaves his genes behind (see Figure 5.4C). ❖

FIGURE 5.4C Forensic science routinely uses tandem array DNA as evidence in criminal cases.

Prokaryotic genomes are concatenations of genes with occasional inserted sequences, whereas eukaryotic genomes have large intergenic regions that play no apparent role in gene replication or function

In the prokaryotes, especially the bacteria, genes are closely packed together, with little intergene DNA. There are also few introns. This genome can be thought of as maximally efficient in the use of DNA. The **prokaryotic genome** is a compact compendium of genetic loci with the occasional transposable element inserted here and there. In such compact genomes, the evolution of the genome is not that different from the evolution of many individual genes combined. Indeed, such genomes are often largely free of introns, making them an even tidier story. This genome structure is sketched in Figure 5.5A.

Some eukaryotes, such as some yeast species, also have very compact genomes. Like prokaryotes, such unicellular eukaryotes have little DNA between genes. They also tend to have relatively few introns. Again, the genome is very compact.

The existence of compact eukaryotic genomes shows that there is no functional requirement for abundant DNA between genes. Nor is there any apparent requirement for introns. These prominent features of genome structure in most eukaryotes appear to be dispensable, at least for some microbial organisms.

Unlike yeast, most eukaryotes have the the type of genomes shown in Figure 5.5B, with large regions of DNA between genic regions, abundant introns, transposable elements, and so on. Most of the DNA of animals and plants has no protein-coding function. The genomes of such organisms are usually a sprawling affair. For long stretches of DNA, there are no genes at all. One way to understand the difference between these genome structures is to think of bacterial genomes as villages, yeast genomes as small cities, and most eukaryotic genomes as megalopolises like Los Angeles (Figure 5.5C). ❖

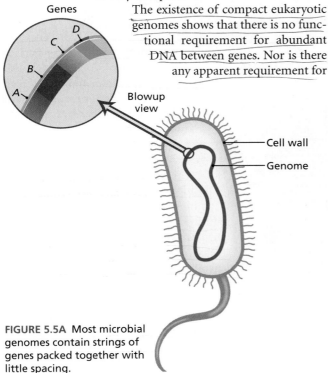

FIGURE 5.5A Most microbial genomes contain strings of genes packed together with little spacing.

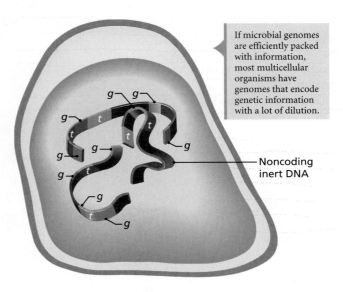

If microbial genomes are efficiently packed with information, most multicellular organisms have genomes that encode genetic information with a lot of dilution.

t indicates a transposable element.

g indicates genes.

FIGURE 5.5B Big eukaryotic genomes have a lot of noncoding intergenic DNA. There is no clear functional purpose for much of this DNA.

FIGURE 5.5C **Two Views of Los Angeles**

Neutral Molecular Evolution

5.6 The neutral theory of molecular evolution is based on genetic drift

In the 1960s, Motoo Kimura and a few other evolutionary theorists proposed that the explanation for molecular evolution was not selection, but genetic drift. They proposed that segregating molecular genetic variation is not important for natural selection. This **neutral theory of molecular evolution** was at first discussed in fairly extreme terms. Many argued against it on the grounds that evolution is full of adaptations, which must therefore be products of natural selection. There were two weaknesses to this criticism. The first weakness was that adaptations can be produced by evolutionary processes other than natural selection specifically favoring their evolution, perhaps as a result of natural selection for very different features. For example, there is some evidence that feathers, which are adaptations for flight, were first produced by selection for some other function in reptiles, perhaps temperature regulation.

The second weakness of this criticism is that evolution has multiple levels. Most molecular variation may have no effect on the organism's phenotypes, even if there is other genetic variation that does affect the organism's phenotypes. This second type of genetic variation can then be shaped by natural selection, even if the first type is not. There is no incompatibility between the neutral theory of molecular evolution and the action of natural selection in adaptive phenotypic evolution.

It is difficult to predict whether or not particular proteins will be favored by natural selection. At the level of nucleotides, however, it is somewhat easier to say in advance how natural selection will be structured, because of the considerable difference between nucleotides in their roles within the genome. There are two levels to this problem: one is the individual gene and the other is the entire genome. Let's consider each in turn.

Figure 5.6A shows a gene, with different regions labeled according to their coding and regulatory roles. The sequences in the middle of the introns are usually free of natural selection—unless a particular intron sequence disrupts the excision of the intron during the processing of the initial RNA transcript, in which case that intron sequence would be selected against. In exons, the third position of some codons is free to vary, because some nucleotide changes at this position are synonymous, as shown in Table 5.6A. For example, all RNA triplets (codons) that start with the sequence UC– code for the amino acid serine. The third nucleotide doesn't matter, in this case.

However, in many cases where third position nucleotides code for the same amino acids, they do not occur at uniform frequencies. Instead, there may be a great preponderance of a particular triplet. This is called **codon use bias.** There is no generally agreed explanation for it. In some cases, it may be an unlikely product of genetic drift. In other cases, it may reflect some distortion in the biochemistry of nucleotides.

An additional possibility for neutral genetic variation arises when nucleotide substitutions result in the use of a different amino acid that is effectively equivalent, perhaps because of similar structure, to the original amino acid.

Far from genes, genetic drift can act with impunity. Note, however, that in such regions there may be transposable elements that evolve on their own, subject to their own natural selection for effective spread through the genome. In addition, there may be stretches of simple repeats, such as ATATATATAT, that expand and contract with unequal crossing over. Therefore, even the seemingly lifeless expanses between genes may evolve by processes more complex than genetic drift on its own.

TABLE 5.6A Synonymous Substitutions

The molecular code is degenerate. More than one triplet of RNA nucleotides may code for the same amino acid.

RNA Triplets	Amino Acid
UUU, UUC	Phenylalanine
UUA, UUG, CUU, CUA, CUG, CUC	Leucine
AUU, AUC, AUA	Isoleucine
AUG	Methionine
GUU, GUC, GUA, GUG	Valine
UCU, UCC, UCA, UCG, AGU, AGC	Serine
CCU, CCC, CCA, CCG	Proline
ACU, ACC, ACA, ACG	Threonine
GCU, GCC, GCA, GCG	Alanine
UAU, UAC	Tyrosine
UAA, UAG, UGA	STOP
CAU, CAC	Histidine
CAA, CAG	Glutamine
AAU, AAC	Asparagine
AAA, AAG	Lysine
GAU, GAC	Aspartic acid
GAA, GAG	Glutamic acid
UGU, UGC	Cysteine
UGG	Tryptophan
CGU, CGC, CGA, CGG, AGA, AGG	Arginine
GGU, GGC, GGA, GGG	Glycine

DNAs also evolve by genomic processes like duplicative transposition (described earlier) and gene duplication by unequal crossing over (described earlier). Many of these events may also be free of natural selection, or close to free from it, especially when they occur in intergenic regions.

There is now little doubt that the evolution of many DNA sequences is not directly determined by the action of natural selection. Instead, it is widely agreed that mutations to DNA sequences often have no effect on the phenotype of the organism, especially because a great deal of the DNA of eukaryotic organisms has no role in determining either amino acid sequences or the regulation of gene expression. When a new molecular variant of no selective significance arises, it is likely to be lost accidentally almost immediately. If that does not happen, the variant molecule will fluctuate in frequency for a time, producing molecular polymorphism in the population. But this polymorphism will have no selective significance. Finally, some neutral DNA sequence variants may rise to fixation in the population, in an accidental *substitution*. Then the population will regain polymorphism only once a new mutation has occurred. ❖

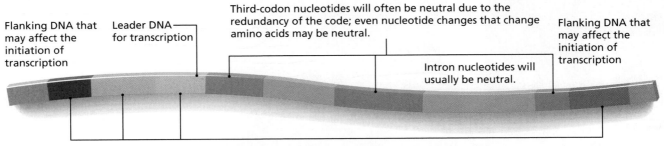

Flanking DNA that may affect the initiation of transcription

Leader DNA for transcription

Third-codon nucleotides will often be neutral due to the redundancy of the code; even nucleotide changes that change amino acids may be neutral.

Intron nucleotides will usually be neutral.

Flanking DNA that may affect the initiation of transcription

The DNA before and after the coding part of the gene may be subject to selection on any nucleotide.

FIGURE 5.6A **Gene Structure and Neutrality**

In the 1960s, the first data on the amino acid sequences of proteins were published. These data were collected from several different species, especially mammals. The kinds of protein that were studied included hemoglobins and cytochrome *c*. Having the amino acid sequences for the same protein in different species naturally led scientists to look at the relationship between the time since the species last had a common ancestor, called **divergence time**, and the number of fixed amino acid differences between any two such species, called the **number of substitutions**.

Emil Zuckerkandl and **Linus Pauling** (Figure 5.7A) pointed out that the number of substitutions per unit of time seems to be roughly constant. There appeared to be a **molecular clock**, which recorded the passage of time by substitutions of amino acids. This finding was puzzling because evolutionary processes scale with the number of generations, not elapsed chronological time. Many species have generation times much less than a year, or much more. The species used in an analysis of molecular evolution might have very different generation times:

This finding was puzzling because evolutionary processes scale with the number of generations, not elapsed chronological time.

rodents and apes, for example. The time unit of calendar years was used in studies of molecular evolution anyway.

A further anomaly arises with multiple amino acid substitutions. An observed difference at a particular amino acid site might have occurred after a sequence of several amino acid substitutions at that site, though only a single difference would be detected in the comparison of two species at that site. Yet even with these problems of timescale and multiple substitutions, the data for molecules like hemoglobin often follow a linear pattern, with amino acid differences accumulating in a clocklike pattern with time.

For a more evolutionary understanding of the molecular clock, look at Figure 5.7B. It is important to bear in mind that evolutionary divergence is a dual process: Two distinct evolutionary lineages are undergoing genetic substitutions through evolutionary time. Therefore, a correct estimate of the **rate of evolutionary divergence** is *not* the ratio of substitutions (K) over time (T), or K/T. Rather, the correct estimate of the rate of evolutionary divergence is as follows:

$$r = K/(2T)$$

There is twice as much evolutionary time as the total time since the last common ancestor suggests, because both of the descendant species diverge from the evolutionary state of the common ancestor.

An interesting scientific maneuver is to use the rate equation to estimate divergence times. If we assume that the rate of protein evolution is constant, then we can use the total number of substitutions to estimate the evolutionary time separating two species. This is done by rearranging the previous formula to obtain

$$T = K/(2r)$$

In one of the most important scientific applications of the molecular clock concept, in 1967 Vince Sarich and Allan Wilson of the University of California, Berkeley, applied this calculation to the evolution of primates. They arrived at the remarkable estimate of 5–8 million years since the last common ancestor

FIGURE 5.7A Linus Pauling, Nobel Laureate

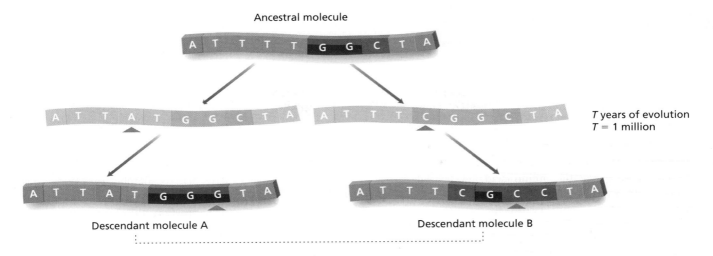

Ancestral molecule

T years of evolution
T = 1 million

Descendant molecule A

Descendant molecule B

Molecules A and B differ by *K* = 4 evolutionary substitutions.
The rate of substitution is $r = K/(2T) = 4/(2 \text{ million}) = 2$ per million years.

FIGURE 5.7B The Concept of Molecular Divergence

FIGURE 5.7C Chimpanzees are our closest living evolutionary relatives.

between chimpanzee and human (Figure 5.7C). At that time, the consensus in the paleoanthropological community was that the common ancestor of chimp and human lived about 20 million years ago.

For this reason, the paleoanthropologists strongly rejected Sarich and Wilson's argument. However, 30 years of further research have led to considerable reductions in nonmolecular estimates of the time since chimps and humans last had common ancestors. (This is discussed further in Chapter 21.) Nonmolecular and molecular clock estimates of the time of divergence separating chimps and humans are now roughly similar. ❖

5.8 Unlike nonsynonymous substitutions, synonymous substitutions proceed at a fairly uniform rate across a wide range of DNA sequences

To set a uniform standard for the molecular clock, scientists wanted data on molecular evolution that would not be affected by natural selection and other variable evolutionary processes. With that in mind, they obtained data on **synonymous DNA substitutions**, changes in DNA sequences that do not change the amino acid composition of proteins. Such synonymous substitutions arise because the genetic code is *redundant:* more than one triplet of RNA codes for most amino acids. For example, serine is coded for by six different RNA triplets: UCU, UCC, UCA, UCG, AGU, and AGC. The coding for serine is thus highly redundant. Tryptophan, on the other hand, is coded for by a single RNA triplet, UGG. It has no redundancy at all. (Note that uracil replaces thymine in RNA molecules, which accounts for the "U" symbol in Table 5.6A, giving the genetic code.) This redundancy seems to allow the evolution of some DNA nucleotides to proceed without any influence from natural selection. However, this assumption depends on the cell using each of the alternative codons uniformly, without bias. This isn't always true. However, synonymous substitutions are far more likely to be equivalent to each other in their phenotypic effects than are nonsynonymous ones.

The scientific interest is this: If DNA evolution proceeds in a clocklike fashion when there are no effects on protein evolution, we should find that the number of synonymous substitutions is uniform, across evolutionary time and among different pro-

> *To set a uniform standard for the molecular clock, scientists wanted data on molecular evolution that would not be affected by natural selection and other variable evolutionary processes.*

teins. Synonymous substitutions should give us the most clocklike data for the process of molecular evolution. Figure 5.8A shows the rates of synonymous substitution among a group of common vertebrate proteins. To a reasonable extent, these rates are uniform: 3.5 to 6.5 substitutions per billion years. Therefore, if we use the number of synonymous substitutions separating two species for these proteins, and an evolutionary rate of about five substitutions per site per billion years, then we should be able to estimate the evolutionary time of divergence fairly accurately. This is probably the most reliable kind of molecular clock to use.

One way that we can test for natural selection in molecular evolution is to compare **nonsynonymous substitution** rates with synonymous rates. Nonsynonymous substitutions involve changes to DNA sequences that *do* change the amino acid sequences of proteins. If amino acid sequences are subject to natural selection, then we expect to find more heterogeneity among rates of nonsynonymous substitutions, as compared to the clocklike rates of synonymous substitutions. What do we actually observe?

Figure 5.8B shows some of the heterogeneity for substitution rates in some common vertebrate proteins, the same proteins that were used to estimate synonymous substitution rates. The rates of substitution are far more heterogeneous for nonsynonymous substitutions—that is, for the DNA changes that result in changes in amino acid sequences. Therefore, even if DNA evolution is fairly uniform when it

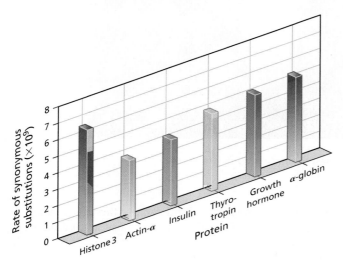

FIGURE 5.8A The Rate of Synonymous Substitutions (per nucleotide) for Genes of Different Proteins

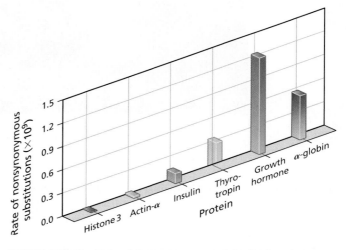

FIGURE 5.8B The Rate of Nonsynonymous Substitutions (per nucleotide) for Genes of Different Proteins

does not make any difference to amino acid sequence (as in the case of synonymous substitutions), DNA evolution is *not* uniform when it changes amino acid sequence. Therefore, the overall pattern of nonsynonymous DNA evolution may involve natural selection.

Is there anything predictable about the rates of evolution of different proteins? One generalization is that the "role" of a protein sometimes suggests what its rate of evolution will be. Consider the evolution of two very different types of protein: structural proteins and the proteins of the immune system. Structural proteins have to fit mechanically with other proteins or cell constituents. For example, **histones** are used to package DNA. **Actin** is a structural part of muscle fibrils. We would expect such structural building blocks to evolve slowly, because they literally have to fit other molecules. As Table 5.8A shows, they do evolve slowly. **Immunoglobulins,** on the other hand, are part of the vertebrate immune system, which generates random variation in antibodies. We would expect some selection for rapid evolution of immunoglobulin amino acid sequence, because this protein must continually evolve in response to the challenge posed by new foreign molecules. And rapid evolution is what we observe, as Table 5.8A shows. In these extreme cases, we can make reasonable guesses about relative rates of protein evolution. However, such guessing is not always so easy, particularly when we are considering the evolution of proteins whose function is not well known. ❖

TABLE 5.8A	Nonsynonymous Substitution Rates in Molecules of Different Types
In mammals, times 10^9	
a. Structural molecules that "fit" closely	
Histone 3	0.0
Histone 4	0.0
Actin α	0.01
Actin β	0.03
b. Immunoglobulins	
Ig V_H	1.07
Ig $\gamma 1$	1.46
Ig k	1.87

SELECTIVE MOLECULAR EVOLUTION

5.9 Natural selection eliminates, substitutes, and maintains specific molecular genetic variants

As described in Chapter 4, natural selection acts on three kinds of genetic variants. The first kind consists of all those genetic variants that are clearly inferior to normal alleles. These inferior alleles undergo purifying selection and are usually eliminated (see Module 4.16). These inferior genetic variants are probably the second most common type of new mutation, after neutral mutations.

The second kind of genetic variant is the class of favored alleles. These variants may be lost due to accidents of sampling, as shown in Figure 5.9A. (Even if an organism has the best genotype, it may still die accidentally.) When that happens, natural selection has failed to recruit a beneficial allele. Otherwise, the favored allele increases in frequency enough so that natural selection seizes hold, taking the favored allele to virtual fixation. A lot of adaptive evolution has involved the occurrence of favorable mutations and their fixation in natural populations, which is a type of *substitution*. This is how many adaptations evolve, even molecular adaptations. Nonetheless, it is often difficult to know which nucleotide substitutions have been selectively favored.

The third kind of genetic variant is made up of those alleles that are not always favored in all genetic combinations. Instead, these alleles are beneficial only in special genotypic combinations. One example of this pattern occurs when genotypes containing two different molecular variants have a fitness advantage over genotypes that have only one of these two variants. This might occur, for example, when a molecule composing the vertebrate immune system leads to more diverse antibodies when it is coded for by two distinct genetic variants, from the same locus. This is *overdominance*, introduced as *heterozygote advantage* in Chapter 4. At the molecular level, an interesting effect of overdominance is that it will foster molecular genetic variation—in principle, at least—as shown in Figure 5.9B. But in practice, convincing examples of overdominance have been very hard to find.

An example of overdominant selection that has already been described is the evolution of the hemoglobin molecule. The hemoglobin genes of northern Europe allow red blood cells (RBCs) to form without sickling. Such RBCs pass through small blood vessels, such as the capillaries, with ease. Unfortunately, these blood cells also leave people vulnerable to infection with malaria, a blood-borne disease caused by a parasitic trypanosome, *Plasmodium*. Hemoglobin evolution is discussed in more detail in Module 4.25.

A variant of the hemoglobin gene causes the RBC to deform. The RBCs take on a sickled shape when two copies of this gene are present, in homozygous combination. This shape makes it difficult for these RBCs to pass through small blood vessels, causing circulatory problems and eventually death.

The heterozygote that combines the alleles for the two kinds of hemoglobin has occasional sickling, but it does not usually cause health problems. The single sickling gene makes it harder for the malaria parasite to establish itself in the circulatory system. For this reason, the heterozygote has the greatest fitness in regions of the world afflicted with malaria. ❖

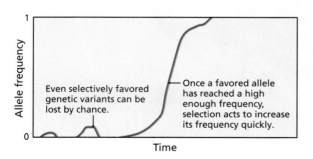

FIGURE 5.9A Evolution of Selectively Favored Genetic Variants

Even selectively favored genetic variants can be lost by chance.

Once a favored allele has reached a high enough frequency, selection acts to increase its frequency quickly.

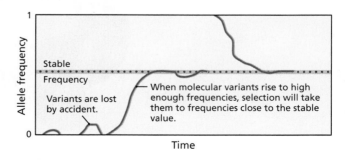

FIGURE 5.9B Evolution with Selectively Balanced Polymorphism

Stable Frequency

Variants are lost by accident.

When molecular variants rise to high enough frequencies, selection will take them to frequencies close to the stable value.

It is uncertain how much nucleotide evolution is due to selection, but there is some evidence for selection on particular nucleotides

In the 1960s a technique was developed to study variation in the amino acid composition of proteins: **protein electrophoresis.** Electrophoresis involves running a current through a gel, then adding proteins to one end of the gel and letting them migrate through the gel for a fixed period of time. The proteins are then stained using a chemical reaction specific to each type of protein. Usually, proteins that have different amino acid compositions migrate to different points in the gel, as Figure 5.10A shows. This allows geneticists to identify at least some of the variant proteins produced by different organisms from a population. Protein electrophoresis was the first relatively unbiased technique that population geneticists had to study genetic variation in natural populations.

When they finally came, the first data on molecular genetic variation were a shock to almost all evolutionary biologists, even the ones who collected the data themselves. There was a vast amount of genetic variation in the organisms first studied—humans and fruit flies of the genus *Drosophila*. Later work, with other organisms, confirmed this pattern. A few species had little genetic polymorphism. Among this monomorphic group, inbred species like self-fertilizing nematodes and cereals were common. But most outbreeding species had large amounts of genetic variation.

How often is selection involved in maintaining molecular genetic polymorphism? One of the best-studied examples is amino acid polymorphism at the alcohol dehydrogenase (*adh*) locus of *Drosophila melanogaster*. In humans this locus is responsible for metabolizing alcohol. There is some evidence that it plays a similar role in the metabolism of alcohol in fruit flies, but it may do other things as well. We don't know.

It has been known for some time that protein electrophoresis detects a protein polymorphism involving two common variants at this genetic locus. One of these variants codes for threonine in exon 4 of the gene, as indicated in Figure 5.10B, while the other variant has lysine in the corresponding site in the protein. This polymorphism is found among *D. melanogaster* populations throughout the world. That it is likely to be subject to selection is also suggested by a north-south gradient in allele frequencies. It is notable that this gradient reverses direction in the Earth's Southern Hemisphere, compared to the Northern Hemisphere. Furthermore, it is known that fruit flies disperse rapidly up and down these gradients. These gradients are therefore unlikely to be ancient relics of migration patterns. Some type of selection must be involved.

The problem is that we do not know what the focus of selection is on *adh*. However, there is every sign that the locus is undergoing selection, selection that maintains genetic variation. This case is an interesting challenge for the next generation of evolutionary biologists. ❖

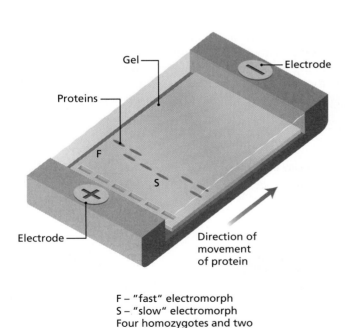

F – "fast" electromorph
S – "slow" electromorph
Four homozygotes and two heterozygotes are shown.

FIGURE 5.10A Protein Electrophoresis

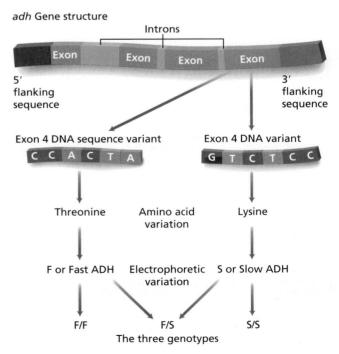

FIGURE 5.10B Genetics of Alcohol Dehydrogenase Polymorphism in *Drosophila melanogaster*

Retrotransposons have a remarkable impact on genome evolution over and above their proliferation within genomes. Because they supply genomes with strong promoters for transcription, as well as reverse transcriptase, they can cause the duplication of genes.

There are several steps in this process, shown in Figure 5.11A. The first is the over-transcription of genes that are located just after retrotransposons. The transcriptional machinery may continue creating mRNA for some distance after the retrotransposons, if the element lacks sequence that would stop transcription. This may cause an overabundance of RNA transcripts from this region of the genome, including complete transcripts of the downstream genes. This can happen because retrotransposons are not normal functional genes, and therefore will not be selected to regulate the transcription that they stimulate.

The next step is that the transcribed RNA is edited, removing introns and other extraneous sequences. At this point, the retrotransposon and the downstream gene(s) may be separated. However, let us assume that they are not. The downstream gene(s) are now physically linked with a retrotransposon structure.

The third step is that the retrotransposon and the downstream gene(s) are reverse transcribed back into DNA. Again,

> *The transcriptional machinery may continue creating mRNA for some distance after the retrotransposons, if the element lacks sequence that would stop transcription.*

the retroposon and the gene may become physically separated at this point.

The fourth step is the reincorporation of the reverse-transcribed DNA, both retrotransposon and regular gene(s). By this point in the process, you can see that the genome size has been increased. Both the retroposon and the downstream gene(s) have made new copies of themselves in the genome. This is like any transposable element, which as a class have the capacity to make many copies of themselves.

But there is a further consequence. There is now a new gene, or genes, in the genome. Its evolution will proceed in one of two directions. The first occurs when there is no useful promoter of transcription located before the gene. As an untranscribed genetic element, which cannot transpose on its own, the new gene is irrelevant where natural selection is concerned. Mutations to its DNA sequence can accumulate, including mutations that stop transcription or translation. Such genes are dead genes, or **pseudogenes.** These genes are detectable by several diagnostic features: close similarity to the exons of another gene, absence of introns due to the processing of the gene as mRNA, and the accumulation of codons that interrupt transcription or translation. Figure 5.11B contrasts a normal gene with a processed pseudogene. Animal and plant genomes are littered with these dead or dying genes.

A new gene takes the second evolutionary direction much less frequently. With considerable rarity, reverse-transcribed genes may reinsert in the genome near an active promoter for transcription—possibly a promoter contributed by a retrotransposon. In this case, the reverse-transcribed gene may still produce a protein, and it may be a target of natural selection.

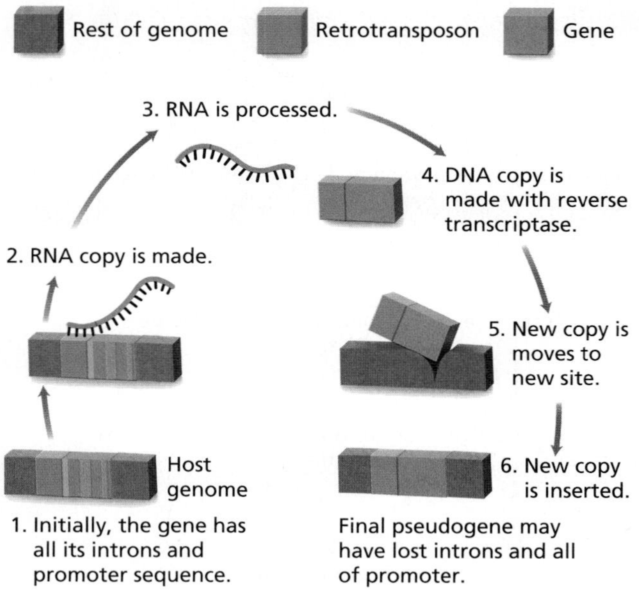

Rest of genome Retrotransposon Gene

3. RNA is processed.

4. DNA copy is made with reverse transcriptase.

2. RNA copy is made.

5. New copy is moves to new site.

Host genome

6. New copy is inserted.

1. Initially, the gene has all its introns and promoter sequence.

Final pseudogene may have lost introns and all of promoter.

FIGURE 5.11A How Pseudogenes Are Made

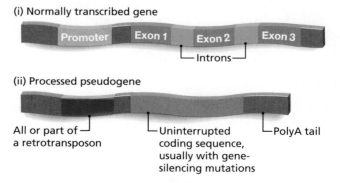

(i) Normally transcribed gene

Promoter Exon 1 Exon 2 Exon 3
Introns

(ii) Processed pseudogene

All or part of a retrotransposon   Uninterrupted coding sequence, usually with gene-silencing mutations   PolyA tail

FIGURE 5.11B Gene Structures of the Quick and the Dead

Such a gene may be called a **retrogene.** This initial transcription does not guarantee the continued "survival" of the gene. The additional protein produced as a result of the transcription and translation of the processed gene may reduce fitness. If so, natural selection will favor mutations that silence the processed gene. Or the additional protein may acquire a new biological function.

An example of this evolutionary process is the autosomal *PGK* gene in mammals, which is homologous to an X-linked *PGK* gene. The autosomal *PGK* gene has no introns, indicating that its DNA came from an RNA intermediate in which introns were excised. Autosomal *PGK* is expressed almost exclusively in the testes, a novel tissue specificity. The maintenance of gene activity by the autosomal *PGK* gene may have happened because the gene located on the X chromosome is normally shut down during spermatogenesis within the testes. In effect, the accidental creation of the retrogene may have allowed mammalian evolution to correct a problem that had limited spermatogenesis before the gene duplication. In such cases, retrotransposons may actively foster adaptive evolution, an ironic side effect of their lives as genomic parasites. ❖

5.12 Genome size is highly variable, perhaps due to the proliferation of useless elements

The total amount of DNA per haploid cell is known as the **C-value,** where "C" stands for *characteristic*. Some aspects of C-value evolution are easy to understand. Bacteria usually have much smaller C-values, about 500 to 13,000 kilobases of DNA. Eukaryotic animals, on the other hand, have C-values of 50 to 140,000 megabases, much greater in size. Figure 5.12A contrasts bacterial and animal genome sizes. This difference in size makes intuitive sense, because animals have many differentiated cell types, so they should have correspondingly more genetic information.

But there are anomalies in the C-value data. Humans have just 3200 megabases of DNA; some lungfish have 140,000 megabases. Why should lungfish need so much more genetic information than humans? Some ferns have 160,000 megabases of DNA. Even a unicellular ameba (*Amoeba dubia*) has 670,000 megabases of DNA, about 200 times more than humans have. Part (ii) of Figure 5.12A shows the relative magnitudes of three of these different genomes. The amoeba genome, however, is too big to fit in the figure and still see the human genome, because the ameba genome is about 200 times larger.

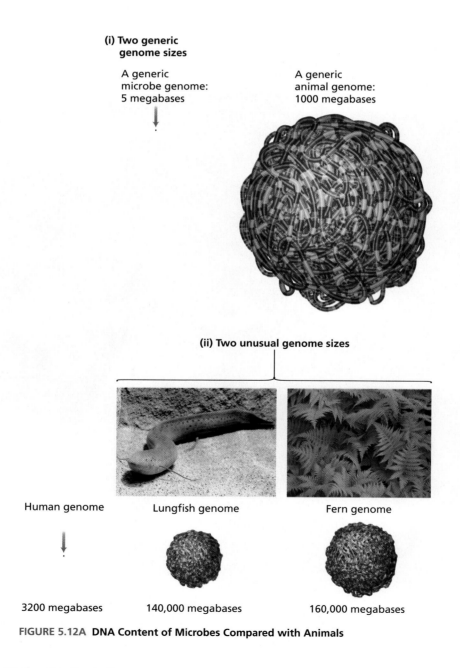

(i) Two generic genome sizes

A generic microbe genome: 5 megabases

A generic animal genome: 1000 megabases

(ii) Two unusual genome sizes

Human genome   Lungfish genome   Fern genome

3200 megabases   140,000 megabases   160,000 megabases

FIGURE 5.12A DNA Content of Microbes Compared with Animals

There are no obvious functional patterns in these DNA data. The amount of DNA that a eukaryotic organism has bears no relation to its morphological or physiological complexity. How are we to make sense of this?

One way to make sense of the wild variation in DNA content is to view it as a by-product of molecular processes like transposition, processes that do not have a simple correlation with the function of the organism, unless they are merely deleterious. That is, instead of viewing genome structure or size as a character comparable to the size or shape of a bone, we might view the genome as a sort of ecosystem, in which genome properties arise via a kind of evolutionary ecology of DNA sequences that copy themselves and proliferate throughout the genome.

In this DNA ecology, some processes eliminate DNA and other processes cause DNA to proliferate, as shown in Figure 5.12B. These processes may spread particular DNA sequences, as well as eliminate specific DNA sequences. In addition to such random mechanisms of loss, it is possible that selection acts against genomes that are overly large. The best evidence for this is the compactness of the genomes of microbes, for which DNA replication is a major part of their overall metabolism. However, this inference is indirect. Duplicative transposition and amplification of tandem arrays are the two obvious molecular mechanisms by which genome sizes can increase. Recent evidence associates an abundance of transposing elements with larger genomes, but this evidence is still indirect. The evolution of genome size is one of the most interesting topics in the study of molecular evolution. ❖

Increasing genome size:
Tandem array amplification; transposable element spread by duplicative transposition, generating both active and inactive elements

Decreasing genome size:
Tandem array loss of genes; loss of DNA during conservative transposition; loss of transposable element sequence by excision from genome

TA – Tandem arrays

TE – Active transposable elements

te* – Defunct transposable elements

g – Translated genes

FIGURE 5.12B **Mechanisms That Change Genome Size**

SUMMARY

1. Biologists have now sequenced the entire human genome, as well as the genomes of other species, from bacteria to mouse. Whole-genome sequencing has clearly revealed patterns of molecular evolution that biologists have been piecing together over the last fifty years.

2. It was once thought that the genome is an organized library packed with information that specified the functioning of millions of genes. We now know that genomes are made up of about 2000–50,000 genes. Prokaryotes have smaller genomes that are reasonably well organized. But eukaryote genomes are generally a muddle. Their genes frequently have useless sequence information inserted at random within them, the introns. There are also large gene deserts between genes, regions that appear to have no function. Some DNA sequences move around genomes, generating mutations and chromosome rearrangements. Large amounts of repetitive DNA may evolve from the repeated unequal crossing over of tandem arrays of repeated DNA sequences.

3. It has been proposed that much of the evolution of DNA sequences within the genome is neutral. That is, it proceeds without control by natural selection, subject primarily to molecular-level processes, like transposition, mutation, and unequal crossing over. Several features of molecular evolution support this model. One is the rough constancy of nucleotide substitutions, called the molecular clock. Another is the relative uniformity of the rate of evolution of DNA sequences that do not affect the amino acid coding of genes, called the synonymous substitution rate.

4. Despite the apparent success of the neutral model of molecular evolution, there must be cases where natural selection intervenes in molecular evolution. Hemoglobin polymorphism in human populations exposed to malaria supplies one case that indicates the action of natural selection. Another example is the polymorphism of alleles at the *adh* locus of *Drosophila melanogaster*. The genome churning of transposable elements also generates new genes, which can be seized on by natural selection to create new genetic functions.

5. Much of molecular evolution is probably irrelevant to the evolution of the visible characters of organisms. But some of it plays a critical role in functional evolution, giving rise to new adaptations at the molecular level.

REVIEW QUESTIONS

1. Transposable elements normally act in what kind of adaptation?

2. Do humans have the largest genome size?

3. Pseudogenes come from what source?

4. Is the genetic code redundant?

5. Are all molecular genetic variants subject to natural selection?

6. When does the molecular clock keep better time?

7. Why does the molecular clock allow us to estimate the times of evolutionary divergence?

8. Offer some explanations for molecular genetic polymorphism.

9. Why is there so much DNA between genes in some eukaryotes?

KEY TERMS

actin
codon
codon use bias
Crick, Francis
C-value
divergence time
DNA-based transposition
eukaryotic gene
exon
genetic transcription
genetic translation

genome
histone
immunoglobulin
intron
Kimura, Motoo
McClintock, Barbara
messenger RNA (mRNA)
molecular clock
neutral theory of molecular evolution
Pauling, Linus

prokaryotic genome
promoter
protein electrophoresis
pseudogene
rate of evolutionary divergence
retrogene
retrotransposon (retroposon)
reverse transcriptase
Sarich, Vince
substitution, nonsynonymous
substitution, number of

substitution, synonymous DNA
tandem array
transposable element
transposase
unequal crossing over
unequal recombination
variegation
Watson, James
Wilson, Allan
Zuckerkandl, Emil

FURTHER READINGS

Alberts, Bruce, Alexander Johnson, Julian Lewis, Martin Raff, Keith Roberts, and Peter Walter. 2002. *Molecular Biology of the Cell*, 4th ed. New York: Garland Publishing.

Gillespie, John H. 1991. *The Causes of Molecular Evolution.* New York: Oxford University Press.

Lewontin, Richard C. 1974. *The Genetic Basis of Evolutionary Change.* New York: Columbia University Press.

Kimura, Motoo. 1983. *The Neutral Theory of Molecular Evolution.* London: Cambridge University Press.

Li, Wen-Hsiung, and Daniel Graur. 2000. *Fundamentals of Molecular Evolution,* 2nd ed. Sunderland, NJ: Sinauer.

Selander, Robert K., Andrew G. Clark, and Thomas S. Whittam, eds. 1991. *Evolution at the Molecular Level.* Sunderland, NJ: Sinauer.

From the origin of species to mass extinction

Speciation and Extinction

The process of evolution is most dramatic when entire species originate or go extinct. Evolution can bring about considerable change within species, as the recent rapid evolution of the human brain illustrates. But the appearance and disappearance of entire species, even groups of species, has had immense significance for life on Earth. The extinction of the dinosaurs about 65 million years ago eliminated the dominant terrestrial animals of the Mesozoic. The evolution of flowering plants transformed the relationship between animals and plants, including groups as diverse as insects and birds, roses and camellias. Most people think of these large-scale events when they think of evolution, even though evolutionary processes are not usually so cataclysmic.

The study of the origin and extinction of species is tremendously challenging. Normally, these events occur over thousands or millions of years, making them impossible to study directly. It is unlikely that a scientist will have enough contemporary data to characterize an ongoing speciation event completely. Fortunately the fossil record provides us with some information about the appearance and disappearance of life-forms, as does molecular evolution. In recent times, the pervasive destruction of habitats and rare life-forms by humans has allowed us to observe the extinction of many species directly, a regrettable scientific benefit of the human impact on the biosphere.

We begin with the study of speciation, the origin of species. A problem that was one of the core mysteries of nineteenth-century biology is now a reasonably understood part of twenty-first-century biology. Punctuated equilibrium is one of the more public evolution controversies of recent years, seemingly a challenge to Darwinism itself. The roots of this controversy, however, lie in mainstream evolutionary biology. The subject of extinction has garnered great attention in recent years, thanks to the fascination of scientists with the possibility that mass extinctions have been caused by the collision of large astronomical bodies with Earth. ❖

ALLOPATRIC SPECIATION

6.1 The biological species concept is based on the reproductive isolation of organisms that are given the opportunity to mate

At the core of the speciation concept is the separation of populations from each other, so that their evolutionary fates become independent. Darwin assumed that this happened, but was usually vague about the biology involved. Ernst Mayr, Theodosius Dobzhansky, and other greats of twentieth-century evolutionary biology wanted to make the crux of speciation clear. They used the **biological species concept** as a way of defining when speciation occurred. This concept makes reproductive isolation the key criterion for the separation of populations into multiple species. **Reproductive isolation** in turn refers to a failure of organisms to reproduce successfully when they are placed together under circumstances in which they would normally mate (Figure 6.1A).

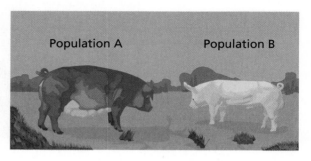

FIGURE 6.1A The Biological Species Concept When members of one population do not mate successfully with members of the other population, no fertile offspring are produced. Mating may never occur, or hybrid offspring may have fitness deficits.

(i) Different habitats or mating seasons

(ii) Lack of sexual attraction

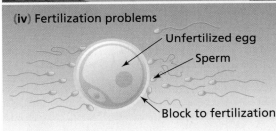

(iii) Coital problems

(iv) Fertilization problems

- Unfertilized egg
- Sperm
- Block to fertilization

FIGURE 6.1B **Prezygotic Reproductive Isolation**

Several features of reproductive isolation need careful thought. Organisms that are being tested should be allowed to develop and produce gametes normally. Mating should not be forced, because part of reproductive isolation is inclination to mate. Indeed, most species are isolated from each other by **prezygotic isolating mechanisms**, as sketched in Figure 6.1B. Even when organisms from two populations are in the same general area, they may not mate because their preferred habitats may be different. These habitat preferences may be specific to the time of mating, with mixing of the populations at other times. Such mating-time preferences can still operate to keep two species isolated from each other.

A common impediment to mating between species is a lack of sexual attraction. This sexual attraction may be mediated by one or several of the sensory modalities: visual appearance, auditory behavior, scent, and so on. Together the senses produce the perception that members of other species are sexually repellent. At least in animals, this may be the single most important factor preventing sex between species.

In a variety of organisms, sex may be attempted between members of different species. In organisms like insects, which use internal fertilization, but have rigid penile and vaginal structures, there may be a lack of "fit" between the genital structures. This can cause a failure of penetration. It can also cause male and female insects to become locked together by their genitals, preventing the reproduction of either. Other species with internal fertilization may achieve ejaculation in hybrid matings, but the sperm of the other species may be expelled after ejaculation.

The final barrier to sex between species arises at the moment sperm come into contact with eggs. Some species have cellular mechanisms that specifically prevent foreign sperm from penetrating their ova. Even in species with external fertilization, like the gamete-shedding sea urchin, mixing sperm from several species will still mostly result in fertilization of eggs by the sperm of the same species. Nature goes to a great deal of trouble to prevent the fertilization of eggs by the gametes of other species. ❖

Not all species are isolated from each other by barriers to fertilization. They are instead reproductively isolated because of **postzygotic isolating mechanisms**, which act after fertilization occurs. For example, horses mate with asses and zebras, and hybrids of these equine species with horses are viable. (One example is the zebroid, shown in Figure 6.2A.) In most animal groups, hybrids are typically inviable, sterile, or have greatly reduced fertility, as shown in Figure 6.2B. The lack of fertility of the offspring of hybrids that are not completely sterile is usually attributed to **hybrid breakdown**. Hybrid breakdown occurs due to a failure of the hybrid's genetic system to produce gametes that can successfully combine with the gametes of the parental species, or with the gametes of other hybrids.

The lack of fertility of the offspring of hybirds that are not completely sterile is usually attributed to hybrid breakdown.

Unlike prezygotic reproductive isolation, postzygotic reproductive isolation usually imposes a considerable fitness cost. Males waste their sperm on unproductive matings. Females may waste energy caring for hybrid offspring that will never produce grandchildren. In the case of mammalian hybrids, like those between equine species, the considerable physiological burdens of gestating and nursing a hybrid offspring will fall on females that mate with members of other species. For all these reasons, organisms should be strongly selected to replace postzygotic mechanisms of reproductive isolation with mechanisms of prezygotic reproductive isolation. In other words, natural selection should normally favor the preemption of the fitness costs associated with interspecific fertilization. This point is schematically summarized in Figure 6.2C. ❖

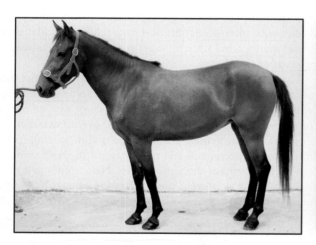

FIGURE 6.2A A Zebroid: A horse/zebra hybrid together with photographs of its parent species

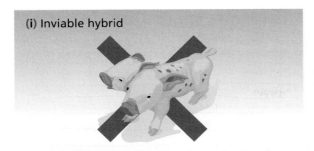

(i) Inviable hybrid

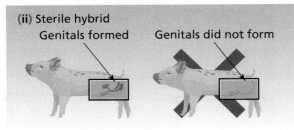

(ii) Sterile hybrid

Genitals formed   Genitals did not form

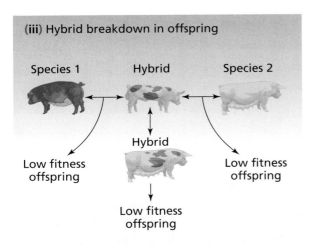

(iii) Hybrid breakdown in offspring

Species 1   Hybrid   Species 2

Hybrid

Low fitness offspring

Hybrid

Low fitness offspring

Low fitness offspring

FIGURE 6.2B **Postzygotic Reproductive Isolation**

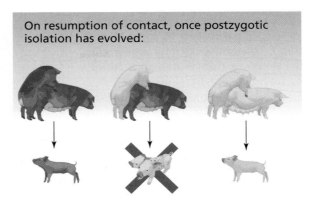

On resumption of contact, once postzygotic isolation has evolved:

With random mating, between-species mating gives no viable offspring and a loss of fitness.

Selection will strongly favor mating only within species, giving additional reproductive isolation.

FIGURE 6.2C **Reinforcement of Reproductive Postzygotic Isolation after Postzygotic Isolation Has Evolved**

The traditional evolutionary scenario for speciation, developed particularly by Ernst Mayr in the 1940s, is based on geographical migration. Over the course of their evolutionary history, species may split in two because of a major geographical barrier to their migration. What constitutes such a barrier will depend on the mobility of the species. Large ocean-flying birds, like the albatross, are able to range over entire hemispheres (Figure 6.3A). But small cave-dwelling worms may be completely isolated from worms of the same species living in a separate cave just a few yards away. Plants might seem more likely candidates for isolation, because they cannot run or fly. But the wind-borne pollen of pine trees (Figure 6.3B) may disperse for hundreds of miles, as may pollen carried by winged pollinators. The important thing is that most species will have barriers to their dispersal, and when populations are located on either side of such barriers they are evolutionarily independent as long as the barriers are maintained. Such geographically separated populations are called **allopatric**.

The evolutionary situation becomes interesting when allopatric populations evolve separately for a long period. As shown in Figure 6.3C, these populations may diverge from each other, because of selection or because of some other evolutionary mechanism. When they diverge enough, they may become so different from each other that they are separate species when the geographical barrier is removed. Note that there is no necessity to this process. The brief splitting up of a species for a short period of time may not result in speciation. Resumption of contact between populations may lead to their evolutionary reunification. Speciation does not have to occur.

If populations have evolved to be quite different during periods of allopatry, they may have postzygotic reproduction isolation before they ever resume contact with each other. However, these same populations may have no sexual inhibition about mating with members of the separate populations before they resume contact. However, on resuming contact, if there is established postzygotic reproductive isolation, there will immediately be selection on members of the two species

FIGURE 6.3A **Albatross**

FIGURE 6.3B **Pine Trees**

to avoid having sex with each other as we have already outlined in the previous module, especially Figure 6.2C. This effect then tends to reinforce the reproductive isolation of the two populations. An initial degree of reproductive isolation leads to the considerable reinforcement of reproductive isolation by selection on sexual and other reproductive behaviors. For this reason, we expect species to have multiple layers of mechanisms that minimize reproductive interactions with other species. Natural selection actively favors the maintenance of reproductive barriers between species. ❖

This graphic shows a fish species split in two by the formation of a land bridge, or isthmus, all the way across a body of water.

Geographical barrier

Once the isthmus has formed, the allopatric populations evolve into separate species.

Evolutionary differentiation of populations occurs

Geographical barrier

These species are reproductively isolated from each other when contact is resumed as a result of the subsidence of the isthmus.

Contact is resumed.

FIGURE 6.3C Allopatric Speciation

In the middle of the twentieth-century, most evolutionary biologists thought the evolutionary changes that bring about speciation arose from adaptation to dissimilar habitats. This basic scenario is illustrated in Figure 6.4A. The idea was that when two populations were separated geographically, they might find themselves in very different habitats, with different kinds of food or, in the case of plants, light levels, among other ecological differences. Many generations in these distinct habitats would lead to contrasting biological adaptation in the separate populations. On resuming contact, members of the two populations might have adaptations that are so different that hybrids between them would have very low Darwinian fitness, and speciation would be achieved.

This is a very plausible scenario. Most evolutionary biologists still suppose that it occurs frequently in evolution. Extensive biological differentiation of allopatric populations has been observed in many organisms. There is a vast body of literature documenting the physiological and genetic differentiation of populations from a number of species, from clines in the frequency of alcohol dehydrogenase alleles in fruit flies to ecological genetic differentiation in clams off the northeastern coast of the United States. Natural selection will

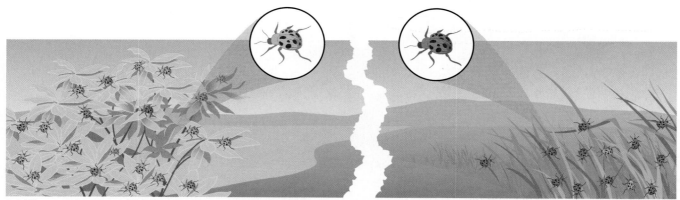

1. Differences between the habitats of the separate areas causes selection for different characteristics.

2. While still separate, the populations become highly differentiated.

3. When the barrier goes away, the populations may now be species.

FIGURE 6.4A **Adaptive Differentiation in Allopatry**

cause some allopatric populations to differentiate genetically. What has been poorly documented is the association between such allopatric differentiation and the evolution of reproductive isolation.

This is not a trivial question, because there are alternative genetic mechanisms that might bring about speciation—particularly at the level of genomic evolution, described in Chapter 5. Transposable elements can bring about hybrid dysgenesis, even when a single transposable element is involved, as has been shown in *Drosophila*. If allopatric populations undergo multiple, but different, invasions of transposable elements, then it is plausible to suppose that hybrids would cause massive genomic disruption from extensive, unregulated transposition, as shown in part (i) of Figure 6.4B.

An alternative genomic mechanism for speciation is one of structural genome evolution, shown in part (ii) of Figure 6.4B. If one genome becomes very different in structure, then hybrids may be subject to a failure of the genetic mechanism, in both gene transmission and gene expression. The Chinese and Indian muntjacs illustrate this pattern. The muntjac is a small deer that barks. The Chinese species has 46 chromosomes, while the outwardly similar Indian species has 6 chromosomes (Figure 6.4C). During allopatry, it appears, the Indian species has undergone numerous chromosome fusions, end-to-end. Hybrids can be made, but they are infertile, no doubt due to structural incompatibility of the two genomes.

Two important points need to be made about genomic mechanisms of speciation. The first is that they are not fully documented, putting them on the same plane as traditional selective scenarios for speciation. The second is that they do not require natural selection. Transposable elements may spread, genomes may expand in size, and chromosomes may fuse without any action from natural selection at the organismal level. Indeed, natural selection on organisms may oppose such evolutionary processes, perhaps ineffectually. ❖

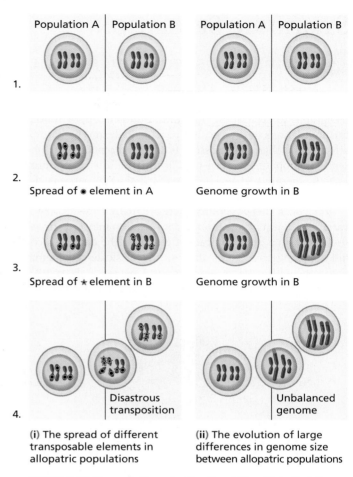

(i) The spread of different transposable elements in allopatric populations

(ii) The evolution of large differences in genome size between allopatric populations

FIGURE 6.4B Genomic Nonadaptive Speciation Note that the opaque bar corresponds to a geographical barrier separating populations.

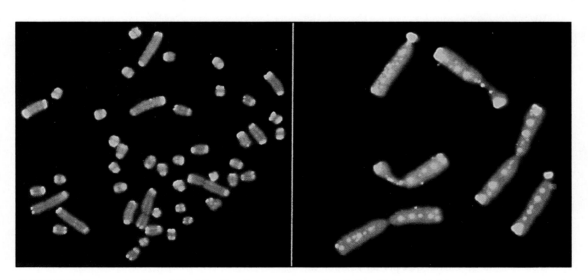

FIGURE 6.4C Reorganization of the Muntjac Genome during Speciation The two staining patterns correspond to two morphologically similar species.

SYMPATRIC SPECIATION

6.5 Sympatric speciation is difficult due to a lack of evolutionary isolation among groups

Geographical separation provides abundant opportunities for populations to diverge, both accidentally and by natural selection. Is there any possibility for divergence in the absence of allopatry, and can such "sympatric" divergence proceed far enough to produce speciation? (Sympatry is the antonym of allopatry).

In favor of this possibility, there are many known cases of variation within populations. Such variation may even be associated with a geographical pattern, as in the cases of alcohol dehydrogenase alleles and latitude, or sickle-cell anemia and areas with malaria, described in Chapters 4, 5, and 22. Such examples tempt biologists to suppose that humans, for example, might evolve isolation between populations exposed to malaria and populations free of malaria. That way people who are free of malaria, and therefore would not benefit from red blood cell (RBC) sickling, would not inherit the sickling allele.

But there are good reasons for supposing that this isolation will rarely occur. The biggest problem is that any allele that might foster the isolation of group A within a population—call it an "A-barrier" gene—would get into the rest of the population. Then all groups would have the A-barrier gene, and isolation could not evolve (Figure 6.5A).

An evolutionary mechanism that might help **sympatric** speciation along is disruptive selection, discussed in Chapter 4. Suppose that disruptive selection acts to eliminate intermediate individuals, leaving only the extreme genotypes. If successful matings with an individual at one's extreme are the only matings that produce viable offspring, then natural selection will be effectively imposing selection against "hybridizations" between the extremes, as well as selection against intermediate individuals. If such selection is sufficiently intense and sustained for enough generations, then prezygotic isolation might evolve, producing speciation without geographical isolation. Such a disruptive selection process is shown in Figure 6.5B.

Although this type of sympatric speciation could operate in principle, it is unlikely to occur in practice. The problem is that natural selection has to enforce the reproductive isolation of extreme groups. If it relaxes for a generation, there would be abundant gene flow between extreme groups. This relaxation would correspond to a breach in the geographical obstacle separating allopatric populations. Because selection is very rarely rigidly sustained, to our knowledge, it is unlikely to produce sympatric speciation in most species.

However, in a few cases it is plausible to infer that sympatric speciation has occurred, and may occur again. We turn to those cases next. ❖

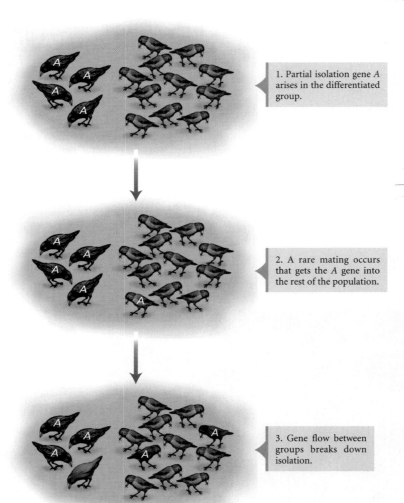

1. Partial isolation gene *A* arises in the differentiated group.

2. A rare mating occurs that gets the *A* gene into the rest of the population.

3. Gene flow between groups breaks down isolation.

FIGURE 6.5A In sympatry it is difficult for genes that give partial isolation to establish full isolation, because there is no external factor establishing evolutionary separation.

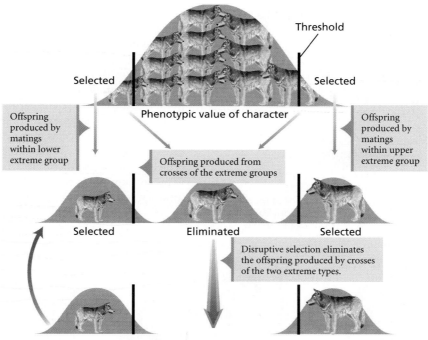

FIGURE 6.5B In principle, disruptive selection could prevent successful reproduction of hybrids between extreme types favored by natural selection, providing that all the hybrid offspring fall between the limits of selection.

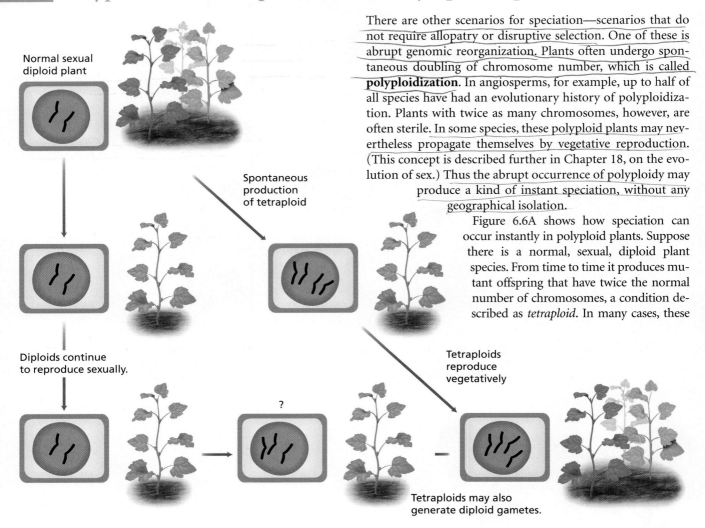

Normal sexual
diploid plant

Spontaneous
production
of tetraploid

Diploids continue
to reproduce sexually.

Tetraploids
reproduce
vegetatively

?

Tetraploids may also
generate diploid gametes.

There are other scenarios for speciation—scenarios that do not require allopatry or disruptive selection. One of these is abrupt genomic reorganization. Plants often undergo spontaneous doubling of chromosome number, which is called **polyploidization**. In angiosperms, for example, up to half of all species have had an evolutionary history of polyploidization. Plants with twice as many chromosomes, however, are often sterile. In some species, these polyploid plants may nevertheless propagate themselves by vegetative reproduction. (This concept is described further in Chapter 18, on the evolution of sex.) Thus the abrupt occurrence of polyploidy may produce a kind of instant speciation, without any geographical isolation.

Figure 6.6A shows how speciation can occur instantly in polyploid plants. Suppose there is a normal, sexual, diploid plant species. From time to time it produces mutant offspring that have twice the normal number of chromosomes, a condition described as *tetraploid*. In many cases, these

FIGURE 6.6A Speciation by Polyploidy in Plants Speciation hinges on whether triploid hybrids of diploid and tetraploid parents are viable and fertile. If they are, they may establish gene flow between the diploids and tetraploids, preventing speciation. If not, speciation will occur.

FIGURE 6.6B Strawberries spread vegetatively using runners.

tetraploid mutants may be inviable or infertile. Very rarely, however, a viable tetraploid is produced—one that can reproduce without outcrossing, perhaps by selfing. Such selfing may occur by self-fertilization, with the production of diploid gametes. Or the plant might reproduce vegetatively, perhaps through the root system or by "runners," lateral stems that produce new plants as they grow away from parent plants. Strawberries reproduce vegetatively using runners (Figure 6.6B). Many trees use their root systems to produce clonal "stands" of trees. An example of clonal reproduction using roots is the trembling aspen (Figure 6.6C).

An important feature of tetraploid derivatives from diploid species is evident when the tetraploids are exposed to the haploid gametes produced by diploid parents. If the tetraploids produce only diploid gametes, the hybrid zygotes will be triploid. In animals, triploids are not usually viable. But plants are more tolerant of varying ploidies. If successful triploids are generated, and then they successfully mate with diploids and tetraploids, there will be a failure of isolation between diploid and tetraploid plants. Speciation will not occur.

But if there is no successful hybridization between new tetraploids and their parent diploids, the tetraploids will be evolutionarily isolated from the diploids. A new species will have formed almost instantly. Furthermore, tetraploids are usually different from diploids in their patterns of growth, metabolism, and so forth. In such cases, ecological differentiation is achieved instantly as well. This pattern of sympatric speciation is generally accepted as an important mechanism of speciation in plants by biologists. ❖

FIGURE 6.6C A stand of trembling aspen, containing many clonal trees.

Another scenario for sympatric speciation involves micro-geographical isolation. Some insects, for example, specialize on particular host plants. If mating is associated with plant type, then there is the possibility that the ecology of plant preference could produce pseudo-geographical isolation on a microscale. Different host plants could thereby play a similar role to geographically separate environments.

The true fruit flies of the genus *Rhagoletis* are divided into different **host races** that feed on different types of fruit. Significantly, mating in such species tends to be located at the site of feeding, making the different host races partially isolated from each other (Figure 6.7A). Therefore, this genus seems to fit the pattern expected for sympatric speciation. But what are the specifics of this organism?

The apple maggot fly, *Rhagoletis pomonella*, feeds on several host plants: hawthorn, apple, and cherry. Mating takes place on the plants where the flies feed, and the eggs are laid on the host plants as well. Potentially, there could be isolation between "host races." Is there? Like other flies, *Rhagoletis* disperse

readily. There is no physical impediment to switching between host plants. On the other hand, this factor would not matter if the adults always chose the plants on which they had grown.

The genetic evidence suggests there are indeed two races of flies. As shown in Figure 6.7B, there are apple-feeding flies and hawthorn-feeding flies. These flies are not visibly different. Electrophoretic studies (the technology described in Chapter 5) of these flies, however, show that the host races have different allele frequencies. The flies of the two races also prefer their "home" type of host plant. The findings suggest that these flies may be undergoing a process of evolutionary differentiation, despite geographical overlap.

This does not mean that speciation has occurred already. Matings have been observed between the host races, and there is no evidence for postzygotic isolation in laboratory crosses. These fly host races may continue to diverge, and so become species. Or they may continue to exchange genes, remaining part of one species. Evolution is not generally compelled to produce new species. ❖

FIGURE 6.7A Hawthorn and apple trees are distributed throughout eastern North America. True fruit flies can feed on either type of tree. The flies that feed on a particular type of tree, apple or hawthorn, also remain on that tree to mate. Note: Flies and trees are not shown to the same scale or with anatomical precision.

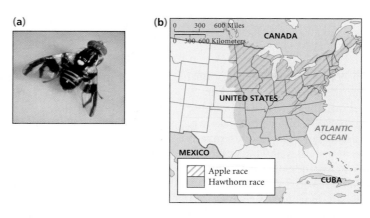

FIGURE 6.7B Apple and hawthorn maggot flies look the same but do not perfectly overlap, because apple trees do not grow in the South.

Laboratory experiments show the feasibility of host-race sympatric speciation

One difficulty with the concept of sympatric speciation is that it rests on processes that are not known to work in nature, except for polyploidization. The differentiation of host races has not been observed to evolve to the point where reproductive incompatibility arises. For this reason, sympatric speciation has needed laboratory studies in which evolution can be shown to drive toward reproductive barriers when there is specialization on different hosts.

William Rice performed an ingenious experiment in which host races were allowed to evolve toward separate species. He used two *Drosophila* eye-color mutants, *vermillion* and *raspberry*. He put these flies into a complex laboratory apparatus in which eight different environments were supplied with food, each habitat having one of the eight possible combinations of height (higher, lower), light (bright, dark), and scent (orange, cocoa). The flies were allowed to move through the apparatus as they wished, as adults. But only *vermillion* females that chose the higher, lighted, and orange-scented habitat (habitat 1) were allowed to contribute eggs to the next generation. In parallel, only *raspberry* females that chose the lower, dark, and cocoa-scented habitat (habitat 2) were likewise allowed to contribute eggs to the next generation. Rice imposed these rules as he collected flies for the next generation. The experimental apparatus and rules of culture are diagrammed in Figure 6.8A. (The apparatus and the rules of selection were changed somewhat partway through the experiment, but the basic experimental strategy was preserved.) Although there were opportunities for flies to mate outside of the eight possible habitats, almost all females did not mate until they had reached a habitat with food. In other words, the two types of mutants were analogous to fly host races in nature.

The experimental question was whether these host races would evolve so that they preferred a particular habitat, with mating confined to that habitat. Rice did not force any particular outcome. He merely contrived conditions in which host-race specialization and reproductive isolation might occur.

Figure 6.8B shows the results. At the start of the experiment, the two eye mutants showed no preference for the habitats that allowed them to successfully reproduce. But over 50 generations, strong preferences developed. As shown, *vermillion* flies of both sexes evolved a strong preference for habitat 1, the habitat in which they successfully reproduced. In parallel, the *raspberry* flies evolved a strong preference for their successful habitat for reproduction—habitat 2. (These data are not shown.)

These results show that strong preferences for habitats can evolve in populations that are ecologically successful only if particular genotypes chose particular habitats—even when these habitats are available to all genotypes, and even if there is no genetic bias in favor of choosing the appropriate habitats to begin with. This laboratory experiment was extremely rigorous in destroying flies making the wrong choice. But it did not directly force the evolution of host-race preferences. Those preferences were produced by natural selection acting on its own. Such preferences are a first step in evolution toward reproductive isolation, and thus sympatric speciation. ❖

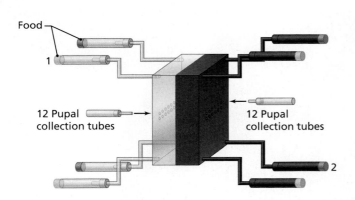

FIGURE 6.8A Habitat Maze Used in the First 24 Generations of the Speciation Experiment Vials with dark medium had cocoa flavoring in the food. Vials with light medium had orange flavoring. Some vials are placed high, others are low. One side of the maze let in light, while the other side was kept dark. This apparatus let flies choose between habitats. The experimenter then selected the flies by their choices.

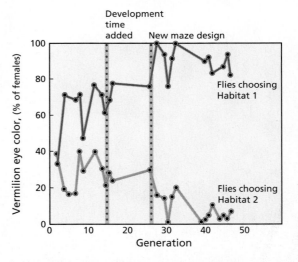

FIGURE 6.8B The results of selection against flies that choose intermediate habitats: Flies make more distinct choices as lab evolution proceeds, resulting in distinct lab species. Flies with *vermillion* eyes become separated from the other flies due to habitat choice.

HYBRIDIZATION

6.9 Hybridization sometimes occurs in nature

Hybridization occurs when members of one species mate with members of another species and produce offspring. This may lead to (1) alleles spreading from one species to another, or (2) a new hybrid form arising from the cross, or (3) little of evolutionary significance occurring, especially if the matings produce only inviable or infertile hybrids.

In any case, hybridization depends, in the first instance, on enough contact between species that sex happens. Reproductive isolation, and the definition of species as evolutionarily separate, seems to imply that species will not hybridize. However, species may be generally isolated from each other most of the time or over most of their geographical distribution, yet hybridization may still occur under unusual local conditions. When local conditions allow hybridization, the locality is called a **hybrid zone**. This scenario is diagrammed in Figure 6.9A. Hybrid zones may arise, for example, when allopatric speciation has recently occurred and formerly separate populations have resumed contact.

Given that hybridization occurs, several further possibilities arise. One we have already seen—selection for enhanced prezygotic isolation when hybrids are inviable or infertile—"reinforcement" of reproductive isolation. The remaining cases involve fertile hybrids. Some of these scenarios are diagrammed in Figure 6.9B and discussed further here. Later in this module, we describe other scenarios.

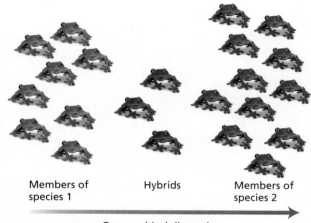

Members of species 1   Hybrids   Members of species 2

Geographical dimension

FIGURE 6.9A Hybrid Zone Arising from Sex between Members of Two Distinct Species

The normal expectation for hybrids between two species is that they will have reduced fitness compared to the two parental species (see Figure 6.9B, Scenario One). Crudely speaking, the hybrids will be dying off soon after they are produced by hybridization. This will tend to produce a narrow hybrid zone, with relatively little evolutionary impact.

Members of species 1 | Hybrids | Members of species 2

Scenario One: Hybrids die off or are infertile.

Scenario Two: Hybrids have sex with parental species, causing an exchange of genes between species.

Scenario Three: The hybrids are superior to their parental species, and outcompete them.

FIGURE 6.9B Three Scenarios for Evolution When a Hybrid Zone forms between Two Species

In some species, particularly plants, hybrids may be comparable in fitness to the parental species. Under these conditions, the hybrid zone will be relatively wide (Scenario Two). In rare cases, the proliferation of the hybrid may lead to frequent additional hybridization between the hybrid group and the parental species. Carried to an extreme, this could lead to the disappearance of the evolutionary barrier between the parental species, and the formation of a larger, heterogeneous species made up of the fusion of the two parental species and the hybrid.

Finally, there could be hybrids that have even greater fitness than the parental species (Scenario Three). This superiority might be specific to the hybrid zone habitat, in which case hybrids might be abundant there, but not spread outward. Alternatively, the hybrids could be generally superior to the parental species. Under these circumstances, many outcomes are possible, depending on whether the hybrids evolve independently of the parental species. If a hybrid does so, it might displace the parental forms, even drive them to extinction. ❖

We have already described instant speciation, when plants produce polyploid offspring. Such polyploids are called **autopolyploids**, because all the genetic material comes from one species. Hybridization allows an additional possibility—**allopolyploidization**, in which the extra chromosomes come from hybridization with members of other species. Furthermore, abrogation of normal meiosis is to be expected if chromosomes that lack DNA sequences required for the process are introduced by hybridization (Figure 6.10A). For this reason, the creation of new species with unusual ploidies by hybridization is expected to occur.

Parthenogenetic vertebrate species have apparently been produced by hybridization. (Parthenogenesis is reproduction with unfertilized eggs.) One example comes from the genus *Poecilia*, shown in Figure 6.10B. The species *P. formosa* is an asexual hybrid of two sexual parent species,

P. mexicana and *P. latipinna*, which live with the asexual species. Other asexual fish are thought to be similar hybrids, particularly species of the related genus *Poeciliopsis*.

An even more interesting example comes from the salamander genus *Ambystoma* (see Figure 6.10B). Two sexual species are known, *A. jeffersonianum* and *A. laterale*, that have two hybrid asexual forms, both successful triploid species. One of the hybrids has two sets of *A. jeffersonianum* chromosomes and one set of *A. laterale* chromosomes. The other hybrid has one set of *A. jeffersonianum* chromosomes and two sets of *A. laterale* chromosomes. In this case, there is no doubt that the new species were produced by hybridization of sexual species that were already established. It is particularly interesting that species formation apparently occurred twice independently, with opposite patterns of chromosomal imbalance. ❖

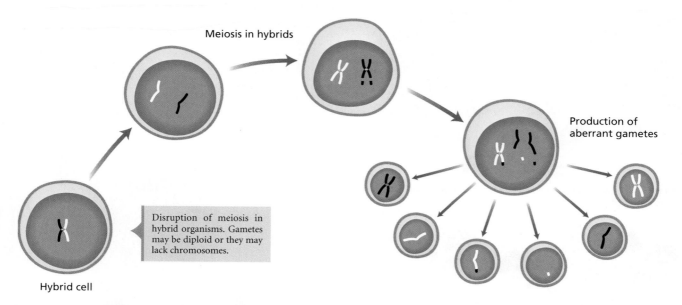

Meiosis in hybrids

Production of aberrant gametes

Disruption of meiosis in hybrid organisms. Gametes may be diploid or they may lack chromosomes.

Hybrid cell

FIGURE 6.10A Aberrant Meiosis in Hybrids Black chromosomal material comes from one species; white chromosomal material comes from the other species.

(i) Mollies and guppies from the genus *Poecilia*

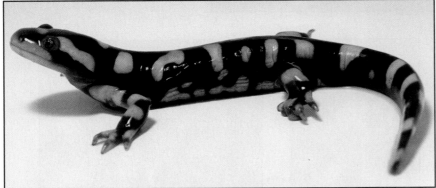

(ii) Juvenile (above) and adult (below) salamanders from the genus *Ambystoma*.

FIGURE 6.10B   Two groups of animals that form triploid hybrid species: (i) killifish (the poecilids); (ii) ambystomatid salamanders.

SPECIES RADIATIONS

6.11 The Cambrian explosion was a spectacular radiation of animal species

From time to time, a large number of species come into existence in a short period of geological time. (A few million years would be a brief period on this timescale.) This type of proliferation of species is called a **species radiation**, or an *adaptive radiation*. The most spectacular of all species radiations is that which occurred during the **Cambrian** period, 510–543 million years ago. At that time, land was colonized by multicellular plant and animal species. Many animal taxa appear in the fossil record for the first time during the Cambrian. Essentially all the major animal body plans evolved in an explosion of diversification, including the chordate body plan that was the basis of human evolution.

The Cambrian explosion is documented in the **Burgess Shale**, one of the greatest of all fossil finds. Located on the side of a mountain in the Canadian Rockies, the Burgess Shale contains fossils of whole aquatic animals, not just fossils of bones and shells. For some reason, soft-bodied animals were trapped in sediment that did not crush them immediately, allowing slow mineralization to preserve their anatomy in some detail. Over hundreds of millions of years of geological churning, the sediment was transported en masse from the bottom of a body of water to the side of a mountain.

Figure 6.11A shows photos of some of the fossils found in the Burgess Shale. Figure 6.11B displays drawings of biologists' best guesses as to what these animals looked like. What is striking is that these animals often do not resemble the fossil species that we know from the 500 million years since the Cambrian period. That is, there appears to have been a Cambrian explosion of diversity, which was then cropped back. The Burgess Shale ecosystem is reconstructed in the illustration used for the cover of this book.

Some of these unusual animals have been placed in modern-day taxonomic groups, especially the arthropods. Other Cambrian fossils do not fit in any known taxonomic group. Indeed, some of these animals seem virtually impossible. When it was first found, it was not clear how to orient a fossil called *Hallucigenia* (shown on the book cover). Which side was the bottom and which was the top? Further research also changed the identification of the head end. Yet this fossil animal has been placed in a surviving taxon—the Phylum Onychophora, the velvet worms. ❖

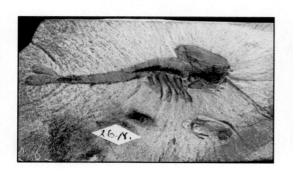

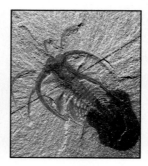
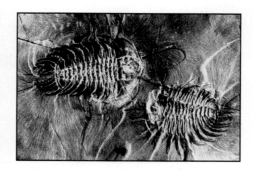

FIGURE 6.11A Fossils from the Canadian Burgess Shale, about 500 Million Years Old See the front endpapers for the names of these fossil life-forms.

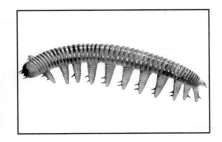

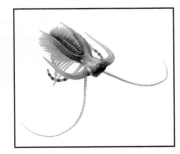
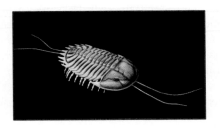

FIGURE 6.11B Reconstructions of the Animals Whose Fossils Are Shown in Figure 6.11A

After the Cambrian explosion, one of the best-known examples of adaptive radiation is the colonization of the **Galápagos Islands** by bird species, especially **finches** and mockingbirds. The thirteen finch species of the genus *Geospiza*, known as Darwin's finches, have been best studied. The diversity of species is shown in Figure 6.12A (see Chapters 1 and 4 also).

Figure 6.12B shows one theory for the evolutionary tree connecting these finch species. The most important idea presented by this tree is that all these species descend from a single founder species. This founder is generally thought to have arrived from the South American mainland some thousands of years ago. Presumably, that founding event was a flock of finches blown west into the Pacific. This is not entirely groundless speculation. Exotic birds and plants still appear on the Galápagos Islands, blown in by wind, or floating in on tides.

Once they got to the Galápagos, Darwin's finches evolved into a number of different species. Some have evolved into insect eaters. Other species eat seeds. Some live on the ground, while others live primarily in trees. The finches are not that different from each other in anatomy. Some are obviously larger. The central feature in their evolution has been their beaks. Some of the finches have beaks twice the size of the

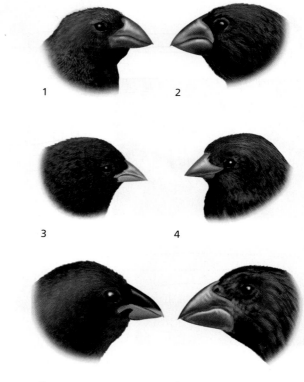

FIGURE 6.12A **Heads of Some Species of Darwin's Ground Finches**

smallest finches, and the shape of the beaks varies a lot. Some finches have beaks that are short but deep; others have pointed beaks that project sharply outward from the bird's face. This variation in beaks corresponds to variation in diet. The birds with large, powerful beaks are able to eat hard-shelled seeds. Cactus finches have long beaks that enable them to forage for food between the spines of cacti. A simple evolutionary inference is that these finches would never have become so diverse if the Galápagos Islands had already been colonized by a full range of bird species before the finches arrived. But because the islands were likely free of most bird species when the first finches arrived, the finches were free to evolve and speciate, filling up a greater number of ecological niches than they would have otherwise. This is a species radiation that occurred to fill an ecological vacuum. ❖

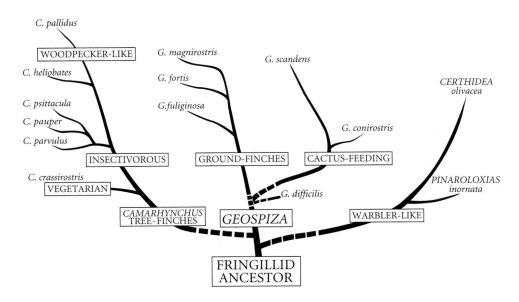

FIGURE 6.12B An Evolutionary Tree for Darwin's Finches

PUNCTUATED EQUILIBRIUM

6.13 Mayr's speciation model implies punctuated equilibrium

Many scientists have thought that Darwin's theory of evolution means that species should gradually change in the fossil record. Sometimes they do. Since the start of the **Tertiary** period, terrestrial mammalian species like horses, bears, and pigs have gotten larger, or smaller, fairly smoothly. This pattern, called **phyletic gradualism**, is shown in Figure 6.13A. The doctrine of phyletic gradualism leads paleontologists to look for "missing links" when there seems to be an abrupt change in the fossil record of evolution. This search for missing links is inspired by the assumption that when there is a missing link, it is only because of a failure to find an intermediate fossil that is in fact out there. The most famous missing link, that between ape and human, is discussed in Chapter 21.

There is an alternative interpretation to phyletic gradualism and missing links. This is the view that there really are abrupt changes in evolution. In this theory, species are normally at evolutionary equilibrium, with little evolutionary change. But when speciation occurs, evolutionary equilibrium undergoes rapid change. This theory is called **punctuated equilibrium**. The evolutionary pattern that arises with punctuated equilibrium is shown in Figure 6.13B. The horizontal branching of the evolutionary process signifies evolutionary change during speciation, while the vertical lines denote unchanging species between speciation events.

How can evolutionists explain punctuated equilibrium? One explanation comes from a theory of speciation developed first by **Ernst Mayr**, but since elaborated by many evolutionary theorists, especially **Stephen Jay Gould** and **Niles Eldredge**. Mayr's original idea was that small **peripheral** populations might be more likely to undergo the kind of **genetic revolution** required to produce a new species. Among other things, such small populations would have a much greater speed of **genetic drift**. This has two consequences: (1) The large populations that make up the bulk of a species may be relatively static evolutionarily; and (2) the processes that produce new species may typically occur in populations so small that they would not be represented in the fossil record. As shown in Figure 6.13C, these processes tend to make the appearance of new species in the fossil record abrupt. Evolution may thereby proceed by normal processes of speciation, but the resultant pattern may resemble punctuated equilibrium more than it conforms to the idea of phyletic gradualism. ❖

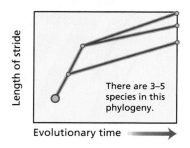

FIGURE 6.13A The Traditional View by Paleontologists of Macroevolution From this viewpoint, the pattern of evolution among species follows that within species with no discontinuities. (New species occur when the phylogeny branches *and* during evolution between branching events.)

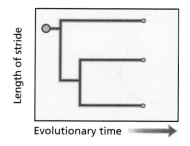

FIGURE 6.13B The Punctuated Equilibrium View of Macroevolution From this viewpoint, the pattern of evolution among species does not follow that within species. It is supposed that most evolutionary change occurs when new species are produced. (New species occur only when the phylogeny branches.)

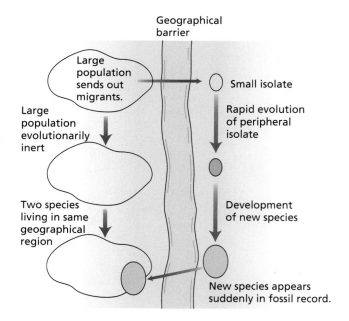

FIGURE 6.13C Mayr's scheme, in which small, isolated populations evolve rapidly and form new species that appear abruptly in the fossil record of evolution. (The scale of the blobs roughly indicates the population size of the organisms they represent.) Time proceeds from top to bottom.

Hopeful monsters can escape evolutionary stasis, in theory

When the idea of punctuated equilibrium was first proposed, it was an application of Mayr's speciation model. Mayr's model in turn was a special case of the type of speciation process that evolutionary biologists have generally accepted. However, in the hands of Gould, the idea became more extreme. Gould went from using Mayr's Darwinian model of speciation to the idea that speciation might occur as a result of almost instantaneous genetic upheavals. Coupled with this was an emphasis on the idea that normal species are locked into **evolutionary stasis**, with little change possible until speciation occurs. Both of these ideas were major challenges to the Darwinian tradition.

There are three main scenarios by which species may show evolutionary stasis. These scenarios are illustrated in Figure 6.14A. The first is a *lack of genetic variation.* If

species are genetically uniform, there is little prospect that natural selection will change their phenotypes. This is a straightforward application of the basic features of natural selection, described at the start of Chapter 4. However, surveys of genetic variability indicate that most plants and animals have abundant genetic variation. A lack of genetic variation is not a viable explanation for the stasis of most species during evolution.

The second scenario is **stabilizing selection**. Stabilizing selection undoubtedly can prevent evolutionary change. But it is not clear that all, or even most, characters of most species undergo stabilizing selection. Evidence for such an extreme assumption is lacking. Furthermore, the ease with which artificial and laboratory selection can change biological characters, a point that Darwin himself emphasized, suggests that the phenotypes of organisms are not generally "locked down" by stabilizing selection.

The third scenario is **fluctuating selection**. If selection has no long-term focus, then evolving characters may move back and forth during the course of evolution, without the kind of trend that is expected from the perspective of phyletic gradualism. The problem with this view is that it is a kind of selective drift, which should eventually lead to a breakdown of stasis within species evolving according to this scenario.

If it is not easy to see why most species should show evolutionary stasis, the problem of abrupt speciation may also be difficult. We have already considered polyploidy and hybridization as mechanisms for instantaneous speciation, in which asexual forms are derived from sexual species. The problem is that speciation by spinning off asexual forms is not common across the evolutionary spectrum. Other species that might undergo instant speciation have a major problem: the **hopeful monster**. A newly tetraploid plant that reproduces vegetatively does not have to find a mate. But most species do not produce asexual offshoots. Therefore a sexual organism that has undergone a radical transformation to become a monster from a new species needs to mate with another of its kind, which is the hopeful part. Frankenstein's monster needed a bride made for him. In a real biological setting, which this cartoon does not show, the genetic upheaval required to produce one monster is rare, so producing two or more monsters is still less likely. This problem is diagrammed in Figure 6.14B. ❖

Scenario One: Lack of genetic variation

> If there is no genetic variation, natural selection cannot act.

Scenario Two: Stabilizing selection

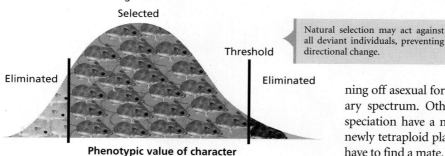

Selected

Threshold

Eliminated

Eliminated

Phenotypic value of character

> Natural selection may act against all deviant individuals, preventing directional change.

Scenario Three: Fluctuating selection

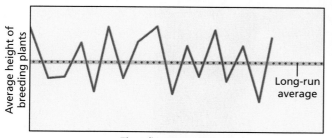

Average height of breeding plants

Long-run average

Time (in generations)

FIGURE 6.14A Three Explanations of Evolutionary Stasis

FIGURE 6.14B The Hopeful Monster Needs a Bride A major mutant (i.e., a monster) can start a new species only if it reproduces asexually, or if a sexually compatible similar mutant is improbably produced.

RETAIL EXTINCTION

6.15 Extinctions destroy unique products of biological evolution

People may complain that things never seem to last anymore—things like marriage or audience attention spans—but one thing is really forever: **extinction**. Parallel evolution does occur, and such parallels can produce rough "repeats" in evolution. One example is the fusiform body shape of sharks, ichthyosaurs, and dolphins. This body shape has evolved repeatedly in diverse phylogenies, and quite independently, as shown in Figure 6.15A.

But we can be confident that these repeats were not complete reincarnations of earlier species. Comparing shark and dolphin DNA would reveal a vast number of evolutionary differences between these forms, differences that instead would ally them with other fish and other mammals, respectively. That is, they are on very different branches of the evolutionary tree of vertebrates, as shown in Figure 6.15B. In the parallel evolution of streamlined shapes for rapid swimming, natural selection has produced a similar result starting from dissimilar evolutionary backgrounds.

Comparing shark and dolphin DNA would reveal a vast number of evolutionary differences between these forms, differences that instead would ally them with other fish and other mammals, respectively.

Therefore, even though dolphins are similar to some aquatic **dinosaurs**, they are not themselves dinosaurs. All aquatic dinosaurs have gone extinct, and they will never come back. Evolution does not supply complete repetitions, only repeated parallels. Once we lose a species, we do not get it back again.

Another way to understand this concept is that the evolutionary tree introduced in Part One directly implies that each species is an historically unique entity. Species are not functional categories like table, knife, or car. You can always make a new knife. Species are like the paintings of Van Gogh—each species a remarkable masterpiece, unique in itself, produced at a singular point in time by a process of speciation that we have yet to understand fully. When such species are lost, they are virtually irreplaceable. The one mitigation is that there may be other, similar species, just as there are several Van Gogh paintings of sunflowers. ❖

FIGURE 6.15A **Body Outlines for Sharks, Ichthyosaurs, and Dolphins, Showing Parallel Evolution for Fast Underwater Swimming**

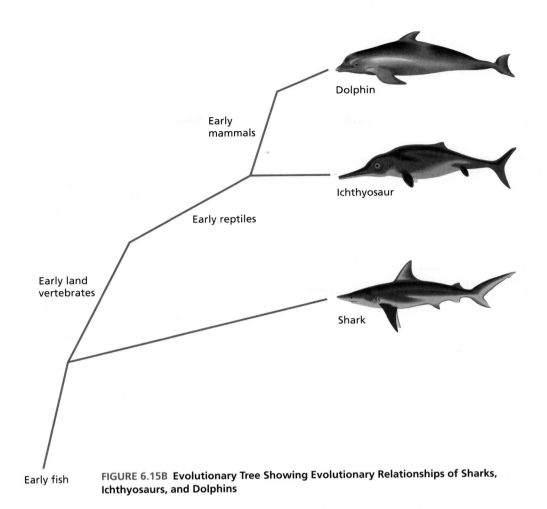

FIGURE 6.15B **Evolutionary Tree Showing Evolutionary Relationships of Sharks, Ichthyosaurs, and Dolphins**

Dolphin

Early mammals

Ichthyosaur

Early reptiles

Early land vertebrates

Shark

Early fish

6.16 Normal biotic diversity is the result of a balance between the processes of speciation and extinction

It is a cliché of evolution that the dinosaurs went extinct 65 million years ago, and then the mammals evolved. This suggests that extinction got the dinosaurs, and the mammals then went from success to success. Such an interpretation, however, falsifies what really happens in evolution. All major taxonomic groups undergo repeated extinctions, as well as repeated speciation events.

One well-studied example of this is mammalian evolution since the extinction of the dinosaurs. During these 65 million years, many successful forms of mammalian life evolved. But even more prominent in the fossil record is the repeated extinction of mammalian species. The evolutionary ancestry of the modern horse, for example, is full of examples of species and genera going extinct, as shown in Figure 6.16A. Usually we do not know why these forms went extinct. It would not be correct to assume that species less like modern horses were always the ones that went extinct, compared to a "horsier" ancestor. Such patterns are occasionally found in mammalian evolution, but are by no means the rule. Figure 6.16B shows some examples of the various mammalian species that went extinct in the last 60 million years. Even though the mammals

were an extremely successful animal in this period, a large number of mammalian species went extinct, even as new mammalian species evolved.

One way to understand this situation is illustrated in Figure 6.16C. Think of species as molecules of water. At any time, species are doing one of three things. First, some of them are evolving from other species, joining the basin of living species. This process is like water dripping from a tap into a basin. Second, there are the species going about their business as living things, increasing or decreasing in numbers temporarily, but otherwise floating in the basin of the living. Third, because the basin is plugged, the water in the basin tends to overflow, letting water drip out. The water dripping out is the normal process of extinction, by which species are being lost at all times. This metaphor is a bit misleading, though: Water in a plugged basin with a dripping faucet will maintain the same level, but the number of species in the world does not stay strictly constant. The important point is that there are three kinds of species: those being created by speciation; those merely existing; and those going extinct. The total number of species at any one time reflects the balance between these three types of species. ❖

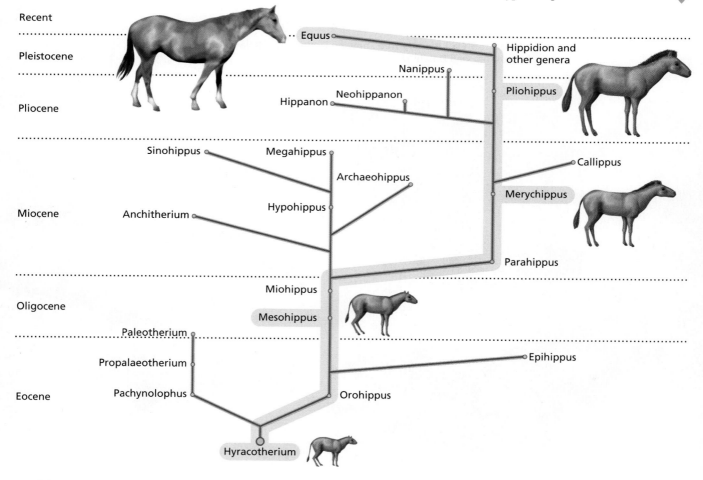

FIGURE 6.16A The Evolution of Horses, Showing Repeated Extinctions

(i) Nesodon of South America

(ii) Giant kangaroo
of Australia

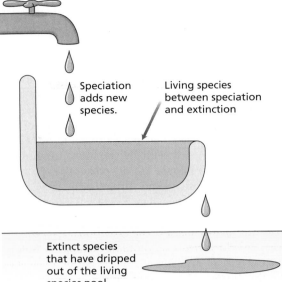

(iii) Woolly mammoth of Eurasia
and North America

Speciation
adds new
species.

Living species
between speciation
and extinction

Extinct species
that have dripped
out of the living
species pool

FIGURE 6.16C Balance Between Ordinary Extinction and Speciation Controls Diversity

(iv) Giant ground sloth of North America

FIGURE 6.16B Extinct Pleistocene Mammalian Megafauna

Retail Extinction **219**

Many factors influence whether or not a species goes extinct. Extinction is primarily a matter of too many deaths among both juveniles and adults. The normal causes of death are also factors in extinction. Disease, predation, or competition are all biotic factors that must kill organisms at all times. A new deadly disease may pose a threat to the survival of an entire species. Potentially, the human immunodeficiency virus (HIV) could kill off our entire species, because the recently successful drugs only slow the progress of the disease rather than curing it. Dramatically virulent pathogens, like the Ebola virus, conceivably could accomplish quickly what HIV is doing only slowly, providing efficient world travel spreads these pathogens faster than they can kill all available victims. The recent rapid spread of the severe acute respiratory syndrome (SARS) virus illustrates the potential for disaster with disease transmission arising from intercontinental travel. Abiotic factors, like excessive heat, desiccation, or salinity, will kill some organisms that are in the wrong place at the wrong time. The aptly named Death Valley in eastern California embodies this principle. Tourists who visit the locale, especially those from Europe, frequently suffer exposure, heat stroke, and death. When there is a major climate change, such as an ice age, many species may freeze to death and thus become extinct. Too much death, however caused, must foster extinction.

The effects of the recent human population explosion have revealed one particularly important point of vulnerability where extinction is concerned—loss of habitat. It is often difficult for humans and other predators to destroy all members of a species, if only because most species are distributed somewhat broadly. But if the habitat of the species is greatly reduced, or cut up by roads, railways, and canals, then it may have few resources with which to produce young or hide from predators. And this vulnerability to loss of habitat is as true of a small plant as it is of a small butterfly. Furthermore, if one species is extinguished, then other species that depend on it for their food or other functions may go extinct themselves. Plants that depend on specific pollinators probably will not survive the extinction of their pollinator species. A number of parasite species are vulnerable to the extinction of just one or two host species: Insect parasitoids are one example; and pathogens like smallpox, which infects only humans, are another. In a sense, the "bumping boxcar" extinction (one extinction causing another, which causes another, etc.) of these additional species is a continuation of the problem of loss of habitat, because the species they depend on are in a sense part of their habitat.

There is extinction data that supports the importance of habitat loss, although it is a bit indirect. Species that have a local distribution, with one or just a few suitable habitats, are inherently more vulnerable to habitat loss

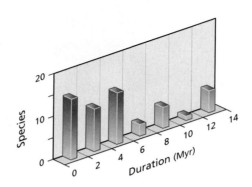

Planktotrophs

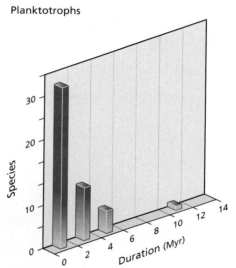

Nonplanktotrophs

(i) Comparison of species duration among marine species

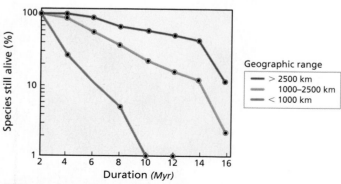

(ii) Comparison of species duration among mollusks

FIGURE 6.17A **(i)** Planktonic species disperse more widely than nonplanktotrophs. Planktotrophic species also last longer before extinction. (Myr is millions of years) **(ii)** Mollusk (snails, clams, etc.) species last longer when they have a wider geographic range. (Myr is millions of years.)

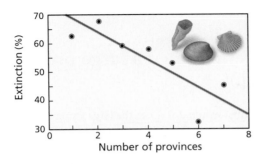

FIGURE 6.17B Marine bivalve genera that inhabited more oceanic regions, or "provinces," were less likely to go extinct at the K-T boundary.

than are species that disperse widely and have a broad geographic range. Therefore, if habitat loss is generally important, then we should see an inverse correlation between breadth of distribution and likelihood of extinction. The broadly distributed species should go extinct less often. For example, marine invertebrates that have a planktonic larval form, which floats long enough to be distributed widely across an ocean, should go extinct less often than forms without a planktonic phase. Figure 6.17A shows that this pattern does exist in studies of planktonic life history and geographic range, at least during normal extinction periods. This pattern also shows up in data from large-scale extinctions, as shown in Figure 6.17B for the extinctions at the end of the **Cretaceous** period. With respect to avoiding extinction, spreading out is better. ❖

MASS EXTINCTION

6.18 Mass extinctions have intermittently eliminated a large proportion of living species

Influenced by Darwin, most evolutionary biologists have thought of the broad evolution of life in terms of the slow production of new species, together with their steady extinction, as discussed in the preceding modules. Most of the time, so far as we can tell from the fossil record, this is correct. More than 90 percent of extinctions occur in this unspectacular fashion. But at a few points of time, this pattern is annihilated by massive levels of extinction.

The most extensive **mass extinction** is the one that happened at the end of the **Permian** period, which was also the end of the **Paleozoic** era, the first of the three major eras comprising the eon of multicellular life known as the **Phanerozoic**. This mass extinction wiped out about 50 percent of all known taxonomic families and an even larger percentage of species—about 90 percent of all species then living. This massively catastrophic event marks the transition between the Paleozoic and the **Mesozoic** eras.

Dinosaurs would dominate the large fauna of the Mesozoic before they too would be wiped out at the end of the Cretaceous, the **K-T boundary**, a point in time that is also the end of the Mesozoic and the start of the **Cenozoic** period. Since then mammals and birds have dominated the large fauna populations of the world. The Cretaceous is the most recent mass extinction event, visualized in Figure 6.18A. The overall sequence of mass extinctions is shown in Figure 6.18B.

Two major scientific issues arise from the phenomenon of mass extinction. First, do species levels recover, or has the Earth been progressively losing species because of mass extinctions? It turns out that species numbers recover from mass extinction events, as if evolution tends to return to a "set point" for species numbers. The causes of this pattern are not fully understood, but some hint of explanation is provided by such events as the mammalian recovery of many large-animal niches, after the dinosaurs went extinct.

FIGURE 6.18A Pictoral Visualization of the Large-Body Impact That is Thought to Have Ended the Cretaceous and Killed Off the Dinosaurs This is explained further in Module 6.19.

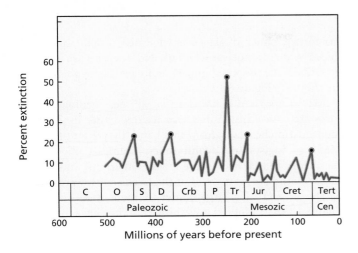

FIGURE 6.18B Patterns of Extinction among Taxonomic Families Since the Cambrian, Showing the Mass Extinctions The rate of extinction of individual species was far higher.

It may be that evolution tends to fill up "vacated" ecological niches after mass extinctions. However, there are many dinosaur features that have not been emulated in the course of mammalian evolution, so this theory is unlikely to be the whole truth.

The second major scientific issue is the cause(s) of mass extinctions. Figure 6.18A depicts one possible cause: the impact of a large astronomical body, such as a comet or large meteorite. This idea is discussed further in Module 6.19. ❖

What causes mass extinctions? Geologists once believed that such normal processes as volcanic activity or climate change caused mass extinction. Some still do.

Since 1980, however, an alternative theory has been proposed that has excited the scientific community. The theory is that moderately large astronomical bodies, either comets or meteors, have intermittently hit the Earth with explosive forces equal to those of many exploding thermonuclear weapons. The best evidence for this **large-body impact** theory has come from the most recent mass extinction, 65 million years ago, at the end of the Cretaceous. At that point in the geological sediment, there are many indications of an extraterrestrial impact: High levels of **iridium**, which is common in

meteorites; "shocked" glass over wide areas, like the effect produced by thermonuclear weapons; and extensive ash and soot deposits in the proposed impact sediment. These phenomena are shown in Figure 6.19A.

Most important is the fact that all these geological features are concentrated in a very narrow band. In the sediment just inches away from this band, they are not found. Many geologists and evolutionary biologists now accept the theory that at least some mass extinctions, and almost certainly the one at the end of the Cretaceous, were caused by the impact of a large comet or meteor. As shown in Figure 6.19B, geologists now think that they have found the site where this large object hit the Earth—Chicxulub, at the tip of the Yucatán peninsula, in Mexico. ❖

(i) Dark band of clay at K-T boundary, suggesting abrupt release of soot and other debris.

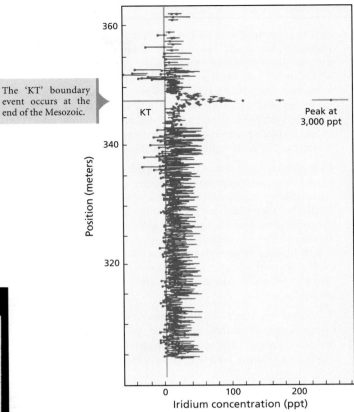

The 'KT' boundary event occurs at the end of the Mesozoic.

(iii) The spike in iridium concentration at the K–T boundary sediment layer.

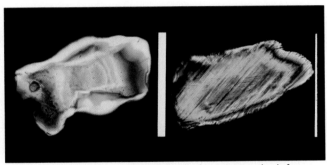

(ii) Shocked quartz on the right, normal quartz on the left; shocked quartz is often found near meteorite strikes.

FIGURE 6.19A   Four pieces of evidence for a large-body impact collected from Cretaceous-Tertiary (K-T) boundary sediment. See Alvarez (1997).

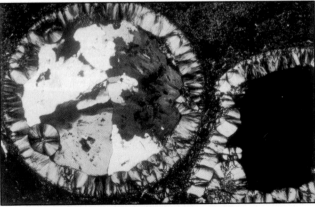

(iv) Microtektites, particles of glass generated by meteorite impacts.

FIGURE 6.19B Location of the Chicxulub crater, which is below ground and partly below water, at the tip of the Yucatán peninsula. This crater may have been produced by the same astronomical body that caused the K-T mass extinction.

SUMMARY

1. Species are defined according to reproductive isolation: Organisms are members of different species if they do not mate when they live in the same area. Speciation is thus the process by which populations develop reproductive isolation. Prezygotic reproductive isolation may be established by a failure to produce zygotes, which may involve a lack of sexual attraction, for example. Postzygotic reproductive isolation may occur when hybrid zygotes do not develop into fertile adults. In some species, both types of reproductive isolation may occur.

2. Geographical separation of populations, called allopatry, plays an important role in speciation. Allopatric populations may evolve distinct mating adaptations that create prezygotic reproductive isolation when they resume contact. Allopatric populations may also develop postzygotic reproductive isolation when they evolve distinct genomic features that render hybrids inviable or infertile. When postzygotic reproductive isolation evolves allopatrically, contact with closely related species may select for prezygotic reproductive isolation, to avoid wasted reproductive activity.

3. Speciation also occurs without allopatry, by a process called sympatric speciation. One evolutionary process that produces speciation sympatrically occurs when genomes are increased in ploidy by a failure of meiosis or by hybridization. Another evolutionary process that may create new species is local mating of specialized fruit eaters.

4. New species sometimes originate in rapid radiations. The most spectacular example of species radiation occurred around 500 million years ago, when most groups of animals and plants evolved. Less dramatic examples of species radiation are known, such as the radiation of finch species on the Galápagos Islands.

5. The fossil record often shows a pattern of intermittently abrupt evolution called punctuated equilibrium. Punctuated equilibrium can be explained in terms of normal Darwinian processes.

6. Extinction occurs frequently. The diversity of living species is the result of a balance between speciation and extinction. Dispersed species tend to go extinct less often.

7. Mass extinctions have occurred at least five times during the last 500 million years. Mass extinctions involve the abrupt loss of the majority of living species—sometimes more than 80 percent. Many biologists and geologists now think that mass extinctions are caused by the collision of large asteroids and comets with the Earth, including subsequent disruptions to the Earth's atmosphere.

REVIEW QUESTIONS

1. What is a biological species?

2. What is the difference between sympatry and allopatry?

3. Does speciation increase the level of adaptation in the new species?

4. Why is allopatric speciation more likely than sympatric speciation?

5. Name one biological process that might produce new species sympatrically.

6. Explain why the evolution of mammals over the last 65 million years could be called an adaptive radiation.

7. Which theory would Darwin have preferred, phyletic gradualism or punctuated equilibrium?

8. What is the normal fate of a species?

9. What kinds of species are more likely to go extinct when mass extinctions occur?

10. What kinds of species are humans driving to extinction?

11. Propose a scenario for the extinction of the human species.

KEY TERMS

allopatry
allopolyploid
autopolyploid
biological species concept
Burgess Shale
Cambrian
Cenozoic
Cretaceous
dinosaurs
Eldredge, N.
evolutionary stasis

extinction
fluctuating selection
Galápagos Islands
genetic drift
genetic revolution
Gould, S. J.
hopeful monster
host race
hybrid breakdown
hybridization
hybrid zone

iridium
K-T boundary
large-body impact
mass extinction
Mayr, E.
Mesozoic
Paleozoic
parthenogenesis
Permian
Phanerozoic
phyletic gradualism

polyploidization
postzygotic isolating mechanism
prezygotic isolating mechanism
punctuated equilibrium
reproductive isolation
species radiation
stabilizing selection
sympatry
Tertiary

FURTHER READINGS

Alvarez, W. 1997. *T. Rex and the Crater of Doom*. Princeton, NJ: Princeton University Press.

Briggs, D.E.G., D. H. Erwin, and F. J. Collier. 1994. *The Fossils of the Burgess Shale*. Washington, DC: Smithsonian Institution Press.

Dobzhansky, T. H. 1937. *Genetics and the Origin of Species*. New York: Columbia University Press.

Mayr, E. 1942. *Systematics and the Origin of Species*. New York: Columbia University Press.

———. 1963. *Animal Species and Evolution*. Cambridge, MA: Harvard University Press.

Otte, D., and J. Endler. 1989. *Speciation and its Consequences*. Sunderland, Massachusetts: Sinauer.

Raup, David M. 1991. *Extinction, Bad Genes or Bad Luck?* New York: W.W. Norton.

Rice, W. R., and E. E. Hostert. 1993. "Laboratory Studies on Speciation—What Have We Learned in 40 Years?" *Evolution* 47: 1637–1653.

Weiner, J. 1994. *The Beak of the Finch: A Story of Evolution in Our Time*. New York: Knopf.

White, M.J.D. 1978. *Modes of Speciation*. San Francisco: W.H. Freeman.

PART THREE

THE DARWINIAN ORGANISM

How organisms function cannot be understood just by studying molecules or cells. The organism is more than the sum of its parts. For instance, regulation of body temperature often depends on behavior. Butterflies spread open their wings in direct sunlight to warm themselves. Lizards raise their body temperatures by resting on rock surfaces that have been warmed by the sun. The integration of physiology and behavior makes thermoregulation possible in these animals. This is only one example of the many ways in which organismal biology is integrated with ecology and evolution.

It is possible to study organismal function and physiology by simply studying how the organism works as a biochemical machine. The effect of temperature on the rate of an organism's biochemical reactions can be studied by itself, as can the biochemical reactions of organisms to extreme heat. But this will not provide us with a complete understanding of how the organism regulates its temperature. We gain deeper insights into organismal biology if we remember that natural selection has had a hand in determining how organisms function. Thus the term the *Darwinian organism*, because the way organisms function is partly the product of natural selection acting in a particular ecological context. Butterflies that have evolved "sunning" behavior are better equipped to survive in cooler environments, and therefore they are favored by natural selection. In general, the process of natural selection molds behaviors and functions in ways that help the organisms regulate their physiology. Scientists are now performing experiments in which they can observe the evolution of important physiological functions. We introduce you to this work in Chapter 9.

Organisms also reproduce, and as *Darwinian* organisms this is their most important function. It should come as no surprise that most aspects of reproduction are under the influence of natural selection. However, the ways organisms reproduce are quite variable. A goal of Chapter 7 is to help you understand why all organisms are not identical with respect to the timing and duration of reproduction. How do we make sense of the fact that bacteria may reproduce every 20 minutes, while large mammals may wait a dozen years before beginning reproduction? Shouldn't there be a single way to reproduce that always has the highest fitness? We can understand the great diversity of life histories by understanding how natural selection acts on reproduction.

The consequences of reproduction take us to the next level of biological organization, the population. The population is part of the individual's environment. The composition of a population will change as organisms reproduce, which in turn sets into motion the conditions for evolutionary change. A change in the composition or size of an

organism's population is like changing temperature, food availability, or any other aspect of the ecological environment. As the number of organisms in a population increases, there will be effects on reproduction. This will alter the effect of natural selection on the traits of organisms, as well as changing the ecology of the whole population. This connection between evolution at the level of the organism and changes in the ecology of whole populations is one of the more fascinating aspects of the Darwinian organism. In Chapter 10 we review how populations grow, and how evolution affects population dynamics.

Populations of animals and plants are not static in space. We see movements of populations, as with the migration of birds during the winter. Even when individuals cannot move, their gametes may. Plants often have elaborate methods for ensuring the wide dispersal of seeds. In Chapter 11 we discuss the importance of dispersal.

After reading Chapters 7 through 11, you should be able to start thinking in Darwinian terms. When interpreting behavior, physiology, or life cycles, you will naturally ask if they have adaptive aspects. These questions help make sense out of the great complexity of biological diversity.

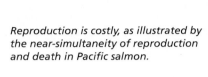
Reproduction is costly, as illustrated by the near-simultaneity of reproduction and death in Pacific salmon.

7

Life History of the Organism

The Pacific salmon leaping up the waterfall is on its way to spawn and then die. For this bony fish, dying and reproducing have a poignant simultaneity. Yet there is much about its life history that is universal. The cycle of birth, reproduction, and death is basic to all organisms, even when the specific features of these life histories vary. Plants disperse pollen. Some animals make their way into the world by cracking open their eggshell, while others struggle to emerge from their mother's body during birth. Some unicellular life-forms and some simple animals just split in two, and the two halves become separate organisms right away. There is great diversity among life cycles.

One thing is unitary to all these life histories: Darwinian fitness, the net reproduction of individual organisms. And fitness in turn is defined by the life history. There are two main ways in which fitness is determined: (1) the product of viability and fertility; and (2) the integration of survival and reproduction over a complex life cycle. We will show how these two kinds of fitness arise, what they mean, and what their consequences are for the life cycle of the organism.

The focus of fitness is reproduction, the key point of every life history. Thinking about life histories in relation to net reproduction gives a useful focus on how life histories evolve. An important thing about reproduction is that, even though it is essential to Darwinian fitness, it may be destructive for the individual organism. Reproduction has costs, and these costs limit its evolution.

If the initiation of new lives is obviously important for any life history, there is also the end of life to consider—death. Death comes in an unrelenting form at advanced ages, through a loss of health and vigor known to us as aging. Aging is a unifying theme of life histories, though it is not universal. ❖

FITNESS AND LIFE HISTORIES

7.1 There are many types of life history

Bacteria have one reproductive episode. Once the cell is ready to divide, it divides exactly once, and then there are two immature cells. Adults do not exist. This is the essential feature of these life histories. Organisms are either developing toward reproduction or reproducing. There is no adult phase between reproductive events.

Bacteria are not the only organisms like this. Many single-celled organisms reproduce by fission—and even some multicellular animals, such as asexual sea anemone species. Figure 7.1A shows some examples.

FIGURE 7.1A Bacteria (left) and Sea Anemones (right)

But having just one reproductive episode does not require fission. Insects that have one generation per year are described as **univoltine**, and they also have one reproductive episode. Examples of univoltine insects are particularly common among the butterflies and moths of temperate North America. In spring, caterpillars hatch out of eggs that have overwintered on a tree branch or on the ground. The caterpillars are quite small at first. Over successive molts, the caterpillars grow substantially. They then form a pupa, in which they undergo a metamorphosis to the adult moth morphology. The adults mate after emerging from the pupa. The females lay their eggs, in a clump or dispersed over host plants, and then die. The cycle then begins again next spring.

Among animals generally, this type of **life history** is called **semelparous.** An interesting point is that semelparous life histories do not have to be annual. The 13-year and 17-year cicadas have a similar pattern of semelparous reproduction, except that in their case the cycle has a length of 13 or 17 years, from the start of one adulthood to the start of the adulthood of the offspring.

Many plants have a growth cycle that is termed *annual*, although some annual plants may survive into the next growing season. *Monocarpic* plants, such as soybean, have a strictly semelparous life history; the adult plants die right after reproducing.

Male marsupial "mice," *Antechinus stuartii*, an Australian species unrelated to mice elsewhere, also have semelparous reproduction. Just before the start of the mating season, the males undergo profound hormonal changes. They become more aggressive and their genitals enlarge. Come mating time, males fight with each other and copulate with females for hours at a time. They lose weight, suffer numerous internal lesions, and lose resistance to parasites. Once the mating season is over, the males die very quickly. None make it to the next mating season. Some females, interestingly, do make it to the next reproductive season, like "annual" plants do. Apparently reproduction is more stressful for males in this species. Figure 7.1B shows a cicada, Figure 7.1C shows shows a field of soybean, and Figure 7.1D shows a marsupial mouse.

Most of the organisms that we are familiar with have a pattern of repeated, or **iteroparous**, reproduction. We have such a pattern ourselves. Most

FIGURE 7.1B A Cicada Molting

FIGURE 7.1C **Field of Soybeans**

FIGURE 7.1E **Snow Geese**

mammals, birds, and reptiles have adult stages that can reproduce more than once, sometimes with long periods of time between each reproductive event.

Just as semelparous species often time their reproduction to the annual cycle of seasons, so do some repeated reproducers produce offspring once a year, at about the same calendar dates. But with iteroparous reproducers, some reproduce again the next year, and possibly the year after that. Snow geese (*Chen caerulescens*), for example, migrate in the thousands to breeding grounds in Alaska and northern Canada in the spring and summer. Once they have reproduced, they fly back south for more hospitable winter conditions. Next spring, they fly north to breed again. Figure 7.1E shows snow geese.

Many trees follow a pattern of annual reproduction, sustained over many seasons. Other trees reproduce less often than once a year. But in either case, the ability of some trees to survive through many rounds of reproduction is probably greater than that of any other organisms. Bristlecone pines (*Pinus longaeva* and *P. aristata*) of the White Mountains of eastern California can survive thousands of years, continuing to reproduce century after century. Trees are

the silent witnesses of lives that are fleeting compared to theirs. Figure 7.1F shows bristlecone pines.

Some repeated reproducers show no annual cycle to their reproduction. Tropical species from plants to insects, like *Drosophila melanogaster*, are more likely to follow this pattern than temperate or arctic species. Humans are a tropical species in their evolutionary origins, and we reveal that in our continuing fertility throughout the year. ❖

FIGURE 7.1D **The Brown Antechinus**

FIGURE 7.1F **Bristlecone Pine**

7.2 The fitness of semelparous organisms is the product of viability and fecundity

Some of the most spectacular "shows" in biology are put on by semelparous organisms—the animals and plants that reproduce a single time. Fields of grain that have grown abundantly suddenly set seed and then die. Mayflies emerge together on summer days, fill the skies with their mating swarm, lay their eggs, and then die in virtual unison, their myriad bodies clogging drains. (There are hundreds of species of mayfly, many of which have vestigial mouths as adults—guaranteeing an early death. A common American mayfly is *Dolonia americana*.)

Among vertebrates, one of the more dramatic semelparous life histories is that of Pacific salmon (genus *Onchorhynchus*). These salmon swim upstream from the ocean, overcoming waterfalls and hungry bears to find their way back to the small streams in which they were born. Schematics of semelparous life-history patterns are shown in Figure 7.2A.

For the scientist, the burst of reproduction that these species show is tremendously convenient. It makes the arithmetic of fitness simple, provided only that all adults are equally successful sexually. This means we assume that partners are so abundant during the brief burst of reproduction that we can assume that every female animal, or female flower, receives sperm or pollen. We also need to assume that all males are equally successful at fertilizing females.

If we make these assumptions, then the fitness of a semelparous organism is simply a product of its *viability* times its *fecundity*. **Viability** can be defined as the chance of surviving from the youngest juvenile stage to adulthood. **Fecundity** is the total output of successful gametes produced by an adult. As shown in Figure 7.2B, the Darwinian arithmetic of **fitness** in these life histories reduces to this product of viability (*v*) and fecundity (*f*), which is fitness (*W*):

$$W = vf$$

Semelparous life is like a pinball game with a single bumper that releases the balls for the next round of the game when it is struck, and at the same time this strike terminates play, so that the parent ball rolls down to its death. ❖

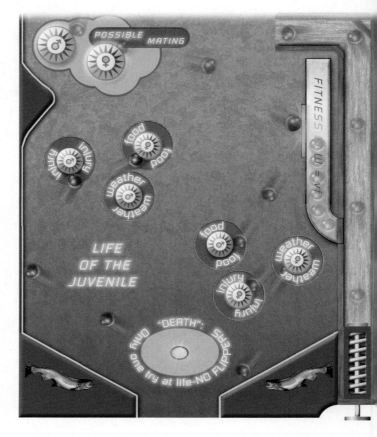

FIGURE 7.2B **Pinball Structure of Semelparous Life Cycle** This structure can be applied to most microbial life cycles, all monocarpic plants, univoltine insects, and many other kinds of organisms.

(i) Semelparous animal life cycle

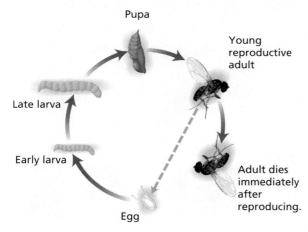

Pupa

Young reproductive adult

Late larva

Early larva

Adult dies immediately after reproducing.

Egg

FIGURE 7.2A

(ii) Semelparous plant life cycle

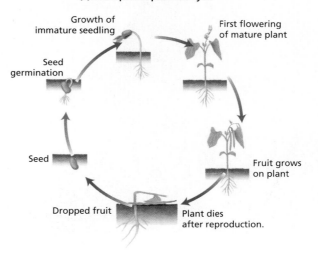

Growth of immature seedling

First flowering of mature plant

Seed germination

Seed

Fruit grows on plant

Dropped fruit

Plant dies after reproduction.

The reproductive capability of an iteroparous organism generally changes with age. Thus an important aspect of a population of iteroparous organisms is its **age-structure:** the numbers and ages of all members of the population. To predict population growth in age-structured populations, we need to understand how patterns of survival and fertility vary with age.

To predict the growth of age-structured populations, we need some estimate of the fertility and survival of individuals as a function of age. Because the age of an individual is continuously changing, some convenient way of recording age must be established. We can usually accomplish this by creating discrete intervals, called **age classes**. All individuals in an age class are treated as equal. The limits of an age class are usually based on the properties of the organism and the amount of data that has been collected. For instance, in human populations, age classes are often broken down into five-year intervals. The first interval would be all individuals from birth to 4 years of age, the second age class individuals from 5 years to 9 years, and so on. This kind of analysis is shown in Figure 7.3A, for human demographic data. For an organism like a fruit fly, the age classes might be single days or two-day intervals. The chances of surviving—or the fertility of—a given age class is then based on the average of all members of the age class.

Age-specific probabilities of survival describe the chance that an individual in a certain age class will survive to become a member of the next age class. In females **age-specific fertility** refers to the number of offspring produced by a female in a particular age class that survives to become a member of the first age class. In males age-specific fertility is the count of offspring fathered by males of the age class.

Two basic techniques are commonly used to estimate age-specific fertility and **survival** probabilities. The most direct technique is to take a large group of identically aged individuals, or a **cohort**, and follow their survival and reproduction as they age. This technique is most easily used with laboratory or cultivated populations of animals and plants.

For some organisms that are very long-lived or that can't be subjected to experimental techniques, we need other methods. These other techniques involve looking at the numbers and ages of individuals in a particular population at a single point in time. After making several assumptions, we can estimate age-specific survival probabilities and fertilities. The product of this analysis is a **static life table**. The most obvious limitation of the ecologist's static life table is that the individuals used to construct it have not experienced the same environmental conditions. These environmental conditions can have a significant impact on current fertility or survival. ❖

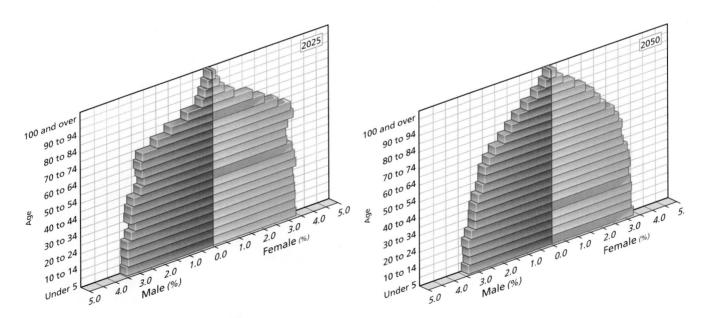

FIGURE 7.3A U.S. Age-Class Estimates for 2025 and 2050

To follow the growth of age-structured populations, we require the fertility and survival probabilities of each age class. With this information, we can predict not only the total population size in the future but also the numbers of individuals in each of the age classes in the future. We will not go into the details of this process, but we will review the basic science of age-structured populations.

Suppose we have an organism with five age classes that we will call newborns, 1-year-olds, 2-year-olds, 3-year-olds, and 4-year-olds. Although the fertility of males could affect population growth, we assume, as is common, that they do not. We will further assume that the age-specific survival of males and females is the same. Table 7.4A summarizes the survival and fertility of the population as a **life table**.

No number is entered in the table for the survival of the 4-year-olds, because they are all presumed to die. If the population initially has 50 individuals in each age class, the growth of the total population and each age class is shown in Figure 7.4A. The total population size is simply the sum of each of the five age classes. The behavior of the age classes over the first five or so years is variable from year to year. This variation reflects the oddities of the initial conditions. After some time we see that each age class starts to exhibit regular increases in size. In fact all age classes are increasing at the same rate, which in this example is about 4.7 percent per year.

When the different age classes begin to grow at the same exponential rate, as seen in Figure 7.4A, the different age classes will always make up a constant fraction of the total population. At this point we say the population has achieved a **stable age-distribution**. The population in Figure 7.4B has

TABLE 7.4A **A Hypothetical Life Table**		
Age Class	Female Fertility	Probability of Surviving
Newborns	0	0.8
1-year-olds	0.5	0.6
2-year-olds	1.2	0.5
3-year-olds	0.3	0.3
4-year-olds	0.02	—

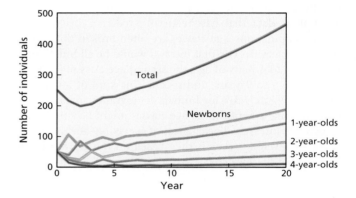

FIGURE 7.4A Population Growth in Age-Structured Populations This population has five age classes, including newborns. The population starts with 50 in each age class. For the first 6 years, the age classes change haphazardly. After 10 years, each age class begins to increase steadily, as does the total population. At this point the total population and each age class are growing at the same exponential rate.

a stable age-distribution with newborns making up 40 percent of the population, 1-year-olds 31 percent, 2-year-olds 18 percent, 3-year-olds 9 percent, and 4-year-olds 2 percent. It is important to remember that even though a population is at a stable age-distribution, it may still be increasing in total size.

The shape of the stable age-distribution can give insights into the relative rate of population growth. In general, the proportion of individuals in the youngest age classes will be greater in faster-growing populations, such as the one in Figure 7.4C. This population has the same characteristics as the population in Figure 7.4B, except the fertility of 1-year-old females is 1.0 and the fertility of 2-year-old females is 2.0 (rather than the fertilities given in Table 7.4A). The resulting population is increasing by 26 percent each year (instead of 4.7 percent), and the fraction of the population in the youngest age classes is greater. ❖

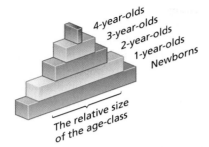

FIGURE 7.4B Population Growth in Age-Structured Populations When all age classes are growing at the same exponential rate, the population is at a stable age-distribution. This means that each age class will occupy a constant percentage of the total population size. For the population in Figure 7.4A, the stable age-distribution is shown here. This type of display is called a population pyramid. The young age classes make up the largest fraction of the total population.

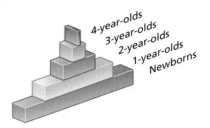

FIGURE 7.4C Population Growth in Age-Structured Populations The relative proportions of individuals in each age class are affected by the rate of population growth. In slowly growing populations, the proportion of the population in the younger age classes is smaller than in rapidly growing populations. In Figure 7.4B, the population is increasing by 4.7 percent each year. In this figure, the population is growing by 26 percent each year.

Age Structure in Human Populations

The differences in age structure of rapidly and slowly growing populations are illustrated with human populations in Figure 7.4D. The rapidly growing Mexican population in 2000 has a sharply rising pyramid, while the more slowly growing U.S. population shows a more even distribution of people in the middle age ranges. Males and females are combined in these pyramids. Compared to the U.S. population, a much larger fraction of the Mexican population consists of very young individuals (less than 20 years old). Likewise, the U.S. population has a much larger fraction of people older than 80 years.

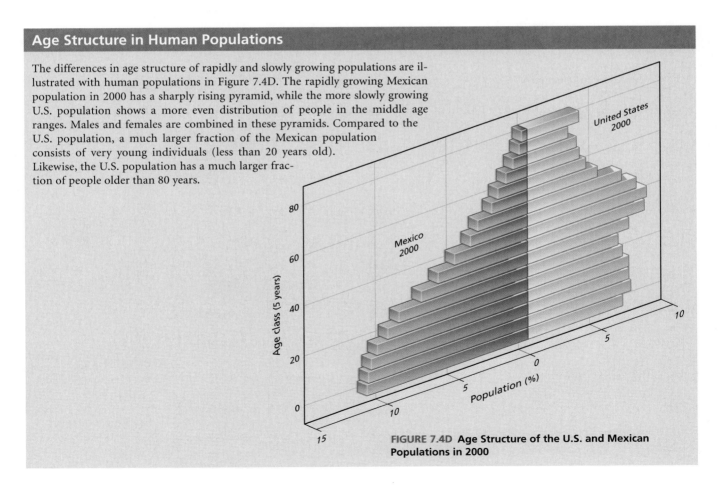

FIGURE 7.4D Age Structure of the U.S. and Mexican Populations in 2000

7.5 In iteroparous organisms, fitness can be calculated from estimates of population growth rates

Most iteroparous organisms are multicellular animals and plants. Their **life cycles** are sketched in Figure 7.5A. One way to think of these life cycles is that they are like semelparous life cycles, but with the first episode of reproduction repeated multiple times in the adults who survive to reproduce again.

(i) Iteroparous animal life cycle

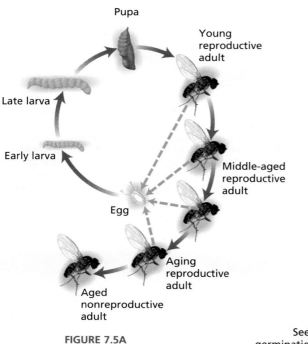

FIGURE 7.5A

Figure 7.5B supplies a schematic for the iteroparous life cycle. An iteroparous life cycle is a more elaborate pinball game. The organism enters the population "at the top," among the juveniles. Then it bounces down through a succession of stages, starting with one or more immature stages of variable duration, followed by adult stages. As the organism proceeds through the adult part of its life cycle, it may produce offspring, like pinball games that release additional balls during play. All new offspring start again at the first stage of the life cycle. In an entire population, many organisms are bouncing to and fro among the bumpers. It is a very complex game—the combination of multiple iteroparous organisms to form a population.

It is difficult to estimate fitness with the iteroparous life history. It is as true of the iteroparous organism as it is of the semelparous that better survival and more reproduction mean greater fitness. The problem is that it is not intuitively clear how much better. For example, increasing fecundity at

(ii) Iteroparous plant life cycle

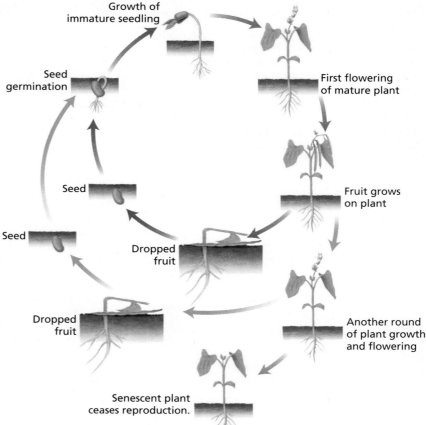

any particular age should be a good thing for fitness; but there are reasons for supposing that increasing early fecundity would be better than increasing later fecundity. After all, in the wild, the organism might not survive long enough to reproduce at the later age, so any genetic capacity that it might have to reproduce then would be wasted.

This accounting problem in estimating fitness can be solved using the concept of population growth in age-structured populations. Repeated reproduction automatically makes populations age structured. The initial growth of an age-structured population is fairly chaotic, with the potential for large swings in the composition of the population—and thus large changes in reproductive output. Over time, population growth tends to produce stable proportions of the different age groups, as explained earlier.

With stable proportions, the population size grows at a stable rate. This stable rate of population growth defines the fitness of the population. There are several different names for this fitness, including **intrinsic rate of increase** and the **Malthusian parameter**. It is usually represented by the lowercase letter r, although sometimes m or $\lambda = e^m$ is used, with m referring to Malthus. This definition of fitness is intuitively appealing because fitness is based on the idea of net reproduction, and net reproduction must determine the rate of growth in population size.

❖

FIGURE 7.5B Structure of Iteroparous Life Cycle

7.6 A small increase in semelparous fecundity may be favored over iteroparity

An amazing fact about life histories is that semelparous animals and plants reproduce in a single burst, and then die. Why don't these organisms continue reproducing? Why does the Pacific salmon usually die right after spawning?

At one level, the answer appears to be that these organisms are killed by the act of reproduction. If the gonads of Pacific salmon are removed before spawning, they can live several years longer than intact, reproductive salmon; this is shown in Figure 7.6A. If the floral structures of a soybean plant are removed every time they grow, then the deflowered soybean plant can live much longer than an intact reproductive plant. Castrated *Antechinus* "mice" live months longer than intact "mice." For these organisms, and many like them, sex kills. Semelparous organisms do not go on reproducing, because they cannot. They are dead or dying.

But at the evolutionary level, this answer is not adequate. *Why don't semelparous organisms make some physiological investment in their continued survival?* This is called Cole's Paradox. The key must be reproduction, because the castration experiments indicate that there is a **trade-off** between early reproduction and continued survival, as the Pacific salmon example shows. Semelparous animals and plants must be shifting resources to reproduction, away from their survival. The suspicion is that, in so doing, they increase their fitness—the product of viability and fertility.

We can calculate the conditions under which fitness is increased by shifting resources from later survival to earlier reproduction. One of the interesting special cases occurs when there is no juvenile or adult mortality. Under these implausible conditions, an increase of as little as one additional offspring during the first reproductive period is enough to make semelparity favored by natural selection over iteroparity. Even when there is significant mortality, semelparity may be favored by natural selection with just a small increase in fecundity during the first bout of reproduction. This quantitative comparison illustrates the extent to which natural selection is biased in favor of reproducing quickly, rather than waiting. ❖

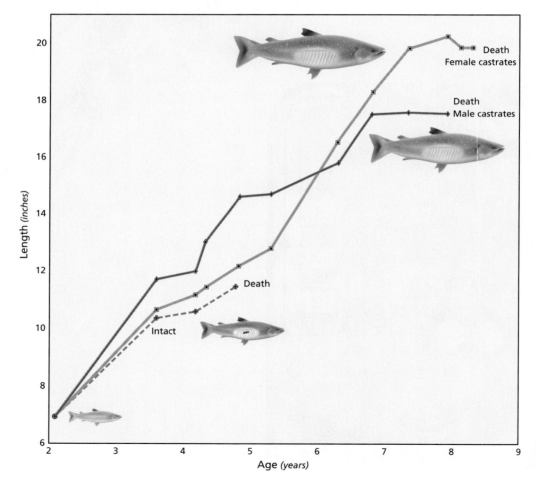

FIGURE 7.6A Pacific Salmon Castration Data Instead of dying within weeks, castrated Pacific salmon live for years. As the graph shows, castrated males and females go on growing for years after the intact fish have died. (From Robertson, 1961.)

TRADE-OFFS IN LIFE-HISTORY EVOLUTION

When evolution increases one life-history character, another life-history character may decrease

Evolutionary costs or trade-offs arise when there are limits to the resources available to organisms—limits to the available food calories, water, amino acids, and so on. Organisms cannot solve their biological problems by getting more of such limited resources. A plant can synthesize only so much material from photosynthesis in a single day. An animal can feed only so much, so the amount of food energy available to fuel its metabolism must be limited. All of this can be summarized by stating that resources are limited.

Given that resources are limited, all organisms have the problem of what to do with them. For multicellular organisms, this problem is partly one of allocation of resources between reproduction and survival of the body. (Other allocation problems are discussed later.) This type of allocation in turn involves the flow of materials between organs. As shown in Figure 7.7A, a hypothetical fish example, the **cost of reproduction** often involves shunting materials to reproductive organs from the other tissues. Thus in the fish example shown, allocation will include the shunting of fats to gonads from the rest of the body. In the female gonad, fat can be used to provision eggs, increasing fecundity. In the male gonad, fat can be used to make more sperm. Or the fat can be retained in the rest of the body, where it can be used as a source of fuel for general metabolism, including the metabolism required for locomotion,

growth, and other functions. A growing organism, providing it survives, will later be able to build a bigger gonad. Resources devoted to reproduction and growth impinge on the animal's survival. The effect on life span of removing the gonads of Pacific salmon reveals this trade-off dramatically.

The evolutionary cost that is best established is the cost of reproduction. Much of the evidence for this cost comes from experiments or situations in which reproductive activity is increased, resulting in decreased survival; or in which reproduction is decreased, resulting in increased survival. We have already seen the example of Pacific salmon castration. A human example comes from institutionalized male patients with gross intellectual impairment. Some decades ago, a number of American male patients were castrated by their doctors. The castrated individuals lived longer than intact males did. Apparently, production of testosterone or related male behavior impairs survival in these institutionalized patients.

One of the best studies of this kind involved a fruit-fly species, *Drosophila subobscura*. J. Maynard Smith reduced the reproduction of female fruit flies by several different methods: maintaining some of them as virgins, sterilizing others by irradiation, and genetically eliminating the ovaries of others. As shown in Figure 7.7B, any manipulation that decreased fecundity increased adult female survival. ❖

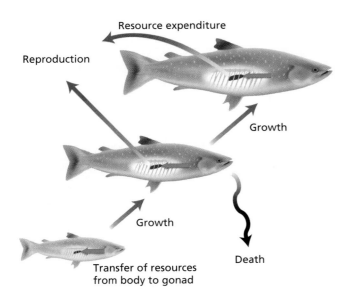

FIGURE 7.7A Shunting of Resources between Body and Gonad in a Growing Fish The growing fish faces trade-offs in its use of food energy. This energy can go to reproduction, growth, or survival. Growth may foster both survival and reproduction later in life.

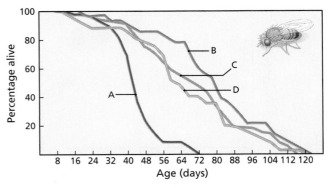

FIGURE 7.7B Effects of Mating and Reproduction on Survival in *Drosophila subobscura* A = normal mated females. B = females that lack ovaries and lay no eggs. C = virgin females, which may lay some eggs, but fewer than normal. D = mated females that have been sterilized by maintenance at high temperatures. The normal mated females lay more eggs and die sooner. (From Maynard Smith, 1958.)

7.8 Trade-offs between survival and reproduction may lead to the evolution of reproductive restraint

We will now consider the consequences of the cost of reproduction in more detail. When the cost of reproduction is considered theoretically, it is often found that natural selection favors producing an intermediate number of offspring. The logic behind this conclusion is shown in the box, shown below. Total fecundity rises with the number of offspring, because that is what fecundity means. If there is a trade-off between parental fecundity and parental survival, the survival of the parent will fall as fecundity increases. If we use net fecundity as a measure of fitness per reproductive season, the curve relating total fecundity to the number of offspring produced in each season of reproduction is humped. If no offspring are produced, individual fitness is zero. If many eggs are laid, the parent will quickly die—approximating the semelparous pattern of reproduction. Somewhere in the middle may be the right fecundity per round of reproduction, the seasonal fecundity at which fitness over the whole life is at a peak.

The best-studied examples of this pattern are in nesting birds, because experimenters can easily add or subtract eggs

from the nest, as noted earlier. Figure 7.8D shows the data from a study of this kind using great tits (*Parus major*). Great tits usually lay an intermediate number of eggs—about eight. If adults are given additional eggs to rear, they almost always succeed in rearing the young to fledging. Why don't they lay more eggs each year? The data of Figure 7.8D reveal that adults laying more eggs have lower survival rates. These findings were obtained in natural breeding populations, so they demonstrate well what evolution actually does. Faced with a trade-off between fecundity and survival, evolution compromises, leading to the evolution of reproductive restraint.

Note that in some cases evolution will not compromise. In the case of semelparous organisms, evolution pushes reproduction to its maximum value, effectively killing the parent. Natural selection is always focused on fitness, and thus on reproduction. For this reason, survival must ultimately lose out against reproduction, when there is a trade-off. We live only in order to reproduce. ❖

Selection Favors an Intermediate Number of Offspring

The figures show the situation birds face within a breeding season. This is like semelparous breeding. As an evolutionary problem, fitness for the breeding season is maximized using the product of fertility and viability, the net number of offspring reared.

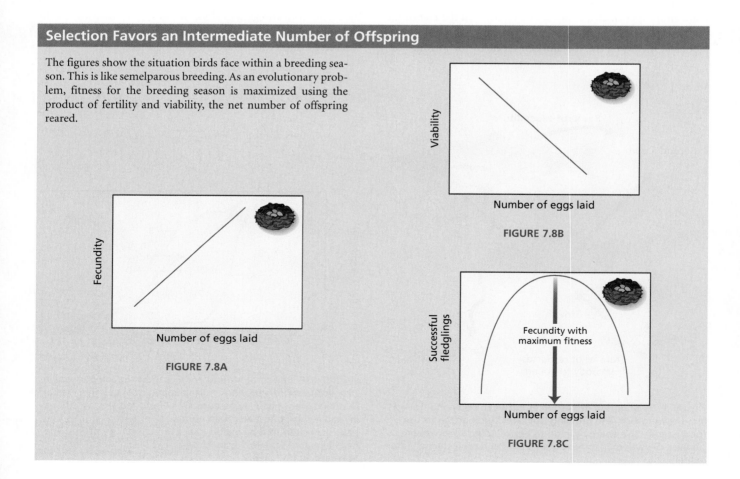

FIGURE 7.8A

FIGURE 7.8B

FIGURE 7.8C

Fecundity with maximum fitness

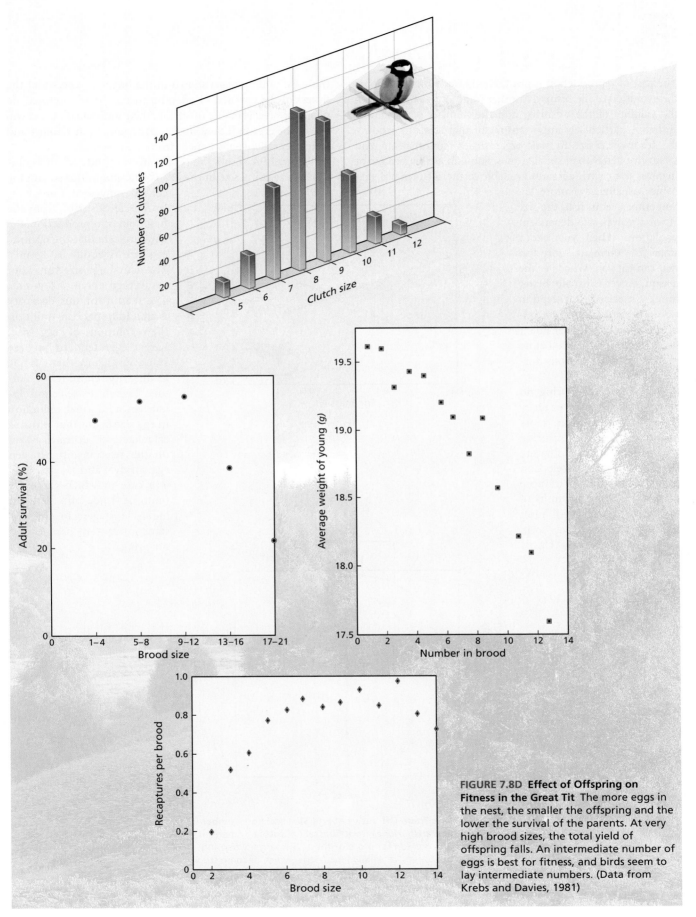

FIGURE 7.8D **Effect of Offspring on Fitness in the Great Tit** The more eggs in the nest, the smaller the offspring and the lower the survival of the parents. At very high brood sizes, the total yield of offspring falls. An intermediate number of eggs is best for fitness, and birds seem to lay intermediate numbers. (Data from Krebs and Davies, 1981)

7.9 Trade-offs between offspring size and offspring survival lead to the evolution of intermediate size

The cost of reproduction is not the only trade-off that shapes the evolution of life history. Another trade-off occurs between the viability of the offspring and the number of those offspring. A particularly important factor that tends to produce this trade-off is size. In most organisms, larger offspring tend to survive better than smaller ones. Sometimes this size factor involves the caloric reserves available to the offspring. Bigger babies usually have more fat. Sometimes, as in fish, the size factor involves size-dependent predation. When you are a small fish, there are more fish that can eat you. Whatever the reason, bigger is usually better in most species. (An exception would be larger babies in humans, which had higher infant death rates in the era before routine Cesarian sections during delivery.)

But if larger offspring do better, it is easier for virtually all parents to produce more offspring when the offspring are smaller than when they are larger. And natural selection favors the production of more offspring, within the limits of their cost in reduced adult survival, as we have seen. Therefore, we have a classic trade-off situation. As shown in the box, we expect that the fitness of the parent will often be greatest at an intermediate offspring size. (For the offspring, their individual fitness will usually be greater if they are larger, because both viability and fecundity will usually benefit from increased size.)

Figure 7.9A shows the results from a study of the evolution of offspring size in the lizard *Uta stansburiana* from the coastal ranges of California. In female offspring from the first bout of reproduction, the data indicate that the optimal egg size is about 0.5 grams. The actual egg size turns out to average about 0.4 grams. The reason for this disparity is an additional constraint on the evolution of egg size. Larger eggs tend to become stuck in the mother's body, leading to reproductive failure. Therefore, selection favors an additional reduction in egg size below that which is calculated as optimal based on the relationship between egg survival and its size. This is a case in which there are multiple trade-offs between life-cycle stages, in this instance between parent and offspring. ❖

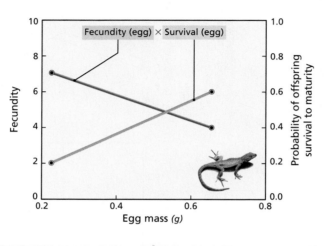

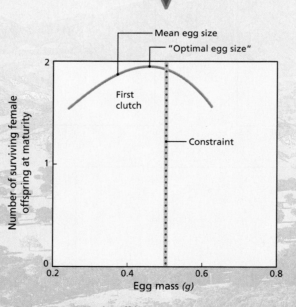

FIGURE 7.9A A Trade-Off between Egg Size and Egg Number in a Lizard (*Uta stansburiana*) from Coastal California Females that lay small eggs can lay more of them. But larger offspring survive better, except for really large eggs, which become stuck inside the mother.

Evolution of Offspring Size

The larger each offspring is, the fewer a typical parent can produce.

Offspring that are too small die off. But once the offspring are large enough, they get no additional benefit from being larger.

Calculating the effect on parental fitness of offspring size gives an intermediate peak.

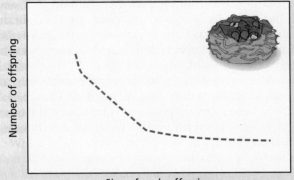

FIGURE 7.9B

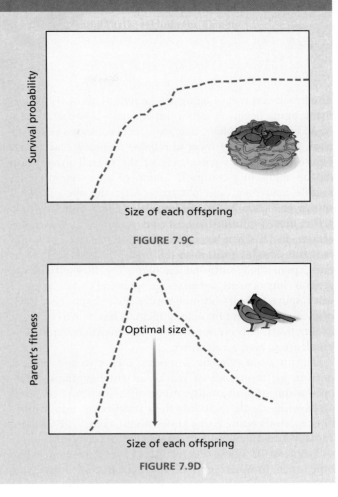

FIGURE 7.9C

FIGURE 7.9D

EVOLUTION OF AGING

7.10 Aging has been studied from very different perspectives, including evolutionary biology

The broad features of aging are known to all of us. In humans, **aging** begins as minor impairments between the ages of 30 and 50: gray hair, hair loss, reduced sexual performance, loss of muscle tone, unreliable memory, and so on. Figure 7.10A shows something of the overall progression during aging, in a composite photo of a boy, young man, middle-aged man, and old man. After the age of 50, as shown in Figure 7.10B, the impact of aging becomes altogether more profound. Almost all degenerative disorders increase in frequency. Cardiovascular disease becomes rampant. Strokes and cancer kill many. In total, death rates rise exponentially until the age of 90. Together with death, general impairment increases progressively. All types of work, sport, and play become harder, if not impossible. Each of the organs, the brain among them, is less able to perform its functions. This is a spectacle of pervasive deterioration—unrelenting, even accelerating.

The puzzle of aging has interested biologists and medical doctors for thousands of years. Ancient Egyptians developed a cult of immortality; much of their culture centered on cheating death, including their practice of mummification. Aristotle wrote a short book about aging. The Chinese **Taoists** built much of their religion around the explanation and control of aging (Figure 7.10C). In Renaissance Europe, such luminaries as **Sir Francis Bacon** worked on aging as a research problem. In modern times, the National Institute on Aging has given millions of dollars to cell biologists who propose to solve the problem using techniques of molecular biology.

Aging is both easy and difficult to study. It is easy to document the changes that occur in organisms during aging, with one exception: It is very difficult to study organisms that live a long time. Animals like elephants are often considered long-lived; and they are, if we compare them to most other animals. But the really long-lived organisms are trees, shrubs, and grasses. Individual trees, particularly the bristlecone pines mentioned earlier, can live thousands of years. Some biologists doubt that such organisms age at all. However, it is hard to have a very strong opinion, simply because we study too small a fraction of their lives. We just don't know. Meanwhile, it is probably more appropriate to confine scientific attention to aging in organisms that biologists can study as a practical matter.

In such organisms, the biology of aging is quite clear. As chronological age increases, well-studied plants and animals tend to suffer pervasive deterioration across the full range of functions and tissues. It is difficult to find organisms that improve biologically with adult age. In humans and rodents, in particular, we have amassed a large amount of data on aging, from the level of biochemistry to that of cognitive function. None of it is encouraging.

In researching many biological processes, it is common to study mutants or manipulated organisms in which the process is stopped. For example, we study vision in fruit flies by using mutants that are blind. The problem with the study of aging is that we cannot create organisms that stop aging completely. Instead, some scientists study aging in organisms that die sooner rather than later. This creates quite a few problems, because dying sooner can occur via many physiological processes that have nothing to do with normal aging. For a lot of organisms, accidents—such as being stepped on—end life sooner. In aging research, it is more appropriate to test for interventions in which life

FIGURE 7.10A **The Aging of the Human Male**

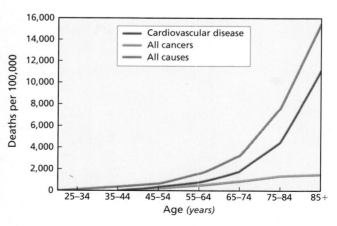

FIGURE 7.10B **Different Causes of Death as a Function of Age**

FIGURE 7.10C **Painting of a Taoist Rite for Achieving Physical Immortalitiy**

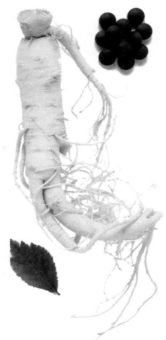

span is increased, because normal aging processes must somehow be alleviated in such longer-lived organisms.

Evolutionary biology has developed a program for the study of the biological aging of whole organisms. The foundation of this program is an evolutionary theory that explains aging based on a decline in the force of natural selection, which we explain in more detail shortly. Using this evolutionary approach, it has been possible to postpone aging in experimental organisms, such as fruit flies and mice. Progress has also been made on the genetics of aging, particularly in nematodes and fruit flies.

Between genetics and evolutionary biology, a lot of headway has been made in research on aging. The evolutionary research is outlined in the remainder of this chapter, in the following sequence. First, we describe aging as a life-history phenomenon. Then we introduce the basic theory underlying the evolutionary biology of aging. This theory leads immediately to a general and important prediction about the occurrence of aging in nature, especially the situations in which it should always occur and those in which it should never occur. We end this chapter by discussing the core of the research program, which consists of experiments in which evolutionary principles are used to substantially postpone aging. ❖

7.11 The survival and fertility of iteroparous plants and animals change with age

In many natural populations, death is often due to environmental factors such as disease, predation, and accidents. However, if a population is largely shielded from these factors, a process called senescence will gradually accelerate mortality and decrease fertility with age. **Senescence** is the gradual breaking down with age of the physiological machinery, and the subsequent increase in rates of mortality.

The effects of age on survival and fertility can be studied in **cohorts**, or groups of identically aged individuals. Typically, a fixed number of young individuals are followed until all have died. At regular intervals, researchers record the number alive and the number of offspring produced by members of the cohort. This life-table information can then be used to estimate age-specific mortality and fertility as in Table 7.4A.

Figure 7.11A, part (i), shows age-specific mortality in fruit flies. Initially there is a dramatic increase in mortality at young ages (5–20 days). However, at approximately 20 days of adult life, the mortality rates reach a plateau. These data were collected for a large cohort (~10,000 individuals of each sex). The mortality rates in the plateau apply only to 1 percent or less of the cohort of flies. This plateau has been seen in many other organisms, including humans. It is likely that the same evolutionary forces that lead to aging are also responsible for these plateaus.

The fertility of males and females also declines with age, as shown in part (ii) of Figure 7.11A. Some organisms do not reach peak fertility immediately after they reach sexual maturity, as with the herb, *Phlox drummondi* (Figures 7.11B and 7.11C).

It is possible for individual organisms to survive past the age when they last reproduce. Today this happens quite often with human females, for instance. Is there is an evolutionary advantage to living past reproductive age? For many species, the answer is no. However, for some species, mostly vertebrates, post-reproductive individuals may still take part in the care of offspring and thus contribute to their fitness.

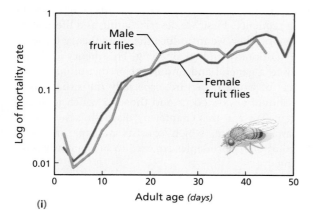

(i)

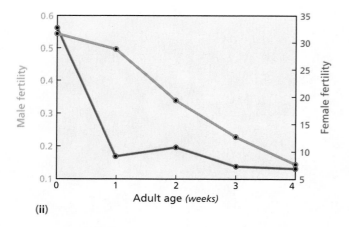

(ii)

FIGURE 7.11A In fruit flies, mortality increases in a linear fashion on a log scale, until late in life, where a plateau occurs. Female fecundity is the number of eggs laid. In males, fertility can measured by giving males multiple females and seeing how many are fertilized. Both fecundity and fertility decline with age in both sexes.

FIGURE 7.11B Fecundity in *Phlox drummondi* Plants exhibit the same general trends in survival and fertility as do animals.

Humans follow these same general patterns (Figure 7.11D). Between 1994 and 2000, females aged 25–29 years produced the most offspring per 1000 females. In the United States the delay in peak female fertility is affected by factors like schooling, starting careers, and other considerations that influence individual decisions to have children. Although these factors also influence to some extent the very low fertility of women aged 40–44 years, physiological infertility is also a significant factor at these ages. ❖

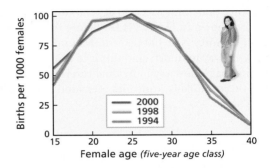

FIGURE 7.11D Age-Specific Fertility of U.S. Females

FIGURE 7.11C In this plant, *Phlox drummondi*, each flower produces three ovules; usually, every ovule becomes a seed.

The core idea of modern evolutionary theories of aging is the decline in the force of natural selection with age. This is not an assumption or a speculation. It is instead a result that has been derived mathematically from the first principle of an age-structured population, as defined at the start of the chapter. Therefore, we know that the **force of natural selection** acting on a fixed percentage change in age-specific survival has a declining effect after the start of reproduction in age-structured populations.

The force of natural selection depends on the age of an organism, as shown in Figure 7.12A. Before the start of reproduction, the force is at 100 percent. Natural selection is very strong. One way to understand this is to think about a dominant gene that kills every individual carrying it. These genes are known in humans, fruit flies, nematodes, and other or-

ganisms. But death must come at a particular age. Suppose that it comes before reproduction, before the age b. When it does, then that gene kills itself off. It will never be transmitted into the next generation. Natural selection has completely eliminated it, thanks to its 100 percent strength.

Now consider the period after the last reproduction, at age d. After d, a lethal gene has no effect on its own reproduction, because its carrier could have already produced offspring. Thus very bad genetic effects become neutral for natural selection, so long as they occur at late enough ages.

Between the start and end of reproduction in the population as a whole (ages b and d), the force of natural selection progressively falls because less and less reproduction occurs at ages after the lethal gene takes effect. With partially deleterious genes, the effect is not as dramatic, but it is qualitatively the same—initial strong natural selection, selection falling in strength during the period when reproduction occurs, and finally an absence of effective natural selection after the end of reproduction.

With natural selection as the ultimate source of most adaptation, this pattern of natural selection as a function of age is expected to produce health during youth and decrepitude at later ages—in a word, aging. ❖

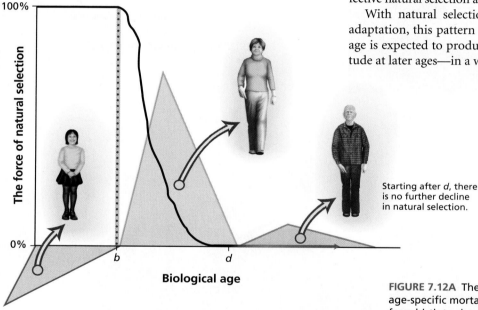

Starting after d, there is no further decline in natural selection.

b = age at which reproduction first occurs
d = last age at which reproduction occurs

FIGURE 7.12A The force of natural selection acting on age-specific mortality has three phases: (1) Childhood, from birth to shortly after maturation (b); (2) aging, the middle phase; and (3) late life, after the end of reproduction (d).

The Aging of Fertility

Fertility ages, too: All organisms from normal age-structured populations that have been followed carefully in the laboratory eventually become less fertile. It turns out that the force of natural selection acting on fertility also eventually declines with age. However, it does not necessarily decline from the start of adulthood. It can even increase into middle age. This may be part of the explanation for the greatly increasing fertility of some fish and trees during their adult period.

A key feature of the evolution of declining fertility in late life is that it is not a mere by-product of the same evolutionary or physiological processes that produce declining survival. There is a sepa-

rate evolutionary pattern that leads to the reduction in fertility. Anatomically and physiologically, this disparity helps explain the aging of fertility when the reproductive organs are relatively autonomous from the rest of the body. In many organisms, insects being one example, somatic cells no longer divide during adulthood; yet germline cells continue to replicate. This may explain the lack of synchrony between reproductive aging and the aging of the rest of the body. This lack of synchrony is obvious in the case of human menopause, but many females in a variety of mammalian species lose fertility long before the rest of their body dies of aging.

An important feature of the evolution of aging is the difference between populations that are expected to age and those that are expected not to age. The critical factor is the nature of reproduction as shown in Figure 7.13A.

Some organisms **reproduce vegetatively**. Part of the organism breaks off, and then both parts develop independent lives. This is common among trees, shrubs, and other plants. It is not as common among multicellular animals, except for the Cnidarians (corals, hydras, sea anemones). Some hydras are shown in Figure 7.13B. *Hydra* species often reproduce by budding a new hydra from their bodies. In tubular sea anemones, some species reproduce only by splitting in two, the two half "tubes" then closing in to form a functionally complete anemone, a process called **fission**. When the mortality rates of budding hydra and splitting anemone species are carefully monitored in the laboratory, they do not increase with age. These species are essentially free of aging. The reason is that these organisms never become adults. The single reproductive act creates two juveniles, so there is no adult to age. If aging were to occur in such organisms, it would lead to the deterioration of the offspring from fission. But since that is the only way reproduction occurs, all members of the population would progressively deteriorate. This universal deterioration would result in the extinction of the population. Therefore, with fissile reproduction, aging does not evolve.

The other extreme includes organisms that reproduce only by eggs or ovules (see Figure 7.13A). Most multicellular organisms reproduce this way—vertebrates, insects, mollusks, most flowering plants, and so on. In all these species, the requirements of the conventional evolutionary theory of aging are met. Aging is the firm expectation.

Some biologists don't think that fish age. However, the fish that they refer to are those that are hard to study, such as long-lived rockfish species. Fish that are studied for their entire lifetimes in the laboratory are always found to age. In general, there are no examples in which nonvegetative species have been shown to be free of aging. The non-aging vertebrate is a myth, even though an attractive one.

Besides the immortal and the mortal organisms, there are some species that can reproduce both vegetatively and sexually. These ambiguous organisms include coral and many plant species. For ambiguous organisms, there are no general predictions; they may or may not age. ❖

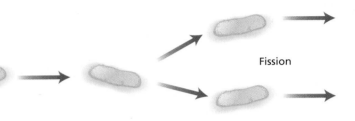

(i) Immortal or non-aging organisms

Fission

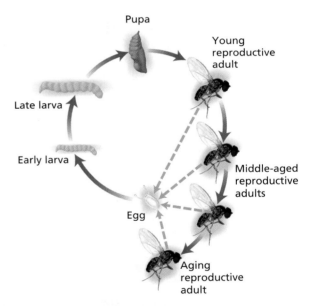

(ii) Mortal or aging organisms

Pupa

Late larva

Early larva

Egg

Young reproductive adult

Middle-aged reproductive adults

Aging reproductive adult

FIGURE 7.13A **Two Types of Organisms** (i) Immortal, or non-aging, organisms include, for example, bacteria, many protozoans, and some coelenterates, all with fissile or fragmenting reproduction. (ii) Mortal, or aging, organisms include plants and animals that reproduce only by means of eggs or seeds.

FIGURE 7.13B **Hydra, a Small Aquatic Invertebrate Animal**

Changing the force of natural selection can produce rapid evolution of aging patterns

The idea of the evolution of aging leaves the impression that aging can be changed only with difficulty, over long periods of evolutionary time. This impression is erroneous. Evolution can change aging patterns very quickly.

This rapid evolution has been shown with fruit flies. The experimental trick involved is to change the first age of reproduction. As we noted earlier, the force of natural selection on survival starts to decline once the population begins to reproduce. In fruit flies, it is easy to change the age of first reproduction experimentally simply by discarding all eggs laid before the preferred age of reproduction. Normally adult fruit flies are kept until the age of 14 days, made to lay eggs, and then discarded. But the adult flies can be kept for a longer period as shown in Figure 7.14A. This regimen changes the force of natural selection by increasing it at later ages. If aging is controlled by natural selection, then this change should produce large changes in mortality patterns, if evolution is given several generations over which to act. As shown in Figure 7.14B, this is in fact what happens: Delayed first reproduction leads to increased survival at later ages within 10–20 generations.

The flies that live longer in these experiments have evolved extensive physiological changes over the course of more than 100 generations of additional selection. They resist stress better. They can fly longer than other flies. They are able to have sex at later ages. The females are more fecund when older. These are not organisms that simply endure decay for a longer period; they are more vigorous at later and middle ages, long after most normal flies are dead. Physiologically, these flies do not prolong aging, they prolong active life.

These experiments have shown that aging is eminently controllable using evolutionary tools. They do not, however, suggest that humans should be selected for delayed aging. Progress with human aging cannot wait the hundreds of years that human breeding would take, if we were to follow the evolutionary pattern of the late-breeding fruit flies. ❖

Selection regime B

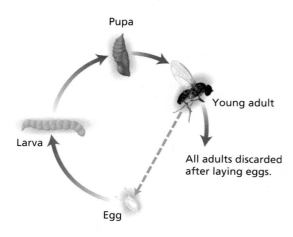

Selection regime O

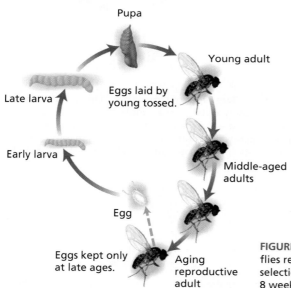

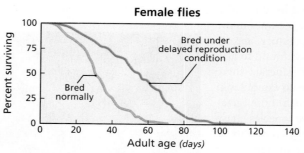

FIGURE 7.14B The evolution of postponed or slowed aging in fruit flies arises from delayed reproduction.

FIGURE 7.14A Lab Evolution of Aging in Fruit Flies (i) In the B regime, lab fruit flies reproduce in the first few days of adult life. (ii) In the O regime, natural selection is prolonged by delaying female reproduction until adults are at least 8 weeks old.

SUMMARY

1. Life history is the central clearing ground of evolution, ecology, and organismal biology. The life history determines fitness, the central parameter of evolution. The life history sums up all the ecological forces that shape an organism's way of life in its environment. The life history also makes manifest the organism's multiple adaptations, the articulation of anatomy and physiology that defines the potentialities and limits of that organism.

2. A basic problem of life history is the arithmetic required to proceed from life history to fitness. In semelparous organisms this arithmetic is relatively simple, when there are no complications due to mating. Under such conditions, the fitness of a semelparous organism is the product of viability and fertility. But many plants and animals reproduce multiple times over their adult life span. These species have age-structured populations. Age-structured populations require different models. Specifically, we need to know the age-specific fertilities and the age-specific survival of individuals. Age-structured populations grow exponentially at a constant rate after they have reached their stable age-distribution, in theory. In these iteroparous organisms, the arithmetic of fitness is far more complex. However, in populations that have stable age-structure, fitness is given by the growth rate of the population.

3. Trade-offs arise when one part of the life history has a functional antagonism with another part of the life history. The best-known trade-off is that between reproduction and adult survival. Reproducing more tends to result in earlier death among the adults of many species, whether that increased reproduction occurs because of genetic, anatomical, physiological, or behavioral manipulation. This cost of reproduction sometimes leads to the evolution of restrained reproduction. There are other life-history trade-offs, including those between fecundity, offspring size, offspring viability, and growth.

4. Aging is not universal. It occurs in nonfissile organisms because the force of natural selection on survival falls with adult age. In fissile organisms, like many Cnidarian and plant species, the force of natural selection does not fall, and aging does not evolve. Patterns of aging are readily shaped evolutionarily by manipulating the force of natural selection, which is itself easy to manipulate by changing the age of first reproduction.

REVIEW QUESTIONS

1. List some life-history characters, and define each one.
2. In semelparous organisms, which two variables define fitness?
3. What kind of semelparous animals are univoltine?
4. If an animal or plant reproduces by fission, can it be iteroparous?
5. List some life-history trade-offs or costs, and define each one.
6. How is it possible to make inferences about the cost of reproduction from manipulating the number of eggs in bird nests?
7. What is the effect of delaying the onset of a population's reproduction for many generations?
8. What do you predict will be the effect on early fertility of delaying the start of reproduction for many generations?

KEY TERMS

age-class	Cole's paradox	iteroparous	survival
age-specific fertility	cost of reproduction	life cycle	Taoist
age-specific probabilities of survival	fecundity	life history	trade-off
	fertility	life table	univoltine
age structure	fission	Malthusian parameter	vegetative reproduction
aging	fitness	semelparous	viability
Bacon, Sir Francis	force of natural selection	senescence	
cohort	intrinsic rate of increase	stable age-distribution	

FURTHER READINGS

Charlesworth, Brian. 1980. *Evolution in Age-Structured Populations.* London: Cambridge University Press.

Cole, Lamont C. 1954. "The Population Consequences of Life History Phenomena." *Quarterly Review of Biology* 29:103–37.

Finch, Caleb E. 1990. *Longevity, Senescence, and the Genome.* Chicago: University of Chicago Press.

Krebs, John R., and Nick B. Davies. 1981. *An Introduction to Behavioral Ecology.* Sunderland, NJ: Sinauer Associates.

Maynard Smith, John. 1958. "The Effect of Temperature and of Egg-laying on the Longevity of *Drosophila subobscura.*" *Journal of Experimental Biology* 35:832–42.

Robertson, O. H. 1961. "Prolongation of the Life of Kokanee Salmon (*Oncorhynchus nerka kennerlyi*) by Castration before Beginning of Gonad Development." *Proceedings of the National Academy of Sciences USA* 47:609–21.

Roff, Derek A. 1992. *The Evolution of Life Histories: Theory and Analysis.* New York: Chapman and Hall.

Rose, Michael R. 1991. *Evolutionary Biology of Aging.* New York: Oxford University Press.

Sinervo, Barry, and Alexandra Basolo. 1996. "Testing Adaptation Using Phenotypic Manipulation." In *Adaptation*, edited by M. R. Rose and G. V. Lauder, 149–85. San Diego: Academic Press.

Stearns, Stephen C. 1992. *The Evolution of Life-Histories.* New York: Oxford University Press.

Plants and animals can exist in physically challenging environments due to a variety of physiological adaptations.

Physical Ecology of the Organism

The environment has two distinct components. The biotic component consists of the living organisms. Much of ecology is devoted to the study of the interactions of coexisting plants and animals. Topics like energy flow, competition, predation, and parasitism are all examples of these types of biological interactions. The abundance and distribution of species must be affected in some way by the numbers and kinds of parasites, predators, and competitors in the local environment. However, these biological interactions cannot tell us the whole story.

Physical aspects of the environment are also important in determining the abundance and distribution of species. Many limitations of the physical environment are obvious; for example, photosynthetic plants need light in order to live. Some places on Earth are simply too cold or too hot for many species to live there. We can learn about adaptation by studying organisms that live in extreme conditions. Even in the harshest environments, we find organisms that have evolved unusual adaptations that permit them to live there. For instance, how is it possible for fish to live in oceans so cold that the temperature is actually below freezing?

The size and shape of an organism also determines the kinds of interactions it will have with its environment. For instance, there is a simple relationship between the surface area of an organism and its size. Because it is in direct contact with the physical environment, usually either air or water, the surface area of an organism is an important location for the exchange of heat and gases. We will see that there are important consequences to organism function that follow as a direct consequence of the surface area to volume relationship. ❖

TEMPERATURE AND LIGHT

8.1 Animals regulate their temperature in various ways

Some of the most conspicuous adaptations of plants and animals are those that help overcome problems created by the physical environment. Animals have a variety of ways to reduce heat loss in cold environments, for example, fur, feathers, and fat layers. Plants in very warm, dry climates have thick, waxy leaves to help prevent water loss. In this module we will review some of the problems that the physical environment—particularly temperature and light—poses to organisms. We will also look at how various organisms deal with these problems.

Temperature   When we consider the possibility of life on other planets, we can immediately eliminate most of them, because we believe they are either too hot or too cold to support life as we know it. Even on Earth there are places where temperatures are too extreme for life. In those places where life can exist, the rate of an organism's chemical reactions—its metabolism—will be affected by its body temperature.

Animals are often classified according to how they maintain their body temperature, but the relevant terminology is often used imprecisely. For example, the terms **cold-blooded** and **warm-blooded** are used to describe animals. These terms refer to how the animal feels when touched. Birds and mammals are considered warm-blooded, and all other animals are called cold-blooded. However, a lizard that has been basking in the sun can feel quite warm to the touch.

Other commonly used terms are *poikilotherm* and *homeotherm*. These refer to whether an animal's body temperature tends to vary (**poikilotherm**) or stay constant (**homeotherm**). However, these terms also tend to be confused, since there are mammals that hibernate, causing their body temperatures to vary significantly. Deep-sea fish show little variation in their body temperatures, because their environmental temperature changes little.

A better set of terms refers to how animals generate body heat. **Ectotherms** require external sources of heat energy, while **endotherms** generate heat internally from their own metabolism. These various definitions, and the animal groups they apply to, are summarized in Figure 8.1A.

If a plant or animal cannot regulate its body temperature, its metabolism will slow down as the temperature cools down. Animals that cannot regulate their body temperature will become less active in cold conditions. Some animals, the endotherms, attempt to avoid these problems by using the chemical energy of metabolism to keep their body temperature constant. As a consequence, endotherms can live and remain active in a far greater range of environments than ectotherms can.

This does not mean that ectotherms are completely at the mercy of the environment. In later modules we will consider the effect of temperature on organismal function in more detail, and we will review some of the short-term responses that ectotherms can make to help them adjust to new environmental conditions.

Light   An important issue in contemporary global ecology is the level of ozone in our atmosphere. This level is important because ozone filters out harmful ultraviolet light (UV) energy. However, UV light is not the only part of the light spectrum that is subject to filtration. Both the gaseous atmosphere and water absorb significant portions of the light energy that arrives from the sun. This filtering has a profound impact on the energy available to plants for photosynthesis. Both the type and amounts of light are affected by atmospheric filtering.   ❖

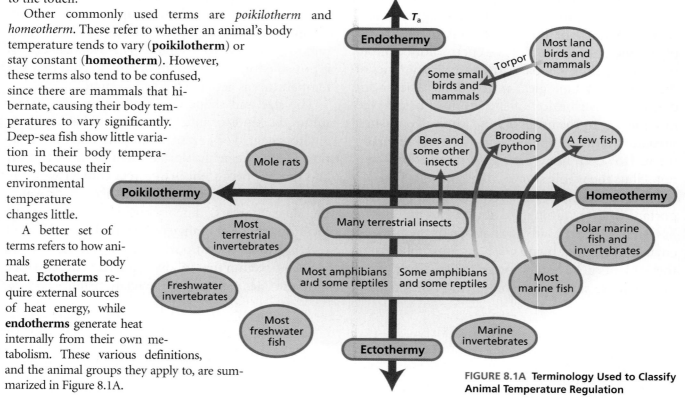

FIGURE 8.1A Terminology Used to Classify Animal Temperature Regulation

Temperatures on Earth vary over a tremendous range. In Antarctica, low temperatures have been recorded down to −89°C; in other environments, high temperatures range from 80°C in deserts to 100°C in hot springs. In the deep sea, hydrothermal vent temperatures may reach 350°C.

The internal temperatures at which plants and animals can function are much more limited. While some unicellular organisms can live at 90°C, few multicellular organisms can live at temperatures much in excess of 50°C. No organism can function once it is frozen, so the lowest body temperature of functioning organisms is determined by the freezing point of seawater, which is −1.86°C. Some animals can live in environments where the temperature dips to −60°C, but these are birds and mammals that maintain a high internal body temperature by using metabolic energy. Some organisms, like nematodes, tardigrades, and certain insect eggs, can be revived after freezing in liquid helium (−269°C). As Figure 8.2A shows, the bodies of most endotherms operate within a fairly narrow range of temperatures. This range is somewhat higher for ectotherms, but it is still narrow compared to the range of environmental temperatures.

Animals and plants lose and gain heat energy in different ways. The flow of energy shown in Figure 8.2B is typical of animals, but many of these energy flows also occur in plants.

Conduction is the flow of heat energy that occurs when two bodies at different temperatures come into direct contact. Heat energy flows from the warmer to the colder object. On the outside surfaces of animals, fur and feathers provide insulating surfaces that reduce heat loss through conduction. Some ectotherms use conduction to warm their bodies. For instance, lizards lie on rock surfaces that have been heated by the sun in order to warm their bodies.

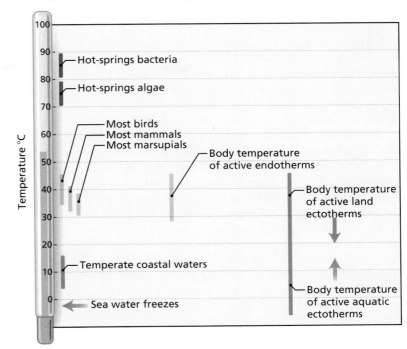

FIGURE 8.2A **The Temperature Range Tolerated by Different Plants and Animals**

Heat may be transferred between two bodies at different temperatures via a third moving or fluid layer of either gas or liquid. This process is called **convection**. Many animals have a small layer of air, surrounding their bodies, that is warmed by the animal. Convection mediates the transfer of heat from the organism through the boundary layer and into the surrounding environment. Heat loss by convection is typically much quicker than by conduction.

Objects that have heat energy also emit **radiation**. The type and amount of radiation depend on the temperature of the object and other properties that are related to its color and reflective surface. Living organisms emit energy in the infrared wavelength range. The amount of energy lost is about 300–500 watts per square meter. A resting mammal with one square meter of radiating surface may only produce 20 watts from their own metabolism. However, most of the energy lost from radiation is gained through the absorption of radiant energy from the surrounding environment.

As water changes state from liquid to gas, it absorbs considerable energy. Such **evaporation** is an important means of dissipating heat on the surface of an organism. The rate of evaporation will be a function of the surface temperature of the organism, the relative humidity of the air, and convective processes. Some larger land animals use evaporation as a means of cooling. However, since evaporation entails water loss, it is usually not an efficient method for cooling very small organisms or for cooling in very dry environments. ❖

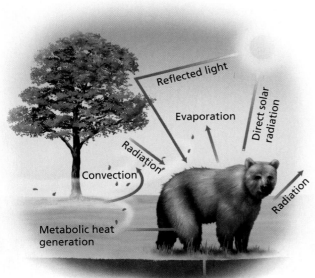

FIGURE 8.2B **Thermal Exchanges between an Animal and the Environment**

8.3 The temperature coefficient, Q_{10}, expresses the effect of temperature on organismal function

The kinetic energy of molecules increases with temperature, which also increases the rate of most chemical reactions. There are some limits to this effect in living things. For instance, temperatures may get so high that the folding of protein enzymes is disrupted, and the reactions dependent on these enzymes then cease altogether.

Over normal metabolic temperatures, there is a simple relationship between temperature and chemical reaction rates. The rate at which chemical reactions accelerate due to temperature is called the **temperature coefficient**, or $\boldsymbol{Q_{10}}$. As shown in Figure 8.3A, the log of the reaction rates at elevated temperatures increases with temperature in a linear fashion.

Q_{10} is used to express the effect of temperature on organismal function. The temperature coefficients for many biological processes are in the range of 2–3. A Q_{10} of 2 means that the rate of the metabolic process doubles for each increase of 10°C. Thus an increase of 20°C will quadruple the rate of the reaction with a Q_{10} of 2.

For organisms that cannot regulate temperature, like plants and ectothermic animals, most of their metabolic processes will increase as they warm up. If we take a typical insect—like the Colorado potato beetle—that has been raised at

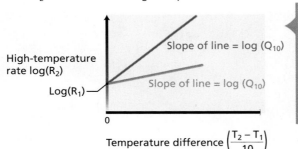

T$_1$: Low temperature
T$_2$: High temperature
R$_1$: Reaction rate at low temperature
R$_2$: Reaction rate at high temperature

High-temperature rate log(R$_2$)

Slope of line = log (Q$_{10}$)

Log(R$_1$)

Slope of line = log (Q$_{10}$)

0

Temperature difference $\left(\dfrac{T_2 - T_1}{10}\right)$

- When the temperature difference gets larger, the difference between the initial and final reaction rates gets larger.
- The reaction described by the red line is more sensitive to temperature than the blue reaction, (Q$_{10}$ > Q$_{10}$).
- When the final and initial temperatures are the same, the final and initial rates are the same.

FIGURE 8.3A Increasing Temperature Accelerates Chemical Reactions

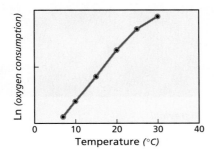

FIGURE 8.3B Oxygen Consumption of the Colorado Potato Beetle

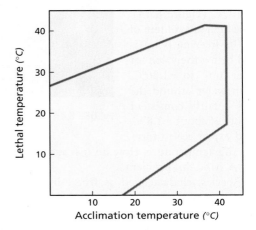

FIGURE 8.3C Acclimation and Lethal Temperatures for Goldfish

8°C, and increase the ambient temperature, the insect's oxygen consumption predictably increases (Figure 8.3B). The elevated oxygen consumption reflects the beetle's increased need for energy to fuel the accelerated metabolism rates brought on by high temperatures.

Often the metabolic performance of an organism is less efficient when its body temperature has just been significantly increased or decreased. Most organisms go through a process called **acclimation** if left in these new conditions. Acclimation may take from days to weeks and involve physiological and biochemical changes that allow the organisms to perform better in the new environment. These changes may include the increased or decreased production of different enzymes, or changes in the molecular components of cell membranes. Sometimes this acclimation process is called *adaptation*; however, this term can be confused with the evolutionary concept of adaptation by natural selection. We will avoid using the word *adaptation* to describe individual physiological responses to the environment.

We can follow the process of acclimation to different temperature regimes in fish by determining the low and high temperatures at which adults can survive. In Figure 8.3C we see that adult goldfish increase both their lower and upper tolerance zones as they become acclimated to higher temperatures. ❖

Life at extreme temperatures reveals how organisms adapt to environmental stress

Very high and very low temperatures present different types of problems for living organisms. At high temperatures, proteins that are essential for life may become unfolded, or *denatured*, and so are unable to perform their structural or enzymatic functions. At very low temperatures, water within the cells begins to freeze. The solid ice crystals that then form may cause irreversible damage to the cells, killing the organism. Yet some fish live in the polar oceans at temperatures of about −1.8°C (Figure 8.4A). Due to the high salt content of seawater, its freezing point is depressed below the value for pure water (0°C) to −1.86°C. The ocean water around the poles has a fairly consistent temperature of about −1.8°C. We would expect ice to form

FIGURE 8.4A The Antarctic fish *Trematomus borchgrevinki*

in fish at this temperature. How do fish such as the one in Figure 8.4A avoid this problem?

Some species, characterized as **freeze-intolerant**, die if frozen but can live in very cold environments by preventing ice formation. Living organisms accomplish this in the same way people prevent water in their car radiators from freezing: with antifreeze. Adding antifreeze solutes to water depresses the freezing point. Certain compounds do this quite effectively. One compound used by a number of insects is glycerol. For instance, the winter gallfly survives winters in Alaska by increasing glycerol concentration to almost 50 percent. The overwintering immature gallflies can survive temperatures down to −60°C.

Fish also use antifreeze, but a different type. Antarctic fish have compounds called **glycoproteins**, composed of protein and sugar molecules, that effectively inhibit ice crystal formation, as shown in Figure 8.4B. Antifreezes have been found in fish from 11 different families, and the chemical structures of these antifreeze molecules are quite different from each other. This information suggests that the use of antifreeze compounds has evolved independently several times.

Other groups, labeled **freeze-tolerant**, can survive periods of freezing. These organisms typically employ two methods of coping with freezing. The first is to maintain high levels of glycerol in their cells; this reduces the temperature required for freezing. Glycerol also can reduce cell damage once freezing has occurred. Secondly, when ice does form in a freeze-tolerant organism, it often occurs in the intercellular spaces, thus reducing damage to cell contents. Ice formation requires a seed or nucleating agent. The nucleating agent can be an ice crystal, or it may be some other compound. Many freeze-tolerant species have high-molecular-weight proteins in their intercellular spaces that act as nucleating agents and thus encourage ice formation outside of the cell.

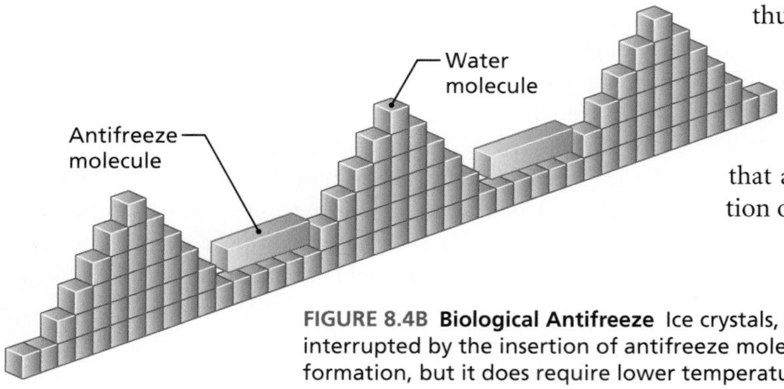

Water molecule

Antifreeze molecule

FIGURE 8.4B Biological Antifreeze Ice crystals, represented as cubes, have their growth interrupted by the insertion of antifreeze molecules. The antifreeze does not prevent ice formation, but it does require lower temperatures for ice to form.

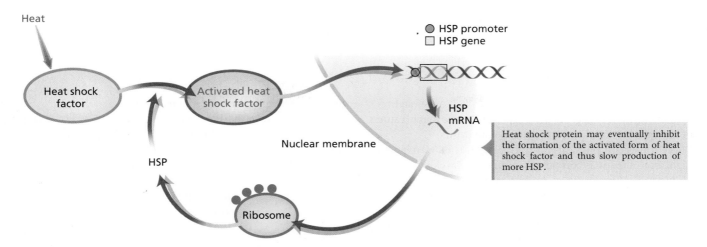

FIGURE 8.4C **The Regulation of Heat Shock Proteins**

High temperatures are also stressful and potentially deadly. Many organisms show a similar response to a sudden increase in temperature, called the **heat shock response**. The heat shock response is rapid, happening in a matter of minutes to hours. It involves the production of several types of proteins called **heat shock proteins** or **HSPs**. The HSPs act as molecular chaperones for proteins that have become unfolded. A molecular chaperone stabilizes protein structure and aids the protein in resuming its correct three-dimensional structure so that it can function properly.

How does the heat shock response work? The sudden increase in temperature causes a monomeric molecule called the *heat shock factor* to become a trimer. In this activated form, the heat shock factor interacts with the cell's nuclear DNA to initiate the transcription of HSP RNA, which is then translated into functional HSP (Figure 8.4C).

Several other stress conditions can trigger the heat shock response. It is a general physiological mechanism for coping with environmental stress. ❖

8.5 The physical properties of light striking the Earth constitute a key environmental factor mediating the physiology, distribution, and abundance of organisms

Almost all biological systems ultimately depend on sunlight for their energy. The properties of light thus influence many features of biological systems, such as where life can exist.

An important physical property of light is its wavelengths. Light from the sun is composed of multiple wavelengths, some visible to the human eye and others not (Figure 8.5A). Energy in the ultraviolet range (not visible to the human eye) is further divided into ultraviolet A (315 nm to 380 nm) and ultraviolet B (280 nm to 315 nm). This high-energy light can damage biological tissues, but fortunately much of it is absorbed by the ozone in our atmosphere (Figure 8.5B). The other wavelengths of light are also reduced in intensity as they pass through the Earth's atmosphere, but the blue and ultraviolet are the most attenuated. Much of the blue light is scat-

Light from the sun is composed of multiple wavelengths, some visible to the human eye and others not.

tered (reflected) in the atmosphere, giving rise to the blue color of the sky. Most of the sun's energy reaches the Earth's surface in the range of visible light. It is these wavelengths that plants use for photosynthesis.

Plant chlorophyll absorbs light in the blue and red wavelengths and reflects green light, giving leaves their usual green color. Light is attenuated very rapidly in water (Figure 8.5C). As a consequence, most water plants are found close to the surface of the water column. The light-limited distribution of plants is determined by the **compensation point:** the intensity of light at which the production of energy by photosynthesis just equals its consumption by respiration. At light intensities below the compensation point, a plant suffers a net loss in energy.

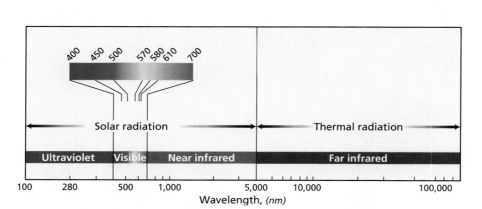

FIGURE 8.5A Properties of Light That Reaches the Earth's Surface Visible light is a small portion of the electromagnetic spectrum.

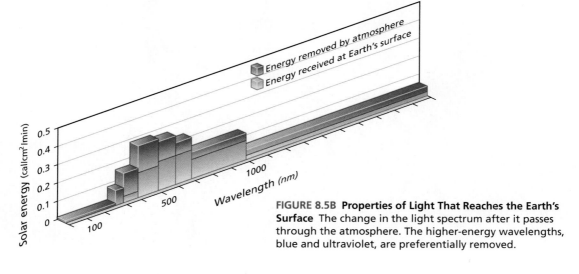

FIGURE 8.5B Properties of Light That Reaches the Earth's Surface The change in the light spectrum after it passes through the atmosphere. The higher-energy wavelengths, blue and ultraviolet, are preferentially removed.

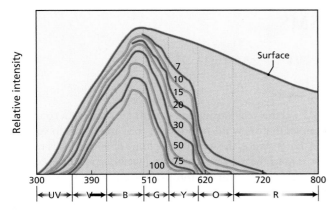

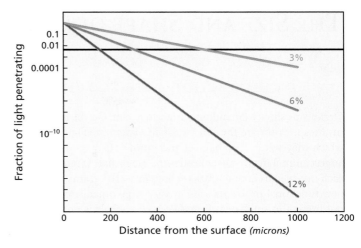

FIGURE 8.5C **Properties of Light That Reaches the Earth's Surface** The absorption of light energy in pure water. Total energy is reduced very quickly. The numbered lines show the intensity of light at different depths, measured in meters, below the water surface. By 100 meters, very little visible light penetrates the water column. The longer, red wavelengths are more rapidly removed.

FIGURE 8.5D **Light Penetration in a Leaf**

Even in bright sunlight, energy is absorbed as it penetrates the surface of a leaf. How deeply can chloroplasts be located in a leaf and still receive enough light to stay above the compensation point? The attenuation of light through a transparent medium can be described by an equation known as Bouguer's Law. This law predicts an exponential decrease in light intensity with distance from the surface of the leaf. Figure 8.5D shows this relationship for three different concentrations of chloroplasts: 3, 6, and 12 percent of leaf volume. The horizon-

tal black line represents the compensation point, assuming bright sun. Even with this favorable assumption, the chloroplasts should not be deeper than a few hundred microns (1 micron = 0.001 mm). Only those chloroplasts at distances above the black line get enough light. Because of this limitation, plants do not have thick round leaves, since light cannot penetrate the interior of such an organ. Leaves must be flat, with much surface area, because their chloroplasts must be concentrated on the surface layer of the leaf. ❖

THE SIZE AND SHAPE OF ORGANISMS

8.6 The surface area to volume ratio of an organism affects its interaction with the environment

Organisms show tremendous variation in size. On the small end of the spectrum are the unicellular *Mycoplasma* (Figure 8.6A), which may weigh less than 0.1 picogram (10^{-13} grams). The largest animal that has ever lived is the blue whale (Figure 8.6A); each one weighs about 100,000 kilograms (10^8 grams). But do new functional problems arise in very large organisms that are not present in small organisms? Perhaps a whale is no different than a *Mycoplasma* in the way it moves, acquires energy or dissipates heat, except it happens on a much larger scale.

Surface Area to Volume Ratio

In fact, new functional problems arise when organisms get larger. Some of the most important problems arise from changes in the surface area to volume ratio of an organism. We can roughly quantify these problems by considering a sphere. From high school geometry, recall that the surface area of a sphere is equal to $4\pi r^2$, where r is the radius of the sphere. The volume of the same sphere is equal to $(4/3)\pi r^3$. The ratio of surface area to volume is therefore equal to $3/r$. The biological meaning of this relationship is that as a sphere gets larger, it has less surface area per unit of volume. Larger organisms will have relatively smaller surface areas.

Since the volume of an organism is related to its overall size or weight, we can plot the surface area to volume ratio as a function of volume or size (Figure 8.6B). The relative surface area decreases for larger spherical objects. Although most plants and animals are not spheres, the same general relationship holds.

Why is this relationship important? Let's look at a couple of examples that show the relevance of the surface area to volume ratio. For instance, the weight or volume of an animal is related to how much heat it generates from metabolism. However, heat is lost to the environment through the animal's surface. Thus, the loss of metabolic heat energy by radiation depends on the surface area of the animal, not its volume. In another instance, the amount of water in a plant or animal is proportional to its size or volume. However, water is usually lost through the exposed surface of the plant or animal, so surface area is also the most relevant parameter in determining water stress.

The consequences of these relationships are often dramatic. Desert plants are under severe water stress; as a result, their photosynthetic surfaces are round and large to reduce the total surface area (Figure 8.6C). Plants that live in humid tropical areas, on the other hand, may have broad, very flat leaves because water loss is less of a concern and maximizing exposed area to the sun is more important.

The Importance of Scale Effects

When we look at morphological or life-history characters in many species, we often see a relationship between the quantitative value of the character and the organism's size. Later in this chapter we consider the problem of support in the context of an organism's size. We will also examine methods for studying the relationship between size and functional characters. Such "scale effects" are some of the most important of all quantitative patterns in biology.

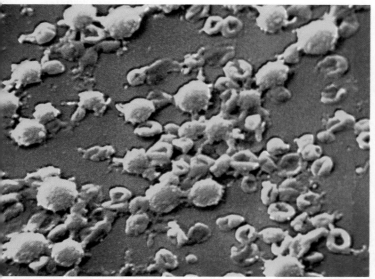

Mycoplasma agassizii

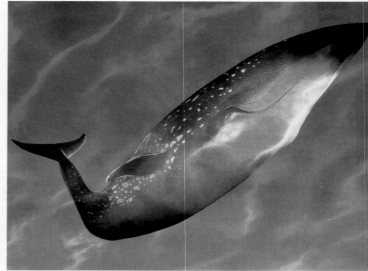

Blue whale

FIGURE 8.6A Organisms Vary Greatly in Size

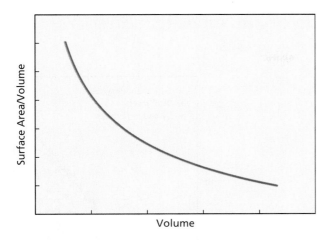

FIGURE 8.6B **Relative Surface Area Decreases for Larger Organisms**

In evolutionary biology, scale effects must frequently be taken into account. For instance, suppose two populations of an organism have evolved in different thermal environments, one hot and the other cold. One hypothesis might be that the metabolic rate of the organism in the hot environment is higher due to genetic changes following evolution. A controlled experiment would raise individuals from both populations to a common temperature and measure their metabolic rates. Before comparing these metabolic rates, the experimenters must also determine that the individuals from each population are the same size, since larger individuals will have higher metabolic rates when all other factors are equal. This is an example of a scaling effect. While it might be interesting that there are size differences between individuals from each population, that is certainly a different effect than is the case where two individuals of the same size have different metabolic rates. ❖

FIGURE 8.6C These plants differ in the amount of exposed surface area. These differences reflect the varying levels of water stress in the tropics and deserts.

8.7 Changes in size have a major effect on organismal structure and function

As organisms change in size, they are likely to encounter a variety of new problems. Some of these problems have been the subject of previous modules in this chapter. For instance, for the very smallest organisms, which are 2 mm or less in diameter, diffusion can supply the cell with all its oxygen needs. 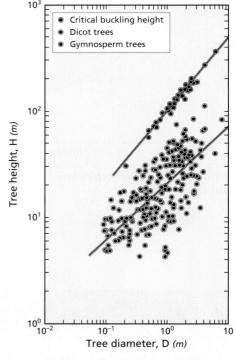 However, larger organisms need some type of specialized delivery system for oxygen. Animals that rely on blood for gas transport have circulatory systems to deliver oxygen and take up carbon dioxide to body tissues. Animals with open circulatory systems have a limited ability to control the velocity and distribution of blood. Closed circulation found in all vertebrates and some invertebrates has blood flowing through a continuous circuit of tubes.

Plants and animals also encounter increasing problems with support as they get bigger. The greater mass of large organisms means greater gravitational forces. Plants eventually bend and break under their weight unless their trunks are thick

FIGURE 8.7A The Diameter of Tree Trunks Determines Buckling Height A tree buckles under its own weight if the trunk is too narrow.

enough to withstand gravity. From the physical properties of wood, the thickness of tree trunks required to prevent collapse can be calculated (Figures 8.7A and 8.7B). The required thickness of a tree's trunk grows exponentially as the size of the tree increases (notice the log scale in Figure 8.7A). Actual trees are not as large as the theoretical curve would permit. This finding is reasonable, since real trees have to withstand other forces, such as wind from storms, in addition to gravity.

Animals face similar problems. Skeletons serve as the support systems of vertebrates. The width of critical bones increases much faster than does the size of the animals in which they are found (Figures 8.7C and 8.7D). Do the requirements of bone size place natural limits on the size of land animals? They might, but before that limit is reached other factors come into play, including the dissipation of heat. Excess heat generated by metabolism must be dissipated from the skin surface. While there are several factors that affect the dissipation of heat in large animals, their small surface area to volume ratio is an important one. It is no coincidence that the largest known animals—whales—live in the ocean. Life in the ocean means that the surface of the whale is in contact with water, which has a much higher heat transfer capacity than air and so can more effectively remove heat. ❖

FIGURE 8.7B Redwoods are one of the largest terrestrial plants; they must have a correspondingly wide base to support the entire structure.

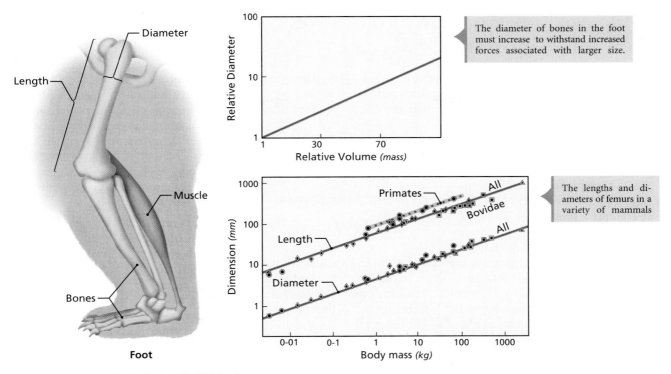

The diameter of bones in the foot must increase to withstand increased forces associated with larger size.

The lengths and diameters of femurs in a variety of mammals

FIGURE 8.7C Larger Animals Require Thicker Bones.

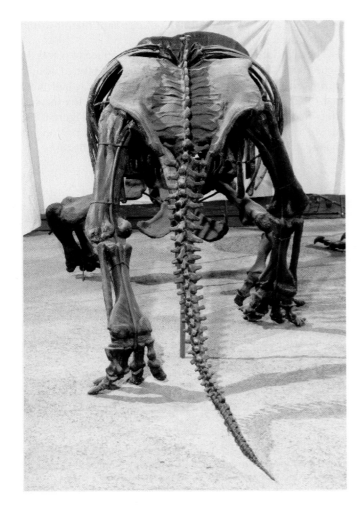

FIGURE 8.7D The triceratops, which is thought to have weighed 6–8 tons, was supported by immense legs.

8.8 Allometric methods are used to quantify changes in form and function associated with size

One way to study and understand a physiological process is to compare that process in many different organisms and see how it changes. However, organisms differ in size; and that alone might lead to differences in the physiological phenomenon under study. How could we determine if the change in a physiological character was simply due to a size difference, or whether some other factors were important?

If we consider a simple cube with edge length L, it is easy to compute its surface area ($6L^2$), and its volume (L^3). Even when the volume is not a cube, the surface area is proportional to L^2 and the volume (V) to L^3. These relationships imply that an organism's length is proportional to $V^{1/3}$ and the surface area is proportional to $V^{2/3}$ (Figure 8.8A). In practice, if there appears to be a relationship between a character and weight raised to the two-thirds power, it is reasonable to suppose that the character is affected by the surface area of the organism.

The study of the relationships between physiological characters and size is called **allometry**. These types of relationships can be written mathematically as:

$$character = aV^b$$

where a and b are constants, and V may be the volume but more often is the weight of an organism.

If we take logarithms of this equation, it looks like this:

$$log(character) = log(a) + (b)\,log(V)$$

Therefore, the slope of this log-linear equation is b.

It would take less energy to support 100 kg of elephant than it would 100 kg of mice.

The metabolic rates of some mammals and birds are plotted against their weights on a log scale in Figure 8.8B. The slope of this line is 0.75. Thus, the metabolic rate of animals increases faster than expected if it were determined by surface area ($b = 0.67$), but slower than expected if it were directly related to weight ($b = 1$).

These allometric relationships are important when comparing animals of different sizes. For instance, if you wanted to compare the metabolic rates of seed-eating birds to the metabolic rates of insect-eating birds, you would first need to account for the sizes of the birds. For example, if the insect-eating birds happened to be larger on average than the seed-eating birds, we would expect the total organismal metabolic rates to be greater due to the size differences alone.

The results in Figure 8.8B can also be used to address whether metabolic rate per kilogram, or **specific metabolic rate**, of an animal is different for small and large animals. When analyzed this way, the specific metabolic rates decrease with size of the organism. In other words, it would take less energy to support 100 kg of elephant than it would 100 kg of mice.

The principles of allometry can also be applied to plants. In Figure 8.8C the total mass of seed per fruit is plotted against fruit biomass. The slope of this line is just about 1, indicating that the mass of seeds changes in proportion to the mass of fruits that contain them. In a similar fashion, in Figure 8.8D, we see that the number of seeds in the cones of conifers changes in direct proportion to the mass of the cones. ❖

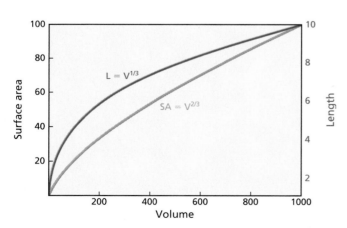

FIGURE 8.8A **Linear Dimensions and Surface Area Are Related to Organismal Volume** Physiological characters that are related to the length or surface area of an organism will also show a relationship to the volume or weight of an organism.

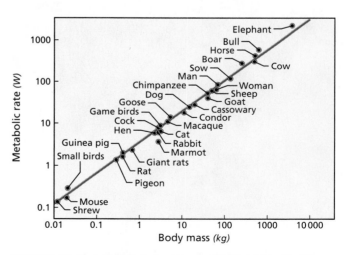

FIGURE 8.8B **Metabolic Rate Increases with Weight to the Three-fourths Power** The metabolic rates of many birds and mammals as a function of size. Each axis has been log transformed. Metabolic rate is proportional to (weight)$^{0.75}$.

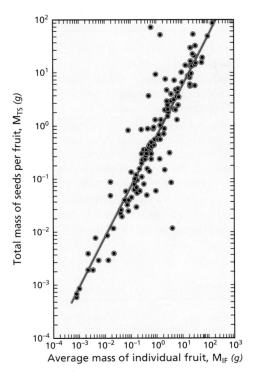

FIGURE 8.8C **Seed Mass vs. Fruit Mass**
The slope of this line is 0.93.

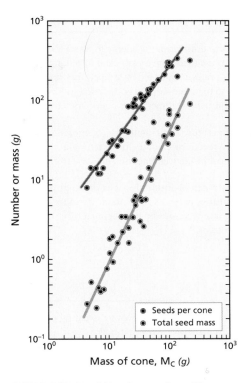

FIGURE 8.8D **Seed Number vs. Cone Mass**
The slope of this line is 1.06.

The Size and Shape of Organisms **269**

SUMMARY

1. Ectotherms utilize external energy to warm their bodies. Endotherms use their own metabolism to generate internal heat. Metabolic rates are affected by temperature, and thus endotherms are able to maintain normal body functions over a broader range of environmental temperatures than ectotherms are.

2. The temperature coefficient, or Q_{10}, measures the rate at which chemical reactions change due to temperature increases of 10°C. Most biological reactions increase from two- to threefold.

3. Organisms can be found over a wide range of environmental temperatures, from −1.86°C to 90°C. Some animals survive freezing temperatures by producing a form of antifreeze from glycoproteins.

4. Our atmosphere filters out much of the high-energy radiation from sunlight.

5. Plants are limited to growing in areas where the light intensity is great enough to at least balance their metabolic requirements. This light intensity is called the compensation point.

6. The surface area to volume ratio of a round organism will decrease with increasing radius of the organism. Since the exchange of heat and the loss of water occurs through the surface of an organism, the size and shape of plants and animals directly affects their ability to exchange heat and retain water.

7. Many characteristics of plants and animals are correlated to the size of the organism. The study of the relationship between physiological measures and size is called allometry.

REVIEW QUESTIONS

1. Many animals cool themselves by sweating or panting. What are the physical processes that make sweating and panting effective means of cooling?

2. Fruit-fly larvae generate 1 mL of CO_2 per hour at 25°C. If we assume that the metabolic rate has a Q_{10} of 2, how much CO_2 would a larva generate at 28°C?

3. Why must chloroplasts be located near the surface of leaves?

4. Discuss three types of problems that confront plants and animals as their size increases.

5. Plant seed mass increases with the size of the plant. Suppose it has been determined that for the standard allometric equation, $b = 0.75$ for seed mass. As plant size is doubled, how much greater would the mass of the seeds be?

KEY TERMS

acclimation	ectotherm	heat shock protein	specific metabolic rate
allometry	endotherm	heat shock response	temperature coefficient
cold-blooded	evaporation	homeotherm	warm-blooded
compensation point	freeze-intolerant	poikilotherm	
conduction	freeze-tolerant	radiation	
convection	glycoprotein	Q_{10}	

FURTHER READINGS

Niklas, K. J. 1994. *Plant Allometry*. Chicago: University of Chicago Press.

Schmidt-Nielsen, K. 1990. *Animal Physiology*. Cambridge: Cambridge University Press.

Willmer, P., G. Stone, and I. Johnston. 2000. *Environmental Physiology of Animals*. Oxford: Blackwell Science.

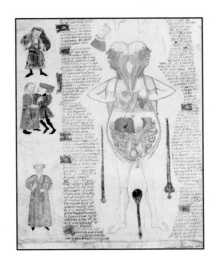

This artwork by John Arderne shows the thirteenth century view of human blood flow.

9

How Organisms Work

Much of ecology and evolution deals with interactions between many individuals and the properties of whole populations. However, the physiological function of individuals may be affected by the environment. Because proper physiological function also affects the survival and reproductive capacity of individuals, these functions are also subject to natural selection.

Some consequences of adaptation to different environments can be seen in the physiology or morphology of different species. For instance, plants like cacti, which live in very dry environments, have thick, round, photosynthetic structures rather than thin leaves. These types of structures help the cacti reduce the loss of water, which is a critical stress in a desert environment.

Plants, animals, and microbes can adapt to their environment in short periods of time. Often this type of microevolution is limited by a species' current morphology or physiology. Thus, in response to an increase in arid conditions, it is unlikely for a rose to evolve thick, waxy leaves in a dozen generations. But this type of short-term evolution is important, because many environments may change over short periods of time. In this chapter we will review how organisms adapt to stresses like desiccation, starvation, and nitrogen waste products. ❖

CHEMICAL TRANSPORT

Whole organisms must cope with regulating solutes, gases, and water

The problems faced by whole organisms are not necessarily the same as those faced by single cells. But there are certainly areas of overlap. They include managing water and salt balance and transporting wastes and gases.

Water and Solutes The factors affecting the flow of water or solutes across cell membranes are also important for the whole organism. If there is a difference in the concentration of a substance between the inside and the outside of a cell, there will be a tendency for net diffusion in the direction of lower concentration. For whole organisms, the relative magnitude of this problem can be determined by comparing the total concentration of solutes in the organism's cells with the total concentration in the surrounding environment. Figure 9.1A shows the range of animals and environments as a function of the concentration of solutes, with land and freshwater environments having lower concentrations of solutes than do brackish (somewhat salty) water and seawater.

Marine invertebrates and elasmobranchs tend to have body fluids that are very close to seawater in their solute concentrations. Accordingly, we do not expect much flux in water and solute concentrations in these animals. Brine mosquito larvae, by contrast, find themselves in environments with very high solute concentrations relative to their body fluids. Without some type of regulation, we would expect water to be lost from the cells of mosquito larvae in these types of environments. Small organisms like fruit flies that live in the desert have great demands placed on their internal

water reserves (Figure 9.1B). We look more closely at water, gas, heat, and ion exchange in whole organisms in Modules 9.2 and 9.3.

Gas and Water Transport Most cells rely on diffusion for their uptake of oxygen and to rid themselves of carbon dioxide. For whole organisms, this process is much too slow for all but the smallest animals. Most animals rely on some sort of circulatory system to bring oxygen from the external environment to cells, which it can then enter by diffusion. There are a wide variety of circulatory systems in nature. We survey some of these systems in Module 9.6.

Plants also require nutrients and fluids. However, without muscles, plants must depend on nonmotile processes for moving fluids. Yet exceptionally large trees, like redwoods, can transport water over distances of up to 100 meters! We consider water transport in plants in Module 9.5.

Physiological Systems Evolve Different modes of blood transportation have evolved in different animal groups, as we will see in Module 9.6. By studying the different circulatory systems of invertebrates and birds, we get some appreciation for the large-scale changes brought about by evolution. But these types of changes happen slowly across many taxa. Within any population or species the types of changes in physiological systems will be much more modest. Later in this chapter, we look at how physiology may evolve within populations in response to specific environmental stress. ❖

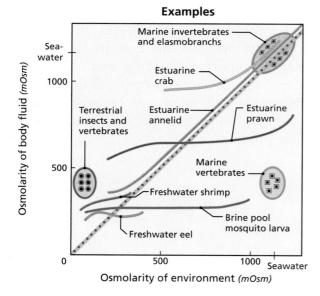

FIGURE 9.1A The relationship between environmental osmolarity and body fluid osmolarity in a variety of animals. Long lines indicate that the animal may live in a range of environmental conditions.

FIGURE 9.1B Small fruit flies live naturally in the rots or wounds in desert cacti.

Animals are mostly made of water. For proper cell function, cells add small amounts of sodium, potassium, calcium, and other ions to their water. The balance of water and ions in animal cells is affected by the outside environment. Many animals live in freshwater, seawater, or brackish (somewhat salty) water. These animals have net flows of water and ions into or out of their cells depending on the concentration of ions in their tissues and in the surrounding water. Terrestrial animals are in direct contact with gaseous air, so they lose water primarily through exposed tissues.

Terrestrial animals are in direct contact with gaseous air, so they lose water primarily through exposed tissues.

The **osmotic concentration** of a fluid depends on the concentration of all solutes, which include ions and nutrients. Seawater typically has 35 grams of salts per liter of water. Organisms with salt concentrations that are higher than the surrounding environment are called **hyperosmotic**. Their cells experience an influx of water and an outflow of ions. Cells with a lower osmotic concentration than the outside environment are **hypoosmotic**. These cells experience an inflow of ions and an outflow of water.

Aquatic animals have two options. They may keep the osmotic concentration of their cells equal to the environment; that is, they can be **osmoconformers**. Or, they may keep the osmotic concentration of their cells different from the environment; that is, they can **osmoregulate**.

Many marine organisms are osmoconformers, although some fish and marine mammals are hypoosmotic. However, many osmoconformers have specific ions that are either more or less concentrated than in the surrounding water. These ion concentrations are actively regulated by the animal cells. Ocean salinity is relatively constant. Not surprisingly, many marine organisms cannot survive large changes in salinity. Such organisms are called **stenohaline**. Some marine organisms, including small crustaceans, live in high-tide pools that undergo large salinity increases as water evaporates in the summer heat and large salinity decreases when pools fill up with rainwater. Organisms in these environments are usually adapted to withstand large salinity changes and are called **euryhaline**.

Skin and gill surfaces are important for the ion and water regulation of aquatic animals (Figure 9.2A). Hyperosmotic organisms such as fish and invertebrates show active uptake of ions at their gills. In worms and amphibians, the skin is used for osmoregulation. Marine teleost fish are hypoosmotic to their environment. They use their gills to remove ions from their cells. This action occurs at sites in the gills called *chloride cells*, where Na^+ and Cl^- ions are exported. Other animals—like birds and reptiles—that live in marine environments do not have gills but use salt glands to excrete excess ions. Marine mammals can produce urine that is more concentrated than their blood. These animals can rid their bodies of excess ions through their urine.

Because the respiratory structures of land animals are moist, they are important sites of water loss (Figure 9.2B). Most water intake for terrestrial animals comes from drinking and eating. Some foods produce water as a by-product when metabolized, although this is not a major source of water for most animals. For instance, for every gram of lipid metabolized, about 1.07 grams of water is produced. The water in exhaled air is recovered by some animals by cooling the air prior to exhalation. This recovery is achieved using the countercurrent circulatory systems that we review in Module 9.3. Cooled air holds less water, so some condensation and water recovery can be achieved by this process. ❖

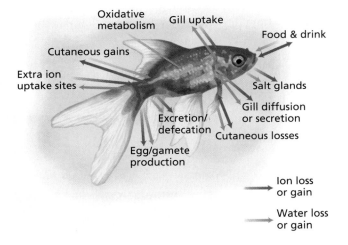

FIGURE 9.2A **Water and Ion Loss in Aquatic Animals**

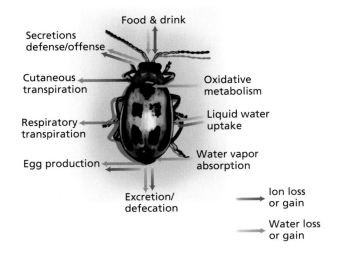

FIGURE 9.2B **Water and Ion Loss for Terrestrial Animals**

Many important physiological processes require the diffusion of ions or molecules in and out of the organism or from one tissue to another. The processes include extracting oxygen from air or water, extracting or eliminating ions between the environment and the organism, and concentrating and eliminating waste products in urine. Many animals employ a common solution to these problems; it is called countercurrent exchange. **Countercurrent exchanges** involve the flow of two liquids, or a liquid and a gas, in opposite directions. Concentration gradients between the two substances cause molecules or heat to move more efficiently between the two flows than would be the case if the flows moved in the same direction.

For instance, in Figure 9.3A we see two flows in opposite directions. The top flow, which moves from right to left, initially has a low concentration of a solute or a low temperature. The bottom flow, moving in the opposite direction, is a source of the solute or heat source and has a very high concentration of the solute or high temperature when it first enters the system, on the left. Because of this difference in concentration or temperature between the two flows, the solute or thermal energy moves from the bottom flow to the top flow. This move-

ment lowers the concentration of the solute in the bottom flow. As this flow moves from left to right, it continues to encounter lower and lower concentrations of the solute in the top flow. This gradient permits the continued movement of the solute from the bottom flow to the top flow. This countercurrent exchange permits the solute in the top flow to reach a much higher concentration than if the two flows had been moving in the same direction.

Fish gills work on the principle of countercurrent exchange (Figure 9.3B). Unoxygenated blood from the heart is pumped into the gills, where a large surface area of capillaries facilitates the diffusion of oxygen from water into the bloodstream. The fish also move water through their gills in a direction that is opposite the flow of blood. The principles of countercurrent exchange are also used in the kidneys of many animals for removing and recovering valuable ions from urine.

Another application of countercurrent flows is heat exchange in tuna. Most extremities of the tuna are at ambient temperatures. However, by using a countercurrent flow of blood to and from their muscles (Figure 9.3C), tuna can keep their muscles at a temperature that is well above ambient. This

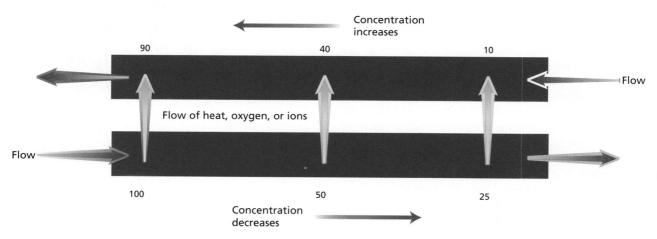

FIGURE 9.3A Countercurrent Exchangers

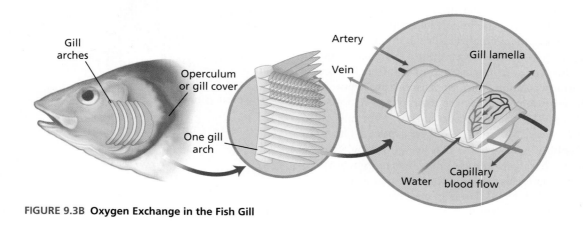

FIGURE 9.3B Oxygen Exchange in the Fish Gill

warmth permits tuna to swim faster and longer than can equivalently sized fish without heat exchangers. The activity of the tuna's swimming muscles produces metabolic heat that warms up the venous blood. As this blood flows away from the muscles, cool arterial blood moving in the opposite direction and in close proximity to the venous blood is warmed prior to reaching the muscles. The heat generated by these muscles can be substantial. Tuna fishermen who have worked especially hard to land a tuna notice that the fish can become overheated and "burnt." The meat of these "burnt tuna" is less valuable as a result. ❖

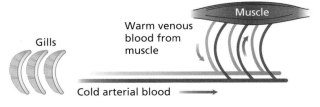

FIGURE 9.3C **Heat Exchanger in Tuna**

9.4 The uptake of oxygen by animals is accomplished by gills, lungs, and occasionally skin

The most efficient means of extracting energy from glucose is by complete oxidation in the Krebs cycle. This process requires that oxygen be delivered to cells constantly as it is consumed and that the waste product of respiration, carbon dioxide, be removed. Oxygen is abundant in the atmosphere, making up about 21 percent of our air. It is reasonable to ask why animals cannot simply let oxygen diffuse from the air into their cells.

The answer is that diffusion is a slow process. Even with the high levels of oxygen in the Earth's atmosphere, diffusion would not deliver enough oxygen for organisms much larger than 2 millimeters (mm) in diameter. As a result, most organisms have organs for extracting oxygen from the atmosphere and then some kind of circulatory system for delivering this oxygen to cells.

In most aquatic animals the extraction of oxygen is done by the **gills,** which are skin surfaces turned outward (Figure 9.4A). These gills may appear as *tufts* similar to those found in marine tube worms or as modifications of tube feet used by echinoderms. *Filamentous* gills have elaborate circulation. They are found on aquatic arthropods and in some salamanders and tadpoles. *Lamellar* gills are arranged as a series of flat plates that can be oriented toward the flow of water to generate countercurrent flows. These gills are found in fish and some crustaceans.

If water simply stands next to the gill surfaces, the available oxygen will be depleted locally, and the animal will find itself in low-oxygen or **hypoxic** conditions. To avoid this problem, the water must be continually moved to bring fresh water with high levels of oxygen into contact with the gill surfaces. Many fish accomplish this through movement (Figure 9.4B). The fish opens and closes its mouth and operculum (the covering of the gills) to create water flow past the gills. The fish first opens its mouth, letting in fresh water, while closing its operculum to prevent water from leaking out. The fish then closes its mouth and reduces the volume of the mouth cavity, while simultaneously opening the operculum. Water then flows over the gills in a single direction, creating the countercurrent flow of water and blood. Certain other fish, such as tuna, can create water flow only by swimming with their mouths open. This technique is called **ram ventilation**. As the fish swim faster, their increased need for oxygen is matched by increased water flow over their gills.

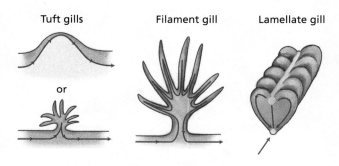

FIGURE 9.4A Different Types of Gills

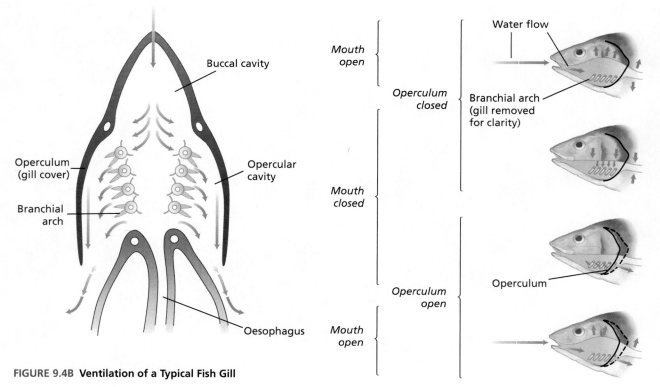

FIGURE 9.4B Ventilation of a Typical Fish Gill

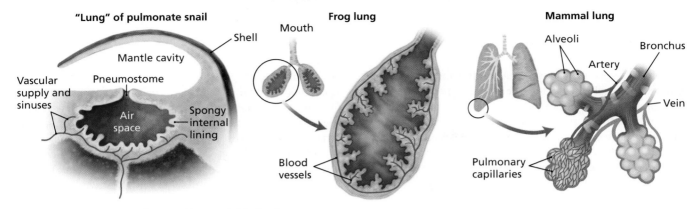

FIGURE 9.4C **The Lungs of Several Terrestrial Animals**

"Lung" of pulmonate snail — Shell, Mantle cavity, Pneumostome, Vascular supply and sinuses, Air space, Spongy internal lining

Frog lung — Mouth, Blood vessels

Mammal lung — Alveoli, Artery, Bronchus, Vein, Pulmonary capillaries

Terrestrial animals typically use lungs to extract oxygen from the air. A **lung** is an invagination of skin to form a respiratory surface (Figure 9.4C). Lungs are most often found in terrestrial animals, although marine sea cucumbers also have lungs. The morphology of lungs can be simple, like those found in land snails. Snail lungs have a simple invagination with some convoluted skin surface to increase the exposed surface area to air. This basic design is more elaborate in frogs and becomes quite complicated in mammals, where small branched alveoli structures create a large surface area for the exchange of oxygen and carbon dioxide. As mentioned earlier, diffusion occurs only over small spatial scales. In the lung the exchange of gases occurs by diffusion through the moist surfaces of thin membranes. In humans, the total surface area of the lungs is 100 m^2, about the size of a tennis court.

For lungs to provide a continuous supply of oxygen, fresh air must continuously replace old air. The bird lung is particularly efficient in this regard. The mechanism of air flow in the bird lung is shown in Figure 9.4D. The anterior and posterior sacs act like the bellows of a blacksmith requiring two inhalations and two exhalations. In the first breath, air is pulled down the mesobronchus into the caudal air sacs. In the first exhalation, the air is forced out of these sacs, primarily into the actual lung, through the parabronchi (in one direction). The air in the lung is pulled into the cranial air sacs upon the second inhalation, and at the second exhalation, it is forced from this second set of air sacs out of the body. The directional flow of air out of the lungs permits birds to have capillaries running countercurrent to the flow of air, achieving efficiencies that other vertebrates cannot. This countercurrent configuration permits the bird lung to be somewhat smaller and lighter, which in turn facilitates flight. ❖

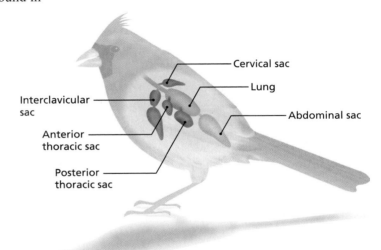

Cervical sac, Lung, Interclavicular sac, Anterior thoracic sac, Posterior thoracic sac, Abdominal sac

Model

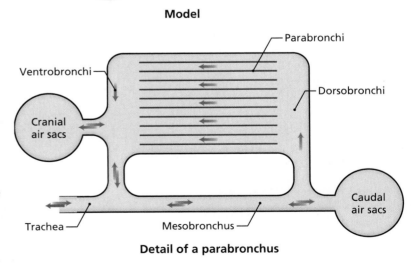

Parabronchi, Ventrobronchi, Dorsobronchi, Cranial air sacs, Caudal air sacs, Trachea, Mesobronchus

Detail of a parabronchus

FIGURE 9.4D **The Operation of a Bird Lung** The cranial air sacs include the anterior thoracic, cervical, and the interclavicular sacs, while the caudal air sacs include the posterior thoracic and abdominal air sacs.

One of the biggest problems facing plants is the management of water. This problem includes the reduction of water loss and the transport of water from the roots to the rest of the plant.

About 90 to 95 percent of a plant's water loss occurs through its leaves. As we saw in Module 8.5, leaves must have a large amount of exposed surface area to effectively capture light energy. This surface area is a potential site for water loss. Many plants layer their leaves with hydrophobic waxes that make excellent water barriers. Yet plants must also capture carbon dioxide (CO_2) from the atmosphere to fuel photosynthesis. They do this by allowing air to pass through openings in the leaf called **stomata**. These small pores also permit water vapor to escape. The diffusion of water vapor through the stomatal pores is called **stomatal transpiration**.

Stomates are very effective at gas diffusion (Figure 9.5A). Typically the amount of gas that can diffuse through a circular opening is proportional to its area. However, for very small openings, the relative diffusion efficiency increases (Figure 9.5A). The reason for this greater efficiency is that the amount of gas passing through an opening is constrained by the size of the opening; but once the gas is through, it can bend around the edges, creating an increased rate of flow. For a circular opening with a radius of r, the area is πr^2 while its circumference (edge) is $2\pi r$. Thus, if the radius is cut in half, the area is reduced by three-fourths; but the edge is reduced by only one-half. This means that several small openings will permit more gas flow than will one large opening of the same area. The efficiency of CO_2 gain also means water vapor is lost quickly.

Although stomates close to conserve water when carbon dioxide intake is not needed, or when the plant is severely stressed, the plant must still have some means of transporting

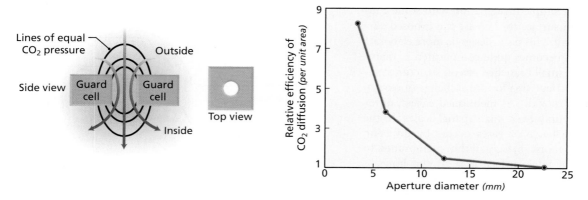

FIGURE 9.5A **CO_2 Gain and Water Loss through Stomates**

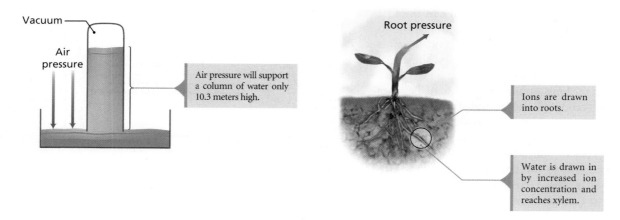

FIGURE 9.5B Air and root pressure cannot account for water movement in the tallest trees.

water to its leaves (Figure 9.5B). For redwoods and Douglas fir trees, water collected at the roots must travel over 100 meters to reach the top of the tree! If a vacuum could be created in the water vascular system of plants, called the **xylem**, then air pressure alone might push the water to a height of about 10.3 m at sea level. Air pressure alone cannot push water higher than this.

There is also a pressure created at the roots of some plants. As plants collect ions in the roots, the concentration difference between the plant cells and the soil can draw water into the cells, creating a positive pressure in the xylem. This pressure can be measured in some plants, but at best could account for about 16 percent of the pressure needed to move water up the tallest trees. In any case, positive root pressures are not always present, so they cannot be a general explanation for water movement in plants.

The best explanation for water movement in plants involves **cohesion,** the tendency of water molecules to stick together. This explanation begins with water being lost from the leaves by transpiration. The loss of water through the stomates creates a negative pressure that begins to pull on the water column in the xylem. Two forces are crucial for transmitting this pull all the way to the roots of the plants. The first is the cohesive forces holding water molecules together. These forces arise from the polar nature of water molecules and the hydrogen bonds that form between neighboring molecules. The second force is an adhesive force between water and the lining of the xylem. This lining is hydrophilic and also forms hydrogen bonds between water and the walls of the xylem cells. The result is an effective transmission of the negative pressure created by transpiration down the entire water column.

How exactly does the loss of water through the stomates create a pull in the first place? Water that is exposed to air loses some molecules that have sufficient energy to leave the liquid phase and become gaseous (Figure 9.5C). Gaseous water may also condense and return to the liquid phase. Dry air in contact with water takes some time to reach an equilibrium in which

the number of molecules leaving the liquid phase equals the number entering it (Figure 9.5C). When this equilibration has occurred, so that condensation equals evaporation, the force exerted by the gaseous water creates **vapor pressure**. In the space inside the leaf just beyond the stomates, there is a great deal of exposed water that is subject to vapor pressure. The rate of transpiration depends on the difference between the vapor pressure in the stomatal spaces and the vapor pressure in the air just outside the stomates (Figure 9.5D).

Under still conditions a thin layer of air called a **boundary layer** can develop, just outside of the stomates, that has a higher vapor pressure than the surrounding air. This boundary layer can reduce the loss of water from the stomates. However, strong winds can increase water loss by removing the boundary layer. Some plants have hairlike structures, called **trichomes** (Figure 9.5D), around the stomates to foster the creation of the boundary layer and thus reduce water loss. ❖

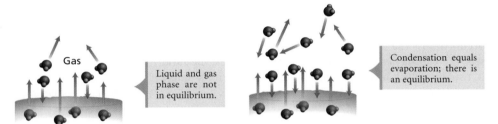

FIGURE 9.5C Vapor pressure is the force exerted by a gas in equilibrium with its liquid phase.

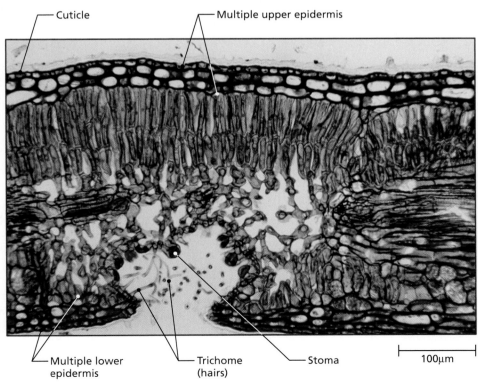

FIGURE 9.5D **Leaf Structure**

Circulating fluids in animals supply metabolites and remove wastes. They carry nutrients and oxygen to tissues, and they remove carbon dioxide and other waste products. This circulation is accomplished by either of two designs: open or closed circulation.

In some **open circulatory systems** there are no specialized vessels for transporting fluids. Transport is accomplished by movement of the body wall. In other open systems, like those found in molluscs and crustaceans (Figure 9.6A), specialized vessels transport fluids through part of the organism; the vessels then open up into a body cavity. These open systems may include hearts and vessels with valves to prevent the backflow of fluid.

Closed circulatory systems do not open directly into body cavities, and the provisioning of nutrients and gases occurs across the thin walls of the smallest circulatory vessels, the **capillaries**. In closed circulatory systems, the fluid is usually called **blood**. Closed circulation permits stable blood composition.

The flow of blood through the major organs differs in invertebrates and vertebrates.

For large, active animals, circulation requires a pump. It may be as simple as a thickened section of one of the principal vessels that conduct fluids. Pumps with nervous system synchronization and internal valves that ensure one-way flow are referred to as **hearts**. Both open and closed circulatory systems may include hearts.

The flow of blood through the major organs differs in invertebrates and vertebrates, as Figure 9.6B shows. Invertebrates have one-way blood flow. After leaving the body tissues, the blood goes to the gills, then to the heart, and finally back to the tissues. In vertebrates, blood leaving the tissues goes first to the heart. In vertebrates with one-way blood flow, like fish, the blood leaving the heart then goes to the gills, receives oxygen, and finally passes directly to the tissues. Other vertebrates have a dual circuit of blood flow. The **pulmonary circuit** carries deoxygenated blood, under low blood pressure, to the lungs for gas exchange and then returns oxygenated blood to the heart. The **systemic circuit** carries oxygenated blood, under

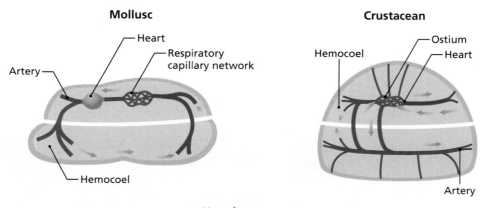

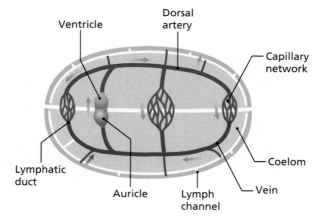

FIGURE 9.6A Open and Closed Circulatory Systems

high blood pressure, to all body tissues for gas exchange and then returns deoxygenated blood to the heart. In mammals about 50% of the blood volume will be found in the veins, 21%-25% in the arteries and capillaries, with the remainder in the heart and pulmonary circulation.

Besides the heart, the blood vessels affect the flow of blood. Figure 9.6C shows how the velocity, pressure, and volume of blood changes as it circulates through the body of a mammal. For instance, the **aorta**, the artery leaving the heart, contains the highest blood pressure. The aorta is connected to numerous arteries. These arteries have thick, muscular walls that prevent them from collapsing as they twist through the body. The narrower arteries are called **arterioles**. They can respond to a variety of stimuli to expand or contract; these processes are called **vasodilation** and **vasoconstriction** respectively. This responsiveness gives arterioles a key role in regulating blood pressure. The capillaries cover a large surface area in the organism (Figure 9.6C). There are more **venules** (small veins) and veins in the body than arteries, and these veins and venules have the capacity to hold large volumes of blood. The walls of veins are elastic, allowing them to expand as blood volume increases. The veins and venules also have valves to prevent the backward flow of blood, which can occur because the pressure in veins is very low. ❖

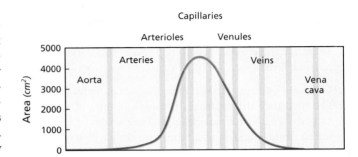

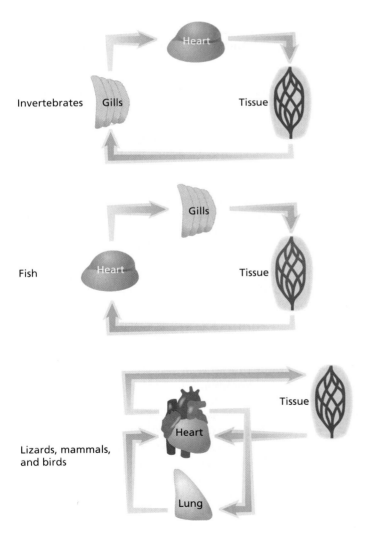

FIGURE 9.6B Comparative Circuitry

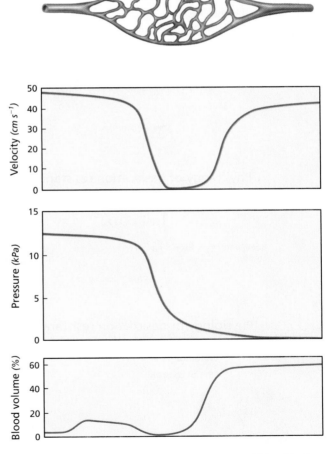

FIGURE 9.6C The panels follow three properties of blood in the body of mammals: blood velocity, blood pressure, and blood volume (as a percent of the total).

EVOLUTION OF PHYSIOLOGICAL SYSTEMS

9.7 Desiccation is a major problem for terrestrial life

The exchange of gases in terrestrial plants and animals typically requires contact between air and moist tissue. This contact inevitably leads to water loss through evaporation. Animals that are covered by moist skin will also lose significant amounts of water through their skin. Many terrestrial climates become hot and dry for at least some periods of time, thereby exposing animals and plants to drying conditions.

There are three basic ways to cope with desiccation at the organismal level:

1. The plant or animal may reduce the rate of water loss.
2. The organism can maintain greater amounts of water in its body. This water can either be bulk water, or it can be stored using special storage compounds like glycogen.
3. The organism can develop the ability to tolerate the loss of water.

All three of these strategies for coping with water loss are employed by some plants or animals to different degrees. We will examine these in a little more detail by illustrating how evolution alters the physiology of fruit flies that have been selected to survive extended periods of water stress (Figure 9.7A). Biologist Michael Rose and his colleagues developed experimental populations of *Drosophila* by keeping adults in large cages without food or water for a period of about 20-60 hours. A population might start with 9,000 adults, and by the end of the desiccation period there would be only about 1,000 survivors. Control populations were derived from the same starting populations and maintained in the same fashion, except that during the period of desiccation these flies received water but no food. The absence of food would kill about two-thirds of the control population (Figure 9.7A); but these deaths were primarily due to starvation, not desiccation. This selection procedure was repeated for more than 100 generations.

After many generations of selection, the flies selected for desiccation resistance became markedly better at surviving periods of desiccation (Figure 9.7A). These same flies also lose water at about half the rate of the control flies. How did they differ from the control flies? The investigators examined the flies' **cuticle,** or outer covering, which has layers of chitin and protein that are relatively impermeable to water. (Cuticles are more effective water barriers than skin is.) The outer layer of the cuticle is called an **epicuticle,** and it has high levels of lipids that act as effective water barriers. It appears that the exoskeletal cuticle of the desiccation-resistant flies has a greater proportion of long-chain hydrocarbons, although the total amount of hydrocarbon is about the same. These long-chain lipids have higher melting points and are better barriers to water. The changes seen in the chain length of lipids in the *Drosophila* populations are also seen in different insect species. Species that are active in the summer have longer-chain hydrocarbons than do similar species that are active in the winter.

Were other differences found between the dessication-resistant flies and the control flies? The desiccation-resistant flies also have a greater amount of bulk water per individual than do the control flies. There appears to be no difference in the dehydration tolerance of the two types of flies. When these flies die, they all have roughly the same amount of water as a percentage of their total size. Dehydration tolerance does differ

In each generation, the experimental flies experience a prolonged exposure to a desiccating environment; the controls do not.

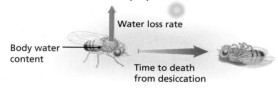

No water or food

9000 → 1000

The survivors are used as parents for the next generation.

Water but no food

3000 → 1000

= 1000 flies

Physiology of desiccation resistance in control populations

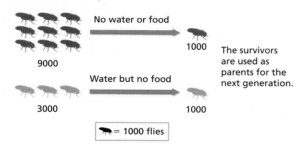

Water loss rate

Body water content

Time to death from desiccation

Physiology of desiccation resistance in desiccation-resistant populations

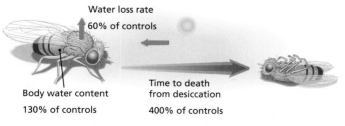

Water loss rate
60% of controls

Body water content
130% of controls

Time to death from desiccation
400% of controls

FIGURE 9.7A Laboratory Evolution of Desiccation Resistance

between species. For instance, earthworms, slugs, and snails can tolerate large (40–80 percent) water loss before dying. Most mammals can only tolerate losses of between 5 and 10 percent of their body water. Camels, which can tolerate losing 30 percent of their water, are an interesting exception.

Animals may reduce their water loss in several other ways. Behavior can be important. Many animals are inactive during very hot periods, to avoid water loss. Other animals seek moist conditions under rocks or leaves, where they are surrounded by water-saturated air. Large animals may also reduce water loss by reabsorbing water in their kidneys before they excrete nitrogen wastes. The ability to excrete concentrated urine varies greatly among animal species, although some desert-adapted mammals can excrete urine that is more concentrated than seawater (Figure 9.7B). ❖

FIGURE 9.7B The Kangaroo Rat Kangaroo rats live in desert habitats and get all their water from their foods, either as bulk water or as a by-product of the metabolism of their food.

9.8 The ability to tolerate nitrogen wastes is molded by natural selection

When animals use proteins and amino acids for energy, they ultimately produce small nitrogen molecules, like ammonia, that are toxic in large concentrations. Consequently, animals have well-developed physiological mechanisms for eliminating nitrogen. Most but not all animals eliminate nitrogen either as ammonia, urea, or uric acid (Table 9.8A). Animals that eliminate ammonia directly are almost always aquatic or live in environments where they are in intimate contact with water. This lifestyle facilitates removal of ammonia, which is very soluble in water. Animals that cannot get rid of ammonia rapidly must convert it to a less-toxic compound, like urea or uric acid, to prevent cell damage. Some animals possess the ability to produce more than one type of nitrogen waste product. Spiders and scorpions excrete guanine. The method of nitrogen excretion also depends on embryonic conditions. In those animals with eggs that are sealed from the environment, a less toxic waste product is required.

The levels of internal or external nitrogen waste products may become high, requiring organisms to develop adaptations to these conditions. For instance, fruit-fly larvae excrete ammonia directly into their larval food environment (Figure 9.8A). In very crowded cultures, the levels of ammonia may become high and contribute to the stress of crowding. Can organisms, like fruit flies, adapt to high levels of nitrogen wastes? To study this problem, Laurence Mueller

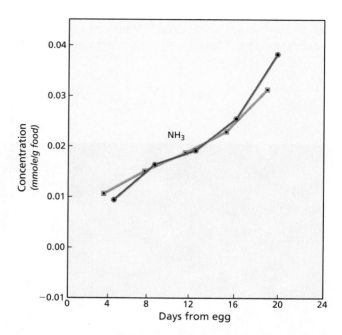

FIGURE 9.8A In crowded cultures of fruit flies, the levels of ammonia increase dramatically due to ammonia excretion by the larvae. The two lines represent two different populations.

TABLE 9.8A	Different Routes of Nitrogen Elimination
Mode	**Major End Product**	**Representative Animal Groups**
Ammonotelism	NH_3, NH_4^+	Freshwater and marine invertebrates, teleost fish, aquatic amphibians, isopods, *Drosophila* larvae
Ureotelism	Urea	Adult amphibians, mammals
Uricotelism	Uric acid	Terrestrial gastropods, terrestrial insects, lizards, snakes, birds

Ammonotelic

Ureotelic

Uricotelic

and colleagues created multiple populations of fruit flies that either were raised as larvae on high concentrations of ammonia or were raised as controls in low-ammonia conditions. Over time the populations of fruit flies raised on ammonia were dominated by genotypes that showed elevated resistance to ammonia, relative to controls. This adaptation was easily determined by raising both the ammonia-resistant and control populations on food with high levels of ammonia added directly to the food (Figure 9.8B). The survival from egg to adult in the ammonia-resistant population was about 35 percent greater than that of the control population (Figure 9.8B).

What are the ammonia-resistant populations doing that permits them to withstand such high levels of ammonia? Fruit flies, as well as many other organisms, may convert ammonia to a less-toxic compound, glutamate, by the following reaction: $NH_4^+ + H^+ + NADH + \alpha\text{-ketoglutarate} \leftrightarrow glutamate + NAD + H_2O$. This reaction is catalyzed by an enzyme called glutamate dehydrogenase. The ammonia-resistant lines make 40 percent more of this enzyme, even when they are raised on normal food. This suggests that ammonia resistance may in part come about by the rapid conversion of ammonia to a less-toxic compound.

Another surprising consequence of ammonia resistance is that resistant larvae feed more slowly and are less vigorous foragers (Figure 9.8C). It is hard to imagine that this behavior could help these larvae survive better. In all likelihood this change in behavior does not directly help develop ammonia resistance, but reflects a trade-off. That is, to marshall sufficient energy to detoxify ammonia, these larvae may have to draw energy from other activities—like their foraging behavior. ❖

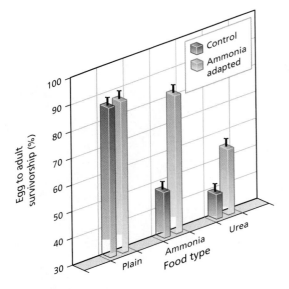

FIGURE 9.8B Populations selected for resistance to ammonia show elevated levels of survival in food laced with ammonia compared with control populations. Interestingly, the ammonia-adapted populations also show increased resistance to the novel compound, urea.

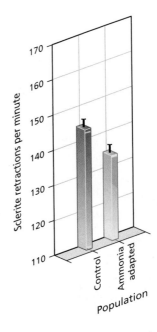

FIGURE 9.8C The ammonia-resistant populations show reduced feeding rates relative to their controls. This may be due to trade-offs in energy requirements for feeding and detoxifying ammonia.

Fat is beautiful when episodes of starvation are a predictable part of life

There is no more basic requirement for an animal than eating. While reproduction is the ultimate goal of the Darwinian organism, without food an animal may not survive to have an opportunity to reproduce. How animals search for food is an important topic in ecology that will be covered in more detail later in the book. But what happens if an animal is unsuccessful in finding food? We know that if we do not eat, we lose weight as our body metabolizes its fat reserves to produce needed energy. If starvation is severe and regularly encountered, how will animals adapt to these types of conditions?

To study this problem, evolutionary biologist Michael Rose has created populations of fruit flies that are regularly subjected to several days of starvation conditions early in their adult life. Over time, genotypes that are able to resist

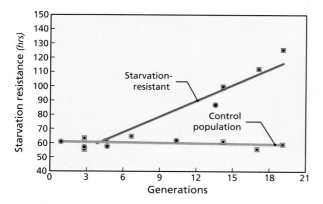

FIGURE 9.9A The average time until death due to starvation for females increases steadily in the starvation-selected populations, but shows no change in the control populations.

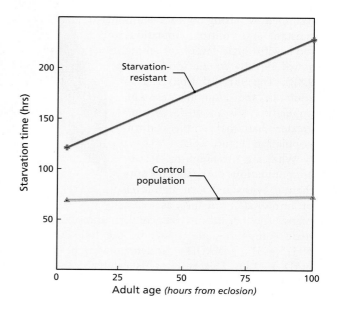

FIGURE 9.9B Very young adult females have poor starvation resistance relative to older adults in the starvation-resistant population.

starvation have become more common in these starvation-resistant populations as compared to controls that never experience starvation (Figure 9.9A). Within 20 generations, the starvation-resistant populations can live almost twice as long (120 hours vs. 60 hours) without food as the controls can. But how are they doing this?

A more careful examination of the starvation-resistant populations shows that the ability of adults to resist starvation is present as soon as the adult has completed metamorphosis (adult age zero hours); this ability increases dramatically over the first 96 hours of adult life (Figure 9.9B). No such change occurs in the control populations. These observations suggest that the flies in the starvation-resistant populations are changing their physiological state during their early adult life.

When the total weight and amount of fats (lipids) are measured in the 96-hour-old adults, we see that the starvation-resistant flies are not only heavier, but have larger reserves of fat (Figure 9.9C). This fat reserve actually increases substantially during the first 96 hours of adult life in the starvation-resistant populations. Thus, the picture that has developed is that starvation-resistant females feed vigorously during the first four days of adult life and rapidly build up their fat reserves. These reserves, which are in fact being built up during the larval period, slow down the development of the starvation-resistant larvae as well as somewhat decreasing their survival rate. These reserves ultimately permit these flies to withstand the inevitable episode of starvation. In a standard laboratory environment—and in many natural environments—such drastic and predictable episodes of starvation do not occur. Consequently, building up large fat reserves is not useful, and the reduction in larval survival would prevent such evolution. However, in environments with regular bouts of starvation, these reductions in viability are tolerated so adults can survive to reproduce.

The evolution of starvation resistance also affects traits other than those that directly affect the ability to resist starvation. We call such effects pleiotropic effects. In this example, selection for starvation-resistance has also increased the longevity of the starvation resistant populations (Figure 9.9D). These results are interesting because it is known that populations of fruit flies selected for longevity and late reproduction also show increased starvation resistance. These congruent results suggest a connection between the genes and traits that affect longevity and starvation resistance. ❖

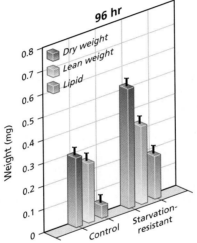

FIGURE 9.9C Starvation-resistant females weigh more than controls, largely due to the storage of lipid.

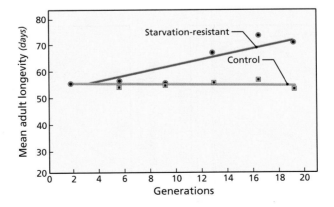

FIGURE 9.9D A correlated response to selection for starvation resistance is increased longevity.

ENERGY PRODUCTION AND UTILIZATION

9.10 Many factors affect energy production and utilization

An animal spends a good portion of its life in searching for and consuming food. Food supplies an animal with water and essential nutrients like vitamins, amino acids, carbohydrates, and lipids. Molecules such as proteins, lipids, and carbohydrates may be broken down or **catabolized** to produce energy. Some of this energy may be used to build large molecules. Collectively these building processes are called **anabolic** processes. Animals typically store energy as adenosine triphosphate (ATP).

Energy Production Animals can use several catabolic routes in producing ATP. In low-oxygen environments, anaerobic metabolism is used to produce energy. Starting with glucose, here is the simple equation for one of the most common anaerobic pathways:

$$\text{glucose} + 2\text{ADP} + \text{P}_i \rightarrow 2\text{ lactate} + 2\text{ATP}$$

where ADP is adenosine diphosphate and P_i is a phosphate molecule. This equation shows that glucose is broken down into lactate, and the energy released is used to make ATP from its precursor ADP and phosphate. No oxygen is consumed. Animals may rely on anaerobic metabolism if they live in low-oxygen environments, such as estuarine mud flats, or if they live as parasites in the guts of vertebrates.

Even animals that normally live in well-oxygenated environments may use anaerobic metabolism for brief periods of time. During periods of extreme exercise, animals' muscles utilize anaerobic respiration and produce high levels of lactic acid. Not all animals produce lactic acid, although this compound is the most common end product of anaerobic respiration. Goldfish produce ethanol as the final product, while swimming cephalopods and marine bivalves produce octopine.

Of course, the most efficient catabolic pathway for energy production is aerobic metabolism, which utilizes the Krebs cycle and oxygen. The overall equation is:

$$\text{glucose} + 36\text{ADP} + 36\,\text{P}_i + 6\text{O}_2 \rightarrow$$
$$36\text{ATP} + 6\text{CO}_2 + 6\text{H}_2\text{O}$$

The equation shows that glucose is broken down to produce carbon dioxide and water, and the energy released is used to make ATP from ADP and phosphate.

This aerobic pathway releases much more energy from glucose than anaerobic metabolism does. We can see this difference by comparing the numbers in the two equations. Thus, for each molecule of glucose we get 18 times more ATP in aerobic metabolism (36 ATP) than we get in anaerobic metabolism (2 ATP).

We also see from the equation for aerobic metabolism that for every molecule of glucose consumed, six molecules of oxygen are consumed and six molecules of carbon dioxide are produced. Thus, rates of aerobic respiration can be inferred from the amount of oxygen consumed or the amount of carbon dioxide produced. We take a closer look at ways of measuring these rates in Module 9.11.

Net Energy Gain The relative presence or absence of oxygen is not the only environmental factor that affects energy production in organisms. The net gain of energy is the difference between the energy an organism takes in, or consumes, and the amount of energy it loses to the environment. Organisms lose energy in the form of excreted wastes as well as metabolic energy, which generates heat that is dissipated to the atmosphere.

For ectotherms, the temperature of the environment can have a significant impact on energy intake. In the graph at the left in Figure 9.10A, for example, we see that for fish the rate of energy intake increases with temperature until it becomes very hot and approaches the lethal range for this species. Energy loss also increases, mostly due to increases in the background metabolic rate. However, as we see in the graph at the right in Figure 9.10A, there is clearly an optimal temperature

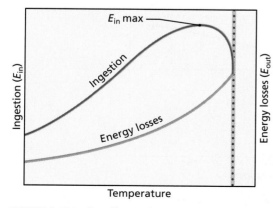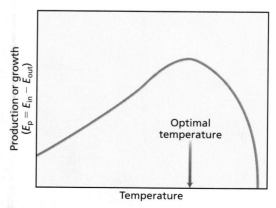

FIGURE 9.10A The effects of temperature on energy ingested (E_{in}) and energy lost (E_{out}) by fish (left graph). The graph on the right shows the effects of temperature on net energy gained (E_p).

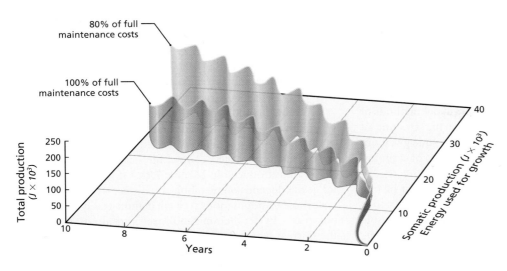

FIGURE 9.10B The effects of reducing maintenance metabolism by 20 percent in the marine mussel *Mytilus edulis*. The amount of energy used for growth is increased by 30 percent, and the total energy available for growth and reproduction is doubled over a 10-year period.

where the net energy gain is greatest. This net energy gain can be used for growth or reproduction.

The allocation of energy is a crucial issue in evolutionary biology. It is often argued that limits are placed on the types of life-history characters that can evolve due to the limited energy any organism has to allocate to different functions. As an example of the possible forces at work, consider Figure 9.10B. Two hypothetical energy budgets were used to model the allocation of energy to growth and all other functions in the marine mussel *Mytilus edulis*. In one budget it was assumed that the normal "housekeeping" metabolic functions could be carried out for 20 percent less energy than the usual amount. Over time this reduction in background energy use permitted the organism to allocate much more energy to growth and reproduction. These types of calculations suggest the important evolutionary role of energy allocation in individual organisms. ❖

Animals use energy to move, to grow, for general maintenance, and to reproduce. To understand the use of energy, we must first determine the total amount of energy used per unit of time, or the **metabolic rate**. The two principal methods of measuring metabolic rates measure either the amount of oxygen consumed or the amount of carbon dioxide given off. These measurements assume aerobic metabolism of carbohydrates, lipids, and proteins for energy.

The amount of energy used in the consumption of one liter of oxygen or the production of one liter of CO_2 depends on the source of the energy. When carbohydrates are the sole source of energy, there are equal amounts of oxygen consumed and carbon dioxide produced, a one-to-one ratio. For every liter of O_2 consumed, 5.0 kilocalories (kcal) of energy are produced. When lipids are the sole fuel, the ratio of CO_2 formed to O_2 consumed, sometimes called the **respiratory quotient (RQ)**, is 0.7. In this case, 4.7 kcal of energy are produced. In most organisms, the RQ is between 0.7 and 1.0, indicating that a mixture of carbohydrates and lipids is being used to produce energy.

Metabolic rates change when animals become active. The extent to which metabolic rates can change varies greatly among species. When an animal is resting and not under any unusual stress, the metabolic rate is referred to as the **standard metabolic rate**. For ectotherms, this rate depends on the ambient temperature, as we saw in Module 8.3. For endotherms, the metabolic rate is relatively independent of the ambient temperature, and the standard metabolic rate is called the **basal metabolic rate**. As animals become progressively active, the metabolic rate rises until it reaches the **maximum metabolic rate**.

The ratio of the maximum metabolic rate to the basal metabolic rate is called the **factorial aerobic scope**. For vertebrates the factorial aerobic scope is 5–12; for invertebrates it is typically 2–10. Insects that fly may show factorial aerobic scopes of 30–50 due to the very high metabolic rates of insect flight muscles. The aerobic scope indicates the capacity for sudden bursts of high activity.

In Module 8.8, we saw that the *specific metabolic rate*, or metabolic rate per kilogram, decreases with increasing size in mammals. Longevity also tends to increase with increasing size (Figure 9.11A). These two relationships suggest that longevity also

increases with decreasing metabolic rates. Obviously longevity is not determined universally by metabolic rate, although there may be particular taxa for which it is important.

How then could we test the idea that increased life span would be associated with decreased metabolic rates? Experimental fruit-fly populations, called O's, live almost three times longer than their control flies, called B's. (These fruit flies were introduced in Module 7.14.) If the hypothesis that longevity depends on metabolic rates were true, we would expect these O-populations to have demonstrably lower metabolic rates. In fact their metabolic rates are almost the same as the normal B-populations (Figure 9.11B). Longevity cannot be attributed to lower metabolic rates.

All energy that animals absorb is excreted as waste, or used for growth, reproduction, or internal metabolism. Metabolic energy in turn can be partitioned to bodily maintenance functions, processing and digesting food, and locomotion or other physical activities. In growing *Drosophila* larvae, for example, a major portion of their energy budget is spent on feeding and movement. As *Drosophila* larvae feed, they extend their mouth hooks, grab food, and then retract their mouth hooks. This movement also pulls the rear of the body

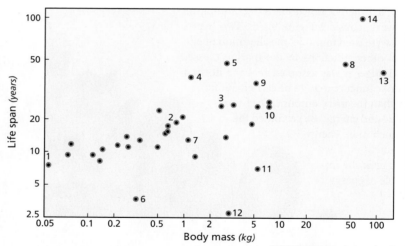

FIGURE 9.11A Is Longevity Related to Metabolic Rates? Larger animals tend to live longer and have lower metabolic rates.

1. *Microcabus murinus*
2. *Carcopithecus jacchus*
3. *Carcopithecus talapoin*
4. *Lemur fulvus*
5. *Cebus albifrons*
6. *Cheirogaleus major*
7. *Hapalemur simus*
8. *Pan troglodytes*
9. *Macaca mulatta*
10. *Carocebus albigena*
11. *Alouatta seiculus*
12. *Propithecus verreauxi*
13. *Gorilla gorilla*
14. *Homo sapiens*

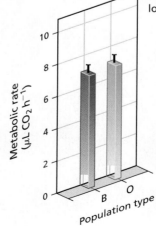

FIGURE 9.11B Is Longevity Related to Metabolic Rates? Populations of *Drosophila melanogaster* selected for increased longevity (O) show no difference in the metabolic rates of active tissue compared to controls (B).

forward, causing the larvae to move as they feed. A good-sized larva performs this feeding extension and retraction movement 120–160 times per minute! These larvae are the epitome of eating machines.

In stressful larval environments, *Drosophila* slow their feeding rates, presumably reducing the amount of energy used in feeding so it may be diverted to cope with the new stress (Figure 9.11C). For *Drosophila*, these stresses include high levels of ammonia or urea in their larval food, larval crowding, and exposure to parasites that attack feeding larvae. Because energy is used for growth and reproduction, it is likely that the energy budgets of organisms have been well tuned by natural selection. ❖

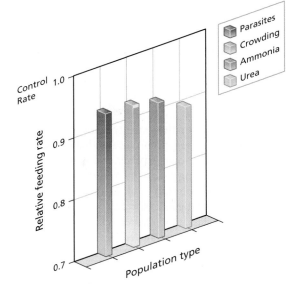

FIGURE 9.11C Activity Decreased in Populations Adapted to Stress The feeding rates of populations of *D. melanogaster* adapted to various stresses are compared to the controls producing a relative feeding rate. The populations adapted to the stress show a reduction in their feeding activity, possibly as a means of conserving energy needed to cope with the stress.

The "cost of transport" is a key metric of the energetic expense of moving in motile organisms

The importance of movement was highlighted in Module 9.11. There we saw that the feeding rates of *Drosophila* larvae slowed as the populations adapted to stressful environments that placed additional demands on their energy budgets. In general, movement requires more energy than resting does. Most of us can appreciate this from our own experience. For instance, it is fairly easy to sustain a walking pace of 15 minutes per mile. For short periods, healthy people can walk at this pace without breathing hard or getting fatigued. However, if we try to run at the speed of a world-class marathoner—5 minutes per mile—we will quickly become exhausted.

As animals increase their speed, they adjust their movements to reduce their energy ex-

penditure. For instance, horses have three different **gaits**: walk, trot, and gallop (Figure 9.12A). As horses move faster, they switch from walking to trotting and then to galloping. When horses are made to move at different speeds using just one gait, their energy use per meter traveled has a minimum at intermediate speeds but increases at lower and higher speeds (Figure 9.12A). The minimum energy consumed per meter traveled is actually the same for all three gaits. The histograms in Figure 9.12A show the speeds that horses themselves choose when moving by each of these gaits. Horses select the speed that consumes the least energy for a given gait.

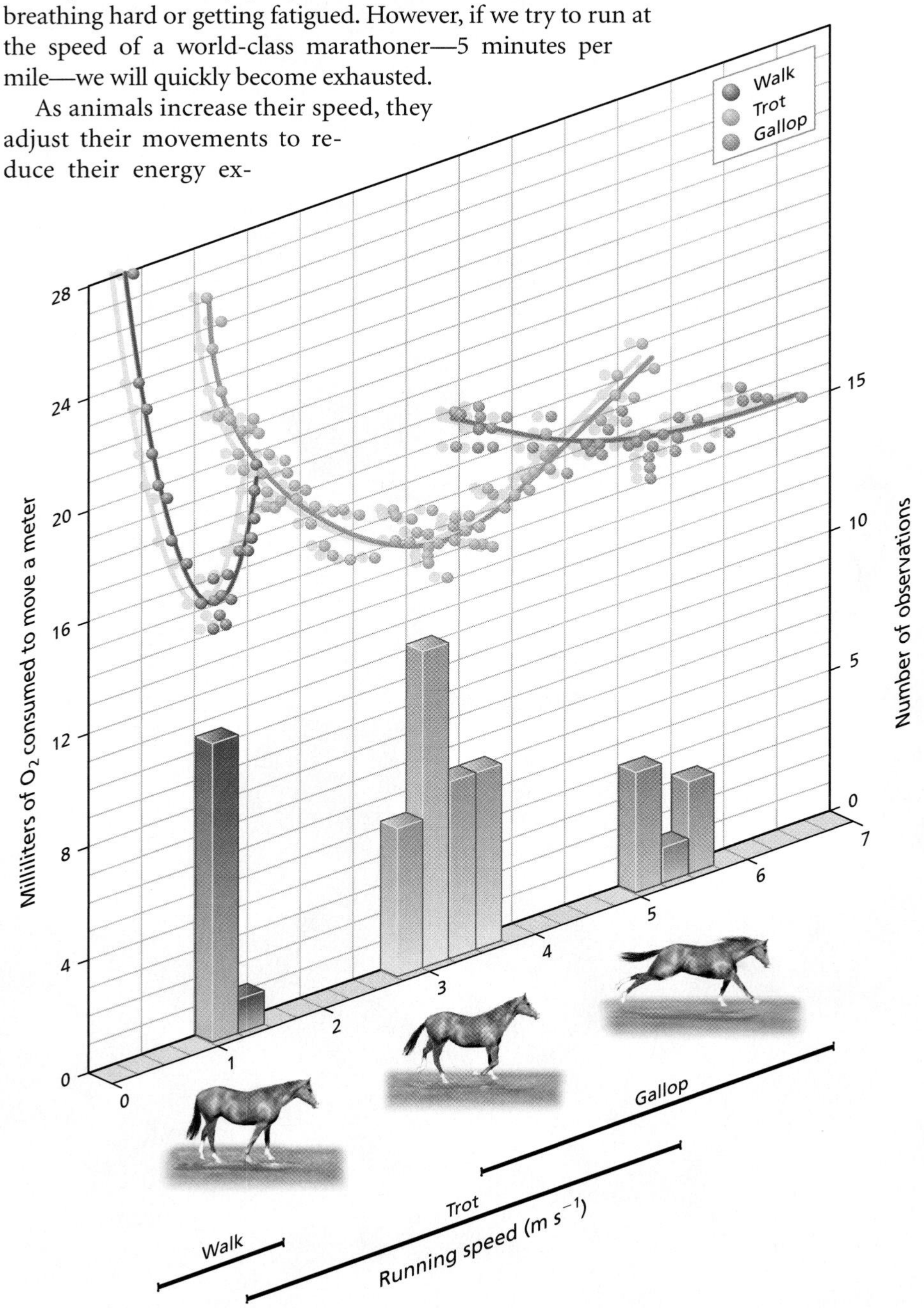

To study different modes of transportation, we need to have some measure of the energetic requirements of movement. If we measure the amount of oxygen consumed by an animal while moving, then we know how much energy is being used. To standardize these energy measurements, we can then calculate how much energy is used per meter traveled per kilogram of animal. For example, in Figure 9.12B the cost of transport is measured in calories per gram-kilometer (cal/g-km). With these units, we can calculate the **cost of transport** to compare the movement of different species.

The cost of transport in different media (air, water, or land) varies because of two factors. First is the energy needed to support the animal. Second is the cost of resistance from moving through the medium.

In water, the energy needed for support is less than that needed in air or on land, due to the buoyancy of animals in water. Land animals need to support their weight, but they receive support from the land while in contact with it. Animals that travel in the air need to supply energy continuously to support themselves.

On the other hand, water offers the greatest resistance to motion due to the higher viscosity of water compared to air. When we compare the costs of locomotion in different

FIGURE 9.12A Metabolic Rate of Horses as a Function of Speed

media, there is a decrease in the energy costs for larger animals. Animals of similar sizes reveal that the cost of transport is greatest on land and least in water, as Figure 9.12B shows. Transport in water can be efficient if the speed of travel is kept low, thus reducing the energy lost to drag, because almost no energy is needed for support. These results also explain why animals that migrate very long distances either fly or swim. ❖

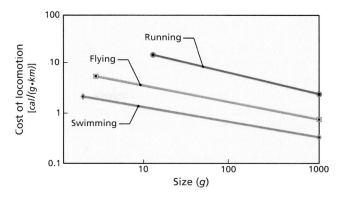

FIGURE 9.12B Costs of Different Modes of Transportation

It seems simple to suppose that natural selection will favor increases in reproductive rates, competitive ability, immune system competence, and many other traits. But we might wonder if there is any limitation to the improvements natural selection can make. A central organizing assumption in evolutionary theory has been that the energy intake of organisms will be limited; and thus adaptations that require energy will often have to get it by diverting energy from some other component of the individual's energy budget.

An early articulation of these ideas was made by Martin Cody in 1966. He was interested in understanding why there was a gradient in the numbers of eggs laid by birds as one goes from high latitudes toward the equator (Figure 9.13A). Specifically, birds of many different species that live in high latitudes tend to lay many more eggs per clutch than do birds that live near the equator. The generality of this trend certainly suggests there must be some common explanation based on fundamental aspects of biology rather than on the detailed peculiarities of the ecology of any particular bird species. Cody's explanation was based on the assumption that each bird will have a limited amount of energy to spend on the activities of competition, predator avoidance, and egg production (Figure 9.13A). In the tropics there are relatively more competing species and predators. Thus, Cody reasoned that these tropical species will have less energy available for egg production and ultimately will produce smaller clutches.

These ideas were developed further by Madhav Gadgil and William Bossert in 1970, in a formal model. The focus of their model was to understand the timing and levels of reproduction during an organism's life. They considered time and energy to be limited and required for maintenance, reproduction, and growth. Thus, expending more energy in maintenance might increase the chance of survival but then would take energy away from reproduction. Evolutionary biologists consider that the key to understanding how different life-histories evolve revolves around understanding how organisms can best allocate energy to these different functions and achieve the greatest gains in fitness.

Another example of the effects of energetic trade-offs was developed by Derek Roff and his colleagues. They studied a species of sand cricket (Figure 9.13B), *Gryllus firmus*, that has two different wing morphologies, long winged and short winged. The long-winged morphs are capable of dispersal but have new additional energy requirements due to the flight muscles that have high respiration rates. Roff measured metabolic rates directly and found female long-winged morphs to have significantly higher metabolic rates (Figure 9.13C). However, as postulated by Gadgil and Bossert, if there is an increase in maintenance energy there should also be a corresponding decrease in some other component of the energy budget. In this case Roff documented substantial declines in the biomass of female gonads of the long-winged forms (Figure 9.13C). ❖

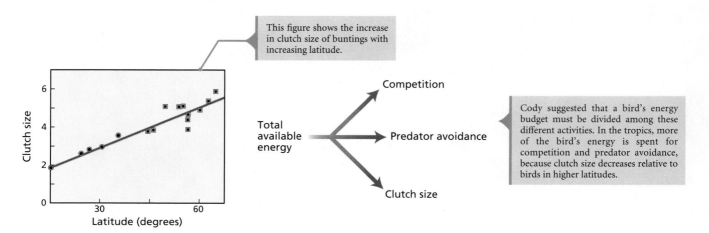

This figure shows the increase in clutch size of buntings with increasing latitude.

Total available energy → Competition

Total available energy → Predator avoidance

Total available energy → Clutch size

Cody suggested that a bird's energy budget must be divided among these different activities. In the tropics, more of the bird's energy is spent for competition and predator avoidance, because clutch size decreases relative to birds in higher latitudes.

FIGURE 9.13A Cody proposed an energetic explanation for variation in avian clutch size.

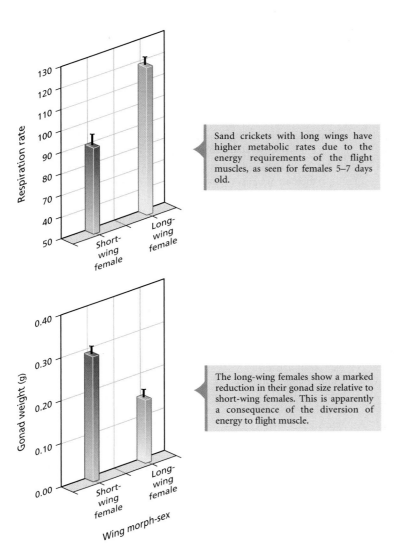

Sand crickets with long wings have higher metabolic rates due to the energy requirements of the flight muscles, as seen for females 5–7 days old.

The long-wing females show a marked reduction in their gonad size relative to short-wing females. This is apparently a consequence of the diversion of energy to flight muscle.

FIGURE 9.13B **The Long-Winged Form of the Sand Cricket, *Gryllus firmus***

FIGURE 9.13C Reproduction and dispersal capability trade-off in sand crickets.

SUMMARY

1. Animals in aquatic environments will experience net flows of water and ions in or out of their cells, depending on the osmotic concentration of the surroundings. Some animals are osmoconformers and do not try to regulate the concentration of solutes of their cells. Others animals maintain concentration gradients.

2. Animals must also take up oxygen from the environment to fuel aerobic metabolism. This is most often accomplished with gills and lungs. These organs typically have large surface areas exposed to oxygenated water or air.

3. Many physiological systems for oxygen uptake, heat exchange, and ion reabsorption use countercurrent exchange to improve efficiency.

4. Plants must transport water over large distances. In some trees the distances from roots to treetop can be 100 meters. This transporting of water is accomplished through the joint effects of transpiration and the cohesive forces of water.

5. Many plants and animals live in hot, dry environments and must cope with high rates of water loss. Dessication, like many physiological systems, is subject to natural selection.

6. Fruit flies adapt to dry environments by storing more water and slowing down their rate of water loss.

7. Fruit flies increase their fat reserves to resist starvation and evolve increased longevity as a correlated response to starvation resistance.

8. Transportation is a major portion of the energy budget of most animals. The standardized cost of transportation is greatest on land and smallest in water.

9. Life-history evolution involves allocating energy between maintenance, growth, and reproduction in a manner that has the greatest effect on fitness.

REVIEW QUESTIONS

1. Review the types of stress placed on the cells of animals that live in hyperosmotic conditions vs. hypoosmotic conditions.

2. Give two examples of countercurrent exchange systems, and explain how they work.

3. Why can't air pressure alone explain the movement of water through plants?

4. What are trichomes, and how do they prevent water loss in plants?

5. Why do veins in blood circulatory systems have valves on them?

6. What physiological changes occur in populations of fruit flies selected for desiccation resistance?

7. What are the major patterns of nitrogen excretion used by animals?

8. Fruit flies selected for starvation resistance change in a number of ways, including the storage of lipids. Why don't all fruit flies employ these adaptations?

9. How can the respiratory quotient be used to infer the type of fuels an organism is using to produce energy?

10. How does energy affect the course of life-history evolution?

KEY TERMS

ammonotelism	countercurrent exchange	maximum metabolic rate	stenohaline
anabolic	cuticle	metabolic rate	stomata
aorta	epicuticle	Mueller, Laurence	stomatal transpiration
arteriole	euryhaline	open circulatory system	systemic circuit
basal metabolic rate	factorial aerobic scope	osmoconformer	trichome
blood	gait	osmotic concentration	ureotelism
boundary layer	gills	osmoregulate	uricotelism
capillary	heart	pulmonary circuit	vasoconstruction
catabolize	hyperosmotic	ram ventilation	vasodilation
closed circulatory system	hypoosmotic	respiratory quotient	vapor pressure
Cody, Martin	hypoxic	Rose, Michael	venule
cohesion	lung	standard metabolic rate	xylem
cost of transport			

FURTHER READINGS

Borash, D. J., V. A. Pierce, A. G. Gibbs, and L. D. Mueller. 2000. "Evolution of Ammonia and Urea Tolerance in *Drosophila melanogaster*: Resistance and Cross-tolerance." *Journal of Insect Physiology* 46:763–69.

Chippindale, A. K., T. F.J. Chu, and M. R. Rose. 1996. "Complex Trade-offs and the Evolution of Starvation Resistance in *Drosophila melanogaster*." *Evolution* 50:753–66.

Cody, M. 1966. "A General Theory of Clutch Size." *Evolution* 20:174–84.

Crnokrak, P., and D. A. Roff. 2002. "Trade-offs to Flight Capability in *Gryllus firmus*: The Influence of Whole-Organism Respiration Rate on Fitness." *Journal of Evolutionary Biology* 15:388–98.

Gadgil, M., and W. Bossert. 1970. "Life Historical Consequences of Natural Selection." *American Naturalist* 104:1–24.

Schmidt-Nielsen, K. 1990. *Animal Physiology*. Cambridge, UK: Cambridge University.

Willmer, P., G. Stone, and I. Johnston. 2000. *Environmental Physiology of Animals*. Oxford, UK: Blackwell Science.

Cannery Row, Monterey, California, 1973

10

Balancing Birth and Death

Cannery Row in Monterey in California is a poem, a stink, a grating noise, a quality of light, a habit, a nostalgia, a dream. . . . In the morning when the sardine fleet has made its catch, the purse-seiners waddle heavily into the bay blowing their whistles. The deep-laden boats pull in against the coast where canneries dip their tails into the bay.

These words from John Steinbeck's *Cannery Row* describe a sardine fishery thriving off the coast of Central California. Yet shortly after *Cannery Row* was written in 1945, the sardine population vanished. Today Cannery Row is a tourist attraction. What happened to the sardines is now an ecological mystery.

Did the sardines off the coast of California simply move to a different habitat? Or did the fishing industry drive the population to extinction? To answer these questions we need to understand how populations grow in number, and we need to uncover the biology that determines growth.

Many factors affect the growth of a population. Some depend on the size, or *density*, of the population, particularly how crowded it is. For this reason, these factors are called *density dependent*. Other factors may not depend on population size. One such factor is dispersal, the movement of individuals between populations. Of course, weather and other features of the physical environment affect the growth of populations, as well as their dispersal.

Population growth affects natural selection, and natural selection in turn shapes population growth. Interrelationships between population growth and natural selection provide fascinating intersections of ecology with evolution. Some of our most important insights into the natural world have come from the study of the subtle ways in which populations grow in number. ❖

THE POPULATION BOMB

10.1 Populations are collections of interbreeding individuals and the basic units of ecology and evolution

In his influential book, *The Population Bomb* (1968), renowned population biologist Paul Ehrlich describes a night he experienced in Delhi, India:

> The temperature was well over 100, and the air was a haze of dust and smoke. The streets seemed alive with people. People eating, people washing, people sleeping. People visiting, arguing, and screaming. People thrusting their hands through the taxi window begging. People defecating and urinating. People clinging to buses. People herding animals. People, people, people, people. ... since that night I've known the *feel* of overpopulation.

The recent history of the human population is one of almost unfettered growth. It is shocking to realize how quickly the planet is becoming populated. In 6000 B.C. there may have been about 5 million humans. Before that, it probably took 1 million years for the population to increase from 2.5 to 5 million. We call the time it takes for the population to double in size a **doubling time**. From 6000 B.C. to A.D. 1650, the doubling time averaged about 1000 years. However, by 1850 the population had again doubled from its level in 1650. By 1930 the doubling time had fallen to 80 years. By the time *The Population Bomb* first appeared in 1968, the doubling time was about 35 years, and the world human population was about 3 billion.

One important prediction of exponential growth is that if births exceed deaths, then—if we are also given unlimited time—the population will get infinitely large. But it is impossible for a planet of limited size and resources to sustain exponential growth of any population forever. Thomas Robert Malthus (Figure 10.1A) was one of the first to articulate this idea in 1789 when he wrote, "The power of population is so superior to the power in earth to produce subsistence for man, that premature death must in some shape or other visit

the human race." The logic behind these conclusions is so simple, we may consider a fundamental principle of ecology to be that *no population will grow exponentially forever.*

The ideas of Malthus were so persuasive to Darwin that they became a crucial component in the development of his own thinking, as we saw in Chapter 1. According to Malthus, populations will tend to produce an excess of progeny that cannot all survive. For Darwin the important question was, which individuals will die and which will survive? He answered the question with his theory of natural selection: Those individuals possessing traits that better adapted them to their environment should be most likely to survive.

The temperature was well over 100, and the air was a haze of dust and smoke.

In evolutionary biology a **population** is a group of individuals that regularly exchange genes. It is this exchange of genetic material that links the evolution of the population's members. Ecologists are often less concerned with the exchange of genes than with the physical boundaries within which individual organisms may be found. To be consistent in this book, we will use the evolutionary definition. However, we need to bear in mind that defining the ecological boundaries of populations may be difficult.

Suppose we wanted to predict the number of people reaching retirement age in the next 20 years. Estimates of these numbers can be most easily made with the help of a mathematical model of population growth. As with most models in population bi-

ology, however, there is seldom a single model that we can apply to all organisms. So before turning to the models, we must address the general life cycle of the organism. There are two major categories of life cycles we could model. Both were introduced in Chapter 7.

The first life cycle is one with **discrete generations**. This means that reproduction is synchronized among the adults: There is a breeding season. We often see manifestations of this life cycle during springtime in temperate climates. Released from the rigors of winter, adults greet each other with a view to participating in sex and other reproductive activities. The offspring from this round of breeding then develop and enter the population of breeding adults in the next generation. Among the many organisms that actually follow this life cycle are many annual plants, insects that produce one generation per year, and salmon. When generations are discrete, time moves in sudden jerks. It does not flow.

The second major type of life cycle involves organisms with continuous reproduction. Instead of discrete pulses of reproduction, reproduction provides a continuous flow of new recruits into the population. Therefore, this type of life cycle is represented by continuous time. There are no jerks in the ecological process, only a seamless procession.

Because discrete time models do not require knowledge of calculus, we will use them exclusively in this chapter. Often the biological conclusions are qualitatively the same with both models. We are not usually sacrificing valuable ecological insights by limiting ourselves to discrete-time models.

Another important aspect of models of population growth is age structure. Organisms whose adults may reproduce multiple times and live for an extended period have an **age structure;** that is, they consist of individuals of many different ages in different proportions. This means we need to know something about how mortality and survival change with age. It is easier to describe the growth of populations without age structure, and so Modules 10.2 through 10.14 focus on these types of populations. ❖

FIGURE 10.1A Thomas Malthus forecast the outcome of exponential growth.

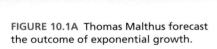

T. R. Malthus (1766–1834)

Thomas Robert Malthus graduated from Jesus College, Cambridge, with honors in mathematics. Although Malthus was concerned about overpopulation, he was apparently not concerned about inbreeding; he married his cousin, Harriet Eckersall, and had three children. His most lasting work—*An Essay on the Principle of Population, as it Affects the Future Improvement of Society with Remarks on the Speculations of Mr. Godwin, M. Condorcet, and Other Writers*—was published in 1798 anonymously. This essay met with success, and Malthus published several revisions and summaries of this work. Malthus was appointed Professor of Political Economy at East India College in 1805, where he worked until his death.

Malthus conjectured that there were two principal mechanisms to halting population growth: *Preventive checks* included mechanisms like postponed marriage, abstinence, homosexuality, birth control, and abortion. *Positive checks* were the more severe forms of population control, which entailed increased death rates from sources like war and famine. Malthus was keenly aware that these positive checks would not be meted out evenly and that the poorest would suffer most. Consequently, he favored the use of preventive checks like postponed marriage as a means of avoiding the devastation of positive checks.

We start with the exponential model of population growth, a model that almost never applies to any real population—or, if it does, will apply for only brief periods of time. This does not mean the model is useless. In fact, by learning the predictions of the exponential model, we are led to more realistic descriptions of population growth.

The development of this model requires some assumptions about the organism's life cycle. For simplicity we focus attention in Figure 10.2A on asexual organisms, which reproduce synchronously. This bout of reproduction can be thought of as a breeding season. Limiting this discussion to asexual organisms

may seem extreme, but in fact the inclusion of two sexes does not substantially alter the predictions of the exponential model.

We also assume that each individual dies after it has reproduced, and the next generation is constituted only from the surviving offspring. Many organisms have this type of life cycle: annual plants, black widow spiders, Pacific salmon, and others described in Chapter 7. If we allowed some adults to survive to the next generation we would get similar results, but the development of the model would be a bit tricky.

The key assumption of the model is that each individual, on average, produces a constant number of offspring,

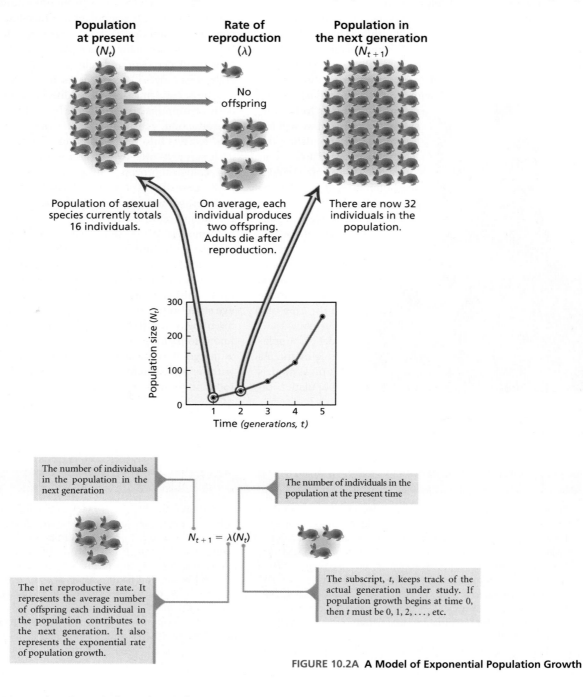

FIGURE 10.2A A Model of Exponential Population Growth

referred to as λ (lambda). We also call λ the **net reproductive rate**. We assume that this rate of reproduction does not change over time or as the population increases in size.

Figure 10.2A shows a model of exponential population growth using the example of an asexual population that starts out with 16 members and grows at a net reproductive rate (λ) of 2. This population increases in size over time, and the numbers that are added to the population each generation get larger. As long as each individual in the population produces a slight excess beyond replacement ($\lambda > 1.0$), the population will grow in this manner. These are the characteristic signs of exponential growth. Not only do the numbers increase by larger amounts in each generation, but as a great deal of time passes, the population size is predicted to get infinitely large. This is the major defect in this model,

since we know from common sense and practical experience that no population on Earth can become infinitely large. There must be some forces at work in natural populations that prevent this growth. We will discuss some of those forces later in this module.

In the very unlikely case that $\lambda = 1.0$, the population will remain at a constant size that equals its initial size. The likelihood of this occurring is effectively zero. If the average individual in the population does not replace itself ($\lambda < 1.0$), then the population will decline in size every generation until it goes extinct.

The following box shows how to predict mathematically what the size of an exponentially growing population will be in the distant future. ❖❖

Population Growth over Multiple Generations

The model of exponential growth in Figure 10.2A shows how population size changes over a single generation. Is there some way to predict the population size at some distant time in the future for exponentially growing populations? In fact, there is an easy way to do this. Remember that the relationship shown in Figure 10.2A holds for any generation. So if we start with some arbitrary time in the future, time t, then we know that

$$N_t = \lambda N_{t-1}. \qquad (1)$$

In other words, the size of the present generation equals the net reproductive rate times the size of the previous generation.

It would be helpful to have N_t on the left-hand side of the equation and some simple relationship involving the initial population size (N_0) on the right-hand side. To do this we must replace N_{t-1} with some other relationship. We know from the figure that if we take the right-hand side of this equation and replace N_{t-1} in equation (1) above with this value we get, $N_t = \lambda(\lambda N_{t-2}) = \lambda^2 N_{t-2}$. If we continue with these substitutions, we get the useful result, $N_t = \lambda^t N_0$.

From this relationship we can easily produce figures, like Figure 10.2B here, showing the long-term behavior of exponentially growing populations. Thus, an initial group of 50 mice that had a net reproductive rate of 2.0 would grow to 200 mice (50×2^2) after two generations, and to 51,200 mice after 10 generations (50×2^{10}).

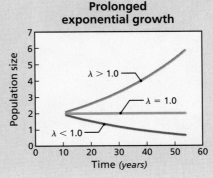

Prolonged exponential growth

$\lambda > 1.0$

$\lambda = 1.0$

$\lambda < 1.0$

Population size / Time (years)

FIGURE 10.2B
Prolonged Exponential Growth

No biological population can grow exponentially forever. Eventually, populations become so crowded that food, shelter, and other essential resources are difficult to find. Individuals that cannot find sufficient resources may die, or if they survive, their capacity to reproduce may be impaired. Accordingly, the net reproductive rate, which we call λ, can be divided into two components as follows:

λ = (*number of offspring produced*) × (*probability that each offspring survives to reproduce in the next generation*)

The two components are given in parentheses above. The first is the total number of offspring produced by each individual—sometimes called fertility—in the previous generation. The second is the probability that each of those offspring will survive to reproduce—sometimes called viability—in the next generation.

Crowding can reduce the number of newborn offspring, reduce offspring survival, or both. The specifics will vary among organisms. For example, larval growth in insects and amphibians can determine the size of the reproductive adult. If food is scarce, larvae may grow slowly and produce small adults. These small adults may then have fewer offspring, as shown in Figure 10.3A. If food levels are very low, death may result, as illustrated in Figure 10.3B. Like-

wise, if adults starve, they may have reduced fertility. For example, human females whose body fat has been reduced due to starvation or intense exercise may fail to ovulate.

Crowded populations may also suffer from increased disease. This could arise from increased transmission of infectious disease from one individual to the next, thanks to more frequent contact. Or it may come from the accumulation of feces, urine, and other waste products that accompany crowded conditions. The black plague in Europe may have been one such example, as the following box describes.

Another consequence of crowding is increased susceptibility to predators. Predators sometimes concentrate their efforts on the most common prey in a given area, all but ignoring acceptable but less common species. Thus the most numerous species may also suffer increased mortality from its popularity with predators.

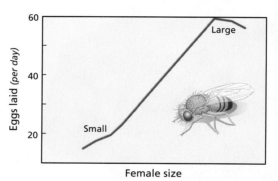

> Fertility may be reduced by:
> • limited food for adults
> • increased behavioral interactions
> • reduced size of adults due to crowding of prereproductive individuals

FIGURE 10.3A   The Relationship between Female Body Size and Egg Laying in Laboratory Fruit Flies

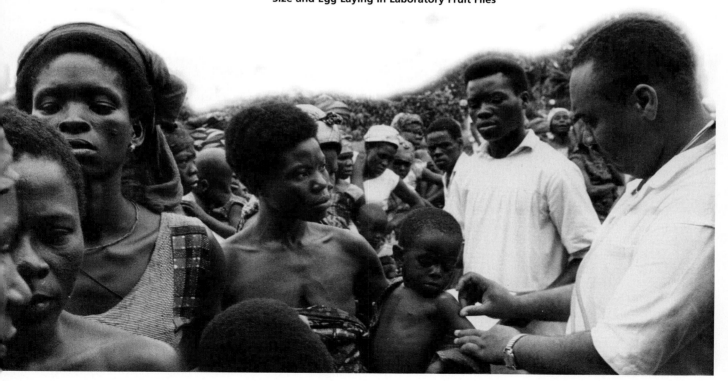

In mammals, crowding can stimulate the formation of steroids such as cortisone, cortisol and corticosterone, which help the organism adapt to stress by creating ready reserves of carbohydrate for quick energy. However, prolonged stress and production of these steroids have adverse side effects, which include the weakening of the immune system and sometimes death.

In general, the deleterious effects of high population size get more severe as the level of crowding increases. This will lead to further reductions in fertility and survival, which in turn will further reduce the value of λ. Ultimately, the net reproductive rate will either equal one or be less than one. At this point the population will cease growing. A goal of Module 10.6 will be to predict when a population will cease growing and thus how many individuals can be reasonably supported by a particular environment. ❖

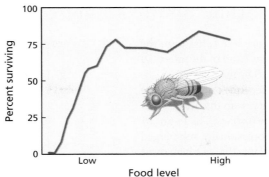

Survival may be reduced by:
* limited food
* increased waste products
* increased susceptibility to predators
* increased stress
* infanticide and cannibalism
* disease

FIGURE 10.3B The Relationship between Food Level and Survival in Laboratory Fruit Flies

The Black Plague

Humankind has experienced three worldwide epidemics of the deadly disease known as the black plague (Figure 10.3C). The second pandemic, which ravaged Asia and Europe for three centuries, may have killed 25 million people in Europe during the fourteenth century—nearly one quarter of the total population. This disease results in death in 50 percent or more of those who are afflicted and untreated. It is caused by the bacteria *Pasteurella pestis*, which infects rats and humans and is primarily transmitted by fleas.

As Eurasian populations grew during the Middle Ages, the lack of sanitation helped maintain large populations of rats. When epidemics of the plague killed large numbers of rats, the fleas would move to their less favored human hosts with devastating effects. Although the cause of the disease was unknown until 1894, ancient civilizations were aware of important aspects of its epidemiology. The fifth-century Indian medical work, *Bhagavata Purana*, implores people to leave their homes, "when rats fall from the roofs above, jump about and die." The Venetian Republic was so concerned about the disease that in 1374 it required all ships coming from plague-infested territories to wait 40 days before landing, so any infections could run their course. This 40-day period, or *quaranti giorni*, is the root of the English word *quarantine*, meaning enforced isolation to prevent the spread of disease.

FIGURE 10.3C *The Plague*, by Arnöld Bocklin

The *rocky intertidal zone*, shown in Figure 10.4A, has been the site of many important studies in ecology. Land in the intertidal zone is splashed and partially submerged by the ocean. The duration of these periods of submersion depends on the tides, the local weather conditions, and the distance from the ocean—criteria that subdivide the intertidal into two zones. The *upper intertidal zone* is mostly splashed, only occasionally submerged during the highest tides. Animals and plants that live here must be able to withstand heat and water loss. The *lower intertidal zone* is only briefly exposed to the air each day. Animals and plants here, such as the barnacles and algae in Figure 10.4A, are constantly pounded by the surf and must be securely fastened to rocks; otherwise they will be washed out to sea.

FIGURE 10.4A The Rocky Intertidal Zone in the Pacific Northwest

Many intertidal animals survive by filtering food brought in by the tides and surf. Many small, often microscopic, plants and animals live near the surface of the ocean. They are called *plankton*. Mucus secreted by *filter-feeding* animals traps plankton, which is then used as food. For a filter-feeding animal, the intertidal can be a luxurious place to live because the continuous movement of water ensures a constant supply of food. In many intertidal habitats, the first resource to vanish when population size is high is not food but open space for attachment. Thus, one limit to plant and animal numbers may simply be the available space. Any environmental factor that limits the distribution or abundance of an organism is called a **limiting resource**.

The relationship between free space and animal numbers was demonstrated by Paul Dayton. He created two experimental regions in an intertidal zone. The control region (Figure 10.4B) contained undisturbed populations of sea

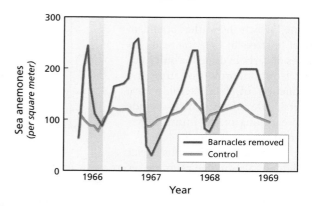

FIGURE 10.4C Paul Dayton removed 45 percent of the barnacles in some areas and none in others (control areas). In the winter and fall sea anemones grew rapidly in the area lacking barnacles, due to the extra free space. In the summer the numbers of anemones declined due to warm, drying conditions.

anemones (*Anthopleura elegantissima*) and barnacles (*Balanus cariosus*). The test region was the same except that 45 percent of the barnacles were removed, creating more free space. As Figure 10.4C shows, the number of sea anemones stayed relatively constant in the control region. But in the test region, the numbers of sea anemones fluctuated widely. Their numbers increased rapidly in the fall and winter as they occupied the newly cleared space. In the summer the number of anemones declined due to the drying heat and wind.

Although space can be a limiting resource in the intertidal zone, as Dayton's experiment shows, it is not always a limiting resource. Other aspects of the physical environment may also determine when space is limiting. For example, although sea anemones can reproduce by a vegetative process of budding, many invertebrates, such as barnacles, broadcast large numbers of larvae out to sea. These larvae must develop and then be washed by tidal currents onto open space. In Oregon and Washington, it appears that there are always large numbers of these invertebrate larvae settling, leaving little free space (Figure 10.4A). In California the nearshore currents carry the larvae farther from shore, increasing the likelihood that the larvae will die before settling. Consequently, there is more free space in the California intertidal zone than in the Northwest, as Figure 10.4D shows. ❖

FIGURE 10.4B On the San Juan Islands in Washington State, sea anemones and barnacles live in the same area of the intertidal zone. The sea anemones are often clumped below large barnacles and thereby protected from drying in the air.

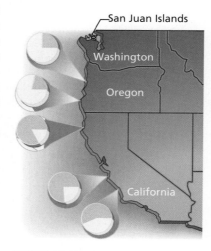

FIGURE 10.4D The amount of free space (orange) in the intertidal zone increases from Oregon to California.

Experimental and theoretical ecology began with investigations of single-species population growth

How do Malthus's "positive checks," such as famine, affect growing populations? Rather than generating catastrophes, they usually act by creating gradual declines in reproductive rates as population size increases. Ecologists refer to the operation of such checks on population increase as **density-dependent population growth**, with *density* referring to the numbers of organisms inhabiting a particular area.

FIGURE 10.5A Raymond Pearl

In fact, the relationship between population growth and density was the first focus of ecological theory. Raymond Pearl (Figure 10.5A) vigorously explored this relationship in the early part of the twentieth century. Pearl championed the use of a mathematical model, called the *logistic equation* (see Module 10.6), to describe density-dependent population growth. Much of Pearl's experimental research was used to support the logistic equation.

Pearl initiated a strong experimental laboratory research program to investigate the theory of density-dependent population growth. The use of experimental laboratory research is still not as common in ecology as it is in many areas of science, despite early laboratory research by Pearl (see the box), Gause (see Module 12.6), and others. Ultimately, ecologists are interested in understanding the forces and principles responsible for determining the abundance and distribution of organisms in their natural environments. Experimental laboratory research involves environments that are different from those in nature, and usually simpler. This difference has led some ecologists to question the validity of inferences made from laboratory systems.

The use of simplified laboratory systems has served other branches of science very well. As an example, a physicist may be interested in understanding the laws that govern the motion of objects in the Earth's atmosphere or on its surface. In these types of "environments," friction affects the velocity and acceleration of objects. However, physicists have found it helpful first to study the laws governing the motion of objects in the absence of friction. Introductory courses in physics still begin instruction with these simple models of motion. After developing and testing these simple theories, physicists then modified their laws of motion to include the complications of friction.

The science of ecology is not nearly as advanced as physics. Ecologists are still developing very basic models. Thus, although we accept the notion that the growth of most populations will be affected by factors other than density, many of our models are focused on the idea of density dependence by itself. However, progress in ecology will require the careful development and testing of additional ideas, which must of necessity start out simply.

Ecology has also advanced by carefully designed experiments using natural populations. In addition, we have learned much through careful observation of organisms in their natural habitats.

Many organisms experience a decline in net reproductive rates with increasing population density. In effect, increasing population size regulates the ultimate number of individuals in the population. However, this regulation may be far from perfect. A basic issue in ecology is the extent to which fluctuations in population size are due either to the environment or reproductive regulatory mechanisms, as seen in Chapter 1.

Some species regularly experience periods of crowding. These organisms are often able to endure severe crowding, managing to survive and reproduce. These abilities are rooted in adaptations of behavior or morphology. As we will see in Module 10.7, these adaptations are important for understanding how organisms live in changing environments. ❖

Raymond Pearl (1879–1940), the Founder of Experimental Ecology

After obtaining his Ph.D. from the University of Michigan, Raymond Pearl studied in Europe with the biostatistician Karl Pearson. Pearson instilled in Pearl an appreciation for mathematical approaches to biology. Pearson also taught a method of achieving scientific generalizations that appealed to Pearl. This method was to take a wide variety of scientific observations and reduce them to a brief formula or a few words, called a law.

With this background Pearl was eager to find the laws that governed population growth. He undertook research at the Institute for Biological Research at Johns Hopkins University, which he had established with support from the Rockefeller Foundation. Pearl did experimental research with fruit flies to test models of population growth.

He was also interested in predicting the growth of human populations with the logistic model. Pearl's work was criticized for his attempts to extrapolate human population growth from the logistic model. Some scientists thought that the only way to control human population growth was through the genetically based improvement of human intelligence. Pearl's theory seemed to contradict this approach by suggesting that a variety of natural factors will predictably and reliably slow down population growth. Ultimately, the long-term predictions of the logistic model were not very accurate. In the 1920s Pearl predicted that the population of the United States would be about 197 million by the year 2000. In fact the real number was closer to 272 million, although it had been boosted by immigration.

The effects of growth-limiting factors increase with population density. If net reproductive rates decline as population density increases, the population will reach a density where the net reproductive rate is exactly one, and each individual in the population simply replaces itself. When this happens, the population will cease growing until population densities decrease and net reproductive rates are again greater than one. It is also possible for the population size to get so large that net reproductive rates drop below one. In that case, the population size will decrease until the net reproductive rates again rise to one.

To make these predictions more quantitative, we need to define the mathematical relationship between population size and net reproductive rate. No single equation will apply to all organisms. Indeed, we expect this relationship to vary from one species to another, and perhaps between populations of the same species. But many important consequences of density-dependent population growth can be understood by examining simple models.

One such model, known as the *logistic model,* is based on a linear relationship between density and net reproductive rate. In contrast to the exponential model that we examined in Module 10.2, which sees the net reproductive rate as constant, the logistic model assumes that the reproductive rate decreases linearly as population density increases, as the graph in Figure 10.6A shows. The properties of the logistic model depend on numerical constants called parameters. The logistic model uses only two parameters. The first is r, the **intrinsic rate of increase**. This parameter determines the maximum rate of growth at low densities. The second parameter, K, is called the carrying capacity. The **carrying capacity** reflects the maximum number of individuals that the environment can support.

In Figure 10.6A we see that as the number of individuals in the population (N_t) gets close to zero, the population size in the next generation is approximately $(1 + r)N_t$. If $r > 0$, then the population is growing at an exponential rate equal to $1 + r$. As the population grows, the

rate of exponential growth slows, reflecting the combined effects of reduced fertility and increased mortality brought about by crowding. (Mathematically, the slowing occurs because $-rN_t/K$ becomes significantly less than zero and this quantity is added to the exponential growth term, $1 + r$.) As the population size gets closer to the value K (the carrying capacity of the environment), the net reproductive rate declines and gets closer to one. When the population size is exactly equal to K, the population ceases to change size. We call this point $N_t = K$ an *equilibrium* of the

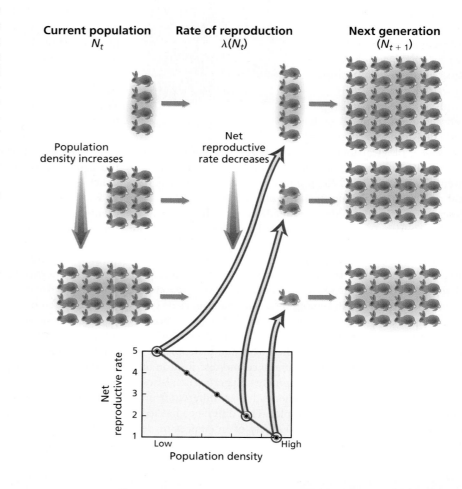

Current population
N_t

Rate of reproduction
$\lambda(N_t)$

Next generation
$(N_{t + 1})$

Population density increases

Net reproductive rate decreases

Net reproductive rate

5
4
3
2
1

Low   High

Population density

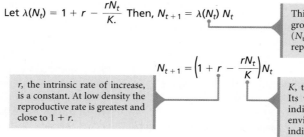

Let $\lambda(N_t) = 1 + r - \dfrac{rN_t}{K}.$ Then, $N_{t+1} = \lambda(N_t)\,N_t$

This looks the same as the exponential growth model, except we have added (N_t) next to the λ to emphasize that reproductive value depends on density.

$$N_{t+1} = \left(1 + r - \frac{rN_t}{K}\right)N_t$$

r, the intrinsic rate of increase, is a constant. At low density the reproductive rate is greatest and close to $1 + r$.

K, the carrying capacity is a constant. Its value will depend on how many individuals can be supported by the environment. When $N_t = K$, each individual just replaces itself and the population size stays constant.

FIGURE 10.6A A Model of Logistic Population Growth

logistic model because population size does not change at that population density.

This simple mathematical result illustrates the general principle that density-dependent population growth causes populations to grow toward their equilibrium density whether they start at low or high densities. Ecologists use this theoretical pattern to explain the maintenance of steady population densities in both experimental and natural populations. In some species, the logistic model does a reasonable job of predicting population growth (see box). ❖

Logistic Population Growth in Real Populations

How does the growth of real populations compare with the predictions of the logistic model? The growth of real biological populations can be studied by examining the total change in population size as a function of density. The total change in population size is simply the present population size minus the size in the previous generation, or in terms of the notation we developed earlier, $N_{t+1} - N_t$. This change can then be compared to the change predicted by the logistic equation. Using the results developed in Figure 10.6A, a little algebra shows that $N_{t+1} - N_t = rN_t - rN_t^2/K$. For the unicellular ciliate, *Paramecium caudatum*, r has been estimated from the observations in Figure 10.6B to be 0.922 and K as 200. Thus, when the density of *P. caudatum* is 100 animals per 0.5 cc, the expected change in population size is $0.922 \times 100 - 0.922 \times 100^2/200 = 46.1$. The solid line in Figure 10.6B shows the predicted changes in population size from the logistic equation. The observed changes, shown by the circles, generally are fairly close to the predicted changes.

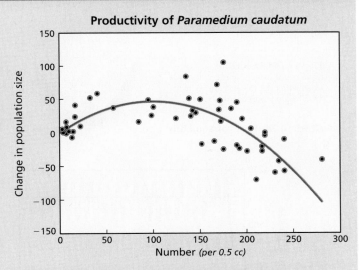

FIGURE 10.6B The growth of *Paramecium caudatum* as a function of population density. The solid line shows the predicted change in population size from the logistic equation; the symbols show the observed change.

For the logistic model of population growth, when net reproductive rates are equal to one, the population size should stay constant. The population size at which the population ceases growing is an equilibrium. Many factors other than population density may affect the population size. Environmental fluctuations, such as those produced by weather, may lead to either additional deaths or enhanced survival, changing population size by small amounts above or below the equilibrium. This type of small change is called a **perturbation**.

Will the population return to the equilibrium size after a perturbation, or will it move to some new equilibrium? The answer to this question hinges on a property of the equilibrium known as its **stability**. If the population returns to the equilibrium after a perturbation, we say the *equilibrium is stable*. If the population moves away from the equilibrium size, this is evidence of an *unstable equilibrium*. Figure 10.7A presents a visual analogy for stable and unstable equilibria.

For the logistic model the stability of the equilibrium, $N_t = K$, is determined by the magnitude of r, the intrinsic rate of increase. If r is greater than zero but less than two (the red curve in Figure 10.7B), the equilibrium at K, the carrying capacity, is stable. Let's try to make sense of this condition. If r is less than zero, net reproductive rates are less than one at all densities and the population will die out (recall that at very low density the net reproductive rate will be approximately $1 + r$). When r is greater than zero but less than two, net reproductive rates are

STABLE EQUILIBRIUM
If the ball is moved up the valley, it is returned by the force of gravity to its equilibrium position at the bottom.

UNSTABLE EQUILIBRIUM
In this case, any slight perturbation of the ball from its resting position causes the ball to continue moving away from its equilibrium position at the top of the hill.

FIGURE 10.7A **Types of Equilibria**

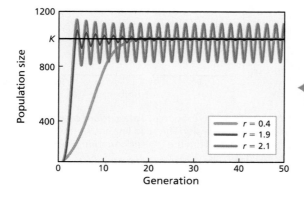

For the logistic model, the stability of the equilibrium point, $N_t = K$, is determined by the value of r.

	$r = 0.4$
	$r = 1.9$
	$r = 2.1$

FIGURE 10.7B **Stability of the Logistic Model**

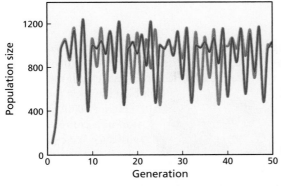

If *r* is very large, population size fluctuates in a seemingly erratic fashion. This behavior is called chaos. Populations that display chaotic dynamics are very sensitive to small displacements.

The graph shows the population size predicted from the logistic equation with $r = 2.7$ and $K = 1000$. The only difference between the two curves is that the red line started at 100 individuals and the blue line started at 101.

FIGURE 10.7C **Chaos**

greater than one and the population size increases at low density. As population size increases, the net reproductive rate decreases, slowly approaching one and allowing the population size to gradually approach the carrying capacity (green curve). As *r* gets larger the population tends to overshoot and then undershoot the equilibrium size, oscillating but eventually approaching the carrying capacity (red curve). Eventually, when $r > 2.0$ (blue curve), the overshoots and undershoots continue indefinitely.

When *r* gets very large (which for the logistic model means $r > 2.57$), the population dynamics become very erratic. This type of behavior is called **chaos**. Chaos is a type of behavior exhibited by many things besides population growth—from electrical circuits to economies. One property of chaos in populations is that population sizes can change abruptly and dramatically in a single generation—as shown in Figure 10.7C, where a theoretical population with a carrying capacity of 1000 occasionally drops to 360 individuals. It

isn't easy to predict the future of a chaotic population. Figure 10.7C shows how different the future population sizes are in two chaotic populations that start at almost the same sizes (100 vs. 101). With real populations, prediction is even more complicated because population size also varies due to random environmental factors.

The box below demonstrates stable and unstable equilibria in two laboratory environments. Laboratory experiments, like the one below, aid the ecologists in dissecting the important events that determine population stability. In flour beetle populations adults often cannibalize pupae. The stability of the flour beetle populations depends critically on these rates of pupal cannibalism. It also turns out that flour beetle larvae cannibalize eggs and occasionally pupae. However, larval cannibalism is less important for stability of the beetle populations. Ultimately, high levels of adult cannibalism and high adult mortality can lead to population cycles and even chaos. ❖

Stability of Laboratory Populations of Fruit Flies

Below are data from laboratory populations of the fruit fly *Drosophila melanogaster*. Figure 10.7D shows five populations kept under conditions that result in a stable carrying capacity. During the first five generations there is a slow and steady increase in the population size. After generation five, the populations fluctuate around their carrying capacity. The populations in Figure 10.7E are in a different environment, where the carrying capacity is not stable. These populations show large increases in size immediately followed by large decreases. For *Drosophila* an unstable environment is one in which adult food levels are high and larval food levels are low. The high food level provided to adults results in substantial increases in egg production, thus effectively increasing the value of *r*.

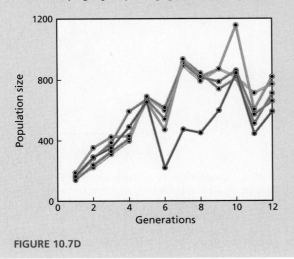

FIGURE 10.7D

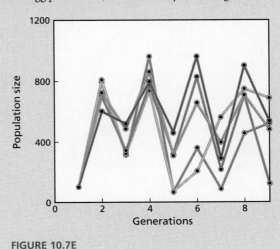

FIGURE 10.7E

The logistic equation is based on the premise that crowding reduces survival and fertility. What are the biological mechanisms that generate these effects? Here, we consider some of the biological mechanisms that occur in organisms with complex life cycles. Some organisms start life as small versions of adults and then grow to their adult size. Other organisms, including some frogs and insects, have juvenile and adult forms that are morphologically quite different. They may even live in different habitats. For example, many insects start life as crawling or burrowing larvae with limited ability to move large distances. After an active period of larval growth, they form a resting phase as a pupa. Usually the pupa is covered with a protective case, inside of which the insect undergoes a dramatic metamorphosis to the adult form. Usually the adult can fly. Likewise, amphibians normally begin life as an aquatic organism with gills. To become adult, their bodies undergo a metamorphosis to

form legs that permit the animal to travel on land (see Figure 10.8A). The combination of two distinct life-forms into one life cycle has led ecologists to call the typical insect and amphibian life cycle *complex*.

Figure 10.8A is a generalized illustration of a **complex life cycle**. Organisms with such life cycles devote the first part of the cycle to growth and the second part to reproduction. For the insect and amphibian life cycles, the adult phase is better able to disperse. But in marine invertebrates the dispersal cycle is often reversed. The pre-reproductive stages live as zooplankton, small floating animals that are dispersed by the ocean currents. To mature, the larval zooplankton settle out and attach to rock, only then beginning reproduction.

Because the different life stages may occupy very different environments, the effects of population crowding may differ between life stages. We have already discussed the problems of limited space faced by many marine intertidal invertebrates. Although less is known about the planktonic phase of these organisms' life cycle, the effects of density must be much less severe among plankton.

Crowded larval conditions are probably common for many amphibians and insects. These organisms have evolved ways of coping with crowding that increase their chances of surviving to become adults. For instance, when some toads (*Bufo americanus*) are crowded as larvae, food is in short supply. The larvae then grow much more slowly and metamorphose into very small adults (see Figure 10.8B). This small body size reduces the number of offspring they can produce as adults. However, had the larva died from starvation, they would not have had offspring at all.

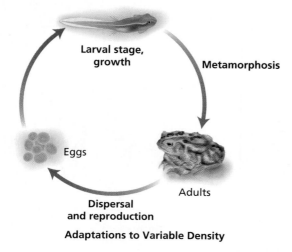

FIGURE 10.8A A Generalized Complex Life Cycle

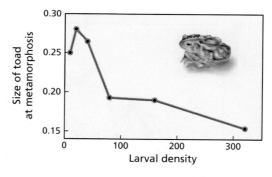

FIGURE 10.8B Adaptations to Variable Density: Toads When toads are crowded, the tadpoles may metamorphose at a small size and thus avoid death by starvation. Interestingly, when the density is very low there are probably too few larvae to stir up all the food and keep it suspended in the water column. As a result, there is a small decline in body size at the lowest larval density.

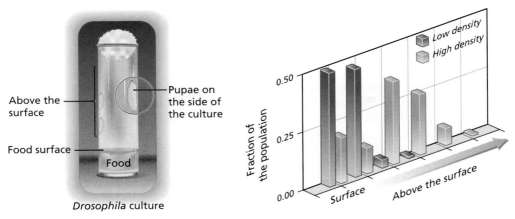

FIGURE 10.8C Adaptations to Variable Density: Flies Larvae of the fruit fly, *Drosophila*, often form pupal cases on the surface of the food medium in uncrowded cultures. If the cultures are crowded, the pupae can sink into the soft food and drown. In such crowded cultures the larvae try to avoid pupating on the surface and tend to crawl up the sides of the culture, away from the food surface.

In laboratory cultures of fruit flies (*Drosophila melanogaster*), the consistency of the food depends on the numbers of larvae that are feeding in it. When there are few larvae, the food may remain relatively hard and dry. With many larvae, the food becomes a fluid mess, much like quicksand. To become adults, the larvae must find a place to form their pupal case and undergo metamorphosis. In uncrowded cultures, they tend to pupate on the surface of the food or close to it (Figure 10.8C). But in crowded cultures pupating on the surface is dangerous because the pupae would sink into the food and die. Under these conditions, the larvae tend to avoid the surface of the food, instead crawling high up the side of the culture as shown in Figure 10.8C.

Tribolium, the common flour beetle, also has a complex life cycle—although all life stages are typically found in a common environment. For these species an important component of population regulation is cannibalism (Figure 10.8D). As more larvae are produced, they consume more eggs. Since eggs become larvae, high numbers of larvae will reduce the numbers of larvae in the future through their consumption of eggs.

The levels of cannibalism may also respond to natural selection. When *Tribolium* are experimentally provided eggs from close relatives, their propensity to cannibalize them decreases over time due to group selection. This type of natural selection is described in more detail in Chapter 20. ❖

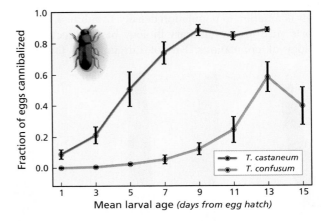

FIGURE 10.8D Cannibalism is an important mechanism of population regulation for flour beetles. At a fixed egg density, *Tribolium* larvae consume more eggs as they get older. As more eggs are eaten, fewer larvae are produced; this relaxes the intensity of regulation on egg numbers.

DENSITY-DEPENDENT NATURAL SELECTION

The early theories of *r*- and *K*-selection were verbal

The importance of the natural environment in shaping the adaptations of organisms is inescapable. Yet evolutionary biology in the first half of this century paid little attention to ecology. In 1930 the great English statistician and population geneticist, Sir Ronald Fisher, proposed that natural selection will always increase the mean fitness of a population. Fisher called this theory the Fundamental Theorem of Natural Selection. In 1962 the American ecologist, Robert MacArthur (see the box), tried to redirect the course of evolutionary thought by developing an ecological analogue to Fisher's Fundamental Theorem of Natural Selection. This attempt of MacArthur's did not result in a lasting theory, but it did illustrate the point that ecology and evolution might be combined, and biologists in both fields have continued to explore this intersection.

Ecologists devote much attention to the response of population growth to density. And some evolutionary biologists study adaptation to population density. Combined, these two efforts yield an evolutionary biology based directly on the ecology of populations. Part of this union is the field of *density-dependent natural selection*. It is the first success of evolutionary ecology. In this module we outline its basic theory and key experiments.

Some of the seminal ideas behind density-dependent selection were first summarized in 1967 by MacArthur and E. O. Wilson. They called the theory *r*- **and *K*-selection**, where *r* and *K* refer to the parameters of the logistic model. Specifically, *r* is the intrinsic rate of increase, or maximum rate of growth at low population densities, and *K* is the carrying capacity, or the maximum number of individuals that the environment can support. Over the last 30 years, various experiments have tested these ideas. The history of this research provides important lessons for research methodologies in evolution and ecology. We review some of these lessons here.

The original development of this theory was qualitative. MacArthur and Wilson argued that for populations living at low density with abundant resources, natural selection would favor traits that contribute to *r*, the intrinsic rate of increase in the logistic equation. It was not clear exactly what traits could make this contribution, so there were many suggested candidates. (These candidate traits are referred to as **life-history characters** since they are associated with the timing and process of reproduction. You will be familiar with these characters from Chapter 7.) For instance, some suggested that decreases in **generation time**, the average time between birth and the production of offspring, and increases in fecundity would be the most likely changes brought about by life at low population densities. Others suggested that increasing *r* could be accomplished by a decrease in body size.

At high densities the opposite suite of traits would evolve, causing an increase in the carrying capacity for individuals with these traits. For example, if individuals became more efficient at extracting energy from the available food, a given environment could support a larger number of these efficient individuals.

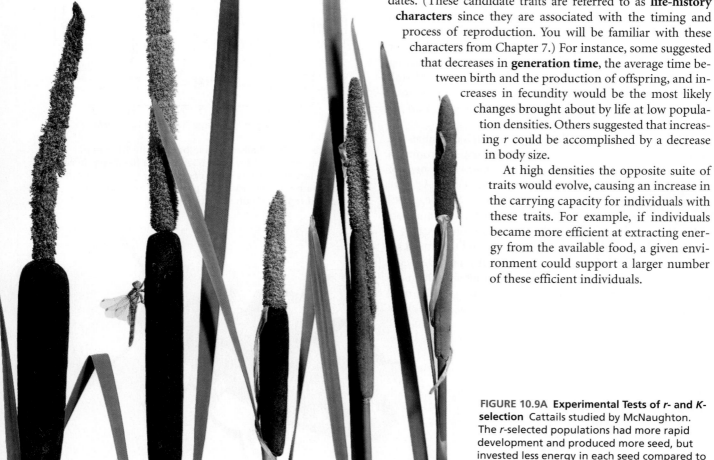

FIGURE 10.9A Experimental Tests of *r*- and *K*-selection Cattails studied by McNaughton. The *r*-selected populations had more rapid development and produced more seed, but invested less energy in each seed compared to the *K*-selected populations.

A central thesis of *r*- and *K*-selection is the notion that evolution cannot simultaneously maximize *r* and *K*, so there are trade-offs. As a result, the suite of traits that are favored at low densities are different from those favored at high densities. So while selection at low density may favor increased fecundity, that increase may be accompanied by a decline in competitive ability at high densities. It is hard to predict what these trade-offs will be like, or if they will even exist. Fortunately, experimental tests have given insights into the nature of these trade-offs.

Some early tests of these predictions compared different species that lived under different density conditions. The idea was that each species represented a different end product of evolution and that the differences between the species were largely due to the different densities at which they lived. We review two of the classic studies in this field, from the early 1970s. McNaughton compared two species of cattail (*Typha*, Figure 10.9A) and concluded that species from the northern United States had traits consistent with *r*-selection, whereas species from the southern United States had traits consistent with *K*-selection. Of course it is hard to be sure that environmental density was the only important difference between these species; or that other factors, such as climate or genetic drift, did not contribute to the observed differences.

Some other early studies focused on different populations of one species. Gadgil and Solbrig studied populations of dandelions (Figure 10.9B). They found that a genotype of dandelion prevalent in an undisturbed site was a better competitor but produced fewer seeds than a genotype common in a disturbed site. The assumption was that population densities should be higher in the undisturbed sites. However, there was no way to know what the past densities had been at these sites, or if there were factors other than density that varied between sites and affected the evolution of competitive ability and seed number. Gadgil and

Solbrig looked at competitive ability and seed number because the measurement of actual rates of population growth in nature is difficult. ❖

FIGURE 10.9B Experimental Tests of *r*- and *K*-selection Gadgil and Solbrig found that the predominant genotype of dandelion in undisturbed areas (*K*-selected) was a better competitor than the predominant genotypes in the disturbed area (*r*-selected). However, the *r*-selected genotype produced more seeds than the *K*-selected genotype did.

Robert H. MacArthur, Innovative Theoretical Ecologist (1930–1972)

**FIGURE 10.9C
Robert H. MacArthur**

Much of the interest in selection at varying densities was inspired by the theoretical work of Robert MacArthur (Figure 10.9C). MacArthur's short career was marked by the development of highly original theories concerning species diversity, life histories, and many other important problems in ecology. MacArthur emphasized theory that could explain general patterns in ecology, as opposed to searching for hypotheses that might apply to only a few species. Critics of MacArthur pointed out that many of his theories were overly simplistic.

The most extensive development of his theory for density-dependent population growth appeared in his 1967 book with E. O. Wilson, *Island Biogeography*. MacArthur and Wilson called their theory of density-dependent selection *r*- and *K*-selection, emphasizing the consequences of selection at extreme densities. At very low density, they argued that the genotypes with highest *r* values would be favored; at very high densities, the genotypes with the highest *K*-values would win. These ideas were attractively intuitive, but at times their application was exaggerated, which caused scientists to dismiss them prematurely. However, the theory of *r*- and *K*-selection has been made more rigorous (as outlined in Module 10.11) and the predictions from this type of theory have been tested, as described in this chapter. The groundwork laid by MacArthur has led to substantial advances in our understanding of ecology and evolution.

Great differences exist within species in their ability to tolerate crowding

A major goal in the field of evolutionary ecology has been to determine the relationship between *life-history characteristics* (traits associated with the timing and rate of reproduction) and *fitness*. The chance of surviving to become an adult, **viability**, must be related to fitness. But viability alone is not equivalent to fitness, because the organism must also leave offspring. Therefore the number of offspring produced will be related to fitness. But fertility alone is not fitness if the organism never finds a mate. There is certainly no single, easily measured character that we can assume is equal to fitness.

Nevertheless, some characters may be very closely related to fitness. For example, the *per capita contribution to population growth* is a plausible measure of fitness. This measure, which can be related to the number of offspring left in the next generation, depends on both survival and fertility. Population growth rates will vary with population density. At high density the carrying capacity of a genotype may be equivalent to fitness. This is a somewhat abstract concept, but in principle refers to the ability of a genotype to withstand crowding. While at low densities, the initial reproductive rate may be more important for fitness. For these reasons, it is attractive to think of *fitness* as an organism's contribution to population growth.

If we utilize this definition of fitness, then natural selection may shape population growth rates if population growth

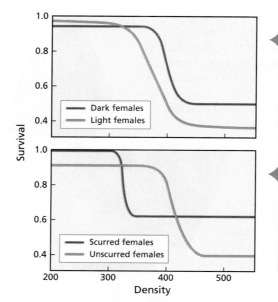

In Soay sheep a single locus determines the color of the sheep's coat. The allele that produces light-colored coats when homozygous is recessive to the allele that produces dark brown coats.

These same sheep also show genetically based differences in horn shape. Some sheep have scurred horns (small and twisted), while others are unscurred. The horn type is determined by one or two loci.

In 1996 Paul Moorcroft and his colleagues analyzed eight years of survival data from natural populations of Soay sheep in Scotland. The survival of females at different densities is shown on the right.

From these results it appears that at low density all phenotypes survive well, but at high density the dark phenotype does better than the light phenotype and the scurred phenotype does better than the unscurred.

FIGURE 10.10A Different genotypes of Soay sheep vary in their ability to survive crowding.

rate varies genetically within populations. Figures 10.10A and 10.10B show two examples of this type of genetic variation. The first example (Figure 10.10A) comes from a natural population of Soay sheep (*Ovis aries;* Figure 10.10C) living in Scotland. In these populations, there is clearly strong selection in favor of sheep with dark coats and scurred (small, twisted) horns during periods of high density. The reason these genotypes survive better is unknown, but they are a remarkable example of density-dependent selection in a natural population.

The second example (Figure 10.10B) involves populations of fruit flies (*Drosophila melanogaster*) in the laboratory. Fruit-fly chromosomes were manipulated to create different populations of flies. All individuals in a population were made homozygous for their second chromosome. When rates of population growth were estimated, large differences were found between genetically different populations. For this collection of genotypes, the differences in fitness, measured as net reproductive rates, were greatest at low densities. These net reproductive rates are the same as those defined in Module 10.2. ❖

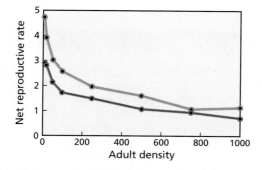

FIGURE 10.10B Net Reproductive Rates Vary among Genetically Different Populations of Fruit Flies Populations of fruit flies (*Drosophila melanogaster*) can be rendered homozygous for large portions of their genome. The graph shows two such populations and their net reproductive rates at a variety of densities. Note that at very low density, one population grows 50 percent faster than the second population.

**FIGURE 10.10C
Soay Sheep**

Because rates of population growth depend on both the survival and fertility of individuals, genetically based differences in population growth rates will affect fitness, and natural selection may change the frequencies of genotypes. As we saw with Soay sheep and fruit flies, the differences between the growth rates of particular genotypes can vary with density. Soay sheep are differentiated at high density, and fruit flies show the greatest growth rate differences among genotypes at low density.

We can determine the outcome of selection with varying population density using a simple genetic model with two alleles (A_1 and A_2) at a single locus. If we assume population growth follows the logistic equation, then we can summarize the growth characteristics of each genotype by its specific values of r and K (Figure 10.11A). In the example in Figure 10.11B, the A_1A_1 homozygotes have highest fitness at low density, the A_2A_2 homozygotes have highest fitness at high densities, and the heterozygotes are intermediate. These fitness differences reflect the fact that the A_1A_1 homozygotes have high values of r but low values of K, relative to the A_2A_2 homozygotes (Figure 10.11C).

What then is the outcome of natural selection? The outcome of natural selection depends on the environment. In crowded environments, the A_2A_2 homozygotes have the highest growth rates and the highest fitness and therefore increase in frequency to fixation (i.e., their frequency approaches 100 percent; Figure 10.11D). In those environments where population density is kept low, the A_1A_1 homozygotes have highest fitness, and they increase in frequency at the expense of the genotypes carrying the A_2 alleles.

The particular density conditions experienced by natural populations vary. In stable environments, populations may have long periods of uninterrupted growth, reach their carrying capacity, and thus experience strong selection for those traits that increase growth rates at high densities. It is also possible that, in some environments, floods, drought, and winter freezes keep population numbers well below the carrying capacity. In these environments, we would expect selection to favor those traits that will increase growth rates at low density.

Natural constraints may prevent any single genotype from being best for both r and K. One explanation for the existence of such constraints may be that organisms have limited energy stores, so that energy devoted to high reproduction at low densities may not be available for surviving stressful conditions at high population densities. For instance, an individual could increase r by producing more eggs. One way to tolerate crowded conditions, and thus periods of reduced food, is to store fat. But eggs and fat both require energy, so they may trade off against each other. This type of trade-off may be fairly common. ❖

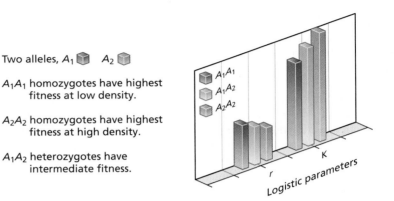

$$W_{A_1A_1} = 1 + r_{11} - \frac{r_{11}N}{K_{11}}$$

- Intrinsic rate of growth for the A_1A_1 homozygote
- Carrying capacity for the A_1A_1 homozygote

$$W_{A_2A_2} = 1 + r_{22} - \frac{r_{22}N}{K_{22}}$$

- Total population size, which is the sum of all three genotypes

FIGURE 10.11A Fitness equals each genotype's net reproductive rate.

Two alleles, A_1 ▧ A_2 ▧

A_1A_1 homozygotes have highest fitness at low density.

A_2A_2 homozygotes have highest fitness at high density.

A_1A_2 heterozygotes have intermediate fitness.

FIGURE 10.11B Genetic variation for population growth.

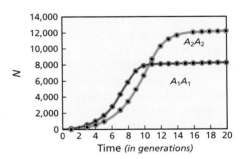

FIGURE 10.11C This figure demonstrates that when the homozygous genotypes are kept in isolation, they grow logistically and attain their carrying capacities. The carrying capacity of the A_2A_2 genotype is greater than that of the genotype.

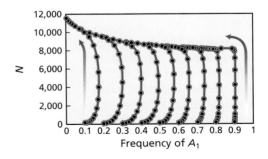

FIGURE 10.11D Changes in allele frequency and population size over time from several different initial allele frequencies. When population density is low, the curves lean to the right, indicating that the A_1 allele is increasing due to the superior fitness of A_1 carrying genotypes at low density. At far left in the figure, the A_2 allele has become fixed while the total population has grown and become crowded.

10.12 Natural selection often cannot increase population growth rates at high and low density simultaneously

If trade-offs exist between the responses to selection at different densities, then the genotypes favored by natural selection in uncrowded environments should be different from the ones favored in crowded environments. One way to test this idea is to create populations that differ only in the density they are exposed to as they evolve over many generations. Will the growth rates of these populations then fit the theory of density-dependent selection?

Just such an experiment has been carried out with fruit flies. The experiment, illustrated in Figure 10.12A, used three populations that had been maintained in the laboratory at low densities for 198 generations. From each of these three populations, a new population was created that was maintained with crowding. After 25 generations of evolution at high densities, the rates of population growth (measured as net reproductive rates per week) were measured in both the low-density and high-density populations, at both high and low densities. Because there were three replicate populations, genetic differences that appear in all three high-density populations must have arisen due to natural selection as opposed to a random process like genetic drift.

We use the net reproductive rates to estimate relative fitness. A generation in these environments takes about three weeks, so relative fitness was estimated as the growth rate of the low-density population divided by the growth rate of the high-density population cubed. From the graph in Figure 10.12A we see that at a test density of 10 larvae, the relative fitness of the low-density populations is greater than 1.0, due to their superior growth rates. At the two higher test densities, 750 and 1000, the average relative fitness of the low-density populations is less than 1.0, indicating the superior fitness of the high-density populations. However, natural selection has been unable to produce a genotype that does best at all densities. This is consistent with the underlying hypothesis that there will be insurmountable trade-offs in an organism's ability to do well at both extreme densities.

These results show that evolution has affected population growth rates. However, individual traits have also evolved in response to crowding. Three of these traits are well-defined larval behaviors. A behavior called pupation height refers to the distance traveled by a larva up the side of its culture prior to settling and becoming a pupa. Some larvae do not travel at all but will pupate the surface of the food. Other larvae may travel to the very top of the culture. The populations that have evolved at high density are more likely to travel far up the side of the culture and less likely to pupate on the surface of the food compared to the populations that have evolved at low density. In a crowded culture, nearly 80 percent of pupae that pupate on the surface of the food die, so avoiding the surface as a pupation site is clearly adaptive in crowded environments.

A second behavior that has changed in response to density-dependent natural selection is a measure of the feeding rate

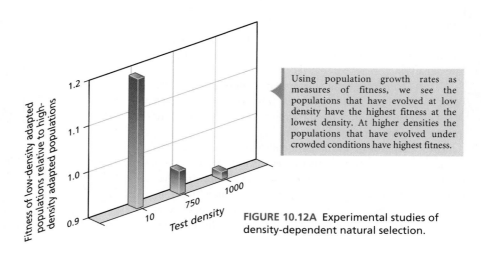

Three populations of fruit flies (*Drosophila melanogaster*) are kept in the laboratory for 198 generations at low larval and adult densities.

Three populations are continued at the low-density conditions.

Three populations are maintained in crowded larval and adult conditions.

After 25 generations of selection, population growth rates are measured at low (10), moderate (750), and high (1000) adult density.

Using population growth rates as measures of fitness, we see the populations that have evolved at low density have the highest fitness at the lowest density. At higher densities the populations that have evolved under crowded conditions have highest fitness.

Fitness of low-density adapted populations relative to high-density adapted populations

Test density

FIGURE 10.12A Experimental studies of density-dependent natural selection.

of larvae. Fruit fly larvae have only one goal—to grow. The larvae feed and grow continuously until they surpass a minimum size required to successfully metamorphose and become an adult. In environments with limited food it turns out that the best competitors have the highest feeding rates (see Module 12.2). Consequently, the populations of fruit flies that have evolved at high population density have evolved much higher feeding rates than their low-density counterparts. A third behavior is called the foraging path length and is described in the box below. ❖

Behavior Evolves in Response to Density

Density-dependent natural selection can even affect the evolution of behavior. Consider fruit-fly larvae, which show genetic polymorphism for foraging. When larvae are put on a flat surface, some—called *rovers*—move a lot, and some—called *sitters*—do not (Figure 10.12B). Larvae from populations that have low densities are mostly sitters, whereas rovers are more common in populations that have been kept at high densities for many generations.

How can we explain the evolution of these foraging behaviors? In crowded environments, rovers are probably better at finding food and avoiding waste. In uncrowded environments, the extra movement of rovers may just waste energy. This sitter-rover polymorphism appears to be controlled by a gene that codes for a cyclic guanosine monophosphate-dependent protein kinase. This protein has been implicated in nervous system function in other organisms.

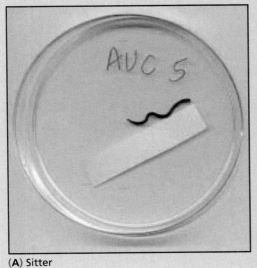

(A) Sitter

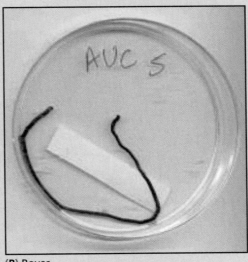

(B) Rover

FIGURE 10.12B
The foraging path length of sitters (left) and rovers (right). The trails are paths left by larvae in a yeast solution.

The logistic model produces sustained fluctuations when r is sufficiently large, as we saw in Module 10.7. Although r is a parameter that reflects genetically based components of survival and fertility, r may also be affected by environmental factors. For example, the number of offspring an individual produces may be affected by either the amount or the quality of food the organism eats. Some environments may be uncrowded but have poor food quality, such as low levels of protein and carbohydrate, which may depress r and enhance population stability. The converse is also true.

Blowflies are insects with a flying adult stage and a crawling larval stage. These insects are a significant agricultural pest in Australia, where they infest grazing animals. In Figure 10.13A we give an example of a blowfly population whose stability is affected by environmental factors. In the 1950s, A. J. Nicholson studied the factors that affect population stability in blowflies. His research was not motivated by recent theoretical results concerning population stability, but it has recently been of great inter-

est to ecologists. In Nicholoson's experiment, the blowfly larvae and adults received food independently. The first 630 days of observations, during which the adults were given large quantities of food, show dramatic population cycles, as Figure 10.13A shows. After that period, when the level of adult food was reduced, the cycles were significantly attenuated.

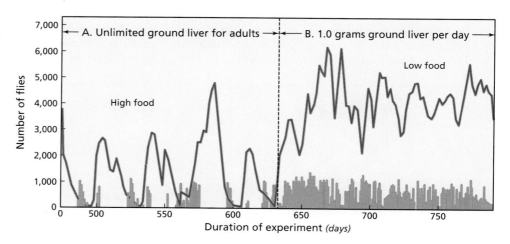

FIGURE 10.13A **Environmental Factors Affect Population Stability** Populations of blowflies show changes in stability when adult food levels are changed. High levels of food cause increases in female fecundity, and the population becomes unstable. The red line indicates the total number of adults and the green vertical lines represent the number of new recruits added to the adult population. During the first part of the experiment almost all new adult recruits are produced only when the total population size is low.

Besides environmental factors, genetic changes in populations may also affect the value of r in the logistic equation, and hence population stability. To investigate this problem, Mueller and his colleagues introduced five populations of fruit flies, which had evolved for 43 generations at low larval

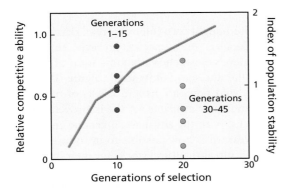

FIGURE 10.13B **Evolution of Population Stability** Over the first 25 generations in environments with crowded larval conditions, there was strong selection for increased competitive ability as the larvae adapted to the cultures (see blue line). However, the stability of these populations seems unaffected by the evolution of life-history traits in response to crowding. An index of population stability shows roughly the same range and magnitude of effects in the first 15 generations of selection (red circles) as it does in the last 15 generations of selection (green circles).

and adult densities, into environments with low levels of larval food and high levels of adult food. This resulted in very crowded larval cultures and population sizes that cycled around an equilibrium but never settled down to stable population densities.

What happened in this experiment? During the first 25 generations, the populations evolved in response to crowding. They adapted to the crowded larval conditions, and this adaptation is reflected in measurable improvements in competitive ability, as the blue line in the graph of Figure 10.13B shows. Changes in population size were recorded every generation and used to compute an index of population stability. The closer this index is to zero the more stable the population. As the red and green circles on the graph in Figure 10.13B show, there was little change in the stability of the populations; the population stability indices overlap extensively when early generations are compared with later generations. These results suggest that over ecological time spans—perhaps from dozens to a few hundred generations—the stability of a population is more likely to be influenced by changes in the local environment than it is to be influenced by changes in the genetic composition of the population.

Ecological interactions between different life stages have important consequences for the stability of flour beetle populations too. This simple experimental system is described in more detail in the following box. ❖

Cannibalism Affects Population Stability of Flour Beetles

The dynamics and stability of flour beetle (*Tribolium castaneum*) populations are affected by adult cannibalism of pupae as well as adult mortality. It is thus possible to study the dynamics of flour beetle populations by manipulating these events. Theoretically, we expect that low rates of adult mortality and cannibalism should result in the population gradually approaching a stable equilibrium, while high rates of adult mortality and cannibalism should destabilize the population. Figures 10.13C and 10.13D show that when

adult mortality was manipulated, the populations behaved as predicted, gaining or losing stability as expected.

Cannibalism may be affected by levels of crowding and by natural selection. It is not clear whether natural selection on flour beetles would have different effects on stability than those we saw in fruit flies.

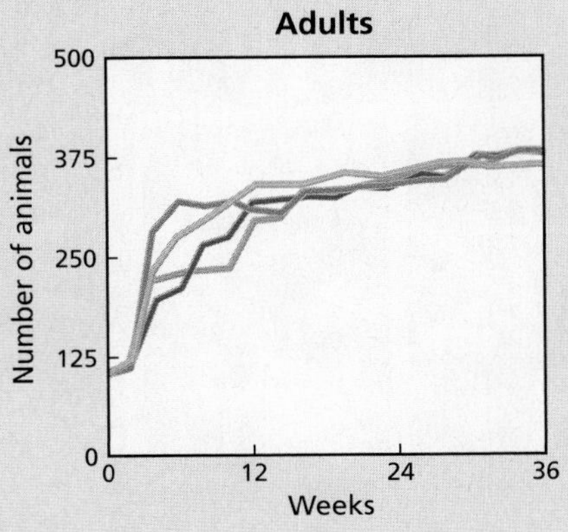

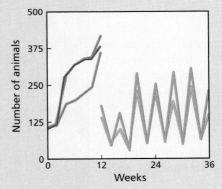

FIGURE 10.13C In this experiment adult mortality was kept low, and the population approaches a stable equilibrium point of about 375 adults. Each line shows an independent replicate population.

FIGURE 10.13D In this experiment, researchers increased the adult mortality rate after 12 weeks. This caused the population to move away from the stable point and enter a two-point cycle. The peak of the cycle occurs at about 225–250 animals, and the trough is at 50–75 animals.

THE BOMB DID NOT BLOW

10.14 A combination of increased food production and changes in demographic patterns has helped humans avert Malthusian catastrophes

If there is one principle we can derive from this chapter, it is that no biological population can grow exponentially forever. This principle also applies to the human population. Is the human population indeed headed for global starvation and ecological catastrophe, caused by our overabundance? In this module we review two major trends suggesting there is hope that human population growth can avoid the worst possible scenarios. This is not to say that some regions of the world will not suffer starvation and severe disease from time to time. Even with no further increase in the human population, its current members will continue to deplete natural resources, and replacements for these must be found. Nevertheless, gains that have been made in food production and in decreasing human fertility might prevent the population bomb from exploding.

Humans practiced animal and plant breeding long before we had any understanding of genetic mechanisms. Indeed, Darwin used the great success that breeders had in creating animal breeds to support his theory of adaptation by natural selection. For Darwin, nature employed the same mechanisms of differential selection, with the most important traits being those that helped individuals survive and reproduce in their environment.

However, as we have added to our knowledge of the genetic basis of natural and artificial selection, improvements in these techniques have become possible. For instance, we now understand that most of the progress achieved in any artificial selection program that uses outbreeding organisms depends on pre-existing genetic variation in the population. Consequently, preserving genetic variation in economically important species is vital to the long-term success of any breeding program. In addition, most scientists now realize that evolution is never static. Today's super corn variety may need to be modified in the future as new insect or microbial pests evolve.

Many agricultural breeding programs now include the preservation of genetically diverse stocks and ancestral populations as important sources of genetic variation. The levels of genetic variation in some of these species is impressive. For instance, store-bought corn (maize) is remarkable for its consistent appearance. As a species, however, maize harbors tremendous levels of genetic variation—some of which is visible in the color and shape of its seeds (Figure 10.14A).

However, wild populations and stocks of maize are not the only avenue now available for the improvement of agricultural stocks. With the advances in genetic engineering of the last 20 years, it is now possible to introduce genes from other species into agricultural stocks for their improvement. In this module we review some of the impressive gains that have been made with genetically engineered crops.

The gradual decline in mortality among developed countries has been associated with a decline in fertility, thus preventing rapid population growth. However, in the first half of the twentieth century, underdeveloped countries experienced very rapid declines in mortality as modern medical advances were introduced. The result was unprecedented population growth in these countries. Are there any indications that these countries have made any progress decreasing fertility?

Although fertility has been generally lower in developed countries than underdeveloped countries, many developed countries—including the United States—had fertility rates above replacement in the middle of the twentieth century. However, as we see in Figure 10.14B, fertility has shown a steady decline in the United States since the 1950s. This trend has also been observed in other developed countries. Female

FIGURE 10.14A The Range of Corn Seed Phenotypes

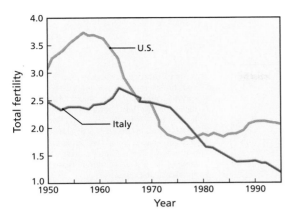

FIGURE 10.14B Observed Total Fertility in Italy and the United States

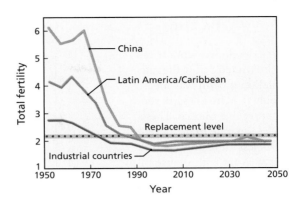

FIGURE 10.14C Past and Projected Total Fertility in Different World Regions

fertility in Italy, for instance, has dropped below replacement levels since the late 1970s.

However, no country has shown a more dramatic decline than China, which has dropped from nearly six offspring per female in 1970 to two in 1990 (Figure 10.14C). More importantly, many other regions of the world have experienced similar declines (Figure 10.14C). The major reason for these declines in the developing countries has been changes in the behavior of people. Many couples have chosen to limit their families to two children through the use of contraceptives. These changes in behavior have been aided by massive international efforts to instill family planning in these countries. In many developing countries, there has also been a postponement of marriage and child bearing. In 1950, most Asian women were married by the time they were 20. In 1980 the average age of Asian women at their first marriage was 20–25 years.

This transition to lower fertility in the more developed world is quite different from events in the less developed countries. In Europe and North America, lower fertility was often reached without the aid of modern contraceptive techniques—and in many instances, despite social and religious forces opposing such practices. Next we will learn more about how these reductions in fertility affect population growth projections for the next century. ❖

In more developed countries, such as the United States, the rate of population growth was much lower than that in a less developed country like Mexico. These differences separate many developed and less developed countries today. What is the cause of these differences?

Very broadly, the differences in population growth rates may be due to differences in birthrates, death rates, or both. To summarize the survival of human populations, the maximum life span is broken into intervals of five or ten years. If the age classes were spaced at five-year intervals, then the first age class would represent newborns and children up to the age of five years, and the second age class would be children just over five years up to ten years, etc. The chance that an individual survives from the first age class to age class x is often represented by the symbol l_x. The estimated chances of survival for females in Mexico and Spain are shown at the top of Figure 10.15A. For the same age classes, fertility of females may be summarized as the number of offspring born to females in age class x. This value is often represented by the symbol, m_x. Female fertility for the Mexican and Spanish populations is also shown in Figure 10.15A (middle). In 1966 the Mexican population had higher death rates, especially at young ages,

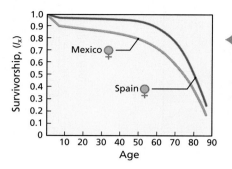

A sharp distinction between the developed and undeveloped countries exists in age-specific patterns of survival and fertility. The figure on the left shows the survival and fertility of females in Spain and Mexico in 1966. The major differences in survival are at very young ages, when Mexican females have a much greater chance of dying. Mexican females have more children than Spanish females do at all ages.

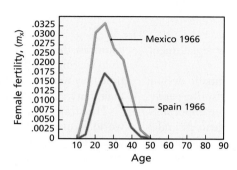

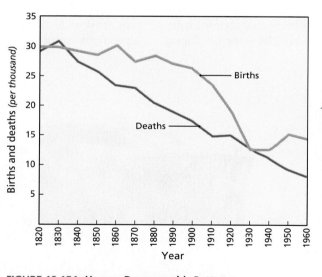

The transition to low rates of population growth in European populations was made gradually by reductions in both fertility and mortality. The patterns shown in the figure are for Sweden between 1820 and 1960.

FIGURE 10.15A Human Demographic Patterns

compared to the Spanish population, but still had higher population growth rates. The reason for this is that female fertility in Mexico was often substantially higher than in Spain at all ages.

Many European populations had much higher birthrates 200 years ago. However, as modern medical technology and hygiene were introduced to Europe, mortality gradually declined, especially among infants. This was accompanied by a gradual decline in the birthrate (Figure 10.15A). Of course, until recently the less developed countries lacked modern medical technology. The rapid introduction of antibiotics and other modern medical practices to less developed countries has suddenly reduced mortality with hardly any change in fertility. The result is rapid population growth.

We can summarize patterns of births and deaths in human populations with a few simple demographic parameters. If the total number of individuals in the populations is N, then the **crude birthrate** of a population in any year is simply the total number of births in that year (B) divided by the population size (N). The **crude death rate** is the number of deaths (D) in a given year divided by the total population size (N). The **crude rate of population increase** is simply births minus deaths divided by the total population size: $(B - D)/N$.

TABLE 10.15A World Demographic Parameters, 1995–2000			
Country or Area	Growth Rate (%)	Crude Birthrate (per thousand)	Crude Death Rate (per thousand)
World Total	1.3	22	9
Developed regions[a]	0.3	11	10
Less developed regions[b]	1.6	25	9
Least developed regions[c]	2.4	39	15

[a] All regions of Europe and North America, Australia/New Zealand, and Japan.

[b] All regions of Africa, Asia (excluding Japan), Latin America and the Caribbean, and the regions of Melanesia, Micronesia, and Polynesia.

[c] According to the United Nations in 1998, these consisted of 48 countries; 33 are in Africa, 9 in Asia, 1 in Latin America and the Caribbean, and 5 in Oceania.

As an example, let's consider Mexico in 2000. The demographic parameters for that country were as follows: $N = 100,350,000$, $B = 2,310,000$, and $D = 502,000$. Thus, the crude birthrate expressed as a rate per thousand is $(2,310,000/100,350,000) \times 1000 = 23$. The death rate is 5 per thousand, and the rate of increase is 18 per thousand, or 1.8 percent.

Based on birth and death rates, the world's populations generally fall into three categories (see Table 10.15A): (1) High birth and high death rates are found in the least developed countries; (2) High birth and low death rates are found in the less developed countries; and (3) Low birth and low death rates are found in the developed countries.

Many of the developed countries have very low growth rates, and in some cases population sizes are decreasing. Nevertheless, the world population is still growing rapidly. The total world population passed 6 billion in 1999. During this period the population was adding 1 billion new people every 12 years. In the early part of the twentieth century, it took nearly 33 years for the world population to add 1 billion people. The United Nations predicts that the world population will reach 8.9 billion by the year 2050. ❖

The use of selection and genetic engineering vastly expanded agricultural productivity—though their long-term ecological effects are not known

The production of food in modern agriculture faces a variety of problems that set limits not only to the productivity of existing farms but also to where farming can be carried out. These limits include insect, bacterial, and fungus pests; lack of nutrients in the soil; weed species that compete for those nutrients; and water. Other problems are more subtle but of growing importance. For instance, in many arid areas of the world, crops can be raised with irrigation. However, the high rates of evaporation result in increasing soil salinity, sometimes to the point that plants are unable to grow. Genetic engineering is now being used or considered as a solution for all the problems we have just reviewed. The result of these efforts is increasing agricultural productivity. Let us review some of these recent advances.

The general methodology of genetically engineering crops is shown in Figure 10.16A. A gene of interest, say one that fixes nitrogen from the atmosphere, is carried in bacterial cells. The gene of interest and the bacterial host are together called the **transgenic construct**. This gene is introduced into a susceptible genotype, or **transformation variety**, of the crop species. The presence of the target gene in any particular plant can be determined by a genetic analysis of tightly linked marker loci. Inbreeding will make the target gene homozygous. Then by a series of crosses with the genotype actually used for growing crops, the so-called **elite variety,** the gene of interest is brought into the useful elite variety.

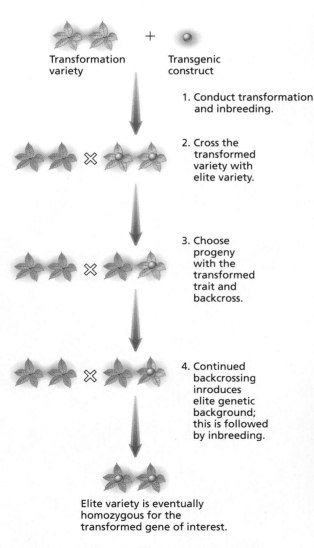

Transformation variety   +   Transgenic construct

1. Conduct transformation and inbreeding.

2. Cross the transformed variety with elite variety.

3. Choose progeny with the transformed trait and backcross.

4. Continued backcrossing inroduces elite genetic background; this is followed by inbreeding.

Elite variety is eventually homozygous for the transformed gene of interest.

FIGURE 10.16A Genetic Engineering of Crop Species

(i) Crown gall infecting a rosebush

(ii) Crown gall infecting (arrows) a grapevine

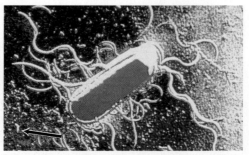

(iii) *Bacillus thuringiensis* spores and Bt toxin crystals (arrow).

FIGURE 10.16B Examples of genetic engineering in crop species.

FIGURE 10.16C Genetic analysis of wild-type, domesticated, and hybrid tomatoes allows identification of important genes that affect color and shape of the fruits. These lines serve as reservoirs of important genetic information.

Crown gall (Figure 10.16B) is caused by a bacterium (*Agrobacterium tumefaciens*) with a small, circular piece of DNA called a plasmid (explained in Chapter 5) that infects many plant species, including roses, grapes, cherries, and peaches. The bacterium invades wounds in the plant and injects the plasmid into plant tissue. The plasmid DNA then integrates into the plant nuclear DNA and causes the plant cells to build the gall. Ironically, the ability to transfer DNA from a plasmid into plant nuclear DNA has made *A. tumefaciens* a useful vector for making transgenic constructs. The growth of crown gall is inhibited by another bacteria, *Agrobacterium radiobacter*. *A. radiobacter* was genetically engineered so that it could not pass on its genetic information to *A. tumefaciens*. These genetically engineered *A. radiobacter*, known as strain K1026, are now used quite effectively to control crown gall.

A soil bacterium, *Bacillus thuringiensis*, makes a protein, Bt toxin, that kills a variety of insect pests (Figure 10.16B). The gene coding for Bt toxin has been introduced into more than 50 different crop species. These plants thus have continuous protection from pests, eliminating the need for costly pesticide spraying. As with man-made pesticides, some insects are now developing a resistance to Bt toxin. In addition, plants that make Bt toxin excrete it into the soil, where it can persist for long periods of time. It now appears that this lingering Bt toxin can have detrimental affects on "good" insect species that naturally consume pest species. More research will be needed to see if there are ways to avoid these undesirable effects of Bt toxin.

The ancestral species (*Lycoperscion pennellii*) of today's cultivated tomato (*Lycoperscion esculentum*) looks very different from contemporary domesticated plants (Figure 10.16C). However, the ancestral species is a reservoir of genetic variability that can be useful for developing new varieties of crops. By making crosses between the cultivated tomato and the ancestral species, hybrids with intermediate traits can be created. By measuring these traits in the hybrids and the two parents, it is possible to map the genes that affect important color, shape, and other tomato characters. These genes can then be isolated in varieties for future use in tomato breeding programs. ❖

We began Module 10.14 with a review of evidence that many countries have undergone a decline in fertility in the last 50 years. This trend is expected to continue in the developing countries of the world well into the twenty-first century. These declines imply that populations that are now growing will eventually stop growing and even start decreasing in size. Because it is impossible to know exactly how quickly fertility will decline in the future, scientists have estimated the most probable changes. These estimates can then be used to predict when certain populations will start declining in size (Figure 10.17A).

From these estimates we see, for instance, that in the European part of the former USSR there is about a 90 percent chance that the populations will start declining in size by 2015. Less developed countries like China and sub-Saharan Africa lag behind, but show an increasing likelihood of population decline as the twenty-first century progresses (Figure 10.17A). For the entire world population, there is a 50 percent chance that numbers will start declining by the year 2075.

We show what this means for total population size in Figure 10.17B. By 2050, the total world population is predicted to be about 3 billion more than the current number of 6 billion. However, growth is expected to slow by the end of the twenty-first century and not to exceed 10 billion. In fact, current estimates suggest there is about a 15 percent chance that in the year 2100, the world population will be less than it is today.

For the entire world population, there is a 50 percent chance that numbers will start declining by the year 2075.

These predictions depend on the projections of fertility decline. What evidence suggests that fertility will not increase in the future? In fact, this could happen; and then these predictions would be of little use. However, recent historical trends suggest it is unlikely for countries that have experienced fertility declines to show a later increase in fertility. In Figure 10.17C, the changes in fertility between 1950 and 1990 have been recorded. The number of countries that fall into one of four categories of fertility in 1950–55 and 1990–95 are listed in this figure. Very few countries show an increase in fertility (e.g., numbers in the orange zone). Most countries show movement to lower fertility (the blue zone) or no change (the purple zone). If these trends continue to hold, then the widespread drop in fertility is not likely to show a reversal in the next century.

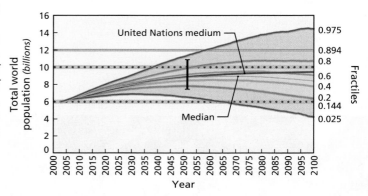

FIGURE 10.17B Projected world population size. Based on information about fertility declines around the world, the best estimates of the total human population are shown here as a dark grey line labeled, "United Nations medium." Due to some uncertainty about this projection, the various shaded bands give some indication of the range of uncertainty in these predictions.

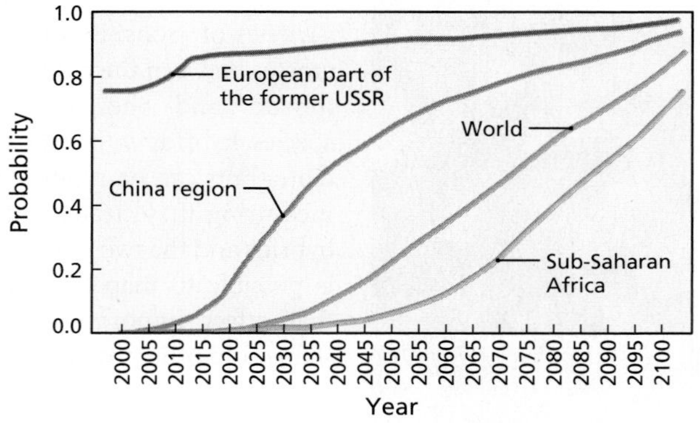

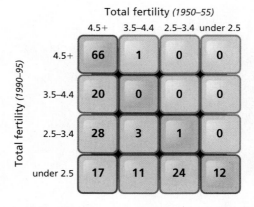

FIGURE 10.17A Probability of population decline. Due to declines in human fertility, populations are predicted to stop growing in the future. This figure shows the probability (y-axis) that a given region in the world will start its population decline at some time in the future. By the year 2100, current predictions suggest there is a very high chance that almost all parts of the world will have started their population decline.

Total fertility (1950–55)

	4.5+	3.5–4.4	2.5–3.4	under 2.5
4.5+	66	1	0	0
3.5–4.4	20	0	0	0
2.5–3.4	28	3	1	0
under 2.5	17	11	24	12

Total fertility (1990–95)

FIGURE 10.17C Fertility change over 40 years. Numbers indicate number of countries in each category.

SUMMARY

1. When birth and death rates are constant, populations grow or decline exponentially.

   a. When births exceed deaths, this means that a population will increase in size indefinitely.

   b. The most obvious factor that will prevent exponential growth is the effect of crowding on fertility and on survival.

   c. Food and space are two resources that will be in short supply as density increases.

2. A simple model of density-dependent population growth is the logistic model.

   a. This model has two parameters, r and K. K is called the carrying capacity, and it determines the equilibrium population numbers that can be supported by the environment.

   b. The second parameter, r, determines the maximum rate of population growth at low densities.

   c. The parameter r of the logistic model also determines whether the carrying capacity is a stable equilibrium.

   d. When r is greater than 2.0 the equilibrium becomes unstable; the population may exhibit cycles or even seemingly random variation called chaos.

3. Natural selection may also affect population growth. If there is genetic variation for traits that affect r or K of the logistic, we expect per capita rates of population growth to increase evolutionarily through a combination of increases in either r or K, or in both r and K.

4. In populations of fruit flies kept at either very high or very low densities, we observe appropriate improvements in population growth rates. However, there are trade-offs such that the genotypes that do well at high densities do not do well at low density, and vice-versa.

5. The human population may avoid large-scale catastrophes from overpopulation. This is a consequence of two major events.

   a. Advances in agriculture in the last 50 years have allowed food production to keep pace with population growth.

   b. There are signs that many countries are undergoing reductions in fertility. The long-term consequence of this decrease is that world population size may stabilize or even decline by the year 2100.

REVIEW QUESTIONS

1. The exponential growth model predicts that population size will be constant when λ is exactly equal to one. Why do you think the exponential model is rarely offered as an explanation for the growth of populations that exhibit stable population size?

2. In a population growing exponentially with $\lambda = 1.5$, how many generations would be required to increase the current population size by tenfold?

3. List some factors that may slow rates of population growth at high densities, and give some specific examples.

4. In a logistically growing population with $r = 1$ and $K = 1000$, there are two population sizes that result in no change in population size over time. What are these population sizes?

5. At a single locus with two alleles (A and a), the genotypic values of r and K are as follows:

Genotype	r	K
AA	1.0	900
Aa	1.2	1100
aa	1.5	1000

   If this population is allowed to grow to an ecological equilibrium, what will happen to the frequency of the A allele?

6. The stability of some populations is affected by aspects of their ecology. Give some examples for which the biological factors affecting stability are well documented.

7. Are the crude rates of human population growth density-dependent or density-independent estimates of future growth?

KEY TERMS

age structure
carrying capacity
chaos
complex life cycle
crude birthrate
crude death rate
crude rate of population increase

density-dependent population growth
discrete generations
doubling time
elite variety
equilibrium
equilibrium population size

exponential population growth
generation time
intrinsic rate of increase
life history characters
limiting resource
logistic population growth
net reproductive rate

perturbation
population
r- and K-selection
stability
transformation variety
transgenic construct
viability

FURTHER READINGS

Bongaarts, J., and R. A. Bulatao (eds). 2000. *Beyond Six Billion: Forecasting the World's Population*. Washington, DC: National Academy Press.

Costantino, R. F., R. A. Desharnais, J. M. Cushing, and B. Dennis. 1997. "Chaotic Dynamics in an Insect Population." *Science* 275: 389–91.

Dayton, P. K. 1971. "Competition, Disturbance, and Community Organization: The Provision and Subsequent Utilization of Space in a Rocky Intertidal Community." *Ecological Monographs* 41: 351–89.

Ehrlich, P. R. 1968. *The Population Bomb*. New York: Ballantine.

Gadgil, M., and O. T. Solbrig. 1972. "The Concept of r- and K-selection: Evidence from Wild Flowers and Some Theoretical Considerations." *American Naturalist* 106: 14–31.

Gallagher, R. 1969. *Diseases That Plague Modern Man*. Dobbs Ferry, NY: Oceana Publications.

Hastings, A. 1997. *Population Biology*. New York: Springer-Verlag.

Lutz, W., W. Sanderson, and S. Scherbov. 2001. "The End of World Population Growth." *Nature* 412: 543–45.

McNaughton, S. J. 1975. "r- and K-selection in *Typha*." *American Naturalist* 109: 251–61.

Moorcroft, P. R., S. D. Albon, J. M. Pemberton, I. R. Stevenson, and T. H. Clutton-Brock. 1996. "Density-Dependent Selection in a Fluctuating Ungulate Population." *Proceedings of the Royal Society of London B* 263: 31–38.

Mueller, L. D., and A. Joshi. 2000. *Stability in Model Populations*. Princeton, NJ: Princeton University Press.

Mueller, L. D., P. Z. Guo, and F. J. Ayala. 1991. "Density-Dependent Natural Selection and Trade-offs in Life History Traits." *Science* 253: 433–35.

Paoletti, M. G., and D. Pimentel. 1996. "Genetic Engineering in Agriculture and the Environment." *Bioscience* 46: 665–73.

Roughgarden, J. 1971. "Density-Dependent Natural Selection." *Ecology* 52: 453–68.

Sokolowski, M. B, H. S. Pereira, and K. Hughes. 1997. "Evolution of Foraging Behavior in *Drosophila* by Density-Dependent Selection." *Proceedings of the National Academy of Science USA* 94: 7373–77.

Wilbur, H. M. 1980. "Complex Life Cycles." *Annual Review Ecology and Systematics* 11: 67–93.

Zamir, D. 2001. "Improving Plant Breeding with Exotic Genetic Libraries." *Nature Reviews Genetics* 2: 983–89.

Dispersal may involve the movement of a single individual or large groups such as these elephants.

11

Dispersal

Populations are not static. That is, their members and gametes are not restricted to staying within the boundaries of the population. Individuals may move to seek a different place to feed or to overwinter. This type of migration is typical for many birds that live in temperate climates. Bird migrations may cover thousands of miles. These movements let birds avoid harsh winter conditions and find new sources of food. Other organisms may travel that far, but not under their own power. Many marine organisms travel thousands of miles while floating on the ocean currents.

These movements can have both ecological and genetic consequences. Dispersing individuals may find new suitable but unoccupied habitats and thereby initiate new robust populations. Even small numbers of new emigrants may carry genetic variation that is not currently in the population. These individuals can have a large effect on population differentiation.

Many organisms have no effective means of movement. These organisms can choose an alternative strategy of dormancy. Either the adults or some juvenile form may simply assume a more robust condition and try to weather out the bad conditions. Plants obviously lack the ability to move, and they exhibit many types of dormancy—especially with seeds. In this chapter we will develop these ideas so that the full ecological, genetic, and evolutionary significance of dispersal, migration, and dormancy can be well understood. ❖

DISPERSAL AND MIGRATION

11.1   Migration and dispersal have a variety of important genetic and ecological consequences

Animals and plants have a wide variety of ways to move from one place to another. Dispersal may be passive or active. In **passive dispersal**, whole organisms, their gametes, or their seed are carried by currents of air and water or attached to moving animals. Passive dispersal is not necessarily haphazard dispersal. Organisms that depend on passive disperal often have elaborate structures to take advantage of wind or air currents. Seeds, for example, may have structures resembling sails or propellers that allow them to travel farther in gusts of wind (Figure 11.1A). We will see in this section that the physical environment can profoundly influence the distribution of animals that rely on ocean currents to disperse their larvae.

In **active dispersal**, organisms, especially animals, disperse under their own power and so may control the distance and direction that they travel. This dispersal can be very dramatic and regular, with animals moving in large cohesive groups between two locations. Such dispersal is called **migration**. Bird migration may cover distances of 10,000 miles or more. Migrations are often seasonal, but they may be more frequent. For instance, animals following the tides may move several times a day, as the height of the tide changes.

Less organized dispersal includes individuals moving away from a local population or group. Individuals settled in a particular area may still move substantial distances within some characteristic area or **home range** just to find food or a mate.

Later in this section we will explore the relationship between the size of the home range and the energy requirements of a particular animal.

Migration between Populations Links Their Fates
Population size may vary, due either to the imperfect action of density-dependence or to changes in the environment. Occasionally these population fluctuations may become severe enough that a population goes extinct. Yet these periods of population extinction may be brief if populations frequently exchange migrants with nearby populations, since migrants may recolonize an extinct population. For this reason, predicting the long-term fate of a population depends not only on the characteristics of the individual population but also on other populations that may communicate with it via the process of migration.

Genetic Polymorphisms for Dispersal
In some organisms the propensity to disperse is under genetic control. Such systems are interesting because they allow us to analyze the consequences of dispersing. For example, to assess the fitness consequences of dispersing, we can keep track of the survival of those individuals in a specific environment that do not disperse versus those that do disperse. If environmental conditions occasionally deteriorate, due either to crowding or to seasonal fluctuations in resources, the ability to disperse can increase an organism's chances of finding a new, suitable environment. The frequency of dispersing forms in many species appears to be highest in the environments that are subject to these types of deterioration.

So why, for example, don't all insects disperse? The energy needed to form wings and wing muscle may actually detract from an individual's ability to produce offspring. It may not always pay to invest in the machinery needed to disperse. This uncertainty in the relative benefits of dispersing or staying put is reflected in some species where only a part of the population has the capability to disperse. In these cases, we say the population is **polymorphic** for dispersal ability.

Two species of *Gryllus*—*G. firmus* and *G. rubens*—have a genetically based wing polymorphism. The wingless form (bottom in Figure 11.1B) does not disperse; and it has less wing muscle, lower

FIGURE 11.1A Some plants produce seeds that are designed to take advantage of wind dispersal.

FIGURE 11.1B The Two Winged Morphs of the Cricket *Gryllus* On the bottom is the wingless form; on the top is the winged morph.

metabolic rate, and greater ovary weight than the winged form (top, Figure 11.1B). Males of the wingless form have higher mating success. The price for dispersing is reduced fertility in both males and females.

Whether a given individual develops wings or not is determined by an insect hormone called *juvenile hormone*. During the last larval stage, the wingless forms have higher levels of juvenile hormone than do the winged, and this prevents the development of wings and wing muscle. The amount of juvenile hormone in an insect is, in turn, finely controlled by levels of juvenile hormone esterase, an enzyme that breaks down juvenile hormone.

Tony Zera and his colleagues have shown that the levels of juvenile hormone esterase can easily be raised or decreased by artificial selection. This observation tells us not only that the level of juvenile hormone esterase in an insect is under genetic control, but that there is genetic variability in populations for this genetic control. Thus, we would expect that natural selection can control the frequency of dispersal by changing juvenile hormone esterase levels. ❖

11.2 A population may consist of many small populations linked by migration

So far we have treated populations as isolated groups of individuals. In fact, many species consist of a large number of populations that are linked through occasional exchange of migrants. This collection of groups is sometimes called a **metapopulation**. What happens to the whole metapopulation may differ from what happens to its constituent groups.

We have already seen how density dependence affects population size (Chapter 10). Density dependence and environmental fluctuations may cause populations to go extinct. Populations may also fluctuate for other reasons, such as the impact of predators. Without migration, those habitats where extinction has occurred would remain unoccupied. But migration can reintroduce a species and thus prevent the progressive extinction of all populations. Figure 11.2A describes an experiment that shows the differing fates of large single populations and metapopulations.

Migration can be both stabilizing and destabilizing. Figure 11.2B shows a hypothetical metapopulation consisting of 10 smaller populations. Each of these has some chance of going extinct in a particular period of time. If there were no migration between these populations, then all 10 populations would eventually become extinct. But if migration occurs at a rate (m) that exceeds the rate of extinction (e), then the metapopulation will approach an equilibrium. What happens if the migration rate gets too large? Then the equilibrium will not be stable, just as the carrying capacity is not stable if r, the intrinsic rate of increase, becomes too large. ❖

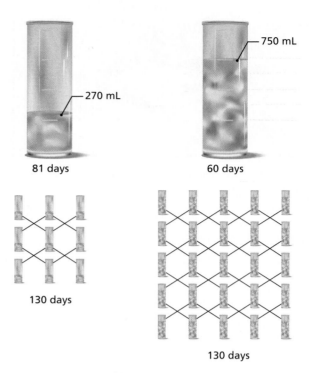

FIGURE 11.2A Holyoak and Lawler (1996) created laboratory populations of single-celled ciliates. The prey species was *Colpidium stritum*, which was eaten by the predator ciliate, *Didinium nasutum*. The predator and prey were kept in two types of environments: (1) a large single culture, which was either 270 mL or 750 mL in volume; and (2) a metapopulation. The metapopulation consisted of either 9 (30 mL) small bottles joined by hollow tubes or 25 (30 mL) bottles joined by hollow tubes. The numbers of predator and prey were determined at regular intervals. The average time until all predators went extinct was calculated from this information. In the metapopulations, the predators persisted until the end of the experiment (130 days). In the large single populations, the predators went extinct well before the end (81 days for the 270-mL culture and 60 days for the 750-mL culture).

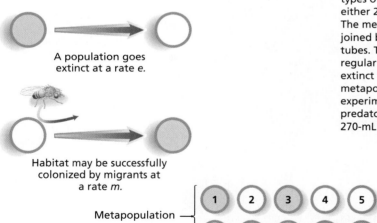

A population goes extinct at a rate *e*.

Habitat may be successfully colonized by migrants at a rate *m*.

Metapopulation

In the example at the left a metapopulation is displayed as consisting of 10 populations. Each population occupies a different geographical site. At some time, four populations are colonized and six are not. In this case the fraction of occupied sites is $p = 4/10 = 0.4$.

After some time has passed we find that populations 1 and 7 have gone extinct; but 2, 6, and 8 have been colonized. Now the new fraction of colonized sites is $p = 5/10 = 0.5$. Over time the average number of sites occupied approaches an equilibrium.

FIGURE 11.2B In a metapopulation, individual populations may go extinct, and unoccupied habitats may be colonized by migration. If the rate of migration is greater than the rate of extinction, the fraction of colonized sites will approach an equilibrium value of $1 - (e/m)$. As shown here, however, the precise sites that are empty and occupied will always be changing.

In the search for food, shelter, or mates, many animals move within characteristic areas called home ranges. The movement that determines the home range is more limited than that of dispersal. An individuals' home range may also change over time, especially when seasons affect the availability of food.

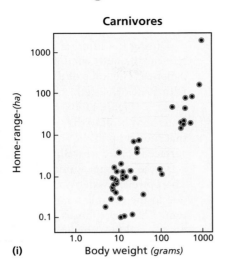

Carnivores

Home-range (ha) / Body weight (grams)

(i)

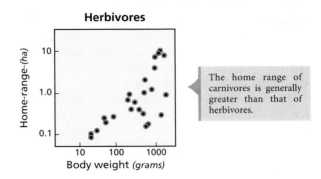

Herbivores

Home-range (ha) / Body weight (grams)

> The home range of carnivores is generally greater than that of herbivores.

While it is usually not difficult to monitor the movements of animals in their home ranges, it is not always obvious why individual animals move. But sometimes we can explain particular patterns of movement. For instance, in species where males mate with multiple females, males often move more widely than females, because they are searching for *both* food and mates.

What factors affect the size of the home range? There are several:

1. *Diet of the animal* Carnivores have to search greater distances than herbivores do, all other things being equal, because their prey is mobile and sparse compared to plants. Part (i) in Figure 11.3A shows these trends in data from carnivores and herbivores.

2. *Size of the organism* All other things being equal, larger animals need more energy than smaller animals do, so they need to search over greater distances to get the required food. Part (ii) in Figure 11.3A shows for both birds and rodents that larger animals require more energy.

3. *Distribution of food resources* If the food is patchy in its distribution, then animals may have to travel greater distances to find sufficient food.

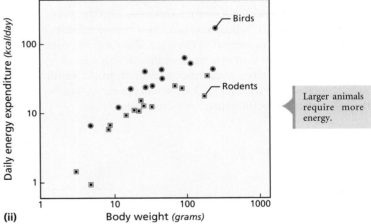

(ii)

Daily energy expenditure (kcal/day) / Body weight (grams) / Birds / Rodents

> Larger animals require more energy.

The relationship between home-range size and energetics can usually be described by a simple mathematical relationship. In the following equation, W is the weight of the organism, a is a constant reflecting the organism's basal metabolism, and b is a constant indicating the rate at which energy requirements increase with size:

$$\text{Energy} = aW^b$$

Animals that require more energy will have larger home ranges; see part (iii) in Figure 11.3A. For example, in both birds and rodents, the constant b is roughly the same, while the constant measuring basal metabolism, a, is greater for birds than for rodents. The higher basal rate for birds reflects the high energy costs of flight.

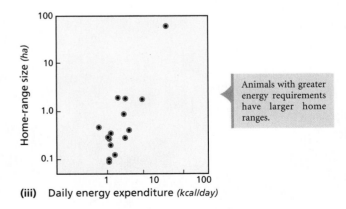

(iii) Daily energy expenditure (kcal/day)

Home-range size (ha)

> Animals with greater energy requirements have larger home ranges.

FIGURE 11.3A Home Range and Energetic Requirements

11.4 The dispersal of many marine organisms is mediated by ocean currents

We have seen the importance of space as a limiting resource in the intertidal environment (Module 10.4). Ocean currents play a crucial role in dispersing juvenile forms of many plants and animals that live in the intertidal zone. These physical factors also contribute to the relative differences in free space along the West Coast of the United States (see Figure 10.4D). These important biological consequences of ocean currents warrant a closer examination of the forces causing ocean currents.

As an example, let's look at the local movement of water off the West Coast of the United States, the area that is home to the barnacles and sea anemones we discussed in Chapter 10. One important force contributing to currents in this area is the **Coriolis effect**, which arises from the Earth's rotation. Viewed from the North Pole, the Earth rotates in a counterclockwise direction. Because of the Earth's rotation, drifting objects appear to deviate from a straight line.

Figure 11.4A shows one way to envision this phenomenon. Imagine you are on a carousel that is moving in a counterclockwise direction (observer X in Figure 11.4A). If an individual in the center of the carousel throws a ball to someone standing directly behind you (observer B in the figure), the ball will appear to veer off to the right. The Coriolis effect is more complicated than this, because the Earth is a rotating sphere instead of a circular carousel. Viewed from the South Pole, it rotates clockwise, so the ball would appear to veer to the left; nevertheless, the general principle holds true.

During the summer months, a strong and fairly consistent north-south wind develops off the west coast of North America. The force of the wind on the surface waters, along with the Coriolis effect, produces a westward, offshore flow of water (Figure 11.4B) called an **Ekman flow**. As the surface water moves off to the west, it is replaced by deep, cold water that moves up to the surface. This flow of cold water to the surface is called **upwelling**. Upwelling is responsible for the very cold summer water temperatures along the west coast. The upwelled water often has higher levels of phosphorus due to the organic matter brought up from the ocean bottom. This phosphorus can support additional growth of **phytoplankton**, the microscopic plants that grow in the ocean waters.

How does the Ekman flow affect the dispersal of intertidal larvae? Figure 11.4C illustrates two ways. First, young larvae are carried away from the shore and out to the ocean by this flow. The maximum distance that intertidal larvae travel away from the shore is proportional to the strength of upwelling, as Figure 11.4C (left graph) shows. (In Figure 11.4C, a positive upwelling index indicates that the current is moving offshore, while a negative index means the water is actually flowing onshore). Second, the Ekman flow will also make it more difficult for larvae to make it back to the shore when they are mature, so the settlement of larvae on the central California coast decreases as the strength of upwelling increases, as Figure 11.4C (right graph) shows. ❖

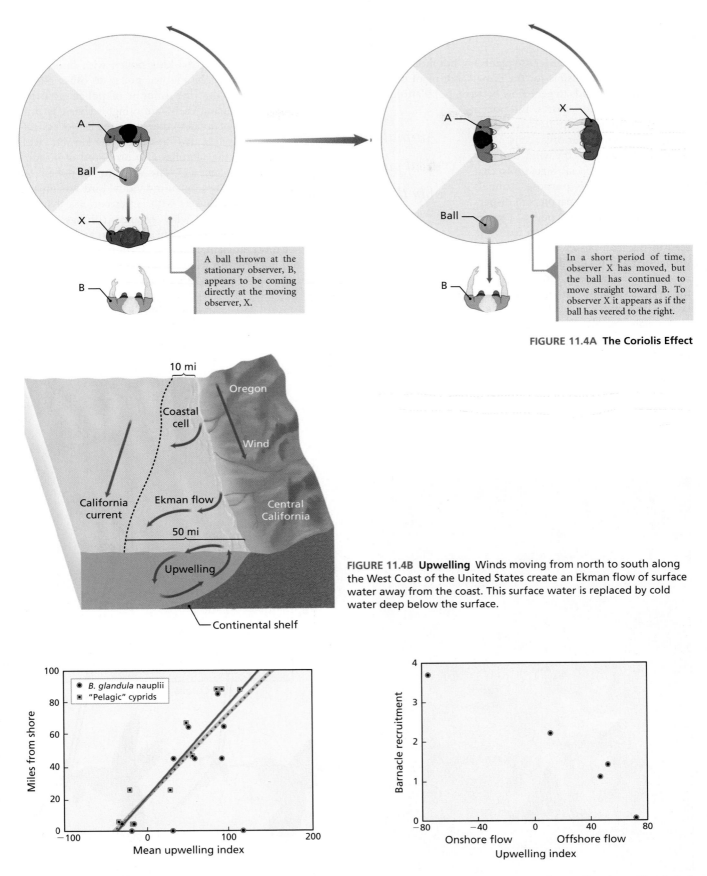

A ball thrown at the stationary observer, B, appears to be coming directly at the moving observer, X.

In a short period of time, observer X has moved, but the ball has continued to move straight toward B. To observer X it appears as if the ball has veered to the right.

FIGURE 11.4A **The Coriolis Effect**

FIGURE 11.4B **Upwelling** Winds moving from north to south along the West Coast of the United States create an Ekman flow of surface water away from the coast. This surface water is replaced by cold water deep below the surface.

FIGURE 11.4C The figure at left shows the distance larvae are transported as a function of the strength of upwelling (upwelling index). The figure at right shows that barnacle recruitment onshore depends decisively on the direction of currents and their strength (after Roughgarden et al., 1988).

11.5 Plant morphology affects the efficiency of passive dispersal

How do plants disperse? Because plants are not mobile, seeds and spores are responsible for the dispersal of plant populations. Pollen also leads to gene flow between populations.

Dispersal of seeds, spores, and pollen often depends on help from environmental factors or other organisms. Many seeds are dispersed by the wind, and they may even have special structures that help accomplish this purpose. Animals are frequently critical partners in dispersal of seeds, spores, and pollen. For example, when an animal eats a fruit, the seed in the fruit passes through the animal's digestive tract undigested and is later deposited in the feces, often quite far from the original plant. Other seeds become attached to the fur, bristles, skin, or feathers of animals, which transport the seeds as they move.

Animals are frequently critical partners in dispersal of both seeds and pollen.

Animals are also instrumental in dispersing pollen of many plants. An interesting example involves the plant *Ipomopsis aggregata* (Figure 11.5A). Individual plants of this species produce both male (pollen) and female (seeds) gametes. However, individual plants cannot fertilize their own seeds, so they must **outcross**, using pollen from another individual. These plants are most often pollinated by hummingbirds (Figure 11.5B). Diane Campbell and her colleagues (1996) have carefully documented the dispersal of pollen from individual plants by marking pollen with fluorescent dye and then following the marked pollen to other plants. The biologists found that the amount of pollen exported during each hummingbird visit is proportional to the size of the flower's corolla, for two reasons. (1) Flowers with wider corollas also produced more total pollen. (2) The amount of pollen removed by an individual bird was greater for flowers with larger corollas than for flowers with smaller corollas. Pollen transfer by plants with wide corollas seems to be more efficient. Hummingbirds inserted their bills deeper into flowers with wide corollas and thereby extracted a greater amount of pollen (Figure 11.5B). ❖

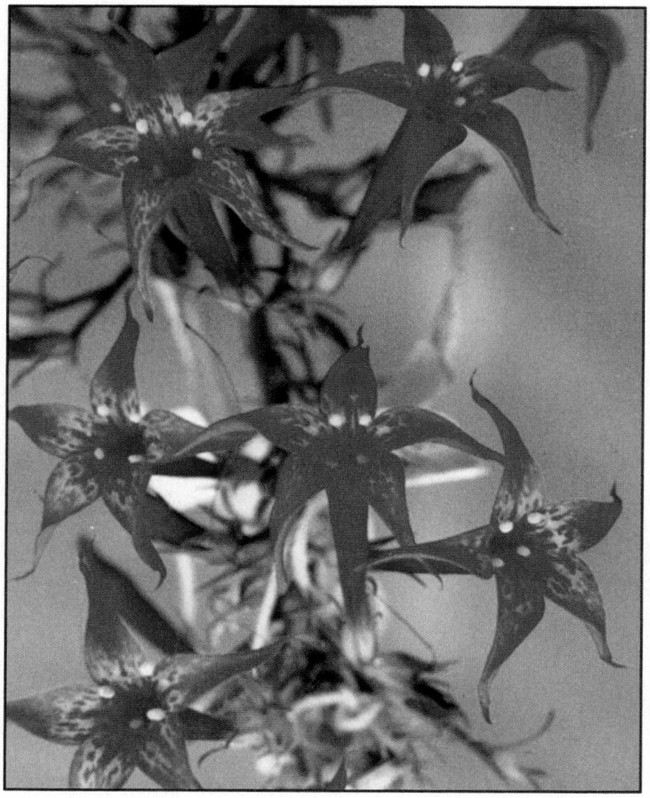

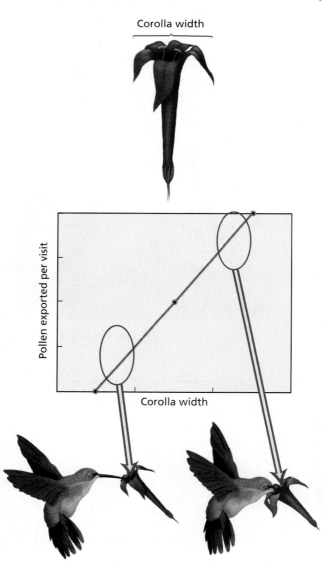

FIGURE 11.5A *Ipomopsis aggregata*

FIGURE 11.5B Hummingbirds place their bills deeper in flowers when the corollas are wide. The deeper the bill, the more pollen is transported.

Some species escape bad conditions by "traveling" through time: dormancy

Dispersal is an adaptation that takes plants and animals from an inhospitable environment to an environment in which survival or reproduction are more likely. But many plants and animals cannot effectively disperse (Figure 11.6A.) Some species cannot disperse because they and their gametes are not significantly motile. This is evidently true of many plants, which obviously lack locomotion and may also lack dispersing pollen or seeds. This is also true of sessile animals, such as sponges and sea anemones. Fissile sea anemones even lack a dispersing life-cycle stage, unlike barnacles, which disperse widely as juveniles before becoming sessile as adults.

Organisms that do not disperse are at the mercy of local conditions. Unfortunately, such local conditions may become inimical to the continued survival or reproduction of the organism. Physical trauma may injure sessile forms. Extremes of temperature, light, moisture, and desiccation may kill or damage the organism. Overcrowding may limit the resources available to an organism, impairing its survival or reproduction. But without dispersal, what can be done?

One solution is dormancy. Dormancy is an adaptation that involves shutting down most metabolism, enabling the organism to survive on little or no food. Dormancy also reduces vulnerability to external stressors, possibly including heat and desiccation. The Darwinian opportunity is that the period of dormancy could be followed by a more active period during which survival and reproduction are feasible. Dormancy becomes a way of "traveling" through time to a better habitat. Dormancy can be an effective alternative to dispersal.

Perhaps the most familiar example of dormancy is the hibernation of bears from temperate climates (Figure 11.6B). The reduction in metabolism enables bears to survive most of the winter, with little foraging or feeding. Other forms of dormancy can be much more radical. Plant seeds, in some cases, are almost completely shut down metabolically.

Rather than a single simple adaptation, dormancy is exhibited over a diverse range of species, with widely varying consequences for the ecology and evolution of those species. For some species, dormancy is brief. For other species, dormancy is part of a seasonal life cycle. For still other species, dormancy may allow the species to travel many years through time, with more impact on population genetics than on dispersal. This temporal range is shown in Figure 11.6C. ❖

FIGURE 11.6B Grizzly Bear Pulls Leaves into its Den

1. Transient dormancy

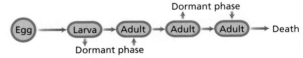

2. Seasonal dormancy

3. Indefinite dormancy

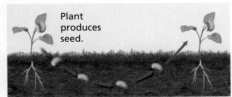

FIGURE 11.6C Patterns of Dormancy

FIGURE 11.6A Plants and Animals with Limited Dispersal Ability: (i) strawberries and (ii) fissile sea anemones

11.7 Plant seeds are some of the longest-lasting dormant life-cycle stages

After fertilization, seeds store nutrients and reduce their water content. A *seed coat* covers the surface of the seed, and the coat is thick and hard in a number of species. The coconut shell is a famous example of the physical robustness attained by some seed coats. (Figure 11.7A shows several seed structures.) In many respects, seeds are among the most invulnerable of all forms of life, resisting heat, freezing, desiccation, water, and so on.

Once the seed has fully developed, it enters into a period of dormancy in which its metabolic rate is extremely low. Growth and development entirely cease. The seed is perhaps the most perfect resting stage exhibited by multicellular life. Plants distribute seeds widely. The soil of many habitats contains a hidden reserve of dormant seeds called a **seed bank**. Seed banks link generations through time: seeds produced decades earlier may lie dormant in soil, although the seeds of many species are viable for just a few years (see Figure 11.7B). Dormancy may thereby connect otherwise isolated biological generations, increasing the breeding population size and reducing genetic drift.

The adaptive significance of the seed is that it allows a plant to disperse dormant offspring through space and time until changed conditions allow *germination*, when the developing plant, or *seedling*, is produced by the seed. The timing of germination is crucial. A very interesting feature of plant biology is the environmental cues that seeds use to trigger germination. Benign conditions for seedling growth often serve as cues, especially reasonable moisture. Because seeds are quite dry, they naturally tend to take up water. This can

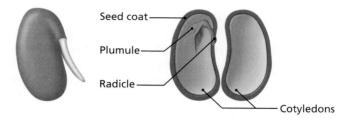

1. Common bean. The fleshy cotyledons of the common garden bean, a dicot, store food that was absorbed from the endosperm when the seed developed.

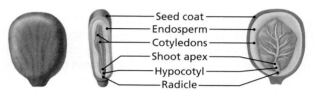

2. Castor bean. The castor bean has membranous cotyledons that will absorb food from the endosperm when the seed germinates.

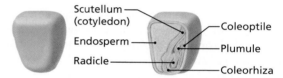

3. Corn. Like other monocots, corn has only one cotyledon (the scutellum.) The rudimentary shoot is sheathed in a structure called the coleoptile.

FIGURE 11.7A Seed Structure

lead to the rupturing of the seed coat, with germination following straightforwardly. Germination occurring in different plants is shown in Figure 11.7C, proceeding to the right. Germination can also be triggered by quite different cues—including fire, an important ecological agent in some habitats, such as the California chaparral where many grasses and shrubs germinate in response to fire. ❖

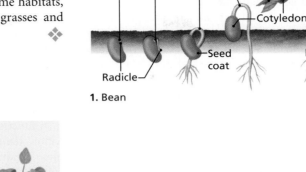

1. Bean

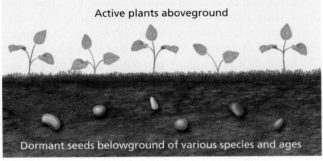

Seeds may germinate whenever conditions are good, as long as they are still viable.

FIGURE 11.7B A Seed Bank

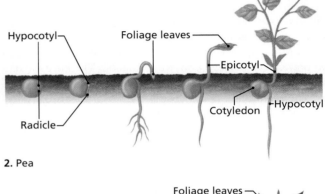

2. Pea

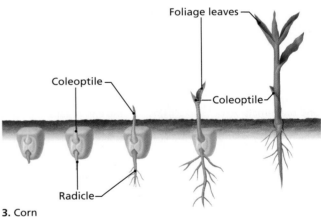

3. Corn

FIGURE 11.7C Seed Germination

Many animals and plants survive in seasonal climates through the use of dormancy

Although equatorial ecosystems have relatively stable environments, toward the poles all habitats are affected by extreme annual fluctuations in temperature. Indeed, the circumpolar environment may entirely forestall completion of the life cycles of most invertebrates, and most plants, for most of the year.

The solution that many species have found is to spend much of the year in a dormant phase. In plants, the primary form of seasonal dormancy is the seed stage. Because of their low water content, seeds are particularly resistant to the formation of ice crystals, the main agent of cell damage at low temperatures. Deciduous trees shed their leaves and cease photosynthesizing. Many univoltine insects overwinter as early larvae inside egg cases. Many overwintering animal larvae make cryoprotectant to prevent cell damage from ice formation. The life cycle of a univoltine moth is shown in Figure 11.8A.

The ecological impact of seasonal dormancy is considerable. While endothermic vertebrates can exploit the winter environment by generating heat metabolically, ectothermic invertebrates—such as insects—are essentially unable to exploit the winter environment. Local sources of heat, such as decaying compost or hot springs, may occasionally allow insects to survive; but these insects are the exception. Thanks to

dormancy, ectotherm species can flourish in northern circumpolar regions, greatly adding to their ecological diversity. Some endotherms, such as bears, exhibit **hibernation**, in which metabolic rate and body temperature fall. In some seasonal habitats, a similar dormancy pattern may occur in dry or hot summer periods, which is called **estivation**. A species does not have to be a ectotherm in order to use dormancy to get relief from climatic extremes.

A functional problem of seasonal dormancy occurs when winter temperatures rise to unusually high levels. Under these conditions, dormancy may be broken. If the dormant animal or plant experiences an increased metabolic rate, but does not feed, it may quickly run out of stored calories and starve to death (see Figure 11.8B). Selection will oppose breaking dormancy in this case, thus dormant overwintering is often cued to environmental signals that are more reliable than ambient temperature. Daylength is a common cue used by both plants and animals to control entry into dormancy as well as emergence from dormancy. ❖

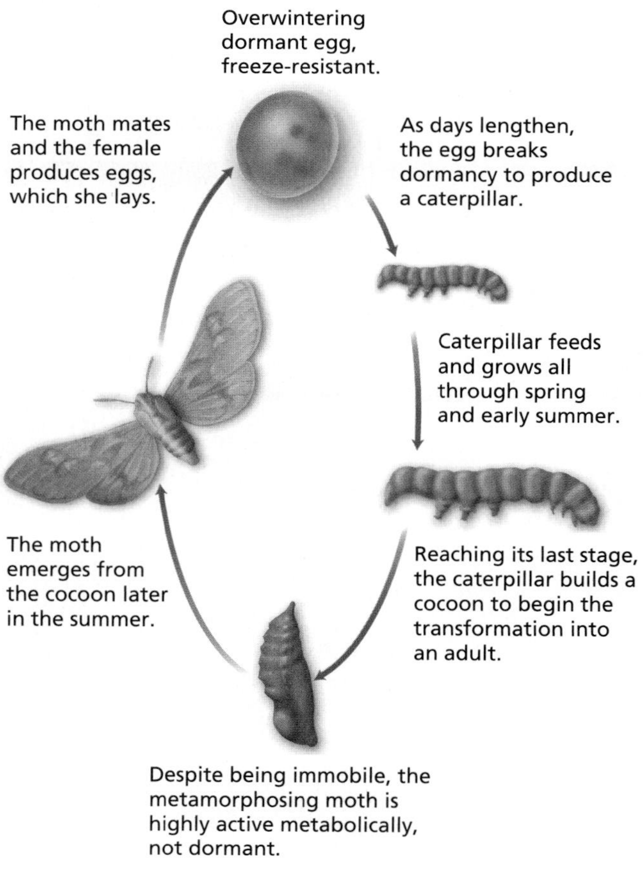

Overwintering dormant egg, freeze-resistant.

As days lengthen, the egg breaks dormancy to produce a caterpillar.

The moth mates and the female produces eggs, which she lays.

Caterpillar feeds and grows all through spring and early summer.

Reaching its last stage, the caterpillar builds a cocoon to begin the transformation into an adult.

The moth emerges from the cocoon later in the summer.

Despite being immobile, the metamorphosing moth is highly active metabolically, not dormant.

FIGURE 11.8A The Life Cycle of a Univoltine Moth

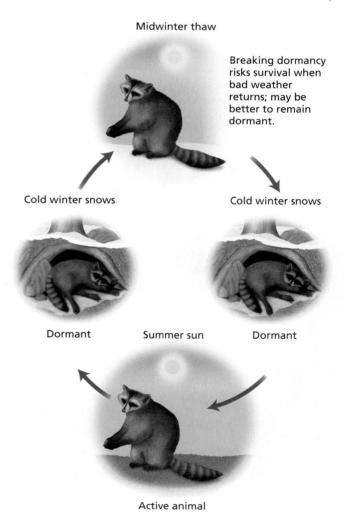

Midwinter thaw

Breaking dormancy risks survival when bad weather returns; may be better to remain dormant.

Cold winter snows

Cold winter snows

Dormant

Summer sun

Dormant

Active animal

FIGURE 11.8B Breaking Seasonal Dormancy

If seeds provide a means to propagate through long periods of time, and seasonal dormancy allows species to utilize environments with extreme fluctuations, there are also temporary opportunities to use dormancy. Starvation is a common stress for animals, and some undergo dormancy or torpor during periods of reduced nutrition. The nematode *Caenorhabditis elegans* has a *dauer* stage that it enters partway through development, when conditions are stressful. As a dauer, the nematode greatly delays maturation to adulthood. An interesting feature of dauers is that they are not immobile; they move actively.

A somewhat similar case is nutritionally restricted rodents, both rats and mice, which greatly reduce their reproduction so long as food is restricted. However, these rodents have metabolic rates that are similar to those of well-fed rodents, in proportion to total body weight. They are also highly active, moving around with great fervor in their cages, unlike the superficially torpid, well-fed rodents. This entire syndrome is diagrammed in Figure 11.9A. It is important to understand that the effect of diet can be reversed. Reduced feeding converts a large ad lib rodent into an underweight healthier rodent, and vice versa.

A similar experiment was done on humans during the Biosphere II project (see Figure 11.9B). Biosphere II was intended as a mock-up of a colony on another planet, dependent on recycling its own resources. Food was grown within Biosphere II using materials within the facility, as well CO_2 and excrement from the humans. Due to an error in setting up the facility, there was not enough plant growth. Access to food was restricted for the Biospuereans, as it has been for experimental rodents. Slowly, the Biospuereans lost weight. As they did so, many indices of cardiovascular health improved. For example, blood pressure fell, the level of triglycerides (fat) in the blood fell, and serum cholesterol levels fell. It became very hard for the Biospuereans to work, and even harder to enjoy life. When the facility was opened, after years of reduced calories, all but one of the participants went off the restricted diet. Their cardiovascular health fell, their weight increased, and it became easier for them to function. In effect, they came out of dormancy.

In a number of species, dormancy may occur intermittently. The phenomenon of sleep in mammals can be seen as an example of temporary, marginal dormancy. Dormancy may take on a variety of forms, each form husbanding resources so that the organism can live long enough to develop or reproduce in good conditions. ❖

Mice are allowed to grow normally and become adults.

Some mice are fed ad lib.

"Restricted" mice are given 30–40% of the ad lib calories.

Ad lib mice become bigger; sperm production and menstruation are normal; they are less active.

Restricted mice are smaller and they have limited fertility; they are much more active.

Ad lib mice die sooner of a variety of ailments, from kidney problems to cancer; they are very inactive later in life and do not learn well.

Restricted mice live 30–40% longer than ad lib mice, with fewer major illnesses; they remain active longer and learn well.

FIGURE 11.9A Restricted diets produce animals that reproduce little or not at all, while living longer. The best-known example of this pattern is caloric restriction in rodents, rats and mice particularly. Mice are used as an example.

FIGURE 11.9B The Biosphere II Facility

CONSEQUENCES OF DISPERSAL

11.10 Novel ecological structures: metapopulations

One way to appreciate the complexities of metapopulations is by studying a particularly good example. One such example is the metapopulations of the Glanville fritillary, *Melitaea cinxia* (a fritillary is just a spotted butterfly). This metapopulation, which has been studied extensively by Ilkka Hanski and his colleagues, is located on an island in southwest Finland (Figure 11.10A). The adult females typically mate once and then lay several large batches of eggs on one of the two host plants that are available. Most of the local populations (solid dots in Figure 11.10A) are quite small, consisting of larvae from just a few egg batches.

Metapopulations can be characterized by several important attributes.

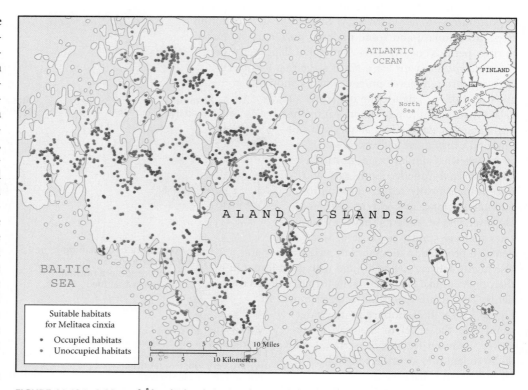

FIGURE 11.10A A Map of Åland Islands in Southwest Finland The circles show suitable habitats for *Melitaea cinxia*. The red circles are occupied habitats; the green circles are unoccupied.

1. They consist of spatially discrete breeding populations. This means that there must be purposeful movement or migration for individuals to get from one population to another.
2. All populations have a high risk of extinction. As a result, we expect to find at any time some suitable habitats that are unoccupied due to a recent extinction.
3. Recolonization of an extinct population is possible. This prevents the rapid extinction of the entire metapopulation.
4. There is asynchrony in local dynamics. While some populations have large numbers, others will be close to extinction. These types of fluctuations are not strongly correlated from one population to the next.

The Glanville fritillary has all four of these important metapopulation attributes.

As the area of patches increases, there should be more resources, and the patch should be able to support more individuals. Consequently, we expect extinctions to be less frequent when larger patches are occupied. This is exactly what has been observed in the Glanville fritillary (Figure 11.10B). In this metapopulation individuals do not move far when they migrate (Figure 11.10C), so it is not at all unusual for some populations not to receive any migrants for some time after they

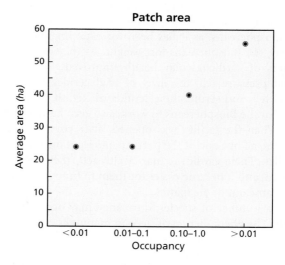

FIGURE 11.10B As patch area increases, the fraction of patches occupied also increases.

have gone extinct. This gives rise to the unoccupied patches, which are fairly frequent (Figure 11.10A).

Under what conditions would a whole metapopulation go extinct? It would seem plausible that if a metapopulation had very few connected populations, or a small **network**, then the chances of all going extinct simultaneously would be greater than if there were very large numbers of connected populations. The relationship between the number of patches in a network and extinction has been examined in the Glanville fritillary (Figure 11.10D). Hanski has concluded from these observations that a metapopulation should have 20 or more well-connected populations in order to be reasonably certain of prolonged persistence. ❖

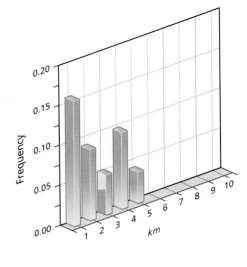

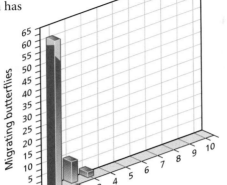

(i)

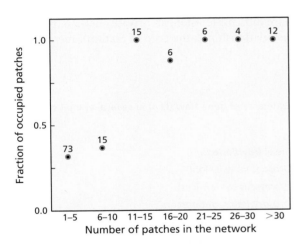

FIGURE 11.10C In the Glanville fritillary, most migrating butterflies are found within 1 km of their home (i). The number of butterflies that successfully migrate to a new population decreases with distance between the two populations (ii).

FIGURE 11.10D Interconnected patches or networks are more likely to be occupied if the number of patches is large. The numbers above each point indicate the number of different networks used to compute the fraction of occupied networks.

We have already seen how migration can prevent population extinction. From a genetic perspective, the equivalent of extinction is the loss of genetic variation. The loss of genetic variation due to genetic drift can also be prevented by migration. In Figure 11.11A we have represented a group of small populations, all with two alleles called A and a. The A allele is shown as red and the a allele is blue. If these populations are completely isolated, then over many generations we expect either the A or a allele to be lost from each population. Which allele is lost is random, but it is certain to happen if we wait long enough. At the point when all populations have become fixed for just a single allele, there will be no heterozygotes in any populations (Figure 11.11A; Module 3.23).

Migration involves the movement of subsequent breeding of individuals from one population to the another. If migrants do not mate and have offspring, then their movement to a population will have no direct genetic consequences. The effects of migration on population differentiation will depend on both the migration rate (m, Figure 11.11A) and the population size, N. The forces of drift and migration work in opposite directions. Drift will tend to eliminate genetic variation; and the smaller the population, the faster the rate of elimination. Migration will generally reintroduce genetic variation, thus preventing its loss. Eventually these two forces will balance each other. An equilibrium level of heterozygosity will be reached that is equal to

$$2pq(4Nm)/(4Nm + 1).$$

As an example, consider the case where the two alleles have the same frequency, 0.5. The Hardy-Weinberg expectation is that heterozygotes will be 50 percent of each population. The drift and no-migration expectation is for 0 percent heterozygotes. In a population with just a 2 percent (0.02) migration rate and a population size of 1,000, the equilibrium frequency of heterozygotes should be 49.4 percent—very close to the Hardy-Weinberg expectation.

Is it possible to estimate the N and m for real populations? In fact, Montgomery Slatkin devised a method to estimate the product of Nm using a concept called **private alleles**. If genetic variation is measured in many subpopulations, private alleles are those found in only one of the subpopulations examined. If there is a lot of migration or drift is very weak, because the population size is large, it will be more difficult to establish private alleles in populations. This relationship is the core of Slatkin's method. In

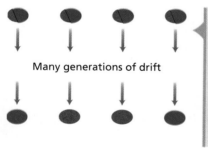

Many generations of drift

In this example the two alleles, A(red) and a (blue) are initially at a frequency of p and q in every population. If there is no migration between any populations, one of the two alternative alleles will be fixed in each population. There are no heterozygotes.

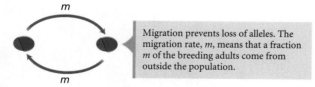

Migration prevents loss of alleles. The migration rate, m, means that a fraction m of the breeding adults come from outside the population.

FIGURE 11.11A Migration and Population Differentiation

Table A. Estimates of gene flow (*Nm*) in several animal species.

Species	Nm
Marine mussel (*Mytilus edulis*)	42
Fruit fly (*Drosophila willistoni*)	9.9
Mouse (*Peromyscus californicus*)	2.2
Fruit fly (*Drosophila pseudoobscura*)	1.0
Pocket gopher (*Thomomys bottae*)	0.86
Mouse (*Peromyscus polionotus*)	0.31
Salamanders (*Plethoden cinereus*)	0.22

Table A we show estimates of Nm for several animal species. We see very high values for marine mussels. This seems reasonable since these organisms distribute their immature larval forms into the ocean, and the larvae may be carried great distances by ocean currents before they settle and become adults. Conversely, the salamander, *Plethodon cinereus*, includes samples from the southern United States, and as far north as Quebec, Canada. It is unlikely that a small terrestrial salamander can traverse even a small fraction of these distances. Therefore we expect small estimates of gene flow.

Another interesting example of migration occurs when there is essentially one-way flow from a large, mainland population to smaller, island populations (Figure 11.11B). In this case the migration affects allele frequencies only in the smaller population. In Figure 11.11B the frequency of an allele, A, is P on the mainland; and at time t, it is p_t on the island. If the migration rate from mainland to island is m, then the frequency of the A allele on the island changes each generation, as shown in Figure 11.11B. One application of this theory is to estimate the rate of gene flow between races in the United States. The U.S. African American population is much smaller than the U.S. Caucasian population and can be considered the island population in this theory. One allele in the Rh blood group is quite different in East Africans (0.63) and U.S. Caucasians (0.028). The allele's present frequency in the U.S. African American population is about 0.446. This suggests the frequency has been falling due to movement of Caucasian alleles into the African American population. If we assume that 300 or so years of African presence in the United States represents about 10 generations, then we can substitute values in the formula in Figure 11.11B to obtain the migration rate as $(0.446 - 0.028) = (1 - m)^{10}(0.63 - 0.028)$. After some algebra, we can find $m = 0.036$. This means that on average, about 3.6 percent of the alleles in the African American population come from U.S. Caucasians. After 10 generations this would mean about 30 percent of the African American alleles are from Caucasian ancestors. Using other genetic markers, estimates of gene flow into the U.S. African American population, or admixture, have been estimated at between 20 and 30 percent. ❖

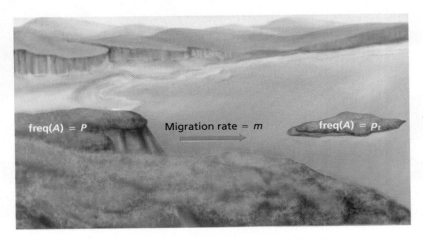

freq(A) = P

Migration rate = m

freq(A) = p_t

After one generation
$$p_{t+1} - P = (1 - m)(p_t - P)$$
After n-generations
$$p_{t+n} - P = (1 - m)^n(p_t - P)$$

FIGURE 11.11B **Migration from a Large Mainland to an Island**

SUMMARY

1. Animals and plants may move move or disperse from one location to another. These movements may be under the power of winds or currents, which is called passive dispersal, or under the organism's own power, called active dispersal.

2. More highly organized movements are called migrations.

3. Metapopulations are large numbers of discrete populations that are linked through migration.

4. The home ranges of animals are related to their environment and food requirements. Larger home ranges are required when resources are patchily distributed. Carnivores will generally require larger home ranges than herbivores will.

5. Organisms that cannot disperse may use dormancy as a means of escaping unfavorable conditions. Some organisms will become dormant for short periods of time. Others have a regular cycle of dormancy that coincides with the seasons. Still other organisms exhibit dormant phases that are released from dormancy by some special cue.

6. Metapopulations of butterflies show several general trends. The chance of patches being occupied increases with increasing patch area. The number of populations in the communicating network increases the chances of at least one patch being occupied. It seems at least 20 or more populations are required for a network to essentially be guaranteed of occupancy.

7. Even small amounts of migration are an effective means of preventing population differentiation.

REVIEW QUESTIONS

1. Discuss the different factors that affect home-range size.

2. How does upwelling and the Ekman flow affect barnacle recruitment?

3. What factors cause or release organisms from dormancy?

4. A mainland population has a common allele for brown eyes in an insect; the allele is at a frequency of 0.92. Migrants from this mainland arrive at a small island population and make up 3 percent of the breeding population in each generation. The brown-eye allele is at a frequency of only 0.1 on the island. What will the frequency of the brown-eye allele be after six generations of migration?

KEY TERMS

active dispersal	home range	outcross	seasonal dormancy
Coriolis effect	indefinite dormancy	passive dispersal	seed bank
Eckman flow	metapopulation	phytoplankton	transient dormancy
estivation	migration	polymorphic	upwelling
hibernation	network	private alleles	

FURTHER READINGS

Campbell, D. R., N. M. Waser, and M. V. Price. 1996. "Mechanisms of Hummingbird-Mediated Selection for Flower Width in *Ipomopsis aggregata*." *Ecology* 77: 1463–1472.

Hanski, I. 1999. *Metapopulation Ecology*. Oxford, UK: Oxford University Press.

Holyoak, M., and S. P. Lawler. 1996. "Persistence of an Extinction-Prone Predator-Prey Interaction through Metapopulation Dynamics." *Ecology* 77: 1867–79.

Roughgarden, J., S. Gaines, and H. Possingham. 1988. "Recruitment Dynamics in Complex Life Cycles." *Science* 241: 1460–66.

Slatkin, M. 1981. "Estimating Levels of Gene Flow in Natural Populations." *Genetics* 99: 323–335.

Swingland, I. R., and P. J. Greenwood. 1983. *The Ecology of Animal Movement*. Oxford, UK: Oxford University Press.

Zera, A. J., and R. F. Denno. 1997. "Physiology and Ecology of Dispersal Polymorphism in Insects." *Annual Review of Entomology*, **42**: 207–31.

PART FOUR

ECOLOGY OF
INTERACTING SPECIES

Species interact in a variety of ways. Some compete for common resources, some species may be food for other species, and some species may benefit one or many other species. In this part, we examine some basic types of interactions between species.

When Darwin wrote about a struggle for existence, the image he conjured up is one of a competition for resources essential to life. Species that have similar ecological requirements will be forced to compete for scarce resources like food, space, or light. In Chapter 12, we review some adaptations used by plants and animals to compete and the impact of competition on population dynamics.

Interactions between species are not always indirect. Some animals eat other animals or plants. These types of interactions are not subtle or inconsequential. A spider that quickly subdues an insect caught in its web makes the struggle for existence obvious. It should not be surprising that predators have evolved elaborate means for successfully capturing prey and that prey have likewise found ways of trying to avoid being eaten. In Chapter 13 we introduce such predators and prey.

Parasites and hosts represent another important feeding relationship. These interactions are determined both by the adaptations of parasites that overcome host defenses and by the adaptations of the host to resist parasites. In Chapter 14 we consider the coevolution of host-parasite systems. Chapter 14 also introduces the notion that species can interact in a way

that mutually benefits each species. Because different species by definition have independent evolutionary histories, understanding the development of mutualistic interactions is a major challenge for evolutionary biology.

When you walk through a forest or field, you see many plants and animals, not just pairs of predators and prey. This ensemble of organisms is called a community. Its members interact in many overt and subtle ways. In Chapter 15, we consider the relationship of community members from several different perspectives. One common thread is energy. Through the feeding relationships of community members, energy flows from plants that fix energy from the sun, up through multiple levels of animals that either feed on plants or feed on other animals. In addition to energy, organisms require nitrogen, water, and carbon. The complex interactions between the environment and the biological community that cycle these nutrients are also examined in Chapter 15.

The local physical environment affects many characteristics of biological communities. We are not surprised by the very different appearance of the flora and fauna in the desert compared with those in a tropical rain forest. We recognize the importance of water for life and accept that great differences in the availability of water will have large effects on the types and numbers of species that can survive. In Chapter 16, we study some of the factors that determine the local climate. We also survey the typical types of communities that appear in particular environments.

Any student of ecology realizes that many biological communities and even more individual species are adversely affected by human activity. Biological communities are an important reservoir of future medicines, genetic information, and recycling resources that should be preserved. In Chapter 17 we consider the principles of conversation biology that may forestall the loss of species and communities.

The growth of human populations often has negative consequences for many natural communities of plants and animals. Understanding how species interact and how communities function can prepare us to make informed decisions in the future. We still have much to learn about biological communities, but the grave impact of human activities requires that our current knowledge be applied now. In Chapters 12–17 you will learn what ecologists know about interacting species.

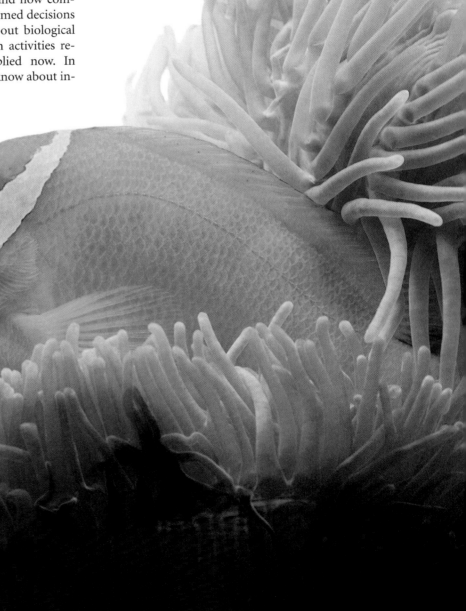

Sea gulls competing for food

12

Competition

The recognition of competition as an important biological process dates back to Darwin. In his *Origin of Species*, Darwin emphasized a struggle for existence that hinges on the presumption that not all individuals can be supported by the existing resources. For Darwin the competition between individuals was the mechanism that led to adaptive traits increasing within a species. Much of the early work on competition was in fact motivated by Darwin's ideas.

In 1934 the great Russian ecologist G. F. Gause presented much of his experimental work on competition between species of protozoans in a book entitled *The Struggle for Existence*. It is clear from reading Gause that he believed research into the mechanisms of competition was critical for an understanding of Darwin's theory of evolution. As our understanding of natural selection and ecology has progressed, our ideas about competition have changed, but competition remains an important process in ecology.

Although the role of competition in natural selection is still considered vital, ecologists recognize that competition between species may be an important factor affecting the abundance and distribution of a species. In this chapter we will review some of the ways that plants and animals compete for essential resources. We will also see that competition may be a mechanism that explains how some species are able to coexist while others are not. Natural selection is also important in this chapter. We find that some circumstances favor an increase in competitive ability, and others favor the evolution of traits that will reduce the levels of competition. ❖

THE ECOLOGICAL AND EVOLUTIONARY PROCESS OF COMPETITION

12.1 Plants and animals compete for resources

Competition between individuals is an inevitable consequence of increasing population size and limited resources. **Competition** is an interaction between individuals of the same species—or two or more species—that reduces survival, fertility, or both. Discussions of competition typically distinguish competition between individuals of the same species, known as **intraspecific competition**, and competition between individuals of different species, known as **interspecific competition**. For closely related species, the limiting resources and mechanisms that give rise to interspecific competition may be quite similar to those involved with intraspecific competition.

There is one important difference between these two types of competition. Interspecific competition may be reduced or completely eliminated when one species evolves traits that effectively prevent the two species from competing for common resources. But this does not happen in intraspecific competition. For sexually reproducing species, at least, different individuals tend to require the same resources and utilize the environment in a similar fashion, so ultimately they are

unable to avoid competing with members of their own species. Asexual organisms, such as bacteria, may evolve clones that become genetically differentiated to the point where their resource utilization results in little competition. It is not uncommon, however, for such differentiated clones to be classified as different species.

While it is impossible to classify all mechanisms of competition into two categories, the ecological literature refers to two major categories of competition. **Contest competition** (Figure 12.1A) refers to the ability of some individuals to monopolize a particular resource in short supply. For many animals, competition for space or territories would be considered a type of contest competition. Animals may go through ritualized behavior that determines a dominance hierarchy and ultimately who will posses certain important resources. An important outcome of contest competition is that the distribution of resources is very uneven. Some individuals may get sufficient resources while others get none.

Scramble competition (Figure 12.1A) occurs when all individuals have access to the limiting resource and their acquisition is like a free-for-all. Some individuals may still get more resources than others because they differ in behaviors or morphologies that are important for resource acquisition. However, individuals are not able to sequester or monopolize resources when competition occurs by a scramble process. To use a concrete analogy, imagine the dinner meal at a prestigious but anonymous private boys' school in England.

The competition for water and nutrients between these trees is an example of **scramble** competition. No one tree can monopolize the resource; rather the amount of resource obtained by each tree will depend on the surface area of its roots and other factors. This picture also illustrates the local neighbor concept of plant competition.

Both the spotted hyenas and vultures are interested in the dead zebra. This leads to a direct conflict or **contest** for the resource. The hyena will most likely be able to monopolize this resource.

Male red-winged blackbirds also engage in a **contest** competition for space. A male defends a particular piece of territory that will be used for foraging and will be important for attracting mates. Males with especially good territories may have several females nest there and thus will father many more offspring than will males with inferior or no territory.

FIGURE 12.1A Different Types of Competition

The rules of the school are that no second helpings are permitted until the boy has finished his first helping. The catch is that there is never enough food for all to have second helpings. In this competition, the winners are those that get a solid grasp on the fork during grace and chew as little as possible. An important lesson here is that success in competition comes with a price—indigestion.

Basic biological differences between plants and animals result in differences in the process of competition. Following are some important basic differences between plants and animals:

1. *Plants are sedentary.* Consequently, the distance over which competitive interactions occur is fixed and usually smaller than for a similarly sized animal. An important factor in plant competition is an individual plant's immediate neighbors, rather than the size of the total population. For those animals that are sedentary, similar patterns arise.

2. *Tremendous size variation in plants.* The size and growth of plants can vary dramatically, even within the same species, depending on levels of nutrition, sun, and water. Thus, when assessing the competitive interactions of plants, the number and size of competitors will be important. Animals also vary in size, although the range of variation is typically not as great as in plants.

3. *Competition for common resources.* Most species of plants require the same essential resources for growth: light, water, and nutrients. These common requirements imply that competition ought to be more common among a wider variety of plant species. While there are important differences between some plant species (for instance, certain plants that can fix nitrogen from the atmosphere as opposed to absorbing it from soil nutrients), we expect that plants will often compete for the same essential nutrients.

Our examples of competition and the definitions of scramble and contest competition suggest that in general, competing species exert a negative effect on the growth and reproduction of another species. However, theoretical ecologists Robert Holt and John Lawton described situations in which two species might have a negative effect on each other, even though they do not compete in a direct sense as we have defined it. Holt and Lawton called these situations **apparent competition**.

As an example, consider two animal species that are prey for a common predator. Or consider two plant species eaten by a common herbivore. Suppose one of the prey species increases significantly in number. This may ultimately increase the numbers of predators. Now the predators may increase their rate of attack on the second prey species, obviously having a negative impact on them. In this way, the increase of the first prey species indirectly has a negative impact on the second prey species.

Although apparent competition is an interesting phenomenon, in this chapter we will focus on examples of direct competitive interactions. ❖

Competition is ecologically important because it is caused by resource limitation. Competition is important to evolution because the ability to obtain adequate resources affects fitness. Fitness may be affected due to changes in survival, fertility, or both. As we will see, the connection between competitive ability and fitness means that genetically based differences in competitive ability can result in the evolution of increased competitive ability.

The exact nature of the competitive process often depends on the details of how a particular organism makes its living. For example, many insects, such as fruit flies, have a larval stage with limited dispersal ability that feeds on food close to the hatching site. If many eggs hatch in a small space, the larvae may have to compete for limited food in order to survive and reproduce. In Figure 12.2A we see that when *Drosophila* larvae are reared on small amounts of food, their chance of surviving to the adult stage can be quite small. Different genotypes of *Drosophila*—in this case, flies homozygous for the white-eye allele and wild-type (normal) flies—appear to respond to limited food in a similar manner. As the line graphs show, each genotype displays an increase in viability with increasing food level until a maximum survival rate (less than 100 percent) is achieved.

What happens when white-eye larvae and wild-type larvae are put together so they compete with each other, at least when food levels are limited? At low food levels, the chance that a white-eye larvae will survive is much greater when it is competing with wild-type larvae than when it competes with other white-eye larvae, as the bar graph at the top of Figure 12.2A reveals. This is a direct consequence of the fact that the white-eye larvae are better competitors and somehow manage to get a disproportionate amount of the food. Conversely, at low food levels the wild-type larvae suffer a

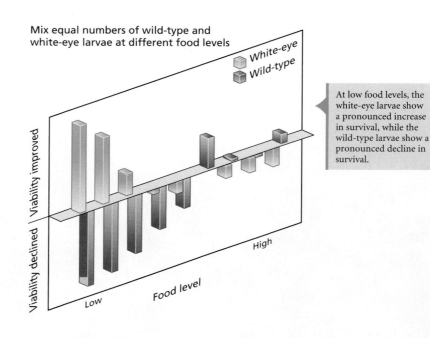

Mix equal numbers of wild-type and white-eye larvae at different food levels

White-eye
Wild-type

Viability improved — Viability declined

Food level

Low — High

At low food levels, the white-eye larvae show a pronounced increase in survival, while the wild-type larvae show a pronounced decline in survival.

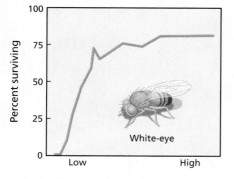

White-eye

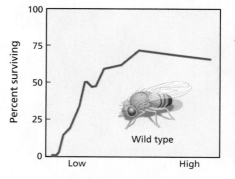

Wild type

When larvae from either the white-eye or the wild-type flies are raised separately, their chances of surviving to the adult stage at different food levels are similar. However, if white-eye and wild-type are raised together, they will compete for the limited food and their chances of surviving will be different.

FIGURE 12.2A **Food levels affect egg-to-adult viability in *Drosophila***

reduction in their chances of surviving when they compete against the white-eye larvae. These wild-type larvae are poor competitors. In these competitive circumstances, the white-eye larvae have greater fitness relative to the wild type due to their relative increase in viability.

How do *Drosophila* larvae compete? In fact, the competitive process here is a simple case of scramble competition. *Drosophila* larvae cannot monopolize resources; but they simply eat the food as fast as possible, before it is all gone. In Figure 12.2B we see that when raised in isolation from each other, the wild-type and white-eye larvae can produce about the same number of surviving larvae as they consume their food. The only difference is that the white-eye larvae consume the food much faster. When the larvae are placed together, the elevated feeding rate of the white-eye larvae gives them an advantage, and they are able to consume a disproportionate amount of the food and thus enjoy an increase in viability. ❖

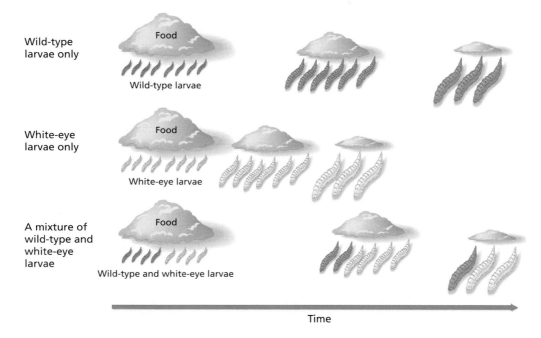

FIGURE 12.2B The Mechanism of Competition in *Drosophila*

The Ecological and Evolutionary Process of Competition **355**

12.3 Plant competition for limited resources may lead to stable coexistence

We have considered one example of how animals compete for food, but how do plants compete for resources? Many resources may limit plant growth in different environments. Some of these resources include water, light, and nutrients such as nitrogen and phosphorus. Some resources, like water and nutrients, are in a constant state of flux. Natural processes such as rainfall and decomposition add these resources back to the local plant environment. As plants grow, they remove these resources, and the level available to all plants is decreased. One way to understand plant competition is by focusing on the dynamics of resource utilization and production. Let's look at plant competition by using graphs to visualize these dynamics.

What happens when two plant species compete in the same environment?

In Figure 12.3A, we consider how plant growth is regulated by the available nutrients. Growth rates increase as nutrient levels increase. At the same time, individual plants lose biomass due to herbivores and other factors. This rate of loss must at least be equaled by growth if the plant is to survive. This leads to the first conclusion: A minimum level of resource is required to sustain a viable population. In Figure 12.3B we show that in fact, more than one essential resource may determine a plant's fate.

Next, we suppose that in an environment with sufficient resources to maintain growth, plants will grow until they drive the resource levels down to a point where plant growth just balances loss of biomass. In Figure 12.3C this

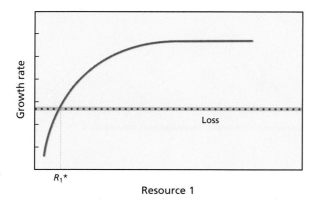
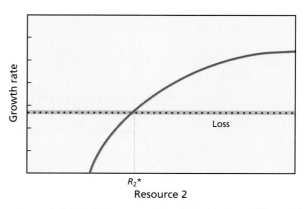

FIGURE 12.3A Plant Competition for Limited Resources An individual plant's growth rate increases with increasing resource level. Loss of biomass is due to herbivores, seed predators, and environmental disturbance. An equilibrium plant biomass is reached when the growth rate equals the loss rate. For resource 1 and resource 2, this occurs at the starred values, R_1^* and R_2^*, respectively.

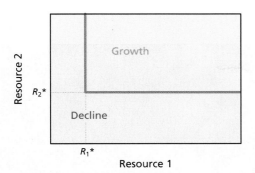

FIGURE 12.3B **Plant Competition for Limited Resources** We can illustrate the combinations of resource 1 and resource 2 that result in sustained growth (green), sustained decline (red), or no net growth (blue lines).

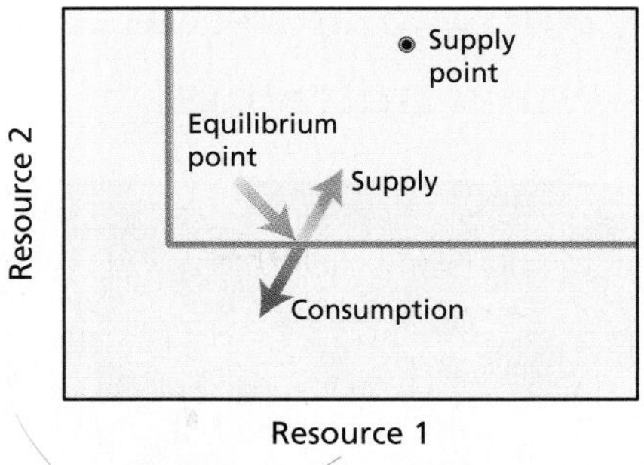

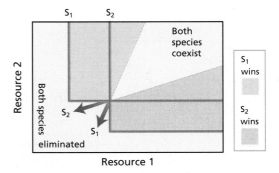

FIGURE 12.3C **Plant Competition for Limited Resources** Resources are continually added to the environment. The supply point represents the level of resources 1 and 2 in the absence of plants. Renewal of resources is indicated by the green vector and will always point toward the supply point. The point on the blue line where the plant's rates of consumption (red vector) are exactly opposite the green vector represents the equilibrium level of resources.

equilibrium is represented by the point on the blue line where the two vectors (arrows)—plant resource consumption (red arrow) and environmental resource supply (green arrow)—point in opposite directions.

Finally, different environments can be represented by different rates of resource renewal. In Figure 12.3D, environment A results in relatively high levels of resource 2 and low levels of resource 1 at equilibrium, as compared with the resources for environment C.

What happens when two plant species compete in the same environment? Two different species will never have exactly the same resource requirements. Figure 12.3E shows the equilibrium resource curves for two different species of plants, called S_1 and S_2. We see that S_1 requires more of resource 2 and less of

resource 1 to maintain a viable population compared with S_2. We also see that the consumption vector of S_1 is close to a vertical line, indicating a high rate of consuming resource 2. Since S_1 requires more of resource 2 to live than S_2 does, we conclude that S_1's high rate of consumption of resource 2 has a greater impact on its own species than it does on species S_2. In a similar fashion, we see that the consumption vector of S_2 is close to a horizontal line, indicating a high rate of resource-1 consumption. Again, this should have a greater impact on other members of S_2 than on members of species S_1. These relationships imply that both species exhibit higher levels of intraspecific competition than interspecific competition, and this ultimately permits both species to coexist if the supply point is in the wedge-shaped region indicated in Figure 12.3E. ❖

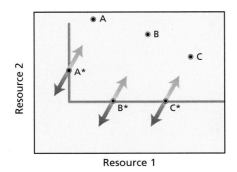

FIGURE 12.3D **Plant Competition for Limited Resources** Three different environments are represented by three different supply points (A, B, C). The consumption by plants is assumed to be the same. The result is that three different equilibrium levels of resources are reached (A^*, B^*, C^*) in each environment.

FIGURE 12.3E **Two-Species Competition** In this figure the equilibrium growth lines for two species, S_1 and S_2, are shown as blue lines. The consumption rates for each species are also shown as red vectors. If the supply point of the environment falls in the green region, S_1 reduces resource 1 to a level where S_2 cannot exist. Just the reverse happens if the supply point is in the dotted region. The wedge-shaped blank area indicates the range of environments where the two species can coexist.

In some ways it is easier to study competition by observing plants rather than animals. Because plants do not move, they can be placed in close proximity to each other, and the nutrients and water they need to grow can be carefully distributed. One technique for estimating competitive ability in plants was developed by de Wit and is illustrated in Figures 12.4A and 12.4B. The technique is based on measuring some characteristic that should indicate the severity of competition, such as plant weight. The total number of plants placed in competition determines the overall level of competition. If we keep the total number constant but vary the relative proportions of competing species (or genotypes), then relative competitive ability can be inferred.

If total weight of one species changes in direct proportion to its relative frequency (black lines on the graphs in Figures 12.4A and 12.4B), then there is no difference between the presence of the alternative species (or genotype), and the two species (or genotypes) are competitively equivalent. When the growth curve (red line) is above the black line (see Figure 12.4B), it indicates that the first species (or genotype) gains weight faster in the presence of a competitor than in the presence of members of the same species (or genotype). In this case, the first species (or genotype) is a superior competitor to the second species (or genotype). When the growth curve of the first species is below the black line (see Figure 12.4B), just the opposite conclusion may be drawn: The first species (or genotype) is an inferior competitor.

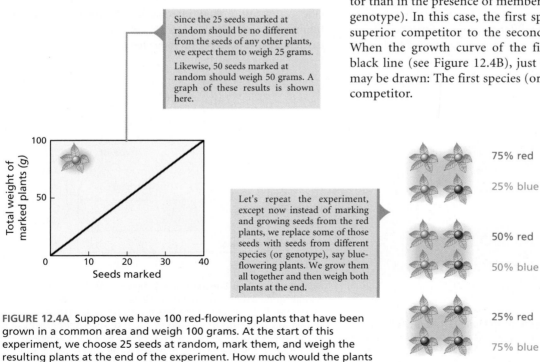

Since the 25 seeds marked at random should be no different from the seeds of any other plants, we expect them to weigh 25 grams.

Likewise, 50 seeds marked at random should weigh 50 grams. A graph of these results is shown here.

Let's repeat the experiment, except now instead of marking and growing seeds from the red plants, we replace some of those seeds with seeds from different species (or genotype), say blue-flowering plants. We grow them all together and then weigh both plants at the end.

75% red

25% blue

50% red

50% blue

25% red

75% blue

FIGURE 12.4A Suppose we have 100 red-flowering plants that have been grown in a common area and weigh 100 grams. At the start of this experiment, we choose 25 seeds at random, mark them, and weigh the resulting plants at the end of the experiment. How much would the plants be expected to weigh?

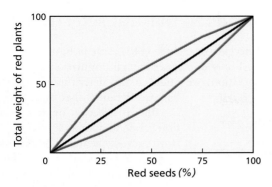

FIGURE 12.4B What's the outcome of the experiment in Figure 12.4A? This figure shows two possible outcomes from the experiment. If the red line is above the black line, then the red plants are better competitors than the blue plants. If the red line is below the black line, the red plants are inferior competitors to the blue plants.

Plants interact with a variety of fungi. In some cases these interactions are beneficial to the plant, and in other cases the fungi are pathogens. Rust fungi are pathogens to many plants, and occasionally plants develop a genetically based resistance to the rust. Burdon and his colleagues studied two genotypes of the skeleton weed, one resistant to rust and the other susceptible (see Figure 12.4C). When the plants were not exposed to rust, the susceptible genotype revealed a slight competitive advantage. However, if both genotypes were exposed to rust during their growth, the resistant genotype had a tremendous advantage in competitive situations.

Mycorrhizae are fungi that infect the roots of many plants and often increase the ability of the plant roots to absorb nutrients such as phosphorus and potassium. The presence of mycorrhizae may significantly affect the outcome of competition. In one experiment, two different species of plants were raised under competitive conditions with and without inoculations of mycorrhizal fungi (see Figure 12.4D). In the absence of mycorrhizae, an index of competitive ability showed little difference between the two plants. However, with mycorrhizae, one plant (*Holcus*) demonstrated a clear competitive advantage. ❖

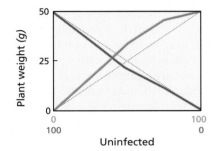
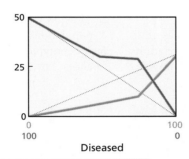

The blue line and blue axis labels indicate % susceptible.
The red line and red axis labels indicate % resistant.

Skeleton weed, *Chondrilla juncea*

FIGURE 12.4C **Competition between Different Genotypes** Burdon competed two genotypes of skeleton weed. One genotype was susceptible to a rust fungus, the other was resistant. When both plants were uninfected, the susceptible type is a better competitor. Not surprisingly, when both plants are exposed to the fungus, the resistant genotype is a much better competitor.

Lolium perenne
Perennial rye grass

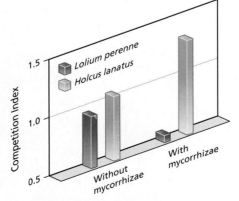

Holcus lanatus
Velvet grass

FIGURE 12.4D *Lolium* and *Holcus*, when planted together, compete for phosphorus and potassium uptake through their roots. Inoculating the roots with mycorrhizal fungi increases the competitive ability of *Holcus* for these limiting nutrients.

The Ecological and Evolutionary Process of Competition **359**

Is competitive ability affected by natural selection? Two important attributes of competitive ability suggest that it is. First, relative competitive ability affects survival and thus fitness. Any genetically based trait that improves competitive ability will increase the fitness of its carriers. Secondly, there appears to be genetically based variation for competitive ability, at least in fruit flies.

To test whether competitive ability responds to natural selection, it would seem best to place one population in an environment where competitive ability is at a premium. This can be accomplished by crowding individuals so the level of food resources per individual is low. A control would consist of similar populations in an environment with abundant resources.

Figure 12.5A outlines an experiment with fruit flies using three replicate experimental populations (K's) where adults and larvae were extremely crowded. The control populations (r's) had abundant food resources, especially among the larvae.

Note that an important feature of experiments in evolutionary ecology is the replication of whole populations. Genetic differences may arise between two populations for at least two very different reasons—natural selection or random genetic drift. If natural selection causes rare genetic variants in the crowded populations to increase in frequency due to their favorable effects, this should happen in each crowded population. Thus, the effects of natural selection are expected to be consistent across replicate populations that are subject to the same environmental treatment. If traits like competitive ability change due to chance genetic changes brought on by drift, then independent populations

Note that an important feature of experiments in evolutionary ecology is the replication of whole populations.

will not necessarily change in the same fashion. If the experiment used only one experimental population and one control population, we could never determine if genetically based differences that arise over time are due to selection or drift. Replication is essential.

After many generations of evolution in the r- and K-environments, the competitive ability of larvae from these populations was measured (using the technique illustrated in Module 12.2) For each set of populations, competitive ability was greater in the populations that evolved with scarce resources (K's) than in their controls (r's). As expected, the populations that have adapted to scarce resources have improved competitive ability.

How do we know that the differences between the r- and K-populations are genetically based? It is certainly possible for phenotypic differences to arise due to environmental differences. For instance, larvae in crowded cultures typically grow slowly and are smaller compared to larvae raised in uncrowded environments. To eliminate phenotypic differences due to environmental difference, all

larvae are raised in a common environment for two generations prior to the measurement of competitive ability. Why two generations? If this was done for just one generation, then the larvae to be tested would come from eggs that might have different levels of nutritional resources because their mothers came from the two very different environments. To eliminate the possibility of this type of *maternal effect*, larvae are raised in a common environment for two consecutive generations. ❖❖

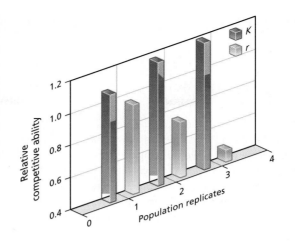

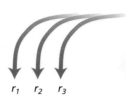

Wild-type population

r_1 r_2 r_3 K_1 K_2 K_3

Six replicate populations were created from a single source population. Replicate populations are used to aid in sorting out genetic changes due to natural selection vs. those due to random genetic drift.

One group of populations, called *r*'s, were kept at low adult and larval densities and served as controls. The other group, called *K*'s were kept at high larval and adult densities. Food was especially scarce for the growing larvae.

After 128 generations in these environments, larval competitive ability for food was measured. Prior to these measurements the *r*- and *K*-populations were cultured for two generations in a common environment. The common environment for the first generation should remove any effects that the different densities may cause. However, since the mothers of the first generation may produce eggs that are affected by their density environment, a second generation is used to remove these maternal effects.

FIGURE 12.5A Evolution of Competitive Ability

Alexander J. Nicholson (1895–1969)

FIGURE 12.5B
Alexander J. Nicholson

A. J. Nicholson (Figure 12.5B) was an Australian entomologist with interests in host-parasite models, population regulation, and experimental population ecology. Nicholson was a strong proponent of the use of theoretical models in ecology. Nicholson thought that ecology would advance through the combination of observations of nature, model construction, and experiments.

In 1935, with physicist Victor A. Bailey, Nicholson published an important paper that developed host-parasite models that are still referred to today. A central concern of Nicholson's was to describe the forces that regulate population numbers. He became convinced that intraspecific competition was one of the most important factors regulating populations. In his writings he coined the terms *contest* and *scramble* competition to distinguish between territorial and nonterritorial animals.

Nicholson also carried out some of the most extensive laboratory experiments with populations of blowflies (*Lucilia cuprina*). Under his culture techniques, these blowflies produced dramatic cycles in population numbers. These blowfly data are still the subject of theoretical analysis 50 years later.

THE CONSEQUENCES OF COMPETITION

12.6 Gause developed his competitive exclusion principle from experiments with *Paramecium*

In the previous module we reviewed the effects of competition on survival of individuals of the same species as well as on individuals of different species. What would happen to the distribution and abundance of two competing species if they competed over a prolonged period of time? Several answers seem plausible. (1) While negative competitive effects may reduce the numbers of each species, they may nevertheless coexist indefinitely. (2) The effects of competition on survival may be so severe that one or perhaps both species would go extinct over time. (3) For many animal species that are mobile, one species may simply leave the area in which competition occurs to avoid the negative effects. In this module we will see that ecological theory may help us answer these questions.

The outcome of two-species competition has been studied carefully in the laboratory as far back as the 1930s. In 1934 Georgii F. Gause published an account of these experiments in his historic book, *The Struggle for Existence*. In these experiments, Gause utilized small ecosystems, called microcosms, to study competition between different species of unicellular organisms called *Paramecium*. In these simple microcosms, different species of *Paramecium* feed on a growing bacterial population.

Gause's experiments documented two of the possible consequences of competition we noted earlier. When *P. aurelia* and *P. cuadatum* were placed in the same culture, both populations initially increased in numbers. At first, when the densities of the two

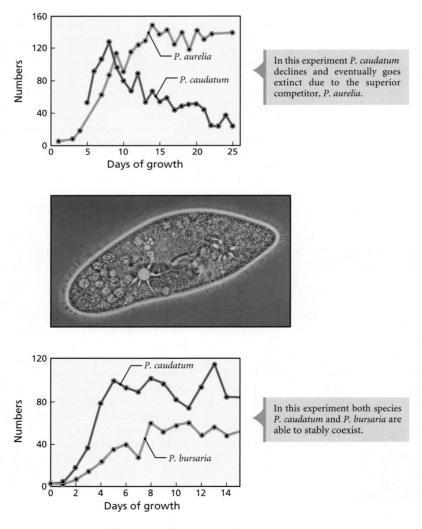

In this experiment *P. caudatum* declines and eventually goes extinct due to the superior competitor, *P. aurelia*.

In this experiment both species *P. caudatum* and *P. bursaria* are able to stably coexist.

FIGURE 12.6A Gause's Experiments

species are low, there should be adequate resources for both species. However, when resources become limiting and competition ensues, we see in the first experiment (at top in Figure 12.6A) that the numbers of *P. caudatum* start to decrease, and *P. aurelia* continues to increase and eventually level off. Although not shown, *P. caudatum* is eventually driven to extinction.

In the second experiment shown in Figure 12.6A, *P. cuadatum* is raised with a third species, *P. bursaria*. In this case, both species increase in numbers and then level off. Both species are able to coexist despite the limited resources.

Gause noted that exclusion of one species often occurred between closely related species when they were placed in highly simplified environments. These simple environments essentially forced the two species to compete for exactly the same resources. The simple environment also prevented the

two species from moving and avoiding competition. This finding led Gause to propose that no two species can coexist if they use the environment in precisely the same fashion. This prediction is sometimes called the **competitive exclusion principle**.

Gause's proposal implies that a key to understanding the consequences of competition is to know the various resource and habitat requirements of a species. A description of all such requirements for a species is sometimes called the **ecological niche**. In this final section of the chapter, we will explore in more detail the concept and evaluation of ecological niches. The important connection we wish to make here is the relationship between the consequences of competition and the idea of an ecological niche. ❖

The logistic model of population growth, discussed in Chapter 10, summarizes the impact of intraspecific competition on population dynamics. Competition between individuals of the same species results in declining survival and fertility. Similar effects may follow from competition between different species. The population dynamic consequences of interspecific competition can be studied with the **Lotka-Volterra model**.

In Figure 12.7A the Lotka-Volterra equation is developed following the same process used to derive the logistic equation in Chapter 10. In fact the Lotka-Volterra model looks just like the logistic equation, but with the addition of a term, $r_1\alpha_{12}N_t^{(2)}/K_1$, that reflects how population growth of species 1 is slowed by competing species 2.

The coefficient α_{12} is called the **competition coefficient**. It is a positive number that measures the effect of species 2 on the growth of species 1. If $\alpha_{12} = 1$, then each individual of species 2 has the same effect on the growth rate of species 1 as an individual from species 1. If $\alpha_{12} > 1$, then each member of species 2 has a more severe effect on the growth of species 1 than on their own members. If species 2 is a very weak competitor, then $\alpha_{12} < 1$. When $\alpha_{12} = 0$, species 2 has no effect on the growth of species 1.

The Lotka-Volterra equations can be used to study the outcome of competition. With two species there are three possible outcomes of competition: (1) Species 1 grows to its carrying capacity (K_1) and species 2 is eliminated (goes extinct). (2) Species 2 grows to its carrying capacity (K_2) and species 1 is eliminated. (3) Both species coexist at population sizes less than their carrying capacity. These different possibilities can be inferred from a graph that plots combinations of $N^{(1)}$ and $N^{(2)}$ that yield no growth of species 1 and species 2. Such a graph appears in Figure 12.7B.

For species 1, the blue line in the graph in Figure 12.7B illustrates these zero-growth combinations. The shape of this line is determined by using the equations in Figure 12.7A and finding values of $N^{(1)}$ and $N^{(2)}$ that make the expression in braces {} equal to 1.0. When this term is equal to 1.0, then every individual of species 1 will just replace itself, leading to no net growth of the population. If $N^{(1)}$ and $N^{(2)}$ are below the blue line, then $N^{(1)}$ will increase. For population size combinations above the line, $N^{(1)}$ decreases. ❖

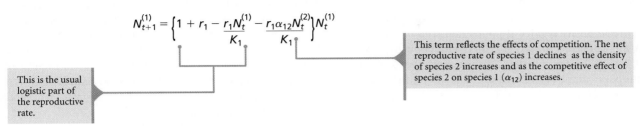

$$N_{t+1}^{(1)} = \lambda(N_{t}^{(1)}, N_{t}^{(2)})N_t^{(1)}$$

This equation looks similar to the equation for the basic model of growth except there are now superscripts next to the N's. These refer to the two different species under study. So $N^{(1)}$ will be the number of species 1 and $N^{(2)}$ the numbers of species 2.

Population growth for species 1

$$N_{t+1}^{(1)} = \left\{ 1 + r_1 - \frac{r_1 N_t^{(1)}}{K_1} - \frac{r_1 \alpha_{12} N_t^{(2)}}{K_1} \right\} N_t^{(1)}$$

This is the usual logistic part of the reproductive rate.

This term reflects the effects of competition. The net reproductive rate of species 1 declines as the density of species 2 increases and as the competitive effect of species 2 on species 1 (α_{12}) increases.

Population growth for species 2

$$N_{t+1}^{(2)} = \left\{ 1 + r_2 - \frac{r_2 N_t^{(2)}}{K_2} - \frac{r_2 \alpha_{21} N_t^{(1)}}{K_2} \right\} N_t^{(2)}$$

This equation for species 2 looks exactly like that for species 1 except every 1 has been replaced by a 2, and vice versa.

FIGURE 12.7A The Lotka-Volterra Competition Model In the Lotka-Volterra model the net reproductive rate (λ) of each species depends not only on its density but also on the density of the competing species.

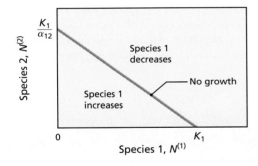

FIGURE 12.7B Find values of $N^{(1)}$ and $N^{(2)}$ that make the reproductive rate of species 1 exactly 1 (that is, where there is no net growth of species 1). The blue line represents combinations of $N^{(1)}$ and $N^{(2)}$ where there is no net growth of species 1. Combinations of $N^{(1)}$ and $N^{(2)}$ above the blue line result in species 1 decreasing in size. Combinations below the line permit species 1 to increase in numbers.

An Application of the Lotka-Volterra Equations

Suppose there are two competing species, and species 1 is characterized by the following parameter values:

intrinsic rate of increase, r_1: 1.2
carrying capacity, K_1: 1000
competition coefficient, α_{12}: 1.25

If $N_t^{(1)} = 500$ and $N_t^{(2)} = 900$, will $N_{t+1}^{(1)}$ be greater or smaller than 500? In other words, are the numbers of species 1 increasing or decreasing?

From Figure 12.7B we can infer that if $N_t^{(2)} > K_1/\alpha_{12}$, species 1 will be in the region where its population size decreases. Since $K_1/\alpha_{12} = 1000/1.25 = 800$, and $N_t^{(2)}$ is greater than this value, the number of individuals in the species 1 population will decrease in the next time interval.

The Lotka-Volterra model of competition predicts competitive exclusion or stable coexistence

What useful ecological predictions can we make from the Lotka-Volterra model? How do these predictions match up with our intuition of competition? Figures 12.8A through 12.8C show three cases of interspecific competition. Each figure shows the lines describing no growth for both species 1 and species 2.

In Figure 12.8A we see that if population sizes of species 1 and species 2 fall below the red line, then both populations increase in size. If their population sizes are above the blue line, then both populations decrease in size. The interesting area in this figure is where the points fall between the red and blue

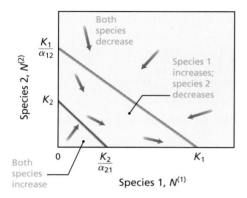

FIGURE 12.8A Predicted Outcome of Two-Species Competition: Species 1 Eliminates Species 2 In this and Figures 12.8B and 12.8C, the blue line is the no-growth curve for species 1 and the red line is the no-growth curve for species 2. The black arrows show the general direction of change in $N^{(1)}$ and $N^{(2)}$ from different starting points.

lines. Here we are above the zero-growth line of species 2, so species 2 will decrease in size. However, we are below the zero-growth line of species 1, so species 1 will increase. The Lotka-Volterra dynamics will push the system into the area between the two lines, no matter where we start. (In other words, if we start with both populations increasing in size, they will eventually grow to the sizes represented by the area between the lines. If we start with both species decreasing, they will eventually shrink to sizes represented by the areas between the lines.) As a result, the numbers of species 2 dwindle down to zero, and species 1 increases to its carrying capacity.

Why does this happen? It helps to look at the competition coefficients α for species 1 and species 2. If we examine the points at which the blue and red lines intercept the y- and x-axis in Figure 12.8A, we see that $K_1/\alpha_{12} > K_2$ and $K_1 > K_2/\alpha_{21}$. Let's assume the two carrying capacities are the same ($K_1 = K_2$), so then the inequalities are reduced to expressions that we can interpret: $1 > \alpha_{12}$ and $\alpha_{21} > 1$. Roughly speaking, this means that species 2 exerts a weak effect on species 1 (since $\alpha_{12} < 1$); in fact, its effect is less than the effect of species 1 on itself. (Recall from Module 12.7 that when the competition coefficient α equals 1, then members of the second species have the *same effect* on the growth of members of the first species as do other members of that first species.) Another way of saying this is that the interspecific competitive effects of species 2 on species 1 are weaker than the intraspecific competitive effects of species 1.

What about the effects of species 1? Species 1 is a strong competitor (since $\alpha_{21} > 1$) and reduces the growth rate of

species 2 more than species 2 individuals reduce their own growth rate. In other words, the interspecific competitive effects of species 1 on species 2 are stronger than the intraspecific competitive effects of species 2. Not surprisingly, the Lotka-Volterra equations are saying that if one species is a very strong competitor and the second is very weak, the strong competitor can drive the weak one to extinction. The meanings of *strong* and *weak* are made explicit by this theory.

Figure 12.8B shows the same process of competitive elimination of one species, but now species 2 is the superior competitor. See if you can derive relationships that the competition coefficients (α's) must satisfy, as we did earlier.

Figure 12.8C shows the very interesting case of coexistence of both species. Because we do find competing species coexisting in nature, it is important to understand the conditions under which this outcome is expected. If we again assume that $K_1 = K_2$, examination of Figure 12.8C reveals that $1 > \alpha_{12}$ and $1 > \alpha_{21}$. This condition suggests that both species exhibit levels of interspecific competition that are weaker than the levels of intraspecific competition. When interspecific competition is symmetrically attenuated in this fashion, both species may coexist. ❖

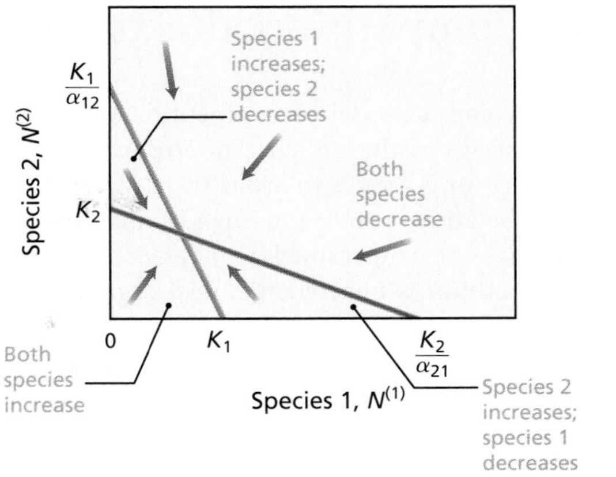

FIGURE 12.8C Predicted Outcome of Two-Species Competition: Both Species Coexist The point where the red and blue line cross is the ultimate equilibrium point. The x and y coordinates of this point are the equilibrium numbers of species 1 and species 2, respectively.

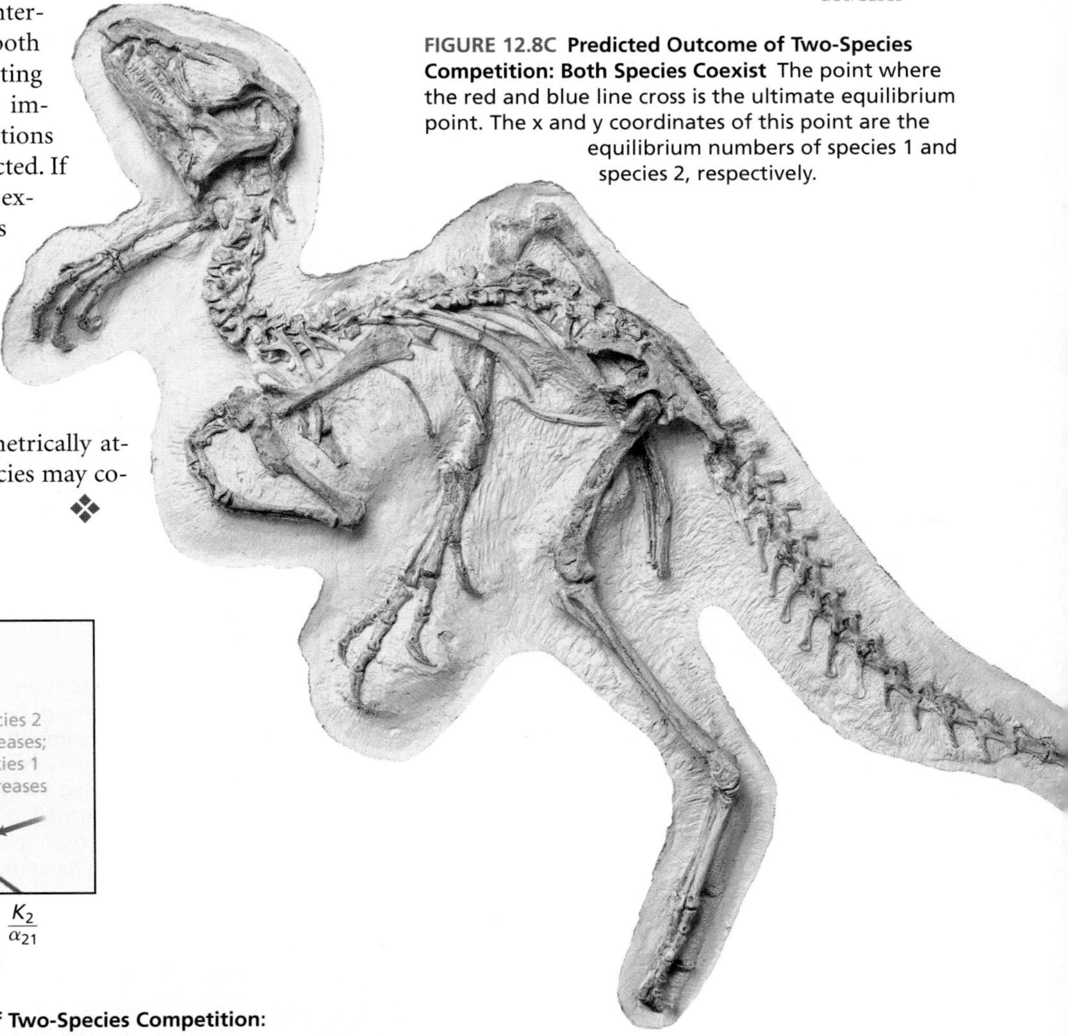

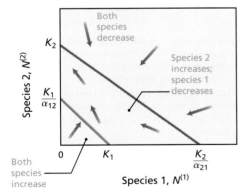

FIGURE 12.8B Predicted Outcome of Two-Species Competition: Species 2 Eliminates Species 1

Predicting the Outcome of Competition from the Lotka-Volterra Equations

Suppose two species grow according to the Lotka-Volterra competition equations and have the following parameter values:

K_1	K_2	α_{12}	α_{21}	K_1/α_{12}	K_2/α_{21}
1000	2000	0.4	1.6	2500	1250

What is the ultimate outcome of competition? By examination of Figures 12.8A through 12.8C, it is clear that the values of r do not affect the outcome of competition. We need to look at the ratios of the carrying capacities to the appropriate competition coefficient and compare these ratios to the carrying capacities. In this case we see that $K_1/\alpha_{12} > K_2$ and $K_1 < K_2/\alpha_{21}$. This corresponds to Figure 12.8C or stable coexistence of both species, as you can determine by looking at the carrying capacity and the ratio on each axis of the graph.

The Lotka-Volterra model of competition predicts the demise of a species in the presence of strong competition. Is there any way for a species to avoid such a fate? The predictions from the Lotka-Volterra model assume those two competing species are constrained to a physical environment where competition is unavoidable. Real organisms may not be constrained in this fashion. The negative effects of competition may be reduced or eliminated if the two species can avoid each other. Joseph Connell (1961) first demonstrated this response to competition for two species of barnacles on the coast of Scotland.

Chthamalus stellatus is found in the upper region of the intertidal zone that is most removed from the splash of waves and most exposed to air and sun. *Balanus balanoides* is found lower in the intertidal zone, less exposed to the drying conditions of air and sun. Both animals make their living by firmly attaching themselves to rock surfaces and then filtering food particles from the waves that splash over them. Observations of young settling barnacles showed that there is extensive overlap in the location of the settling of both species (Figure 12.9A). By the time the barnacles have become adults, however, there is very little overlap in the distribution of the two species (see figure).

Connell observed that when an individual *Balanus* came into contact with a neighboring *Chthamalus*, the *Balanus* barnacle shell would actually grow underneath the *Chthamalus* shell and dislodge it. In other instances *Balanus* would simply grow over the shell of *Chthamalus*. This appeared to be competition, but another possible explanation was that *Chthamalus* were weakened because they cannot live in this middle region of the intertidal zone, so the direct contact with and apparent removal of *Chthamalus* by *Balanus* was not the ultimate reason *Chthamalus* did not live there. Connell reasoned that he could directly test the effects of competition by physically removing *Balanus* from the middle of the intertidal range. *Chthamalus* larvae settled in the middle range, but would they persist in the absence of *Balanus*? Con-

nell found that indeed they did. The absence of *Chthamalus* adults in the middle intertidal zone is a direct consequence of the competitive effects of *Balanus*. The environment in this middle range is perfectly suitable for *Chthamalus*, but competition restricts their distribution to the high end of the intertidal.

Balanus, on the other hand, is unable to persist in the high intertidal due to environmental conditions. The prolonged periods of exposure to air, wind, and sun require that plants and animals living in the high intertidal resist desiccation, or loss of water. *Balanus* is unable to survive in these conditions, and in effect provides *Chthamalus* with a refuge from competition. ❖

The distribution of larval *Balanus* and *Chthamalus* along the intertidal is shown by the colored lines that match the colors of these words in this label.

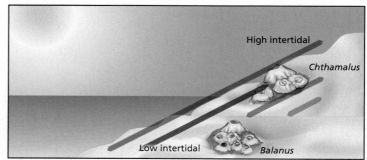

The distribution of adult *Balanus* and *Chthamalus* match the color of these words in this legend.

FIGURE 12.9A Competition between Barnacles in the Intertidal Zone When the larvae of *Balanus* and *Chthamalus* settle in the intertidal, there is extensive overlap in the distribution. As the larvae develop and become adults, their distribution along the intertidal changes. The adult *Balanus* are not found at the very high end of the intertidal, due to dry conditions. Adult *Chthamalus* retreat from the middle of the intertidal zone due to competition with *Balanus*.

Social Darwinism

Charles Darwin had described evolution by invoking powerful images of competition between species and a struggle for survival. These powerful images and the logic of Darwin's ideas had very far-reaching consequences. For example, G. F. Gause's research in 1934 on competition was clearly motivated by Darwin and an interest in understanding the mechanisms of evolution.

Darwin's ideas spread, however, and were embraced by fields other than biology. Many economists and anthropologists could not resist using Darwinian phrases like "survival of the fittest" and the "struggle for existence" to support and lend credence to their own theories. This misapplication of Darwin's ideas to fields for which they were unintended has been called **social Darwinism**.

Some economists, for instance, tried to argue that there is a natural order to economic systems, just as there is in biological systems, and that governments should not interfere with this natural order. They predicted that economic systems left to their own devices would experience competition between companies, and the victors of this competition would improve the overall quality of the economy. We can get the flavor of these arguments from the words of the classical French economist, Yves Guyot, who wrote in 1881 that "Darwin speaks of the struggle for existence. It is the struggle for economic existence that has been the cause of all material progress. Competition rouses from the apathy of content, and unceasingly stimulates the effort to improve. It is the grand agent of evolution. Competition fixes the natural level of prices."

THE ECOLOGICAL NICHE

12.10 Several ecologists contributed to the development of the ecological niche concept

Long before Gause used the Lotka-Volterra theory of competition to understand competition between *Paramecium,* ecologists were collecting information on the habitats and feeding relationships of many plants and animals. This purely descriptive activity gave ecologists an appreciation for the ways in which many organisms differed in their requirements for space, food and light. In 1913 Joseph Grinnell (Figure 12.10A) was the first to use the word **niche** to describe the specific habitat, requirements, and ecological role of a species. From Grinnell's writing it is clear that he considered the niche as a description of both the spatial and dietary dimensions of a species. For instance, many plants and animals cannot live in desert environments, because it is too hot and dry. Some animals may have very specific food requirements. For example, not only are Koala bears herbivores, but they eat only the leaves from a few species of eucalyptus trees.

Grinnell also recognized that aspects of a species' ecological niche were not only related to its nutritional requirements but could be related to its need to avoid predators. Although Gause is typically given credit for formulating the competitive exclusion principle, Grinnell expressed a similar idea in 1917 when he said, "It is, of course, axiomatic that no two species regularly established in a single fauna have precisely the same niche relationship." Gause developed this notion more thoroughly by connecting the niche relationships to species extinction through his experiments.

Charles Elton also used the concept of niche extensively, but in a somewhat different fashion than Grinnell did. In Elton's 1927 book, *Animal Ecology,* he emphasized the feeding role and activity of animals in defining their niche. Thus, according to Elton, it is important to know what an animal eats and who eats that animal.

There are many similarities between the two ecologists' concepts of niche. (1) Both considered the niche to be a constant, immutable aspect of species ecology. (2) They included dietary requirements as an important aspect of the niche. (3) They saw avoiding predators as an important aspect of the niche for many animals. Yet neither Grinnell nor Elton provided a conceptually simple, precise definition of the niche.

George Evelyn Hutchinson was next to tackle the problem of defining a species'

niche, with the goal of being more precise and concrete. The details of Hutchinson's theory of the niche were outlined in a paper entitled "Concluding remarks," delivered at a symposium at Cold Spring Harbor, New York, in 1957. He suggested that every important environmental and ecological aspect of a species be numerically quantified. Each of these variables would represent a different axis of a geometric figure. The range of variables for each axis would be the limits required for the species to successfully survive and reproduce.

If we consider just three axes, we could in fact draw the three-dimensional shape that represents what Hutchinson

Joseph Grinnell (1877–1939). Grinnell obtained his undergraduate degree from Throop Polytechnic Institute, now known as California Institute of Technology, and his graduate degree from Stanford University in 1913. He was appointed Assistant Professor at the University of California, Berkeley and served as Director of the Museum of Vertebrate Zoology until his death. He was especially interested in the natural history of birds and mammals and published over 500 papers during his career.

Charles Elton (1900–1991). A British ecologist who helped start the Nature Conservancy Council in 1949, had concerns about the impact of introduced species on natural populations. He completed two important books in ecology, *Animal Ecology* and The *Pattern of Animal Communities.* He also developed the concept of the pyramid of numbers, which stated that large predators were rare (top of the pyramid) and small animals were abundant and at the bottom of the food chain.

George Evelyn Hutchinson (1903–1991). Born in Cambridge, England, Hutchinson was interested in ponds and nature at an early age. He published his first scientific paper at the age of 15. He spent most of his academic career at Yale. His interests were in limnology. However, he is best known for his papers synthesizing important concepts in ecology including his formulation of the ecological niche.

FIGURE 12.10A Major Contributors to the Concept of the Niche

called the **fundamental niche**. Thus, the x-axis might represent the range of temperatures at which the organism can survive. The y-axis could be the range of food sizes an animal might be able to feed on. Finally, the z-axis could be the range of altitudes over which the animal can live. Figure 12.10B shows an example.

The **realized niche** differs from the fundamental niche in that it reflects the impact of other species. The realized niche is the part of the niche volume that does not overlap the fundamental niche of any other species; it also includes the areas that overlap, but in which the first species is able to survive anyway. The realized niche is a subset of the fundamental niche; that is, the realized niche is totally included within the volume of the fundamental niche. Using Hutchinson's definition of *niche,* we would restate Gause's competitive exclusion principle in this way: No two co-occurring species may have exactly the same realized niche.

Following up on Elton's concept of niche, ecologists often group together species that make their living in a similar fashion. These groups are referred to as **guilds**. Seed-eating birds might form a guild, for example, as might insects that live in decaying fruit. There is a good deal of latitude in the use of the word *guild,* and the common thread that defines a group may vary from one ecologist to the next. In this book, the context in which a group is considered a guild should be clear from the discussion. ❖

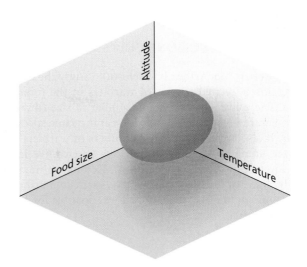

FIGURE 12.10B A Fundamental Niche

12.11 Determination of the realized niche can reveal how different species avoid competition

In 1956 Robert MacArthur had set out to study the ecology of five closely related species of warblers in the forests of Maine. These five species were all similar in size (three are shown Figure 12.11A) and tended to eat the same species of insects. It also appeared that the numbers of warblers were resource limited. Not only did all five species live in the same forest, but they could be found feeding on the same tree. In light of Gause's competitive exclusion principle, MacArthur could not explain how it was possible for all five species to coexist.

He decided to collect detailed information on the feeding behavior of these birds. This involved long and tedious observations of the birds in their natural habitat. MacArthur carefully timed the duration, and location within a tree, of the foraging behavior of each species.

The picture that emerged is shown in Figure 12.11B for three of the five species studied by MacArthur. Although the birds would eat the same species of insects, they foraged in different parts of the tree. Thus, the myrtle warbler (now called Audubon's warbler) spent most of its time

Bay-breasted warbler
Dendroica castanea

Cape May warbler
Dendroica tigrina

Myrtle warbler
Dendroica coronata

FIGURE 12.11A **Species of Warblers**

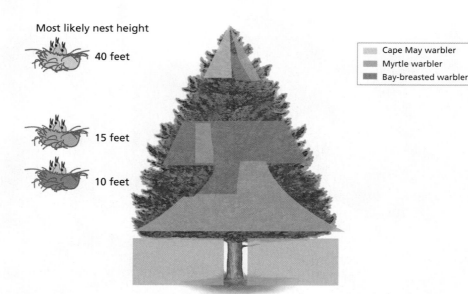

Most likely nest height

40 feet

15 feet

10 feet

Cape May warbler
Myrtle warbler
Bay-breasted warbler

FIGURE 12.11B **Foraging Position of Three Species of Warblers in Maine** The different colors show where each bird species spent most of its time searching for food and feeding. While all three species can be found in the same forest and trees, Cape May warblers tend to feed near the top of the trees, Bay-breasted warblers in the center, and myrtle warblers at the bottom or underneath the tree.

feeding at the bottom of the tree—or actually on the ground underneath the tree. When the myrtle warbler fed higher up on the tree, it tended to forage close to the trunk of the tree. The bay-breasted warbler would feed at the middle heights, both close to the trunk and at the tips of branches. Finally, the Cape May warbler usually fed at the top of the tree and on the outer branches. Even the location of each bird's nest was roughly related to its preferred area of foraging.

Since the species of insects were unlikely to move between these different regions, each bird had its own food resource pool that overlapped little with the resource pools of the other species. In this fashion, the magnitude of competition and the negative impact of one species on the next was reduced.

By today's standards, MacArthur's study is lacking several important components. For instance, he never showed directly that if two species foraged in the same location, there would be insufficient food for both; and thus they would suffer from the effects of this competition. However, MacArthur's interpretation of the distribution patterns of birds was novel, and it inspired many ecologists to more thoroughly study the effects of competition in many other natural systems. ❖

12.12 The number of species that exist in a particular environment may be determined by competition

When we think about the observations made by MacArthur, a natural question is: How many species of warblers could coexist in the same forest? Even very large trees have a limit regarding how finely space can be divided among different species. Thus, there will be a limit to how many species can coexist in—or be packed into—a particular ecological space. One way to think about this problem is to consider competition at the simplest level—one essential resource. We will assume that this resource varies on a scale that we can quantify, for instance, the size of seeds or insects eaten by birds. It seems reasonable that a given species would have a preferred—or most commonly used—resource value, and that its use of larger or smaller resources might decrease.

We show this type of relationship in Figure 12.12A. Each of three species has a bell-shaped utilization of the single resource, but the mean or preferred resource value is different for each species. The degree of overlap of each of these curves roughly indicates the level of competition that is expected between species. The first graph shows weak competition, as indicated by the small colored areas. The second graph shows much higher levels of competition. Not only is the colored area greater, but the difference in the preferred resource level from one species to the next is smaller. Because the three species have only a small difference in their preferred resource level, we say these species are *tightly packed* in this environment.

If there were only two species in a particular guild, could a third species be packed between them on this single resource axis? Another way of posing this question is to ask if a third species could successfully invade this guild. This question can be answered theoretically by using the Lotka-Volterra competition equations, and by using the resource utilization curves to estimate competition coefficients. This is fairly complicated theory, so we will simply review the possible outcomes. As Figure 12.12B shows, there are basically three different possible outcomes of this invasion: (1) The third species successfully invades, and all three species coexist. (2) The

third species successfully invades, but one or both of the resident species is driven to extinction. (3) The invading species cannot displace either of the residents and is itself driven to extinction.

If this type of invasion were to occur repeatedly, we might expect that after awhile, the existing array of species would be packed on the critical resource axis as tightly as possible within the limits prescribed by competition. Is there evidence that resource utilization by organisms is structured in this fashion by competition?

For birds, the types of foods consumed is related to the bird's size. In local environments where two species of sandpipers are found, the difference in the size of birds can be expressed as the ratio of the larger species to the size of the smaller species. The observations show that the size ratio of the two species varies greatly, but there is a marked peak of size ratios at 1.2–1.3 (Figure 12.12C). We can see that if the pool of species were simply placed together at random, there should be no peak at 1.2–1.3. These results suggest that some process is preventing two competing species from being too similar or too different. This is exactly what we would expect if competition were determining these size differences. ❖

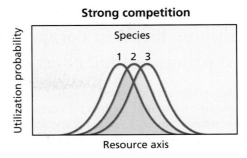

Weak competition

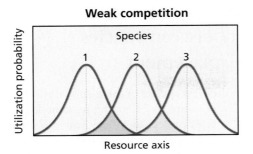

Strong competition

FIGURE 12.12A How Tightly Can Species Be Packed? The blue area indicates the relative intensity of competition between species 1 and species 2. The yellow indicates the relative intensity of competition between species 3 and species 2. The green indicates the relative intensity of competition between all three species.

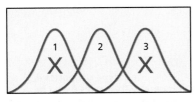

In a two species-guild can a third species (2) invade and successfully coexist if its resource utilization is between the two existing species?

Possible outcomes

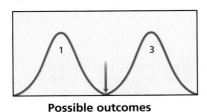

YES. Now all three species are tightly packed on the resource axis.

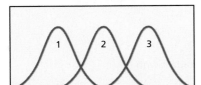

YES. But species 1, or species 3, or both species 1 and 3 go extinct.

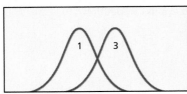

NO. If the two species are already tightly packed, the invading species will be driven to extinction.

FIGURE 12.12B Can a Third Species Invade?

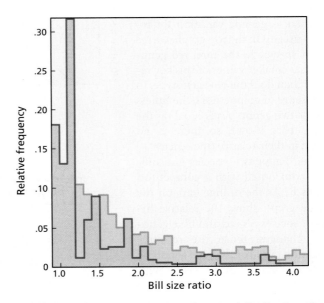

FIGURE 12.12C The red bars show the actual distribution of bill size ratios among North Dakota sandpipers. The green bars show the expected ratios if these pairs of species were randomly placed together. The observed peak ratio at 1.2–1.3 is much higher than expected and may indicate the influence of competition between the bird species.

12.13 Important morphological or behavioral traits may evolve, reducing levels of competition between species in a process called character displacement

Are competing species forever doomed to suffer reduced fitness if their competitors cannot be driven to extinction? The answer is not necessarily. One way for a species to escape the negative effects of competition is to avoid it altogether. One way to accomplish this is to relocate to someplace without competitors. However, if this is not possible, evolution may over time cause changes in morphology or behavior in one or both competing species. Such changes will ultimately reduce the level of competition. If they were to occur, we would say that the morphological or behavioral character has been *displaced* due to the competitive interactions.

Figure 12.13A shows how **character displacement** might happen. As with other models of natural selection, we start by assuming that there is genetic variation in the population. In this case we assume, for simplicity, that competition depends on a single resource and that there is genetic variation for resource usage. On the top graph, we see that in species A, the rare, red genotype uses smaller values of this key resource than does the blue genotype. In the absence of competitors, the fitness of these two genotypes is equal (as the orange bars show), so there is no change in their relative frequencies.

What happens if species A is suddenly confronted with a competitor, species B? As the orange bars on the middle graph show, the relative fitness of species A's red and blue genotypes changes. Since the resource utilization curve of the blue genotype overlaps more extensively with species B, the blue genotype suffers disproportionately. As a result, the relative fitness of the red genotype is now greater than that of the blue genotype. The result of this change is the gradual increase in the frequency of the red genotype at the expense of the blue genotype. After sufficient time has gone by, the average resource utilization of species A is now much lower than it had previously been because of the increased frequency of the red genotype, as the bottom graph shows.

In small coastal lakes of southwestern British Columbia, several different species of three-spined sticklebacks are found. Two species are co-occurring in some lakes. In these lakes, one species prefers to forage near the surface of the lake; that species is referred to as limnetic (Figure 12.13B). The second species forages near the bottom of the lake and is called benthic. One genetically based character that can be used to distinguish these two species is the length of the gill raker (see Figure 12.13B). The limnetic species has longer gill rakers than those of the benthic species. In other lakes there is just a single stickleback species whose gill raker size is between those of the benthic and limnetic species. By examining the stomach contents of these fish, ecologists can infer the diet of each species. In the two-species lakes, the benthic species has a diet dominated by invertebrates that are

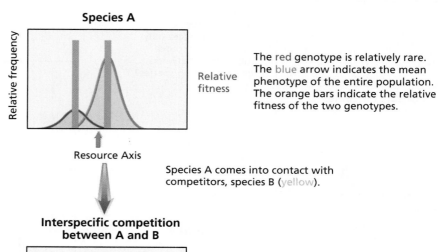

Species A

Relative frequency

Relative fitness

The red genotype is relatively rare. The blue arrow indicates the mean phenotype of the entire population. The orange bars indicate the relative fitness of the two genotypes.

Resource Axis

Species A comes into contact with competitors, species B (yellow).

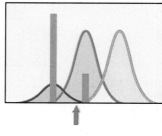

Interspecific competition between A and B

Now the blue genotype of species A must compete with species B and its fitness drops relative to the red genotype, which avoids competition.

Many generations later

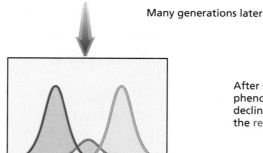

After sufficient time, the mean phenotype of species A has declined as the frequency of the red genotype has increased.

FIGURE 12.13A Character Displacement

found only on the lake bottoms, while the limnetic species derives almost all of its diet from species that live near the lake surface. In the solitary lakes the single stickleback species has a diet that is almost equal in its distribution of bottom- and plankton-derived species.

These observations suggest the hypothesis that the ancestral species of the benthic and limnetic species had gill rakers and feeding preferences similar to those of the solitary species. However, when the species were placed together, the levels of competition favored the differentiation of the two species such that they specialized on feeding in different localities of the same lake. ❖

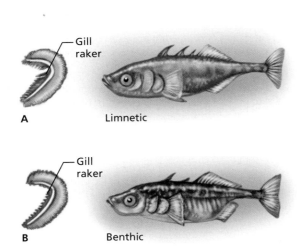

A   Limnetic

B   Benthic

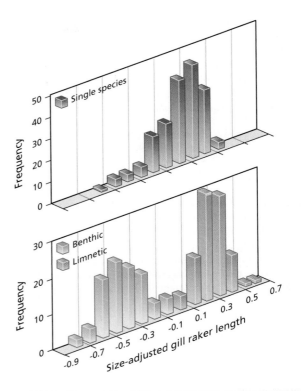

FIGURE 12.13B **Character Displacement in the Sticklebacks**

SUMMARY

1. The mechanisms of competition are typically different in animals and plants.

   a. Animals compete through two mechanisms, called contest and scramble competition. Contest competition often involves competition for territory.

   b. Plants are affected most strongly by their nearest neighbors. This is due to several factors common to plants: They are sedentary, they vary tremendously in size, and they often compete for common resources.

2. When resources are limited, it is expected that natural selection will favor increased competitive ability. One of the few direct demonstrations of this simple prediction has been made with populations of fruit flies.

3. Early studies of competition between different species of *Paramecium* showed that extinction of one species was one possible outcome of competition. When two species coexisted, it was due to differential use of resources in the environment; this led G. F. Gause to suggest that no two species using the environment in exactly the same manner may coexist. This prediction is referred to as Gause's competitive exclusion principle.

4. A description of the ecological requirements of a species is called its niche.

5. The Lotka-Volterra model is a mathematical summary of the interactions of competing species. For two species with the same carrying capacity, the Lotka-Volterra model predicts extinction of one species when the effects of interspecific competition on it are stronger than the effects of intraspecific competition. The two species may coexist when each species competitively affects the other to a lesser degree than it affects itself.

6. The effects of competition can also be used to understand the number of competing species that can stably coexist in a particular environment.

7. In some situations, competition between different species may lead to the evolution of behavioral or morphological characters that will ultimately diminish the intensity of competition between the species. This process, called character displacement, appears to account for morphological differences among three-spined sticklebacks.

REVIEW QUESTIONS

1. In this figure, the blue lines show the equilibrium growth lines for a rose and a weed species. The red arrows show the consumption rates for water and nitrogen for each species. What do you predict the outcome of competition will be if the supply point for these resources is located at points A, B, or C?

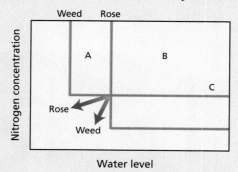

2. In this figure, seeds from a weed and a rose are grown in different proportions, from 0 to 100 percent of each. The weights of the final plants are determined and recorded by the colored lines. Which plant appears to be the better competitor? Why?

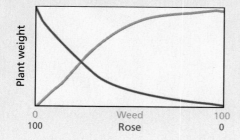

3. In Module 12.8 we considered several possible outcomes of competition according to the Lotka-Volterra model. In this graph we show a set of conditions not considered previously. Try to predict the outcome of competition in this case. Hint: There is more than one outcome.

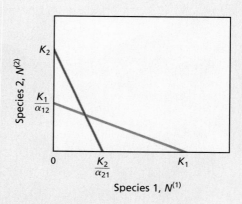

4. Review and explain how the meaning of the niche concept has changed over time.

5. In Module 12.12 we described how character displacement would proceed when a genetically variable species encountered a strong competitor. How would you modify this description if the second species also contained genetic variation similar to the variation present in the first species?

KEY TERMS

apparent competition
character displacement
competition
competition coefficient
competitive exclusion principle

contest competition
ecological niche
fundamental niche
guild
interspecific competition

intraspecific competition
Lotka-Volterra model
mycorrhizae
niche
realized niche

scramble competition
social Darwinism

FURTHER READING

Bakker K. 1961. "An Analysis of Factors Which Determine Success in Competition for Food among Larvae of *Drosophila melanogaster*." *Archives Neerlandaises de Zoologie* 14:200–81.

Burdon, J. J., R. H. Groves, P. E. Kaye, and S. S. Speer. 1984. "Competition in Mixtures of Susceptible and Resistant Genotypes of *Chondrilla juncea* Differentially Infected with Rust." *Oecologia* 64:199–203.

Connell, J. H. 1961. "The Influence of Interspecific Competition and the Other Factors on the Distribution of the Barnacle *Chthamalus stellatus*." *Ecology* 42:710–23.

Eldridge, J. L., and D. H. Johnson. 1988. "Size Differences in Migrant Sandpiper Flocks: Ghosts in Ephermal Guilds." *Oecologia* 77:433–44.

Gause, G. F. 1934. *The Struggle for Existence*. Baltimore: Williams and Wilkins. Reprint, New York: Dover, 1971.

Holt, R. D., and J. H. Lawton. 1994. "The Ecological Consequences of Shared Natural Enemies." *Annual Review of Ecology and Systematics* 25:495–520.

MacArthur, R. H. 1958. "Population Ecology of Some Warblers of Northeastern Coniferous Forests." *Ecology* 39:599–619.

Mueller, L. D., 1988. "Evolution of Competitive Ability in *Drosophila* Due to Density-Dependent Natural Selection." *Proceedings of the National Academy of Science USA* 85:4383–86.

Schluter, D., and J. D. McPhail. 1992. "Ecological Character Displacement and Speciation in Sticklebacks." *American Naturalist* 140:85–108.

Schoener, T. W. 1989. "The Ecological Niches." In *Ecological Concepts*, edited by J. M. Cherrett, 79–113. Oxford, UK: Blackwell Scientific.

Tilman, D. 1988. *Dynamics and Structure of Plant Communities*. Princeton, NJ: Princeton University Press.

Great White Shark

13

Predation

Understanding the ecology of biological populations usually comes from a consideration of the physical and biological environment. The biological environment includes other organisms with which members of a population interact. Some of these other organisms may be competitors for important resources. Others may be an important food source for some species. **Predation** occurs when one animal species feeds on another and, in that process, kills it or consumes most of the organism. An animal that is killed is called the **prey**. Animals that feed on plants are called **herbivores**. There are many important similarities between predation and herbivory, but here we will treat each separately.

One obvious effect of predator-prey relationships is that the numbers of predators will be affected by the numbers of prey, and vice versa. Prey numbers decline as a result of predation, but predators must consume prey to survive and reproduce. One important goal of this chapter is to develop an understanding of the numerical effects of predation on population sizes of predators as well as their prey.

It is clear that animals will try to avoid being eaten. In fact, we would expect natural selection to favor genotypes that possess adaptations to reduce or avoid risks of predation. In this chapter we examine the ways in which many species have attempted to avoid the clutches of hungry predators.

Predators must also be able to overcome the adaptations of prey. In some cases this might occur through the evolution of morphological characters that aid in being an effective predator, like the massive jaws of the shark. However, more subtle aspects of predator adaptation may affect how predators search or hunt for prey. This chapter also reviews the different strategies that are used by effective predators for capturing food. ❖

PREDATOR-PREY DYNAMICS

13.1 The dynamics of predator-prey populations are intimately connected

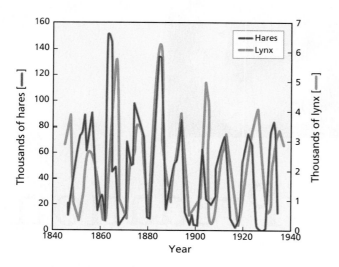

FIGURE 13.1A **Predator-Prey Cycles** The cycles of lynx (green lines) and hare (red lines) across Canada (from Hudson Bay Company's records). Notice the regularity of the cycle and that rises and declines in lynx numbers come after rises or declines in hare numbers.

In nature, populations do not exist in isolation from other populations. In any physical area, we find many different species that interact in a variety of ways. This collection of species is usually referred to as a **community**. A community would require the existence of at least two different species. In this chapter we will consider one of the simplest types of community, one consisting of a predator and prey. This type of community also has a very special relationship between its two members. The major source of food for the predator population is the prey species. Thus, the prey have a positive effect on the survival and population growth of the predators, while the predators have a negative effect on the prey species.

In Chapter 14 we consider a related two-species interaction called **host and parasite**. The host-parasite interaction is also marked by the host serving as food for the parasite. The major difference between the host-parasite and predator-prey relationships is that predators usually kill and consume a whole prey individual. In contrast, many hosts may live indefinitely, supporting large numbers of parasites. Of course these parasites drain energy from their host and make them more susceptible to death from other causes.

Much of the great interest in predator-prey dynamics is a consequence of extraordinary observations made in natural populations. One of the best-documented examples is of the Canadian lynx and its prey, the snowshoe hare (Figure 13.1A). The estimates of numbers come from the records of the Hudson Bay Company's fur trade. Trappers were paid a fixed amount for lynx and hare hides during much of this period, and thus the fluctuations in number of hides are thought to reflect variation in numbers of lynx and hare. This assumption is certainly not precisely correct. For instance, it is known that when hares are common, it is more difficult to trap lynx than when the hares are rare. This trapping bias would tend to make the troughs in the cycles higher than they should be, and the peaks smaller. A more important source of bias is trapper effort. Because the Hudson Bay Company paid the same amount every year, one would expect trappers to intensify their efforts during the peaks of the cycles and perhaps to be discouraged during the troughs.

In any case, more recent information on lynx-hare biology and careful population size monitoring suggests that the basic impression of cycles is certainly correct. In fact the data are remarkable for the consistency of their fluctuations, which take about 10 years per cycle. In addition, the cycles appear in many areas of Canada. This fact led early investigators to suggest that the cycles were a consequence of extrinsic factors like sunspot cycles, ozone cycles, weather cycles, forest-fire cycles, or plant-nutrient cycles. Some of these extrinsic factors show a correlation with the lynx-hare cycles for short periods of time, but none does well for the entire duration of the series in Figure 13.1A. More important, there are no reasonable connections between the lynx-hare cycles and these extrinsic factors. As we review models of predator-prey dynamics, we will see that the observed cycles may be a natural by-product of predator-prey interactions and density-dependent population regulation.

Observations like those made for the lynx and hare are not always possible. For instance, there are very good records of the numbers of voles, a small mammal, in many parts of Europe. It is known that many animals prey on voles, but there are no good records of the predator numbers. There is a consistent north-south transition in vole population dynamics (Figure 13.1B). In more northerly locations, vole dynamics may be chaotic; but toward the south, more stable dynamics appear. One hypothesis for these observations revolves around the predator community that feeds on voles. In the north, predators are more likely to be specialists on voles; in the south, generalists are more common. A generalist predator is one that will feed on a wide variety of prey. The generalists are believed to stabilize the dynamics of vole populations in the south. ❖

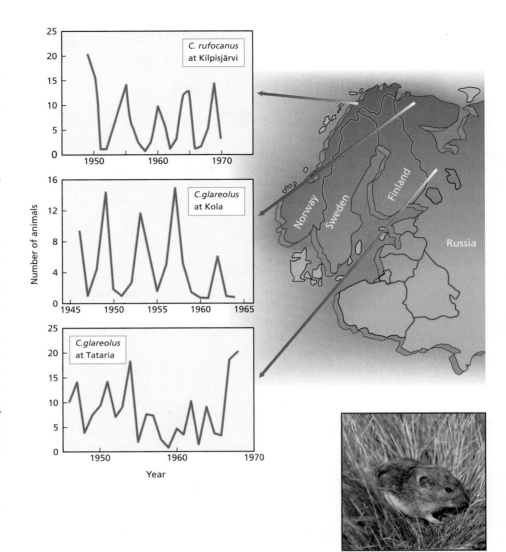

FIGURE 13.1B Population Size Variation in Three Vole Populations Voles of the genus *Clethrionomys* display highly variable population sizes in Northern Europe. However, detailed examinations of the numbers from the most northern populations, Kilpisjärvi and Kola, reveal that they are less stable than the population in Tataria.

13.2 The Lotka-Volterra model of predator-prey dynamics predicts cycles, although for reasons that probably do not apply to natural populations

Observations of natural populations of predators and prey have revealed some dramatic cycles. The case of the lynx and snowshoe hare is certainly one of the best known and most dramatic. Ecologists attacked this problem theoretically in the 1920s. Alfred Lotka was the first to develop a theory of predator-prey dynamics in 1920. Vito Volterra independently derived these results in 1926, so today this ecological theory is referred to as the **Lotka-Volterra predator-prey equations**.

The predator-prey theory makes some very simple assumptions (Figure 13.2A). From our earlier discussion of population growth, we will recognize that some of these assumptions are overly simplistic (Module 10.1). However, it will be easier to understand the more complicated models if we start first with this very simple formulation. Since the words *predator* and *prey* both start with the letter *P*, we will refer to the prey as *victims* and use the letter *V* to represent the number of prey in the population. Likewise, the number of predators in the population is represented by *P*. Unlike some of our previous models, the life cycles of predator and prey are not broken into discrete generations. Instead, the Lotka-Volterra model follows the changes in the size of the predator and prey population over very short time periods. In this model, change in the size of the prey population over short periods is symbolized by ΔV (Figure 13.2A). A positive value for ΔV means the prey are increasing over time; a negative value means the prey are decreasing in size over time. When ΔV is 0, that means there is no change in the population size. In a similar fashion, the change in the predator population size is symbolized by ΔP.

The model is developed by considering what happens to each population in the absence of the other. The prey population is assumed to grow exponentially, at a rate *r*, in the absence of predators. You may wonder how the prey grow exponentially since the equation for ΔV in Figure 13.2A suggests a simple linear increase in ΔV. Recall that ΔV is the size change occurring in a short time period. The principle that gives rise to exponential growth is the same as that regarding the growth of money in your bank account. Your bank adds a small, constant fraction of your money to your account every day as interest. Over a long period of time, because you are adding interest onto interest, the total amount of money in the account grows at an exponential rate. Since the only source of food for the predators is assumed to be the one prey species in our model, in the absence of prey, the predators will die out (Figure 13.2B). For simplicity, the Lotka-Volterra model assumes that the predators also die out at an exponential rate, given by $-d$, where *d* is a positive number.

When both predator and prey are together, prey will be caught and eaten by the predators (Figure 13.2C). The Lotka-Volterra model assumes that the number of prey caught by each predator increases in direct proportion to the number of prey in the population. Thus, if the number of prey is doubled, the number of prey caught per unit of time by a predator will also double. The relative efficiency of the predator at catching prey is given by the parameter *c*. The higher *c* is, the more efficient the predator. The loss of prey per unit of time due to predation is equal to $-cPV$. This term, sometimes called the **functional response**, reflects important aspects of the predator's hunting capabilities. The conversion of captured prey into new predators is determined by the parameter *k*.

The Lotka-Volterra model inevitably predicts that the predator and prey population will continuously cycle (Figure 13.2D). The problem with these cycles is that they are not stable (see Module 10.7). So, if some outside force slightly perturbs the predator and prey away from their current cycle, they simply move to a completely new cycle (Figure 13.2D). This is odd behavior that is not exhibited by real populations; it suggests that certain assumptions of the Lotka-Volterra model need to be changed. We address this issue next. ❖

Instantaneous change in prey population size $= \Delta V$

In the absence of predators, prey population size grows exponentially at rate r.

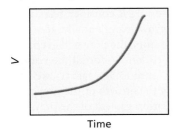

$$\Delta V = rV$$

FIGURE 13.2A **Prey (V = Victims) Alone**

Instantaneous change in predator population size $= \Delta P$

In the absence of prey, predator population size declines exponentially at rate $-d$.

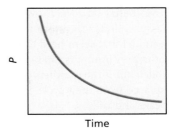

$$\Delta P = -dP$$

FIGURE 13.2B **Predators (P) Alone**

The functional response = number of prey captured per predator

$$= cV$$

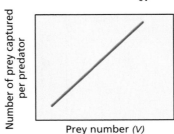

Total number of prey lost per unit time $= cVP$.

The captured prey are converted to new predators at a rate $= kVP$.

So,
$$\Delta V = rV - cVP$$
and
$$\Delta P = kVP - dP$$

FIGURE 13.2C **Prey and Predators Together**

FIGURE 13.2D **Cycles in the Lotka-Volterra Model Are Not Stable.**

13.3 More realistic models incorporate density-dependent prey dynamics and predator satiation

The Lotka-Volterra model was based on several overly simplistic assumptions. These assumptions lead to some predictions by the Lotka-Volterra model that seem hard to defend. For instance, cycles in predator-prey numbers can be generated by the Lotka-Volterra model; but if the numbers of predator and prey are slightly displaced from their cycle, they start a new cycle with different peaks and valleys.

The first assumption that we might consider relaxing is the assumption that the prey grow exponentially. It would seem that in the absence of predators, the prey will eventually feel the effects of density dependence—for all the reasons considered in Module 10.1. In Figure 13.3A, part (i), we have modified the equation reflecting the growth of the prey population to include logistic population growth of the prey. The

We can assume that the prey will eventually be density regulated in the absence of predators. If the carrying capacity of the prey is *K*, then logistic growth of the prey results in:

$$\Delta V = rV\left[1 - \frac{V}{K}\right] - cVP$$

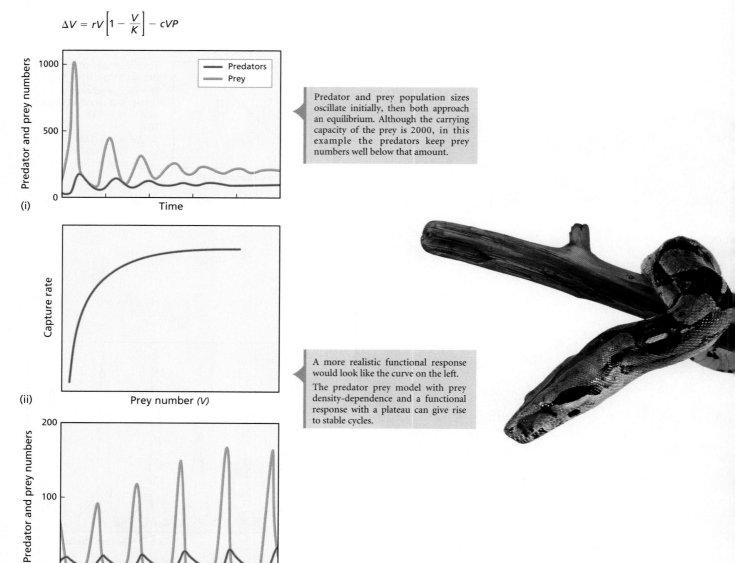

Predator and prey population sizes oscillate initially, then both approach an equilibrium. Although the carrying capacity of the prey is 2000, in this example the predators keep prey numbers well below that amount.

A more realistic functional response would look like the curve on the left.

The predator prey model with prey density-dependence and a functional response with a plateau can give rise to stable cycles.

FIGURE 13.3A Refinements to the Lotka-Volterra Predation Model

logistic part of the red equation in Figure 13.3A is written as $rV[1 - V/K]$. This form of the logistic was encountered earlier in the box at Module 10.6. The addition of this density dependence to the Lotka-Volterra model in Figure 13.3A results in the predator and prey populations approaching a stable population size. It is worth noting that even though the carrying capacity of the prey is 2000, in this example the predators keep the equilibrium number of prey well below this value, at about 200 (Figure 13.3A).

Some assumptions about the behavior of the predators in the Lotka-Volterra model are also overly simplistic. The Lotka-Volterra model assumed that the predators would continue to catch more prey in direct proportion to the number of prey in the population. We know that this cannot always be true, for at least two different reasons: (1) There will be a point at which the number of prey that a predator has captured and eaten is so great that the predator simply cannot eat any more prey. At that point we say the predator is **satiated**. Satiation should produce a leveling off of the functional response, as shown in part (ii) of Figure 13.3A. When we say "level off," we mean that adding more prey to the population will not result in more prey being caught, since the predators are already eating as much as they can. (2) The predator must consume the prey it catches. The consumption of prey will take some time because it involves biting, chewing, and swallowing the prey. The time to complete these activities is called **handling time**. Again, we expect that eventually the handling time for a large number of prey will be so great that the predator is unable to catch more prey. At that point the functional response should again level off.

In part (iii) of Figure 13.3A we see the changes in predator and prey numbers that are predicted from the model with logistic population growth of the prey and a functional response that shows the effects of handling time and satiation. In this case we see there are

cycles (although that is not always the case), and these cycles are stable, unlike those for the Lotka-Volterra model. So if the predators or prey populations are perturbed away from the cycle they will, over time, return to the same original cycle. Most cycles observed in nature are believed to be stable because they regularly experience perturbations due to random environmental fluctuations.

Do real populations show these types of cycles? In fact they do. Luckinbill (1973) has studied a predator-prey system consisting of unicellular flagellates. The prey species in this system is *Paramecium aurelia*, and the predator is *Didinium nasutum*. *Paramecium* grows in an approximate logistic manner in the absence of *Didinium*. In Luckinbill's laboratory cultures the equilibrium number of *Paramecium* is about 850 per milliliter. When both species are present, they show cyclic fluctuations (Figure 13.3B). The predator always reaches its peak numbers slightly later than the prey species. Likewise, the valley—or lowest number—of predators is also always slightly later than the valley of the prey. This is the same behavior we see in the theoretical model shown in Figure 13.3A. ❖

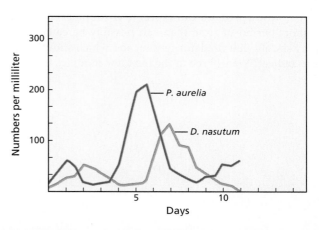

FIGURE 13.3B Numbers of Predator (*Didinium nasutum*) and Prey (*Paramecium aurelia*) per Milliliter. In the absence of predators, *Paramecium* grows logistically to an equilibrium of about 850 per mL.

HOW TO BE A PREDATOR

13.4 A variety of factors determine how predators forage

Predators are faced with many decisions that do not always have simple answers. However, the way predators deal with these decisions may have a substantial impact on their long-term survival and the survival of their offspring (Figures 13.4A through 13.4C). For instance, it is worth recalling that even predators have enemies. Many birds will feed on insects, but themselves may be prey for other birds or small mammals. Consequently, predators may have to evaluate the relative danger of foraging in particular places or for extended periods of time. All other things being equal, many predators may seek to minimize the time they spend foraging in order to reduce their exposure to predators or other hazards. If this is the case, then it may turn out that some species are better to include in the diet than others. For instance, insects that need to be dug out of the bark of a tree may take much longer to find and eat than insects that live on the surface of leaves.

Suppose a predator can expect to encounter more prey in a short period of time than can possibly be eaten. Which prey should that predator ignore, and which should it pursue and eat? We will see in the next few modules that several factors are important in determining the answer to this question. What is clear is that factors such as the nutritional value of the prey, ease of capture, and number of prey all may have an impact on the behavior of predators. Many of the decisions that predators make concerning which prey to consume are also factors for nonpredatory animals that consume resources like seeds or plants. Thus, although the following modules refer to predators, most often many of the problems we discuss are also encountered by a wide range of foraging animals.

Do Animals Forage Optimally?
One approach to the study of foraging behavior has been to assume that the rule used by foragers to make decisions leads to optimal solutions. In this context, *optimal* can mean several things. If it appears that it is most important for an animal to minimize the time it spends in foraging, then the optimal solution would be the one that leads to the minimum time to gain a certain level of nutrition. Alternatively, *optimal* could refer to achieving the maximum return of energy from a fixed foraging effort.

FIGURE 13.4A Lions will cooperate to capture prey.

FIGURE 13.4B Birds of prey often must rely on their keen sense of vision for hunting.

Why Would We Expect These Behaviors to Be Optimal?

One view on this question is that behaviors are the product of natural selection, and thus the animal with the optimal behavior ought to have the highest fitness, all other things being equal. While this idea seems reasonable, several other factors may prevent behavior from being optimal.

One fact about natural selection is that it is not always able to maximize fitness or produce the best phenotype. This can be due to complications in the way the genetic system works, or in aspects of the way fitness is determined. For example, when a heterozygote at a single locus is the most fit genotype, the optimal solution would be to have a population composed entirely of heterozygotes. Due to Mendelian segregation, we know that in every generation we will continue to see less-fit homozygotes in the population; there is no way to get rid of them.

Another complication is that fitness is determined by many things, not just how well an organism forages. Thus, the evolution of foraging behavior must occur along with the evolution of many other traits—like reproductive behavior and competitive ability. Optimal foraging behavior may not evolve due to the conflicts with the direction of evolution of other traits. What use, then, are models that determine the optimal pattern of foraging? If the model has been constructed correctly, then no animal should be able to do better than the optimal. For these reasons, the optimal behavior can be used as a standard against which real behavior can be compared. The construction of these models can also suggest the importance of different factors that can affect rates of energy intake, for instance. This in turn can help the experimental ecologist in designing appropriate experiments to test which factors are most important for determining the foraging behavior of real organisms. As we will see in the next few modules, many animals forage in a manner that is consistent with predictions based on some very simple models. ❖

FIGURE 13.4C Web-building spiders wait for prey to come to them rather than moving to find them.

13.5 Foragers may optimize energy gain per unit of time, or minimize time spent foraging

Predators employ a diverse set of strategies to capture prey. Some predators, like spiders, sit still and wait for prey to come to them and become tangled in their webs. Large predators, like lions, may move great distances to find prey and then often need to run at great speeds to capture prey. Can we detect any patterns to these foraging strategies? Ecologists have addressed this question by trying to determine what the best foraging behavior might be and then seeing if animals forage in this manner. However, determining the best foraging behavior requires that we know how to measure how well a forager is doing.

It may be that the time spent foraging exposes the individual to potential danger, perhaps to other predators, or alternatively takes away from time the individual could use for reproduction or watching and caring for offspring. In this case the best predator would be one that can get sufficient food in the shortest period of time. Another possibility is that the forager simply tries to gain the most energy possible per unit of time spent foraging. These two views of what foragers are attempting to do in many cases will yield the same prediction of the best strategy. We first review the strategies used by time minimizers.

There are two components to the time spent in finding and consuming prey. The first component, usually called the **waiting time**, refers to the average time between encounters with prey. For these calculations we will assume that a foraging predator will catch and consume each prey encountered. Once a prey is caught, the predator must consume it. The time spent catching and consuming the prey is called the handling time. Obviously, a predator cannot be pursuing or catching another prey while it is still handling the first. It may turn out that several prey are available for predator consumption. In Figure 13.5A we have examples where the predator has a choice of two prey. If each prey is equally nutritious, then the preferred prey would be the one with the smaller handling time. The small red prey (A) takes only a half-hour to handle, while the large green prey with thorns (B) takes an hour to handle.

It would seem reasonable for the predator to take the preferred prey; but are there circumstances where the predator should take both types of prey? In Figure 13.5A we compute the average time between prey consumption under two different circumstances. To do this calculation we need to know the encounter rate, P_i, of each prey. The encounter rate is the number of times per hour a particular prey will be encountered. If this rate is five times per hour, then the average time between encounters, or the waiting time, is the reciprocal of five, or 12 minutes. In case 1 and case 2 the preferred prey (A) is five times more likely to be encountered than the less-preferred prey (B). However, in case 1, both prey are 10 times more common than they are in case 2.

In case 1 we see that if the predator takes only prey A, the average foraging time is 0.7 hrs. If the predator takes both prey, then it will encounter six prey per hour, so the waiting time is 1/6 hr. The handling time is the average over both prey. Since prey A is five times more common, it will represent five-sixths of all prey caught, resulting in an average foraging time of 0.75 hrs. For case 1 we see that the predator will do best by taking only prey A and ignoring prey B.

In the second case of Figure 13.5A, the absolute frequencies of both types of prey have decreased. Doing calculations like those in case 1, we see that the predator would now do best to take both prey. The reason for this conclusion is that now the predator spends most of its time in waiting for a prey as opposed to handling it. Therefore, even if a prey is the less-desirable type, the predator should take one because it will be a long time before another prey of either type comes along. ❖

Take only 🥦 = (waiting time) + (handling time)
= (1/5) + (1/2) = 0.7 hrs

Take 🥦 and 🐛 = (1/6) + [(5/6)(1/2) + (1/6)(1)] = 0.75 hrs

Case 2:

Take only 🥦 = (waiting time) + (handling time)
= (1/0.5) + (1/2) = 2.5 hrs

Take 🥦 and 🐛 = (1/0.6) + [(0.5/0.6)(1/2) + (0.1/0.6)(1)] = 2.25 hrs

	P_A	P_B
Case 1	5.0	1.0
Case 2	0.5	0.1

FIGURE 13.5A Foraging Strategies That Minimize Time

Predators may search for and capture prey by methods that maximize the amount of energy that they take in per unit of time. This would certainly seem to be an efficient way to be a predator, but are animals capable of making decisions that can result in such efficient use of their time? There are no first principles in evolution that will allow us to claim that this must be the case, so we need to make direct observations of foraging behavior to test this idea. N. B. Davies (1977) has studied this issue by observing the foraging behavior of small insectivorous (insect eating) birds called wagtails (*Motacilla alba yarrellii* and *M. flava flavissima*). Davies found that the various insects eaten by these birds often differed greatly in size (Figure 13.6A). The very small prey could be eaten immediately, while larger prey were often held and bashed against a perch prior to eating. This behavior might result in a handling time of 5–10 seconds. We see that the energy ingested per second has a maximum for wagtails at about size 7 mm (Figure 13.6A). Even though larger prey will have more energy per individual, the increased handling time reduces the rate of energy return.

FIGURE 13.6B *Parus major,* **the Great Tit**

The wagtails also show a preference for prey in the 7-mm size category. Of course, this result could simply reflect that the insects in the 7-mm size class were the most common. However, it turns out that insects in the 8-mm size class were most common. These results suggest that wagtails are modifying their selection of prey toward those that give the greatest energy return per unit of time.

We previously reviewed how predators might change their foraging behavior, going from specialization to generalization as prey become scarce. This prediction followed from a time-minimization perspective. If predators instead try to maximize the rate of energy gain, then the most efficient behavior will be to specialize on the highest-energy prey when these are common, but accept multiple prey when they are rare. John Krebs and colleagues (1977) studied the foraging behavior of birds called great tits (Figure 13.6B). These experiments were carried out in the laboratory. Birds were exposed to two "prey" types that were made up of pieces of mealworms. Large prey had about twice as much energy per prey item as the small prey did. The small prey was constructed in a fashion that resulted in both large and small prey having about the same handling time. As a result, the energy return per unit of time was much higher for the large prey. A conveyor belt was used to run these prey past the birds, who were then free to forage. The conveyor belt permitted the scientists to control precisely the relative frequency of encounters with both prey types. When both prey types were equally abundant and encountered frequently, the birds chose the larger, higher-energy prey more often (Figure 13.6C). However, when the frequency of both prey was reduced, the birds became less choosy and were equally likely to select small and large prey (Figure 13.6C).

Thus, from these two studies, we see support for the idea that predators can assess the relative quality and abundance of prey. With that information, predators can then make relatively rapid changes in their foraging behavior that will help them maintain a high level of energy intake. This ability to alter foraging behavior is certainly advantageous, since most predators will over their lifetime experience a range of environmental conditions that will not be best handled by a single strategy. ❖

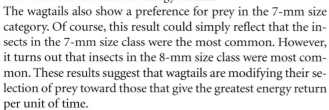

FIGURE 13.6A Maximization of Energy Intake

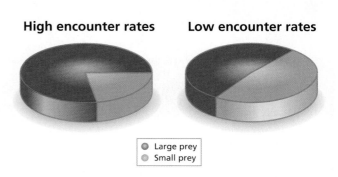

FIGURE 13.6C The Fraction of Two Prey Types in the Diet of Great Tits

13.7 Central-place foragers should recover more food the farther they travel

Some animals forage from a single location. Thus, all trips to recover food are followed by a return trip to the central place. For instance, many birds that are caring for young will hunt and catch prey, and they bring them back to the nest in order to feed their young. A crucial component of this type of foraging is the time spent traveling from the central place to the foraging site. Based on our previous examples, it is clear that if the forager attempts to minimize the time spent foraging, or to maximize the rate of energy capture, an important component of this calculation will be the travel time. The predator in Figure 13.7A may forage in two locations. The nearby location has a travel time of T_1, whereas the distant location has a travel time of T_2, which is greater than T_1. Suppose that the time to capture prey were effectively zero (of course, this is not true; see Figure 13.7B). Then, if T_2 is twice as large as T_1, the forager must bring back twice as much food from the distant location simply to equal the return that is possible from nearby foraging.

But are animals able to take these factors into account? Krebs and Avery (1985) attempted to address this question by studying the feeding habits of the European bee-eater (*Merops apiaster*). The researchers observed adult bee-eaters that were actively foraging to feed their young. A number of different insects were available to the bee-eaters, including small bees with a dry weight of about 25 mg and large dragonflies of about 315 mg. The amount of energy provided by these insects is roughly proportional to their dry weight. As we can see in Figure 13.7C, the number of small prey in the diet of bee-eaters decreases the farther they have to travel. One possible alternative explanation for these observations is that there are just fewer small prey at the distant sites. Field surveys demonstrated, however, that the different prey types did not vary in abundance as a function of proximity to the bee-eater nesting sites. It is reasonable, then, to conclude that the birds are purposefully changing their diet. ❖

T_1

T_2

Tree

FIGURE 13.7A Central place foragers make many trips to and from one location. The time between foraging sites may vary substantially and therefore have an important impact on the foraging behavior in each site.

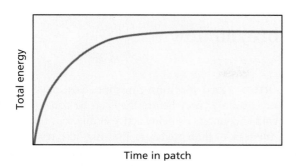

FIGURE 13.7B Total Energy Gain in a Single Patch We expect that initially food will be easy to find and the energy gain will be rapid. As time goes on, the easy food items will be gone; more effort and time will be required to get additional energy. Eventually no additional energy will be found, and the total energy gain will reach an asymptote. At this point the energy return per unit of time is zero.

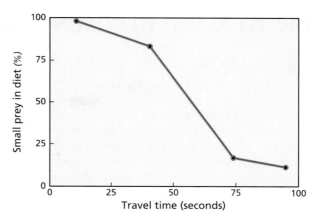

FIGURE 13.7C This figure shows the results of a study with a bird, the European bee-eater. Bee-eaters make many trips from their nest and back to feed their young. Usually they bring back only one insect-prey item at a time. It is clear that they bring back fewer small insects when they travel farther from the nest.

HOW TO AVOID BECOMING PREY

13.8 The process of prey capture can be broken down into multiple stages

Predation can be a major source of mortality for many animals. Consequently, any behavior, morphology, or physiological traits that would help prey avoid predators should be favored by natural selection. There are many stages of predation where prey may employ adaptations to avoid being eaten. In this module we review some of these stages.

FIGURE 13.8A Flounder on the Sandy Bottom of the Ocean

Encounter Predators must come into close physical proximity or encounter prey before they can be consumed. Prey may avoid encounters by being active at different times of the day or the seasons than predators. Prey may also rest in inconspicuous places to avoid encounters (Figure 13.8A). Some prey may develop more acute senses than their predators and be able to leave an area before being encountered by predators.

Detection Prey may have morphological structures or colors that make them blend into the background. We review this possibility in more detail in the next module. Prey may attempt to confuse prey by making sudden, unpredictable movements. Large schools of fish may overwhelm and confuse predators (Figure 13.8B).

Identification Predators need to be able to identify prey, once detected, as something that is worthwhile eating. Some animals may manufacture or consume toxic compounds that make them unpalatable (Figure 13.8C). In these situations the prey will take on colors or morphological patterns to advertise that condition. As we discuss in the next modules, some palatable animals may try to take advantage of this adaptation by looking like unpalatable species.

FIGURE 13.8B Large School of Fish

Approach Once a predator has encountered and identified a prey, it must attack and capture the prey. The prey may still escape capture. It may simply be faster than the predator and be able to escape, or it may be near a hiding place to which it can escape. The prey may startle the predator by assuming an aggressive stance that momentarily delays the predator and allows escape (Figure 13.8D).

Subjugation After the predator has captured the prey, it must either gain control of it or kill it before it is eaten. Some prey may be strong enough to simply escape from a predator. Other prey may have physical properties, like shells or mucus and slime, that can prevent predators from gaining control. Some prey may detach body parts to help them elude predators (Figure 13.8E). Many salamanders and lizards can detach their tails when caught by a predator. Some prey are simply noxious and cause predators to release them due to spines, stings, or bad taste.

FIGURE 13.8C **Monarch Butterflies** These butterflies incorporate toxic and distasteful cardiac glycosides from their milkweed food plants.

Consumption By the time a prey is about to be consumed, it has few options for escape. However, if the consumption of a prey causes the predator to become sick, the predator may avoid this type of prey in the future. If this experience prevents that predator from consuming relatives of the prey, then the production of the toxins that cause these types of negative reactions in predators may nevertheless be favored by natural selection.

❖

FIGURE 13.8D **Puffer Fish, Blown Up**

FIGURE 13.8E **Lizard with Detached Tail**

There are many ways for an animal to avoid being eaten by a predator. One of the most obvious is simply to avoid being detected by predators. Many animals have been extremely successful at blending into the background to avoid detection. These animals often have **cryptic coloration** that resembles a random sample of the visual background they are most likely to be found in. The insects shown in Figures 13.9A through 13.9C illustrate the range of remarkable adaptations to predator avoidance.

Cryptically colored animals do not have to be drab in appearance. Many birds—like parrots, orioles, and tanagers—are brightly colored, but not conspicuous in their natural habitat. Some animals may have coloration that is conspicuous to other members of the same species but cryptic to important predators. This effect can occur when the visual acuity of the prey and predators differs.

Some animals have bright coloration to warn predators that they are innately distasteful. Coloring that serves as a warning is sometimes called **aposematic coloration**. This is a very different strategy than that of cryptic coloration because the goal is to be seen by the predator. The colors used for these warnings are often black and either yellow or red. It appears that many predators have a general aversion to these colors. Animals may communicate their unpalatability in other ways. Some animals, like bees or rattlesnakes, make noises that serve as warnings. Others, like skunks and stinkbugs, have odor signals. Some species of arctiid moths are distasteful to bats. These moths emit ultrasonic pulses that bats can sense and use to avoid them.

Bright colors appear to help predators avoid distasteful prey. In an experiment with chickens, two types of distasteful bait were offered: cryptically colored or conspicuously colored (Figure 13.9D). Initially the conspicuous bait was found more easily and consumed at a greater rate, but over time the chicks quickly learned to avoid this bait. They had a more difficult time learning to avoid the cryptically colored bait. ❖

FIGURE 13.9C A geometrid larva appears to be an extension of the branches of this plant. If you were a predator, how well would you do finding this larva?

FIGURE 13.9A A Walking Stick, *Diapheromera femorata* The color of walking sticks varies from green to brown. To assist with their disguise, they walk slowly and may stay motionless for extended periods.

FIGURE 13.9B A Branch Covered with Thornbugs (*Umbonia crassicornis*) These insects cling to the stems of plants and are especially abundant in subtropical habitats including Florida.

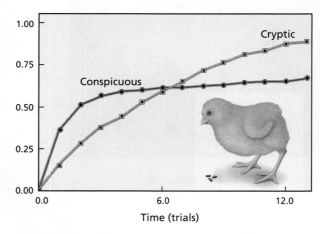

FIGURE 13.9D Chickens learn to avoid conspicuously colored, distasteful bait more quickly than they do cryptic, distasteful bait.

We have seen that predators can learn to avoid distasteful prey, but this requires several trials on the part of the predator. If several different species of prey are all distasteful, a common predator may learn to avoid those prey more quickly if they look similar. Thus, two distasteful species that have a similar appearance will both benefit from the negative reinforcement they exert on a common predator. This type of process can lead to different species looking remarkably similar. These similar-looking species are called **Müllerian mimics**. Müllerian mimics are most likely to evolve when both species are equally unpalatable and equally common.

Mimicry can also evolve between two species—one that is palatable and one that is not. This type of mimicry is called **Batesian mimicry**. Some examples of Batesian mimics are shown in Figure 13.10A. The unpalatable species is called the model, and the palatable species is the mimic. In this case the mimic is benefiting from the model's negative effects on predators. However, the protection enjoyed by the model is negatively affected by the mimic. This is because, as the mimic becomes common, predators learn to associate its appearance with palatability. As a result, natural selection should favor the model to evolve a different appearance from that of the mimic; meanwhile, the mimic should always try to look as similar as possible to the model. In some species of Batesian mimics, there appears to be a polymorphism in coloration (Figure 13.10B); one morph appears to mimic a distasteful species, while a second is cryptic. ❖

FIGURE 13.10A Wasps and Their Fly Mimics *Xylocapa latipes* (1) and the fly mimic *Hyperechia fera* (2). Note that the wasp has two pairs of wings, while the fly has only one. The model *Collyris emarginata* (3) and its mimic *Sepedon* sp. (4). The model *Mesostenus* (5) and its mimic from the family *Stratiomyidae* (6). the mimic *Xylophagus* (8) shows a great elongation of the antennae to match its model species *Mesostenus* (7). The *Solias* wasp (9) and its fly mimic *Laphria* (10). The wasp *Macromeris violacea* (11) hunts spiders. Their mimics from the genus *Midas* (12) have an enlarged wing that makes it appear as large as the pair in *M. violacea* . The fly mimic *Milesia vespoides* (13) has a striking resemblance to the wasp *Vespa cincta* (14), including yellow wing color.

FIGURE 13.10B Species of Swallowtail Butterflies Have Both Mimetic and Cryptic Forms

PLANT-HERBIVORE INTERACTIONS

13.11 Plants show immediate and long-term reactions to herbivory

Unlike animals, plants do not possess many of the same options for avoiding their predators, the herbivores. Plants are not mobile, so they cannot escape from herbivores. They have no behavioral mechanisms for avoiding herbivores. However, this does not mean that plants are at the mercy of herbivores. Two major strategies employed by plants to defend against herbivory are (1) to be resistant to herbivores, and (2) to tolerate the damage caused by herbivores. Plants often defend themselves against herbivores by making toxic chemical compounds that cause herbivores to completely avoid feeding on the plant. An alternative to resistance is tolerance. Plants may still be able to produce flowers and seeds, even after herbivores have removed substantial portions of their leaf area. As we shall see, it may not be possible for a plant to both defend itself and be tolerant of herbivory.

Tolerance and **resistance** are both aspects of plant defense. Resistant plants will show little or no reduction in fitness or the ability to reproduce as a consequence of herbivore attack. **Tolerance** is a relative measure of the fitness reduction caused by a particular level of damage from herbivores. The greater the tolerance, the smaller the fitness reduction. The difference between the fitness of a plant damaged by herbivory and an undamaged plant is called **compensatory ability**. The smaller this difference, the greater the tolerance. In some cases this difference may actually be positive; that is, the plant's fitness is greater after herbivory. For different genotypes of the plant *Asclepias* (Figure 13.11A), compensatory ability increases with increasing root-to-shoot biomass ratio. This means that plants that store more energy in their roots

are better able to reproduce after herbivore damage, perhaps because of their ability to replace lost biomass from the energy reserves in their roots.

Other plants tolerate herbivory due to structural aspects of the plants. For instance, wild tomatoes are better able to tolerate herbivory damage than domesticated tomatoes are. This is because the canopy structure of wild tomatoes can better exploit light resources after damage than can the canopies of the domesticated tomato.

As with many life-history traits, the abilities to resist herbivory and tolerate herbivory appear to trade off. The production of defensive chemical compounds is likely to be energetically costly and thus compromises a plant's ability to tolerate damage from herbivores. Kirk Stowe (1998) selected for high levels of toxic mustard glycosides in the wild mustard, *Brassica rapa*. He did this by employing the classic procedures of artificial selection. He then compared the ability of these high-level defense plants to tolerate herbivory of specific levels. Compared to control plants, which were not selected for resistance, and plants that were selected for low levels of glycosides, the plants with high levels of glycosides showed greater declines in fitness with increasing herbivory levels (Figure 13.11B). These experiments demonstrate that resistance comes with a cost—reduced tolerance.

Herbivory may have effects on the entire herbivore community, because many plants will produce defensive chemicals in response to herbivore damage. Robert Denno and colleagues (2000) have studied this phenomenon in two species of planthoppers, *Prokelisia marginata* and *P. dolus*.

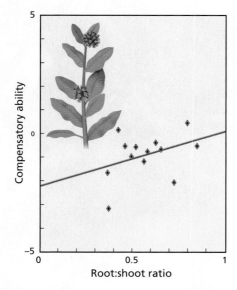

FIGURE 13.11A The compensatory ability of *Asclepias syriaca* becomes greater as more resources are stored in roots; for example, the root-to-shoot ratio becomes greater.

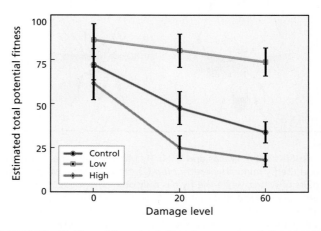

FIGURE 13.11B The wild mustard, *Brassica rapa*, can defend itself against herbivory by producing toxic mustard-oil glycosides, or glucosinolates. Plants selected for high levels of chemical defense show a greater reduction in fitness than do either controls or lines selected for low levels of defense, when they are actually damaged by herbivory. This study shows that high levels of defense have a cost in reduced tolerance.

The development time and adult size of female *P. marginata* were determined in three experimental treatments (Figure 13.11C). In one case, the control—host plant—was fresh and not previously exposed to either planthopper. An intraspecific competition treatment allowed *P. marginata* to feed first on the host plant. Then, that plant material was given to experimental *P. marginata*, whose development time and size were measured. This type of intraspecific competition increased the development time and decreased adult size (Figure 13.11C). The interspecific competition treatment allowed host plants to be fed on by *P. dolus* first, and then the plants were fed on by *P. marginata*. Interspecific competition increased development time to about the same extent that intraspecific competition did. However, adult size was more severely affected by interspecific competition. The overall results suggest that herbivory has direct effects on the plant community and indirect effects on other intraspecific and interspecific competitors. ❖

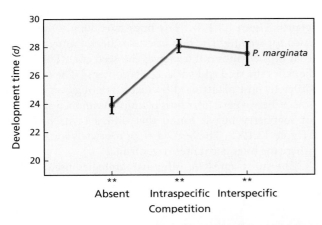
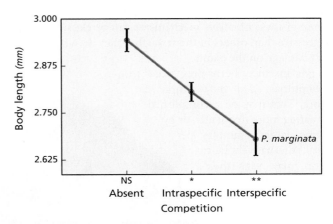

FIGURE 13.11C Development Time and Size of Female Planthoppers Following Three Experimental Competition Treatments In the *absent* treatment, planthoppers fed on fresh material. In the *intraspecific* treatment, planthoppers fed on plants previously fed upon by other members of the same species. In the *interspecific* treatment, planthoppers fed on plants previously fed on by members of a different species. Previous feeding has negative effects on both development time and body size. The effect on body size is most pronounced when there is interspecific competition.

13.12 Herbivores employ various of strategies to overcome plant defenses

Many defenses mounted by plants to ward off herbivores are quite effective. However, this does not mean that herbivores will not adapt to these defenses. In fact, we can think of herbivores as employing a variety of offensive mechanisms to exploit host plants. These offensive mechanisms may be any behavioral, physiological, or morphological trait that increases the performance and reproductive success on host plants. Some of these offensive mechanisms can be classified as less aggressive than others in the sense that they do not inflict direct damage on the plants.

For instance, herbivores faced with suboptimal food may simply eat more. This may be accomplished by eating more frequently, or by eating larger amounts per meal. Alternatively, herbivores may vary their diet to obtain their required calories and nutrition.

Slightly more aggressive mechanisms of herbivore offense include various physiological and morphological adaptations. Many of the toxic compounds produced by plants will be detoxified by enzymes that are part of the cytochrome P-450 system. The importance of these enzymes was demonstrated by Anurag Agrawal and colleagues (2002) on a species of spider mite (*Tetranychus urticae*) that feeds on a variety of host plants. Tomatoes produce a number of toxic compounds. Female spider mites raised on beans lay more eggs than do females raised on tomatoes (Figure 13.12A; compare the two control treatments). Agrawal demonstrated the importance of the P-450 enzymes by feeding the spider mites compounds that inhibited the protective P-450 enzymes, and then raising the spider mites on beans and tomatoes (Figure 13.12A, "enzymes inhibited" bars). This treatment severely reduced the number of eggs produced by females raised on tomatoes. Without these protective enzymes, spider mites would find it difficult to survive on tomatoes.

Herbivores may also sequester toxic compounds in their own cells. The herbivore then gains the benefits of these toxins for protection against their own predators. The variety of sequestered compounds is large and includes cannabinoids, cardenolides, cocaine, and mustard oils.

Herbivores may also evolve morphological traits to better exploit host plants. Soapberry bugs live on a variety of host plants (Figure 13.12B). The bugs have long, tubular beaks used for feeding. The bug inserts its beak through the outer coat of the fruit until it pierces the seed coat. The bug then liquefies the seed and sucks up its contents. The fruits of the different host plants used by soapberry bugs vary widely in size. Where a particular host plant is common, a specific race of soapberry bug is found that differs in its beak length (Figure 13.12C). The beaks are appropriately longer in areas where the bugs encounter larger fruits.

The most agressive offensive strategies used by herbivores result in physical damage or alteration of plant tissue. Some herbivores will secrete substances when they lay eggs on host plants that cause the plant to produce new plant tissue, called a gall (Figure 13.12D). These structures surround the developing larvae and usually provide access to nutrients within the gall. In addition, the gall may serve as protection from pathogens, predators, and parasites.

Some herbivores cut the large veins at the base of plant leaves. These veins serve as canals to transport plant-defensive chemicals. If the insect then feeds on leaf material beyond the veins, it will avoid the harmful effects of the plant's chemical defenses. ❖

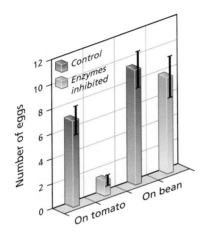

FIGURE 13.12A Number of Eggs Laid by Spider Mites Raised on Either Tomato or Bean Plants The two control treatments show that the spider mites do better on beans, because the tomato produces toxic compounds that the spider mites must detoxify. When the detoxification enzymes are inhibited, the performance of the mites is more severely affected on tomato, indicating the importance of these enzymes for using tomatoes as a host plant.

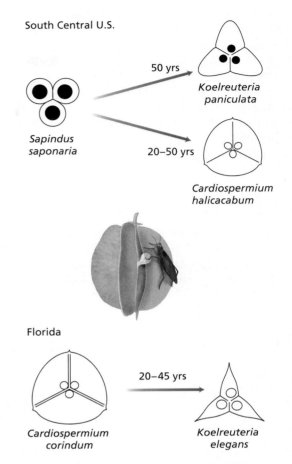

FIGURE 13.12B The soapberry bug feeds on a variety of host plants whose fruits vary widely in size. Different races of soapberry bugs are found on each of these host plants, with different-sized beaks that appear to have evolved in response to the host plant morphology.

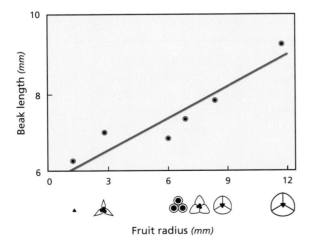

FIGURE 13.12C Beak Length of the Local Soapberry Bug as a Function of the Diameter of Fruit on Its Host Plant

FIGURE 13.12D The bumps on these maple leaves are maple bladder gall.

SUMMARY

1. Many natural populations of predators and prey show pronounced cycles.

2. One of the first and simplest models of predator-prey dynamics was developed by Vito Volterra and Alfred Lotka. This model produces cycles, but they are not stable and thus unlike the cycles of real populations.

3. The Lotka-Volterra predator-prey model can be made more realistic by including density-dependent prey growth and predator satiation.

4. The behavior of predators can be evaluated by comparing it to models of optimal behavior. Although ecologists have conflicting ideas about which properties of behavior should be optimal, two candidates are optimizing energy gain per unit of time spent foraging and minimizing time spent foraging.

5. Birds like wagtails and great tits appear to adjust their diets in a manner consistent with the maximization of energy return.

6. Prey may escape predators by avoiding any one of the stages of predation: encounter, detection, identification, approach, subjugation, or consumption.

7. Some prey sequester toxic or distasteful compounds that effectively dissuade many predators from eating them. These species are often conspicuously colored to warn predators.

   a. Different species that employ these defenses may evolve similar warning coloration. These Müllerian mimics each benefit from their effects on common predators.

   b. Some species evolve to look very similar to distasteful species, even though they lack that type of protection. These Batesian mimics thereby benefit from the protection afforded by the truly distasteful species.

8. Plants show a variety of adaptations to herbivores, including resistance and tolerance. Plants that evolve high levels of resistance may become less tolerant of herbivore damage.

9. Herbivores have a variety of offensive options to make them better at using plant resources. These mechanisms include behavioral changes, morphology, physiology, and active destruction or alteration of plant tissue.

REVIEW QUESTIONS

1. Why do the rises and declines of lynx always follow the rise and decline of hares?

2. Why is it likely that the cycles resulting from the Lotka-Volterra predator-prey model are not a complete explanation of predator-prey cycles in nature?

3. Why should the numbers of prey captured by predators eventually level off at very high prey densities?

4. Suppose a predator can choose from two prey that are equally nutritious. Prey A takes 0.9 hour to handle, and prey B takes 1 hour. The encounter rates per hour are 3 for prey A and 2 for prey B. If the predator minimizes the time spent capturing and consuming prey, should it choose a diet of just prey A, or should it choose both species? Explain your choice.

5. Give two examples of how prey avoid predators.

6. What is the difference between Müllerian and Batesian mimicry?

7. Under what conditions do we expect Müllerian and Batesian mimicry to evolve?

8. What is the difference between plant resistance to herbivores and plant tolerance of herbivores?

9. What evidence suggests that high resistance may lower tolerance?

10. What is meant by herbivore offense? Give three examples.

KEY TERMS

aposematic coloration
Batesian mimics
community
compensatory ability
cryptic coloration

functional response
handling time
herbivores
host and parasite
Lotka-Volterra predator-prey equations

Müllerian mimics
predation
prey
resistance

satiated
tolerance
waiting time

FURTHER READINGS

Agrawal, A. A., F. Vala, and M. W. Sabelis. 2002. "Induction of Preference and Performance after Acclimation to Novel Hosts in a Phytophagous Spider Mite: Adaptive Plasticity?" *American Naturalist* 159:553–65.

Davies, N. B. 1977. "Prey Selection and Social Behaviour in Wagtails (*Aves: Motacillidae*)." *Journal of Animal Ecology* 46:37–57.

Denno, R. F., M. A. Peterson, C. Gratton, J. Cheng, G. A. Langellotto, A. F. Huberty, and D. L. Finke. 2000. "Feeding-Induced Changes in Plant Quality Mediate Interspecific Competition between Sap-Feeding Herbivores." *Ecology* 81:1814–27.

Endler, J. A. 1991. "Interactions between Predators and Prey." In *Behavioural Ecology*, edited by J. R. Krebs and N. B. Davies, 3rd ed., Oxford, UK: Blackwell.

Hochwender, C. G., R. J. Marquis, and K. A. Stowe. 2000. "The Potential for and Constraints on the Evolution of Compensatory Ability in *Asclepias syriaca*." *Oecologia* 122:361–70.

Karban, R., and A. A. Agrawal. 2002. "Herbivore Offense." *Annual Review of Ecology and Systematics* 33:641–64.

Krebs, J. R., and M. I. Avery. 1985. "Central Place Foraging in the European Bee-eater, *Merops apiaster*." *Journal of Animal Ecology* 54:459–72.

Krebs, J. R., J. T. Erichsen, M. I. Webber, and E. L. Charnov. 1977. "Optimal Prey Selection in the Great Tit (*Parus major*)." *Animal Behavior* 25:30–38.

Luckinbill, L. S. 1973. "Coexistence in Laboratory Populations of *Paramecium aurelia* and *Didinium nasutum*." *Ecology* 54:1320–27.

Roughgarden, J. 1998. *Primer of Ecological Theory*, chapter 2. Upper Saddle River, NJ: Prentice Hall.

Stowe, K. A. 1998. "Experimental Evolution of Resistance in *Brassica rapa*: Correlated Response of Tolerance in Lines Selected for Glucosinolate Content." *Evolution* 52:703–12.

Stowe, K. A., R. J. Marquis, C. G. Hochwender, and E. L. Simms. 2000. "The Evolutionary Ecology of Tolerance to Consumer Damage." *Annual Review of Ecology and Systematics* 31:565–95.

Wickler, Wolfgang. 1968. *Mimicry*. New York: McGraw-Hill.

The parasitoid from the movie Alien

14

Parasitism and Mutualism

Biological organisms can both help and harm each other. In this chapter we will explore some of these relationships. On one hand are a vast array of organisms called *parasites*. These organisms complete all or part of their life cycle within or on another organism called a *host*, which suffers some reduction in survival or fertility as a result.

The imaginary parasites in the classic movie *Alien* completed their development within a human host, with results that many of us consider worse than death. One unusual feature about the parasites in *Alien* was that they had no prior history with human hosts, at least not in the first movie. In reality, a fascinating aspect of the biology of parasites is the extreme specialization that is the normal accompaniment of their successful development in particular hosts. As you might expect, hosts often have adaptations that make life difficult for parasites. This type of reciprocal evolutionary change in interacting species is sometimes called *coevolution*. We look at host-parasite interactions in the first part of this chapter, beginning with Module 14.1. We add coevolution to our discussion beginning with Module 14.4.

On the other hand, not all interactions between species are negative. Some species provide benefits to one another. These beneficial interactions are called *mutualisms*. Because mutualisms involve interactions between different species, it is important to understand how such interactions can evolve. If one species provides a behavior or resource that benefits another species, but incurs a cost for doing so, there must be a high likelihood of getting something in return. Otherwise, natural selection would quickly weed out such traits. We find that only under particular ecological conditions do we see mutualistic interactions evolving. Nevertheless, mutualisms present some of the most interesting, complex, and important types of ecological interactions. Beginning with Module 14.6, we consider mutualistic interactions. ❖

HOST-PARASITE INTERACTIONS

14.1 The specialized life cycle of parasites makes them useful for controlling certain pest species

A **parasite** is any organism that feeds in or on another individual organism and is dependent on that organism to complete its development. The organism that the parasite feeds on is called a **host**. The fitness effects of the parasite on the host are negative and certainly may result in death. (The difference between a parasite and an *herbivore* is that the herbivore usually feeds on many plants in the course of its lifetime.) Some parasites are very small, such as bacteria and viruses; others, such as worms, flies and fungi, may be quite large.

Small parasites may be transferred from one host to another directly. We are all familiar with many human diseases, from HIV to the common cold, that are typically transferred directly from one person to another. Small parasites may also be transferred via an intermediate species called a **vector**. Vectors are not usually adversely affected by the parasite, but have a life cycle that ensures transfer to a suitable host. For example, malaria, a severe disease caused by the protozoan *Plasmodium*, employs mosquito vectors to move from one host to another. The mosquito is relatively unaffected by this process.

Life Cycle of a Parasite   Many parasites have specialized and complex life cycles. There may be more than one host species. The host in which the parasite reproduces is called the **definitive host**, while other hosts are called **intermediate hosts**. Often parasites not only have specific species they use as definitive and intermediate hosts, but they also specialize on certain parts of the host organisms for feeding and development. An important question we address in this chapter is: Why have parasites evolved such extreme specialization?

Some insects use other arthropods, usually insects, for the development of their larvae. These insect parasites are called **parasitoids**. Their larvae usually develop within the body of the host. The successful development of the larvae always results in the death of the host (Figure 14.1A). This characteristic makes parasitoids valuable as a means of controlling insect pests. For example, the oriental fruit moth (*Grapholitha molesta*) is a pest of several deciduous fruit crops, such as peaches. The braconid wasp, *Macrocentrus ancylivrous*, is a parasitoid that has been used to effectively control the oriental fruit moth.

There are other examples of using biological species to control pest species. These techniques for pest control are called **biological control**. In this chapter we will examine another example of biological control—the use of a virus to control rabbits that had become pests in Australia. There has been great interest in biological control, which was used in California as far back as 1889 to control the cottony-cushion scale that threatened to destroy the state's citrus industry. The benefits of biological control

FIGURE 14.1A A Moth Caterpillar with Emerging Parasitoid Pupae

relative to pesticides are their reduced cost and lack of toxic environmental effects.

Effects of Parasites on Population Dynamics

Can parasites affect the population dynamics of their hosts? In theory, they certainly can. In this chapter we will review some of the requirements for host-parasite coexistence.

There is also experimental evidence for the impact of parasites on their host population dynamics. For example, red grouse are popular game birds in England. As a result, the numbers of red grouse bagged by hunters have been recorded for some time. These bag numbers are thought to be closely related to the total numbers of birds in the population. The numbers are often highly variable from year to year, as the graph in Figure 14.1B shows. Red grouse are often host to a parasitic nematode that is thought to reduce their fertility. In an experiment, on two different occasions scientists treated grouse in one population with drugs that kill the nematodes. The result of reducing the parasite burden was a marked decline in the magnitude of population fluctuations, as Figure 14.1C shows. This change directly implicates parasites in the population dynamics of these birds. ❖

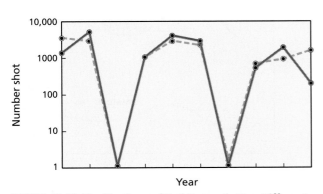

FIGURE 14.1B **The Numbers of Red Grouse in Two Different Control Populations**

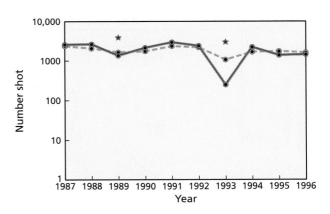

FIGURE 14.1C **The Numbers of Red Grouse in Two Experimental Populations** Antiparasitic drugs were administered to the birds in the two years marked with asterisks.

14.2 Parasitoids cannot be too effective at finding hosts if they are to avoid extinction

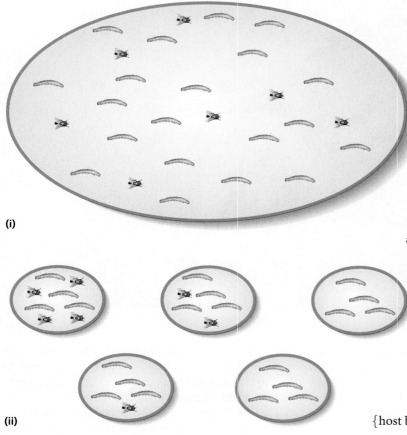

(i)

(ii)

FIGURE 14.2A The hosts and parasitoids are evenly distributed throughout the environment (i). The mean number of hosts that are attacked is equal to aP. The chance that a single host avoids attack is equal to e^{-aP}. If parasitoids show a clumped distribution (ii), then, even though the mean number of hosts attacked is unchanged, the chance of avoiding attack is equal to $[1 + (aP/k)]^{-k}$.

In many ways the dynamics of hosts and parasitoids are similar to those of predator and prey. However, the relationship between parasitoid reproduction and host death is more direct. One of the first models of host-parasitoid interactions was developed by Nicholson and Bailey in 1935.

This particular model was very simple; it assumed that there was no density dependence in the population growth of either the host or the parasitoid. Each parasitoid was assumed to have an identical chance of attacking and laying eggs in a host. If we let that probability be a, and the number of parasitoids be P, then the total number of attacks is simply aP. In some cases a single host might be attacked multiple times, and some lucky hosts would not be attacked at all. Because the larvae of a parasitoid generally kill the host, the Nicholson-Bailey model assumed that all attacked hosts die and give rise to new parasitoids. It further assumed that only the unattacked hosts are able to reproduce.

We can develop a simple model of the number of new hosts and parasitoids in each generation as follows:

$$\{\text{number of new hosts}\} = \{\text{host birthrate}\} \times \{\text{number of hosts not attacked}\}$$

$$\{\text{number of new parasitoids}\} = \{\text{parasitoid birthrate}\} \times \{\text{number of hosts attacked}\}$$

In the original formulation of the Nicholson-Bailey model, the parasitoids and hosts were assumed to be uniformly distributed in the environment, as in part (i) of

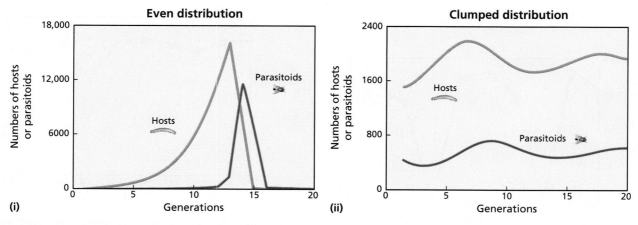

FIGURE 14.2B When the parasitoids are evenly distributed, coexistence is not possible (i). However, when the parasitoids showed a clumped distribution enough hosts avoid parasitism that coexistence is now possible (ii).

Figure 14.2A. With these assumptions, the hosts and parasitoids can never coexist, because either the parasite or both the host and parasite eventually go extinct, as in part (i) of Figure 14.2B. These predictions cannot be correct, because many hosts and parasitoids are found to coexist in nature for long periods of time.

The Nicholson-Bailey model can predict stable coexistence with small modifications. Suppose that the parasitoids are not evenly distributed in the environment, but instead occur in patches as shown in part (ii) of Figure 14.2A. In some patches there are many parasitoids, and in others very few or none. The degree of patchiness is measured by a parameter we call k. If k is less than infinity (∞), then there is some patchiness. If $k = \infty$, then the parasitoids are evenly distributed.

The net effect of patchiness is that there are always more hosts that avoid parasitism than is the case when parasitoids are evenly distributed. For instance, if $aP = 0.5$, then the chance of a host avoiding attack when parasitoids are evenly distributed is 0.6. If parasitoids are clumped ($k = 0.5$), then the chance of avoiding attack is higher: 0.7. It is as if the parasitoids have been unable to find some of the hosts [part (ii) of Figure 14.2A]. Sufficiently high levels of patchiness prevent the host population size from being driven to very low levels and thus going extinct [part (ii) of Figure 14.2B]. In fact, if the patchiness parameter (k) is less than one, then both host and parasitoid can stably coexist.

Is the distribution of parasitoids patchy in nature? In some cases it is. For example, consider the distribution of the parasitoid *Cyzenis albicans*, which often attacks winter moths. Figure 14.2C shows the actual number of parasitoid larvae per host (dots) compared with the number expected if the parasitoids were distributed evenly (green bars) or in a clumped fashion (red bars). The actual observations are more consistent with a clumped distribution (red bars). In addition, the value of the parameter k for these data is 0.6; this value is consistent with stable coexistence, which can occur with clumped distributions. ❖

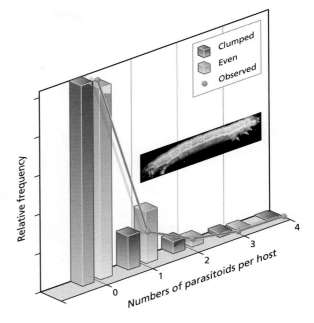

FIGURE 14.2C The number of larvae of the parasitoid *Cyzenis albicans* found in winter moth hosts. Most hosts have no parasitoids, and a few have up to four. The observed number of parasitoids per host is closer to the predicted values of the clumped distribution than it is to the even distribution.

Parasites are often very specialized in their feeding habits and life cycles, to match those of their hosts

A characteristic of many parasites is a complicated life cycle that is closely tied to the life cycles of one or more hosts. As an example consider trematodes, a class of parasitic flatworms from the phylum Platyhelminthes. These worms often have elaborate mouth morphologies for attaching to their *definitive* host, the host in which they reproduce. The *intermediate* hosts need to be organisms that are common in the habitat where the definitive host is usually found. Likewise,

the intermediate host must permit the trematode to find its way back into the definitive host. This transfer can be accomplished if the intermediate host serves as food for the definitive host, or if the intermediate host is often near food the definitive host will ingest, as Figure 14.3A shows.

The type of parasite life cycle shown in Figure 14.3A is considered highly specialized because the parasite interacts with a very limited number of species. Many nonparasitic organisms

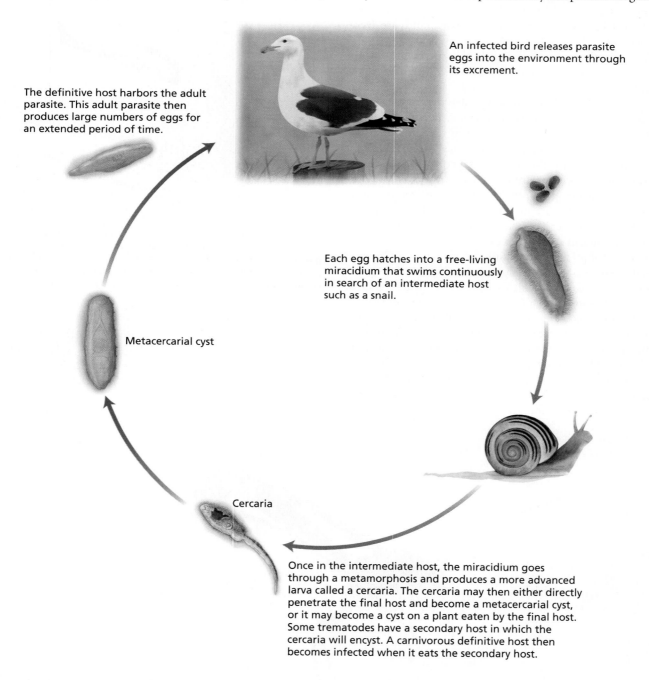

The definitive host harbors the adult parasite. This adult parasite then produces large numbers of eggs for an extended period of time.

An infected bird releases parasite eggs into the environment through its excrement.

Each egg hatches into a free-living miracidium that swims continuously in search of an intermediate host such as a snail.

Metacercarial cyst

Cercaria

Once in the intermediate host, the miracidium goes through a metamorphosis and produces a more advanced larva called a cercaria. The cercaria may then either directly penetrate the final host and become a metacercarial cyst, or it may become a cyst on a plant eaten by the final host. Some trematodes have a secondary host in which the cercaria will encyst. A carnivorous definitive host then becomes infected when it eats the secondary host.

FIGURE 14.3A The Life Cycle of a Trematode Parasite

have less specialized life cycles.
For instance, many seed-eating birds may feed on a variety of seeds from different plants. Carnivorous mammals may eat many different small mammals or birds. Is there something special about parasites that leads to the extreme specialization we often see? The answer lies in the parasitic life cycle. The fact that the parasite completes all or a major part of its development within another organism makes it more likely that parasites will be specialized compared to carnivores or herbivores.

For example, to complete its life cycle, a parasite must clear a number of hurdles. First, many internal and external parasites must have an effective means of attaching to their hosts for long periods of time. Parasites often have elaborate mouth parts or other morphologies for accomplishing this attachment. Some closely related parasites are so specialized that they are unable to attach to the host of their close relatives.

The parasite also needs to be able to withstand the host's defensive responses. Animals have elaborate immune system responses that the parasite needs to withstand. Plants also mount chemical defenses when attacked. Plant parasites eventually have to cope with these chemical defenses. Herbivores, on the other hand, may simply nibble on a plant and then move on to a different part of the same plant or a new plant altogether, thereby avoiding any chemical defenses that take time to build up.

Finally, parasites may also have to contend with enemies. These enemies may be predators, competitors, or parasitoids. Some hosts may make the parasite more vulnerable to natural enemies than other hosts.

It is probably very difficult for any single parasite to clear all these hurdles by attaching itself to any more than a few host species. Recent research has in fact suggested that many parasite species that had formerly been thought to be generalists actually consist of genetically differentiated populations that are themselves specialists. For example, mallards in England have a trematode parasite, *Echinoparyphium recuvatum*. This parasite was thought to be capable of developing on two different species of snails as intermediate hosts—*Lymnaea peregra* and *Valvata piscinalis*. It now appears that there are two morphologically identical but different species of trematode, one that uses *L. peregra* exclusively as its intermediate host and another that uses *V. piscinalis*. Neither of these sibling trematode species can develop on the other's secondary host. Furthermore, as adults these parasites distribute themselves to different locations in the mallards' intestines, as shown in Figure 14.3B. ❖

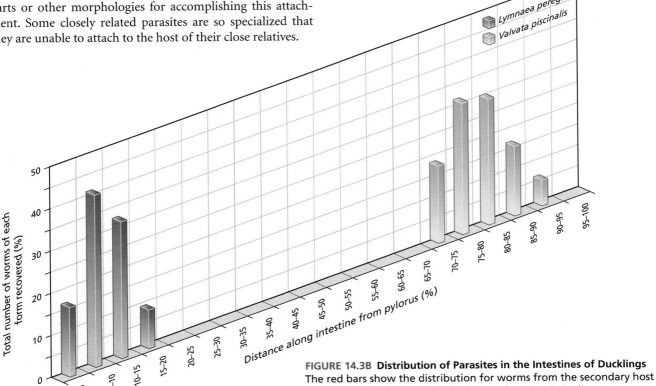

FIGURE 14.3B Distribution of Parasites in the Intestines of Ducklings
The red bars show the distribution for worms from the secondary host *L. peregra*. The green bars show worms from *V. piscinalis*.

As hosts evolve genetic resistance to parasites, the parasites evolve means of overcoming this resistance

The stakes are very high for both parasites and hosts. On one hand, a host may suffer severe reductions in fitness or death when infected by parasites. The parasites, on the other hand, must successfully complete their life cycles in a host if they are to survive. We would expect that evolution would favor any host genotypes that resist attacks by parasites, while parasites would be strongly selected to overcome these host defenses. There is good experimental evidence suggesting that hosts and parasites have evolved elaborate genetic systems that function in these ways.

One such genetic system is the **gene-for-gene system** of host-parasite resistance. Under this system, many genes in the parasite genome might be used by hosts as a means of detecting parasites and destroying them. For instance, each of these genes code for a protein that is involved in the production of some chemical compound that is needed for the parasite physiology but is recognizable by the host as foreign. We will call the alleles of these genes **avirulence alleles**, since they provide the host with a means of controlling the parasite. In

Figure 14.4A, step 1, the avirulent allele at locus 1 is designated V_1, at locus 2, V_2, and so on.

Likewise, suppose that the host has a defense system corresponding to each of the parasite genes. The host is resistant to parasite attack if it has a **resistance allele** (R) corresponding to any one of the parasite's avirulence alleles (V). Thus, a host with the R_1 resistance allele would be resistant to a V_1 parasite, but an r_1 host would be susceptible to a V_1 parasite. If the host is resistant, then the parasite may evolve a **virulence allele** (v) at the appropriate locus. For instance, in Figure 14.4A, step 3, we see a v_1 virulence allele appear in the parasite population, effectively overcoming the resistance of the R_1r_2 host. At this point, the host must then mount its defense by recognizing the parasite at a different avirulent locus, such as V_2 (Figure 14.4A, step 4).

It might seem that the best strategy for a parasite would be to evolve virulence alleles at all relevant loci. Yet this does not appear to happen. Why not? Apparently, maintaining these virulence alleles has a fitness cost that puts a pathogen at a disadvantage if it maintains more alleles than it needs to. Suppose that one pathogen had virulence alleles at three different loci, but the local host could be infected if the pathogen had a virulence allele at only one locus. If a second pathogen genotype appeared with just the single needed virulence allele, then this new genotype could infect the host and also maintain a competitive advantage over the more virulent pathogen.

One example of the gene-for-gene model of host-parasite interactions is the relationship between the wild flax plant and a fungal rust pathogen. One study of natural populations in Australia found 8 different pathogen genotypes and 15 different host genotypes. The most virulent of the rust pathogens had one of the most restricted distributions, as Figure 14.4B shows. This limited distribution shows that virulence is not equivalent to success among pathogens. Likewise, as Figure 14.4B also shows, the host plants often had resistance only to a subset of all possible pathogenic rusts, and one host that was susceptible to all pathogens was quite common in some areas. ❖

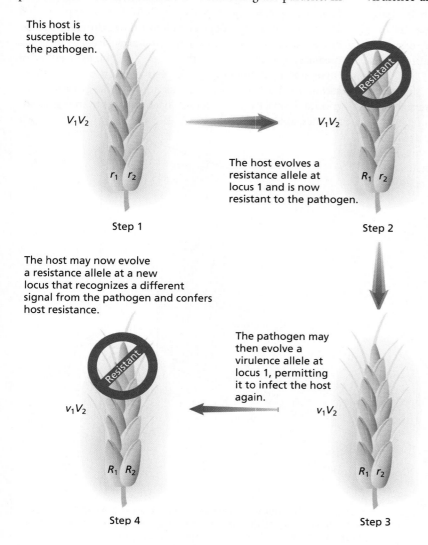

This host is susceptible to the pathogen.

V_1V_2

r_1 r_2

Step 1

The host evolves a resistance allele at locus 1 and is now resistant to the pathogen.

Resistant

V_1V_2

R_1 r_2

Step 2

The pathogen may then evolve a virulence allele at locus 1, permitting it to infect the host again.

v_1V_2

R_1 r_2

Step 3

The host may now evolve a resistance allele at a new locus that recognizes a different signal from the pathogen and confers host resistance.

Resistant

v_1V_2

R_1 R_2

Step 4

FIGURE 14.4A Gene-for-Gene System of Host Parasite Resistance

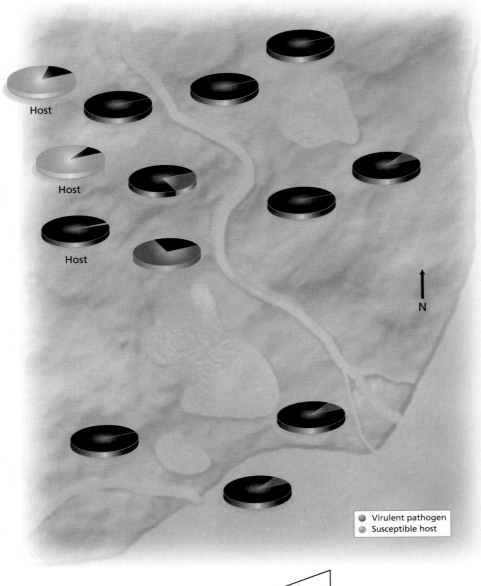

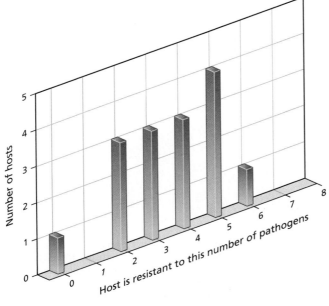

FIGURE 14.4B **Distribution of the Most Virulent Rust and Its Host (above)** The figure above shows the distribution of virulent rust fungus and susceptible host (wild flax) in a 1560-square-kilometer area of New South Wales, Australia. All pie diagrams indicate the fraction of the population that are virulent pathogens except for the three labelled Host. These show the frequency of the most susceptible host in those localities. The graph on the left shows the frequency of resistance among the hosts. Most hosts are resistant to only 2–5 of the 8 common rusts.

14.5 The coevolution of hosts and parasites also depends on ecological factors

When the European rabbit was introduced to the Australian continent, it became a major pest. The lack of any serious competitors or predators allowed the rabbit population to become so large that it depleted resources on grazing lands for sheep and cattle. In an attempt to control the rabbit population, in 1950 scientists introduced the myxoma virus, the cause of myxomatosis disease in rabbits. The virus is naturally found in populations of South American rabbits and produces only a mild disease in them. However, in European rabbits myxomatosis is often fatal. In Australia the main vector for disease transmission is the mosquito. If a mosquito bites an infected rabbit, then the mosquito may carry the disease to the next rabbit it bites.

The ease of disease transmission depends not only on the vector but also on how virulent the disease is. For instance, if the disease is very virulent and results in rapid death of the host, then there may be little opportunity for the infected rabbit to transmit the disease, as part 1 of Figure 14.5A shows. Although many rabbits may die, the virus with high virulence may also die out with them. On the other hand, if the virus produces only mild effects and is rapidly controlled by the host immune system, then the levels of virus in the rabbits' blood may be too low to effectively transmit the disease, as Figure 14.5A, part 3 also shows. Viruses with intermediate levels of virulence may be the most successful. These viruses can multiply to high levels in the host bloodstream, and they persist for a prolonged period because the host does not usually die quickly. This set of conditions increases the chances of successful transmission of the virus to a new host. These factors ultimately favor the evolution of intermediate levels of virus virulence. In fact, samples of virus from the field in Australia show exactly this pattern (Figure 14.5B).

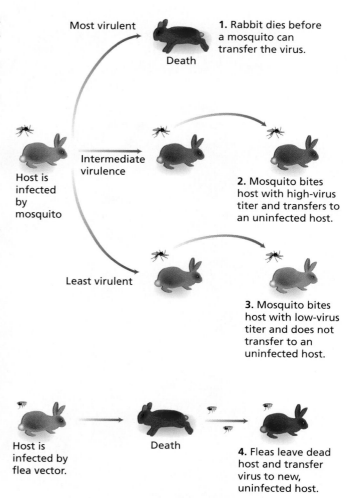

1. Rabbit dies before a mosquito can transfer the virus.

2. Mosquito bites host with high-virus titer and transfers to an uninfected host.

3. Mosquito bites host with low-virus titer and does not transfer to an uninfected host.

4. Fleas leave dead host and transfer virus to new, uninfected host.

FIGURE 14.5A **Effects of Parasite Virulence on Disease Transmission**

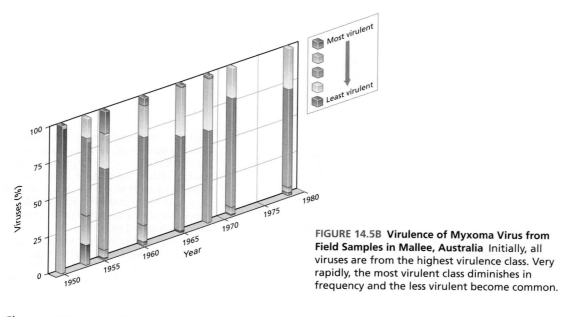

FIGURE 14.5B **Virulence of Myxoma Virus from Field Samples in Mallee, Australia** Initially, all viruses are from the highest virulence class. Very rapidly, the most virulent class diminishes in frequency and the less virulent become common.

At about the same time that the myxoma virus was introduced to Australia, it was also introduced to France and England. In England, fleas rather than mosquitoes are the major disease vector. Fleas tend to stay on their host for prolonged periods of time. If a host rabbit dies from myxomatosis, the resident fleas may then move onto a new host, as shown in part 4 of Figure 14.5A. The life cycle of the flea facilitates the transmission of virus with high virulence. As a result, the level of virulence of the myxoma virus in England is greater than in Australia.

However, the hosts are also evolving. There is, of course, strong selection for increased resistance to the virus as rabbits with low resistance die off. We see in Figure 14.5C a dramatic decline in the virus-related mortality rates among rabbits in England over time. ❖

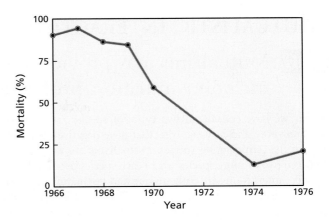

FIGURE 14.5C Percent Mortality of Rabbits in Norfolk, England, Due to a Standard Strain of Myxoma Virus

MUTUALISTIC INTERACTIONS

14.6 Mutualisms may provide several benefits to participating species, including nutrition, protection, and transportation

So far we have considered two types of species interactions, predator-prey and host-parasite, that have negative effects on one of the participating species. **Mutualisms** are interactions between two or more species where each species benefits. Some of these interactions are truly unique and fascinating. Because mutualisms involve genetically isolated species, it is sometimes difficult to understand why one species would evolve behaviors or morphological structures to help another species. To address this question, we will review some of the mechanisms that are thought to be important in the evolution of mutualisms.

In some interactions between species, only one species receives benefits while the second is not affected positively or negatively. These interactions are called **commensalisms**.

Another term that is frequently used is *symbiosis*. This word has been used in many different ways, so we need to define it carefully. We use **symbiosis** to mean a long-term, intimate association between two species. With this definition symbiotic relationships may be mutualistic, parasitic, or commensalistic.

Types of Mutualisms Most mutualisms fall into one of three categories. **Transportation mutualisms** are interactions in which one member of the mutualism has gametes or individuals transported by the other mutualist. **Nutrition mutualisms** involve the exchange of nutrients. These nutrients may be carbon sources or some limiting nutrient for growth, like nitrogen. Some species attack or remove competitors or predators that impinge on another species. These relationships are called **protection mutualisms**. Some species interactions may fall into more than one of these categories. For instance, bees transport plant pollen in exchange for nectar. This type of mutualism would be a combination of transportation and nutrition mutualism.

Do Some Species Exploit Mutualisms? To understand mutualisms, we also need to understand under what conditions species will exploit mutualisms. If a plant produces nectar, an exploiting insect would gather nectar without transferring pollen. Exploiters are able to gain the benefits of the mutualism without incurring any of the costs. We can find examples of potential exploitation in all the major categories of mutualism. For instance, cleaner fish scour the surface of larger fish

FIGURE 14.6A A Small Cleaner Fish Browsing

for parasites (Figure 14.6A). This is an example of a protection mutualism from the large fish's perspective. On occasion, the cleaner fish feed on host tissue. The host fish sometimes consumes the cleaner fish.

Transportation mutualisms can be subverted by plants that mimic nectar-producing plants, but actually supply no nectar to their visitors. Many orchids (Figure 14.6B) have nectarless flowers. Nectar robbers (Figure 14.6C) are insects that chew through the corolla of plants and take nectar without pollinating the flower. ❖

FIGURE 14.6B Orchids rely on insect pollination, but often do not provide nectar rewards.

FIGURE 14.6C Nectar robbers eat through the flowers of plants and take nectar without pollinating the plant.

14.7 Mutualisms may involve the reciprocal exchange of essential nutrients

Some of the most important and widespread mutualisms involve the exchange of nutrients. One of the most important is the association between plants in the pea family, the "legumes," and bacteria in the genus *Rhizobium* (Figure 14.7A). The bacteria live in the soil and, through biochemical manipulation, make the legumes produce nodules on their roots, where the bacteria can live. In these nodules, the bacteria receive protection and carbohydrate. In return, the bacteria take nitrogen from the atmosphere and convert it into ammonia, which the plant uses as a source of nitrogen. Because nitrogen is often a limiting resource for plants, the ability to get nitrogen from the atmosphere is a great advantage for legumes.

Mycorrhizal fungi also form close associations with the roots of many plants (Figure 14.7A). Ectomycorrhizal fungi send hyphae into the plant root tissue, growing between individual root cells (Figure 14.7A). These fungi break down proteins in dead plant matter and thereby supply their host plant with nitrogen. In return, the plant supplies the fungi with carbon compounds. Arbuscular mycorrhizae are some of the oldest mutualistic associations, originating over 400 million years ago. They are found in over 80 percent of land plants. These mycorrhizae have hyphae that actually penetrate the cells of the plant root. Arbuscular mycorrhizae provide their plant host with phosphorus and receive carbon. Plants with arbuscular mycorrhizae are common in the tropics and grasslands, where phosphrorus is often in short supply.

Lichens result from a close association of fungi with either algae or cyanobacteria. This association is so critical that neither species can live on its own. While the algae and cyanobacteria provide carbon by photosynthesis, the fungi provide protection. Lichen are robust colonists of bare rock and dead wood surfaces.

One of the most interesting mutualisms is that between ants from the family *Attini* and fungi from the family *Lepiotaceae*. The ants harvest plant leaves in small disks that they chew to a pulp. They then innoculate the pulp with some of the fungus that they keep growing in underground gardens (Figure 14.7B; see white tufts in the rightmost figure). The fungus is able to digest cellulose in the plant tissue and use it for growth. The ants then harvest fungal tissue for food. This mutualism is thought to have evolved once—over 50 million years ago—and since then, many ant species derived from this common ancestor have continued to cultivate fungus

Nitrogen-fixing bacteria. Bacteria of the genus *Rhizobium* cause legumes to form nodules in their roots, where the bacteria live. The bacteria provide ammonia, a source of nitrogen, to the plants, while the plant provides sugar and protection to the bacteria.

Mycorrhizal fungi. These fungi form associations with plant roots. The fungi supply the plants with nitrogen and phosphorus, while the plant supplies the fungus with carbon.

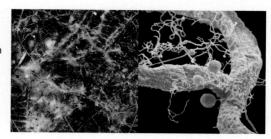

Lichens. These are associations between fungi and single-celled algae or cyanobacteria. The algae provide carbohydrate for the fungus, while the fungus provides protection. Neither the fungus nor the algae can live on their own. Lichens are abundant in the tundras of the Arctic and Antarctic.

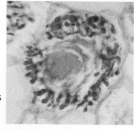

FIGURE 14.7A Nutrient Mutualisms

gardens. These gardens are sometimes invaded by a specialized parasitic fungus, *Escovopsis*, which can inhibit the growth of the favored fungi.

Scientists have long noticed a crust coating the cuticle of the *Attini* ants (Figure 14.7B, left figure). This crust is a filamentous bacterium of the genus *Streptomyces*. All species of ant that raise fungal gardens appear to harbor *Streptomyces* on their bodies. Secondary compounds produced by *Streptomyces* have antibacterial properties and are the source of many antibiotics used in medical practice. *Streptomyces* growing on the ants inhibits the growth of *Escovopsis*, but not the growth of many other fungi.

FIGURE 14.7B **Fungus Farming Ants**

Thus, it appears as if the ants use *Streptomyces* as a means of controlling the pest species of the fungus *Escovopsis*. *Streptomyces* is faithfully transmitted from parental ants to their offspring by contact, thus maintaining a close and beneficial relationship with these bacteria. ❖

14.8 Mutualisms may involve the transportation of individuals or gametes

Pollinators and the plants they visit make up one of the most conspicuous groups of mutualists. Most of us have seen bees visit flowers in the spring and summer months (Figure 14.8A). The role of the bees is to extract pollen and nectar from flowers. In the course of doing this, they transport flower pollen, sometimes over great distances, and fertilize other flowers. The plant avoids inbreeding and, if it attracts many pollinators, may fertilize large numbers of seeds. Flowers are pollinated by many different animals, including butterflies and hummingbirds (Figure 14.8A). Thus, pollinators and plants receive different but equally important benefits from mutualisms.

The transfer of pollen from one plant to another is carried out by a variety of animals including bees (left), hummingbirds (right), butterflies, and other insects. The plant is provided a means of outcrossing and gamete dispersal, while the pollinator receives a nutritional reward from the plant nectar.

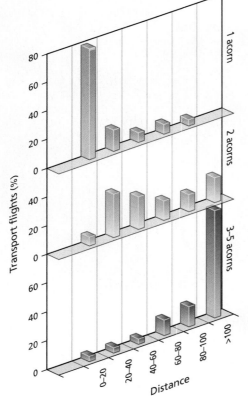

The jay (*Garrulus glandarius*) collects acorns in its mouth and buries them in the soil. The number of acorns taken on any trip varies. To carry 3–5 acorns, the bird will hold one acorn in its beak and the rest in its throat.

The graphs show that the distance the acorns are moved from the parental tree varies. The jay travels farther, on average, when it carries many acorns.

FIGURE 14.8A Transportation Mutualisms

Fertilized seeds may also be dispersed by animal mutualists. The transporting animal often uses the seeds for food. For instance, the jay (*Garrulus glandarius*) can transport up to five acorns at a time (Figure 14.8A). The birds bury these acorns until they are ready to eat them at some time in the future. Individual birds may bury 4600 acorns in a season. Remarkably, the birds appear to recover the seeds from memory, not smell. Acorns that are completely consumed are not dispersed, but many acorns will have sprouted by the time the jay comes back to them. At that point, the jay will eat the soft shoots, but the plant may continue to grow.

Plants that produce large, heavy seeds tend to drop them in the immediate vicinity of the parent tree. This will lead to high levels of competition and low reproductive success for the plant. With the aid of animals that can travel large distances, however, the seeds of a single individual can be distributed to many different locations, some of which are less likely to be as crowded as the immediate vicinity of the parent plant.

Transportation may also involve whole organisms. The carrion beetle, *Necrophorus humator,* carries many small mites on its body (Figure 14.8B). Carrion beetles work cooperatively to bury dead animals that will be used later as food for their young. Although a group of beetles may bury one dead animal, only a single pair of beetles deposits eggs on the buried carcass. The mites hop off the beetles onto the carrion and search out the eggs of the beetle's chief competitor, the fly *Calliphora*. When they find these eggs, the mites pierce their shells and eat the contents.

In experiments where mites were excluded from carrion, beetle larvae were outcompeted by the fly larvae and few beetles survived. Thus, the mites eliminate competitors for the beetles in exchange for transportation to new sites of food. ❖

Necrophorus humator

FIGURE 14.8B Carrion Beetles in Their Underground Brood Chamber

14.9 Mutualism may involve providing protection from predators or competitors

From the examples of mutualism considered so far, we see that the type of benefit gained by each participant in a mutualism may be different. The carrion beetles' mites gain transportation from the mutualism, while the beetles receive protection from a major competitor. For the beetle, this is a substantial benefit. Here we consider some additional examples of mutualisms in which one member receives some sort of protection.

Cowbirds and the oropendula birds of central Panama are part of a mutualism that includes four different species. Cowbirds use oropendulas to raise their young. This type of dependence is called **brood parasitism**. Cowbirds lay their eggs in the oropendula nest, and the oropendula then feed and care for the young cowbirds. Ordinarily oropendulas do not benefit from cowbirds in their nest, because the cowbirds take food that would otherwise go to the young oropendulas. As we might expect, in certain areas of Panama the adult oropendulas are very discriminating, and remove any strange-looking eggs. In these areas, cowbirds often produce eggs that closely mimic the coloration patterns of the oropendula eggs (Figure 14.9A).

In other areas of Panama, cowbirds do not produce mimetic eggs, and the oropendulas are very tolerant of the extra eggs added to their nests by cowbirds. What could explain these two very different behaviors in the same species of bird? It turns out that the discriminating oropendulas almost always occur in areas with large numbers of bees and wasps, while the nondiscriminating oropendulas occur in areas with few bees and wasps. Apparently the presence of bees and wasps keeps away botflies, a parasite of the oropendulas. In areas lacking bees and wasps, there are numerous botflies, and the young oropendulas are infested with these parasites. Fortunately for the oropendulas, the young cowbird chicks actively feed on the botfly larvae they find crawling on their nestmates.

Thus, in botfly-infested areas, the oropendulas improve the chances of their young surviving by raising a few cowbird chicks along with their own. Consequently, these oropendulas do not try to remove cowbird eggs from their nest. But in those areas with large numbers of bees and wasps, cowbirds do not provide benefits to the oropendulas. Instead, they are

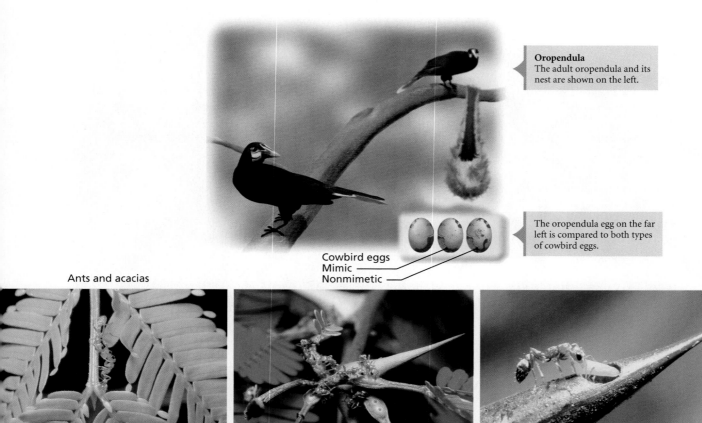

Oropendula
The adult oropendula and its nest are shown on the left.

The oropendula egg on the far left is compared to both types of cowbird eggs.

Cowbird eggs
Mimic
Nonmimetic

Ants and acacias

On the left are acacia leaves with the Beltian bodies on the tips.
On the right are nectaries at the base of the leaf.

The greatly enlarged thorns of the acacia. The interior is a pithy material that the ants excavate prior to occupying the thorn.

FIGURE 14.9A Protection Mutualisms

a drain on parental resources. In these areas, the adult oropendulas attempt to keep their nests free of cowbirds.

Acacia trees in Central America have evolved a number of morphological structures that directly benefit the ants that make their homes in the giant thorns of the acacia. In return, the ants defend the acacias against insect herbivores. The ants also prune neighboring plants that encroach on the acacias' space. At the tips of its leaves, the acacia has specialized structures, called Beltian bodies, that are protein-rich; they are used for food by the ants (Figure 14.9A). The ants also make use of sugar secretions from nectaries at the bases of acacia leaves (Figure 14.9A). But do the ants really benefit the acacia? In one experiment, ants were excluded from some of the branches of acacia trees. Over time, the antless branches were smaller and had fewer leaves than the branches with ants did.

Many plants have small hairs or depressions on the bottoms of their leaves, called domatia (Figure 14.9B). Domatia are found on over 2000 species of plants and seem to be refuges for predatory insects. This suggests that plants may produce these structures to attract insects that will consume herbivorous insects. To test this idea, cotton wool was used to make artificial domatia. Over the course of the growing season, the number of herbivorous spider mites was recorded on plants with added domatia and on those without (see controls, Figure 14.9B). The numbers of predatory bugs on these leaves were also recorded. The results showed a striking decrease in the number of herbivorous insects on the plants with added domatia. Similarly, there was an increase in predatory insects on leaves with added domatia (Figure 14.9B). These results are consistent with the notion that plants have evolved domatia to attract predators and reduce the negative effects of herbivorous insects. ❖

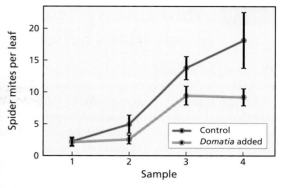

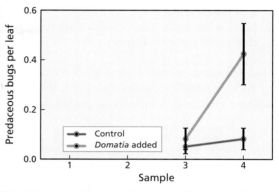

FIGURE 14.9B Adult Big-Eyed Bug on Leaf Domatia The top figure shows the number of mites that appeared on leaves on four successive sampling periods. The bottom figure shows the numbers of predatory bugs on these same leaves. These bugs did not appear until the third sampling period.

14.10 Mutualisms often evolve as a direct consequence of negative interactions between two or more species

The examples of mutualisms that we have reviewed are remarkable for their diversity of species and mechanisms. These interactions must develop due to the action of natural selection, so it would seem reasonable to suppose that natural selection has left some common thread among these mutualistic associations. In fact, at least two principles serve as common threads among these relationships:

1. Many mutualisms evolve out of initially antagonistic pairwise interactions.
2. Many other mutualisms involve three or more species, including an antagonistic pair.

Now let's consider these ideas in more detail.

We have already reviewed typical negative ecological interactions, including competition, predation, herbivory, and parasitism. Many of these interactions reduce the fitness of at least one species substantially. Thus evolution has often favored traits that help organisms avoid, or reduce the impact of, these negative interactions. The development of alleles that confer resistance to parasites is one example of this type of evolution.

In a similar fashion, as our first principle suggests, many mutualisms are thought to have evolved from an initially negative interaction as a way of reducing the negative effects on fitness. For instance, many insects consume plant spores and seeds. Some plants have reduced the impact of these insects through the evolution of floral nectaries, which provide a food source separate from the plant's pollen. Some insects have also evolved behaviors that help ensure the fertilization of plants (Figure 14.10A). Because a fertilized plant produces more seeds, this translates into more food for seed predators.

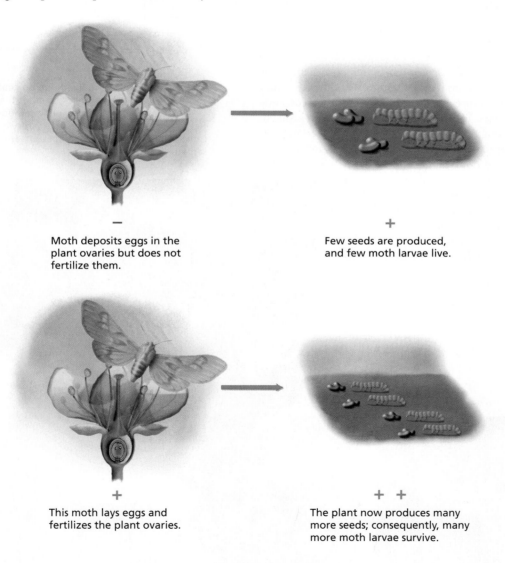

−

Moth deposits eggs in the plant ovaries but does not fertilize them.

+

Few seeds are produced, and few moth larvae live.

+

This moth lays eggs and fertilizes the plant ovaries.

+ +

The plant now produces many more seeds; consequently, many more moth larvae survive.

FIGURE 14.10A Moth-Yucca Mutualism

Often, the larvae of seed predators consume only a portion of all the seeds produced by a plant. Thus, the net impact of insect fertilization can be positive for both the insect and plant. A good example of this type of relationship is the yucca and yucca moth mutualism (Figure 14.10B).

We have already considered two examples of our second principle in this chapter. Recall that acacias benefit from their interaction with ants because of the beneficial impact of the ants on the negative effects of insect herbivores and plant competitors. The ants effectively reduce the negative impact of other species. We also saw that oropendulas can benefit from raising cowbird chicks that feed on parasitic botflies. However, the relationship between oropendulas and cowbirds can turn from mutualistic to negative if botflies are not present in the local habitat. . ❖

(i)

(ii)

FIGURE 14.10B (i) A yucca plant in bloom in Eastern Colorado. (ii) The female yucca moth visits the yucca flowers and collects pollen, which it rolls into a ball. The female then visits another yucca flower, where it drills a hole in the ovary wall of the flower and lays its eggs. The female moth then uses the pollen ball to fertilize the flower. The yucca is effectively fertilized, and the moth larvae are guaranteed a good source of seeds.

14.11 The evolution of mutualisms should be facilitated when the reproduction of host and symbiont coincides

If mutualisms are beneficial to both participants, shouldn't they always evolve when the opportunity presents itself? The answer is not necessarily. Here we consider a theory about the evolution of cooperation (mutualism) between two organisms that formally have a host-parasite relationship. The interesting aspect of this theory is that, depending on the mode of reproduction, cooperation may or may not evolve.

In Figure 14.11A we show a simple host-parasite system that initially consists of a genetically variable parasite population, indicated by different-colored cells. Two modes of reproduction are considered. With indirect transmission, the parasite infects the host and reproduces or feeds off the host and then leaves. Once free of the initial host, the parasite is free to infect any other suitable host and begin the cycle of growth and reproduction (Figure 14.11A). Direct transmission implies that the reproduction of the parasite is accomplished simultaneously with the reproduction

How does the symbiont evolve? The answer will depend on the mode of reproduction.

of the host. So all daughters of the host contain replicates or progeny of the original parasite (Figure 14.11A).

We next consider the fitness consequences on host and parasite when the parasite cooperates with the host. By cooperate, we mean that the parasite does not kill or debilitate the host and may do favorable things to its host, like produce a source of nitrogen. For the host, cooperation means not trying to kill the symbiont and providing resources that the symbiont may require for its reproduction. In Figure 14.11B, part (i), we show the evolution (arrows) of the host strategy (cooperate or attack). If the symbiont cooperates, then it will also benefit the host to cooperate and continue to reap these benefits from the symbiont. In this case, the host fitness (green bar) is greater when it cooperates than when it attacks (red bar). Thus, evolution improves fitness when cooperating genotypes of the host become more common. When the symbiont attacks, the host must defend itself. In

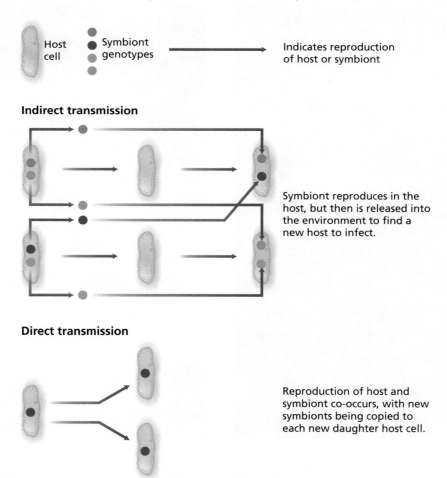

Indirect transmission

Symbiont reproduces in the host, but then is released into the environment to find a new host to infect.

Direct transmission

Reproduction of host and symbiont co-occurs, with new symbionts being copied to each new daughter host cell.

Host cell Symbiont genotypes → Indicates reproduction of host or symbiont

FIGURE 14.11A **Different Modes of Symbiont Transmission**

this case evolution will force any host population that cooperates to evolve an attacking phenotype.

How does the symbiont evolve? The answer will depend on the mode of reproduction. Let's first consider indirect transmission, as shown in part (ii) of Figure 14.11B. In this case the symbiont uses a host cell to grow and reproduce temporarily, and then may move on to other hosts. Even if the symbiont cooperated with the host, a second infection from a noncooperating symbiont might kill the host anyway. Thus the symbiont can never be sure of gaining any benefits from cooperation. Because cooperation usually has a cost, the symbiont's fitness will always increase if it adopts an attack strategy, no matter what the host does.

When transmission is direct, cooperation is favored; see part (iii) of Figure 14.11B. In this case, cooperation by the symbiont improves the host's chances of survival, which then directly improve the chances of the symbiont surviving. The symbiont stands to lose only if it attacks its host. Once the symbiont has evolved a cooperative strategy, then based on the previous discussion, we expect the host to evolve cooperation as well. As we will see next, these theories are amenable to experimental tests. ❖

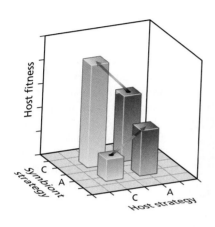

(i) Host fitness. The arrows show the direction of evolutionary change. The host and symbiont can assume a strategy to cooperate (C) or to attack (A). If the symbiont cooperates, it will be in the best interests of the host to reap these and additional benefits by cooperating. If the symbiont attacks, then the host must attack to prevent additional damage.

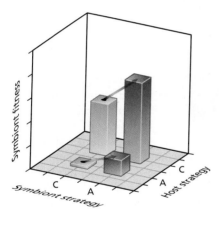

(ii) Symbiont fitness with indirect transmission. Because each host may be occupied by multiple symbionts, cooperation by one symbiont genotype will be in vain if the host will eventually be killed by a second invading genotype. In this instance, the symbiont will evolve to attack even when the host cooperates.

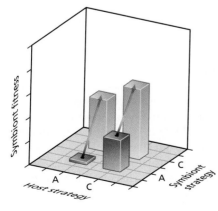

(iii) Symbiont fitness with direct transmission. In this case, the symbiont is transmitted to the daughter host directly. Thus, the fate of the symbiont is more directly related to the fate of the host. Even when the host attacks, the fitness of the symbiont will increase if it cooperates. As we saw in part **(ii)**, once the symbiont adopts a strategy to cooperate, the host will also evolve to cooperate.

FIGURE 14.11B The Evolution of Mutualism

14.12 Levels of antagonism between hosts and parasites may depend on the frequency of opportunities for horizontal transfer

We have seen that species interactions may evolve over varied levels of antagonism, at least in theory. But is there evidence for the evolution of different levels of antagonism? In fact, such evidence exists, and it comes from studies of hosts and parasites.

In one type of transmission, called **horizontal transfer**, parasites move from one host to another within the same generation. **Vertical** or **direct transmission** occurs between different generations, with parasites moving from host parents to their offspring. Investigators hypothesized that parasites that depend on vertical transmission would be much more benign than those with many opportunities for horizontal transmission. The reason is that the host needs to survive and reproduce if parasites that rely on vertical transmission are to propagate themselves. Thus, the parasite's fitness is directly related to its host's fitness with vertical transmission.

The hypothesis has been tested with a system of bacteria (*Escherichia coli*) and bacteriophages (R208). (Bacteriophages or simply phages are viruses that infect bacteria.) These phages reproduce continuously in the bacteria, and new phages can pass through the bacterial membrane without killing the bacteria. This particular phage carries a gene for resistance to the antibiotic ampicillin, which makes bacteria infected with phages resistant to ampicillin. Once infected, a bacterium cannot be reinfected by a different phage. On the other hand, these phages typically reduce the fitness of the bacteria by slowing their growth rate.

To test the impact of different types of transmission on parasite virulence, two experimental treatments were set up, and the organisms were permitted to evolve under them. The first treatment, shown in part (i) of Figure 14.12A, was called "high fidelity," because the reproductions of the phage and bacteria were closely linked. These cultures contained antibiotic along with the food medium. As a result, the only bacteria that could survive were those already infected with the phage carrying the resistance gene. Any phage that left bacterial cells could not infect other bacteria, because the other bacteria were already infected. Consequently, the only way for the phage to reproduce was by producing new bacterial cells—in other words, by vertical transmission.

In the second, "low fidelity" treatment, shown in part (ii) of Figure 14.12A, the cultures contained no antibiotic. Uninfected bacteria grew along with the resistant phage-infected bacteria. The phage could reproduce readily by horizontal transmission as well as by vertical transmission.

What were the results of these experiments? The phage-carrying bacteria that evolved under the high-fidelity conditions (which favor vertical transmission) had growth rates 7 to 40 times greater than the growth rates for the phage-carrying bacteria that evolved under low-fidelity conditions (which permit both horizontal transfer and vertical transmission). These results strongly support the notion that the deleterious impact of parasites on the host is greatly reduced when reproduction is vertical.

A second line of evidence for the evolution of different levels of antagonism between host and parasite comes from the study of 11 species of fig wasps (genus *Pegoscapus* or *Tetrapus*) and their nematode parasites (genus *Paradiplogaster*). Let's first review some relevant life history of these wasps and parasites. A fertilized female wasp will enter the part of the flower that will eventually ripen into the fig fruit. This female wasp is called a **foundress**. With pollen she has collected, she pollinates the flower, lays her own eggs, and then dies in the flower. As the fruit and seeds ripen, the wasp offspring mature, mate inside the fig, and then disperse to start the next generation. Some species of wasps have only a single foundress per fig. Other species often have two or more females laying eggs in the same fig. In a single-foundress fig, all offspring are siblings. With multiple foundresses, both full siblings and unrelated wasps develop in the same fig.

Each species of fig wasp has a distinct species of nematode parasite. Within the fig, these nematodes crawl onto newly emerged fig wasps and enter their bodies. They begin to consume each wasp's body and develop into adults. Obviously, the nematodes can have a negative impact on the reproductive capacity of their host wasp. After a female wasp has reached another fig and died, the adult nematodes emerge from her body, mate, and lay eggs that develop alongside the wasp eggs.

In single-foundress figs, transmission of parasites is entirely vertical, from parents to offspring. However, in multiple-foundress figs, nematodes have opportunities for horizontal transfer by infecting wasps that are unrelated to their host. An examination of the number of wasp offspring produced by different species of nematode-infected wasps showed that the greatest numbers of offspring occurred among wasps that usually reproduced as single foundresses, as Figure 14.12B shows. This result is consistent with the idea that parasite virulence will be reduced if parasites are largely dependent on vertical transmission. ❖

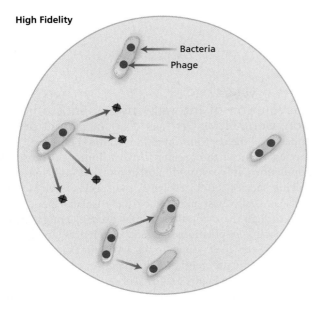

High Fidelity

Bacteria

Phage

(i) In these cultures bacteria are grown with antibiotic, so only those carrying a phage can survive. Phages that emerge from the bacteria have no hosts to infect. Thus, the only way for the phage to reproduce is via reproduction of the bacterial host.

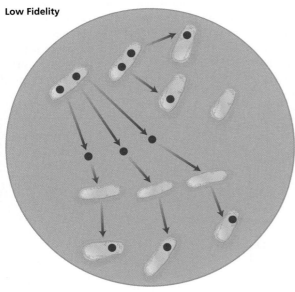

Low Fidelity

(ii) In these cultures there is no antibiotic. Phages can emerge from their host and find many other uninfected hosts. The phages can also reproduce via reproduction of their host, but they are not dependent on this mode of reproduction.

FIGURE 14.12A Experimental Control of Phage Transmission

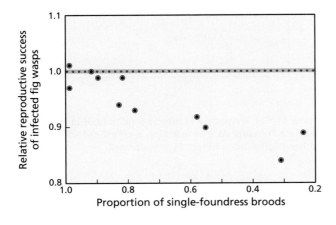

FIGURE 14.12B Reproductive Success as a Function of the Frequency of Single-Foundress Broods

THE COEVOLUTIONARY PROCESS

14.13 Coevolution is a complex process that may depend on selection, migration, and genetic drift

We can define **coevolution** as the reciprocal evolutionary change of interacting species. Paul Ehrlich and Peter Raven first used the term *coevolution* in 1964. In their paper, they discussed the evolution of interactions between plants and insect herbivores.

Coevolution and Speciation: Ehrlich and Raven noted that many plants produce chemical compounds that are toxic to many but not all insects. They suggested that when a new mutation permits a plant population to produce a toxic compound that no insect can tolerate, the plant can then expand its range into new territories and habitats. The plant population may then form a new species. Sometime later, mutations in an insect population may allow the insects to tolerate the newly evolved plant toxins, permitting the insects to follow the plants into these new adaptive zones. New insect species would eventually appear in those areas where the new plant species appeared.

This is only one way that coevolution may accompany speciation. Bacteria have often become associated with plant and animal cells and are transferred to their host's offspring through the mother's egg. Occasionally these bacteria can cause a reduction in the viability of zygotes, if both egg and sperm are not from individuals that possess these bacterial associates. This type of incompatibility can lead to the reproductive isolation and potential speciation of populations with the bacterial symbionts. For instance, some populations of the fruit fly *Drosophila simulans* are infected with a rickettsia bacterium called *Wolbachia*. These parasites are passed on from the mother to her offspring, no matter what type of male she mates with (infected or uninfected). However, uninfected females that mate with infected males show a severe decline in the viability of their zygotes. Although the sperm of the infected males does not carry the parasite, the parasite alters the sperm in some fashion that makes it incompatible with the uninfected female's egg.

The presence of the *Wolbachia* parasite was first detected in fruit-fly populations of southern California, and through the mid 1980s it spread rapidly to central and northern California (Figure 14.13A). Uninfected females would be nearly reproductively isolated if they moved to an area with a high infection rate.

In the modules that follow, we will see that many bacterial parasites and mutualists have a long coevolutionary history with their hosts. Molecular genetic markers and modern methods of phylogenetic reconstruction reveal the close associations of these bacteria and their hosts.

Coevolution of Interactions Depends on Selection, Gene Flow, Drift, and Local Extinctions To understand the coevolution of interactions between different species, we need to consider the distribution of interacting populations. It is easy to think that interactions between coevolving species should evolve at the species level. However, species usually exist as many partially isolated populations. Even though these populations may exchange migrants with neighboring populations, genetic differences are likely to exist between these populations. We saw examples of this type of genetic differentiation in the different populations of flax and rust (see Figure 14.4B). The course of evolution may change as a result of these genetic differences. Local populations may also go extinct and later be recolonized.

The process of recolonization may be important for the long-term evolution of interactions between competing species. Initially, interactions between two close competitors may lead to the local extinction of one of two competing species. However, the opportunity for recolonization of the

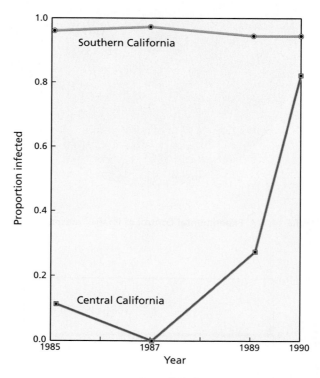

FIGURE 14.13A Frequency of Infected Individuals in Two California Populations The infection was first detected in Southern California and rapidly spread north.

habitat from neighboring populations permits continued interaction and perhaps coevolution of these competing species.

The distributions of interacting species may differ. Thus, in part of the range of each species, there may be no coevolution due to the absence of the other species. A good example that we have already discussed is the behavior of oropendulas, which is quite different in regions with bees than in regions without bees (see Module 14.9). Consequently, the coevolution of traits within a species can vary from population to population.

In addition, the physical environment may vary over a species' distribution and lead to different evolutionary outcomes. For instance, in 1954 Thomas Park showed that in hot-moist environments the flour beetle, *Tribolium castaneum*, was competitively superior to its close relative *T. confusum*, but in cold-dry environments *T. confusum* was always competitively superior. It is reasonable to expect that the types of interactions that evolve may be greatly influenced by such variation in the physical environment.

If this geographic view of coevolution is valid, we ought to see examples of local populations within a species that have specific adaptations that permit interactions with a local second species. One example is bees of the genus *Rediviva* and plants of the genus *Diascia* that the bees pollinate in southern Africa. One bee species, *Rediviva neliana*, is widespread and appears to be the primary pollinator of 12 different species of *Diascia*. However, local populations of *R. neliana* usually occur with only one species of *Diascia*. The bees are attracted to the flowers of *Diascia* by an oil that they use for food. To retrieve the oil, the bee must insert its forelegs down the length of the flower's spur, as Figure 14.13B shows. Different species of *Diascia* have different spur lengths, and there is corresponding variation in the length of the foreleg of the local population of *R. neliana*. In fact, the correlation between foreleg length and spur length is greater than 90 percent. Thus, it appears that as a species *R. neliana* does not specialize on any single species of *Diascia*, but at a local population level there has been close coevolution between plant and pollinator. ❖

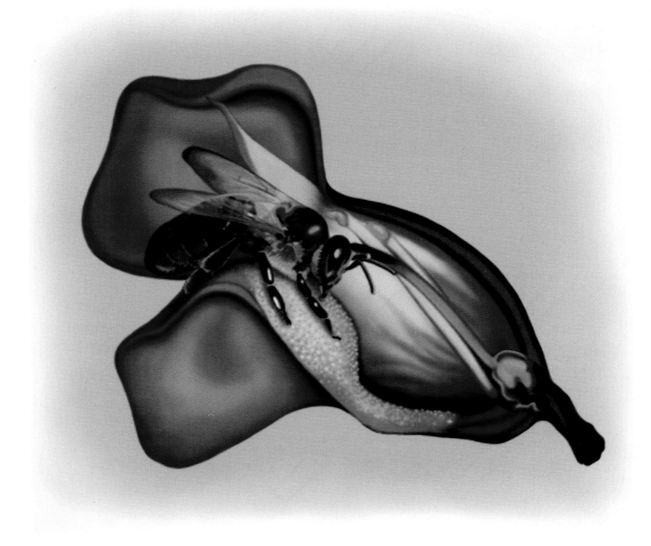

FIGURE 14.13B *Rediviva neliana* **Removing Oils from** *Diascia capsularis* The flower is cut away to show the foreleg of the bee as it is inserted into the spur of the flower.

The tight association between the life cycles of hosts and their parasites suggests that their evolution has been closely linked. If a host moves to a new habitat, the parasite will either have to adapt to these new conditions or go extinct. If these types of movements and the subsequent isolation and adaptation are sufficient to create a new host species, it is not unreasonable to suppose that the parasites might undergo similar evolutionary transitions.

These conjectures can be studied by modern techniques of phylogenetic reconstruction. Suppose that molecular genetic information is available for a group of parasites, and that information is used to construct a phylogenetic tree as described in Chapter 2. In Figure 14.14A we show five parasite species labeled by letters *a, b, c, d,* and *e*. Their four hypothetical ancestors are identified with uppercase letters F, G, H, and I.

Each of these parasites can survive on one or more hosts. The parasite tree can be used to develop numerical estimates of similarity among the hosts. As an example, parasites *d* and *e* share ancestors F, G, H, and I, so their hosts should be more similar than the hosts of parasites *e* and *a*. Using only the parasite phylogeny, a host phylogeny can be constructed and then compared to an independently derived host phylogeny. If these two host phylogenies are congruent, as in the example in part (i) of Figure 14.14A, then there would appear to be coevolution of hosts and parasites. That is, as the host speciated, the associated parasites speciated in tandem.

Of course, some parasites may recently have switched to a particular host. In such cases of recent host switching, we would not expect the phylogenies to be congruent, because the parasite is not a recent descendant of parasites from this host lineage. In

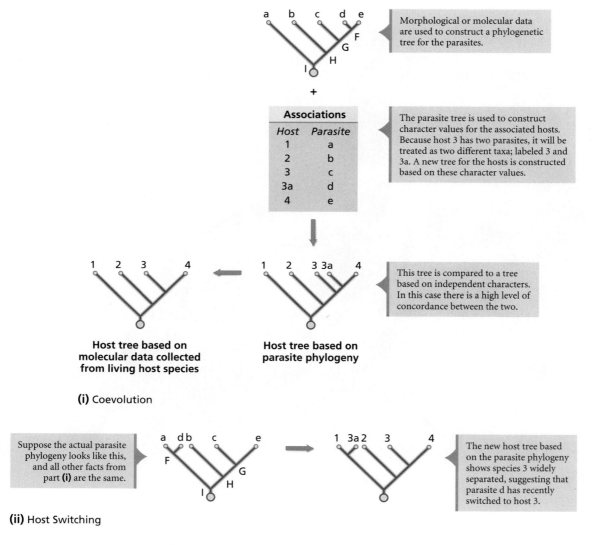

FIGURE 14.14A Phylogenies may be used to reveal patterns of host-parasite evolution.

part (ii) of Figure 14.14A, we show a different parasite phylogeny (with the position of species *d* changed) that produces a different host phylogeny. The host phylogeny has host 3 in two very different locations. Because the host labeled 3a is out of its expected position, we would infer that parasite *d* has recently switched to host 3, whereas parasite *c* has coevolved with host 3.

These methods have been used to study six primate species, as shown in Figure 14.14B: humans (H), Old World monkeys (OW), gorillas (Go), orangutans (O), chimpanzees (C), and gibbons (G) and their associated pinworm and tapeworm parasites. One possible phylogeny of these primates is shown on the right side of Figure 14.14B. The phylogeny based on the parasite phylogeny is shown on the left. Because most primates have several parasites, subscripts are used to show information from different parasites. The black lines in the parasite-derived phylogeny show connections that are congruent with the established phylogeny. The parasites that gave rise to these branches would be candidates for coevolved species. The orange lines show links that indicate possible host-switching events. ❖

FIGURE 14.14B **A Phylogeny of Primates and Their Parasites**

14.15 Coevolution of bacteria and eukaryotic hosts shows little switching between pathogenic and mutualistic lifestyles

Many bacteria have developed close associations with eukaryotic hosts. These bacterial **symbionts** live within the host's body, often within the host's cells. They may have negative or positive effects on the hosts. We have noted previously that many mutualistic interactions develop from previously antagonistic interactions. Is the same true for bacterial symbionts? An analysis of the phylogeny of many pathogenic and mutualistic bacteria (Figure 14.15A) shows that pathogens and mutualists tend to cluster. This clustering suggests that mutualists are more likely to evolve from bacteria that have already established a mutualistic interaction or at least not a pathogenic one. There would appear to be little switching between pathogenic and mutualistic lifestyles.

Why should this lack of switching be the case? We outline a possible explanation in Figure 14.15B. In the course of adapting to life within a host, many genes needed by free-living bacteria are lost. For pathogenic bacteria, this adaptation may often involve the loss of genes for biosynthetic pathways that are unnecessary for bacteria that derive their nutrition from a host. The close adaptation to the host cells may in turn create very small populations of bacteria that rarely come into contact with other bacteria. Many mutualistic bacteria are highly compartmentalized in their host cells and are transmitted maternally by their host from parent to the host offspring. These life cycles reduce the effective population size of the bacteria as well as the opportunities for recombination, making the loss of bacterial genes even more likely. After such severe specialization, the chance that a pathogenic bacterium could become mutualistic is very unlikely, since it now lacks many important genes that might be needed to benefit a host.

As an example, consider aphids, the insect pests of many plant species. Within the body cavity of many aphids is a special structure that holds bacterial symbionts from the genus *Buchnera*. These bacteria are found only in aphids and cannot grow outside of the cells of aphids. The bacteria have the ability to make the essential amino acid tryptophan, which is found in short supply in the aphids' diet. In fact, the *Buchnera* overproduce tryptophan for use by their aphid hosts.

There are many species of aphids with bacterial symbionts that can be compared in a phylogenetic analysis. These analyses show the *Buchnera* symbionts to be closely related to each other, having a common ancestor 150–250 million years ago. ❖

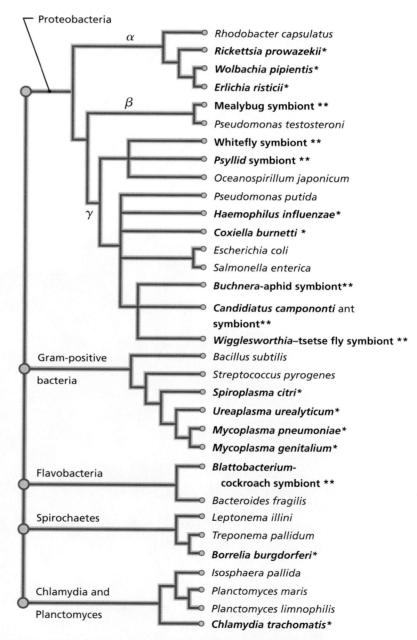

FIGURE 14.15A Pathogenic and Mutualistic Clades within the Eubacteria The single asterisk (*) identifies pathogens; the double asterisk (**) identifies mutualists. Symbiotic bacteria are in bold.

Genetically variable ancestor

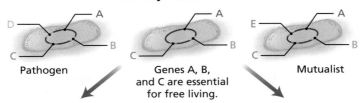

Pathogen   Genes A, B,   Mutualist
and C are essential
for free living.

Adaptation to the host environment

D gene helps
pathogen
adapt to host.

E gene helps
adaptation to
mutualism.

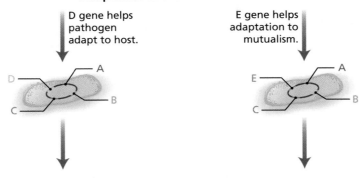

**Bacteria sequestered in special habitats
results in loss of genes and opportunities
for recombination.**

Loss of genes makes a
transition from pathogen
to mutualist unlikely.

**Pathogenic bacteria will often lose biosynthetic
pathways that are important for the development
of mutualistic interactions.**

FIGURE 14.15B Bacterial adaptation to a host environment makes the
transition from pathogen to mutualist unlikely.

SUMMARY

1. Parasites reduce the survival and fertility of their host organisms. In some instances, parasites kill their hosts.

   a. Virulent parasites can drive their hosts to extinction if they are very effective at finding hosts.

   b. Simple models of host-parasite dynamics show that when the distribution of hosts is clumped, the host-parasite populations are more likely to reach a stable population equilibrium.

2. Many parasites have complicated life cycles wherein they alternately live in a definitive host, where they reproduce, and an intermediate host.

   a. The parasite can find its way to new definitive hosts via the intermediate host.

   b. Given the reduction in fitness caused by a parasite, there is strong selection for host genotypes that can effectively attack parasites. Likewise, the parasite will experience strong

natural selection for genotypes that are able to overcome the host's defenses.

   c. Evidence of this evolutionary arms race can be found in natural populations of hosts and parasites today.

3. Not all interspecies interactions involve negative effects on one species. Mutualisms are multiple species interactions wherein all participants experience positive effects from the interactions.

4. There are three basic types of mutualism.

   a. In nutrition mutualisms, one or more species receive some nutritional benefit from the interaction.

   b. Protection mutualisms involve one species protecting a second species from competitors or predators in exchange for some reward.

   c. One species in a transportation mutualism will distribute the gametes or individuals of a second species while receiving some benefit.

5. Models of the evolution of mutualisms suggest that they are more likely to evolve when the reproduction of a symbiont occurs together with the reproduction of the host. Experiments with bacteria and phages have lent support to this view.

6. Species that have close symbiotic relationships are expected to show some concordance in patterns of speciation. Modern molecular techniques can be used to study this type of close relationship, which is observed in primates and their parasites.

REVIEW QUESTIONS

1. Why are hosts and parasitoids more likely to stably coexist when the hosts have a patchy distribution?

2. Suggest some reasons to explain why not all flax plants are resistant to all possible genotypes of rust.

3. What aspects of the life histories of mosquitoes and fleas have an important impact on their ability to be disease vectors?

4. Give an example of each of the following: (i) transportation mutualism, (ii) protection mutualism, (iii) nutrition mutualism.

5. What is the difference between direct and indirect symbiont transmission? How do these different modes of transmission affect the evolution of cooperation?

6. Explain and give examples of how coevolution may vary over the range of a species.

KEY TERMS

avirulence allele
biological control
brood parasitism
coevolution
commensalism
definitive host
direct transmission

foundress
gene-for-gene system
horizontal transfer
host
indirect transmission
intermediate host
mutualism

nutrition mutualism
parasite
parasitoid
protection mutualism
resistance allele
symbiont
symbiosis

transportation mutualism
vector
vertical transmission
virulence allele

FURTHER READINGS

Agrawal, A. A., and R. Karban. 1997. "Domatia Mediate Plant-Arthropod Mutualism." *Nature* 387:562–63.

Bronstein, J. 2001. "The Exploitation of Mutualisms." *Ecology Letters* 4:277–87.

Brooks, D. R., and D. A. McLennan. 1993. *Parascript*. Washington, DC: Smithsonian Institution Press.

Bull, J. J., I. J. Molineux, and W. R. Rice. 1991. "Selection of Benevolence in a Host-Parasite System." *Evolution* 45:875–82.

Currie, C. R., J. A. Scott, R. C. Summerbell, and D. Malloch. 1999. "Fungus-Growing Ants Use Antibiotic-Producing Bacteria to Control Garden Parasites." *Nature* 398:701–4.

Ehrlich, P. R., and P. H. Raven. 1964. "Butterflies and Plants: A Study in Coevolution." *Evolution* 18:586–608.

Herre, E. A. 1993. "Population Structure and the Evolution of Virulence in Nematode Parasites of Fig Wasps." *Science* 259:1442–45.

Hudson, P. J., A. P. Dobson, and D. Newborn. 1998. "Prevention of Population Cycles by Parasite Removal." *Science* 282:2256–58.

Jarosz, A. M., and J. J. Burdon. 1991. "Host-Pathogen Interaction in Natural Populations of *Linum marginale* and *Melampsora lini*: II.

Local and Regional Variation in Patterns of Resistance and Racial Structure." *Evolution* 45:1618–27.

May, R. M. 1978. "Host-Parasitoid Systems in Patchy Environments: A Phenomenological Model." *Journal of Animal Ecology* 47:833–43.

Moran, N. A., and J. J. Wernegreen. 2000. "Lifestyle Evolution in Symbiotic Bacteria: Insights from Genomics." *Trends in Ecology and Evolution* 15:321–26.

Nicholson, A. and V. Bailey. 1935. "The Balance of Animal Populations. Part 1." *Proceedings of the Zoological Society of London* 3:551–598.

Thompson, J. N. 1982. *Interaction and Coevolution*. New York: Wiley & Sons.

———. 1994. *The Coevolutionary Process*. Chicago: University of Chicago Press.

Turelli, M., and A. A. Hoffmann. 1991. "Rapid Spread of an Inherited Incompatibility Factor in California *Drosophila*." *Nature* 353:440–42.

The feeding relationships between species can often be complicated.

15

Communities and Ecosystems

When scientists first began studying biological communities, they were so fascinated with the interactions and dependencies between species that they saw the biological community as a superorganism. Whole species were viewed as organs that performed specific functions for the complete ecological superorganism. The integration and communication between these "organs" was thought to be deliberate and well tuned. One way to think of this idea is to imagine a stitched-together Frankenstein, each sewn-on body part a distinct species.

Today biologists find the analogy between biological communities and organisms superficial. To be sure, there are populations within communities that are highly dependent on each other. And it is also true that biological communities and their physical environments support all life on Earth by such processes as recycling nutrients. The impact of this recycling can be profound. For instance, atmospheric levels of carbon dioxide depend on plant photosynthesis and the respiration of all aerobic organisms. Global temperatures and weather are in turn dependent on atmospheric carbon

dioxide levels, which are covered in Chapter 16 (The Biosphere and the Physical Environment). The coordination and integration of biological communities has vast implications for the Earth.

For this reason, there are few biological topics as important for the future of life on Earth as the functioning of ecosystems. In this chapter, we survey how ecosystems function, from the flow of energy in Module 15.1 (Energy Flow) and the recycling of nutrients in Module 15.15 (Ecosystems) to the portentous problem of the fragility of ecosystems. In Modules 15.8 (Community Organization) and 15.4 (Equilibrium and Nonequilibrium Communities), we consider the factors that determine the number of species in a community. Surprisingly, in some communities predation and environmental disturbance may promote increased species diversity. Islands represent interesting communities, because virtually all species on an island must travel there from some larger mainland. Species diversity on islands is a consequence of dynamic processes. Understanding these forces has important practical applications for the design of ecological preserves, to be covered in Chapter 17 (Conservation). ❖

Energy Flow

15.1 The flow of energy is a central organizing theme in community ecology

We have all strolled through forests or walked along the seashore or lakeside. Even the untrained person will notice a variety of plants in a forest or the many insects and birds near lakes and oceans. These interacting plant and animal populations are part of a biological **community**. The members of such a community will be apparent from their associations or their geographic location. As we have seen in the previous chapter, some plants and animals may interact very closely and affect each other's evolution. While the details of processes such as coevolution were unknown to early ecologists, there was a strong sense that there was a mutual interdependence among the members of a community.

Communities Early in the twentieth century, F. E. Clements developed some of the first ideas about communities. If a tract of land is cleared but then left undisturbed, it will be recolonized by plants over time. This recolonization, or **succession**, may follow a predictable pattern, with some species appearing early in the sequence of recolonization, but later giving way to different species. (Figure 15.1A shows one example of succession.) In studying ecological succession, Clements thought that the species that appeared during succession made up a superorganism, with strong interdependencies much like the organs of a single plant or animal. We will cover succession in more detail in Module 15.7.

In the 1920s, Charles Elton developed a more sophisticated view of communities, one that still persists today. He studied a tundra community on Bear Island in the North Atlantic. Elton's focus was on feeding. Which species feeds on which is one of the most important interactions in an ecosystem. Figure 15.1B shows some of the results from Elton's study. These feeding relationships also reveal a directional flow of energy. Moss captures energy from the sun. Energy in the mosses is then consumed by herbivorous rotifers that are ultimately eaten by ducks. Diagrams that show energy flows are called **food chains**.

The nature of an ecological community is not solely a function of the organisms that make up the community. The physical environment also influences the numbers and types of organisms in a community. Likewise, photosynthesis, respiration, and decomposition affect the physical environment. In 1935 the English plant ecologist A. G. Tansley coined the term **ecosystem** to describe ecological communities and their associated physical environment. In Module 15.15, we will discuss the interactions between biological communities and the physical environment.

Trophic Levels In 1925 A. J. Lotka published his book *The Elements of Physical Biology*. Influenced by his training in chemistry, Lotka advocated the study of communities from a thermodynamic perspective, emphasizing the transfer of energy. This thermodynamic perspective and the importance of food chains were both embraced by Raymond Lindeman in 1942. For his Ph.D. thesis, Lindeman studied the feeding relationships in a bog community in Minnesota. He simplified the analysis of energy flow in this community by focusing on organisms that were at a similar position in the food chain. Such positions are referred to as **trophic levels**.

FIGURE 15.1A The Initial Stages of Succession in a Temperate Forest

In most ecosystems, the lowermost or first trophic level is made up of the primary producers or plants, such as the mosses in Figure 15.1B. These organisms depend on sunlight for their energy. The next trophic level up consists of the herbivores, organisms that consume plants, such as the herbivorous rotifers in Figure 15.1B. The biomass or energy available to herbivores comes directly from the primary producers. It is also affected by the efficiency of conversion of energy. Consumers of herbivores—such as the ducks in Figure 15.1B—are at the next trophic level, and so on.

Lindeman noted that the dependence of each trophic level on the one below it suggests that the amount of energy contained in each level (for example, as plant or animal biomass) should decline as one moves from the lower to the higher trophic levels. Lindeman called this natural progression the Eltonian pyramid. In Modules 15.2 and 15.3, we will study in more detail what is known today about the energy relationships within communities and the factors that affect energy transfer from one trophic level to the next. ❖

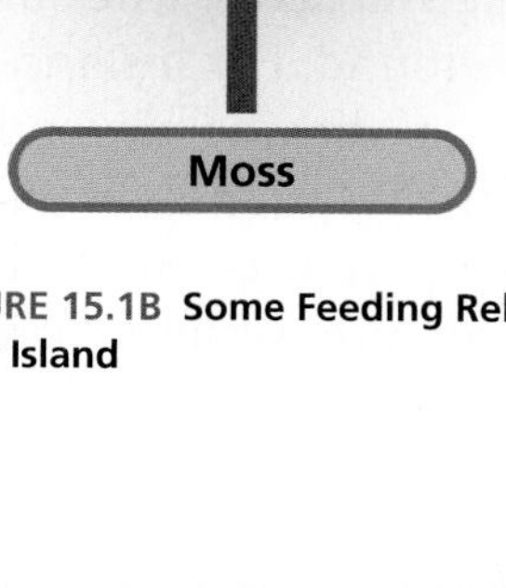

Long-Tailed Duck

↑

Rotifers

↑

Moss

FIGURE 15.1B Some Feeding Relationships from Elton's study of Bear Island

Raymond Lindeman (1915–1942)

During his short life and even shorter academic career, Lindeman managed to write six scientific papers. One of these appeared in the journal *Ecology* after his death, with the title, "The Trophic-Dynamic Aspect of Ecology." This paper is credited with influencing many ecologists to look at the energy and feeding relationships among organisms as an important aspect of community structure. Lindeman received his Ph.D. from the University of Minnesota in 1941. Shortly afterward he moved to Yale University, where he began postdoctoral work with G. Evelyn Huchinson. The original version of Lindeman's trophic-dynamic paper was rejected by the journal *Ecology*. It was only after an appeal by Hutchinson that the editor of *Ecology*, Thomas Park, agreed to publish the paper.

FIGURE 15.1C Raymond Lindeman

In most biological communities, all energy comes from the sun

Biological systems are complicated, but they must follow the same laws of thermodynamics that physical systems obey. The first law of thermodynamics tells us that energy can be neither created nor destroyed. Energy can be changed, however, from one form into another. In biological communities, almost all energy originates from the sun. Green plants capture solar energy and turn it into chemical energy. Because of this special function, green plants are called **primary producers**. The chemical energy is stored by plants as bonds holding organic molecules together. Not all the captured energy from the sun is stored as chemical energy. Plants use some energy for metabolic maintenance, and some is lost as heat.

All trophic levels above plants gain energy by feeding on members of other trophic levels. Herbivores feed on plants and derive their energy from the energy stored in plant tissue. The flow of energy goes in one direction, from plants to herbivores—not the other way. Consequently, the energy content of all the herbivores in a community cannot exceed the energy contained in the primary producers. In fact, it will often be much less, for at least two reasons. (1) The herbivores cannot consume all the plants, or there would be no future source of energy for the herbivores. (2) The conversion of chemical energy in the plants to chemical energy in herbivores is not perfectly efficient. Energy is lost as heat or is unused due to incomplete digestion. Thus the total amount of energy in an ecosystem is determined by the primary producers, though much will be lost to the biological community.

A rough indication of the amount of energy in the primary producer level is their **biomass** produced per year. Figure 15.2C shows the Earth's **biomes** (major communities classified according to their predominant vegetation) and gives an indication of their typical biomass. As we can see, the amount of energy located at the level of the primary producers varies substantially from one type of community to the next.

What causes this variation? The variation seen in Figure 15.2C is a function of environmental factors. Tundra has low primary productivity due to the short growing season near the Earth's North Pole. On the other hand, severe water shortage keeps the productivity of deserts low. The open ocean has plenty of light near the surface and mostly benign temperatures, but nutrients such as phosphorus are in short supply, limiting plant growth. Plants and animals on the ocean surface die and settle to the bottom of the ocean, where their decomposition does not immediately return nutrients to the surface. In some parts of the ocean, currents carry water from great depths up to the surface. These upwelling currents (see Module 11.4) are important for supplying the surface waters of the oceans with nutrients.

Given the dependence of higher trophic levels on lower trophic levels, we expect that the biomass of herbivores would be positively correlated with the plant biomass. For freshwater lakes, this predicted relationship is generally obeyed (Figure 15.2A). When the biomass or the number of species in a community is controlled by the amount of primary production, the community is **bottom-up** regulated. Conversely, if species biomass at most trophic levels is controlled by predation, the community is regulated **top-down**. There is nothing that prevents a single community from experiencing both bottom-up and top-down effects.

In lakes, the effects of top-down regulation by predation can be studied by artificially increasing the numbers of fish that feed on lake zooplankton. In Figure 15.2B, we see that increases in fish numbers lead to a decrease in zooplankton biomass and an increase in plant biomass. Because many zooplankton species feed on plants, reductions in their numbers benefit plants. ❖

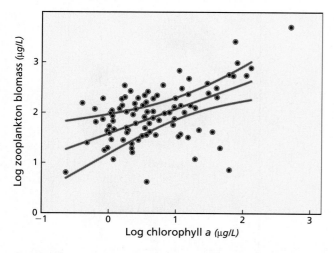

FIGURE 15.2A Biomass of Herbivores (Zooplankton) versus Biomass of Plants (Estimated from Chlorophyll *a*) from Several Different Freshwater Lakes

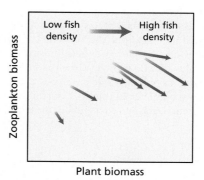

FIGURE 15.2B Changes in Zooplankton and Plant Biomass Following Manipulation of Fish Numbers The arrows show how zooplankton and plant biomass changes in going from low to high fish numbers.

Tropical Rainforest (1800)

Swamp and Marsh (2500)

Boreal Forest (800)

Temperate Grassland (500)

Tundra and Alpine (140)

Desert (70)

Open Ocean (125)

Lakes and Streams (500)

FIGURE 15.2C Primary Productivity in Major Biomes (grams per square meter per year)

15.3 The efficiency of energy transfer from one trophic level to the next varies among communities

What happens to the energy that is captured by primary producers? Some of the energy that is present in the plants is transferred to the next higher trophic level by herbivory and thus the conversion of plant biomass into herbivore biomass. This process continues up the energetic pyramid, as herbivores are eaten by carnivores, and so on.

Insights into community processes can be gained by analyzing such energy flows from one trophic level to the next. No chemical or physical process of energy conversion can be 100 percent efficient. Energy is lost in a variety of ways. In Figure 15.3A, we show how energy is lost as it flows from lower trophic levels to higher trophic levels.

The green arrows in Figure 15.3A represent the energy that successfully makes it from one stage to the next. The red arrows represent the energy that is lost in these transfers. At each of these steps, we can compute the efficiency of energy transfer if we know how much energy from one stage makes it into the next stage. In our example, a fox feeds on birds. Not all birds will be captured and eaten by the foxes, so not all the energy present in the bird trophic level can be converted to fox biomass, for this reason alone. The efficiency of this part of the energy transfer is called the **exploitation efficiency**.

Once a bird is eaten, the fox must convert the energy of its prey to energy it can use. Plants and animals consist in part of materials, such as cellulose and bone, that contain energy but cannot be digested and assimilated by most of the consumers that eat them. Thus, only a portion of the total energy devoured is chemically assimilated. The fraction of consumed energy that is assimilated is referred to as the **assimilation efficiency**.

Some of this assimilated energy will be used for work and maintenance. The rest will be used for growth and reproduction, adding to the biomass of the foxes. The fraction of the assimilated energy that is made into new biomass by foxes determines the **net production efficiency**.

The efficiency of the entire process of energy transfer between trophic levels is called the **ecological efficiency**. Ecological efficiency is the energy content of the higher trophic level divided by the energy content of the lower trophic level, as Figure 15.3A shows.

Energetic efficiencies vary across communities, depending on the lifestyles of the organisms that make up these communities. Figure 15.3B shows the net production and assimilation efficiencies of three categories of organisms. There is no general pattern of assimilation efficiency, with ectotherms and endotherms showing a mixture of high and low values. But the net production efficiency of ectotherms is consistently higher than that of the endotherms. This makes sense, because endotherms must spend a larger fraction of their energy budget maintaining their body temperature and thus have less energy to devote to growth. ❖

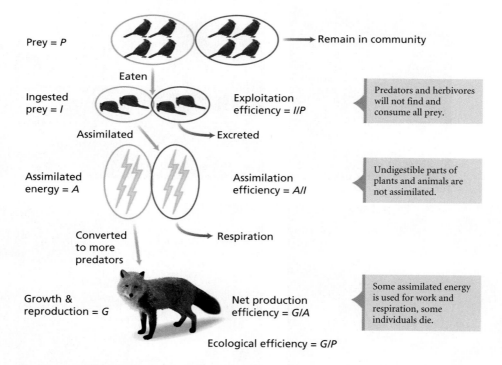

FIGURE 15.3A This figure shows the flow of energy from one trophic level (prey) to a second (predator). The amount of energy is represented by symbols in the leftmost column. The fraction of energy that passes through each step in this process is called efficiency and is shown in the middle column.

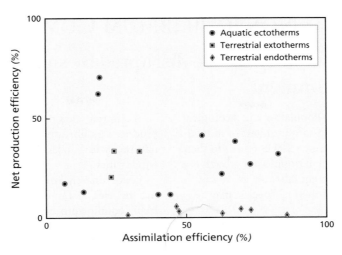

FIGURE 15.3B **Net Production Efficiency versus Assimilation Efficiency of Several Groups of Animals**

Equilibrium and Nonequilibrium Communities

15.4 Community stability can be disrupted by sudden changes in the physical environment

What Is an Equilibrium? Populations in **ecological equilibrium** maintain relatively constant numbers of individuals and, if displaced from these numbers, will return to their equilibrium levels. Density-dependent population growth is a mechanism that can help maintain equilibrium, as we saw in Chapter 10. Communities of competing species may also achieve equilibrium.

One concept of a community in equilibrium is much like the concept of an evolutionary equilibrium of species numbers, maintained by extinction and speciation, introduced in Chapter 6. The metaphor used there was of water dripping into a plugged sink from a faucet, counterbalanced by water dripping over the side of the sink. At an equilibrium population level, some factors tend to increase population size, while others tend to decrease it. If an equilibrium is stable, any change in the numbers of one or more species will be followed by a return to population sizes from before the perturbation. When the community ecology maintains equilibrium, the number of species in a community depends on the strength of competition between close competitors and the food-chain structure. The populaton sizes of all species reach a stable equilibrium size in a stable community.

A different view of community structure is that it does not produce equilibrium. **Nonequilibrium** theory proposes that natural disturbances prevent populations from reaching an equilibrium.

Unfortunately, it is hard to choose between these alternative theories. Just to determine if natural populations are at a stable equilibrium or not is time consuming. One must get accurate census data from populations over many generations. Even then, a community may maintain roughly constant population densities only because of a lack of perturbations. This would be like the "stability" of a tall boulder that would roll or fall over if it were pushed, but has not yet been pushed.

Population Disturbance Populations can be prevented from reaching an equilibrium due to environmental disturbances. These disturbances can come in many forms: storms, fires, drought, floods, even volcanoes. The disturbance may cause relatively small changes in the population size or resource levels or, as in the case of the Mount St. Helens volcano (Figures 15.4A through 15.4C), it may wipe out entire communities over large areas. Large disturbances, like the Mount St. Helens volcano, effectively sterilize an area and then open it to the process of succession that we will review in Module 15.7. The community structure during succession is in constant change. Depending on the community, it can take many years for an equilibrium to be reached after disturbance. In some instances, an equilibrium may never be reached before the next major disturbance.

FIGURE 15.4A Mount St. Helens, Washington, during Its Most Recent Eruption

FIGURE 15.4B Area Surrounding Mount St. Helens Immediately After the Volcano Erupted All signs of life are gone, and the ground is covered with ash.

FIGURE 15.4C **Destruction of Vegetation Following the Eruption of Mount St. Helens**

FIGURE 15.4D **Destruction of Forest from Fire** Although fires can be man-made, they are a regular disturbance in many ecosystems.

Other disturbances are less dramatic and may be important determinants of the species diversity in the community. For instance, forest fires have been a regular disturbance in some ecosystems well before humans arrived (Figure 15.4D). As humans began to manage forests, they naturally tried to avoid the destruction of forests by fires, including fires of human origin and those due to natural factors, such as lightning. It was soon realized that the health and composition of some forests actually required occasional fires and that human intervention to stop all fires was ill advised. We will see in Module 15.5 that, for some communities, moderate levels of disturbance can actually increase species diversity compared to communities with no disturbance.

Island Biogeography
Islands are very special communities that rely on colonization from distant locations to populate communities with plants and animals. Some of the basic forces determining species diversity on islands were first described more than 30 years ago in a book by Robert MacArthur and E.O. Wilson. Their keen insights have proven to be generally accurate and form the core of present-day theory in island biogeography. We will review some of these important ideas in Module 15.6.

Succession
We have already introduced the concept of biological succession in Module 15.1. There are several circumstances that create opportunities for biological succession. Organisms die, snakes shed their skin, and large mammals leave substantial amounts of organic matter in their feces. In all these examples, the sudden appearance of organic matter creates a new habitat that can be colonized through a serial replacement of organisms that we call **degradative succession**. Of course this succession process ends once the organic material has been consumed. We will consider this process in detail in this module and examine how this process of succession is used to produce forensic evidence in criminal cases.

There are also natural processes that rapidly create unoccupied habitat for colonization and succession. The volcanic eruptions already mentioned are one such example. Glaciers may retreat and expose bare soil. Fires and tree falls may create conditions for succession to begin. These processes are referred to as **autogenic succession**. Unlike degradative succession, autogenic succession usually culminates in a mature community, referred to as the climax community. Oftentimes the species that appear early in the successional process may alter environmental conditions in a way that will permit the later successional species to invade and survive. For instance, when the glaciers in Alaska retreat, the exposed soil is first colonized by mosses and shallow-rooted herbs. A little later, alder plants invade. Both the herbs and alder have the ability to fix nitrogen. In time they will increase the nitrogen content of the soil substantially. Alder also acidifies the soil. The result is that the soil becomes more fertile and supports more rapid growth of larger trees, like Sitka spruce.

In some places there may be a gradual change in the species composition as a result of externally changing physical or chemical conditions. These types of changes are sometimes referred to as **allogenic succession**. An example would be the changes that occur as silt accumulates at the mouth of a river system. This is a gradual process that transforms brackish water to soil. As a result, terrestrial species may gradually colonize this new land and displace species adapted to the brackish water conditions. ❖

15.5 The diversity of species in a community may depend on environmental disturbance

In 1961 the idea of competitive exclusion (see Chapter 12) presented a problem for G. E. Hutchinson. His work with marine and freshwater plankton demonstrated that many species of phytoplankton can coexist in the same top layers of water. However, in these top waters, several resources, such as nitrogen and phosphorus, were often in limited supply. Why didn't competition drive all but one species of phytoplankton to extinction?

In a paper entitled "The Paradox of the Plankton," Hutchinson suggested an answer. He argued that the competitive relationships among plankton species were specific to a particular set of environmental conditions. If these environmental conditions changed before the superior competitor could achieve numerical dominance, then many species might persist in a locality for some time.

The role of environmental variation in determining species diversity has been important in ecological thought. In the intertidal zone, for example, many plants and animals compete for space. In an undisturbed environment, species diversity will decrease as the competitively dominant species eliminates other species. However, the intertidal zone is subject to disturbance. Waves or storms may turn over rocks. When this happens, some of the plants or animals on these rocks may be displaced or killed. As the graph in Figure 15.5A shows, when disturbance is very low, we have little displacement of species (green line) and high levels of competition (red line). The result is low species diversity, with the competitively dominant species being most numerous. On the other hand, when disturbance levels are high, there is little competi-

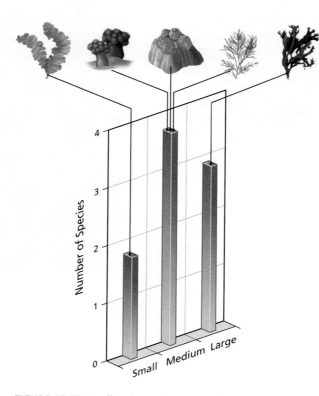

FIGURE 15.5B Small rocks tend to be dominated by the green alga *Ulva*. Large rocks are dominated by the red alga *Gigartina*. The medium-sized rocks have a much larger array of species including *Ulva*, barnacles (*Cthamalus*), sea anemones (*Anthopleura*), and additional species of algae (*Gelidium* and *Rhodoglossum*).

tion (red line), but species are constantly displaced (green line), so only a few good colonizing species are found on the rocks. Therefore, ecologists expect the highest levels of species diversity on intertidal surfaces with intermediate levels of disturbance. This idea is known as the **intermediate disturbance hypothesis** of species diversity.

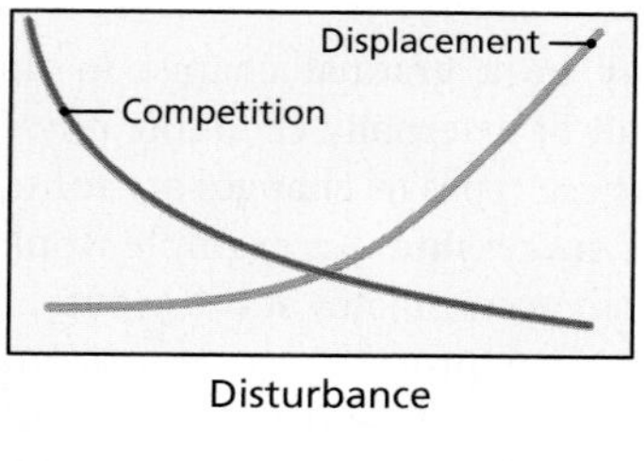

The intermediate disturbance hypothesis was tested in the intertidal zone by Wayne Sousa. Sousa studied the species diversity on intertidal rocks of different sizes. He reasoned that large rocks would be moved only by greater forces, so plants or animals on these rocks would be displaced less often than those on small rocks. The number of species of plants and animals on rocks in three size categories were measured. As Figure 15.5B shows, the most species were found on rocks of intermediate size. Thus, species diversity on intertidal rocks is consistent with the intermediate disturbance hypothesis.

Similar forms of disturbance can be found in other communities. For instance, in forests strong winds or attacks by insects may result in large trees falling (Figure 15.5C). The fallen tree no longer shades the soil surface, and many new plants may find opportunities to grow in these types of open areas. ❖

FIGURE 15.5A As environmental disturbance increases, displacement of organisms from rocks increases. However, competition is greatest when there is little disturbance.

FIGURE 15.5C Openings in the forest may promote new growth. Small gaps in a forest may be created by mature trees being blown down during storms or falling down after being attacked by insect pests.

The number of species on islands represents a balance between extinction and immigration

Biogeographers study the distribution of species. Their point of view is that "islands" may be small parcels of land surrounded by water, or they may be mountain peaks surrounded by valleys. The important requirement is that suitable habitat is surrounded by inhospitable habitat. Plants and animals will occupy the "island" only if they travel from some "mainland" location to the island, or from another island. When a new plant or animal species successfully establishes itself on the "island," we say it has successfully immigrated. The persistence of species on an island requires its successful reproduction and the growth of its population to substantial numbers. In general, the larger the population, the longer its persistence. Eventually all populations go extinct. The only way for an extinct species to become reestablished is by immigration. Ultimately the number of species on an island will represent a balance between immigrations and extinctions (Figure 15.6A).

What factors affect immigration and extinction? The number of species already on an island is one factor. Rates of successful immigration are high when species numbers on an island are low, for several reasons. When species numbers are low, most newly arrived plants and animals represent species not currently on the island. Extinction rates will be low when species numbers are low, since few species are at risk of going extinct, because there is little competition.

Rates of extinction and immigration also depend on factors other than species numbers. Immigration rates are affected by proximity to mainland sources of plants and animals. Islands that are near the mainland have higher immigration rates than do more distant islands (Figure 15.6B). Extinction rates should be sensitive to island size. All other things being equal, small islands should support smaller populations of any particular species, making them more vulnerable to extinction (Figure 15.6C).

These theoretical expectations can be put together to make predictions about the relative number of species on islands as a function of their size and distance from mainland. For instance, if the useful habitat of a particular island were substantially reduced, extinction rates should increase. However, immigration rates should remain the same, so the number of species on the island would be expected to fall.

Daniel Simberloff conducted an experiment to test this prediction on mangrove islands. Over a period of three years, species counts were made on these small islands and then portions of the habitat were cleared and removed from the island (Figure 15.6D). In all cases, reductions in habitat area resulted in observable reductions in the numbers of species. ❖

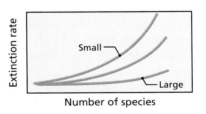

FIGURE 15.6C Extinctions will be less likely in large populations than in small populations. Small islands will generally support smaller populations and thus have higher extinction rates than large islands will.

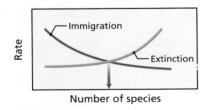

FIGURE 15.6A The number of species on an island will represent the balance between immigration and extinction. The arrow shows this equilibrium.

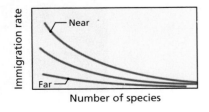

FIGURE 15.6B Immigrants from the mainland are more likely to reach nearby islands than distant islands.

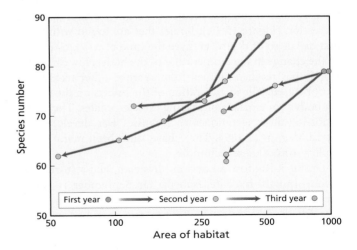

FIGURE 15.6D **Changes in Species Numbers on Small Islands** Each island was followed for three years. The occupied part of the island was reduced, usually in years two and three. The reduction in island area should be accompanied by increased extinction rates and lower numbers of species on the island. This general expectation was observed in each case.

15.7 Habitats go through predictable changes in species composition over time

Most of us have observed the biological changes that occur when a new piece of habitat becomes available for species to occupy. This can happen when a wooded area is cleared for construction or farming, or when fruit falls to the ground and is then left undisturbed. In each case, we see certain organisms making use of the new resources in these habitats. For instance, cleared land might first be occupied by small weedy plants. If the land is left undisturbed for a sufficiently long time, these weeds are displaced by larger bushes. And after very long periods of time, the species of tree that originally occupied the land takes over as the dominant species.

These transitions in community structure over time are called ecological **succession**. Geologic events such as volcanoes and glaciers may leave areas with no existing vegetation or soil. The development of communities on these sites is called **primary succession**. Other places may already have established vegetation and leave seeds and well-developed soil after a disturbance. The changes in these habitats are called **secondary succession**. The end of succession often yields a community of stable species composition. This community is called a **climax community**. As we mentioned earlier, degradative succession does not produce a climax community, because it ends with the exhaustion of some resource. We consider an application of this successional process next.

The sequences of species that characterize degradative succession depend on the habitat available. These species sequences are sufficiently reliable that they can be used in criminal investigations to determine the time of death of corpses. The field of forensic entomology uses the principle of ecological succession to determine when a dead body began to decompose. As a body decays, the moisture content and pH change in such a way that the types of insect species living on the corpse go through a predictable sequence of changes or succession (Figure 15.7A). Bodies that are found within the first day or so of death can have the time of death estimated by the change in the temperature of the body. However, once a body has reached ambient temperature, other techniques must be used. The identification of the insects on the decaying body is one means of making these estimates. The application of these techniques to forensics was developed by Pierre Mégnin in 1894. There have since been many useful applications of these techniques.

Dr. Buck Ruxton, a London physician, had a stormy relationship with his wife, Isabella. On September 15, 1935, Ruxton's wife and their housekeeper, Mary Rogerson, were seen for the last time. Dr. Ruxton said they had gone on holiday to Scotland. On September 29, the parts of two female bodies were found floating in the River Annon. Fingerprints identified one body as Mary Rogerson. The presence of third instar larvae of the blowfly *Calliphora vicina* (Figure 15.7A) was used to determine that the two

Time	State of Corpse	Insect Fauna
First 3 months	Initially fresh, the corpse is subject to bacteria decay and bloating from gases. Larval forms of these insects feed on blood and tissues, first focusing on the heart and digestive system.	 *Calliphora vicina*  *Sarcophaga carnaria*
3–6 months	After fermentation of fats, corpse continues to dry.	 *Piophilia casei*
4–8 months	Remaining body fluids are now absorbed.	 *Ophyra leucostyoma*
1–3 years	The corpse is completely dry.	 *Derestes lardarius*  *Tineola biselliella*

FIGURE 15.7A Forensic Entomology

women had been dead for 12–14 days. Based on this and other evidence, Dr. Ruxton was convicted of murder and executed on May 12, 1936.

Much of the scientific study of insect succession in decaying corpses has been done not with humans, but with other vertebrates such as pigs and dogs. Because it is not clear if the results on these smaller animals would be similar to those on a human corpse, Dr. William Bass has created an experimental site to study this problem (Figure 15.7B). On a three-acre wooded lot near the University of Tennessee campus, he has established an outdoor laboratory known informally as the "Body Farm." At this location, human cadavers have been placed under different conditions to monitor their progressive change over time. Some bodies are embalmed, others not. Some are buried under a carpet of leaves, while others lie on the surface, exposed to the elements. This research will help law enforcement officials determine date of death, often the most important piece of information in determining guilt or innocence in murder cases. ❖

FIGURE 15.7B   The Body Farm at the University of Tennessee

COMMUNITY ORGANIZATION

15.8 The diversity of a community may be affected by competition, predation, or primary productivity

We have already seen how the members of a community depend on each other for energy. This dependence structures their ecological relationships. The most obvious pattern is reduced biomass at higher trophic levels.

Communities can also be understood based on the diversity of the species of which they are composed. **Species diversity** is the number and relative abundance of species in a community. Among the factors that shape species diversity are competition and predation, which we reviewed in Chapters 12 and 13. These processes can affect the number of species coexisting on the same trophic level or on different trophic levels.

Interspecific Competition   In Chapter 12, we examined the conditions for the coexistence of competing species. We saw that if intraspecific competition is stronger than interspecific competition, two species can coexist. The most detailed examination of the Lotka-Volterra competition equations (see Module 12.7) was carried out by Vandemeer.

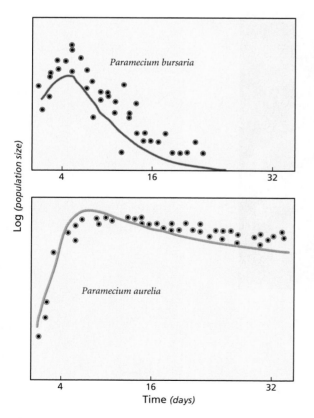

FIGURE 15.8A The numbers of *P. aurelia* and *P. bursaria* when these species are in competition with themselves and *P. caudatum* and *Blepharisma*. Solid lines are the predictions from the Lotka-Volterra equations; circles are the observed numbers.

In his experiment, three species of *Paramecium* and a species of *Blepharisma* were raised separately to estimate their carrying capacity and intrinsic rates of increase, when growing on their own (see Module 10.6). Then pairs of species were raised together to estimate their competition coefficients (see Module 12.7). Finally, all four species were placed together and allowed to grow. Figure 15.8A shows the observed numbers of *P. aurelia* and *P. bursaria*. The population sizes predicted from the Lotka-Volterra equations are also shown as a solid line. The Lotka-Volterra theory not only correctly predicted the extinction of *P. bursaria*, but it also did reasonably well at predicting the actual numbers of all species of *Paramecium*.

Interspecific competitors usually occupy the same trophic level. In some communities, however, an important limiting resource may affect species at several trophic levels. For example, space in the intertidal zone is a limiting resource for species at many different trophic levels, as described in Chapter 7.

Predation   Predation and herbivory involve feeding relationships between species on different trophic levels. However, the ecological factor controlling the numbers of predators or herbivores is not necessarily their food source, as we will see in Module 15.9. For instance, most terrestrial herbivores live in environments with an enormous amount of available plant material. This suggests that the numbers of herbivores in a biological community is not related to the amount of food in any simple way. In 1960 Hairston, Smith, and Slobodkin suggested that it is more likely that the numbers of herbivores are regulated by predators that reduce herbivore numbers well below what can be supported by the primary production. This hypothesis is sometimes called "why the Earth is green." It is an example of top-down regulation and is only one of several reasons why the Earth is green. However, not all herbivores and predators have this type of top-down regulation. Later in the module, we consider additional possibilities.

Food Webs   **Food webs** show the feeding relationships between organisms. A species or group of species is represented by a **node** (point) in the food web. A **link** (line) connects two nodes, indicating a predator-prey or plant-herbivore relationship. Links may be either undirected or directed, as Figure 15.8B shows. A **directed link** shows the flow of energy between two species, while an **undirected link** only indicates that a feeding relationship exists (Figure 15.8B). A **cycle** exists when two nodes feed on each other. When two or more species make a closed circuit, this is called a **loop**. A **chain** is a series of directed links starting from a species that feeds on no other species and ending in a species that is not fed on by any other species. The number of links in a chain is called its **length**.

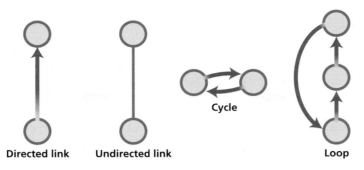

FIGURE 15.8B Components of Food Chains See text for details.

There are several types of food webs. A **source web** arises when a group of species derive all their energy from a single food source. A **sink web** arises when the feeding relationships direct all energy ultimately to a common top carnivore. A **community web** shows the feeding relationships of all members of a community. The mean chain length of a web is the arithmetic average of the lengths of all its component chains.

Understanding the factors that affect the mean chain length of food webs has been an active area of research in ecology. Several hypotheses have been proposed, and we will review evidence for some of them in this module. The **energetic** or **productivity hypothesis** suggests that the mean chain length will be proportional to the amount of energy at the primary producer level. Because energy will be lost with each link in the chain, the total number of links should depend on the amount of energy at the base level.

The **dynamical stability hypothesis** suggests that long chains will be inherently less stable. Thus, longer chains are more likely to be found in benign environments where populations are not subject to large fluctuations in population size.

The **ecosystem-size hypothesis** suggests that chain lengths will be greater in ecosystems with greater physical volume, all other things being equal. This is a complicated theory, but basically a greater number of different species can be supported in larger areas. It happens that average chain length increases with increasing species numbers and thus average chain length also increases with increasing ecosystem size. ❖

15.9 The number of species in a community may depend on predation

Some of the most spectacular examples of the effects of predators involve *biological control.* In Chapter 14 we learned about the explosive growth of rabbits introduced to Australia at the end of the nineteenth century. The introduction of a viral parasite was effective at reducing the numbers of rabbits. A similar example is the aquatic fern, *Salvinia molesta,* that is native to waterways in Brazil. This fern has become a pest in much of the tropics. Its dense populations block waterways and prevent fishing. Effective control has been achieved using a small weevil, *Cyrtobagus singularis.* The adult weevil feeds on the buds of the fern, while the larvae feed on the plant's roots and rhizomes. The specificity of this interaction was demonstrated by the failure of the first control efforts. A weevil from a closely related fern, *Salvinia auriculata,* was tried but failed to control *S. molesta,* even though these two species were thought to be the same for some time.

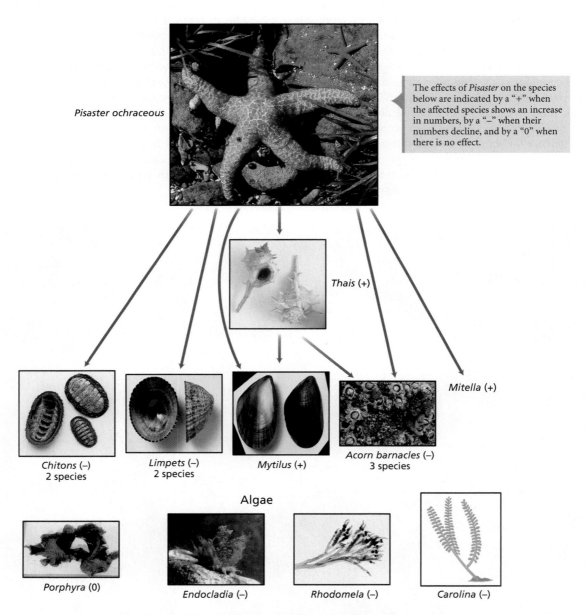

Pisaster ochraceous

The effects of *Pisaster* on the species below are indicated by a "+" when the affected species shows an increase in numbers, by a "–" when their numbers decline, and by a "0" when there is no effect.

Thais (+)

Mitella (+)

Chitons (–)
2 species

Limpets (–)
2 species

Mytilus (+)

Acorn barnacles (–)
3 species

Algae

Porphyra (0)

Endocladia (–)

Rhodomela (–)

Carolina (–)

FIGURE 15.9A *Pisaster,* **a Keystone Predator in the Intertidal Zone**

These examples demonstrate the potential for predation and herbivory to affect the numbers of a prey species, but not the numbers of species. One of the earliest and most influential demonstrations of the importance of predation on species composition was a study by Robert Paine on an intertidal marine invertebrate community. In this study, the ecology of a carnivorous starfish (*Pisaster ochraceous*) was observed at two different research sites. In the experimental site, the starfish was removed and kept out of the area. In the control site, the starfish were allowed to forage as usual. A total of 15 different species were monitored. In the experimental site, the **species richness,** or number of species, dropped from 15 to 8, while in the control area it remained unchanged.

What explains these results? The decline in species numbers in the experimental area was due to increased numbers of the competitively dominant mussel *Mytilus*. Other species of animals and plants were eliminated because the removal of the starfish led to the explosive growth of *Mytilus*. In Figure 15.9B, the species that were negatively affected by the removal of starfish have a negative sign next to their names. A positive sign indicates a benefit to the species due to starfish removal, whereas a 0 indicates no effect. Paine called the starfish a **keystone predator**, because of its central role in maintaining species diversity in this community.

Species numbers in terrestrial communities may also be affected by keystone predators. Wade Worthen studied three species of mushroom-feeding *Drosophila*. These fruit flies serve as food for the predatory rove beetle *Ontholestes cingulatus,* which eats adult *Drosophila*. In the absence of the beetle, there is strong interspecific competition between the three species of *Drosophila*, with *D. tripunctata* often eliminating the other two species. However, when the beetle was added to experimental cultures, the level of competition among *Drosophila* larvae was reduced, and all three species coexisted. As Figure 15.9A shows, beetle predation affected the competitively dominant species *D. tripunctata* most. With fewer *D. tripunctata*, the other two species were able to increase their numbers and so stably coexist. ❖

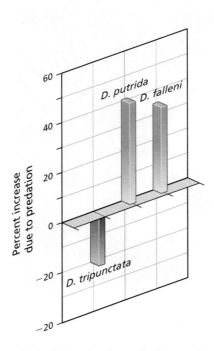

FIGURE 15.9B Effects of a Predatory Beetle on the Composition of a *Drosophila* Community In the absence of the beetle, *D. tripunctata* is competitively superior to *D. putrida* and *D. falleni*, often eliminating them. When the beetle—which feeds on all three species—is added to the community, the numbers of *D. tripunctata* fall and the numbers of the other two species increase. All three species coexist at similar population sizes in the presence of the predator.

The starfish study by Paine demonstrates the potential for predation to shape the diversity of species in the trophic level directly below it. The theory of food webs also predicts that in long food chains the qualitative effects of removing top predators should alternate as one goes down to lower trophic levels. Thus, this theory would predict that in a community with four trophic levels (producers-herbivores-small carnivores-top carnivores) removal of the top predator would have a beneficial effect on the small carnivores. However, once the number of small carnivores increased, that would cause a negative effect on the herbivores. The reduction in herbivore numbers would then have a positive effect on the primary producers. This predicted series of positive and negative effects is sometimes called a **trophic cascade**.

To investigate whether communities showed such cascades, Mary Power undertook a study of a river community (Figure 15.10A). This community also has four trophic levels, as does the example used in the previous paragraph. At the bottom are green algae, *Cladophora* and *Nostoc*. These are consumed by herbivorous insect larvae called chironomids. Chironomids are related to mosquitoes. The chironomids are eaten by predatory insects and juvenile fish, like the stickleback. The rivers that Power studied in northern California also have large fish, steelhead and roach, that feed on the small predators (Figure 15.10A).

Power introduced replicate cages covered with a small mesh that permitted small insects to pass through freely, but not fish. At random, some cages were designated as enclosures and others as exclosures. Each enclosure was stocked with 20 steelhead and 40 roach fish. The exclosures were kept free of large fish. After five weeks, samples were made of all the enclosures to determine the numbers of predators and the biomass of algae. The results were consistent with the predictions of a trophic cascade. The small predators all showed increased numbers in the exclosures relative to the enclosures. However, the next lower level in the food web, chironomids, showed a dramatic decline in the exclosures. This decline of chironomids was accompanied by an increase in algal biomass. Thus, the change in numbers of top carnivores was followed by a series of changes that were alternatively positive and negative at different trophic levels. ❖

Are Aquatic Food Webs Different from Terrestrial Food Webs?

The two examples we have just considered show how strong the effects of top predators in a marine and freshwater community can be. The evidence for such effects in terrestrial communities is less compelling, however. This finding has led some to suggest that there may be a difference between aquatic and terrestrial communities that will naturally give rise to aquatic communities experiencing more top-down regulation. Possible reasons for such differences include the following: (1) Terrestrial food webs are more complex than aquatic food webs. This complexity might make it less likely that the effects of top predators would work their way down to primary producers. Terrestrial plants are also more likely to protect themselves, compared to phytoplankton, from herbivores by producing toxic secondary chemical compounds. This would also make such plants insensitive to changes in numbers of herbivores. (2) Timescales and turnover rates are faster in aquatic systems. Because the plankton at the top layer of the water column contain many small algae, their growth rates are rapid. This may lead to fundamental differences between terrestrial and aquatic systems, or it may simply make it easier to detect strong food-web interactions in aquatic ecosystems.

However, other scientists have argued that an examination of large numbers of studies, not just a few prominent ones, suggests that terrestrial interactions are not substantially different from aquatic communities. There is a lot of heterogeneity within aquatic and terrestrial communities, making it difficult to reach general conclusions about the average behavior. At this time, more work is needed before these questions can be settled.

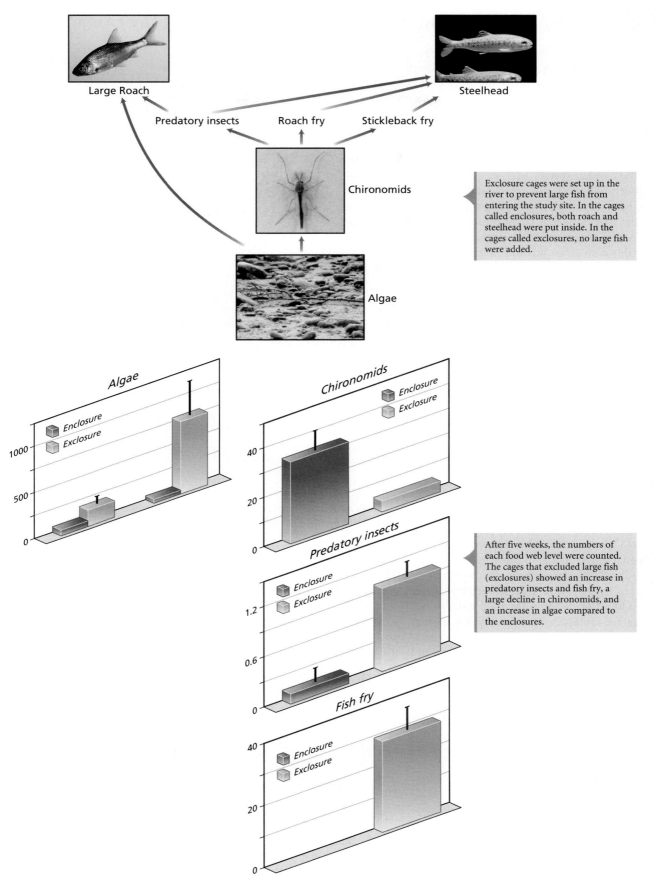

Large Roach

Steelhead

Predatory insects   Roach fry   Stickleback fry

Chironomids

Exclosure cages were set up in the river to prevent large fish from entering the study site. In the cages called enclosures, both roach and steelhead were put inside. In the cages called exclosures, no large fish were added.

Algae

Algae

Enclosure
Exclosure

1000

500

0

Chironomids

Enclosure
Exclosure

40

20

0

Predatory insects

Enclosure
Exclosure

After five weeks, the numbers of each food web level were counted. The cages that excluded large fish (exclosures) showed an increase in predatory insects and fish fry, a large decline in chironomids, and an increase in algae compared to the enclosures.

1.2

0.6

0

Fish fry

Enclosure
Exclosure

40

20

0

FIGURE 15.10A **Effects of Predation in River Food Webs**

15.11   Many features of food webs can be described by the cascade model

Several theories have been proposed concerning the properties of food webs, such as the mean chain length, maximum chain length, and total number of links. One simple theory, called the cascade model, provides numerical estimates of all these quantities. Before describing this model, let's look at how food web data are summarized.

In Figure 15.11A, the food web for the river community considered earlier is shown. A community food web matrix lists the trophic species that are prey as row entries (bold numbers) and those that act as predators in the column entries (bold numbers). If a 1 appears in the matrix, it indicates that the species in that column feeds on the species of that row. A zero means no feeding relationship exists. Quantitatively this is an easier way to summarize a food web than the pictures that show connections between species. A basal species is one that is prey for one or more species and does not eat any other species. In the food web matrix, a basal species will have a column of zeros under its number. In Figure 15.11A, the algae are the only trophic species meeting this definition. A top species is one that feeds on one or more species but is not itself prey for any other species. A top species should have a row of zeros next to its number in the food web matrix. Both the steelhead and large roach are top species in Figure 15.11A. The diagonal elements in this matrix are the cells where the same-numbered row and column intersect. A 1 at these positions would imply that the species can eat itself. In this food web none of the species are cannibals, because all diagonal elements are zero.

Joel Cohen and Charles Newman have developed a simple model to study the properties of food webs. We review the assumptions of the model in Figure 15.11B. It is assumed that there are no loops or cycles in the food web. So no species are cannibals. This assumption means that all the diagonal elements will be zero and all the elements below the diagonal will also be zero. This also means that the species in the community can be organized as a cascade (see Figure 15.11B). Trophic species may feed on species with lower numbers, but are never fed on by species with lower numbers.

The model then assumes that feeding relationships are created at random and that the chance of any allowable relationship forming is determined by a common probability, p. From this simple formulation, the model predicts with some accuracy the mean chain length in many different communities (see Figure 15.11B). This may seem surprising, because community food webs are not put

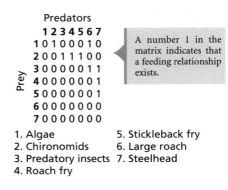

A number 1 in the matrix indicates that a feeding relationship exists.

1. Algae
2. Chironomids
3. Predatory insects
4. Roach fry
5. Stickleback fry
6. Large roach
7. Steelhead

FIGURE 15.11A A Food-Web Matrix

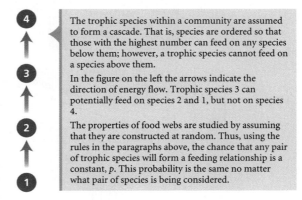

The trophic species within a community are assumed to form a cascade. That is, species are ordered so that those with the highest number can feed on any species below them; however, a trophic species cannot feed on a species above them.

In the figure on the left the arrows indicate the direction of energy flow. Trophic species 3 can potentially feed on species 2 and 1, but not on species 4.

The properties of food webs are studied by assuming that they are constructed at random. Thus, using the rules in the paragraphs above, the chance that any pair of trophic species will form a feeding relationship is a constant, p. This probability is the same no matter what pair of species is being considered.

Consider a community with the four trophic species illustrated above. All possible food chains having two links with species 1 as the basal species and species 4 as the top species are shown below.

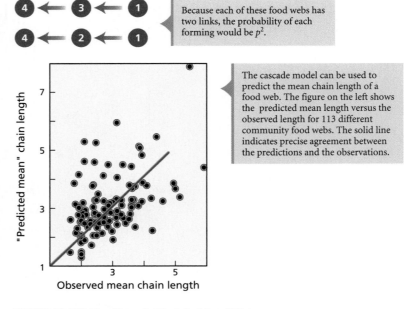

Because each of these food webs has two links, the probability of each forming would be p^2.

The cascade model can be used to predict the mean chain length of a food web. The figure on the left shows the predicted mean length versus the observed length for 113 different community food webs. The solid line indicates precise agreement between the predictions and the observations.

FIGURE 15.11B The Cascade Model of Food Webs

together at random. The ability of the cascade model to do such a good job of predicting mean chain length suggests that while food webs are not created randomly, the details of how they are created are not needed to make accurate theoretical predictions.

This is not unusual in science. It is possible to create a detailed theory of the kinetics of a coin placed in motion by the flip of our finger. We could take into account the forces that our finger generates and the atmospheric conditions—particularly the presence of wind—as well as the position at which the coin is typically caught, and then predict how often we should get a head or tails. If we do this calculation correctly, we would see that the coin will come up heads about 50 percent of the time. Or we can, and often do, use a simple statistical model to predict the chances of getting heads without worrying about all the details of the forces that affect the trajectory of coins. One such model is simply heads half the time, tails the rest.

One interesting result from the theoretical cascade model is that the mean chain length increases slowly with total number of species in the food web. This result can be combined with the theory of island biogeography to arrive at a prediction about the mean chain length and size of the habitat. From the theory of island biogeography, we know that the rate of extinction decreases with increasing island size. If all other variables remain constant, then the equilibrium number of species should increase on larger islands. This result, in combination with the conclusions that mean chain length increases with species numbers, suggests that mean chain length increases as community size or volume increases. ❖

In the last module we noted that the cascade model and the theory of island biogeography predict increasing chain length in larger communities. To test this idea requires samples from many communities that vary in size. There would also have to be some way to rapidly collect information on the mean chain length of each community. David Post and his colleagues (2000) have been able to collect this sort of information for lake ecosystems. They also collected information on the productivity so that they could simultaneously compare mean food-chain length to productivity.

If mean chain length depends only on ecosystem size, we would expect the data collected by Post to look something like Figure 15.12A. Mean chain length would increase with the size of the ecosystem independently of the productivity. Likewise, ecosystems that varied in productivity would show no consistent trend in mean chain length. So what did the results look like?

In Figure 15.12B, we summarize the results of Post's study. Ecosystem size was estimated from the volume of the lake, which is fairly easy to do. The lakes studied were in the northeastern United States. In these lakes, primary productivity is limited by the amount of available phosphorus. Levels of total phosphorus (TP) are highly correlated with the primary productivity and may be used as an estimate of primary productivity. A detailed study of the feeding relationships of the entire community of each lake would take an enormous amount of time. To estimate the mean chain length of each lake, an ingenious method that utilizes radioisotopes of nitrogen was used (Figure 15.12C). This technique is objective and allows information to be collected on a large number of communities.

When the data are analyzed, we can see that there is a very strong positive relationship between ecosystem size and mean chain length (see Figure 15.12B). This relationship holds for

The ecosystem size hypothesis predicts that the food-chain length of a community will increase with increasing ecosystem size, no matter what the productivity.

If productivity was unimportant, then there should be no change in food-chain length with increasing productivity, no mater how large the ecosystem.

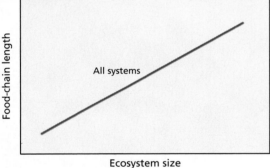
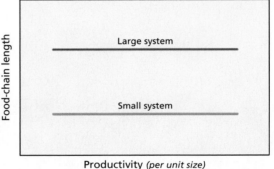

FIGURE 15.12A Ecosystem Size and Food-Chain Length

The mean chain length was estimated by the stable isotope technique (Figure 15.12C) in several lakes that varied in size and productivity.
The mean chain length increases in direct porportion to size, no matter what the productivity of the lake is.

For these same lakes examined in the left graph, there was no consistent relationship between mean chain length and productivity.

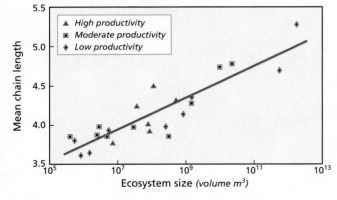
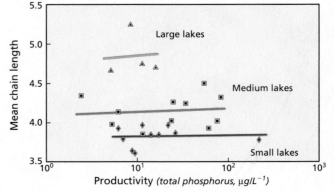

FIGURE 15.12B Tests of the Ecosystem Size Hypothesis

lakes that varied in productivity. On the other hand, there is either no relationship or a very weak relationship between productivity and mean chain length. These observations provide strong empirical support for the importance of ecosystem size in the structure of food webs. ❖

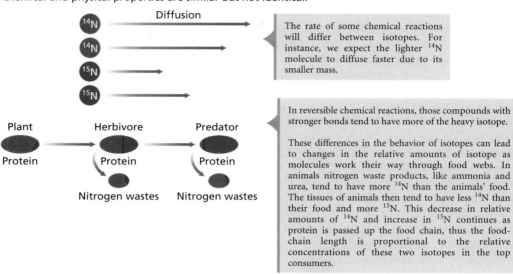

^{14}N and ^{15}N are naturally occurring isotopes of nitrogen. Their chemical and physical properties are similar but not identical.

The rate of some chemical reactions will differ between isotopes. For instance, we expect the lighter ^{14}N molecule to diffuse faster due to its smaller mass.

In reversible chemical reactions, those compounds with stronger bonds tend to have more of the heavy isotope.

These differences in the behavior of isotopes can lead to changes in the relative amounts of isotope as molecules work their way through food webs. In animals nitrogen waste products, like ammonia and urea, tend to have more ^{14}N than the animals' food. The tissues of animals then tend to have less ^{14}N than their food and more ^{15}N. This decrease in relative amounts of ^{14}N and increase in ^{15}N continues as protein is passed up the food chain, thus the food-chain length is proportional to the relative concentrations of these two isotopes in the top consumers.

FIGURE 15.12C Radioisotope Measurement of Food-Chain Length

15.13 Increased productivity can increase food-chain length but decrease stability

In this chapter, we have already seen evidence that increases in primary productivity lead to increases in the biomass of herbivores. This observation leads naturally to the hypothesis that increases in primary productivity might also lead to increases in the biomass of predators that feed on herbivores, and so on. Furthermore, it is conceivable that the total number of trophic levels—that is, food-chain length—might respond to changes in primary productivity.

This idea has been investigated experimentally by Jenkins and his colleagues (1992). They studied communities of bacteria and insects that live in treeholes in the Australian tropics. These communities can be found in a variety of plants that collect water, like bromeliads (Figure 15.13A). Decaying leaves and animals provide the primary energy supply to these communities. Jenkins was able to replicate these communities artificially by placing plastic containers under trees. Each container initially started with some water and decaying leaves. The amount of decaying leaves was varied over three levels: high, medium, and low. The numbers of species and their feeding relationships were determined at regular intervals over a 48-week period. Figure 15.13B shows the results. As the levels of primary energy to the community were increased, the food-web structure became more complex. That is, there were more species in the community and more trophic links. On average, the maximum food-chain length was greatest at the highest productivity. Thus, this study supports the idea that food-chain length is positively correlated with primary productivity.

Are there other factors that might limit the length of food chains? Is it reasonable to suppose that food chains will become longer and longer as primary productivity is increased? One problem that must be considered is the stability of long food chains. If small changes in the numbers of primary producers cause large changes in

FIGURE 15.13A The leaves and central tank of the bromeliad support a community of bacteria and insects.

the numbers of top carnivores, then even modest variation may cause top predators to go extinct.

Lawler and Morin (1993) examined the stability of small communities of protists (Figure 15.13C). In the simplest community there were two species, a bacteria and a bacterivore (bacteria-eating) protist, *Colpidium*. In this simple community, *Colpidium* rapidly reaches its carrying capacity and then changes little in population size over time (panel I). When an additional *Actinosphaerium* carnivore is added, *Colpidium* persists, but its numbers fluctuate (panel II). In a second community with bacteria, *Colpidium*, and the omnivore *Blepharisma*, the numbers of *Colpidium* fluctuate wildly (panel III), and in one replicate the *Colpidium* population goes extinct (panel IV).

Together these two studies show that food-chain length may depend on both primary productivity and the ecological stability of top carnivores. Increasing primary productivity may increase food-chain length up to a point. Very high levels of primary productivity may not result in longer food chains due to the extinction of top carnivores. ❖

Over the first 24 weeks, the number of species increases in all treatments. However, the number of species is always greater in the treatments with higher productivity.

The pattern for the number of trophic links is essentially the same as for the numbers of species.

There is essentially no difference in the maximum food-chain length between the low- and medium- productivity treatments. However, the high-productivity treatment shows a greater food-chain length at all sample points.

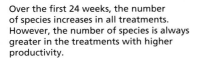
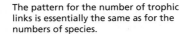
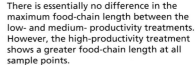
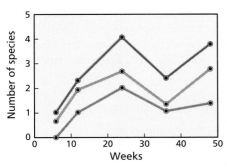
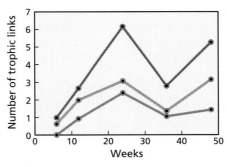
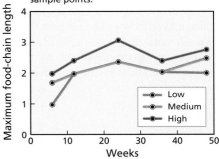

FIGURE 15.13B Community Structure as a Function of Primary Productivity

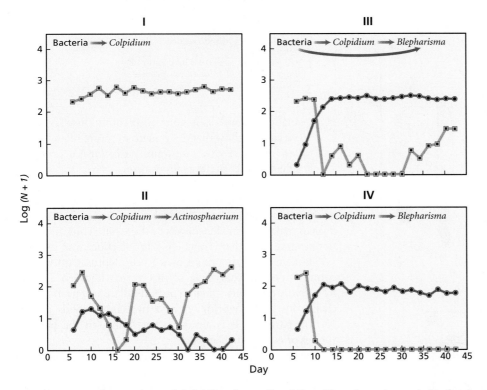

FIGURE 15.13C The Numbers of *Colpidium* (green lines) Over Time In each panel, the food chain of the introduced species is shown. Panel I—a two-species community. II—three species, two links. III and IV—the same three species are shown in both these panels but in different replicate cultures. In III, *Colpidium* persists but fluctuates wildly. In IV, *Colpidium* goes extinct by day 12.

15.14 The structure of communities is also affected by the genetic structure of its members

We have already seen that many predator-prey and host-parasite relationships can be species specific. It is not unreasonable to think that some interactions may even depend on the genotypes within a population. Certainly, for some parasites we have already considered the idea that some genotypes may be susceptible to parasites while others are resistant. Although most species consist of many genotypes, plants possess the ability to generate large levels of genetic variability in a small area by hybridization with other species. It is not unusual to find closely related species of plants that can form viable hybrids. In fact, 30 to 80 percent of all plant species may arise through hybridization.

The areas where different species meet and form hybrids is called a hybrid zone. The progeny that result from the cross of two different species are called the F_1 generation. The F_1 individuals may then mate with one of the parental species. The progeny from this type of cross are called backcross progeny.

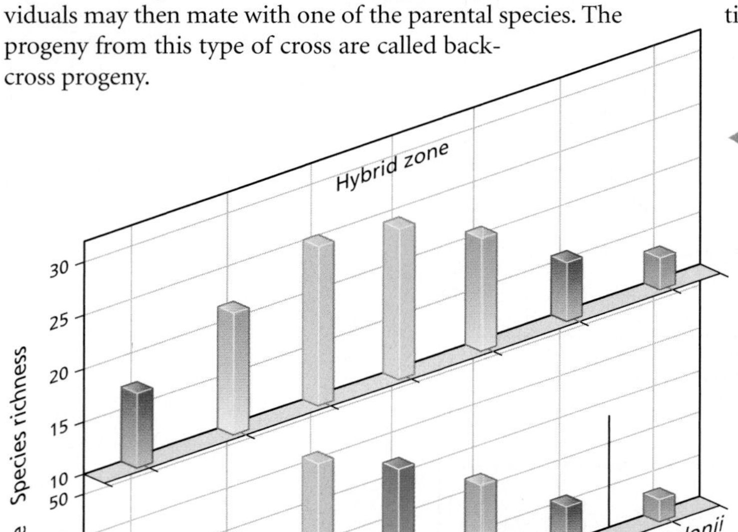

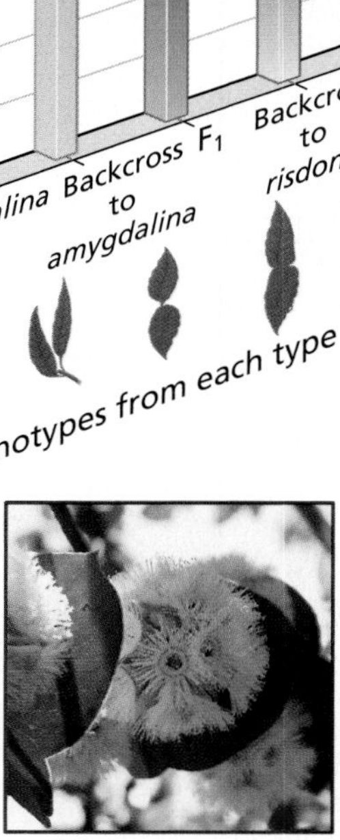

In Australia, two different species of eucalyptus trees (*E. amygdalina* and *E. risdonii*) hybridize. These hybrids have very different leaf morphologies than those seen in either parent (Figure 15.14A), and different physiology. These trees harbor a number of insect and fungal species. Thomas Whitham and his colleagues have examined the insect and fungal communities of each parental species of eucalyptus and of trees found in the hybrid zone. They measured the species richness of these insect and fungal communities, which is a measure of the number of different species. In general for a community, the more species there are and the more even their distribution, the greater that community's species richness will be.

This study revealed that the species richness and relative abundance of insects and fungi was much greater in the hybrid community than in either of the parental eucalyptus populations (see Figure 15.14A). This effect is seen in natural pop-

> Two species of *Eucalyptus* (*E. amygdalina*, *E.risdonii*) intermate and form hybrids in parts of Australia. In these hybrid zones the characteristics of the plants range from mostly *amygdalina* to mostly *risdonii*, and the F_1 hybrids have intermediate traits.
> The species richness and relative abundance of 40 different insect and fungal taxa increase dramatically in comparing the parental strands of *Eucalyptus* to those consisting of mostly hybrids (F_1s). A backcross is a cross between a hybrid and one of the parental species.
> The bars of the same color have statistically the same values of either species richness or relative abundance.

ulations as well as in controlled conditions. These observations show that the genetic structure of an important member of a community can have a profound effect on the species composition.

It is not entirely clear what is causing this effect among the eucalyptus populations. However, these plants produce a large number of oils that deter attacks by insects. The hybrids appear to make intermediate levels of these oils. Thus, they may not have sufficient levels of chemical protection from insects.

These findings are not peculiar to eucalyptus. Cottonwood communities in Utah also form hybrids between *Populus freemontii* and *P. angustifolia*. The bud gall mite appears to specialize on the F_1 hybrids of these two cottonwoods [Figure 15.14B, part (i)]. It is almost never found on the parental species or even on backcrosses.

The effects of these hybrids are not limited to insects and fungi. The hybrid cottonwoods also have a very different morphology than either of these parents [see Figure 15.14B, part (i)]. These hybrids are apparently more attractive to a variety of bird species and, as a result, many more bird nests are found in the hybrid zone than in either of the parental populations [Figure 15.14B, part (ii)]. ❖

FIGURE 15.14A Biodiversity in a Eucalyptus Hybrid Zone

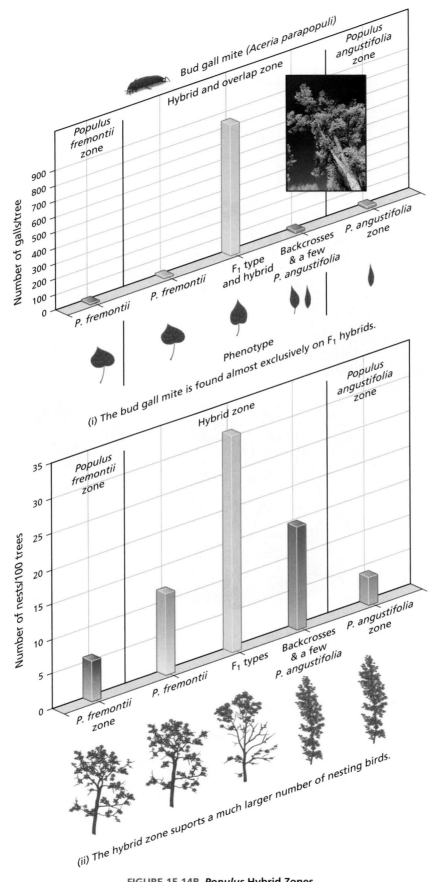

(i) The bud gall mite is found almost exclusively on F₁ hybrids.

(ii) The hybrid zone suports a much larger number of nesting birds.

FIGURE 15.14B *Populus* Hybrid Zones

ECOSYSTEMS

An important feature of ecosystems and their biological communities is their interaction with the physical environment

Life on Earth is almost entirely sustained by energy from the sun. Plants form the essential link between the sun's energy and the energy that is used by virtually all biological life. The Earth is an open system with respect to energy. Energy comes from outside the boundaries of the Earth and its immediate atmosphere (Figure 15.15A).

One consequence of this dependence on the sun is that certain thermodynamic laws that apply specifically to closed systems are violated, because the Earth is an open system for energy. In closed thermodynamic systems, entropy (or disorder) should always increase. However, on Earth the organization of chemical elements into living organisms represents a decrease in entropy. This decrease in entropy is possible only because of the flow of energy into the Earth from the sun.

Nutrient Cycles Life depends on other components in addition to energy. All organisms are composed of various essential elements, such as carbon, nitrogen, sulfur, phosphorus, and others. In addition to these essential elements, all life requires water. We will collectively refer to these required substances as **essential nutrients**. Because the Earth gets no significant amount of these nutrients from outside its boundaries, the Earth is a closed system with respect to these essential nutrients.

Does life use up these nutrients? Is it possible that we will run out of these essential nutrients just like we may one day run out of fossil fuels? The basic answer is no, because life on Earth recycles these essential nutrients between living organisms and nonliving components of the Earth such as the atmosphere and oceans. These cycles of nutrients are called **biogeochemical cycles**, since they depend on both living and nonliving components. We will review three important biogeochemical cycles in Module 15.16.

FIGURE 15.15A
The Earth is an open system for energy, but it is a closed system for nutrients essential to life.

Large pools of nutrients reside in reservoirs on Earth. For nutrients that have a gaseous stage in their cycle, such as water, nitrogen, and oxygen, the atmosphere and oceans serve as important reservoirs. For nutrients with sedimentary cycles, such as phosphorus, rocks and soil serve as the main reservoirs.

The movement of nutrients from one component of the biogeochemical cycle to another is called **flux**. The flux is measured in the amount of nutrient per unit of time. The parts of the cycle with large fluxes are key to understanding the dynamics of the nutrient. These parts of the cycle, if perturbed, would be expected to have the greatest impact on the availability of the nutrient.

Ecosystem Function

Understanding ecosystems requires that we understand the interaction between biological communities and the physical environment. Indeed, it is impossible to understand some processes completely without looking at the interaction of the environment and organisms. Many biogeochemical cycles have important biological components. The cycling of nutrients such as nitrogen depends critically on the ability of microorganisms to carry out important chemical reactions. Physical processes such as weather also depend on the activities of living organisms. Plants and animals have played an important role in the cycling of CO_2 in the atmosphere. However, over the last century humans have significantly increased atmospheric CO_2 through the burning of fossil fuels. This rise in CO_2 is continuing to this day (Figure 15.15B).

As we will see in Chapter 16, CO_2 plays an important role in the global climate, along with water vapor and nitrous oxide. A significant amount of heat energy that would otherwise radiate from the Earth back into space is captured by these molecules,

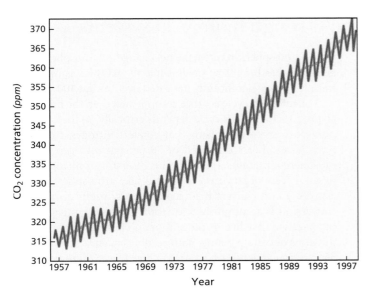

FIGURE 15.15B Change in Atmospheric CO₂ Concentration
These measurements, made in Hawaii, show a steady increase in the levels of CO_2 over a 20-year period. Within a single year there are also small rises in CO_2 concentrations during the winter months. This change corresponds to the decline in CO_2 uptake by plants during the dormant winter months.

keeping the Earth warm through the **greenhouse effect**. This effect refers to the action of our atmosphere that lets visible and ultraviolet light through to the Earth's surface but absorbs much of the heat energy that the Earth radiates back to the atmosphere. There is currently concern that human production of CO_2 may be leading to increases in global temperature. There may also be a positive feedback loop between temperature and atmospheric levels of CO_2. In the remaining modules we will consider evidence that elevated temperatures may accelerate the decomposition of soil carbon and its release into the atmosphere. If these changes were to continue, local climates could change in a significant fashion.

Atmospheric CO_2 levels also change as Earth's ecosystems change. In Module 15.18, we will see that changes in the diversity of biological communities may also affect a community's ability to take up CO_2. Thus, preservation of species diversity may ultimately be critical to preserving the physical environment on which all life depends. ❖

15.16 Essential nutrients are recycled through biological systems

Here we review the cycles of three important nutrients: water (Figure 15.16A), carbon (Figure 15.16B) and nitrogen (Figure 15.16C). In the hydrologic cycle, the largest reservoir of water is the Earth's oceans. The oceans are also the components with greatest flux of water, through evaporation and rainfall, as the numbers in Figure 15.16A show. On land, water is lost by evaporation from the Earth and from plants and animals. This flux is called **evapotranspiration**.

The Earth's atmosphere is composed of 78 percent nitrogen, 21 percent oxygen, and only 0.03 percent carbon dioxide along with other trace gases. Nevertheless all plants depend on atmospheric CO_2 for the carbon used in photosynthesis (Figure 15.16B). In the oceans, carbon dioxide is dissolved in water, where it exists as carbonate ion (HCO_3^-). The carbon fixed by plants either stays in the plant or is consumed by animals. Ultimately, both plants and animals die, and then they decompose through the action of microorganisms. Much of the carbon content then returns to the atmosphere after decomposition.

Some of the carbon in dead organisms, however, is lost to the cycle in sediments. Most of these sediments have very low carbon concentrations, but occasionally there are high concentrations of carbon in fossil fuel deposits. Before the advent of human civilization, these reservoirs contributed little to atmospheric CO_2 concentrations. However, over the last 100 years these reservoirs have been returned to the carbon cycle through human burning of fossil fuels. Although this burning also con-

sumes atmospheric oxygen, the net change in atmospheric oxygen levels has been very small, while the CO_2 levels in the atmosphere have significantly increased over the last 100 years.

The nitrogen cycle is more complicated than the other two (Figure 15.16C). This cycle depends critically on the action of numerous microorganisms that convert nitrogen from one form to another. Composed of 78 percent nitrogen, the atmosphere represents a tremendous reservoir of nitrogen; but relatively few organisms are able to take atmospheric nitrogen (N_2) and convert it to a biologically useful form. This conversion is accomplished for some plants by nitrogen-fixing bacteria that live in the soil or in close association with the roots of certain plants. Both aerobic bacteria such as *Azotobacter* and anaerobic bacteria such as *Clostridium* fix nitrogen. Animals get their nitrogen from the consumption of plant or animal proteins. As plants or animals die, their proteins and amino acids are converted to ammonium ions by microorganisms that derive energy from this process. Ammonium can be taken up by plants for their nitrogen needs, or it can be oxidized to nitrate by a process known as **nitrification**. The first step in nitrification is the conversion of ammonia to nitrite (NO_2^-) by the bacteria *Nitrosomonas*. Nitrite is then oxidized to nitrate (NO_3^-) by another group of bacteria, the *Nitrobacter*. Nitrate can then be taken up by plants or returned to the atmosphere as nitrogen (N_2) by the process of **denitrification**. This last process is carried out by bacteria of the genus *Pseudomonas*. ❖

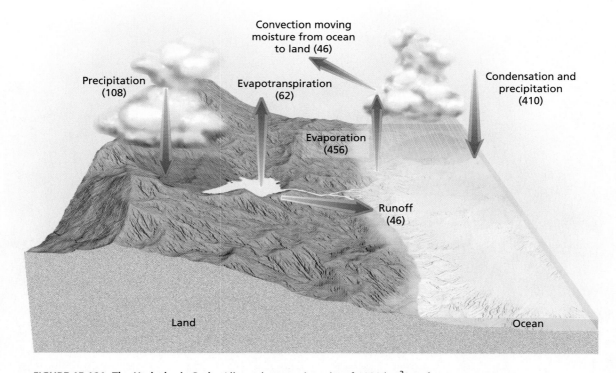

FIGURE 15.16A The Hydrologic Cycle All numbers are in units of 1000 km³/yr of water.

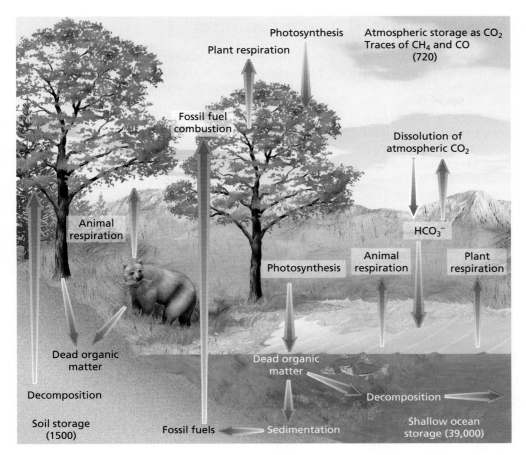

FIGURE 15.16B **The Carbon Cycle** Storage units are in billions of metric tons of carbon.

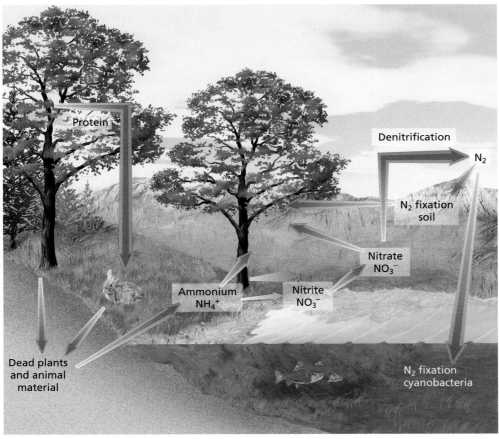

FIGURE 15.16C
The Nitrogen Cycle

15.17 Soil carbon levels are affected by temperature

The levels of atmospheric carbon dioxide are important for their greenhouse effects. The amount of carbon that is stored in soils is two to three times greater than the amounts in the atmosphere. Much of this carbon is in the form of soil organic matter. Because soil organic matter can be decomposed and its carbon released as CO_2, it is a potentially important dynamic source of atmospheric carbon dioxide.

Factors that affect the decomposition of organic matter may also affect the levels of atmospheric carbon. One obvious factor is temperature. Because the metabolic rates of decomposing organisms will increase with increasing temperature, the rates at which soil organic matter is recycled into the atmosphere may also depend on temperature.

In an odd twist of fate, nuclear testing in the western United States has provided an opportunity to explore the important ecological relationship between temperature and soil carbon levels. Nuclear testing between 1958 and 1963 roughly doubled the levels of carbon-14 near the Sierra Nevada mountain range. About the same time, soil samples from the Sierra Nevada were taken and stored as archive samples for later testing.

After 1963, the organic matter in the soils of the Sierra Nevada would be expected to start showing elevated levels of carbon-14 as this isotope became incorporated into plants, and these plants died or shed leaves into the soil (Figure 15.17A). Of course the relative amounts of carbon-14 in the organic matter of the soil would depend on how fast the old organic matter was decomposing and moving out of the soil carbon reservoir.

The Sierra Nevada range is also an interesting study site because the mean annual temperature declines steadily and substantially as one moves from the base of the range to the summits. Using soil samples taken in 1992, Susan Trumbore and her colleagues were able to compare these soils to the archive samples. From these comparisons, the rates of turnover of soil carbon could be estimated at many different elevations (temperatures). Their study showed dramatically that soil carbon turnover is much higher at higher temperatures (see Figure 15.17A). These findings suggest that global warming will increase atmospheric carbon dioxide levels still further.

Based on their studies, Trumbore and colleagues estimated the effects of a 0.5°C increase in temperature on carbon levels in various ecosystems (Figure 15.17B). The effects are greatest in the tropics. In just a single year, all forests would release nearly 1.4×10^{15} grams of carbon, which is nearly 25 percent of the amount released by all fossil fuel consumption in a year. These findings show the complicated dependence of global nutrient cycles on many factors. ❖

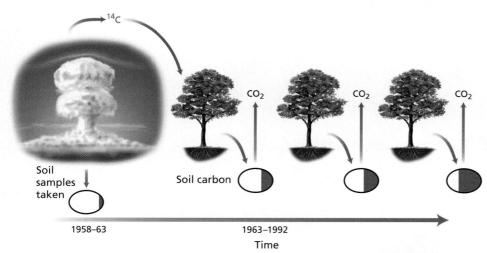

After nuclear testing, the levels of ^{14}C from these tests will increase in the soil carbon at a rate that depends on the rate of carbon turnover in the soil. The red indicates the level of soil carbon-14 at different times. This increase can be documented by comparing soil samples in the 1990s with those taken in the period of 1958–1963.

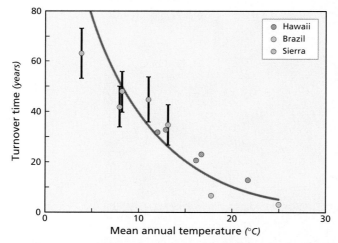

These measurements from the Sierra Nevada and other locations show a rapid decline in turnover time with mean annual temperature.

FIGURE 15.17A Soil Carbon Levels and Temperature

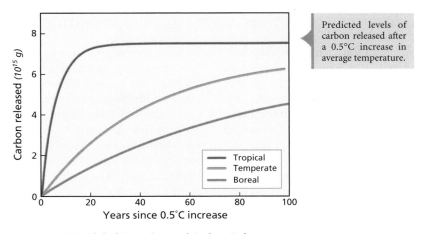

FIGURE 15.17B **Global Warming and Carbon Release**

15.18 Species diversity affects ecosystem performance

We have just seen how ecosystems function to recycle essential nutrients. Does the ability to recycle nutrients depend on properties of the community? Many have suggested that species diversity may have multiple effects on ecosystems, including increased productivity and lower loss of nutrients.

This problem has been examined by Shahid Naeem and colleagues with replicated artificial communities. A large environmental chamber called the *ecotron* was used to create replicate communities with four trophic levels and different numbers of species (Figure 15.18A). The high-diversity community had 31 species, while the low-diversity community had 9 species. These communities were followed for a total of 206 days.

As the plants in each community grew, researchers recorded the fraction of surface area covered by them. The largest changes in percentage of cover were observed in the more diverse community (see Figure 15.18A). This result is not an unavoidable consequence of having more plant species. However, in the more diverse communities, the available space was filled more densely than in the low-diversity communities.

The net consumption of CO_2 was used as a measure of overall photosynthetic rates. These rates were also higher in diverse communities (Figure 15.18B). Given these results, it is perhaps not surprising that the primary productivity was also greater in the high-diversity communities (Figure 15.18C). One possible explanation for these results is that the diverse communities of plants include species that vary in height and leaf shape. This variety may result in the community more effectively making use of the available energy from sunlight.

More diverse communities also reduce available nitrates to lower levels (Figure 15.18D). This means that less nitrogen is leached from the soil. Ultimately this would have a positive effect on sustaining nutrient cycling and soil fertility.

An important implication of these studies is that reductions in species diversity might have a negative impact on ecosystem function. Consequently, we have additional reasons for being concerned about the loss of species due to human activity. The ecosystems that humans rely on for recycling the nutrients that we need may be impaired by the loss of species. ❖

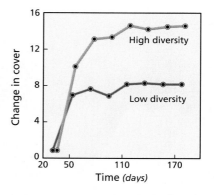

FIGURE 15.18A Replicate communities with either 9 (low diversity) or 31 (high diversity) species were followed over time. In the more diverse communities, there was greater plant coverage of the soil surface.

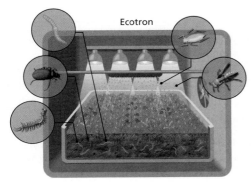
Ecotron

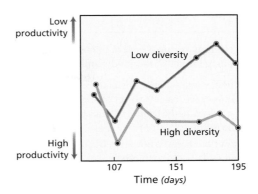

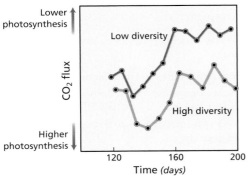

FIGURE 15.18B Overall rates of photosynthesis were also greater in the more diverse communities.

FIGURE 15.18C The amount of plant material (productivity) was greatest in the more diverse communities

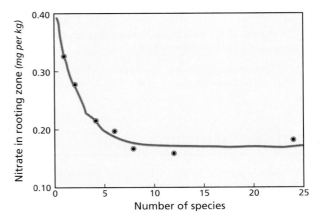

FIGURE 15.18D Diversity vs. Nitrate Concentration

SUMMARY

1. Ecological communities are bound to follow the other laws of science, including the law of thermodynamics.

    a. Because total energy cannot normally be created or destroyed, important insights about communities can be gained by following the transfer of energy from one trophic level to the next.

    b. This transfer of energy is not very efficient, although some communities do better than others.

2. Communities in change are quite common.

    a. Some of the earliest ecological work was motivated by the process of ecological succession.

    b. There is enough regularity in the succession process that it has been used to date the time of death of decaying human bodies.

3. The number of species and the number of trophic levels in a community depend on many factors.

    a. Predators may prevent a superior competitor from eliminating less-effective competitors.

    b. Occasional environmental disturbances can also have a similar effect.

    c. The numbers of species in a community may also have positive effects on the functioning of the entire ecosystem.

4. The cascade model predicts many features of real food webs while making few assumptions.

    a. This model can also be used in conjunction with the theory of island biogeography to predict that the average chain length in food webs will increase with ecosystem size.

    b. Lake communities show increases in food-chain length with ecosystem size.

5. Ecosystems perform many functions that are essential for all life on Earth.

    a. The recycling of important nutrients is one of these functions.

    b. Atmospheric gases like carbon dioxide are produced by ecosystems, and these in turn have effects on global climates.

    c. Ecosystem function appears to be improved by increasing species diversity.

REVIEW QUESTIONS

1. Why does Figure 15.1B support the conclusion that lake communities are top-down regulated?

2. For each of the following terms, explain how energy is lost, making the measure of efficiency less than 100 percent: (i) exploitation efficiency, (ii) assimilation efficiency, (iii) net production efficiency.

3. What did G. E. Hutchinson find paradoxical about plankton communities? How did he explain this paradox?

4. Would you expect the equilibrium number of species on an island far from the mainland to be greater, smaller, or equal to the number on a nearby island? Draw a graph to show how you reached your conclusion.

5. What is a trophic cascade? Are the results in Figure 15.1B consistent with a trophic cascade?

6. In this food-web matrix, indicate which species are primary producers, herbivores, and top carnivores:

	1	2	3	4
1	0	0	1	0
2	0	0	1	0
3	0	0	0	1
4	0	0	0	0

7. Review the roles of different species of bacteria in the nitrogen cycle.

KEY TERMS

allogenic succession
assimilation efficiency
autogenic succession
biogeochemical cycles
biomass
biome
bottom-up regulation
climax community
community
community web
degradative succession
denitrification
directed link
dynamical stability hypothesis
ecological efficiency
ecological equilibrium
ecosystem
ecosystem-size hypothesis
essential nutrients
evapotranspiration
exploitation efficiency
flux
food chains
food web
food-web chain
food-web cycle
food-web length
food-web link
food-web loop
greenhouse effect
intermediate disturbance hypothesis
keystone predator
net production efficiency
nitrification
node
nonequilibrium
primary producers
primary succession
productivity hypothesis
secondary succession
sink web
source web
species diversity
species richness
succession
top-down regulation
trophic cascade
trophic levels
undirected link

FURTHER READINGS

Cohen, J. E., 1978. *Food Webs and Niche Space.* Princeton, NJ: Princeton University Press.

Jenkins, B., R. L. Kitching, and S. L. Pimm. 1992. "Productivity, Disturbance and Food Web Structure at a Local Spatial Scale in Experimental Container Habitats." *Oikos* 65:249–55.

Lawler, S. P., and P. J. Morin. 1993. "Food Web Architecture and Population Dynamics in Laboratory Microcosms of Protists." *American Naturalist* 141:675–86.

Morin, P. J. 1999. *Community Ecology.* Malden, MA: Blackwell Science.

Naeem, S., L. J. Thompson, S. P. Lawler, J. H. Lawton, and R. M. Woodfin. 1995. "Empirical Evidence that Declining Species Diversity May Alter the Performance of Terrestrial Ecosystems." *Philosophical Transactions of the Royal Society London* B 347:249–62.

Paine, R. T. 1966. "Food Web Complexity and Species Diversity." *American Naturalist* 100:65–75.

Post, D. M., M. L. Pace, and N. G. Hairston Jr. 2000. "Ecosystem Size Determines Food-Chain Length in Lakes." *Nature* 405:1047–49.

Power, M. E. 1990. "Effects of Fish in River Food Webs." *Science* 250:811–14.

Smith, K. G. V. 1986. *A Manual of Forensic Entomology.* London: Trustees of the British Museum of Natural History.

Sousa, W. P. 1979. "Disturbance in Marine Intertidal Boulder Fields: The Nonequilibrium Maintenance of Species Diversity." *Ecology* 60:1225–39.

Trunbore, S. E., O. A. Chadwick, and R. Amundson. 1996. "Rapid Exchange between Soil Carbon and Atmospheric Carbon Dioxide Driven by Temperature Change." *Science* 272:393–96.

Worthern, W. B. 1989. "Predator-Mediated Coexistence in Laboratory Communities of Mycophagous *Drosophila* (Diptera: Drosophilidae)." *Ecological Entomology* 14:117–26.

The physical environments of biological communities vary tremendously. For instance, rainfall in terrestrial habitats varies from the hot, dry desert to the hot, wet tropical rain forest (shown here).

16

The Biosphere and the Physical Environment

Organisms are victims of their environments. If it rains or gets very hot, they must find shelter. When it is cold, they must protect themselves from freezing. The physical environment is a major determinant of life and death.

For this reason, the physical environment is important ecologically and evolutionarily. Some environments, like the desert, may simply be too hot and dry for many plants and animals to survive there. Therefore those species will never be found in a desert climate. Many organisms that can survive in the harsh desert climate have adaptations that conserve water, permitting them to flourish in the extreme dryness of the desert.

Extreme environments are not randomly distributed across the Earth. Global patterns of air movement and ocean currents contribute to their creation. In the discussion of "Global Climates", which begins with Module 16.1, we will see how climates are a natural product of such global patterns.

Global forces are not the only determinants of climates. Local geography may also influence climate. Even living organisms may influence the temperature, moisture, or light intensity experienced by other plants and animals, as we will see in the topic "Local Climates," beginning with Module 16.6.

The biological communities that are found in different physical environments vary dramatically in their species composition and biological characteristics. In "The Ecology and Evolution of Biomes," which begins with Module 16.9, we review some of the major biomes on Earth.

Of course, the single species that has had the greatest recent impact on the environment is our own, *Homo sapiens*. Humans often have adverse effects on the survival of plants and animals. Many environmental changes caused by humans happen so fast that the affected populations may not be able to adapt to these changes before they go extinct. We will review some of the more disastrous environmental changes caused by humans in a discussion of "Global Change," beginning with Module 16.13. ❖

GLOBAL CLIMATES

16.1 Global climates are not static, but show major cycles every 100,000 years

Anyone who is considering two job offers—one in San Diego, California, and the other in Fairbanks, Alaska—will think about the great climate differences between the two locations. Fairbanks has a seasonal climate with very cold, long winters; San Diego has very little seasonal variation in temperature, with increased rainfall in the winter being the major seasonal distinction.

Weather and Climates The climates of Fairbanks and San Diego reflect the weather that has been experienced in these locations over long periods of time. **Climate** is defined as the long-term average weather of a particular locality. On any given day, the **weather** may deviate from long-term averages, sometimes by a lot. We typically cannot predict weather accurately more than 3–5 days in advance.

Does Climate Change? The question whether climate can change is very relevant to evolution and ecology, given the pressure that natural selection exerts on organisms to adapt to their environment. Because climate is the long-term average weather of a particular locality, departures from the average that last just a few years are not considered a change in climate. However, there is evidence of long-term changes in temperature that are associated with climate changes. Figures 16.1A through 16.1E document cycles in global temperature changes, each cycle lasting about 100,000 years. Although the average temperature changes by no more than 6°C, these cycles mark major global changes. During the coolest periods, there were

large-scale advances of glaciers southward. **Glaciers** are ice sheets that may have been more than 1 km deep. During these glaciation events, the average surface temperature of the Earth was 10°C. Between glaciation events, the average temperature rises to 15°C. Glaciation events have occurred about every 100,000 years over the last 700,000 years.

We see that over the last 90 years (Figure 16.1D), there has been a gradual warming trend. Because this change is well within the limits of Earth's historical variation in temperature, it is unclear if this marks a natural cycle or is perhaps due to human influence. We will consider this issue in more detail later, in Module 16.5.

Factors That Affect Climate Many factors influence climate. Some factors are global in nature and lead to predictable patterns. We know that the *distance from the equator* is one way to predict climate. Because the equator gets the

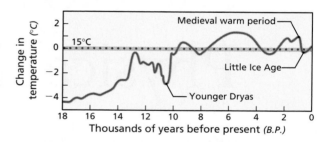

FIGURE 16.1C Long-Term Global Temperature Change over the Past 18,000 Years

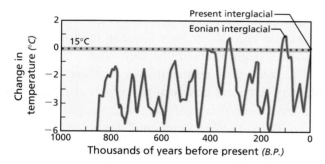

FIGURE 16.1A Long-Term Global Temperature Change over the Past Million Years

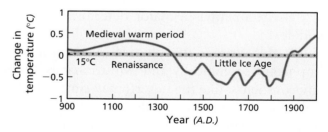

FIGURE 16.1D Long-Term Global Temperature Change over the Past 1000 Years

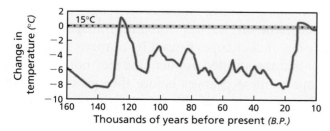

FIGURE 16.1B Long-Term Global Temperature Change over the Past 160,000 Years

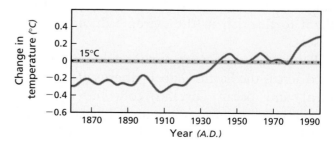

FIGURE 16.1E Long-Term Global Temperature Change over the Past 120 years

most direct energy from the sun, the *temperature* there is typically very warm. As we move to localities north and south of the equator, the average weather should be cooler. However, as we will see in Module 16.2, other global patterns appear as one moves north and south of the equator, such as the air currents that produce desert and rain-forest ecosystems.

Besides temperature, another factor that varies predictably with latitude is **seasonality**, or variation in weather according to the seasons. Seasonality is related to the tilt of the Earth's axis relative to the sun. Thus, as we move north and south of the equator, we also expect to encounter habitats that vary over a yearly cycle. Organisms must not only be able to tolerate the average environment, but they must be able to cope with the extremes. We discuss the basis of seasonality in Module 16.3. Local climate may also depend on conditions other than those that vary with the distance from the equator. For instance, Seattle, Washington, is farther north than Bangor, Maine. Yet Seattle has fairly mild winters in which snowfall is unusual, while in Bangor snowfall is quite common. Clearly, factors other than latitude affect climate (Figure 16.1F). Some of these other factors include proximity to large bodies of water and the prevailing air and ocean currents. We consider these factors in Module 16.4.

Global Climates Affect Ecosystem Composition

By understanding the forces that give rise to these global climate patterns, we start to understand the factors that affect the biological communities that are found in different regions of the world. In Module 16.2, for example, we review global patterns of air flow that are responsible for the hot, moist conditions near the equator and the dry climates found just north and south of the equator. As we will see, many of the great deserts of the Earth are created by the same airflow patterns responsible for the rain forests. These climates largely determine the types of biological communities that are found there. ❖

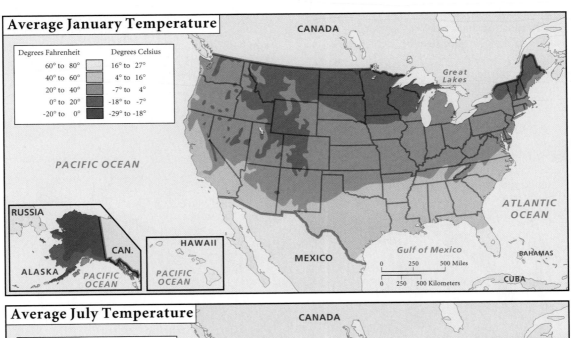

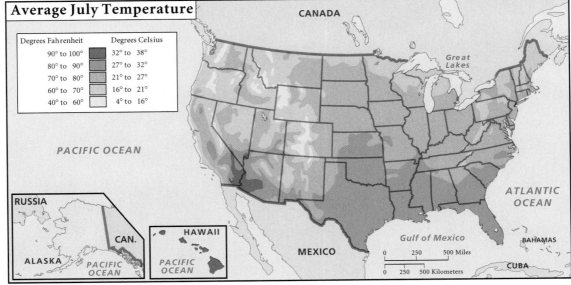

FIGURE 16.1F **Average Temperature Profile of the United States**

16.2 The sun's energy and air currents are responsible for rain forests and deserts

Life on Earth depends on the sun's energy. The most important biological use of the sun's energy is photosynthesis. But the Earth's climates are also dependent on the sun's energy. As Figure 16.2A shows, about 50 percent of the incoming energy from the sun actually hits the surface of the Earth. Much of the sun's energy is reflected back into space. Some is reflected from the surface, but most of it is reflected higher in the atmosphere of Earth. About 70 percent of the incoming solar energy is radiated out to space as longwave (infrared) radiation. Although a lot of energy is radiated from the surface of the Earth, most of that energy is not lost due to the greenhouse effect, which we discuss later in this chapter.

The sun's heating of the surface of the Earth causes air to flow in predictable directions that have important conse-

The Sun's heating of the surface of the Earth causes air to flow in predictable directions that have important consequences for climates near the equator.

quences for climates near the equator. A cycle of airflow called a **Hadley cell** is responsible for both rain forests and deserts in certain areas. Figure 16.2B diagrams the air circulation in a Hadley cell. As a starting point, note that because it receives the most direct sun rays, the equator gets more energy from the sun than does any other place on Earth. This energy heats up the air near the Earth's surface, and the air then starts rising. Air near the equator is also quite moist: and as the air rises, it cools. Eventually the air becomes sufficiently cool that water condenses and is released as rain. This rainfall keeps the Earth's surface moist near the equator. In fact, the major rain forests of the world are mostly found 10° north and 10° south of the equator. The largest of these rain forests is in the Amazon basin of Brazil. The other major rain forests are in Indonesia, Malaysia, and New Guinea and in the central African region of the Congo.

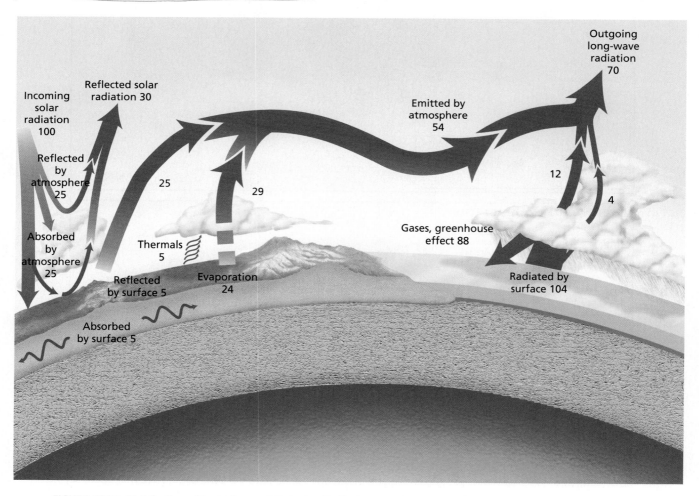

FIGURE 16.2A Distribution of Incoming and Outgoing Radiation as a Percentage of the Total Incoming Solar Radiation

After releasing water, the air above the equator continues to rise and cool; but it is now dry. This dry air eventually stops rising and flows north and south from the equator. This air from the equator then starts pushing back toward the Earth's surface. As this air descends, it is compressed and becomes hot, although it is still quite dry. When this hot, dry air hits the Earth's surface, it absorbs water, creating deserts roughly 20–30° north and south of the equator. This pattern of airflow accounts for some of the great deserts of the world, including the Sahara in North Africa, the Kalahari and Namib in South Africa, the Great Victoria desert in Australia, the Great Indian desert in India, and the Atacama desert in South America. Plants and animals with very specialized adaptations to extreme aridity make up these desert communities.

These airflows tend to result in high air pressure near 20–30° north and south of the equator, and low pressure at the equator. Air moves from high-pressure areas to low-pressure areas. Thus the hot, dry air at the desert moves back toward the equator, picking up moisture along the way and completing the cycle of airflow in a Hadley cell. ❖

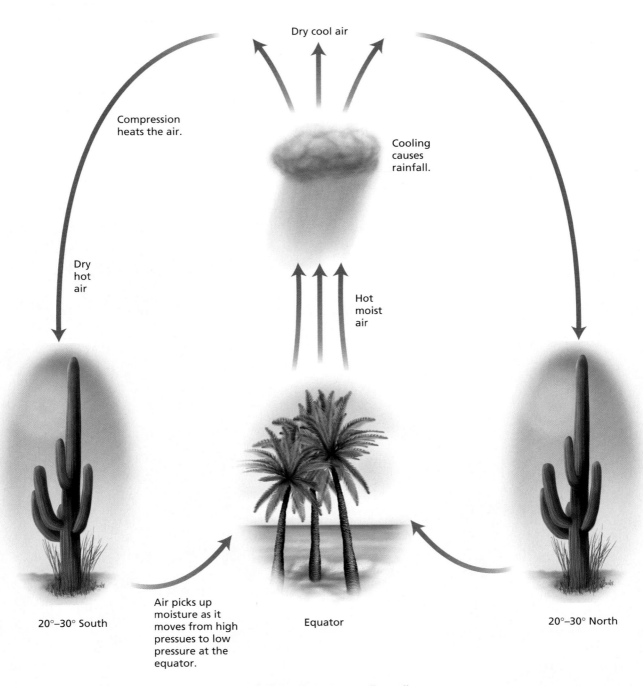

Dry cool air

Compression
heats the air.

Cooling
causes
rainfall.

Dry
hot
air

Hot
moist
air

Air picks up
moisture as it
moves from high
pressues to low
pressure at the
equator.

20°–30° South

Equator

20°–30° North

FIGURE 16.2B Air Circulation in a Hadley Cell

16.3 The tilt of the Earth on its axis results in seasonal cycles in temperature and daylight

The air currents produced in Hadley cells create global climate patterns that persist all year long. Other global patterns are not only persistent and dramatic, but seasonal. The degree of seasonality is related roughly to how far from the equator you are. The seasons experienced by any locality also depend on whether it is north or south of the equator. In Figure 16.3A we show how the Earth's tilt affects the seasons in the Northern Hemisphere. Places south of the equator will experience the opposite set of seasons shown in this figure. For example, while it is winter in the Northern Hemisphere, it will be summer in the Southern Hemisphere.

The energy delivered by the sun to the surface of the Earth can be attenuated, for two reasons. First, because the Earth's atmosphere absorbs energy, light passing through more atmosphere will have less energy when it strikes the Earth's surface than will light passing through less atmosphere. In the spring and fall, for instance, sunlight passes through more of the Earth's atmosphere on its way to the poles than when it hits the equator (see Figure 16.3A).

Second, the energy per unit of area will be greater when light hits the Earth's surface directly, compared with when light hits the Earth's surface at an angle. To demonstrate this point to yourself, take a flashlight and shine it directly on a wall about six inches from the flashlight. Then tilt the flashlight up toward the ceiling, so the light hits the wall at a severe angle. Although the light covers a greater area, it is not as bright as it was when you pointed it directly at the wall.

Because the Earth tilts on its axis, the portion of the Earth that receives the direct rays of the sun varies over a yearly cycle. In the fall and spring, the equator receives the direct rays. These direct rays also pass through less atmosphere, and thus deliver more energy to the Earth's surface (see Figure 16.3A). During the summer months, the point on the Earth that receives the direct rays moves north of the equator, resulting in increased temperatures in the Northern Hemisphere along with longer days. During the winter months, the Northern Hemisphere receives the indirect rays of the sun, like the top part of your wall in the flashlight experiment. Consequently, the Northern Hemisphere cools off and has shorter days.

The Earth's tilt on its axis also affects the length of days. Figure 16.3B shows the distribution of light during the summer in the Northern Hemisphere. Each of the views shown in Figure 16.3B is looking down on either the North or South Pole. As the Earth goes through its daily east-to-west rotation, land in the Northern Hemisphere (top illustration in Figure 16.3B) is shielded from the sun for a much shorter time than land in the Southern Hemisphere (bottom illustration in Figure 16.3B). In fact, areas near the North and South Poles experience 24 hours of light during the summer and 24 hours of darkness during the winter months. These areas are designated as the Arctic and Antarctic circles at 66°30′ for the North and South poles respectively. ❖

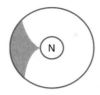
View from the North Pole. The small green circle is the Arctic circle. The Earth's tilt means that this half of the Earth receives more sunlight during the Northern Hemisphere's summer.

View from the South Pole. The small green circle is the Antarctic circle. While it is summer in the Northern Hemisphere, the Southern Hemisphere of Earth receives less sunlight.

FIGURE 16.3B During summer, days are longer.

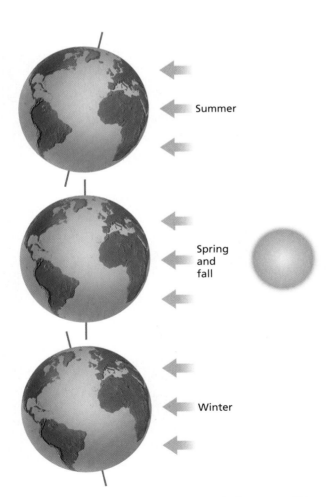

FIGURE 16.3A The Earth's tilt on its axis results in seasons. This figure shows the seasons in the Northern Hemisphere.

Summer

Spring and fall

Winter

16.4 The ocean currents modify land climates

Local climates are affected not only by air temperature, but by the temperature of nearby bodies of water. The critical difference between air and water is the very large heat capacity of water compared with air. It takes 1 calorie of energy to raise 1 mL of water 1°C; it takes only 10^{-4} calories to heat a similar volume of air. As a result, air temperatures fluctuate more than water temperatures. In addition, when the land and air near an ocean are at a different temperature, the ocean can exchange heat with them, causing the air and nearby land either to warm up or cool down, depending on the temperature differences.

Directed flows of water in the oceans are called **currents** or **gyres**. These currents are caused by several factors, including winds that move the surface waters and the Coriolis forces that result from the Earth's rotation on its axis (see Module 11.4). An important source of cold-water currents is the Antarctic gyre (Figure 16.4A) that flows in a clockwise fashion around the South Pole. This current gives rise to the cold Peru current in the Pacific Ocean and the Westwind Drift in the Indian Ocean (Figure 16.4B).

Two important warm currents are the Gulf Stream in the North Atlantic and the Japan current in the North Pacific. Both of these currents are warmed in the southwest portion of their cycles, and they are responsible for warming landmasses in the northeast portion of their cycle. The Japan current warms the west coast of Canada and the United States, while the Gulf Stream provides significant warmth to Northern Europe and Iceland. It is estimated that in the depth of winter, Iceland may receive half of its heat energy from the Gulf Stream.

Another type of current is created by a process called **upwelling** (Figure 16.4C). Offshore winds cause warm surface waters to move away from the land. The water is then replaced by cold, nutrient-rich water that rises from the depths. This process not only cools off the surface waters near the land but also leads to increased productivity of plankton and fish populations. Upwelling is found, for example, off the west coast of Mexico and the coast of Peru. ❖❖

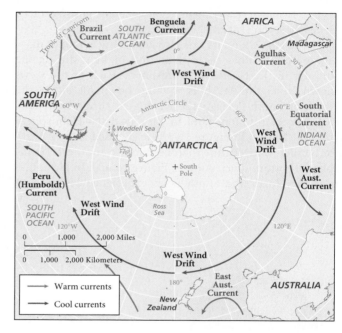

FIGURE 16.4A The Antarctic Gyre

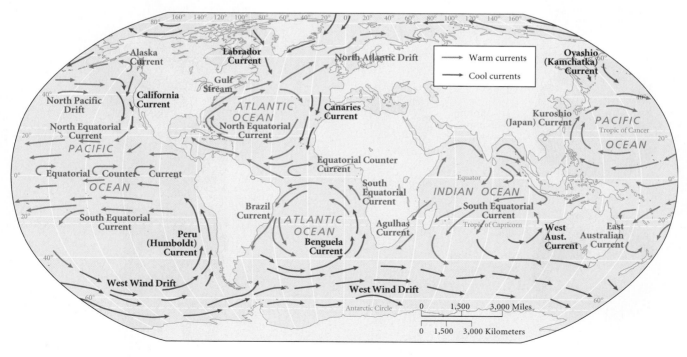

FIGURE 16.4B Ocean Currents

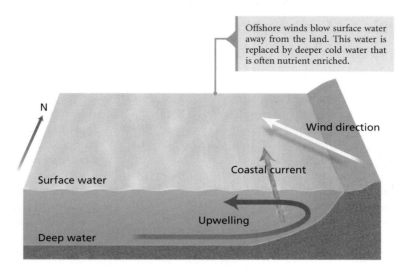

Offshore winds blow surface water away from the land. This water is replaced by deeper cold water that is often nutrient enriched.

Wind direction

N

Surface water

Coastal current

Upwelling

Deep water

FIGURE 16.4C **Upwelling in the Southern Hemisphere**

16.5 Atmospheric CO₂ and water vapor trap much of the sun's energy by a process called the greenhouse effect

We saw in Module 16.2 that only some of the sun's energy strikes the surface of the Earth, and that the Earth also radiates infrared radiation (heat) back to the atmosphere. This energy may then be absorbed by atmospheric gases (Figure 16.5A) and radiated back to the Earth's surface, conserving heat. Collectively these conserving atmospheric gases are called **greenhouse gases**. If the Earth had no atmosphere, its average temperature would be −18°C. So the effect of the atmosphere is critical to maintaining the temperatures necessary for life.

Not all gases play a role in heat capture. The most important greenhouse gas is water vapor. Gases other than water vapor play a fairly small role in capturing heat energy. One of the most important of these gases is carbon dioxide (CO_2). Human activities such as burning fossil fuels and using refrigerants add to the pool of greenhouse gases.

Some of the greenhouse gases contributed by human sources are listed in Figure 16.5A. Chlorofluorocarbons (CFCs) have been used as propellants in spray cans and as refrigerants. Nitrous oxide (N_2O) is produced by the combustion of fossil fuels and the application of fertilizers. Carbon dioxide, produced by burning fossil fuels, has been steadily increasing in the atmosphere for the last 50 years or more. Methane (CH_4) is produced from human activities like raising cattle and burning fuels.

So the effect of the atmosphere is critical to maintaining the temperatures necessary for life.

Will the human introduction of greenhouse gases inevitably cause an increase in world temperatures? Many scientists think so; global temperatures have already begun to rise and the only uncertainty is how much more they will increase. It is clear that there is a close correlation in historical data between atmospheric CO_2 concentrations and global temperatures (Figure 16.5B). The concern is sufficiently great that the world's industrialized countries have attempted to reach agreements on the reduction of greenhouse gases. International agreements are essential, because no country can keep its emissions of greenhouse gases within its own political borders. ❖

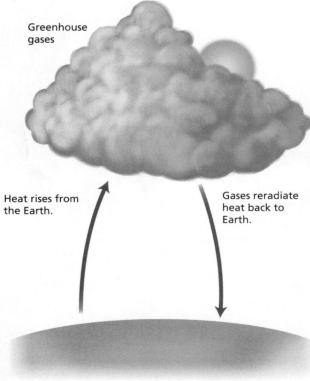

Greenhouse gases

Heat rises from the Earth.

Gases reradiate heat back to Earth.

Fraction of all heat captured by different greenhouse gases		Relative contribution to other gases by human sources	
Water vapor	85%	CFC	15-25%
Small particles of water	12%	CH_4	12-20%
All other gases	3%	O_3	8%
		N_2O	5%
		CO_2	50-60%

FIGURE 16.5A **The Greenhouse Effect**

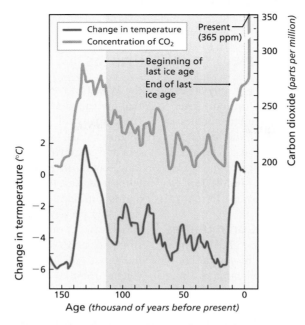

FIGURE 16.5B **Historical Values of CO_2 Concentration and Global Temperatures**

LOCAL CLIMATES

16.6 Many factors may affect local climates

We have seen that local climate may depend not only on latitude and global patterns of airflow but also on local conditions. These local factors include proximity to large bodies of water, such as an ocean, and the prevailing air and ocean currents. Another important factor influencing local climates is local topography, or the position of local geographic features, such as the location of mountain ranges. In Module 16.7 we will see how mountain ranges can create deserts.

Local Climates Affect Biological Communities

Why should we be interested in local climates? One reason is that they can have a substantial impact on biological communities. For example, there are similar local climates along the West Coast of the United States and the coast of Chile that are called **Mediterranean** because of their similarity to the climates of regions bordering the Mediterranean Sea (Figure 16.6A). Mediterranean climates have mild, wet winters and summer droughts that may be one or two months long. Living in this type of climate requires the ability to withstand these long, dry periods. Consequently, many of the plants found in Mediterranean climates have similar adaptations that prevent water loss, such as small leaves, thickened cuticles (waxy outer coverings), glandular hairs, and sunken stomata (pores). The biological communities found along the coasts of California and Chile are called **chaparral** (Figure 16.6B).

Many of these chaparral plants are evergreens. This is advantageous, because it permits the plants to grow in the winter, when most of the rainfall occurs. The exact timing of the winter rains is variable. By having green leaves all year long, the plants can start growing as soon as the rains begin. These plants also have deep, extensive root systems that extract water during dry periods. The root system of one chaparral plant may extend over a fairly large area, and it can prevent competing plants from becoming established—an important influence on ecological competition within these communities.

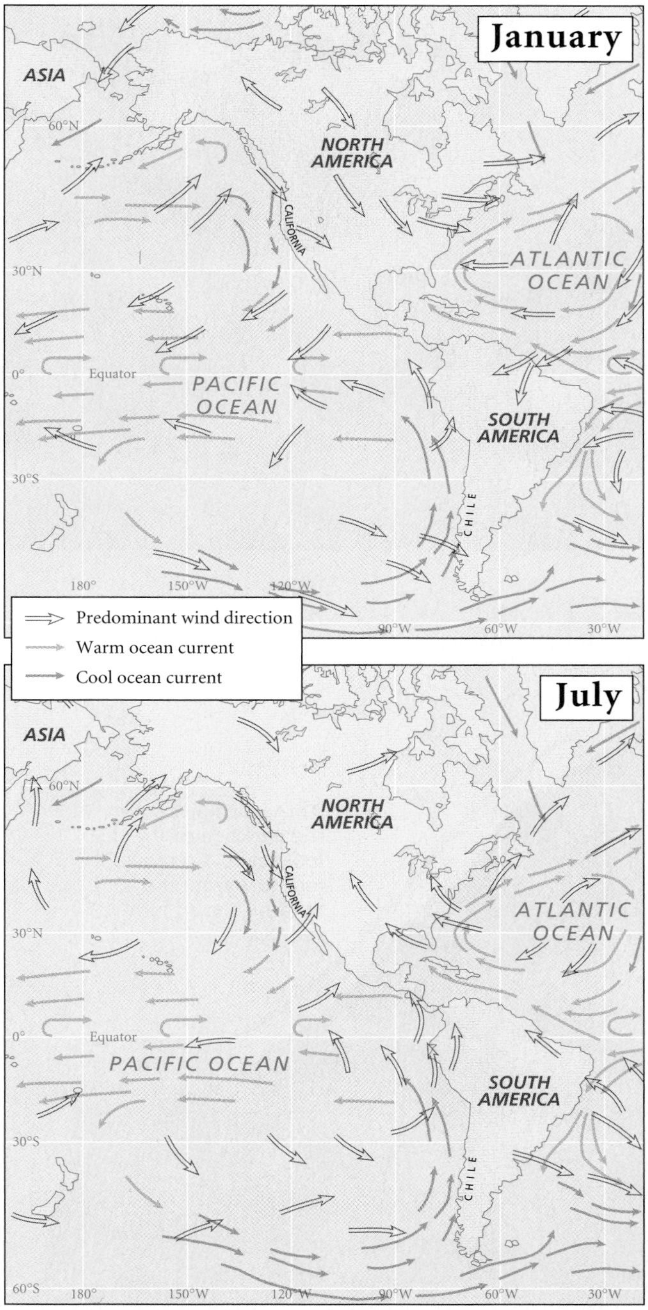
January
ASIA
60°N
NORTH AMERICA
CALIFORNIA
ATLANTIC OCEAN
30°N
0°
Equator
PACIFIC OCEAN
SOUTH AMERICA
30°S
CHILE
180°
150°W
120°W
90°W
60°W
30°W
Predominant wind direction
Warm ocean current
Cool ocean current
July
ASIA
60°N
NORTH AMERICA
CALIFORNIA
ATLANTIC OCEAN
30°N
0°
Equator
PACIFIC OCEAN
SOUTH AMERICA
30°S
CHILE
60°S
180°
150°W
120°W
90°W
60°W
30°W

FIGURE 16.6A Ocean currents off the coasts of California and Chile have a substantial moderating effect on local climates and are responsible for the Mediterranean climate in these areas.

These effects on competition can result in a patchy distribution of plants, with some areas bare of plants.

Plants Affect Local Climate For plants, sunlight is an important aspect of climate. It affects photosynthesis, which is fueled by sunlight. But sunlight also affects the plant temperature, and thus all aspects of metabolism, including growth. In many communities, large trees affect the penetration of light to the lower levels of the forest. Thus these trees modify the local climate of lower-level vegetation, with secondary impacts on the animals living in the understory. In Module 16.8, we review these effects of plants on the light and heat profile of a community. ❖

FIGURE 16.6B Plants adapted to the climates of California (top) and Chile (bottom) show many similar morphologies.

Local topographies can result in predictable effects on climate. One example is found along coastal areas of the world, where there is a predictable flow of moist air from the ocean (Figure 16.7A). If there is also a sizable mountain range near the coast, then this moist air will be forced to climb up the side of the mountains. Air cools as it moves up the side of a mountain, until it reaches the **dew point**, which is the temperature at which the air is saturated with water and condensation begins. Water is less soluble in cold air, just as solutes like sugar and salt are less soluble in cold water than in hot water. The cooling of the air thus results in rainfall and a reduction of the air's water content.

As the now-dry air mass passes over the mountain and begins to go down the other side, it warms and absorbs moisture from the land. For this reason, the side of the mountain that faces away from the prevailing onshore airflow is often dry. The deserts created in this fashion are often referred to as **rain-shadow deserts**, due to close proximity of a mountain range in whose "shadow" they form.

This pattern of climate is observed along the West Coast of the United States in several places. Air flows reliably in a west-to-east direction along the West Coast of the United States. In Oregon, the coastal Cascade mountain range comes into contact with this airflow, and the Great Sandy Desert is found just east of this mountain range (Figure 16.7B). In a similar fashion, in California the Sierra Nevada mountains run north to south and are responsible for several deserts interior to the range, including the Mojave Desert. ❖

The air cools as it moves up the side of the mountain and releases much of its water content as rain.

As the air flows down the backside of the mountain, it removes moisture from the land because the air is now very dry.

Moist air, often from the ocean, travels toward and then up the side of a mountain.

FIGURE 16.7A Creation of a Rain-Shadow Desert

FIGURE 16.7B Rain-Shadow Deserts in the Western United States

Climatic conditions can vary on a local level over distances as small as a few meters. These small or **microclimatic** effects may be very important for the affected organisms. This is especially true of organisms that are not mobile, such as plants.

Often plants themselves may determine the local microclimates. Figure 16.8A illustrates the changes in light intensity in a yellow-poplar stand over a year. There is a predictable seasonal change in light intensity, thanks to the tilt of the Earth. For instance, light intensities of 400–450 lumens occur only during the summer months. Also in the summer, the leaves on the poplar trees filter out much of the light, so that the light intensity drops to 50 lumens at 10 meters above the ground. The shading effects of the poplar trees mean that small shrubs growing on the forest floor have to be adapted to low light levels. During the winter and spring there are fewer leaves on the trees to filter out light, so that the light intensity at the forest floor is greater during the spring than at any other time.

In addition to the size of the trees, other factors such as leaf shape and tree density affect the local microclimate.

Plants also intercept and dissipate heat in a complicated fashion. In a study of meadow vegetation, roughly 45 percent of the incoming radiation and heat were absorbed by the upper layers of plants, as Figure 16.8B shows. Smaller amounts were absorbed by the understory and finally by the ground (Figure 16.8B).

Heat is lost during the day by two routes, evaporation (V) and convection (L). Evaporation is the major source of heat loss during the day. Heat is lost from the plants or ground when water changes state from liquid to gas. Convection depends on the movement of air over the plants and the ground surface. At night the flow of heat energy is in the opposite direction, with radiant energy (Q) accounting for all losses from plants and the soil (Figure 16.8C).

The profile in Figure 16.8C is over a distance less than 1 meter. These types of gradients in energy absorption will be even greater in forests with large trees. In addition to the size of the trees, other factors such as leaf shape and tree density affect the local microclimate. ❖

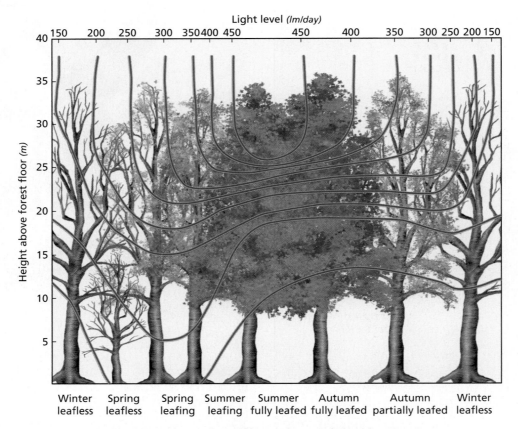

FIGURE 16.8A Light Levels at Different Places and Times in a Forest

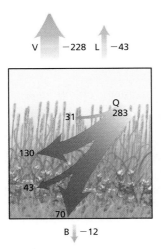

FIGURE 16.8B **Energy Transfer in a Meadow (Day)** The numbers indicate relative amounts of energy coming into or leaving the meadow. The arrow at the bottom of the figure represents energy transfer to the soil and the arrows leaving the top of the figure represent energy exchange with the atmosphere.

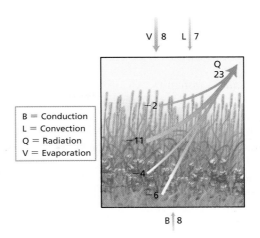

B = Conduction
L = Convection
Q = Radiation
V = Evaporation

FIGURE 16.8C **Energy Transfer in a Meadow (Night)** The symbols have the same meaning as in Figure 16.8B.

THE ECOLOGY AND EVOLUTION OF BIOMES

16.9 The ocean biomes cover 70 percent of the Earth's surface

Plant and animal communities change almost continuously across environmental gradients. Some obvious exceptions are where very different environments meet, such as the land and sea. Ecologists have nevertheless sought to simplify the description of communities by noting some very general categories into which most communities can fit. These categories, called **biomes**, are characterized by their plant and animal communities as well as their geographic location. The plants and animals that are typical residents of a biome are expected to have physiological and behavioral adaptations that make them well suited to the physical properties of the biome. In the next three modules, we consider some of the most important biomes and their distinguishing features.

The oceans are the most conspicuous biome on Earth. They cover 70 percent of the Earth's surface. Life originated in the oceans, and today there are many entire phyla (like Echinoderms) whose members are found only in the oceans. Many of the physical properties of oceans have profound effects on the types of communities of plants and animals that can be supported. We review some of these properties now.

Oceans are characterized by saline water, which has a fairly constant concentration of salts—equal to 35 parts per thousand. Of these salts, 86 percent is composed of sodium and chlorine. While the salinity of the oceans is usually very constant, it can vary—especially near the surface. Rainfall causes local reductions in salinity near the surface. Rivers that flow out to the oceans will create local gradients of salinity where the river mouth meets the ocean. These brackish waters are characterized by very different communities than are found in either the ocean or river.

Despite the many dissolved salts in ocean water, it has typically very little nitrogen and phosphorus. As a result, the levels of primary productivity in oceans are quite low. The primary reason for the low levels of nitrogen and phosphorus is that when plants and animals die in the ocean, they fall to the ocean floor to decompose. There is usually very little mixing of the water near the surface of the ocean and the ocean floor, thus plants that must live near the surface are deprived of these nutrients. In temperate regions, currents occasionally bring nutrients to the surface; these are called upwelling currents (Module 16.4). Temperate regions are relatively more productive than, say, tropical oceanic areas. In the tropics, there are layers of water that differ in their temperatures. These layers, called thermoclines, almost never mix in the tropics. Consequently, tropical waters are some of the least productive ocean regions.

Different regions of the ocean are given names to help in discussing them. The surface areas of the oceans are referred to as **pelagic**, while the bottom is called the **benthic** region (Figure 16.9A). The pelagic region is further subdivided into the **neritic**, referring to the regions over the continental shelf, and the **oceanic** for the remaining area. Other regions of the ocean are characterized by their depths. The **photic** zone refers to the region where light can penetrate, which includes about the top 200 meters. However, the region that can support plant growth is much more limited—from the surface to perhaps 30 meters or so, depending on local conditions.

In the open ocean, plants and animals may be found in the surface water in a community referred to as the **plankton**. The plants, called **phytoplankton**, are mostly unicellular plants. Many invertebrates have larval stages that are part of the animal plankton community; they are called **zooplankton**, and they may feed on the phytoplankton. Many other animals then feed on these small animals. In some especially productive areas, large populations of small crustaceans may live in the zooplankton and then serve as food for large animals like whales.

The region where the ocean meets land is referred to as the **intertidal zone**. Intertidal areas are often capable of supporting very productive communities of plants and animals. The unique feature of the intertidal is that areas within it are exposed to the air for varying periods of time (Figure 16.9B). Consequently, the organisms that occupy the intertidal show varying adaptations to desiccation resistance. Those in the highest reaches of the intertidal may be exposed to air and high temperatures for hours at a time, those in the lowest regions are rarely exposed, and then only briefly. ❖

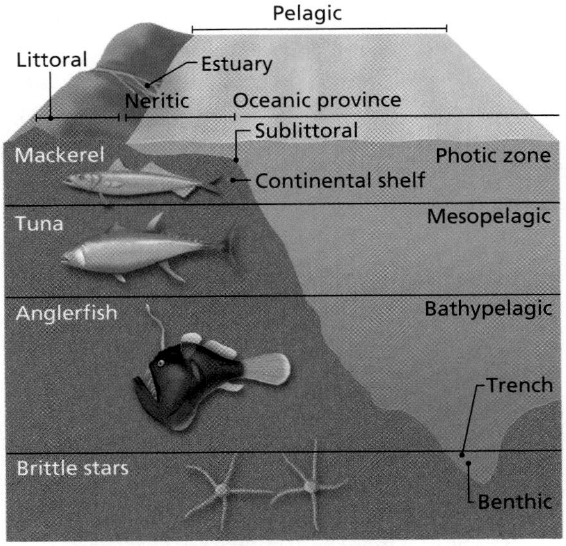

FIGURE 16.9A Major Regions of the Ocean

ZONE 1
Uppermost Horizon. Mainly a region of bare rocks, sometimes with green algae (*Enteromorpha* or *Cladophora*). Characteristic rock-shore animals: the pill bug *Ligia*, the barnacle *Balanus glandula*, the snail *Littorina planaxis*, and the limpet *Acmaea digitalis*. On sand the beach hoppers *Orchestia* and *Orchestoidea*.

ZONE 2
High intertidal. Typically with rockweeds (*Pelvetia*), but bare toward upper limits. Rocky shores: *Balanus glandula*, *Littorina scutulata*, *Tegula funebralis*, and several species of limpets. *Pachygrapsus* very common under rocks. *Aletes* in the south. The chiton *Nuttallino* and the red barnacle *Tetraclita* in high surf. *Hemigrapsus* on mud flats.

ZONE 3
Middle Intertidal. Many algae. Animals predominant here are rarely taken subtidally. Rocky shores: *Hemigrapsus nudus*, the seastar *Leptasterias*, the hermit *Pagurus hemphilli*, many others. *Ischnochiton* under rocks. In high surf: the snail *Thais emarginata*, the common seastar *Pisaster ochraceus*, the goose or leaf barnacle *Pollicipes*, the California mussel (*Mytilus*), and the black abalone. The chiton *Katharina*. *Balanus cariosus* in Puget Sound.

ZONE 4
Low intertidal. Laminarians and corallines. Predominant animals also occur subtidally. Rocky shores have a great variety of animals, but not the hosts of individuals per species that characterize the upper zones. Probably these are animals that have only recently learned to tolerate a slight exposure.

Characteristics of zones

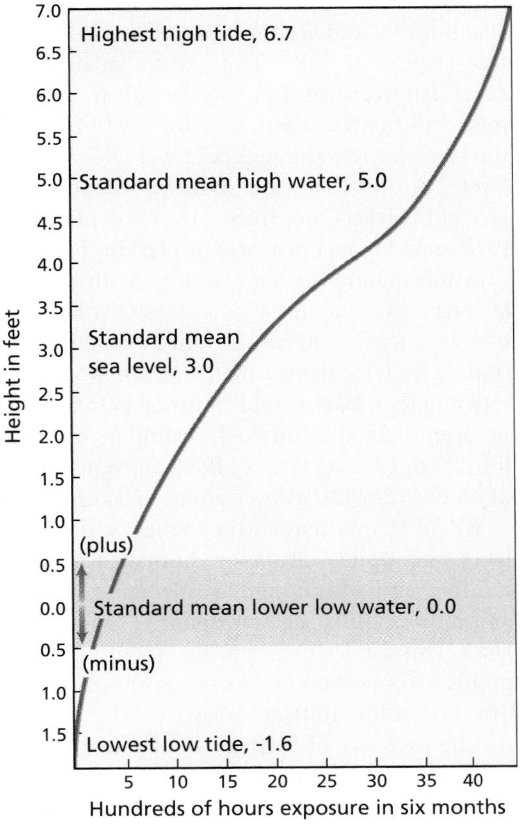

FIGURE 16.9B Intertidal zone experience varying levels of exposure to air.

16.10 The physical properties of water have important consequences for life in freshwater lakes and ponds

Lakes and ponds are inland bodies of freshwater that can vary tremendously in size and biological composition (Figure 16.10A). Their depths may vary from 1 meter to over 2000. Lakes and ponds receive runoff from rainfall that drains off the surrounding land. The characteristics of the local area can then affect the quality of inorganic ions and organic matter that the lakes and ponds receive.

However, three properties of water still determine many of the important physical properties of lakes and ponds: (1) The high specific heat of water means that its temperature will change much more slowly than does air temperature. Diurnal cycles in air temperature will typically have no appreciable effect on the temperature of lakes and ponds. (2) Water is most dense at a temperature of 4°C. (3) Ice floats on water.

During the summer, the top layer of a lake warms and forms a layer of water, called the **epilimnion**, that mixes very little with the lower layers, or **hypolimnion** (Figure 16.10B). This creates a gradient of temperatures, or **thermocline**, that can be quite dramatic depending on the size of the lake. Phytoplankton and plants that live near the surface will keep the epilimnion well oxygenated. However, because organic matter falls to the bottom of lakes, its decomposition depletes oxygen at the bottom of lakes. These conditions lead to what is called the summer stagnation. By the fall, air temperatures have dropped and solar insolation decreases, causing the surface waters to cool. The cooler, denser water then falls to the lake bottom, and warmer water from below rises to the surface (Figure 16.10B). This creates mixing within the lake, called fall overturn. In climates with freezing weather, water may cool to 4°C, at which point it will fall to the bottom of the lake; as the water gets colder, it will stay above this dense layer. Thus the first water to freeze will be water near the surface of the lake. Once frozen, the ice continues to float on the surface and in fact provides insulation from the very cold air, thus forestalling further freezing. At this point, most of the lake will stay at about 4°C, and there will be little movement of water, leading to the condition called the winter stagnation. If ice were denser than water, it would sink to the bottom and then lakes would continue to freeze from the bottom up. Eventually the entire lake would be frozen and all animal life killed. The fact that ice floats prevents the mass mortality of many animals that overwinter in lakes and ponds.

Many factors may affect the levels of nutrients reaching lakes and ponds. However, human activities like logging, mining, agriculture, and construction can add significant amounts of nitrogen, phosphorus, and organic matter to lakes. This can lead to a process of eutrophication. Lakes and ponds without these extra sources of nutrients are called oligotrophic or low-nutrient lakes.

The presence of high levels of nitrogen and phosphorus will stimulate the growth of plants and algae near the surface of the lakes. As the large biomass of plants dies and falls to the bottom of the lake, the oxygen is depleted by aerobic decomposition to the point where aerobic organisms can no longer live [Figure 16.10C, part (ii)]. Although the biomass of a eutrophic lake is greater due to the high primary productivity, the species diversity typically falls. The accumulation of organic matter on the bottom of the lake makes it shallower and may eventually change it to a bog or swamp.

An oligotrophic lake will be deep and have much higher levels of oxygen near the bottom [Figure 16.10C, part (i)]. While the biomass of plants is much lower, the overall species diversity—especially of animals—is much higher in oligotrophic lakes. ❖

FIGURE 16.10A Lakes and ponds come in a variety of sizes and shapes.

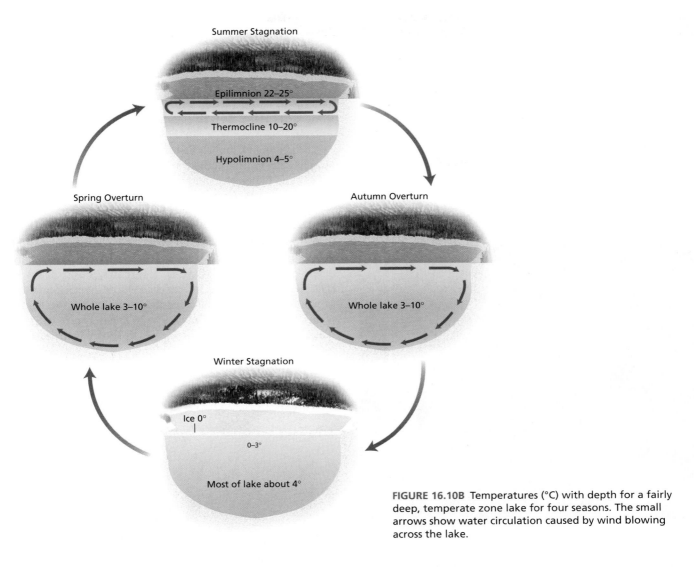

FIGURE 16.10B Temperatures (°C) with depth for a fairly deep, temperate zone lake for four seasons. The small arrows show water circulation caused by wind blowing across the lake.

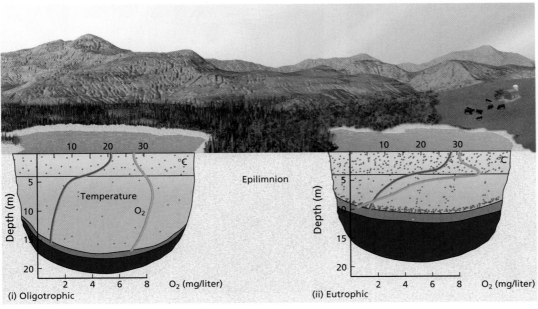

FIGURE 16.10C Comparison of Oligotrophic and Eutrophic Lakes

The Ecology and Evolution of Biomes   **495**

16.11 Adaptations to reduce water loss characterize the plant and animal life found in deserts

Deserts are environmentally stressful biomes that challenge their inhabitants to survive on a daily basis (Figure 16.11A). The hallmark of deserts is low rainfall, or more specifically where potential evapotranspiration is greater than precipitation per year. There is no precise value of rainfall that defines a desert, but rather a continuum, so areas that receive 150–400 mm of rainfall per year are called semideserts. Much of the land area of California would fall in this range. True deserts receive less than 120 mm of rainfall, and extreme deserts less than 70 mm of rain. As we have seen in Module 16.2, global patterns of air circulation in Hadley cells create many of the world's deserts at 20–30° latitude north and south of the equator.

The plants and animals that make deserts their home are almost without exception characterized by unusual adaptations to the dry conditions of the desert. Examining how evolution has adapted these organisms is fascinating. We consider only a few examples here.

Plants lose much of their water through their leaves while their stomates are open. Desert plants have altered designs and physiology to reduce this water loss as much as possible. Many desert plants, like cacti, have no leaves and consist of thick, stalk-like branches. This effectively reduces the surface-area-to-volume ratio of the plant, greatly reducing water loss. Some plants, like the ocotillo (*Fouquieria splendens*, Figure 16.11B), reduce water loss through their leaves by shedding their leaves

FIGURE 16.11A Deserts are typically sparsely covered with plants. To reduce water loss, these plants are often small with thick branches and few if any leaves.

FIGURE 16.11B Ocotillo Cactus, Anza-Borrego Desert State Park, California.

FIGURE 16.11C **Plains Spadefoot,** *Spea bombifrons*, **Kansas**

during periods of drought. After a rainfall, the ocotillo regrows its leaves; this may happen four or five times a year. Many succulents separate the chemical reactions for fixing CO_2 from those that capture light energy. As a result, the plant does not need to open its stomates during the day, when light energy is captured; rather, this is done during the night to reduce water loss.

Animals also show a variety of adaptations to dry conditions. Some of these adaptations are relatively simple, like being active only during the night. Other adaptations are more elaborate. The spadefoot toad, which lives in the desert Southwest of the United States, will bury itself underground and go into a state of reduced metabolic activity called estivation (Figure 16.11C). When rain returns, the toads become active again. Some animals are able to withstand large losses of body water. Desert rabbits can withstand water losses of up to 50 percent of their body weight (Figure 16.11D).

The kangaroo rat is probably the most accomplished water-conserving mammal (Figure 16.11E). These animals never drink water, but gain all their required moisture by metabolizing carbohydrates. During the day, the kangaroo rat remains in underground tunnels to reduce water loss. To reduce the loss of water through respired air, the kangaroo rat has a countercurrent system that cools the air and then reabsorbs much of the moisture. Another major source of water loss for mammals is excretion of urine. Kangaroo rats have extremely efficient kidneys that reabsorb much of the water in urine. As a result, the urine of kangaroo rats is 30 times more concentrated in dissolved solutes than their blood is. No other animal produces such concentrated urine. ❖

FIGURE 16.11D **Black-Tailed Jack Rabbit**

FIGURE 16.11E **Kangaroo Rat**

The Ecology and Evolution of Biomes **497**

16.12 Forests are important terrestrial biomes often characterized by their dominant tree species

On land, forest biomes constitute a heterogeneous but important group of biological communities (Figure 16.12A). Forests are typically characterized by their physical location and their dominant plant species, which are often large trees. The worldwide distribution of some of these biomes is shown in Figure 16.12B. The importance of these biomes is evident from just the extent of the area covered by forests. In this module we review the characteristics of a few of these forest biomes to give you some flavor of their diversity.

Northern Coniferous Forest Northern coniferous forests are found across North America and northern Eurasia. The climate of northern coniferous forests typically consists of cold winters and short, mild summers. Although the rainfall is low (40–100 cm per year), the cool temperatures keep evaporation low; thus, the climates are often humid. These biomes often receive very heavy snowfall.

The forest is usually dominated by one or two tree species. Some typical species are white spruce, balsam fir, black spruce, and white cedar. Coniferous trees are especially well adapted to the environments of northern forests. Their needles shed snow effectively and thus prevent damage that can occur from large accumulations of snowfall. Because the trees do not shed their needles, they can begin photosynthesis as soon as it is warm enough in the spring. This adaptation is especially advantageous in climates with short growing seasons.

Temperate Deciduous Forest Temperate deciduous forests are found in eastern North America, western Europe, Japan, eastern China, and Chile. The climate of these forests typically includes a warm summer and a cool to cold winter. There is also fairly high precipitation. The types of trees that dominate these forests are quite variable. However, some typical species are oaks and maples.

Tropical Rain Forests Tropical rain forests receive more than 200 cm of rainfall per year; in some cases, it can approach 1000 cm. Many rain forests are found in equatorial regions of the world. These areas are prone to becoming rain forests due to the weather patterns of Hadley cells (Module 16.2). One month out of the year may be relatively dry, during the cooler months. However, temperature in the rain forests changes little over a year; it averages about 27°C.

The very high rainfall leaches the rain-forest soils of most inorganic nutrients. The moisture and warm temperatures also mean that organic matter decays quickly. Rain-forest soils do not do well under traditional agriculture either. The low cation exchange capacity prevents these soils from holding nutrients when added as fertilizer, since fertilizer binds to cations. In some locations, when the land is cleared the soil turns to a hard red substance called laterite. The laterite can be used to make bricks, but it supports only scrubby growth.

Tropical rain forests are notable for their very high diversity of plants and animals. Many species in the rain forests remain to be discovered and described. However, there has been increasing pressure to harvest wood from rain forests and convert land to agriculture. As a result, the amount of land surface covered by rain forests is dwindling. It is almost certain that this activity is also leading to the extinction of species, many of which have never been described. ❖

FIGURE 16.12A The top figure shows a pine forest. A deciduous forest in New England is shown below.

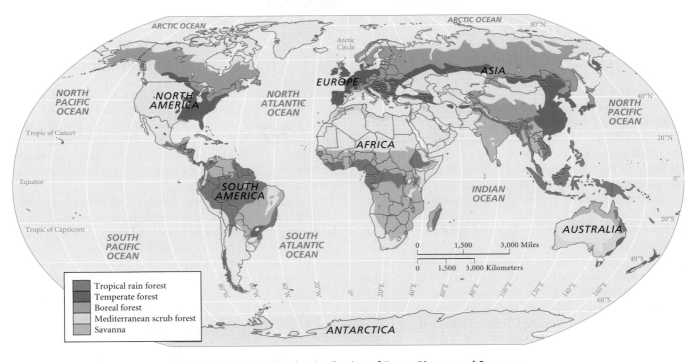

FIGURE 16.12B Worldwide Distribution of Forest Biomes and Savannas.

Legend:
- Tropical rain forest
- Temperate forest
- Boreal forest
- Mediterranean scrub forest
- Savanna

The Ecology and Evolution of Biomes **499**

GLOBAL CHANGE

16.13 Human activities can quickly cause global environmental change

Humans have been aware for some time that their activities can have effects on the environment. Indeed, the industrial pollution of England deposited soot on trees and was partly responsible for the change in frequency of color morphs of the moth *Biston betularia*, as we saw in Chapter 4. These changes had been documented by collections of moths in the late nineteenth century, although natural selection was not implicated as the cause of these changes until the 1950s.

Atmospheric Pollution   The real danger of air pollution was dramatically illustrated in London in 1952. In December of that year, a temperature inversion caused the air to stagnate, and cloud cover blocked the sunlight. Temperatures dipped to below freezing, and the people of London turned up their furnaces. At that time many homes were heated by coal, and the increased coal burning caused the air to become even thicker with pollution. In a seven-day period, nearly 4000 people died of pollution-related complications.

Smog in London and Los Angeles (Figure 16.13A) has effects primarily at a local level, but we are now learning that human pollution can have a truly global effect. That is, pollution in the United States affects everyone on the planet. Burning fossil fuels continues to cause problems. Many of the by-products of coal and gas combustion undergo chemical changes in the atmosphere and can become strong acids, giving rise to acid rain (Figure 16.13B). The acid rain will not necessarily be localized to the area where the pollution was first created. Pollution at the global level means that pollution control will require cooperation from many countries to be effective.

Another form of air pollution is produced by chemicals used as propellants and refrigerants. While the total amount of these gases is not that large, their impacts on levels of atmospheric *ozone* (O_3), an unstable and highly reactive form of oxygen, can be significant. But isn't ozone a type of pollution? In fact, ozone levels can increase close to the Earth's surface due to car emissions; but ozone in the upper part of the atmosphere plays a crucial role in filtering out damaging radiation. Without this filter, humans will experience much higher levels of skin cancer. We discuss the dangers of acid rain and ozone depletion in more detail in Module 16.14.

Agricultural Practices   Civilization would not be what it is today without the development of modern agriculture. The ability of a small number of people to produce food for many has allowed modern civilizations to create a workforce that is not tied to the land and does not need to hunt and gather food. Ecology and evolutionary biology have much to offer agricultural science. Some agricultural pests can be controlled by natural predators and parasites. Other agricultural pests have been controlled by chemical agents. We have seen repeatedly, however, that insects evolve and can become resistant to these chemicals.

Modern agriculture has potential downsides as well. Many plant crops remove nitrogen from the soil. Because crops are often planted in large expanses of *monocultures* (one-species crops), there are no natural means of replenishing the soil's nitrogen supply. However, soil nitrogen can be replenished through *crop rotation* rather than by adding

FIGURE 16.13A Los Angeles Smog Can you find the clear day?

synthetic fertilizer. In crop rotation, a nitrogen-using crop like corn is followed by a crop like soybeans, which fixes nitrogen in the soil. Crop rotation is a means of trying to restore the balance of soil nutrients that becomes disrupted in highly artificial agricultural ecosystems.

In many of the less developed parts of the world, more severe problems face populations that use marginal land for agriculture and grazing. Overuse of these areas can cause land to convert to desert (Figure 16.13C). In Module 16.15 we review this problem and the many factors contributing to it. ❖

FIGURE 16.13B A statue in Chicago damaged by acid rain, before (left) and after (right) restoration.

FIGURE 16.13C Poor agricultural practices in the United States led to soil erosion and the Dust Bowl in parts of Oklahoma during the 1930s.

16.14 Human activities add gases to our atmosphere, leading to acid rain and ozone depletion

In this chapter we have already reviewed the impact of fossil fuel combustion on atmospheric CO_2 and the resulting greenhouse effect (see Module 16.5). But the combustion of fossil fuels has consequences beyond the greenhouse effect. For instance, coal is still a common fuel in industry, especially in power plants. Coal contains sulfur compounds that, when burned, produce sulfur dioxide (SO_2), which is released into the atmosphere. Burning coal and natural gas also produces nitrogen dioxide (NO_2) and nitrous oxide (NO) during their combustion (Figure 16.14A). These compounds may then undergo additional chemical reactions in the atmosphere to give rise to the derivative compounds sulfuric acid (H_2SO_4) and nitric acid (HNO_3). These acids are very soluble in water. They dissolve in rain and snow, returning to Earth in a solution called **acid rain**.

Natural rainfall is slightly acidic (pH = 5.6) because the carbon dioxide normally present in air produces a mild acid called carbonic acid. However, nitric and sulfuric acid are strong acids that can lower the pH of rainwater substantially. Rainfall with a pH as low as 1.5 has been recorded in Wheel-

ing, West Virginia. Acid rain may be responsible for killing forest trees directly and substantially lowering the pH of lakes. A substantial drop in pH has been documented in the Adirondack Lakes region of New York over the past 45 years (Figure 16.14B). These effects of acid rain then cause additional cascading effects on the organisms that live, or depend on life, in these lakes or forests. The effects of acid rain can be seen in the faces of public marble statues that have been eroded by atmospheric acids over time (see Figure 16.13B).

Ozone is an unstable and highly reactive form of oxygen. Unlike the stable diatomic oxygen molecule (O_2) that makes up 21 percent of our atmosphere, ozone (O_3) has three atoms of oxygen and is relatively rare. However, in the portion of our atmosphere called the stratosphere (Figure 16.14C), ozone concentrations reach high values. They also serve their most important function there. Sunlight reaching the outer atmosphere of the Earth contains high-energy radiation, including three types of ultraviolet light (UV). The lowest-energy ultraviolet light is called **UVA**. It can cause some damage to biological cells and typically is not filtered by the atmosphere. **UVB**

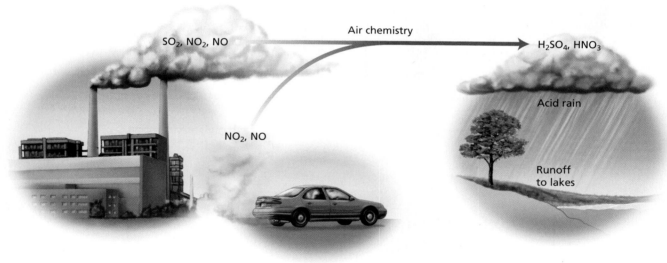

FIGURE 16.14A **Acid Rain Production**

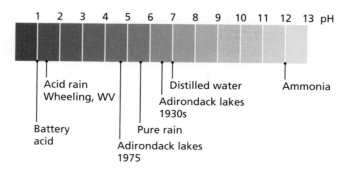

FIGURE 16.14B Acid rain lowers pH in natural lakes.

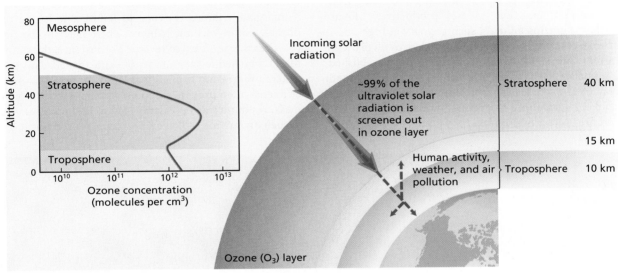

FIGURE 16.14C Ozone absorbs UV light in the stratosphere.

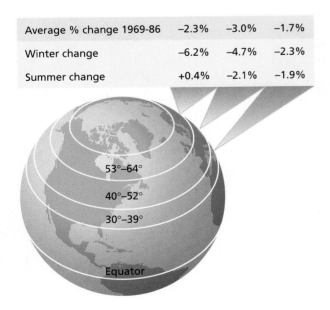

Average % change 1969-86	–2.3%	–3.0%	–1.7%
Winter change	–6.2%	–4.7%	–2.3%
Summer change	+0.4%	–2.1%	–1.9%

FIGURE 16.14D Global Changes in Ozone Levels

has more energy than UVA, and can cause substantial damage to living cells. Ozone plays a crucial role by absorbing UVB radiation, which splits the ozone into one molecule of diatomic oxygen and one molecule of atomic oxygen. This reaction prevents much of the UVB from reaching the Earth's surface. The highest-energy UV is **UVC**. It is strongly absorbed by the atmosphere, so almost none of it reaches the Earth's surface.

In the absence of human activities, the atmospheric ozone reached an equilibrium between its destruction by UV energy and its creation by a variety of other chemical processes. Humans, however, have been adding chemicals to the atmosphere that can dramatically destroy ozone. In so doing, we are increasing our exposure to damaging UVB energy. The chemicals that destroy ozone include chlorofluorocarbons (CFCs), used in aerosols and refrigerators; carbon tetrachloride and methyl chloroform, both used as solvents; and halons, used as refrigerants and in fire extinguishers. These compounds produce reactive chlorine molecules that can destroy ozone. Just one of these chlorine molecules may destroy up to 100,000 ozone molecules before it is removed from the atmosphere. Reduced ozone levels in our atmosphere are now well documented (Figure 16.14D). ❖

16.15 Human agricultural practices have increased the spread of deserts

Deserts are characterized by very low rainfall and sandy soil with little organic matter. Desert soils are also unable to hold much water. There is typically little organic matter in deserts because so few plants can grow in dry sandy soils. Human activity can be a major contributor to **desertification,** the process of becoming a desert. When marginal lands are converted to agriculture or grazing, a number of changes occur that may accelerate the conversion of land to desert. Conversion of land to human use involves removing native plants that may be important in maintaining soil integrity. Often large trees in these areas are the first to go, being harvested for firewood. With the loss of native plants and the planting of commercial crops, there can be a loss of organic matter from the soil, erosion, and loss of soil productivity (Figure 16.15A). Some grazing animals, such as goats, at high densities can be so efficient at feeding that they destroy all plant life and prevent any new growth from becoming established. Larger animals, such as cows, physically trample new growth and break up the soil, fostering erosion. As the land becomes less useful for agriculture or grazing, humans move to new areas and the process is repeated.

In northern Africa, for example, the Masai people have adopted a nomadic lifestyle to make a living on marginal lands. The Masai travel over great distances to find new food for their herds of cattle. The Masai have increasingly faced famine in recent years as a result of desertification and crop failures. These problems have been caused by many factors. Political problems have restricted the movement of the Masai, leading to overgrazing by their cattle. In addition, this region of Africa has been subject to historically severe drought for the last 30 years.

Northern Africa is not the only region of the world facing problems of desertification (Figure 16.15B). Solutions to these problems are hard to come by. Maintaining agriculture in marginal habitats is prone to failure due to the unpredictable nature of these habitats. There may also be cultural practices that are hard to reverse. For instance, the status of a Masai is related to the number of cattle owned. This practice fosters overgrazing in marginal lands. Although world governments often provide aid to prevent mass starvation, these measures in no way necessarily lead to lasting improvements in such underlying ecological problems as desertification. ❖

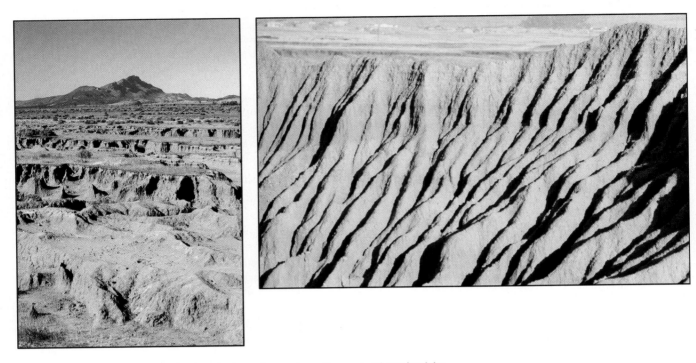

FIGURE 16.15A Soil erosion can lead to the loss of organic matter and soil productivity.

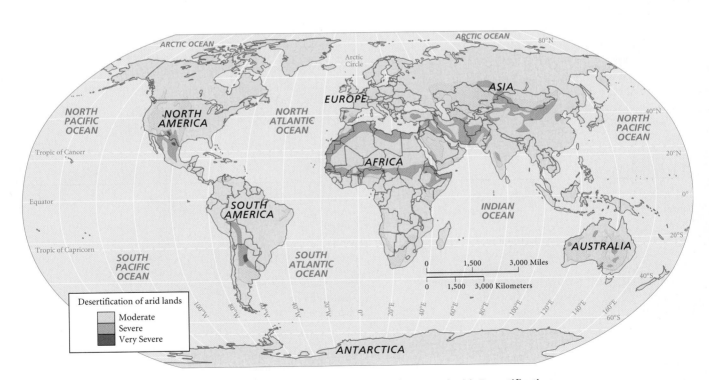

FIGURE 16.15B **Areas of the World Currently Threatened with Desertification**

SUMMARY

1. The Earth's environments vary tremendously. Many patterns of climatic variation are due to physical properties of the Earth and its movement about the sun.

2. Seasonal and daily variation in duration of daylight and average temperature are strongly affected by the rotation of the Earth on its axis and its orbit around the sun.

3. The intense energy of the sun that strikes the equator sets up a predictable flow of air called Hadley cells, which contribute to the existence of rain forests on the equator and deserts just north and south of the equator.

4. Climate is affected by local conditions.

   a. Deserts can form on one side of a mountain as a consequence of the rain-shadow effect.

   b. These types of deserts form only when the right combinations of topography and airflow exist.

   c. Plants can also alter the microclimatic effects. Large trees can substantially reduce the penetration of light and heat to the forest floor.

5. Communities are classified into biomes that have characteristic vegetation and occupy specific geographic and climatic regions.

   a. Oceans cover 70 percent of the Earth's surface and are characterized by low primary productivity.

   b. Deserts form as a consequence of local climates or global conditions like Hadley cells. The plants and animals that characterize the desert community show unusual adaptations to the low water and high temperatures of deserts.

   c. Lakes and ponds often show seasonal cycles of temperature and dissolved oxygen content that have important consequences for their animal communities.

   d. Forests are a heterogeneous terrestrial biome occurring in the tropics and the cold northern latitudes. Forest communities are often identified by their dominant tree species.

6. The global environment is not immune to human activity.

   a. Burning fossil fuels has contributed to elevated levels of carbon dioxide. This in turn may be warming the Earth.

   b. Other by-products of fossil fuel oxidation, like SO_2 and NO_2, contribute to acid rain production.

   c. The protective layer of ozone in our atmosphere is also under attack from certain chlorinated compounds. The potential harm from the loss of ozone is not as immediate and obvious as that from acid rainfall, but in the long run it could be much more dangerous.

REVIEW QUESTIONS

1. Explain how the airflow within a Hadley cell contributes to rain forests at the equator and deserts 20—30° north and south of the equator.

2. Describe two specific ocean currents and their general effects on the climates of landmasses they pass by.

3. What is the greenhouse effect? How do human activities influence this process?

4. How are rain-shadow deserts formed?

5. Why don't most lakes in temperate climates become a solid block of ice in the winter?

6. Describe two adaptations used by plants to cope with the dry conditions of the desert.

7. Describe how acid rain is formed.

KEY TERMS

acid rain	glacier	oceanic	upwelling
benthic	greenhouse gases	ozone	UVA
biome	gyres	pelagic	UVB
chaparral	Hadley cell	photic	UVC
climate	hypolimnion	phytoplankton	weather
currents	intertidal zone	plankton	zooplankton
desertification	Mediterranean climate	rain-shadow deserts	
dew point	microclimate	seasonality	
epilimnion	neritic	thermocline	

FURTHER READINGS

Botkin, D. B., and E. A. Keller. 2000. *Environmental Science.* 3rd ed. New York: John Wiley.

Karl, T. R., N. Nichols, and J. Gregory. 1997. "The Coming Climate." *Scientific American* 276:78–83.

Rowland, F. S. 1990. "Stratospheric Ozone Depletion by Chloroflurocarbons." *AMBIO* 19:281–92.

The endangered Siberian tiger

Conservation

Conservation biology usually invokes thoughts of rare and endangered species like the beautiful Siberian tiger on this page. Certainly over the last half century, the industrialized world has become more concerned with its role in accelerating species extinction. As a result, international cooperation has developed to limit the harvesting of whales, and for some whale species, to prevent it completely; to stop the trade of animal products like ivory from elephants and rhinoceros; and to encourage less developed countries to set aside land for the protection of endangered species and rain forests.

Conservation biology attempts to develop a solid scientific understanding of the forces affecting the long-term maintenance of biological populations and genetic variation within populations. As we will see in this chapter, this understanding often comes from the principles we have already explored in ecology and evolution. Endangered species are not the only focus of conservation biology. Many natural populations are important economic resources. However, we need an understanding of how human harvesting from these natural populations will affect their long-term survival. Simple economic forces cannot always properly weigh the future impact of species extinction. Developing rules for harvesting natural resources is complicated by the different currencies we choose in valuing these resources.

Human activity does not always result in extinction. Humans have transported species to new habitats, where they have done exceptionally well—in some cases too well. Introduced species may become pests and can become the agents of extinction for native species. Another aspect of conservation biology, then, will be to examine how we can control recently introduced species that may change the nature of the ecosystems they have invaded. ❖

BASICS OF CONSERVATION

17.1 Conservation biology requires an understanding of the genetics, ecology, and physiology of managed populations

Biodiversity The variety of plant and animal species on Earth, the genetic variability that exists within each of these species and the variation in communities and ecosystems is referred to as **biodiversity**. In this book we have reviewed many processes that affect genetic variation within species. These processes include genetic drift, inbreeding, and natural selection. We have also learned about ecological factors that affect the numbers of species in a community or ecosystem. Some of these processes are competition, predation, environmental variation, and energy flow. **Conservation biology** is the scientific study of biodiversity and its management for human welfare. As an applied science, conservation biology relies on theories and principles from many other disciplines. We have already noted the importance of ecology, genetics, and physiology to understanding biodiversity. However, other applied fields—like fisheries biology, forestry, and range management—have similar concerns.

The successful application of conservation biology often requires a detailed understanding of the ecology of endangered populations and their community. As an example, we review the decline of the large blue butterfly in England. Creating protected areas for this butterfly was not sufficient to prevent its extinction. A more detailed understanding of the butterfly's ecology was ultimately required.

Our understanding of biodiversity is partially due to our understanding of basic principles of ecology and genetics.

Maculinea arion The large blue butterfly (Figure 17.1A) was common in Southern England in the late 1880s, but its decline was already being forecast. The number of colonies underwent a continual decline for the next 100 years. By 1974 there were only two known colonies and perhaps about 250 butterflies. Two successive droughts caused the extinction of the last colonies in Britain in 1979. Although protected areas had been established in the 1930s for this butterfly, they did not prevent the large blue's ultimate extinction. Detailed studies of the ecology of the large blue ultimately provided the clues needed to design an effective recovery program.

Caterpillars of the large blue feed on thyme plants, but gain little weight. The caterpillar ultimately leaves the thyme and travels a short distance to an ant colony. The caterpillar offers the ant some sugar, and through a variety of tactile and smell signals, cons the ant into bringing the caterpillar back to the ant colony. Once there, the caterpillar feeds on ant larvae. The species of ant that large blue does best with is *Myrmica sabuleti*. These ants prefer close-cropped turf. When the turf is not grazed or burned down, the *M. sabuleti* colonies are replaced with another species of ant, *M. scabrinodes*, which results in a 2 percent decline in the survival of the large blue caterpillar.

Many of the **reserves** set off for this butterfly excluded humans and small mammals that might graze on the grass. Thus the butterfly continued to do poorly because its preferred food resource, *M. sabuleti*, was driven from the habitat.

FIGURE 17.1A *Maculinea arion*, **the Large Blue Butterfly**

Samples of the large blue caterpillar from Northern Europe have now been successfully reintroduced into England.

The idea of setting aside tracts of land has grown and is now applied to large areas with the goal of preserving many species—not just one, like the large blue butterfly. These types of reserves have become necessary as more forest land is lost to agriculture, deserts, and other human activities. In the next modules we provide more detail on the negative consequences of habitat loss and fragmentation.

In the course of creating these ecological reserves, many practical questions arise concerning the size and number and connectedness of the reserves. Some of these questions can be addressed with knowledge we have gained from studies of species loss on islands. We will review some of these ideas from the theory of island biogeography.

Harvesting Conservation biology also deals with the conservation of species that are currently not endangered, but might be due to human activity. Many species are hunted or harvested by humans for their economic value. Marine fisheries are an example of this. These fish are valuable food, but due to their ecology they cannot be easily raised in human captivity, so we rely on natural populations to provide the fish we need. There are a number of difficulties concerning these natural resources. In the case of ocean fish populations, there is no single owner of these fish—and thus no easily recognized authority to determine how many should be harvested at any time. It is also difficult to

determine the number of fish in these populations, so the impact of human removal of fish may not be evident until it is too late. Simple ecological theory can be used to begin our analysis of this problem and provide some insights about how we should limit our harvesting efforts.

Risk Analysis Another important component of conservation biology is the study and preservation of endangered species. One aspect of the study of endangered species is **risk analysis**, the development of quantitative estimates of the likelihood of extinction. This may often require many pieces of information, like the variability of the environment, the reproductive biology of the endangered species, and the dynamics of other species that provide food for the endangered species or feed on the endangered species. For instance, in Figure 17.1B we show information that has been developed to predict the dynamics of a population of red kangaroos in Australia. These animals feed on grasses and forbs that fluctuate widely in available biomass due to highly variable rainfall. The fluctuations in available food level have led to fluctuations in births and deaths that ultimately affect the numbers of kangaroos. The model in Figure 17.1B can be used to determine the chances of the extinction over fixed periods of time. Different scenarios of weather and total available land are used when making these predictions. Together, this information will provide concrete estimates of the vulnerability of kangaroos to extinction. ❖

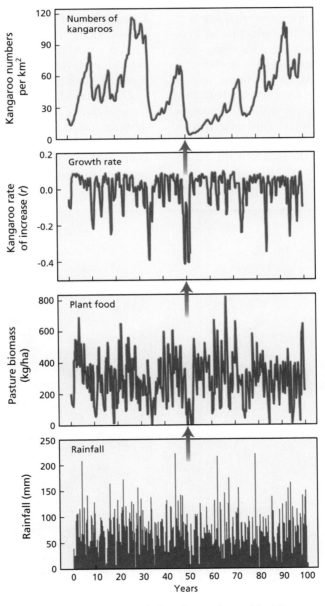

FIGURE 17.1B Predicting Population Fluctuations of Red Kangaroos

17.2 Ecological principles can be used to design reserves

There is an ever-increasing concern for the long-term fate of many plant and animal species, as well as of certain rare ecosystems. In most cases these organisms are threatened due to the negative impact of human activities on their habitat. In an effort to combat these trends, many countries are protecting areas of land or marine habitat as ecological reserves. A reserve may aim for at least three important goals: (1) The reserve may aim at preserving an entire ecosystem with all its important services. For instance, the area that constitutes a watershed may be preserved for the important needs of flood control and water recharge. (2) The reserve may have the general goal of preserving biodiversity. Some reserves might encompass regions in the tropics that are especially rich in the number of different species living there. Other areas may be converted to reserves because they contain many endemic species, that is, species found only in that area. Finally, the reserve may represent a particularly unusual environment that will be lost without protection. (3) Reserves may be set up to protect particular species. Many reserves have as a goal the preservation of particularly conspicuous species that are threatened with extinction—like rhinoceroses, pandas, and tigers.

The numbers of species on an island will stop changing when extinctions exactly equal immigrations.

One branch of ecology, called island biogeography, offers important theoretical concepts that can guide in the design of refuges. **Island biogeography** is concerned with understanding the number of species that are found on isolated habitats or islands. Obviously, an ecological preserve is similar to an island, since it may often be surrounded by inhospitable territory for the organisms within the preserve (Figure 17.2A). One general observation of the species composition of islands has been that the larger the island, the more species are typically found (Figure 17.2B, part i). While the data presented in Figure 17.2B are for birds, similar trends are seen for many different taxa of plants and animals. A guiding principal for understanding these species-area relationships is that, at any time, the numbers of species on an island represent a balance between extinctions and immigration events. We consider these in more detail next.

On any island, we expect that the number of species extinctions per unit of time will increase when there are more species on the island. This follows from the obvious fact that with increasing species, there are more possibilities of an extinction event. However, we also expect that for any fixed number of species, extinction is more likely on a small island than it is on a large island (Figure 17.2B, part ii). This is because all things being equal, large islands will typically have more habitat and resources for any single species, making extinction less likely. On small islands, populations will typically be smaller and more prone to extinction as we detail in Module 17.6.

Immigration is typically from a large source population, like a mainland continent, over inhospitable territory to an island. If everything happens as expected, a migrating plant or animal from the mainland is more likely to hit a nearby island than a distant island. If the island already had many species on it, there is less chance that a migrant represents a new species. Thus rates of immigration decline with increasing numbers of resident species.

FIGURE 17.2A The clear-cut hill in the foreground overlooks the Gifford Pinchot National Forest in Washington. On the right is private land clear-cut by a logging company. The idea of refuges as islands is visually supported by this picture.

The numbers of species on an island will stop changing when extinctions exactly equal immigrations. We can determine the equilibrium number of species on any island by combining the extinction and immigration curves and finding their point of intersection (Figure 17.2B, part iii). We see from this that the expected number of species is much greater on the near island than on the distant island.

At the very least, then, we can conclude that a refuge that aims to prevent extinction of species should be as large as possible. Another problem that is more difficult to answer is, What is better—one large refuge, or several small ones that add up to the same size as the large one? The answer will depend on how much higher the extinction rates are in the small reserves. Suppose their extinction rates were essentially the same as those in the single large refuge. Then it would be much better to have several small refuges because, if a species went extinct in one, you could always recolonize it with the same species from one of the other refuges. On the other hand, if extinction rates were so high in the small refuges that all species would be expected to go extinct in a single generation, then it would be much better to have a single large refuge. ❖

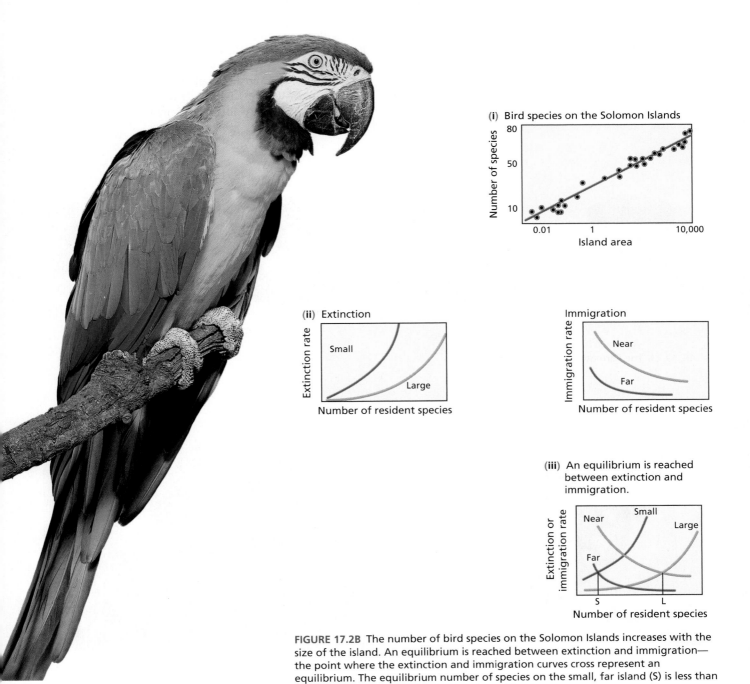

FIGURE 17.2B The number of bird species on the Solomon Islands increases with the size of the island. An equilibrium is reached between extinction and immigration— the point where the extinction and immigration curves cross represent an equilibrium. The equilibrium number of species on the small, far island (S) is less than the equilibrium number (L) on near, large islands.

17.3 The loss of habitat and habitat fragmentation leads to species extinctions

Many natural processes may lead to a break in the continuity of a habitat or **fragmentation**. Some of these breaks may create small gaps. For instance, a tree may fall in a forest and knock down several other trees. This will create a small gap in the forest that is no longer shaded and receives direct sunlight. A storm at sea may wash up large pieces of driftwood that smash against a dense covering of sea anemones and mussels in the intertidal zone. This creates a small opening of space in an otherwise densely packed community. Humans also create gaps in habitats; sometimes these can be quite large. The settlement of humans in forested areas of this country was followed by clearing of trees and initiation of agriculture. At first this created small gaps in the forest habitat (Figure 17.3A); but with time these gaps grew, until only small patches of forest remain. The forest had now become fragmented.

Sometimes we may not even recognize that populations are fragmented in smaller populations. Paul Ehrlich and his colleagues have studied populations of the checkerspot butterfly, *Ephydryas editha bayensis*, near the campus of Stanford University for the last 40 years (Figure 17.3B). The total numbers of butterflies at the Jasper Ridge site appears to be high over a 25-year period of observations (Figure 17.3C). However, the Jasper Ridge site is made up of several small populations—called C, G, and H—whose numbers show very different changes. In fact the smallest population, G, has gone extinct twice during the census period and has once been apparently recolonized (see Figure 17.3C). Thus, fragmentation may create small populations that are simply more vulnerable to extinction. If the fragmentation is severe, there may be no close neighbors to recolonize an extinct population.

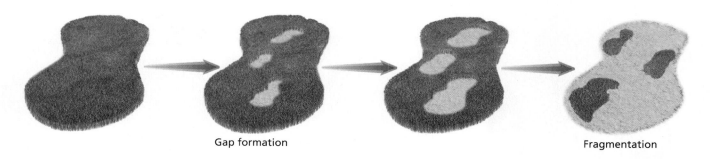

Gap formation

Fragmentation

FIGURE 17.3A The Formation of Habitat Fragments

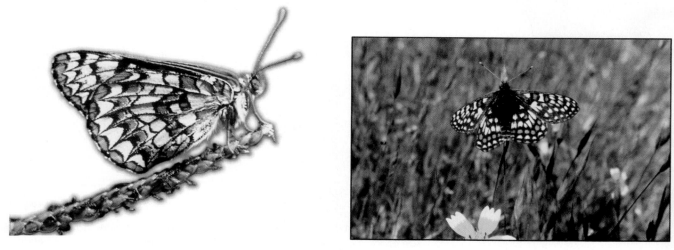

FIGURE 17.3B Euphydryas editha bayensis These butterflies are the subject of the study in Figure 17.3C and is an endangered species known to occur in only a few localities in California.

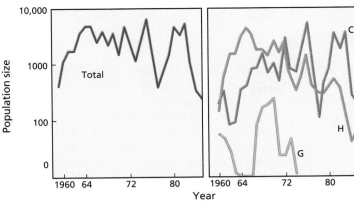

FIGURE 17.3C **Populations of Checkerspot Butterflies at Jasper Ridge, California**

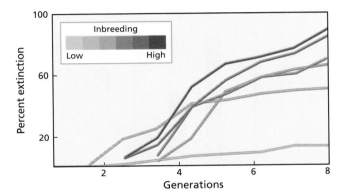

FIGURE 17.3D **Extinction of Laboratory Populations of Fruit Flies Inbred to Different Levels (low to high)**

Fragmentation of habitats may also facilitate species extinction because suitable habitat is destroyed for species with narrow distributions and specialized habitat requirements. Other species may perish because fragmentation is accompanied by barriers to essential movement. Some animals may need to move regularly from one geographic location to another to find new food resources or breeding sites. Some barriers created by humans may be quite subtle. For instance, many small mammals will not cross open roads. Others that try to cross these roads—like the endangered Florida panther—may be hit by cars so often that this becomes a major source of mortality.

Even if small populations persist, they become inbred due to their small size. For organisms that do not typically inbreed, mating with relatives can have devastating consequences. Biologists have suggested for a long time that inbreeding should make populations more vulnerable to extinction, but experimental evidence has only recently been collected. Kuke Bijlsma and his colleagues (2000) studied populations of the fruit fly, *Drosophila melanogaster*, that had been inbred to various degrees. They then made many small populations for each of the inbreeding treatments and kept track of how many populations went extinct over time. The results show that inbreeding demonstrably increased the chance of population extinction (Figure 17.3D). This effect was even more pronounced in stressful environments. The results in Figure 17.3D are for populations exposed to ethanol vapors. ❖

The last century has been marked by the loss of terrestrial forests and the acceleration of species extinctions

There may be anywhere from 10 to 50 million species on the planet today. About 90 percent of these are terrestrial, and about 80 percent of the terrestrial species live in the tropics. It is not surprising, then, that there is heightened concern about the loss of natural habitats in tropical regions. Although many species go extinct without notice, there are also well-documented cases of extinctions. The Rift Valley in East Africa is home to three lakes—Victoria, Tanganyika, and Malawi—that have about 1000 different species of cichlid fish. Most of these species are endemic to these lakes. Lake Victoria has already lost about 200 of its 300 cichlid species. Many of these extinctions were a direct consequence of the introduction of the Nile perch to these Lakes (Figure 17.4A). As humans have become more mobile due to technical advances in travel, they have made it easier for weedy species of plants and animals to move large distances. Some of these transplanted species can seriously disrupt the local communities and in the worst case, like the Nile perch, lead to the outright extinction of resident species.

The major tropical forests of the world are found in Central and South America, Asia, and Africa. The forested regions of these continents have been declining at an alarming high rate recently (Figure 17.4B). For instance, in Central America forest cover has vanished at a rate of nearly 2 percent per year. At this rate, by the year 2019 the forests in Central America and Mexico will be half the size they were in 1981. Much of this loss of forest has been due to the conversion of land to agricultural uses (Figure 17.4C). The result has been a tremendous loss of species.

The best estimates today are that the world loses about 27,000 species per year, mostly as a consequence of the human activities just reviewed. Historically, this rate of species extinction is unprecedented. Prior to the arrival of modern humans on Earth, data from the fossil record suggests that species were disappearing at a rate of about 0.25 per

year. The current rate of extinction is thus about 100,000 times higher than the historical levels. The extinction of a species is not the only negative outcome of habitat loss. Many organisms that suffer reductions in available habitat will, prior to extinction, lose genetic variation, subspecies, and many local populations. Consequently, the genetic variability of endangered organisms is often severely reduced even if they are not driven to extinction.

What are the negative outcomes of species extinctions? Even if we focus on only the direct consequences to humans, the loss of species can be great. Many important agricultural

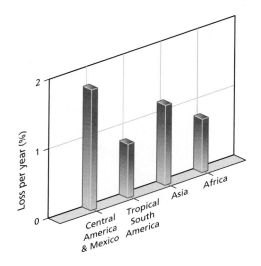

FIGURE 17.4B Annual Rates of Deforestation in Tropical Regions from 1981 to 1990

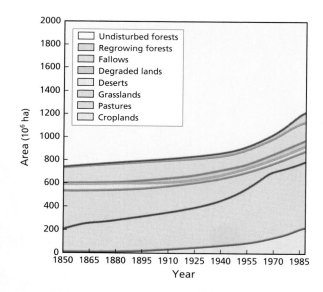

FIGURE 17.4C Land Use in Central America between 1850 and 1985 There is a general loss of undisturbed forests due to increases in pastures and croplands.

FIGURE 17.4A After its introduction, the Nile perch eliminated many endemic species of cichlids from Lake Victoria.

species of plants have been bred for specific characteristics and today appear quite different from their ancestral species. However, we find that agricultural species are constantly challenged with new diseases and insect pests. Having access to the genetic variation of the ancestral species of these crop plants provides opportunities to meet these new challenges by selective breeding or even through recombinant DNA technologies. The loss of these reservoirs of genetic information puts the future of agriculture at risk and makes it more difficult to respond to new challenges.

In the past we have found that plants and animal species harbor important chemical compounds that can be used for important biomedical advances. Clearly, the loss of species at the rates we are now witnessing will ultimately translate into lost opportunities for advances in medical research. Because these discoveries will never be made, we can only guess about their number and possible significance. ❖

17.5 Ecological theory can be used to guide harvesting from natural populations

It comes as no surprise that many species of plants and animals are economically quite valuable. The agricultural industry is an indication of how important these organisms are. However, the focus of modern agriculture is primarily on plant and animal species that can be easily cultivated or domesticated for human use. There are some economically valuable species, mostly animals that cannot be easily raised for human consumption. These species are then **harvested** from natural populations for human consumption. Examples include many species of fish, whales, marine mollusks, and marine crustaceans. These animals are not easily domesticated, because they need too much space, grow too slowly, or must be raised under conditions that are difficult to replicate outside of the natural environment. Often the exact numbers of animals in these natural populations is unknown. However, humans can be very effective harvesters and can cause natural populations to go extinct if they are not careful.

For instance, consider a population with a carrying capacity of 100,000 animals that grows according to the logistic equation. If a constant number of animals are removed from the population every year for human consumption, what will be the effect on the numbers of these animals? In Figure 17.5A we show several curves predicting the outcome of several levels of harvesting on this population. If the harvesting is not severe (removal of 7500, 10,000, or 12,500 animals), the population size declines; but it eventually reaches a stable point where new births equals natural deaths plus harvesting. However, if too many animals are removed (15,000 or 17,500 animals), the population will go extinct fairly rapidly.

Population density affects both birthrates and death rates. If we examine the total number of births in a population, we expect that birthrate will increase as population size increases, but that its rate of increase will slow down—and may even decrease—at very high densities due to resource limitations and their effects on female fertility (Figure 17.5B). Meanwhile, we expect the total number of deaths to increase at an accelerating rate with total density (Figure 17.5B) due to all the increased sources of mortality at high density. When these two curves meet, the population is at its carrying capacity (K): Total births equal total deaths. At densities below the carrying capacity, there is an excess of births over deaths, and some or all of these excess births may be harvested and not cause the population to decline (Figure 17.5B). The **maximum sustained yield** is the density that yields the greatest number of excess births. For populations that follow logistic growth, the maximum sustained yield occurs at $K/2$ (Figure 17.5C).

An interesting prediction for populations growing logistically is that there are two densities that will produce the same yield (see Figure 17.5C). However, the practical effects of harvesting at these two densities may be quite different. For instance, because density N_1 is less than N_2 (see Figure 17.5C), we would expect the animals to be less concentrated if we harvest at density N_1. This means that all things being equal, it will require more effort to harvest the same number of animals at density N_1 compared with density N_2. From an economic perspective, this means it will be more costly to harvest animals at N_1 compared with N_2. However, since the animals are less crowded at N_1, they should have more resources; this may mean they will be larger, and hence more valuable, than animals harvested at N_2. The details depend on the biology of the particular organisms in question and their

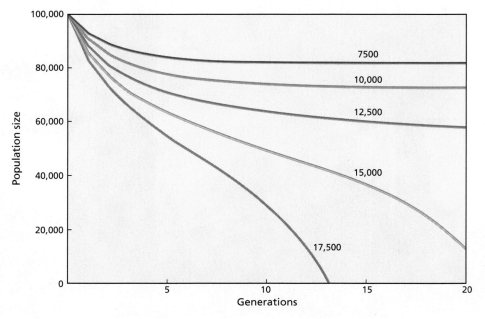

FIGURE 17.5A Population size decreases with increasing harvesting level.

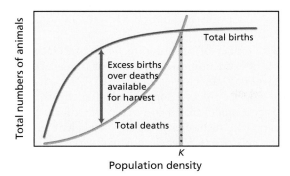

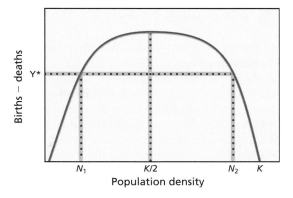

FIGURE 17.5B What density yields the greatest harvest potential?

FIGURE 17.5C Logistic population growth produces the same yield at different densities. Harvesting all excess births at N_1 or N_2 will produce the same yield (Y^*) because the excess number of births over deaths is the same.

response to different levels of crowding. The practical application of maximum sustained yield methods requires accurate information about the species of interest. Unfortunately this information is not available for many species of economic interest.

It is important to realize that the maximum sustained yield may not be the most profitable yield. In fact the most profitable course of action might be to harvest all the animals at once, even if it means extinction of the population. For this reason, many scientists believe that several factors must be used in determining the allowable harvesting levels of vulnerable biological populations. In addition to immediate economic forces, the importance of preserving biological populations for future use must also be considered. ❖

17.6 Risk assessment

How do species go extinct? Although we have abundant evidence from the fossil record of extinctions, it is difficult to determine the actual dynamics of the populations just prior to extinction. Unfortunately, we have well-documented examples of species extinctions over the last few hundred years. Some of these examples are discussed in the next module. Here we simply note that most species typically have their ranges severely restricted prior to extinction. Additionally, the species lives for some period of time at extremely low numbers before vanishing completely. When populations are very small, it stands to reason that all members of the population may fail to reproduce due to some sudden environmental catastrophe. So a virulent disease may kill the few remaining members of a population. Or unusual weather, like a drought or severe winter chill, could wipe out a small population. However, even in the absence of these types of unusual environmental scenarios, populations may go extinct due to natural variation in reproductive success.

Earlier in this book we have treated members of a population as equivalent units, all producing exactly the same number of offspring. In a population with equal numbers of males and females where every female produces 2.2 offspring, we would expect this population to sustain positive growth rates. Although the average number of offspring (λ) is 2.2, the actual number of offspring produced by each female may look more like the data shown in Figure 17.6A. An individual female may produce 0, 1, 2, or more offspring—even though the average is 2.2. Consequently, it is certainly possible that a small number of females may produce no offspring, by chance alone. This type of variability in offspring numbers is called **demographic stochasticity**. A small group of semelparous organisms will go extinct if all females fail to produce any offspring. Even though the chances of this happening in any one generation may be small, the chances over longer periods of time may be substantial. For instance, for small populations we can estimate from Figure 17.6A the chances of at least one year out of 10 with no offspring production, and hence extinction. For very small populations, 1–4 females, these probabilities can be substantial even when the average number of offspring is greater than 2. If some years are characterized by bad environments that substantially reduce the

average number of offspring, to 0.5 for instance, then the chances of extinction over any group of 10 years can be quite high (Figure 17.6A).

To assess the risk of a population going extinct, we need to construct a model that incorporates the information about demographic stochasticity, density-dependent survival and fertility, and effects of environmental variation on survival and fertility. However, before doing this we must decide how

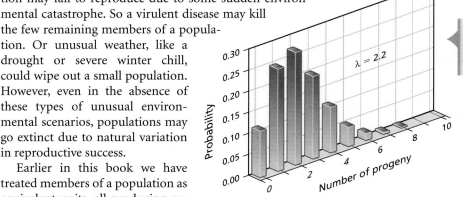

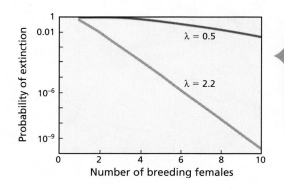

The average female may produce 2.2 offspring, but an individual female may produce no offspring. In this example, 11 percent of the females have no offspring

This figure shows the chance of extinction over 10 generations due to all females producing zero progeny in just one of those generations.

Extinction is greater when there are fewer females and when the average number of offspring produced is low.

FIGURE 17.6A Demographic Stochasticity

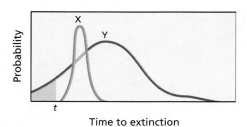

Time to extinction

FIGURE 17.6B Time to Extinction In this figure the mean time to extinction is greater in population Y than in population X. However, the chance of extinction at time t or earlier is greater in population Y. This is due to the large variance in population Y.

we will characterize the risk of extinction. One value that could be estimated is the average time to extinction. Obviously, populations with short average chances of extinction over any group of ten years can be quite high. In Figure 17.6B the possible times to extinction are shown as bell-shaped curves. Because the time to extinction depends on random events, we will not know its value exactly; but we can fix certain probabilities that it will have a particular value. In Figure 17.6B, population X has a lower average time to extinction than does population Y. However, it might not be the population at greatest risk. If we ask what the chances are that the population will go extinct soon (e.g., before time t), we see the answer is essentially zero for population X, but it is equal to the fraction of the Y bell curve that is shaded in red. Population Y is at greater risk of going extinct in the short term because there is a much higher variance in the random factors that affect it—meaning that if all things went really badly, population Y might go extinct quite rapidly. Such rapid extinction cannot happen in population X.

The factors we considered in Figure 17.6B suggest that another way to assess the risk of a population would be to estimate the chance that a population will go extinct before a fixed time period. This can be a useful statistic for managing populations because the ability to sustain certain policy decisions for extended periods of time is limited. If policy decisions like setting aside land or banning agriculture from certain areas are believed to be sustainable for a period of 50 years, then it makes sense to focus such efforts on those species most likely to go extinct in the next 50 years. This might mean that for our example in Figure 17.6B, population Y would be deemed at greater risk than population X and thus deserving of first priority in conservation efforts. ❖

APPLICATIONS

17.7 Applications of conservation biology include designing reserves, reducing species extinctions, and managing exotic populations

Relative to other fields of ecology conservation biology is still in its infancy. Nevertheless, many practical conservation problems require immediate attention. We do not have the luxury of waiting until the science of conservation biology acquires a more rigorous empirical foundation. We next provide an overview of some of these problems. In the modules that follow we pursue these problems in more detail.

An important application of conservation biology is the design of reserves to protect endangered species and ecosystems.

Reserves An important application of conservation biology is the design of reserves to protect endangered species and ecosystems. The use of reserves has in fact predated the development of conservation biology as an academic discipline. One of the earliest reserves also serves as an endorsement of the utility of reserves for preserving endangered species. A French missionary, Pere Armand David, discovered deer living in a walled game park just outside of Beijing, China, in the 1860s (Figure 17.7A). Bones of relatives of these deer can be found in China, and the natural populations appear to have been extinct for 2000–3000 years. It is not known how the captive population in China was established.

Pere David arranged to have a pair of deer shipped to an English park in 1898. At about the same time, the remaining deer in China had escaped from the park and did not survive the social disorder of the Boxer Rebellion. Thus by 1918 there were only 50 remaining Pere David's deer in an English park. These deer were descendants of the original pair shipped to England and some others gathered from around Europe. These deer have been maintained in England. The deer have now been reintroduced in a reserve near Shanghai, China, and appear to be doing well. Without the hunting reserve in China, the species probably would have gone extinct 2000 years ago. Without the establishment of the protected population in England at the turn of the twentieth century, the species would be extinct today. The establishment of several vigorous populations of Pere David's deer around the world should reduce the chances that all populations will go extinct.

As we will see, not all species have been as fortunate as Pere David's deer. Human activity over the last 400 years has been directly implicated in the extinction of many species of plants and animals. Although we are now aware of the fragile nature of many species, that knowledge has not prevented new species from becoming endangered. Many endangered species are located in poor countries that often lack the resources to devote to effective maintenance of reserves. An example of this is the Tsavo Parks in Kenya.

In 1948 the Tsavo National Parks were established in Kenya, with elephants as a primary benefactor of these reserves (Figure 17.7B). However, to create these reserves, many of the native people living in this area were moved out to the borders of the park. From 1957 to 1958, the Kenyan government carried out an intensive antipoaching campaign that was so effective it allowed the park to rely on a

FIGURE 17.7A Pere David's Deer

FIGURE 17.7B African Elephants

very small security force to cover the park's enormous area—21,800 km². The elephant population increased in numbers to about 50,000 just before a severe drought in 1971. Over 9000 elephants died from the drought. The people in the surrounding countryside were also adversely affected by the drought. People who had lost their crops then entered the park to search for dead elephants and ivory. The numbers of people entering the park were far beyond what the security force could contend with, and the psychological deterrent the force represented was lost. After the ivory from dead elephants was gone, poaching ensued. The elephant population dwindled and was estimated to be only 5400 in 1987. If the economic problems of the local people had been effectively dealt with, the poaching could have been prevented.

Extinction All the species shown in Figure 17.7C have shared the fate of extinction. Furthermore, the major culprit of these extinctions was human activity. In Module 17.8 we will go into more detail about some specific cases of extinction. This is a crucial topic because as we all know, extinction is forever.

Exotic Species Human trade, agriculture, and commerce may create other types of problems. In some cases plants and animals are transported by humans to novel environments, where they become serious economic pests. In Module 17.10 we review this general problem and learn what principles from conservation biology can be used to help control or eradicate these exotic species. ❖

Snail-eating coua

Caribbean monk seal

Flying fox

FIGURE 17.7C Extinct Species

Human activity has led to the extinction of many species in recent history

Compared with speciation, even compared with natural selection, extinction is inherently less mysterious than other evolutionary processes. Extinction occurs when species can no longer sustain themselves, which in turn means there are not enough individuals to continue the production of off-

spring at levels high enough to prevent the loss of all individuals of the species. Sometimes the exact moment of final extinction, when the last member of a species dies, is known.

Many cases of extinction caused by human activity are well documented. Take, for example, the dodo (Figure 17.8A). The dodo was a member of the same group of birds as the common pigeon. It was gray, like many pigeons, but its body was radically different—30 to 50 pounds in weight, with small wings of no value for flight and a large beak. The dodo was a single species found only on Mauritius, an isolated island in the Indian Ocean, east of Africa. Portuguese sailors first came upon the dodo in the sixteenth century. It is thought that the name for the dodo is related to the Portuguese slang for *stupid*. The dodos were completely tame and did little to evade capture and slaughter. This is not as paradoxical as it might seem, because the island of Mauritius had no large predators on it before humans arrived, so there would have been little selection for fear of predation in dodos. Although it might be supposed that all the dodos were simply killed and eaten by sailors, two other factors were probably also important. First, humans cut down the trees making up the wooded habitat that the dodo inhabited. Second, dogs and pigs introduced by humans also hunted and ate the dodo. The pigs also competed with dodos for similar food and destroyed their nests. No living dodos have been sighted since 1681. A few museums have preserved pieces of dodos, but no complete specimens remain. The entire extinction process took only about a century.

This one example could be amplified many times over. It is perhaps relatively notable because it was one of the first cases where humans knew that they were responsible for the extinction of an entire species. The expression "dead as a dodo," which was once commonplace, bespoke an awareness of the irretrievable loss that extinction represents—its finality.

The extinction of the dodo might be considered unsurprising, given that it was a single, docile species isolated on a unique island. One step up from the dodo might be the Tasmanian tiger (see Figure 17.8A), or thylacine, a striped marsupial that had a

Dodo

Passenger pigeon

Tasmanian tiger

FIGURE 17.8A Gone but Not Forgotten

Recorded extinctions since 1600

Taxa	Continental	Island	Oceanic	Total
Mammals	30	51	2	83
Birds	21	90	2	113
Reptiles	1	20	0	21
Amphibians	2	0	0	2
Fish	22	1	0	23
Invertebrates	40	48	1	98
Vascular plants	245	139	0	384

morphology and an ecological role generally similar to those of a wolf or coyote. This was a less docile animal than the dodo, with a wider distribution. Europeans were first made aware of its existence by reports published in 1805. It was not long before the animal was classified as a danger to livestock, with a bounty placed on each pelt. It was hunted to virtual extinction by 1930, at about the same time it received legal protection. The last thylacine in captivity died in 1936. Although human hunting again played a role in the species' demise, it is also thought that an outbreak of disease in 1910 may have been a significant part of the extinction process. Unlike the dodo, this species was photographed many times during its last fifty years of existence.

The passenger pigeon is a very different case (see Figure 17.8A). This bird species was hugely abundant in the forests of eastern North America, a broad habitat that should have left it less vulnerable to extinction. At its peak, it was estimated to have a total population of more than a billion individuals. Some observers considered it the most abundant bird in world, which is no longer a decidable question. The bird could reach high speeds, in excess of 60 miles per hour. It was highly gregarious, forming dense flocks in flight and on the ground. Its decline in numbers was partly due to the destruction of its forest habitat by farmers. But another major factor was its extensive use in hunting, both for food and for sport. The slaughter of these birds could be extensive, with individual hunters killing a thousand or more in one day. No other factor has been proposed as an explanation for the bird's extinction. The last known bird of the species, named Martha, died in Cincinnati on September 1, 1914.

This is but a small list of the known extinctions. The total number of known extinctions in recent history is in the hundreds of species (see Figure 17.8A). These numbers must also be viewed as very conservative, because we are probably unaware of many extinctions. Even today we continue to find new species that have been previously unknown. It stands to reason that there must also be many species whose existence and extinction are unknown. ❖

17.9 Some of the most prominent endangered species live in terrestrial ecosystems

Many species are currently facing the fate of passenger pigeons and dodos. Often the organisms recognized as being in danger of extinction are large, prominent, terrestrial organisms. This does not mean that there are no marine organisms facing possible extinction, nor that small insects aren't also in danger, because they almost certainly are. However, it is much more difficult to study these species, especially if they are rare and hard to find. Although the list of endangered species is large, we review a few specific examples here to illustrate the ecological and economic forces involved.

The northern spotted owl lives in old-growth forests in the Pacific Northwest of the United States. Old-growth forests are mature forests that have a relatively stable plant composition, and the dominant trees are often very large and growing little. Old-growth forests are dwindling due to economic pressures to harvest their valuable wood. The area of old-growth forest in Oregon, for instance, is very small and declining (Figure 17.9A). Only about 17 percent of the original old-growth forest in the Northwest remains. In 1987–1989 the total population of spotted owls was estimated to be 2500–3000 pairs. A pair of these birds needs an area of 800–2000 hectares (ha) to find sufficient food (flying squirrels) to survive. However, 800 hectares may contain lumber worth $8 million. Thus, setting

aside sufficient land for 500 pairs of owls is a $4 billion decision. How do you weigh the immediate economic effect due to losses in the lumber industry with the consequences of extinction of an animal that has no immediate economic benefit? This is a difficult question. Of course, the lumber is still in the forest and will be available to harvest for many years to come, but extinction is an irreversible process. Undoubtedly, setting aside these old-growth forests will protect species other than just the spotted owls.

The fate of several species of rhinoceroses is affected by similar pressures—dwindling habitat and economic pressures. The area occupied by several African populations of rhinoceroses is shown in Figure 17.9A along with their former ranges. The growth of human populations, accompanied by the conversion of land to agriculture and harvesting trees from forests, has contributed to the long and continuous decline in the numbers of rhinoceroses. The horns of rhinos are also highly valued as a powdered aphrodisiac, as medicines, and as a ceremonial knife handle. In the early 1990s the wholesale price for Indian rhinoceros horn was $62,400 per kilogram, and $8000 per kilogram for African horns. Thus, even though many rhinos live in protected areas, there are frequent occurrences of poaching. Between 1984 and 1990 an estimated 782 rhinos were killed in Zimbabwe. In Northeast India between 1985 and 1989, poachers killed 243 rhinos. The current estimates of rhino populations are shown in Table 17.9A.

❖

Species	1979	1988	1993
TABLE 17.9A Population Size Estimates of Wild Rhinoceroses			
African black	14,875	3780	2550
African white	3841	—	6784
Indian	—	1200–1500	1900
Sumatran	—	500–900	500
Javan	—	65–70	50

Old growth forests in Oregon

Northern Spotted Owl

The distribution of rhinocerus in Africa.

Black Rhinocerus

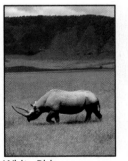

White Rhinocerus

FIGURE 17.9A Endangered species are associated with fragmented and small habitats.

For a conservation biologist the word *exotic* does not refer to some extraordinary characteristic of color, size, or physiology of an organism. An **exotic species** is one that has recently been placed in an environment or ecosystem in which it has not evolved. So while a rabbit is certainly not exotic in the United States, it would be considered exotic in Australia. These introductions are almost always the result of purposeful or inadvertent human activity. Not all species introductions cause problems. However, there are many instances of introduced species becoming serious pests. These successful exotics may do well in their new habitat, for several reasons. They may be superior competitors to the native species. Many of the marsupial animals in Australia are inferior competitors to ecologically similar placental mammals. The introduced exotic often finds itself in an environment without natural predators or diseases. This has of course led biologists to introduce predators and diseases in an attempt to control introduced exotics (see Module 14.5). A small sampling of exotic species pests are shown in Figure 17.10A.

Historically it is now clear that the unchecked introduction of exotic species can cause serious damage to agriculture and the environment. There are several ways to prevent introductions and to minimize their potential damage (Figure 17.10B). One obvious method is to prevent their introduction in the first place. Anyone who has traveled overseas may recall that upon returning to the United States, you are required to answer questions concerning any agricultural products you are bringing back with you. This is a direct attempt to prevent the inadvertent introduction of pests from foreign countries. Sensitive ecosystems and species may be protected by setting aside habitats for them. This management approach is also useful to guard against encroaching human development and activity, like logging.

Despite these efforts, exotics may still gain entry into a new habitat and require additional attention. There may be several different approaches to attempting to eradicate or severely reduce the population of exotics. Controlling exotic insect species may involve the use of pesticides. For animals like rabbits, the introduction of biological disease agents has been tried. Biologists can attempt to limit the spread of exotics even if their numbers cannot be regulated. In 1980 in Northern California, small numbers of Mediterranean fruit flies were discovered in local citrus trees. These introduced pests have the potential to cause tremendous damage to the citrus crops in California. As a result several control methods were utilized, including aerial spraying with the pesticide malathion to reduce or eradicate the established populations. Roadblocks were also set up on the major highways leaving Northern California, and passengers in all vehicles were asked if they had any homegrown fruit in their car. If they did, they were required to dispose of it. These efforts were aimed at limiting the spread of the Mediterranean fruit flies. ❖

European rabbits introduced to Australia grew to tremendous numbers, and the species became a serious pest.

The paperback or cajeput tree was introduced to Florida from Australia. It is tolerant of flooding and recovers well from fires. It is replacing the native cypress, *Taxodium ascendens*.

The South American nutria was brought to the United States for its fur, but escaped to establish wild populations. Nutrias live in marshes in the Southeast and damage the natural vegetation by grazing.

FIGURE 17.10A A Sampling of Introduced Exotics

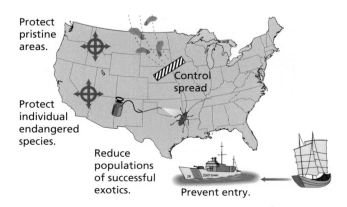

Protect pristine areas.

Control spread

Protect individual endangered species.

Reduce populations of successful exotics.

Prevent entry.

FIGURE 17.10B Methods of Controlling Exotic Species

SUMMARY

1. Small populations of endangered species share characteristics of island populations.

2. Ecological theories of island biogeography then shed light on the factors that affect the diversity of isolated populations.

   a. All things being equal, larger islands or habitats should support a greater number of species.

   b. Diversity will also be increased if it is easy for species to migrate onto an island.

3. Human activity has lead to the fragmentation of many habitats, thus creating the island-like structure of endangered species.

   a. Habitats have not only been fragmented by human activity, but some have often been destroyed.

   b. Tropical rain forests that are home to many endemic species of plants and animals are being rapidly destroyed.

4. Some natural populations are valuable economic resources.

5. Simple models of population growth can be used to determine the maximum sustainable harvest rate for a population. For populations that grow according to the logistic equation, this rate occurs at half the carrying capacity.

6. When populations become very small, they are in increased danger of extinction.

   a. Animals may fail to reproduce by chance alone, and extinction can follow from this type of demographic stochasticity.

   b. The ability to make predictions about chances of population extinction helps guide conservation policy efforts.

7. Extinction is more than just a theoretical concept. In recent history, there are numerous examples of species driven to extinction by human activity.

   a. Just a few examples include the dodo, the passenger pigeon, and the Tasmanian tiger.

   b. The known examples of human-aggravated extinctions number in the hundreds.

   c. The true total number of these extinctions is almost certainly much greater.

8. An important practical problem is the identification and preservation of living species that are currently threatened by extinction—like rhinoceroses and northern spotted owls. In these two examples, we also see the complicated interaction between economic forces and species survival. The solution to these problems is often difficult.

9. Human activity has also introduced species into novel environments, where they have become pests.

   a. The economic costs of these harmful exotic species can be enormous.

   b. A variety of general techniques can be used to deal with exotics. The most successful solutions often require a detailed understanding of the ecology of the introduced species, so that control methods can be tailored for that particular pest.

REVIEW QUESTIONS

1. The large blue butterfly was provided a protected habitat but failed to thrive. Why?

2. How do you think the distance from an island to the mainland will affect extinction and immigration rates on the island? Consider two cases, near islands and far islands that are otherwise the same.

3. Suppose that the dynamics of a fish population are described by Figure 17.5C. Suppose the harvest policy is to catch exactly Y^* fish every year. Describe the different consequences of this policy for the fish population if the actual number of fish is around N_1 vs. N_2.

4. Review the different methods for protecting habitats against the introduction of exotic species.

KEY TERMS

biodiversity	exotic species	island biogeography	risk analysis
conservation biology	fragmentation	maximum sustained yield	reserve
demographic stochasticity	harvesting		

FURTHER READINGS

Bijlsma, R., J. Bundgaard, and A. C. Boerema. 2000. "Does Inbreeding Affect the Extinction Risk of Small Populations? Predictions from *Drosophila*." *Journal of Evolutionary Biology* 13:502–14.

Caughley, G., and A. Gunn. 1996. *Conservation Biology in Theory and Practice*. Cambridge, MA: Blackwell Science.

Cox, G. W. 1997. *Conservation Biology*. Dubuque, IA: Wm. C. Brown.

Ehrlich, P. R., and D. D. Murphy. 1987. "Conservation Lessons from Long-Term Studies of Checkerspot Butterflies." *Conservation Biology* 1:122–31.

Meffe, G. K., and C. R. Carroll. 1994. *Principles of Conservation Biology*. Sunderland, MA: Sinauer.

PART FIVE

DARWINIAN BIOLOGY IN EVERYDAY LIFE

If you want to understand your personal life or your health scientifically, you need to develop a Darwinian perspective. Darwinian biology does not supply the whole truth about your health or social problems, but it reveals some of their deeper meanings, even sometimes their origins.

There are few things in life as confusing as sex. This is true not only for perplexed teenagers, but for Darwinian biologists as well. Sex is an unsolved problem of biology. There is no agreed-upon explanation for the evolution of sex, one of the more widespread and important phenomena in the history of life on Earth. This is more than a mere oversight. Biologists have struggled for decades to explain sex in ecological and evolutionary terms, as we show in Chapter 18. Nobody is particularly happy with the result of their efforts. If nothing else, the failure to explain scientifically the vexed thing suggests that everyday human difficulties with sex may be rooted in the very complexity of the ecology and evolution of sex. At least you will see here that you are not alone in your confusion.

If the inner mystery of sex has resisted Darwinian analysis, more success has been achieved with mating. Many aspects of choosing the gender of your offspring, or even your own gender, turn out to be readily explained in relation to Darwinian adaptations to understandable environmental conditions. Even such a mundane thing as the balance between the numbers of males and females in a population turns out to have a meaningful Darwinian explanation. Problems like mate choice and promiscuity are also eminently resolvable, if not perfectly predictable, using Darwinian logic. It

is not just media images that make teenagers promiscuous; chimpanzees have been practicing group sex for many years without any encouragement from the Fox network or E! Channel. We raise some of these delicate questions in Chapter 19, but be forewarned that we are barely scratching the surface of the Darwinian study of mating strategies.

Why are people ever nice? From Chapter 1, you probably realize that altruism was one of the earliest issues in the controversies about Darwinian evolution. Alfred Lord Tennyson epitomized this issue when he referred to "Nature, red in tooth and claw." In the nineteenth century, a common reaction to Darwin's theory of evolution was that it gave educated people the impression that life in nature was nothing but rapacity. Indeed, this impression led a wide range of political figures to recast history as a struggle, between classes in the case of Karl Marx and other communists, or between nations in the case of Adolf Hitler, Benito Mussolini, and many others. Yet the overwhelming finding of behavioral biologists in the early twentieth century was that most animals are not perpetually warring with each other. Instead, animal behavior is often surprisingly passive, acquiescent, and self-sacrificing. Three main Darwinian ideas have been put forward to explain the degree to which nature is peaceful: group selection, kin selection,

528

and strategy selection. We introduce these theories in Chapter 20.

Human evolution is one of the most absorbing topics in biology. It is a multifaceted area of research, from fossil hunting to determine the timeline of the evolution of the hominid brain to molecular analysis of the differentiation of contemporary human populations. A great deal of information has been collected about human evolution. It is absolutely untrue to say that human evolution is a complete mystery. We know a remarkable amount about how our ancestors evolved, dating back more than 45 million years ago. We also know a great deal about the molecular genetic differentiation of present-day human populations. Indeed, our knowledge of human molecular genetics is so great that we can now readily identify people from tiny amounts of DNA, whether they are fallen soldiers, impostors, criminals, or indeed crime victims. In Chapter 21 we lay out these considerable achievements in the study of human evolution.

Chapter 21 also raises a still more delicate issue—the significance of Darwinian biology for the understanding of our behavior. A lot of people, even many biologists, want Darwinism left out of the analysis of human behavior. Others, especially physical anthropologists, want Darwinism brought into all scientific discussions of human behavior. Given the range of strongly held feelings on this issue, we can do no better than offer several alternative points of view. But it would be irresponsible to hide this issue from a new student of Darwinian biology. Among other things, it surfaces in a variety of media outlets, from drugstore magazines to public interest journals. This situation arises naturally from the long-standing public interest in the significance of Darwinism for people's everyday lives, an interest often expressed in the nineteenth century.

The practice of medicine is dominated by two main components—the clinical experience of physicians and biological knowledge derived overwhelmingly from the molecular biology disciplines, from biochemistry to cell biology. A new movement that rejects this historical tradition has arisen among Darwinian biologists—Darwinian medicine. It has had very little impact on the practice of medicine anywhere in the world, but Darwinian medicine is forcing many people, even a few medical doctors, to rethink their health-care practices. In some cases, patients have themselves taken on Darwinian perspectives and strategies regarding their medical care. Whether you do so or not is up to you. We are not trying to convince you to seek advice from an evolutionary biologist instead of a physician. Our point of view is that medical practice and education should incorporate insights from Darwinian biology that might save lives or reduce suffering.

It may seem inconceivable that ideas used in studying millions of years of evolution might be useful in our everyday lives, but we think that they are—at least at the level of understanding. In the future, we may see a range of societal practices—from laws regulating our sex lives to the treatments offered by medicine—substantially reformed by the application of Darwinian biology. You can get in on the ground floor of this transformation by reading Chapters 18 to 22.

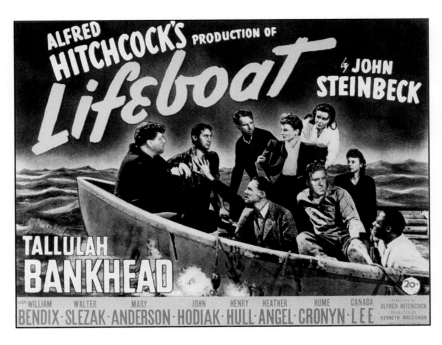

Sex is the Queen of evolutionary problems.

18

Evolution and Ecology of Sex

Sex is a huge and obvious part of the living world, from blooming flowers to copulating horses. Furthermore, sex is a preeminent part of the human experience. Yet sex confounds evolutionary biologists. They do not have a widely accepted explanation for sex, the exchange of genetic material between organisms. Indeed, some very good arguments can be made that sex should not exist. This makes sex a big problem for Darwinian biology, because evolutionary biologists are in the business of explaining why organisms do the things they do.

In the last three decades of the twentieth century, few evolutionary puzzles received more attention than sex. Many theories were proposed, and some were even tested experimentally. None of them worked very well—or at least they didn't work as well as they were supposed to work.

In this chapter we survey some of the mysteries of sex. We begin by showing why sex is a major problem from an evolutionary and ecological point of view. The existence of the male, especially, is puzzling. It sometimes seems, at least as far as evolutionary theory is concerned, as if the world should be inhabited only by asexual females. But it is not, for better or worse. We then explore some possible advantages of sex.

If sex is such an oddity, how did it come to exist in the first place? As we'll see later in the chapter, some surprising explanations have been offered for the origins of sex. The story of sex does not get any simpler as we go further back in time. ❖

WHY IS SEX A PROBLEM?

18.1 Many species do not have sex

One powerful argument against sex is that many organisms do not have sex at all. Some species, such as bdelloid rotifers, have not had sex for millions of years. Some sea anemones do not have sex; they reproduce by splitting in two, as described in Chapter 7. Even more common is **asexual reproduction** by fragmentation—particularly in plants, where it is practiced by strawberries, mint, aspen, juniper, and creosote, among many other species. This tells us that sex cannot be explained just by saying, "Birds do it, bees do it." This statement does not work, because many rotifers *don't* do it. The common occurrence of asexuality, even though it is a minority choice evolutionarily, means that we must explain why sex predominates. It helps to begin by considering what constitutes asexuality. Fortunately, in thinking about this, we can be guided in part by the real asexual organisms that live around us.

There are two kinds of asexual organisms. First is the type whose recent ancestors never had any form of sex. That is, these organisms are not recent offshoots of sexual organisms. A number of protozoa are like this—especially, so far as we know, some species of amoeba. Amoebas are the proverbial simple unicellular "animal" lifeforms. They are recognizable by their shapeless form, with projecting cytoplasm radiating out from an ill-defined center.

To the best of our knowledge, there are amoeboid species that have not had sex in millions of years. Other amoeboid species are sexual, however. Among the asexual species, each cell is a clone made from the same DNA as its parents. In these species, only mutations supply genetic variation. This type of species is pretty rare. Most organisms are sexual in some way.

The multicellular animal that has not had sex for the longest time is the **bdelloid rotifer**. These small aquatic invertebrates can be seen only under light microscopes (Figure 18.1A). They have few cells in their bodies, and they are not very complex physiologically. They have lived without sex for tens of millions of years.

> *To the best of our knowledge, there are amoeboid species that have not had sex in millions of years.*

The second kind of asexual organism is a recent evolutionary offshoot from sexual species. Many organisms are asexual in this sense, from octaploid asexual plants to triploid fish. We have already mentioned these polyploid species in Chapter 6.

One of the most interesting vertebrate examples of asexual reproduction is the whiptail lizards of the genus *Cnemidophorus* (Figure 18.1B). Sexual and asexual species of this genus are found in the desert Southwest of the United States and Mexico. The asexual species originate from hybridizations of two of the sexual species. The asexual lizards are highly heterozygous, because heterozygotes breed true in these species. They have few functional advantages over their sexual congeners, except for their ability to reproduce asexually, which doubles their reproductive fitness.

Sometimes asexual organisms depend on sexual species in order to reproduce. For example, in some asexual organisms, copulation is used to start reproduction, even though fertilization does not occur. In asexual species of the genus *Poeciliopsis* (Figure 18.1C), the initiation of development requires penetration of the egg by a sperm cell from a male of a sexual species belonging to the genus. These sexual species originally produced the asexual species by hybridizing with each other, which probably explains the continued dependence of the asexual form on sex with males from these particular species.

There are many variations on this theme, but they all reveal a common pattern. These recently derived asexuals usually do not have complete evolutionary liberation from the sexual life cycle. They are often halfway between sexuality and asexuality.

What is the general scientific importance of the various asexual forms? Whether long-standing or evolutionarily recent, they show that there is nothing impossible about the evolution of asexual reproduction. Sex cannot be explained by asserting that "Asexual reproduction just doesn't happen." Asexual reproduction happens all the time. ❖

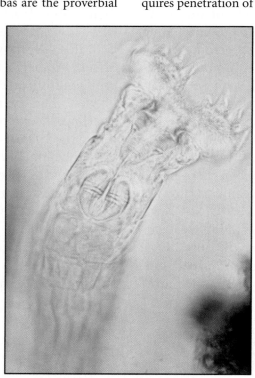

FIGURE 18.1A A bdelloid rotifer, a small aquatic animal that is asexual.

FIGURE 18.1B *Cnemidophorus* lizards are common in the desert southwest.

FIGURE 18.1C An asexual fish, *Poeciliopsis 2 monacha-lucida*, is in the foreground of this photo.

Understanding why sons make sex costly is essentially an exercise in arithmetic. The argument is structured as a comparison between sexual and asexual forms that were recently derived evolutionarily from a common ancestor. To make the argument as simple as possible, it is assumed that these two forms are identical except for their method of reproduction. The arithmetic of this evolutionary situation is further simplified by making additional convenient assumptions. The first is that females are limited to a fixed number of offspring, whether those offspring are male or female. We also assume that a solitary female raises the same number of offspring as a female who is accompanied by a male. With many bird and fish species, this assumption is not correct. But we can get around this problem by considering the sex problem in terms of regular sexual females *versus* asexual females that nonetheless copulate normally with males who keep them around despite their asexuality. In the second case, the asexual females are "pretending" to be sexual; but they only have daughters that are clones of their mother.

From the evolutionary standpoint, the issue is the number of copies of her genes that a female gets into the next generation. In sexual species, each reproductive act involves a genetic contribution from a mother and a father. In asexual derivatives of these species, each reproductive act involves a genetic contribution from a mother only. The asexual mother does not have to share her daughter genotypes with a father; she is essentially cloning herself.

The most concrete way to visualize the situation is to imagine sexual and asexual females reproducing in parallel within the same population, as shown in Figure 18.2A and the following box. We will also assume that sexual females produce equal numbers of sons and daughters, but that asexual mothers have daughters only. If we make the total number of offspring of both genders equal, then sexual mothers have half as many daughters as the asexual mothers

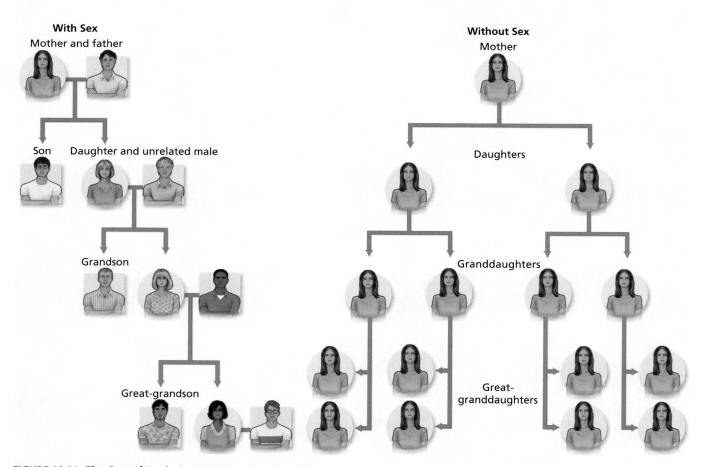

FIGURE 18.2A The Cost of Producing Males In this simple theoretical model, we assume that each female can produce only two children and that asexuals produce daughters exclusively. Asexual females swamp the sexual females numerically, driving out sexuality.

do. Continued generation after generation, this pattern will cause asexual females to outnumber sexual females by greater and greater numbers. Evolutionarily, the asexual females are outcompeting the sexual females in the production of daughters. All the sons are sexual, which might seem like an advantage for the sexual mothers. But sons do not limit the total productivity of offspring. As the number of asexual females becomes greater and greater relative to the sexual females, the sexual females are bred out of the population. Among other things, the sexual males will have a harder time finding a sexual female with whom to mate. ❖

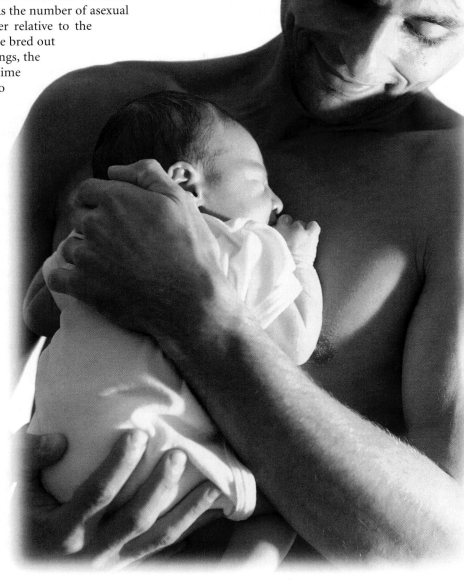

The Arithmetic of Sexual and Asexual Reproduction

To keep things simple, let's assume that both sexual and asexual mothers produce two offspring each. We will also assume that all children grow up to become adults. Sexual offspring are half males and half females. We start with "a" asexual mothers and "s" sexual mothers.

Let us go through the life cycle step-by-step for both types of reproduction:

In each generation, the number of asexual females doubles, while the number of sexual females holds steady. With time, the asexual females will increase in number relative to the sexual females. The asexuals have a selective advantage.

Asexuals: Adult females: a ⟶ eggs: 2a ⟶ Adult females: 2a ⟶ eggs: 4a

Sexuals: Adult females: s ⟶ eggs: 2s ⟶ Adult females: s ⟶ eggs: 2s

Sex requires sexual anatomy and exposure to predators or venereal diseases

Sex causes ecological problems. One of the most unavoidable is that sex requires organisms to develop structures and, in the case of animals, behaviors for fertilization. In species with two sexes, the adaptations of males to achieve fertilization are often obvious. Penises and aggressive male sexual behavior are examples of sexually related adaptations, common in two important taxa, arthropods and vertebrates. But females often need structures for receiving sperm or pollen. With respect to both male and female functions, flowering plants have elaborate structures for sexual reproduction. Growing structures like these can have a significant metabolic cost. In asexual organisms, all of these costs can be avoided (Figure 18.3A).

Another problem of sex is exposure to predation. In animals with protracted fertilization, the period of copulation may bring increased risk of being detected by a predator, or simply a greater risk of being captured due to the distractions of copulation. Some insects, for example fly species, copulate for dozens of minutes at a time, the male remaining mounted on the female. Although some insects can copulate "on the wing"—like dragonflies of the order Odonata—other insects, such as laboratory fruit flies of the genus *Drosophila*, have their flight impaired by copulation. Organisms that do not copulate may nevertheless face an increased risk of predation due to sex. Many species of animals, from fireflies to frogs, use signals to attract mates. Usually it is males who signal, attempting to attract females. But other animals may be paying attention. Bats home in on male frogs chirping at night, diving on them in the darkness and snatching them away for food. The power of sexual selection is so strong that frogs still make a great deal of noise to attract a mate, even when that same noise attracts predators. Yet none of this risk would be necessary if there were no **sexual reproduction**.

At its most intimate, animal sex is a wonderful opportunity for pathogens to infect new hosts. Sex is in some ways a feature of biology that seems to be specifically designed for the care and comfort of disease organisms. In animals, sex brings members of the same species close to one another—when fertilization is **external**, as is the case with most fish. With **internal fertilization**, males and females bring genitalia into intimate contact. Not only are normal dermis and its lubrications brought together, internal fertilization may carry ejaculatory fluid well into the body. Under these conditions, pathogens have excellent opportunities to infect new hosts. The box "Venereal Diseases in Humans" lists some of the pathogens that can infect humans.

Asexual organisms are generally free from this kind of intimate contact with members of the same species, except in such situations as combat over disputed territory. Mothers might infect their daughters, but the absence of sex would forestall a variety of infections of the mother. ❖

Venereal Diseases in Humans

One of the best examples of a species bedeviled with sexually transmitted diseases (STDs) is the human. Listed here are a few of the pathogens transmitted primarily during human copulation.

Name	Organism	Pathologies
Gonorrhea	*Neisseria gonorrhoeae*	Infertility, blindness
Syphilis	*Treponema palliidum*	Infertility, dementia
Chlamydia	*Chlamydia psittaci*	Infertility
AIDS	Human immuno-deficiency virus (HIV)	Immune failure, death
Genital herpes	Herpesvirus	Lesions, blindness
Genital warts	Papilloma virus	Warty tissue, cancer

Note that these **venereal diseases** will, on average, reduce the fitness of the individuals infected with them, due to reduced fertility or death. In asexual organisms, diseases that require such close physical contact for infection would be acquired only from the mother, if then.

(i) Development of sexual organs

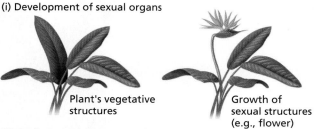

Plant's vegetative structures

Growth of sexual structures (e.g., flower)

(ii) Predation risks from sexual acts

Exposure to predation from seeking a mate

(iii) Sexually transmitted disease

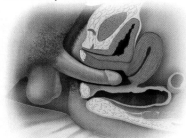

Genital contact transmits disease efficiently

FIGURE 18.3A Other Costs of Sex

Leaving aside problems such as the cost of producing males and the risk of venereal disease, there is a general disadvantage to sexual reproduction that arises from the way in which it handles genetic information. Sex is inherently a gene shuffler. It takes the two parental genotypes and combines them in novel ways. A fish that produces millions of offspring sexually may produce no two that are exactly the same. This novelty occurs because the combinations that can be made from the alleles residing at the genetic loci of two parents are astronomically large in number. (See the box, "Genetic Combinations with Sex.") It is somewhat like holding a bridge hand with thousands of cards. The chances that you will ever have the same such hand again in your card-playing career are essentially negligible. Therefore, if there were any particularly good reason to want to get the same genotype again, sexual reproduction would not be the right way to organize the life cycle.

Yet there is indeed a very profound reason to reconstitute the same genotypes. The genotypes of reproducing adults are the successful genotypes. Those are the genotypes that created phenotypes that could survive to reach adulthood. And when these phenotypes also reproduce, we know that they are successful in terms of fertility as well. This may seem like no big achievement, when it is viewed from the perspective of Americans growing up in suburbs. Your chance of becoming a reproductive adult is better than 4 out of 5. But in most organisms the odds are much worse. Only a few of the million eggs produced by a cod will become a reproducing adult. Only a few of the hundreds of eggs laid by a moth or a fly will become a reproducing adult. These odds suggest that the successful fish or insect may have a genotype that is fairly special. Therefore, why should this genotype be broken up by sexual reproduction? Wouldn't it be better if it was transmitted to the next generation as is?

With many genetic loci contributing to fitness, sexual reproduction chops up good combinations with efficiency. It takes only one locus to generate problems with sexual reproduction. As shown in "One Last Cost of Sex," the second box, in the case of heterozygote superiority at one locus, sex is unable to preserve the best genotype. This means that the average fitness of the population will be much lower than the average fitness of an otherwise identical asexual population. In the case of sickle-cell anemia, tens of thousands die because people in malarial regions cannot give all their children the heterozygous genotype for sickling that is resistant to malaria, as described in Chapter 4. This is a perfect example of the genetic problems that the human species faces because of our dependence on sexual reproduction. ❖

Genetic Combinations with Sex

The magnitude of genotypic variation that sex generates is astronomical. Let's take a simple example and suppose that there is free recombination between 10 genetic loci. Let us also suppose that we have 2 different alleles at each genetic locus. In sexual species, this means that we can have

$$2^{10} \text{ different gametes, or } 1,024.$$

If we consider both parents, each of which contributes a gamete, this allows 1,048,576 different genotypes in the population, assuming that the parental origin of each allele is followed. This with just 10 genetic loci.

But ordinary animals, such as insects and worms, have about 10,000 genetic loci, each with some dozen or more alleles. Suppose there are 10 alleles per locus. The total number of gamete genotypes is then $10^{10,000}$. The number of diploid zygotes that can be created from this many gametes is the square of this number: $10^{20,000}$.

One Last Cost of Sex

Suppose A_1A_2 has fitness 2, but A_1A_1 and A_2A_2 have fitnesses of 1.

In an asexual diploid population, the asexual A_1A_2 genotype would increase in frequency, finally fixing.

In a sexual population, the frequency of the A_1 allele would evolve toward a frequency of 0.5.

In the asexual case, mean fitness would evolve toward a value of 2.

In the sexual case, mean fitness would evolve toward a value of

$$p^2(1) + 2pq(2) + q^2(1) = 0.25 + 1 + 0.25 = 1.5$$

Thus the condition of sexuality results in a decrease in mean fitness.

The basic reason for this result is that selection with asexuality selects on the full genetic variance, while selection on sexual populations acts only on the breeding value—the impact of an allele averaged over all the genotypic combinations in which it occurs.

IS SEX A GOOD THING DESPITE ITS PROBLEMS?

18.5 Sex cannot be explained by evolutionary history

Evolutionary research has turned to the task of explaining the existence of sex. This task has long-standing historical roots. One of the traditional problems of medieval philosophy was explaining evil and other imperfections of the world. Why are some babies born dead? Why does the plague ravage Europe? Why is there sin? The attempt to explain these paradoxes of what was then seen as God's Creation was taken very seriously. The people who did this kind of work were called "apologists." In an analogous way, evolutionary biology has had many apologists for sex. Their arguments, and the counterarguments against them, will be the concern of this set of modules.

Sometimes these arguments get fairly complicated. In reading about them, you should always bear in mind what is going on. Sex seems anomalous, on evolutionary grounds. Yet evolutionary biology tries to explain life on evolutionary grounds. With the problem of sex, this field of science has run into serious trouble. Like flies caught on a spider's web, evolutionary biologists have been struggling with the paradox of sex.

There is an important argument that biologists like to use about sex. Several features of life (besides sex) are problems for organisms. For example, whales and dolphins have lungs instead of gills. They must come to the water surface to breathe, even if they dive thousands of feet between breaths. What's the explanation? The evolutionary "apology" for this feature of whale respiration is that whales evolved from terrestrial mammals in the last 50 million years. They have lungs due to an accident of evolutionary history, not because lungs are beneficial in their aquatic lifestyle. In this sense, the historical element of evolution—which was a major focus of Part One, "Introduction to Darwinian Biology"—leaves many organisms with failures of adaptation.

A phylogenetic feature that such **historical imperfections** of evolution tend to share is that they crop up sporadically in evolutionary trees. Most of the mammalian evolutionary tree is terrestrial, not aquatic. Lungs are a limitation only for one isolated branch of this tree. From this, we can infer a general

pattern. Evolutionary imperfections tend to be sporadic in their occurrence. This is shown in Figure 18.5A.

If sex is an evolutionary problem only rarely, like lungs are for whales, then asexual reproduction should occur at most sporadically in an evolutionary tree dominated by sexual life-forms. Most species of eukaryotes are sexual, and asexual forms are usually recent derivatives from sexual ancestors, as shown in Figure 18.5B. This pattern suggests that sex is of general benefit, while asexuality is a sporadic derivative of sex—and perhaps asexuality is usually disadvantageous.

But there are counterarguments to this kind of broad comparative apology for sex. Some features of life may be conserved even if they are not necessarily beneficial. The classic example of this kind is the fact that almost all adult insects have six legs—yet it is extremely unlikely that six legs is exactly the right number for the hundreds of thousands of insect species. Four legs might be better for some of them.

Characteristics may be preserved in evolutionary trees for reasons that are not related to evolutionary conservatism. Of particular importance are the deleterious effects of pathogens and genetic parasites on organisms. These effects may be generally sustained simply because such parasites are hard to get rid of. Most vertebrates have large quantities of DNA that does not encode protein or regulate protein synthesis, as described in Chapter 5. This DNA is not apparently there because it is beneficial. It is there because it tends to accumulate, and it is hard for cell evolution to eliminate it. Likewise, the fact that sex is widespread does not show that it is good for the organism. Sex may simply be hard to get rid of. ❖

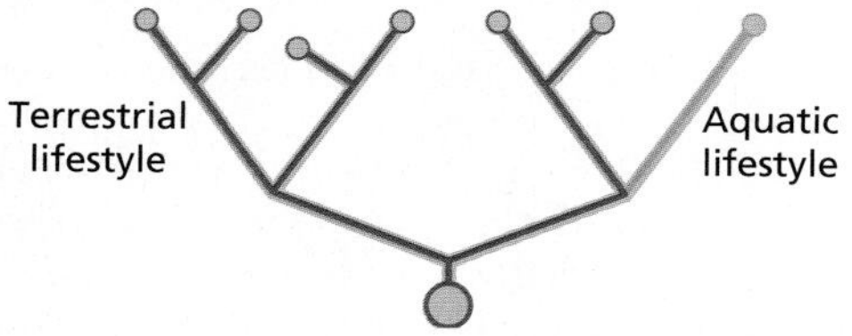

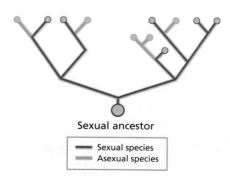

FIGURE 18.5A Partial Phylogeny of Mammals with Lungs
Though lungs are not ideal for aquatic life, they are retained in recently evolved aquatic mammals.

FIGURE 18.5B Partial Phylogeny of Sexuality Most taxonomic groups that have asexual species are primarily sexual, with a few exceptions. This suggests to some evolutionary biologists that sex normally improves fitness.

18.6 With moderately frequent beneficial mutations, sex can speed up the rate of adaptation

The traditional textbook argument in support of sex as an adaptation concerns the substitution of favorable new mutations. In large asexual populations, as we saw in Chapter 4, selection effectively "picks" the mutation that gives the highest fitness. That mutant then increases in frequency until it is fixed, with a frequency of 100 percent, or nearly so. If several new mutations are beneficial, only one of them can be fixed at a time, even when these mutations are at different genetic loci. This occurs because mutants at different loci in asexual populations are equivalent to alleles at the same locus in sexual populations. The asexual genome is a unified evolutionary entity, the whole thing succeeding or failing in selection as one unit. In principle, this makes selection in asexual populations very inefficient. As shown in part (i) of Figure 18.6A, if there are two beneficial mutants in an asexual population, then selection will fix one at a time—the best first usually—and then continue on to fix each of the other beneficial mutations in succession.

An example of this pattern was supplied by the *Escherichia coli* experiments described in Chapter 4. The asexual populations of those experiments underwent one favorable **sequential substitution** at a time, which made the evolutionary improvement in fitness a process of discrete "steps."

Sexual populations that have beneficial mutations at different locations in the genome can undergo substitution of all the beneficial mutations at the same time (Figure 18.6B). It is not surprising that sexual populations can do this. Sex recombines genetic loci, producing many combinations of the alleles at different loci, as described in Chapter 3. At least some of the time, sex will allow selection to produce genomes that have all or most of the beneficial mutations. And such genomes will be strongly favored by selection.

A useful analogy might be professional sports teams. In a league where trading and free agency are banned, it will be hard to assemble a strong championship team. But with trading and free agency, a team that has plenty of money should be able to combine players from other teams into a star-laden powerhouse. Sex is no different; it can combine good alleles. This verbal and graphical argument makes it seem as if sex should normally be of great evolutionary benefit.

But there is a major flaw in this argument—the rate at which beneficial mutations occur. If beneficial mutants are rare, as will be the case in smaller populations, multiple mutations will not be undergoing substitutions. Both sexual and asexual populations would fix beneficial mutants one at a time. There is no need to recombine lineages with different beneficial mutants. This scenario is shown in part (ii) of Figure 18.6A.

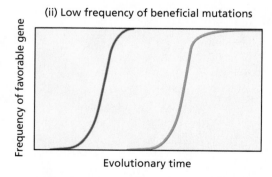

(i) High frequency of beneficial mutations

The second mutation cannot spread at the same time as the first.

(ii) Low frequency of beneficial mutations

FIGURE 18.6A Beneficial mutations are substituted one at a time in asexual populations. A mutation that increases fitness less than an earlier mutation can spread in an asexual population only when the later mutation has occurred in carriers of the first gene, which normally happens only when the first mutation is nearly fixed.

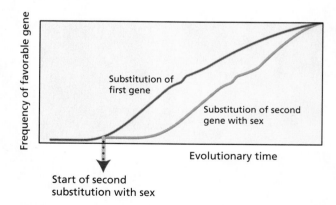

Substitution of first gene

Substitution of second gene with sex

Start of second substitution with sex

FIGURE 18.6B With sex, beneficial mutations at multiple loci can increase in frequency at the same time, because the process of gene frequency change at multiple loci is independent when there is no linkage disequilibrium between alleles. Furthermore, sex reduces such linkage disequilibrium.

On the other hand, if beneficial mutants occur at a very high rate, both sexual and asexual populations will rapidly evolve increased fitness. In particular, asexual genomes would then quickly fix multiple beneficial mutations without sexual recombination, just because such mutations are occurring so often (this case is not shown in Figure 18.6A).

This **accelerated evolution** theory does not require that sexual and asexual forms compete directly against each other. Instead, it is usually offered as an explanation of the greater proliferation of sexual forms over the entire range of macroevolution. ❖

Time Required for Substitutions of Favorable Mutations

The key to the rate of evolution is the amount of time it takes for a mutation to occur plus the amount of time it takes for such a new mutation to be fixed by selection. When there is a moderate rate of favorable mutations, sex may be beneficial compared to asexuality. But this is not always true. There may be few possible mutations that can increase fitness for a particular population evolving in a particular environment. Even if there are many possible beneficial mutations, small populations will not receive such mutations, because the number of new mutations is given by the mutation rate times the population size.

18.7 Sex may reduce competition between siblings, increasing the fitness of sexual parents

Suppose you were playing one of those state lotteries where you pick the number of your ticket. If you had the money to buy four lottery tickets, would you pick the same number four times? No, you wouldn't, because you wouldn't increase your chance of winning. If you were buying four tickets for a lottery, you would pick four different numbers.

Asexual reproduction is somewhat like picking the same lottery ticket number again and again, because all asexual offspring are genetic clones of their parent. The environment is occasionally the equivalent of the lottery. Sometimes only one or a few organisms will survive in a particular environment—probably those with genotypes that produce the phenotypes that fit this environment best. Sexual reproduction inherently produces more diverse offspring. Therefore, when there is intense competition, sexual reproduction may be more likely to "win," because it produces the genetic equivalent of more lottery tickets. Note that in this **lottery model**, sexual and asexual forms compete with each other directly.

But as plausible as this argument seems, it has flaws. Suppose you were buying tickets to four different lotteries, one ticket per lottery. Then it doesn't matter if you pick the same "lucky" number each time, because these tickets are not competing against each other. In the competition between forms of reproduction, a key factor is the competition of sibs, and other close relatives, with each other. It is only if there is such **sib competition**—that is, when the genotypes are "tickets" in the *same* lottery—that the increased diversity of sexual offspring might give an evolutionary advantage.

In ecological terms, sibs will be grouped into the biological equivalent of the same lottery only when they tend to inhabit the same locally discrete unit of habitat. These are sometimes called patches. Examples of such patch competition include pathogens competing within hosts, maggots in animal carcasses, small islands, isolated trees inhabited by insects, and so on. The pattern of such patch-based competition is presented in more detail in Figure 18.7A.

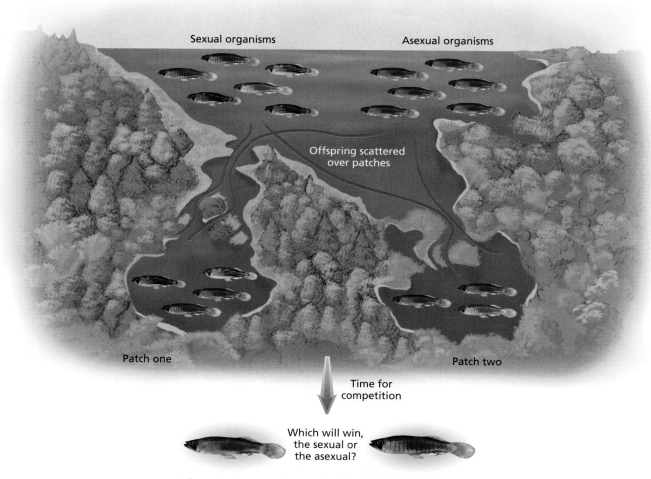

FIGURE 18.7A This illustration shows a competition between sexual and asexual fish when the species has a "patchy" ecology, in which only one competitor in a patch survives to produce offspring.

The relationship between the ecology and the genetics of the evolutionary situation is illustrated in Figure 18.7B, in which each habitat selects for a particular genotype. The idea is that each particular environment that progeny grow up in has a particular set of requirements, which are so exigent that only the organism with the best genotype will survive to reproduce.

This theory makes a general association between ecological variability and genetic variability due to sex. How plausible is this association in the real world? There are many examples of organisms that can reproduce either sexually or asexually. If sexual reproduction is beneficial in variable environments, then organisms should reproduce sexually before environmental change or dispersal to new environments, and they should reproduce asexually when exploiting a stable environment. This pattern is indeed exhibited by some species of plants and insects, among them aphids. Therefore, it is reasonable to view sex as a problem in which evolution, genetics, and ecology are commingled. On the other hand, this association between sex and ecological variation does not show that any particular model for the evolution of sex in a variable environment is correct. ❖

Patch-Structured Ecology

The sib-competition model for the evolution of sex is a complex one. It brings ecology together with evolution. Several requirements have to be met—localized and intense competition; joint dispersal of siblings; and very careful matching between genotype and ecology.

It is reasonable to ask how often ecology stringently sorts genotypes into the successful and unsuccessful. One special case is disease, which we discuss in Module 18.8. Another possibility is that organisms may undergo stringent selection for particular genotypes when there are extreme environmental fluctuations: drought, flood, cold winters, hot summers, and so on. But an additional possibility arises when there is an interaction between environment and competition between individuals in a patch. It could be that direct competitors make the ecology stringent, perhaps by denying each other food in the case of animals, or by denying each other sunlight in the case of plants. This range of possibilities was described in detail in Parts Three and Four, earlier in this book.

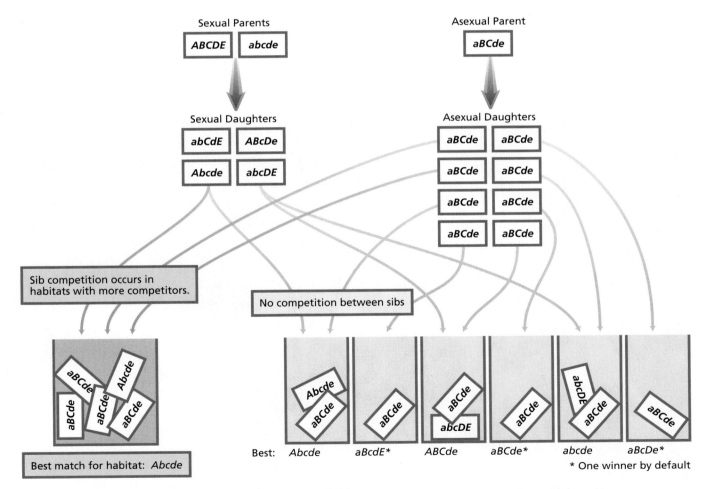

FIGURE 18.7B The Sib Competition Lottery Model The haploid competitors vary at five genetic loci, with large letter and small letter alleles for each locus. The coding for the genetic loci is matched to a parallel coding for the environment. The winner has to have a match with the environment at all five loci. Sib competition occurs when a parent sends multiple offspring to the same patch.

Sex may generate variability required for hosts to evolve faster than their diseases and parasites

The coevolution of pathogens and their hosts is a special case of the sib-competition scenario. Like the sib-competition model generally, hosts are patches for their pathogens, and pathogens compete with each other to exploit the host. In many cases, only one pathogen genotype may win the competition for exploitation of the host (Figure 18.8A).

But there is a major difference from other cases of sib competition: *Both* the pathogen and its host are evolving. This makes their coevolution into a kind of arms race. And in particular, this is an arms race in which there is a selective advantage to being different. Pathogens will be selected to infect the more common hosts; likewise, hosts will be most selected to resist infection by the most common pathogen. A rare type of host may be able to evade infection by the pathogen, simply because the pathogen is relatively less selected to exploit that host type. Conversely, a rare type of pathogen may be able to infect the host readily compared to the common types of pathogen. This is a perfect situation for selection to give an advantage to the generation of variable offspring, as in sexual reproduction, as opposed to unvarying clonal offspring, as in asexual reproduction. This competition will proceed within hosts that receive both sexual and asexual pathogens as well as throughout entire populations of hosts.

The benefits of producing diverse progeny will be sustained from generation to generation. Both pathogen and host will be continually evolving in response to each other. When one species is relatively more successful, it will increase the intensity of selection on the other species.

How can this **pathogen-host arms race** end? One or the other species may go extinct. Another scenario for this arms race ending is when coevolution makes it possible for the pathogen to infect the host with little effect on fitness. The cold viruses, for example, have reached this point in coevolution with humans, although the otherwise vulnerable hosts (the elderly, infants, and those with immune compromise) can still die even of relatively benign respiratory infections. (See Chapter 22 for more discussion of human contagious diseases.)

A criticism of this pathogen-host arms race model is that many species may face little risk of significant fitness consequences from pathogens. But even when this is true, such species may have evolved their freedom from infection. Furthermore, that freedom from serious infection may be undermined by the development of a new deadly pathogen. After the discovery of HIV, we must concede that deadly diseases may never leave any species untouched forever. ❖

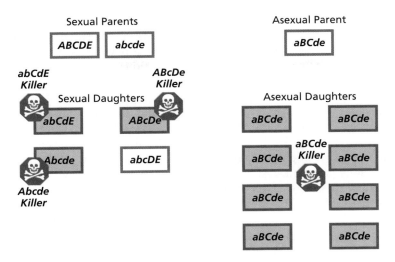

Sexual Parents

ABCDE   abcde

Asexual Parent

aBCde

abCdE
Killer

ABcDe
Killer

Sexual Daughters

abCdE   ABcDe

Abcde   abcDE

Abcde
Killer

Asexual Daughters

aBCde   aBCde

aBCde   aBCde
Killer   aBCde

aBCde   aBCde

aBCde   aBCde

FIGURE 18.8A Escaping from Predators and Disease The genotype required to resist disease is *abcDE*. Again, the sexual offspring have four chances to win, the asexual only one.

Coevolutionary Races between Pathogens and Hosts

A common image of evolution is that of organisms adapting to their physical environment, perhaps with selection favoring resistance to drought in a desert or to extreme cold in the arctic. This type of evolutionary ecology is a major concern of Part Three. But another form of evolution is also important—adaptation to other living species. Nowhere is the importance of this form of adaptation more acute than selection on pathogens and their hosts. Pathogens are selected to infect and replicate within their hosts. But the hosts are selected to resist these pathogens.

An important feature of this evolutionary interaction is that pathogens usually have many more generations than their hosts do, during the same period of time. The importance of this feature to human life is illustrated by HIV, a virus that evolves within each patient with extreme rapidity (Figure 18.8B.) Almost everyone who is infected with HIV eventually dies of the infection, because it out-evolves our immune response. As fast as our immune system generates antibodies to the proteins at the surface of HIV, the virus evolves new proteins. But it is significant that a few individuals appear to be able to resist HIV indefinitely, demonstrating resistance to becoming infected as well as resistance to disease progression toward AIDS. Humans are sexual. This gives human populations great genetic variability. This variability may help us to overcome the threat posed by HIV, despite the irony that sex helps to spread HIV.

There is evidence that chimpanzees suffered a deadly epidemic over the past 2 million years—an epidemic that wiped out all chimpanzees except those bearing a particular allele for disease resistance. The epidemic may have been caused by a pathogen related to HIV.

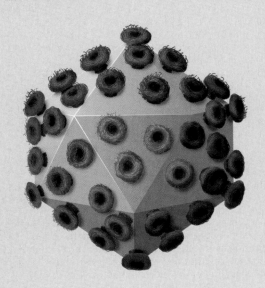

FIGURE 18.8B Drawing of the Human Immunodeficiency Virus (HIV)

Sex may help get rid of deleterious mutations over the entire genome

Deleterious mutations happen all the time. We tend to imagine that natural selection will get rid of these deleterious mutations as soon as they occur, but that is not true. The problem for asexual populations is that small asexual populations will tend to lose the individuals that are entirely free of deleterious mutations by accident. This is a special case of genetic drift, though the alleles involved are not neutral. (See Chapter 3.)

As shown in Figure 18.9A, an asexual population that starts with a spectrum of deleterious mutation is at some risk of losing the mutation-free lineage. All clones will then

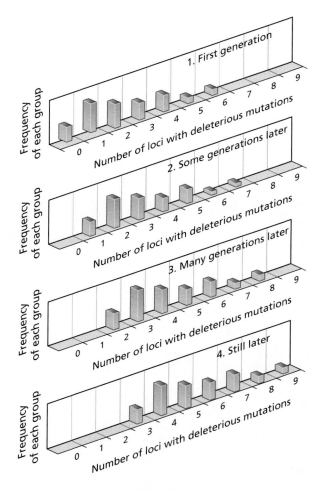

FIGURE 18.9A Muller's Ratchet in Asexual Organisms As genetic drift proceeds, first genotypes that are completely free of any deleterious mutations are lost, then genotypes that have only one mutation, genotypes that have only two, and so on. Eventually a small asexual population has low fitness.

have at least one deleterious mutation at some locus in their genome—although it won't always be the same locus. Then the same thing can happen again, and the lineages with only one mutation may be lost. In this way, the independent clonal lineages of an asexual population will tend to deteriorate with time by a ratchet effect, as the evolutionary process produces a progressive reduction in mean fitness. (A *ratchet* is a type of wrench with an internal gear that allows it to be turned in one direction only.) This deterioration may ultimately lead to the extinction of the clonal lineage, as the level of deleterious mutations becomes very high. This process is called **Muller's ratchet**, after the Nobel Laureate H. J. Muller, who first pointed out this problem for evolution. In the following box, "The Mysterious Deaths of Laboratory Protozoa," we discuss what is likely a real-life example of Muller's ratchet.

Muller's ratchet is made worse as new deleterious mutations are introduced into the asexual population. One thing that molecular biology guarantees for us is the continued occurrence of deleterious mutations scattered through the genome, thanks to low-frequency errors of DNA replication. Therefore, this problem of deleterious mutations is one that we can assume will always arise during evolution.

Now consider what will happen if these lineages can undergo sexual recombination. This recombination will produce offspring that vary in the number of loci that have deleterious mutations. Some offspring may have very few of these mutations over their entire genome, and these offspring will be favored by natural selection. In this way, selection combined with sex can prevent the progressive accumulation of deleterious mutations over all loci. It may even be able to keep the burden of deleterious mutations at low levels.

This ratchet model assumes that the effects of deleterious mutations over many loci are *additive*. That is, having two loci with deleterious mutations is twice as bad as one locus with a bad mutation, having three loci with bad mutations is three times as bad as one, and so on. When this assumption is not correct, the advantage of sex is not automatic. Things change if combinations of loci with deleterious mutations are not as bad as the effects of individual loci with bad mutations added together. In such cases, getting three or more loci with deleterious mutations might be no worse than having two. Then large asexual populations will do better than sexual populations at removing deleterious mutations from their loci. Therefore, the problem of deleterious mutations does not necessarily require the evolution of sex. ❖

The Mysterious Deaths of Laboratory Protozoa

An interesting puzzle of classical experimental biology was the tendency of cultures of protozoa to deteriorate and then die out. For a long time, such culture death was considered an aging phenomenon, perhaps due to confusion about the fact that these protozoa cultures were populations, rather than individuals. In an interesting historical reconstruction, McGill University biologist **Graham Bell** showed that these cultures were probably subject to Muller's ratchet. They tended to be fairly small in number, sometimes just a few individuals, and they were cultured over many generations. These conditions are ideal for the accidental loss of clones with fewer deleterious mutations, as shown in Figure 18.9A.

However, the action of Muller's ratchet in these laboratory populations is not a reliable guide to the accumulation of deleterious mutations in nature. Natural populations will usually be larger than the small numbers maintained in a biologist's laboratory.

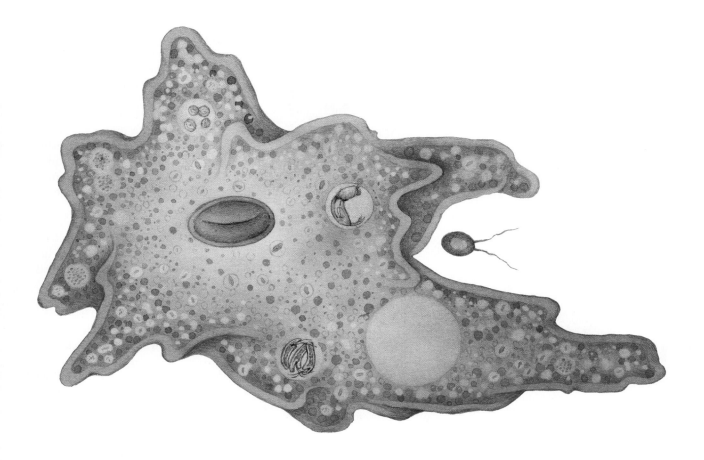

Sex may be maintained because newly asexual females have depressed fitness

The scientific literature on the evolution of sex is a kind of game. Sexual females are compared theoretically with asexual (or "parthenogenetic") females, as if the asexual females are exactly the same as the sexual females, other than the lack of sex in their lives. But this is not always true.

First, newly **parthenogenetic** females usually do not have a reproductive system as efficient as that of sexual females. This is not because asexual females necessarily must have inefficient reproduction. Instead, the evolution of any new structure or function is likely to be inefficient at first. Evolu-

tion usually takes many generations to shape adaptations so that they are highly efficient. It takes time for mutations that enhance adaptations to occur, and then each of these mutations takes some time to increase in frequency.

For example, recently evolved types of "flight," such as flight in gliding squirrels, are very awkward compared to flight in animals that have had it for millions of years, like bats and birds. Likewise, a newly asexual female usually has much reduced fertility. For example, when new asexual lineages of *Drosophila* species are isolated in the laboratory, they usually have fertility that is only a few percent of the normal sexual level. When this occurs, asexual forms will be selected against relative to sexual variants.

Second, newly asexual females face a particularly pertinent hazard that comes from the environment—males. Sexual males will continue to be attracted to asexual females, before these females have evolved distinctive morphology and behavior. Therefore, sexual males will attempt to mate with asexual females. If these females resist, they may be damaged. If they do not resist, then they face the problem of fertilization. As shown in Figure 18.10A, fertilization can have various consequences. One possibility is that the asexual female's egg may respond to the fertilization event and initiate meiosis, so that fertilization can proceed normally. In this event, asexual reproduction is terminated. This result occurs normally in new asexual females of *Drosophila*.

Another possibility is that the asexual female's diploid egg is fertilized by haploid sperm, producing triploid offspring. In most cases, triploid offspring will not be viable. And if they are viable, then they will have reduced fertility. Finally, triploid offspring may be viable, but reproductively isolated. This result can lead to a new triploid species (see the box, "Triploid Asexual Species Reveal the Struggles of Parthenogens").

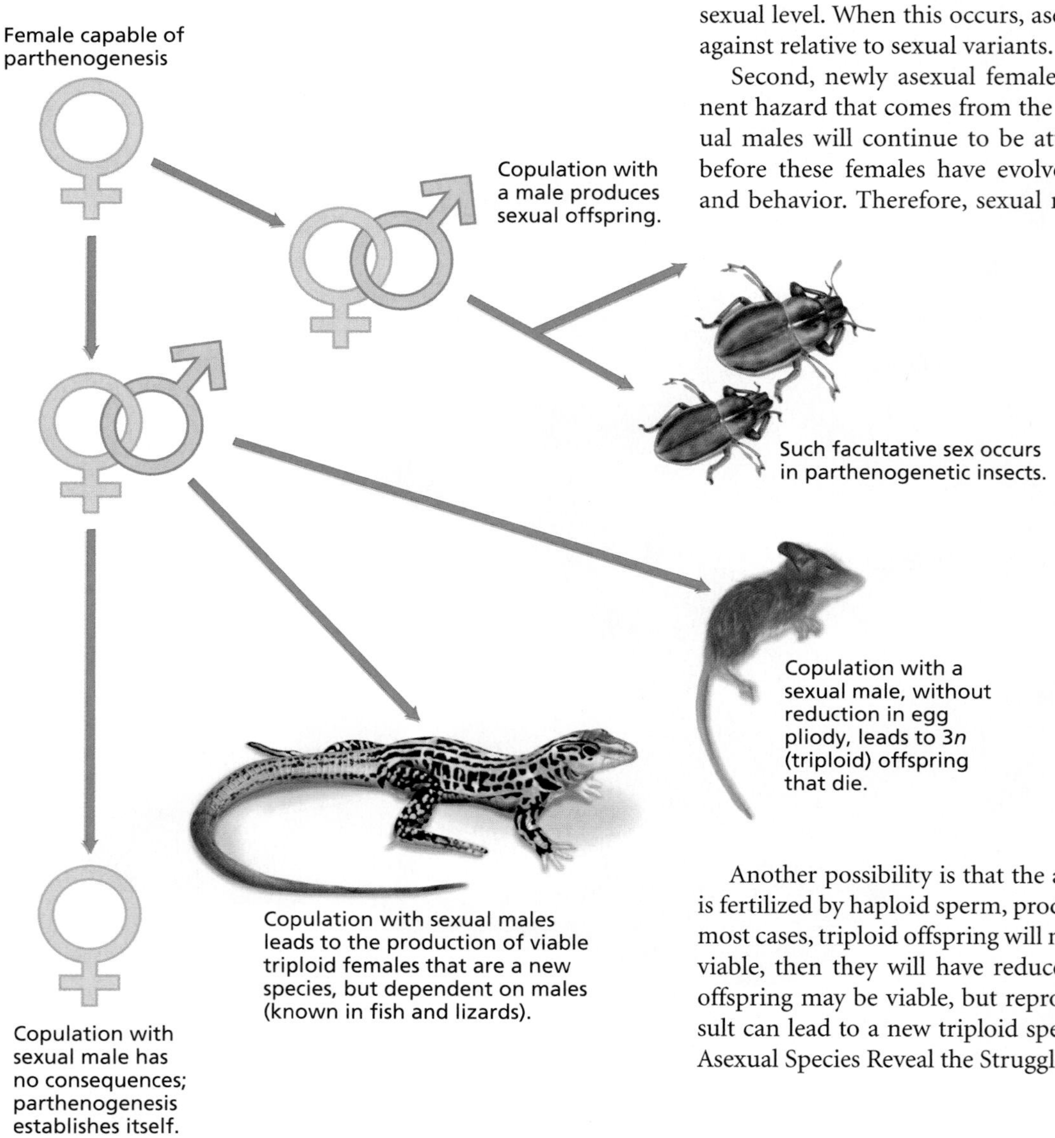

Female capable of parthenogenesis

Copulation with a male produces sexual offspring.

Such facultative sex occurs in parthenogenetic insects.

Copulation with a sexual male, without reduction in egg pliody, leads to 3*n* (triploid) offspring that die.

Copulation with sexual males leads to the production of viable triploid females that are a new species, but dependent on males (known in fish and lizards).

Copulation with sexual male has no consequences; parthenogenesis establishes itself.

FIGURE 18.10A It may be hard for parthenogenetic females to escape from males. The problem is that females with the capacity to reproduce parthenogenetically often have sex anyway, resulting in the production of offspring that may be sexual, dead, or triploid. This prevents the establishment of parthenogenesis.

Triploid Asexual Species Reveal the Struggles of Parthenogens

Triploid asexual species have evolved multiple times, suggesting that they originated from the sexual fertilization of asexual females that have diploid eggs. There are even vertebrate examples. The whiptail lizards of the genus *Cnemidophorus* (Figure 18.1B) have triploid asexual female races in the desert Southwest of the United States. Fish of the genus *Poeciliopsis* have races of triploid asexuals derived by hybridization of two sexual species, *P. lucida* and *P. monacha* (Figure 18.1C). Salamanders of the genus *Ambystoma* are triploid hybrids of two sexual species, *A. laterale* and *A. jeffersonianum*. See Module 6.10 for more on *Poeciliopsis* and *Ambystoma*.

On the other hand, it should not be assumed that these asexual triploids evolve very often. There are no known triploid asexual mammalian species, for example. What these rare examples show is that males continue trying to fertilize females that are evolving asexual methods of reproduction, possibly even with the cooperation of the females, who will not know that they have the evolutionary opportunity to escape from males. Only human females would know that.

In all these cases, the evolutionary dynamic is not one of "fair" competition between sexual and asexual females. There is nothing fair about sex. The difficulty of making the transition to asexuality, particularly in the potentially "hostile" environment of males, may prevent the females' escape from sex altogether. In this case, sex is maintained whether a population would do better without it or not. ❖

ORIGIN OF SEX

18.11 The origin of sex is even more complicated than its maintenance

It is not obvious that sex is beneficial in general. As we have seen throughout the discussion "Is Sex a Good thing Despite Its Problems?" beginning with Module 18.5, there are specific situations where sex might be selectively favored. But these situations are not universal. So why is sex so common?

One possible answer is that if sex is difficult to escape from, then sex may be evolutionarily "sticky." Once a group of organisms has it, then it may be hard to get rid of—because of males or because of lower fertility among parthenogenetic females. Therefore, sex could be as common as it is because it "sticks" evolutionarily, like tar to a shoe on a hot day.

But this leaves the problem of how groups of organisms evolve sexual reproduction in the first place. Sex is an elaborate reproductive adaptation. It is doubtful that it could be acquired in an accidental manner. Again, this seems to indicate that sex must be a beneficial adaptation produced by natural selection. But this argument is not quite as solid as it first appears.

It is important to compare sex with asexuality in a way that reflects evolutionary history. The evolution of sex does not depend on an abstract or perfect competition of sex with asexuality. Instead, it depends on the evolutionary forces that determine the **origin of sex** and its maintenance. We can think of sex as a giant box for evolution, as shown in Figure 18.11A . Sex originates when a population enters this box. Asexuality evolves from sex when a population leaves the box. Selection maintains sex when evolutionary pressures push populations back into the sex box.

What we have seen to this point is that sex might be maintained either by selection favoring sex or by selection evolutionarily punishing females that try to escape from sex. Given this ambiguity about the maintenance of sex, can we get a better understanding of sex by considering its origin?

In considering the origin of sex, note that there are at least three very different processes involved: syngamy, recombination, and gametogenesis. We will consider each of them in order.

Syngamy, or *fertilization*, is the fusion of genomes to produce a genome with doubled ploidy, as shown in Figure 18.11B. There are advantages to cells fusing and doubling gene number. For example, if either haploid genome has one or more recessive deleterious alleles, then syngamy would rescue the haploid genomes from the full deleterious effects of these alleles. But note that if this is the advantage of syngamy, then it is hard to understand the advantage of later reducing ploidy during gametogenesis.

Recombination occurs when strands of DNA that have similar sequences of nucleotides physically touch each other, unravel their helices, and create new combinations of DNA segments, as shown in Figure 18.11C. While the genome size is doubled, cells have the opportunity to recombine their chromosomes.

One puzzle regarding the origin of recombination is that it is likely to be very inefficient in its early evolution, leading to the production of unbalanced combinations of genes. These might be chromosomes that have lost large sections, thanks to improper resolution of the physical crossing of DNA strands that occurs during recombination. Efficient structural resolution of strands of DNA is normal in present-day recombination. But during the evolutionary origins of recombination, it must have been haphazard. Why would cells undergo an initially inefficient process of recombination?

Gametogenesis occurs when sexual cells produce gametes with half the ploidy by some type of reductive division, as shown in Figure 18.11D. The famous instance of gametogenesis is meiosis, but the process can differ from the typical textbook formulation. From an evolutionary standpoint the question is, why would gametogenesis be useful? One scenario is that larger

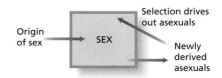

FIGURE 18.11A Sex can be thought of as an evolutionary box. When sex originates, a species evolves into the box. When selection maintains sex, females that abandon sex are selected against by some process, such as sib competition or the effects of males.

genomes are too costly to maintain, compared to the haploid genomes of gametes. But if that is true, why would syngamy evolve in the first place?

In the following modules we will consider some facts of molecular and cell biology, to see if they can help to make sense of the origin—and perhaps the later evolution—of sex. One fundamental point is that in order to spread, sex does not have to benefit the cells that have it. Instead, it may spread as a result of selection for the spread of parasitic DNA sequences. Sex and recombination may also evolve as a result of selection for seemingly unrelated molecular adaptations, the foremost being DNA repair (see Module 18.13).

These additional possibilities for the evolution of sex underscore the likelihood that sex is not a simple evolutionary adaptation. It has been affected by many different selection pressures, not all of them consistent with each other. This makes understanding it something of a nuisance for evolutionary biologists—but still kind of fun. ❖

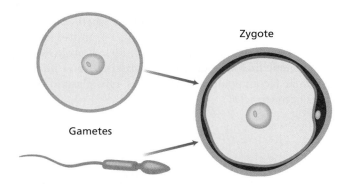

FIGURE 18.11B Syngamy takes place when gametes of two types meet to produce a zygote. Usually these gametes are an egg and a sperm.

FIGURE 18.11C Recombination occurs when one chromosome swaps genetic material with another chromosome of the same type.

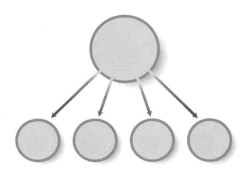

FIGURE 18.11D Reductive division produces gametes, which are usually haploid.

18.12 Simple forms of sex can originate from mobile genetic elements

At least one scenario for the origin of sex has no qualitatively difficult problems. This scenario supplies an origin that does not depend on sex being generally beneficial. Yet it does not preclude the possibility that sex is intermittently beneficial.

The general idea is as follows. We know that bacteria often have closed loops of DNA called **plasmids**. Plasmids replicate independently of the host genome. Sometimes plasmids encode the formation of *pili*, long bridging structures that enable plasmids to pass from one cell to another—as shown in Figure 18.12A, part (i)—by a process called **conjugation**. Plasmids are sometimes beneficial to their bacterial hosts. For example, some plasmids encode genes for resisting antibiotics such as ampicillin. The evolution of resistance to antibiotics in bacteria has come about in part from the spread of plasmids. But other plasmids spread despite measurably deleterious effects. An interesting thing about plasmids is that they can spread whether they are beneficial or not, because they establish a primitive form of genetic exchange that could be considered proto-sexual. Such genetic exchange does not necessarily result in chromosomal recombination, but it does mix plasmid-borne genes between cells.

(i) Bacterial conjugation

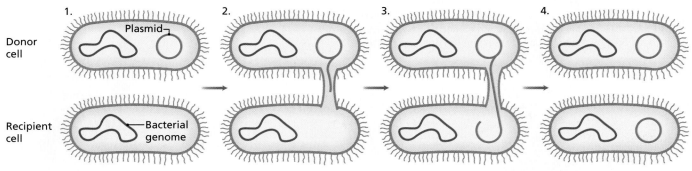

(ii) Transient cell fusion

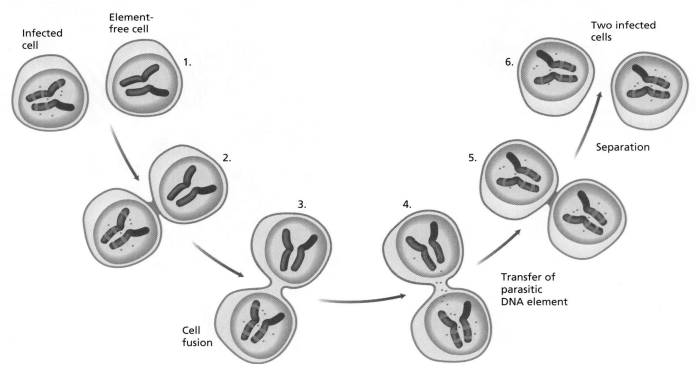

FIGURE 18.12A Primitive Transfer of Genes Plasmids are transmitted from bacterium to bacterium by a process known as conjugation. Conjugation is coded for by plasmids. **(ii)** Parasitic DNAs that float in the cell *and* foster transient cell fusion can foster their own transmission. They might also establish primitive sex.

Eukaryotic cells do not have conjugative plasmids like those in bacteria. But a somewhat similar process might have been involved in the origin of eukaryotic sex. Consider the cell fusion scenario in Figure 18.12A, part (ii). Suppose we have a mobile genetic element that can replicate in the host eukaryotic genome. If this mobile element contains a gene that alters the cell membrane so that cells are more likely to fuse on contact, and this element then "infects" any cells that undergo transient fusion with a carrier of the element, then it will transmit itself through a population of cells by such transient cell fusion. In this process, there is a substantial advantage to such elements from transient **cell fusion**. The transient cell fusion scenario easily explains both proto-syngamy and proto-gametogenesis, in that its spread requires a *cycle* of cell fusion and disconnection.

In addition, this type of mobile genetic element can spread through a population without being beneficial. It can do so as long as its deleterious effects are not too great relative to the probability that the element will soon infect a new cell host. In this model, sex spreads because it is contagious, not because it is beneficial. However, beneficial effects are not precluded, like those of antibiotic-resistant genes carried on conjugative plasmids.

This model for the origin of sex shows that sex does not have to be generally beneficial to spread. Sex might have been a by-product of a mobile genetic element, evolving only for the benefit of the element, but then retained because males or low asexual fertility made it difficult for females to escape sex evolutionarily. Or sex could have always been generally beneficial for some reason that is not now known. ❖

Two More Forms of Sex

We have already considered two basically different forms of genetic exchange—conjugation and eukaryotic sex. There are two other forms of genetic exchange in the living world. The first is **transformation**, the genetic exchange that occurs when bacteria absorb the DNA secreted by other living cells or by dead cells, and then recombine it into their genome. It is not clear whether transformation is beneficial. One possibility is that the absorbed DNA is just a metabolically cheaper source of nucleotides. In other words, transformation could be DNA cannibalism.

The second additional type of genetic exchange is **transduction**, the genetic exchange that occurs when viral infection of bacteria produces viral capsules that bear host DNA and that then find their way to other bacterial cells. Transduction may be entirely accidental. It is not as well known as the other forms of genetic exchange.

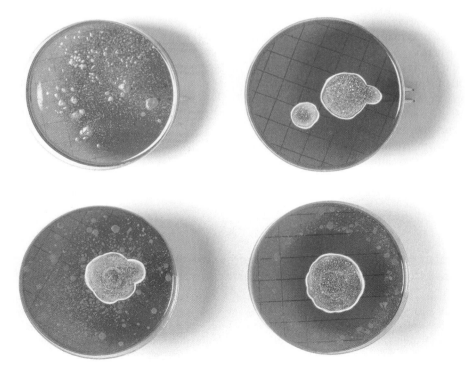

18.13 Recombination may have evolved as a by-product of selection for DNA repair

A preeminent fact about the evolution of life is that DNA has to be repaired . Other types of molecules, such as protein, can be discarded when damaged. But DNA is the information repository of the organism, and it has to be repaired if at all possible. Thus all organisms have **DNA repair** machinery, the details of which are part of the study of molecular biology.

One form of DNA damage is potentially grievous—double-strand breaks. Such breaks completely interrupt the DNA sequence of a chromosome. In diploid eukaryotic cells, however, double-strand breaks can be repaired using the DNA template provided by the homologous DNA sequence on the matching homologous chromosome, as shown in Figure 18.13A. A side effect of such double-stranded break repair is that the resolution of the repair process can lead to recombination, in which the sequence from one chromosome is swapped with the sequence of the other chromosome.

This side effect raises the important possibility that recombination is not itself favored by natural selection. Instead, it might be an incidental by-product of a process that must be favored very strongly—DNA repair. This model is somewhat controversial, and the details of DNA repair do not always fit this theory. For example, male *Drosophila* do not undergo recombination, yet their DNA repair processes seem to work just fine. The DNA repair theory of recombination supplies a completely different perspective on the evolution of recombination. Recombination may not really be part of the evolution of sex, as such. It may instead be a by-product of the evolution of DNA repair.

To put this discussion in perspective, consider that evolutionary geneticists often construct theoretical models in which selection favors, or opposes, recombination. Often these models depend on such ecological patterns as the correlations between different environmental features (e.g., temperature, humidity, etc.), environmental variation through time, and so on. Under some conditions, these models predict the evolution of frequent recombination. Under other conditions, they don't.

Recombination may not really be part of the evolution of sex, as such.

Yet recombination, while not quite universal, is one of the most widespread features of genetic systems. How could something so close to universal depend on ecological particulars? The generation of recombination by DNA repair seems to be a more appropriate solution to the occurrence of recombination, because DNA repair is a nearly universal molecular process, vital to the continuation of life. The DNA repair theory for recombination may not be correct. Cells could be doing something else that leads to the evolution of recombination. But this theory does seem to be at the right level of generality. ❖

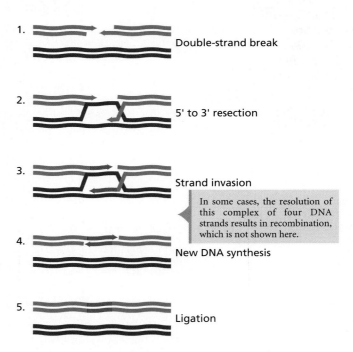

1. Double-strand break

2. 5' to 3' resection

3. Strand invasion

> In some cases, the resolution of this complex of four DNA strands results in recombination, which is not shown here.

4. New DNA synthesis

5. Ligation

FIGURE 18.13A Double-strand break repair is a process that maintains the integrity of DNA. At the same time, under some conditions, it may foster recombination.

Resolving Contradictions in the Evolution of Sex

Now that we have gone through the range of theories about the evolution of sex, one obvious pattern is that many aspects of biology have been connected to the evolution of sex: ecological advantage, accumulation of mutations, mobile genetic elements, destructive males, and DNA repair. Still, it may not be too early to come to some general conclusions.

First, it is doubtful that sex evolved and is maintained because of a single universal evolutionary mechanism. Second, the evolution of sex has probably been driven by multiple evolutionary mechanisms acting at different levels—molecular, organismal, and population. It would be convenient if the entire phenomenon of sex could be explained by just one hypothesis, but that now seems very unlikely .

SUMMARY

1. Sex is the most challenging mystery in biological evolution. At the core of this mystery is the evolutionary advantage that females would derive from reproducing without a male genetic contribution. Such "parthenogenetic" reproduction would free females from a 50 percent dilution of their genes. Sex also breaks up successful genotypes, which the genotypes of all parents must be. There are other problems with sex, problems less related to genetics. The growth and operation of sex organs imposes a physiological burden, as does mating behavior in animals. Sex exposes organisms to greater risks of predation and disease, especially diseases spread during fertilization.

2. There are many adaptive theories of sex, proposing that sexual organisms have an advantage over asexual organisms under certain environmental conditions. Most of these advantages turn out to be dependent on specific ecological conditions that are unlikely to be universal, or even common. When populations are large, but not too large, and beneficial mutations common, but not too common, sex can accelerate evolution. When sibs commonly compete against each other in harsh, changing environments, sexual mothers may have more reproductive success. Contagious disease may eliminate all but rare genotypes that only sexual parents can produce.

3. A different kind of theory is based on the problem of deleterious mutations. Under some conditions, asexual lineages go extinct from the ratchet-like accumulation of such mutations. Under other conditions, large asexual populations may endure deleterious mutations better than sexual populations do. It is not clear whether deleterious mutations help explain sex.

4. One way to explain the maintenance of sex involves the difficulties faced by newly asexual females: (a) They have been observed to have low fertility, and (b) fertilization by males may terminate asexual reproduction or produce triploid offspring of low viability.

5. It is difficult to explain the origin of sex using adaptive models. The origin of sex can be explained instead in relation to the advantage that mobile genetic elements would receive from transient, inefficient cell fusion. Recombination might be explicable as an incidental side effect of DNA repair, one of the most important molecular adaptations of the cell, though there are some problems with this theory.

REVIEW QUESTIONS

1. Why don't males evolve asexual reproduction, as females do?
2. Explain how sex can be viewed as a generator of variation to help organisms adapt to a varying environment.
3. Explain how molecular recombination might help fitness in organisms with damaged DNA.
4. Describe the scenarios that might arise if a fertile male found his way to a newly established colony of asexual females.
5. Why is sex most useful if siblings compete against siblings in small ecological patches?
6. Construct a scenario in which sex is useful to humans in our reproduction.
7. Why do bacteria have sex without syngamy, while sex in plants and animals involves syngamy?
8. Which ecologies favor the evolution of sex, and which favor the evolution of asexuality?
9. Over the entire period of evolution, do you think that sex has generally increased fitness or decreased it?

KEY TERMS

accelerated evolution
asexual reproduction
Bell, Graham
cell fusion
chlamydia
conjugation
Cnemidophorous
DNA repair
external fertilization

gametogenesis
genital herpes
genital warts
gonorrhea
historical imperfections
internal fertilization
lottery model
Muller's ratchet
origin of sex

parthenogenesis
pathogen-host arms race
plasmids
Poeciliopsis
recombination
reductive division
rotifer, bdelloid
sexual reproduction
sequential substitution

sib competition
syngamy
syphilis
transduction
transformation
triploid asexuals
twofold cost of males
venereal disease
zygote

FURTHER READINGS

Bell, Graham. 1982. *The Masterpiece of Nature: The Evolution and Genetics of Sexuality*. Berkeley: University of California Press.

———. 1988. *Sex and Death in Protozoa: The History of an Obsession*. Cambridge, UK: Cambridge University Press.

Bernstein, Carol, and Harris Bernstein. 1991. *Aging, Sex, and DNA Repair*. San Diego: Academic Press.

Halvorson, Harlyn O., and Alberto Monroy, eds. 1985. *The Origin and Evolution of Sex*. New York: Alan R. Liss.

Hamilton, W. D. 2002. *Narrow Roads of Gene Land*. Vol. 2. *Evolution of Sex*. Oxford, UK: Oxford University Press.

Maynard Smith, John. 1978. *The Evolution of Sex*. Cambridge, UK: Cambridge University Press.

Michod, Richard E., and Bruce R. Levin, eds. 1988. *The Evolution of Sex*. Sunderland, MA: Sinaver.

Williams, George C. 1975. *Sex and Evolution*. Princeton, NJ: Princeton University Press.

How do you love?

19

Mating Strategies

For sexual organisms, mating is more important than adult survival, because for them a failure to mate guarantees a failure to reproduce. This is underscored by the amazing biochemical, morphological, and behavioral contrivances that animals and plants use to mate. It is literally true that nature would be less colorful if not for selection to mate, because some of the brightest coloration found in the wild is supplied by the mating plumage of birds and the flowers that plants use to attract pollinators.

The evolution of mating also generates Darwinian deviousness. Mates often have different goals. One party to a mating may seek multiple sex partners, while the other seeks a mate that is faithful. Mates may also attempt to deceive each other about their desirability.

We have sorted the evolution of mating strategies into three issues. The first issue revolves around the multiplicity of sexes, including the mere existence of different sexes and the ratio of sexes within populations. If there were no differences between gametes, many of the complexities of mating strategy would go away.

The second issue is the choice of sex. For some animals and plants, the determination of sex is relatively flexible. Both animals and plants may switch sexes during their lives. Or they may be both sexes at the same time, and thus hermaphroditic.

The third issue is the degree of promiscuity. Some organisms, like nesting bird species, are usually monogamous. Other species mate by squirting their gametes into a pool of swarming gametes produced by many individuals, the ultimate in orgiastic behavior. Plants with wind-dispersed pollen are effectively promiscuous, even though there are no copulations to count. A further complication is that the sexes may differ substantially in their promiscuity. ❖

GAMETES AND SEXES

Most sexual animals have two types of gamete: sperm and eggs

Why are there ever two sexes? The answer turns out to be a complicated story, as you might expect. The simplest scheme for sex would be for any gamete to fertilize any other gamete, and any gender to mate with any other, a situation that is hard to imagine. Bacterial *transformation*, described in Module 18.12, does not involve gametes. It is a type of protosexuality in which DNA is exchanged without regard to the type of DNA donor or DNA recipient. This tells us that sex does not absolutely require gametes or sexes.

Bacterial *conjugation*, also described in Module 18.12, does involve the rough equivalent of mating types. For example, some bacterial cells that have conjugative **"F" plasmids** can donate DNA to cells that lack F (Figure 19.1A). There is also some evidence that cells that possess F plasmids inhibit other F-bearing cells from conjugating with them.

The simplest form of eukaryotic sex is **isogamous** sex— sex with two gametic types that are not distinguishable at the level of cell size, structure, and so on. This is a common type of sex among algae. A much-studied example is sex in *Chlamydomonas*, a mobile alga in which the two mating types are identical in morphology (Figure 19.1B).

Because sex can happen without any differences between the **gametes**, why do most sexual animals have sex with sperm and eggs? One answer is that primitive gametes may have been more like sperm. **Eggs** may have evolved later as *immobile gametes* that were found by *mobile gametes*, the **sperm**, so the eggs could put all their resources into provisions, instead of mobility (Figure 19.1C). In this theory, eggs provision the offspring and sperm find the eggs.

A second theory for the evolution of sperm and eggs is that the original gametes were more egg-like: relatively large and less mobile. Sperm might then have evolved as small, mobile gametes that could be produced in massive numbers, in order to fertilize as many of the large, less mobile gametes as possible. In this scenario, eggs might be the original type of gamete, with sperm being the type of gamete that contributes less cell volume, but fertilize eggs more often. Possibly both scenarios have elements of the truth. ❖

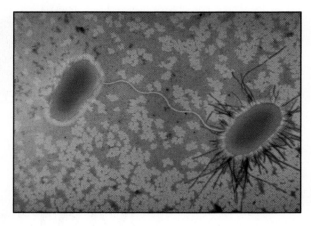

FIGURE 19.1A Conjugation between bacterial cells results in the transfer of plasmids.

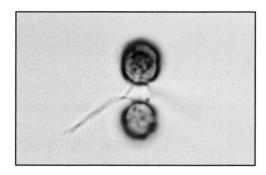

FIGURE 19.1B Sex in *Chlamydomonas* and other algae involves the fusion of two similar haploid cells.

FIGURE 19.1C One theory for the evolution of sperm and egg cells is that the egg provisions the zygote with nutrients while the sperm is responsible for finding the egg.

A traditional puzzle in the evolution of mating is that the ratio of males to females, the **sex ratio**, is usually about one to one, or 50:50. That is, the number of males is usually about the same as the number of females. This ratio is convenient in monogamous species, because everyone should be able to find a mate. But this does not explain how evolution maintains even sex ratios.

The solution to this evolutionary puzzle lies in the consequences of deviation from one-to-one sex ratios, as shown in Figure 19.2A. The iron law of sex is that each sexual offspring has one mother and one father, regardless of the sex ratio. Consider the consequences of having more females than males among the breeding adults, as shown in the figure. Under these conditions, males have more offspring than females do, on average, because the males have to do extra duty. Therefore, if mothers can control the sex of their offspring, they should have sons. (For example, insects of the Order Hymenoptera, in which fertilized eggs are diploid females and unfertilized eggs are haploid males, control fertilization to control the sex of their offspring. See Module 20.6.) Their sons are going to have more offspring, on average, compared to daughters. To maximize the number of grandchildren that they will have, mothers should bias the sex ratio of their offspring in favor of the rarer sex.

Sex ratio fluctuates in nature, due partly to differences in rates of predation on males and females and partly to differ-ent disease risks, among other ecological processes that are sex-biased. If females can adjust the sex ratio of their offspring, we should see a pattern in which deficits of males are met with increased production of males, while deficits of females are met with increased production of females. When mothers choose the sex of their offspring according this rule, deviations from one-to-one sex ratios would be eliminated by the kind of damped fluctuation of sex ratio shown in the lower half of Figure 19.2A.

However, many animals and plants cannot control the sex of their offspring. Sometimes this occurs because of chromosomal sex determination. In mammals and flies, there are X and Y sex chromosomes, which do not recombine with each other. Two XX chromosomes make a female, one X chromosome makes a male. Under these conditions, unbiased Mendelian segregation of chromosomes will automatically produce one-to-one ratios among X and Y sperm. Fertilization of X-bearing eggs by these gametes will almost always produce unbiased sex ratios at birth. In such cases, evolution has produced chromosomal machinery for sex determination that is automatically unbiased.

Other organisms have sex ratio determination that is dependent on temperature, on many genes, or on cytoplasmic factors. These methods of sex determination generate even sex ratios in some species, and uneven or variable sex ratios in other species. ❖

Say you are a mother. Which sex should your child be? Say the population is mostly female.

15 females for every 8 males

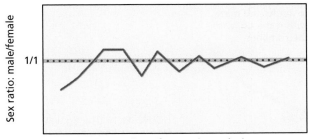

Sex ratio: male/female

1/1

Generation of sex-ratio evolution

This situation gives the following pattern of sex-ratio evolution.

FIGURE 19.2A Sex Ratios in Normal Populations In sexual populations, every offspring has a mother and a father. If there is a shortage of fathers, each male has to do extra duty. This increases the fitness of males over females, and thus sons over daughters. Therefore, individual mothers should produce sons only, in such a population. If there is a shortage of mothers, a similar argument shows that it is then better to produce daughters.

19.3 The hymenopteran sex ratio system is often used to bias sex ratios when mating is incestuous

Fig wasps inject their eggs into figs (Figure 19.3A). In some species, the eggs then hatch, grow into adults, and mate within the fig, before winged adult females leave the fig and disperse. When there is only one mother laying eggs in a fig, the sex ratio is skewed toward females, because only one or two sons are needed to fertilize all the mother's daughters, and the number of daughters is the primary determinant of the mother's fitness.

With multiple wasps laying eggs in a fig, there is competition between unrelated males for access to females. In this situation, females lay more unfertilized eggs, which will become males. The inside of the fig becomes a dark, closely packed arena for competition between males, with strong selection on mating success among males, not so different from the typical singles bar. Figure 19.3B is a cartoon of the fig wasp life cycle.

For species in which winged, mature males emerge from the fig, sex ratios are closer to even. In such cases, mating does not necessarily take place within the fig; it occurs outside as well. Mating might even occur on a different fig tree. Male dispersal undermines the incestuous pattern of a few males mating with more numerous sisters. ❖

FIGURE 19.3A This fig wasp is standing on one of the figs that it uses as an incubator.

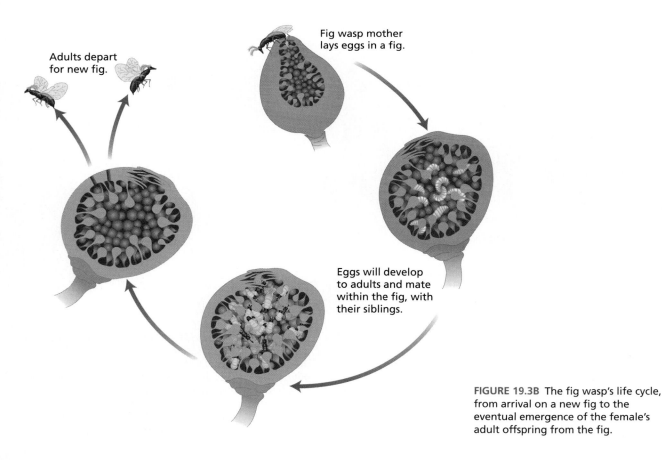

Adults depart for new fig.

Fig wasp mother lays eggs in a fig.

Eggs will develop to adults and mate within the fig, with their siblings.

FIGURE 19.3B The fig wasp's life cycle, from arrival on a new fig to the eventual emergence of the female's adult offspring from the fig.

WHICH SEX SHOULD YOU BE?

Separate sexes evolve when it is hard to combine male and female sexual functions

Why should there *ever* be two separate sexes? **Hermaphrodites** are able to mate with any member of the species, regardless of their gender. In some species, hermaphrodites can even mate with themselves, if they cannot find any other mate. This seems like sexual utopia.

But there is a quantitative evolutionary problem with hermaphrodites. Consider a simple model for the evolution of hermaphrodites in a population that also contains males and females. Suppose that males produce N sperm each generation, and females produce n eggs. We can calibrate the reproductive output of hermaphrodites relative to separate-sex reproduction, with individual hermaphrodites producing aN sperm and bn eggs, where a and b will each almost always be less than one. Under these conditions, evolutionary theory tells us that populations with males and females can prevent the evolution of hermaphrodites if $a + b < 1$. What this means is that hermaphrodites cannot evolve in sexual populations if the total production of gametes by the hermaphrodites is too small.

Consider the special case where $a + b = 1$ and $a = b$. Then $a = b = 1/2$. This means that the hermaphrodite is equally good at being a producer of sperm and a producer of eggs. Under these conditions, the hermaphrodite is on the edge of being able to invade. This is logical, because the hermaphrodite's male and female sexual functions are each half those of the corresponding sexual individual, male or female. This result suggests that, if there is symmetry between the fertility of male and female functions in hermaphrodites, then hermaphrodites will evolve if they can do better than the average fecundity of males and females. They will fail to evolve if they do worse than that.

The key factor is whether a hermaphrodite *gains* fertility, due to some advantage of combining male and female sexual functions, or *loses* fertility, due to some disadvantage of combining male and female sexual functions. In other words, what matters is whether there is synergy or interference when the sexual functions of two different genders are combined. Under these conditions $a + b$ will be greater or less than 1, respectively.

It is easy to imagine that it would be disadvantageous for an animal with internal fertilization to combine male and female sexual functions. Among other problems, who would stick what into whom? Some species, including slugs and earthworms, have solved this problem (Figure 19.4A). Plants do not seem to have many difficulties with anatomically combining both sexual functions (Figure 19.4B). A vast number of plant species are hermaphroditic. ❖

FIGURE 19.4A **Hermaphroditic Slugs Copulating**

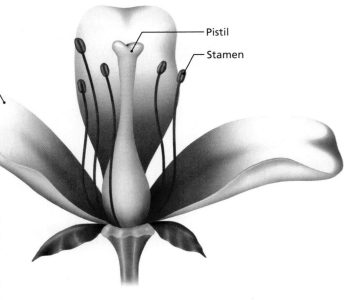

Petal

Pistil

Stamen

FIGURE 19.4B **A Typical Flower with Both Male and Female Parts**

19.5 The evolution of hermaphrodites also depends on the genetics of self-fertilization and the reproductive ecology of mating

The hermaphroditic lifestyle might seem like a good Darwinian option. If the hermaphrodite is sufficiently agile, it can mate with itself. This overcomes one of the most profound costs of the sexual lifestyle—the zero fitness of celibacy.

But there are problems with **self-fertilization**, or having sex with oneself. The obvious one to an evolutionary biologist is that selfing will tend to make recessive deleterious alleles homozygous. On the other hand, the gametes of a selfing hermaphrodite will all be fertilized.

We have already shown how the fertility of hermaphrodites may limit their evolutionary success, especially if it is difficult to combine male and female functions in one body. We can analyze the problem of hermaphroditic selfing in a similar manner using the parameters shown in Figure 19.5A. Let V be the depressed fitness resulting from selfing. If the fitness of organisms that do not self is 1, then we will normally assume that V is less than 1. On the other hand, there is the loss of gametes that occurs when a male or female fails to mate. Let the proportion of gametes that are fertilized in a self-incompatible organism (all males and all sexual females are self-incompatible; hermaphrodites may be self-compatible) be given by f. With these parameters, we

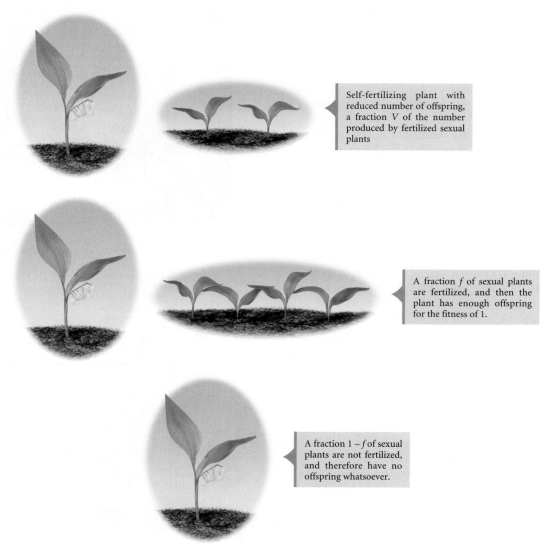

Self-fertilizing plant with reduced number of offspring, a fraction V of the number produced by fertilized sexual plants

A fraction f of sexual plants are fertilized, and then the plant has enough offspring for the fitness of 1.

A fraction $1 - f$ of sexual plants are not fertilized, and therefore have no offspring whatsoever.

FIGURE 19.5A The Evolution of Self-Fertilization with Inbreeding Depression
Self-fertilization can cause inbreeding depression compared with sexual fertilization. But sexual plants may not be fertilized at all, leaving them with zero fitness.

FIGURE 19.5B Some hermaphroditic snails do not fertilize themselves.

can say that a gene for self-incompatibility will spread if $V < f/2$. This means that the greater the cost of selfing for fitness, and the easier it is to find a mate, the more readily self-incompatibility will spread. And since males and females are self-incompatible, this is also a theoretical condition for the evolution of males and females.

What is the actual situation? There are some hermaphroditic species in which selfing is rare or entirely absent. In the snail genera *Helix* and *Cepaea*, hermaphrodites do not fertilize themselves (Figure 19.5B). Plants are more often hermaphroditic than animals, and they have evolved a wide range of characteristics that reduce selfing. In some cases, female and male flowers from the same plant are mature at different times. In other cases, the male and female parts of flowers that are hermaphroditic are positioned so that the flower's pollen is unlikely to fertilize its own ovules. Some plants have even evolved genetic systems that recognize their own pollen and prevent it from fertilizing their own ovules. There is abundant evidence for the evolution of mechanisms that reduce or prevent selfing in plants, whether they are her-

maphroditic or not. To amplify this point further, there are numerous mechanisms that species with separate sexes use to avoid inbreeding. For example, it is common for animals of one sex to disperse from the groups in which they were born, usually males. This is common in primates, such as baboons, among other mammals. Animals go to a lot of trouble to avoid inbreeding, because it is usually so deleterious. And selfing is the most extreme form of inbreeding.

But some hermaphrodites still self. A lack of heterozygotes in a population is a characteristic sign of selfing and other forms of inbreeding. It is common to find such a lack of heterozygotes in hermaphroditic plant species, especially among the cereal plants. The soil nematode, *Caenorhabditis elegans*, normally self-fertilizes, though it occasionally produces males. It is significant that this nematode shows few signs of **inbreeding depression**. In terms of the parameter V, we can say that it has a high value in this nematode species. Therefore, it is probably the case that $V > f/2$ for this species, which means that evolution has not favored self-incompatibility among nematodes. ❖

The puzzle that hermaphrodites present in turn leads to another aspect of mating strategies—the nature of male and female functions in mating.

It is the cellular essence of mating that males supply tiny gametes with relatively little material for provisioning the zygote. There are some exceptions to this pattern, where fertilizing males supply materials that nourish the zygote. But generally male sexual function comes down to supplying a haploid genome for combination with the egg's genome.

Females, by contrast, supply an egg that is far more provisioned with resources than the male gamete is. In addition, the female also supplies a haploid genome for the zygote. As a by-product of provisioning the eggs that they produce, females almost always produce far fewer gametes than the males of the species.

This difference in cellular contribution to fertilization need not, however, be continued at later stages of reproduction. In a number of fish species with external fertilization, the male takes on the role of caring for the offspring of his mating. In emperor penguins, the male incubates the egg on his feet, holding still for months during the dark Antarctic winter (Figure 19.6A).

On the other hand, males in most species contribute absolutely nothing to the care or nurturing of their offspring. The strategic aspect of this situation is that the

What is striking is how little resistance some males display to being eaten. In species where the male usually mates only once, being eaten by the female sex partner may increase Darwinian fitness.

ideal reproductive strategy is partly a function of what the other sex does, during fertilization and afterward.

One twist on the mating issue is supplied by females who eat their mates. **Cannibalism** is fairly common in the animal world, and most cannibals are females. Female spiders and some female insects are frequently cannibals. Females often eat their young, but that is not of concern here. The interesting point for mating strategies is that females often eat their mates, but males do so less often.

The black widow spider and the praying mantis are the popular examples of female cannibalism, but some arthropods are cannibalistic more often than those species (Figure 19.6B). Cannibal females eat their mates, especially. Because the males may be good sources of nutrition, and mating often renders them vulnerable, this practice by females is not so surprising. What is striking is how little resistance some males display to being eaten. In species where the male usually mates only once, being eaten by the female sex partner may increase Darwinian fitness. It is a simple way for the male to contribute to the nutrition of his offspring,

because the female may still be provisioning eggs in her body, eggs which she will later fertilize with stored sperm from the male(s) that she has eaten. It does raise the risk, however, of being seduced by a female that has already had sex, but wants to eat another male. Offering food to a female may be more than seduction, a common mating behavior. It may help a male survive, because the female may be too busy eating the food offering to eat the male during or immediately after sex. There is evidence that arthropod males who are devoured by females during copulation have longer mating times, increasing the transfer of sperm to the female, and thus increasing the number of offspring that the male has. Yet another twist is supplied by species in which the male body is broken off as a result of the devouring female, leaving the male genitalia behind to seal off her vagina, preventing other males from fertilizing the female. The degree to which the semelparous males of some arthropod species evolve to give up their lives in order to reproduce with cannibal females is remarkable.

Mating is full of risks, and these risks generate issues of strategy. How best to find a mate, have sex with them, and then perhaps help with the offspring—when your sex partner may eat you, give you a disease, or abandon you to take care of the offspring on your own? These are difficult—but interesting—questions. ❖

FIGURE 19.6A Emperor penguins show a high degree of paternal involvement in the rearing of offspring.

FIGURE 19.6B Some species of arthropods—both spiders and insects—eat their mates.

19.7 Some species switch from one sex to another in order to increase fertility

Several species of animals and plants change from one sex to the other during the course of their lives. In animals, this change usually occurs only once, and the direction is consistent. Either the species is male when young and small, and later becomes female, or the other way around. Plants are less predictable in their sex changes. The timing and circumstances of this transformation help reveal the underpinnings of mating strategies.

The most basic question is, why change sex at all? **Michael Ghiselin** suggested that natural selection favors sex change when reproductive success depends on age or size, but the two sexes have marked differences in this dependence on age and size. Sex change arising from such factors is known in a wide range of animal species, from polychaete worms to shrimp to mollusks to fish. In plants, sex change is even more ubiquitous.

Consider **sex switching** in fishes. There are fish species where males change into females as they get older, and species where females change into males as they get older. Sex change can occur in as little as six weeks. Sex change occurs in species that live in a wide variety of habitats—freshwater swamps, the waters of continental shelves, coral reefs, and the deep sea. An important feature of fish biology is that unlike most terrestrial vertebrates, fish continue to grow significantly as adults. Associated with that growth is a progressive increase in the fecundity of females. Therefore, if everything else were equal, we would expect sex-changing fish to start life as males, and then become females later. On the other hand, larger males may have a mating advantage over smaller males, perhaps from greater success in fighting, which would tend to favor a sex change from female to male, as the fish grows. Both patterns are known to occur in sex-changing fish

Male

Female

FIGURE 19.7A A school of *Anthias*, a genus of coral reef fish, with both males and females.

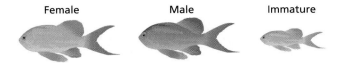

Female   Male   Immature

1. Initial condition

2. Removal of male

3. Sex change by largest female

FIGURE 19.7B Sex Switching in the Coral Reef Fish *Anthias*
Unlike most animals, this fish can switch sexes as an adult. When a school has lost its adult male, the largest female usually becomes male and takes over the mating duties.

species, so both factors may be involved in the evolution of sex change in fishes.

Some of the most interesting patterns of sex change occur in tropical reef fish. One such species is the reef fish *Anthias squamipinnis* (males and females are shown in Figure 19.7A). This is a common fish with a wide range—from the waters of the Middle East to the Philippines. *Anthias* is a group-living fish that feeds at coral reefs. Each group has about 10 percent males, the rest being mature females and immature juveniles. There are no small males. Males compete with each other for access to breeding territories. If a male is removed from a group, females switch to male. The largest females are most likely to switch to male. The female fish appear to respond to the loss of a male by having their largest mature adult female switch to male. This pattern is shown in Figure 19.7B.

In plants, sex change is not always so clear-cut. Some species have male and female flowers. Shifts between male and female sexual functions can occur simply by changing the relative proportion of male to female flowers. Both directions of sex change are known in plants—first male and then female, or first female and then male. A wide range of environmental factors lead to sex change among plant species. One pattern in plants' sex change is that stressful environments (drought, cold weather, poor soil, and lack of light) tend to elicit a change from female to male. There is some evidence that males survive such environments better than females, but an additional factor could be that environmental stress impairs female fertility more than male fertility. ❖

HOW MANY PARTNERS?

19.8 There are three main mating patterns: promiscuous, monogamous, and polygamous

If you have decided on your sex, you have another fundamental choice: whether to seek one or many partners. To limit the complexity of this issue, in the remaining modules we will discuss species that have only two sexes—male and female. For the vast majority of animal species, this assumption is usually correct. However, it is not a particularly accurate assumption to make concerning plants, in which a minority of plant species have only male and female plants. But for the sake of introducing the evolutionary issues, the two-sex animal model is the simplest case.

Many species adopt a **promiscuous** mating pattern, in which any individual that clearly (when females are choosing) or roughly (when males are choosing) is a member of the same species is accepted as a mate. And then—an hour, day, or month later—a different mate may be chosen. Under these conditions, males may be selected to produce a lot of sperm, because large numbers of sperm should give an advantage when the female's reproductive tract is already full of sperm from other males. All things being equal, producing more sperm under these conditions will give rise to more offspring for the male.

Chimpanzees are very promiscuous, and male chimpanzees have very large **testes**. Larger testes typically produce more

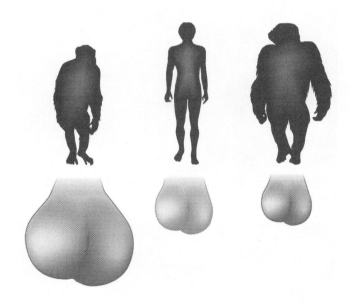

FIGURE 19.8A Chimpanzees have large testes due to the evolutionary advantage that accrues to producing a lot of sperm in promiscuous species. Gorillas are not promiscuous, and have small testes. Humans are intermediate in testis size.

sperm. Gorillas, which have very little promiscuity, have quite small testes. Chimpanzee males often find that the vaginas of their mates contain sperm from other males, but gorilla males rarely do. Humans have testes of intermediate size for our body weight (Figure 19.8A). This suggests that we are moderately promiscuous as a species.

But in a sense, complete promiscuity is no mating strategy at all. The truly strategic settings for sexual selection are monogamy and polygamy. Within these settings, however, individuals may be more or less promiscuous. The details of monogamous and polygamous roles will be taken up in the next two modules. For now we will outline the strategic contrasts of these very different mating strategies.

In **polygamy**, an individual of one sex has sexual access to two or more individuals of the other sex. Usually it is the male that has access to multiple females, but it can be the female that has access to multiple mates. Sometimes this sexual access is exclusive, whereby other individuals cannot have sex with the polygamist's mates. In other cases, interlopers may sneak copulations with the polygamist's mates. But even when this occurs, the polygamist will usually achieve the most copulations with its mates.

In **monogamy**, the male and female usually live together and rear offspring together. They may even approximate equality in their investment of resources—as in nesting birds, where males may incubate eggs as much as females, and then later

feed the nestlings (Figure 19.8B). The fidelity of monogamous animals has been a matter of great interest among scientists. Recent research using molecular markers to assign paternity has found that **adulterously** produced offspring are common among many animal species that had previously been regarded as strictly monogamous.

The critical mistake that naturalists and biologists made in the past was to confuse functional partnership with sexual fidelity. There are many bird species, for example, where the male and the female divide the labor of incubating and feeding their offspring. But this connubial harmony does not mean that the female, or the male, will not have sex with the bird from the next tree when their partner is off looking for food for the young ones.

This finding raises the question as to whether "monogamy" is really about sex at all. Is it better viewed as cooperative rearing behavior, cooperation proceeding despite the dubious paternity of some of the offspring? For the time being, we will take monogamy at face value here. But beware—in its strict form, monogamy may be quite rare. There is genetic evidence that the Canadian snow goose is strictly monogamous, but its case is apparently remarkable. ❖

FIGURE 19.8B Male birds often incubate and feed the offspring produced by their mates.

19.9 Sexual selection favors individuals that are sexually attractive, combative, or territorial

If the mating system is not monogamous, then there may be **sexual selection** to acquire as many mates as possible, especially among males. The question is then, what is the focus of sexual selection? The answer is several things, sometimes in combination.

There are two main types of beautiful organisms—flowering plants that attract pollinators and male birds that attract females. (Some fish are beautiful, too.) Mammals are generally drab, by comparison with birds. The thing that gives away what bird evolution is doing is the difference between the sexes. From peacocks to birds of paradise, male birds are often beautiful; and the female of the species is a drab, camouflaged creature (Figure 19.9A). A great deal of thought has gone into evolutionary models in which such attractiveness is produced by subtle genetic mechanisms, like **runaway selection**. In runaway selection, there are accidental genetic correlations between mating preferences and anatomy—which fuel greater and greater extremes of sexual preference for strange colorations or shapes. An additional hypothesis is choice based on the awkwardness or bright coloration of prospective mates, on the grounds that those who

survive such a **handicap** must have excellent genetic qualities. But recent research has shown that some female birds just want to mate with males decked out in finery, even if that finery has been added by experimenters. There is no need for an elaborate evolutionary history explaining why female birds go for flashy males. Note that pollinating birds and insects also often prefer strong colors, which seems to have led to the evolution of brightly colored flowers. Could there be something about the evolution of vision that tends to produce such sensory preferences incidentally? Some fish, like swordtails, also exhibit an inherent preference for certain types of odd morphology and coloration.

If females do generally prefer very colorful males, why have colorful appearances evolved so rarely? A partial answer might be that sexual selection for prettiness is opposed by other forms of natural selection. In particular, in most terrestrial organisms, a brightly colored male is advertising his availability as lunch for a predator. (A different evolutionary pattern arises among poisonous animals, like gila monsters, that advertise their dangerousness by bright colors. These animals are not facing as much predation risk.)

FIGURE 19.9A Peacocks are extravagantly decorative compared to peahens.

Birds are expected to be less vulnerable to predators, because they can fly away at speed, and the flying bird species have the most colorful males of any vertebrate group. In a sense, flight may allow bird species to indulge the taste for gaudiness that many females apparently have.

There is a more widespread means by which males, and sometimes females, acquire an abundance of mates—contests with other males (or females). These contests are ubiquitous among animal species, from insects like horned beetles to such mammals as elephant seals and zebras. However, for reasons discussed in Chapter 20, these contests are often much less violent than might be supposed. The loser is usually not killed, or even injured. Instead, the loser of these contests tends to slink away, accepting that the victor gets the mates. However, in some animals, like lions, there are fights to the death over access to groups of females. Such savage contests are among the most dramatic instances of conflict within species (Figure 19.9B).

The third way to pick up multiple mates is by acquiring a territory that is useful to prospective mates. Typically, this territory might provide an abundance of food for feeding both mates and their offspring. The red-winged blackbird is an example of this behavior pattern. The females want the nice territory, and maybe the territorial male as well. This diversity of selection pressures is summarized in the box "Sexual Selection for the Promiscuous."

One last topic is the difference in sexual selection acting on the monogamous and the polygamous. When an organism has numerous mates, each of these mates must supply most or all of the offspring care. Another issue, especially among spiders and insects, is that the polygamist has to avoid mates that will attack, kill, and eat them. For monogamous species, this threat may not matter as much; but polygamous animals have the potential to live and mate again later. ❖

FIGURE 19.9B Male animals fighting with each other for access to females.

Sexual Selection for the Promiscuous

- Promiscuity may be found:
  - in both sexes, when neither supplies much parental care
  - in neither sex, when both are monogamous and faithful
- The promiscuous seek:
  - mates that will do all required incubating, if they are available
  - mates that will not attack or damage the promiscuous mate (Note: cannibalism of courting males by females is common.)
  - mates that may supply good genetic material for the offspring
- The promiscuous are selected to:
  - find multiple mates
  - be found attractive

The incubator is selected to find sexually attractive and helpful mates

Incubators are animals that take care of their offspring once they are born live or laid as eggs. Many species do not "incubate" in this sense. The pattern of sexual selection for the incubator strategist is qualitatively unlike the pattern of selection for promiscuity and polygamy. Incubators have two basic mating situations: (1) with a promiscuous/polygamous mate, or (2) with another incubator. We will consider these scenarios separately.

With a promiscuous or polygamous mate, the incubator usually does not get much help with **parental care** of the offspring, at least in relation to the direct efforts of the mate. There may, however, be some transfer of resources if a polygamous mate defends a territory that offers more shelter, food, and so forth. But in many species, this is not the case. When it isn't, the incubator may be choosing its mate on the basis of some genetically influenced qualities that are beneficial when they recur in the offspring of the incubator. But it is also possible that incubators are simply attracted to their mate due to some form of sensory stimulation that they supply. In such cases, the mate is little more than a provocation to reproduce. This may be all there is to the many beautiful male birds, as well as the other attractive male vertebrates and invertebrates that have evolved sporadically. See the box, "Two Role Reversals," for cases where the male is the incubator sex.

With a mate that is also an incubator, the pattern of sexual selection changes radically. Most species in which both male and female are incubators are monogamous, at least socially. Much like partners in human marriages, monogamous pairs of incubators share work and resources, although there may be some gender-based specialization. Therefore, incubators in these species will seek mates that are good co-workers and not abusive. (Physical abuse is common between animal mates.) But it is not clear how incubators can be assessed for these qualities in advance. These qualities are not likely to be associated with showy plumage or orange skin. If the species is territorial, then incubators may seek mates that have already acquired good territory for feeding or protecting offspring.

The box, "Sexual Selection for the Incubator," summarizes the selective situation. ❖

Sexual Selection for the Incubator

- Both sexes may be incubators, when both care for the young.
- Incubators seek:
  - mates that will help care for the young
  - mates that will not interfere with the incubator's task (Note: cannibalism)
  - mates that will supply good habitat
  - mates that may supply good genetic material for the offspring

  Note: These goals may conflict.
- Incubators are selected to:
  - find a mate
  - be found attractive
  - care for their offspring

Males often mate promiscuously, leaving the females to take care of the offspring. But there are a number of exceptions in which females take this role.

In some sandpiper species, the male is smaller than the female, and incubates eggs that are laid by its sex partner. Females tend to be promiscuous, going from male to male (Figure 19.10A).

In many fish species, males take care of their young offspring, while the females do not. The ultimate example of this is the seahorse male, which keeps its offspring in a pouch until they have developed. At that time, they are expelled in a burst (Figure 19.10B).

FIGURE 19.10A Some females reverse roles and compete with other females for additional mates. The spotted sandpipers have such a role reversal, the female having a sort of harem of males that she guards.

FIGURE 19.10B The male seahorse incubates his offspring in a special pouch at the front of his abdomen. When the offspring are mature, the male "gives birth" to them as shown here.

SUMMARY

1. Sex is complicated in itself, but it becomes still more complicated with the evolution of multiple types of gametes and multiple sexes. A fundamental question about animal mating is why there are usually two different types of gametes—the small, motile sperm cell and the much larger, immobile egg. One possibility is that this differentiation reflects a division of labor, with the egg supplying nourishment for the zygote and the sperm taking on the role of finding the other gamete. An alternative interpretation is that fertilizing sperm cells first evolved merely to hijack eggs by getting there before other eggs could.

2. When there are different types of gametes, the evolution of sexes that produce different gametes is possible. When both male and female offspring are produced, 50:50 ratios are favored by natural selection, because the sex in short supply has a fitness advantage. With incestuous mating, sex ratios evolve toward a female bias, because males can usually fertilize numerous females.

3. Mating can get even more complex, however, because female and male reproductive functions can be combined by hermaphrodites. Hermaphrodites will evolve when male and female reproduction can be combined efficiently, as they are in many plants. Separate males and females evolve when the two sexual functions cannot be efficiently combined, which is apparently true of most animals. An advantage of separate sexes is that self-fertilization is not possible, and self-fertilization can cause inbreeding depression. This gives a reliable benefit to the evolution of separate sexes.

4. Even when sexes are kept separate, some animals and plants switch from one gender to the other. Sometimes organisms start out female and switch to male, and sometimes the opposite pattern occurs. Usually there is only one such switch. In some species, larger body size conveys increased female egg production. In other species, larger body size gives males an advantage in competing for mates.

5. One of the deep divides in mating strategy is between polygamy and monogamy. Usually males are the polygamous sex, but sometimes females are. The frequency of strict monogamy is probably overestimated, in that many monogamous species commit adultery. The selection pressures on promiscuous or polygamous individuals revolve around competing for mates and successfully fertilizing them. For monogamous species, a major feature of mating strategy is finding a mate that will be a good partner in producing and, perhaps, rearing offspring.

REVIEW QUESTIONS

1. Why are there sperm and eggs in animals and not just one type of gamete?

2. Why do sex ratios tend to evolve toward even numbers of males and females in species that do not inbreed normally?

3. Why isn't every organism hermaphroditic?

4. Why doesn't every organism have both males and females?

5. Why aren't all males promiscuous?

6. Why are some females promiscuous?

7. What should a monogamous female seek in a mate?

8. What should a polygamous male seek in a mate?

9. Why are male birds the most brightly colored animals of all?

KEY TERMS

adultery	handicap	parental care	sex switching
cannibalism	hermaphrodite	polygamy	sexual selection
egg	inbreeding depression	promiscuity	sperm
F plasmid	incubator	runaway selection	testes
gamete	isogamous	self-fertilization	
Ghiselin, Michael	monogamy	sex ratio	

FURTHER READINGS

Charnov, E. L. 1982. *The Theory of Sexual Allocation*. Princeton, NJ: Princeton University Press.

Cronin, H. 1992. *The Ant and the Peacock: Altruism and Sexual Selection from Darwin to Today*. Cambridge, UK: Cambridge University Press.

Ghiselin, Michael. 1974. *The Economy of Nature and the Evolution of Sex*. Berkeley, CA: University of California Press.

Judson, O. 2003. *Dr. Tatiana's Sex Advice to All Creation*. New York: Owl Books.

Maynard Smith, J. 1978. *The Evolution of Sex*. Cambridge, UK: Cambridge University Press.

Michod, R. E. 1995. *Eros and Evolution: A Natural Philosophy of Sex*. Reading, MA: Helix Box.

Michod, R. E., and B. R. Levin, eds. 1988. *The Evolution of Sex*. Sunderland, MA: Sinauer Associates.

Why are animals ever nice?

20

Social Evolution

When Darwin first published *Origin of Species* in 1859, a widespread reaction among intellectuals and religious figures was that his theory of evolution undermined morality. In the place of a divinely ordered world featuring the retribution of the deity who created all forms of life, Darwin's theory of life was entirely materialistic, with no important connection to a deity or a spiritual realm. In the nineteenth century, it was widely supposed that such materialism provided a justification for utilitarianism, socialism, and even communism. Notably, Karl Marx agreed with this assumption. He regarded Darwinism as a valuable support for his dialectical materialism. Charles Darwin himself held no such view. Although liberal in his politics, he was very far from being a revolutionary.

The central issue was whether evolution was corrosive or supportive of morality, law-abiding behavior, and political stability. At first it seemed as if Darwinism was an excellent justification for un-bridled competition, a laissez-faire society of dog-eat-dog, and conflict between economic classes. That was the type of social behavior that was expected to evolve with Darwinian selection. But decades of evolutionary and ecological research have shown that there is no universal tendency to rapacious social behavior; it is a possible outcome, but not a necessary one. In many situations, the evolution of social behavior tends to produce what humans consider moral behavior. This surprising reversal of intuitive expectations for the evolution of behavior is the concern of this chapter.

Biologists have used three main ideas to explain why evolution has favored the evolution of restrained social behavior. These ideas are group selection, kin selection, and strategy selection. Our main theme is *why evolution has favored seemingly ethical behavior in so many animals,* despite the Darwinian benefits that seem to come from selfish behavior, on first appraisal. ❖

GROUP SELECTION

20.1   Biological altruism is critical for social evolution

At the core of many debates about social evolution is the phenomenon of **altruism**. In evolutionary biology, unlike human politics or psychology, altruism is conceived in relation to the consequences of behavior for the Darwinian fitness of defined individuals. Typical examples of biological altruism include one animal aiding another—supplying food, perhaps—or helping another animal fend off an attack from a predator. Scientists have often noticed this type of behavior among both birds and mammals. Groups of younger male chimpanzees sometimes cooperate to harass couples that are mating. Once they have chased off the copulating male, they take turns mating with the female left behind. (Note from this example that biological altruism is not necessarily "nice" behavior as humans see it.) In scrub jays *(Aphelocoma coerulescens)*, older offspring help the parents feed their younger siblings. Cooperative behavior like this is often reminiscent of humans helping each other, within families and within combat units during wartime.

Some of the most dramatic forms of altruism are found among the **social insects**: group-living insect species such as honeybees, colonial wasps, ants, and termites. In these in-

> *Typical examples of biological altruism include one animal aiding another—supplying food, perhaps—or helping another animal fend off an attack from a predator.*

sects, most members of the group do not reproduce. Honeybee **workers** live out their lives supplying resources to the small subset of males and females that do get to reproduce—especially the **queen** of the hive, which may be the only reproductive female (Figure 20.1A). In social insects, the sterile castes of worker and **soldier** behave as if their lives are of no importance. They do not hesitate to put their bodies in harm's way to stop predators from threatening the hive. They will also attack potential prey with abandon, even when those prey might be dozens of times larger than the ants or termites that attack them.

Altruism is a problem for evolutionary biology because we expect that natural selection will oppose behavioral tendencies that might lead to death or sterilization. The logic of natural selection suggests that individual animals are selected to survive and reproduce. Yet some individuals behave in ways that predictably reduce their fitness. The problem of altruism is sketched in Figure 20.1B for the issue of restrained aggression in conflicts between members of the same species. It is not only sterile honeybees that are a puzzle. The lack of aggression between animals that need the same resources to survive and reproduce is also anomalous. ❖

FIGURE 20.1A A Honeybee Queen Attended by Sterile Female Workers

We might expect nature to be violent, as animals attack each other for food, shelter, or mates, to increase their fitness.

Yet animals instead are mostly peaceful in their dealings with members of their own species, even when doing so reduces their fitness.

FIGURE 20.1B The Problem of Altruism

In **group selection,** entire populations are sometimes selected in the same way that individuals are selected in individual selection. Group selection then works by the extinction and propagation of populations. The extinction of a population is the equivalent of the death of an individual in individual selection. Propagation occurs when new populations are founded by emigration from another "reproducing" population. When the populations of a species are well defined, many features of group selection are comparable to selection acting on individual members of an asexual species. This does not mean that populations are always structured this way, nor does it mean that group selection can work only under such conditions. But it is one scenario in which group selection might work.

How might the concept of group selection explain altruism? The connection between group selection and social behavior is that **selfish behavior** might foster individual fitness, yet undermine the survival of the group as a whole. Group selection might act to prevent overgrazing by deer, for example,

even though each deer might benefit from eating more. If group selection did not act in this way, then herbivores like deer might be at risk of eating all their food, causing the population to go extinct. Individual selection should, at first sight, favor just such selfish behavior on the part of deer and other animals. The more the deer eat, the better their health and the more they should be able to reproduce, ignoring the consequences of overgrazing for the ecology of the local habitat.

When the process of group selection for altruism is modeled using computer simulation or mathematics, the results show that group selection can work under some conditions, even when individual selection favors selfish behavior. But the conditions under which altruism is increased in the population are extreme. For instance, selfish behavior must be very disruptive in its effects on group, or population, survival. Groups must go extinct frequently when selfish behavior is common, or they must have reductions in their ability to colonize new habitats.

(Module continues on next page.)

Michael J. Wade has argued that an important determinant of the power of group selection is the pattern of population propagation. The more propagation is itself a "group" process, as in the ocean crossing of the *Mayflower* colonists who founded New England, the more likely it is that group selection will succeed in increasing altruistic behavior. This effect can be characterized in terms of two extreme alternative scenarios for propagation—the migrant pool scenario and the propagule scenario.

Figure 20.2A shows the **migrant pool** scenario. In this scenario, the migrants that found new populations are drawn from all surviving populations. Somewhat as in the melting pot of latter-day American immigration, all migrating individuals are mixed together to found new populations. These populations then grow in numbers, so long as the number of selfish individuals is not too high. Populations with a lot of selfish individuals tend to go extinct. In this type of group selection, populations are not formed by well-defined groups,

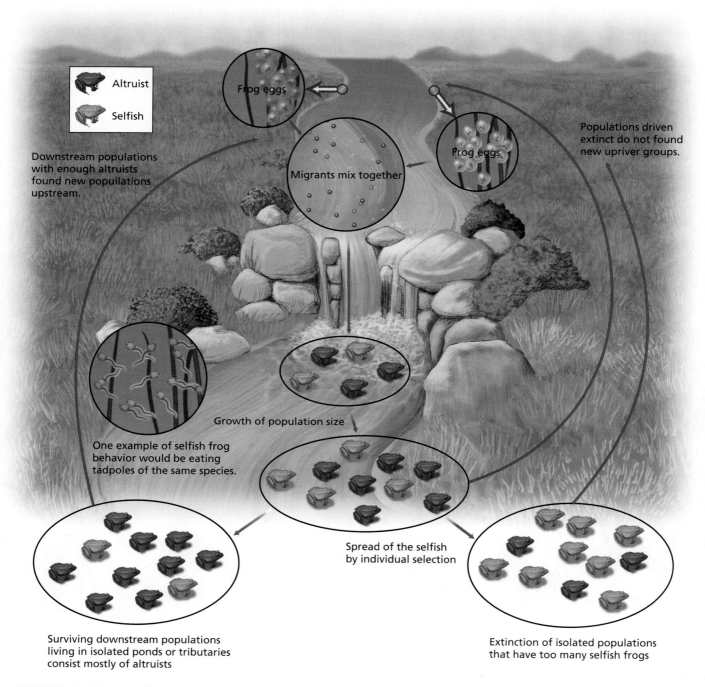

Altruist

Selfish

Frog eggs

Frog eggs

Populations driven extinct do not found new upriver groups.

Downstream populations with enough altruists found new populations upstream.

Migrants mix together

One example of selfish frog behavior would be eating tadpoles of the same species.

Growth of population size

Spread of the selfish by individual selection

Surviving downstream populations living in isolated ponds or tributaries consist mostly of altruists

Extinction of isolated populations that have too many selfish frogs

FIGURE 20.2A Migrant Pool Model for Group Selection In this model, new populations are founded by mixing individuals from different populations.

though the fate of a population depends on its mix of altruistic and selfish individuals.

Figure 20.2B shows the **propagule** scenario. In this scenario, the migrants that found new populations all come from the same ancestral population. There is no mixing of migrants to found new populations. However, once the new groups are founded, their ecology and evolution are the same as they are in the migrant pool scenario.

Theoretical work shows that the propagule scenario (Figure 20.2B) is much more likely to lead to successful selection for behavior that benefits the group. This occurs because the mixing of migrants in the migrant pool scenario (Figure 20.2A) dilutes the impact of the group selection process. Group selection can work with either model, but it is far more difficult for group selection to overcome individual selection with migrant pool mixing. This may occur because, unlike the propagule scenario, the group structure of the migrant pool is partly broken down by the mixing of migrants. ❖

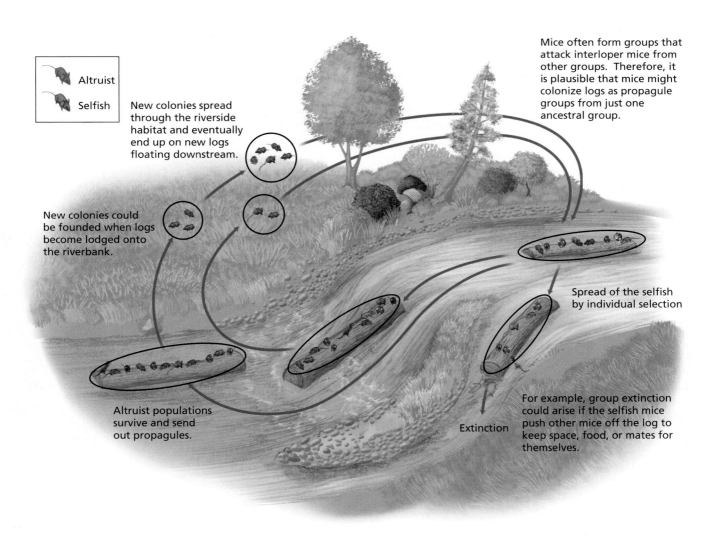

Altruist

Selfish

New colonies spread through the riverside habitat and eventually end up on new logs floating downstream.

New colonies could be founded when logs become lodged onto the riverbank.

Altruist populations survive and send out propagules.

Mice often form groups that attack interloper mice from other groups. Therefore, it is plausible that mice might colonize logs as propagule groups from just one ancestral group.

Spread of the selfish by individual selection

Extinction

For example, group extinction could arise if the selfish mice push other mice off the log to keep space, food, or mates for themselves.

FIGURE 20.2B Propagule Model for Group Selection In this model, each new population is founded by individuals that come from only one established population.

20.3 Group selection may be the best explanation for some cases of biological altruism

Theoretical analysis of group selection shows that it is possible for altruism to evolve through group selection, as we have just seen. But are actual animal populations likely to evolve altruism as a result of the process of group selection? Two kinds of information help us to answer this question. The first consists of laboratory experiments that mimic theoretical models. The second is the natural history of populations, insofar as that natural history can shape group selection in nature.

Figure 20.3A summarizes an experiment testing the power of group selection in a laboratory population of the flour beetle, *Tribolium*. Flour beetles are one of the major agricultural pests, primarily consuming stored grain. If you eat baked goods regularly, you have eaten a fair amount of flour beetle. Flour beetles eat more than grain. They also eat each other—the adults eat eggs and juveniles, the larvae eat eggs, and so on. This cannibalism is obviously selfish behavior, raising the question of whether it is possible to devise a group selection experiment to reduce its impact.

Suppose that flour beetles are kept together in vials, as groups of several dozens in size. If they are given a fixed period for population growth, the more altruistic beetles should reach larger population sizes, because they will not be eating each other. Following our previous discussion of patterns of group selection, Michael J. Wade propagated group-selected populations in discrete propagules, without migration, so there was no mixing of migrants. When Wade imposed group selection, only the vials with large populations were allowed to start additional populations, as shown in Figure 20.3A.

When this procedure was performed over multiple generations, the results were clear. Group selection led to the maintenance of higher population sizes, compared to controls that were not group selected. This experiment verifies our theoretical analysis: Group selection *can* work, when the conditions are right. One of these conditions is that groups found

new populations using discrete propagule groups, without mixing of migrants.

But how likely is it that conditions favoring group selection will occur in nature? A key point is that most species are very unlikely to undergo group selection. Species that do not live in groups are unlikely to undergo group selection. Many species of animals, plants, and microbes occur sporadically in the wild, having little contact with others of their species. Species that have unstructured populations are also unlikely to undergo group selection. Some species live in groups; but these groups exhibit little subdivision for group selection to act on, and they do not propagate themselves as groups. Large colonies of communally nesting birds, large herds of ungulates such as deer and antelope, and human populations all fit this pattern. These points are summarized in the box, "Group Selection in Nature: How Common?"

But certain species might well undergo group selection. Pathogens that infect hosts as single propagule groups, with-

Group Selection in Nature: How Common?

Unlikely to show group selection

- Animals with large, randomly mating populations without local breeding (e.g., most herd mammals; flying insects)
- Plants that distribute pollen over a wide area (e.g., most conifers and grasses)

Possible cases with group selection

- Pathogens that infect as propagule groups (e.g., lethal viruses)
- Animals that migrate as small social bands with mating restricted to the band (e.g., some of the social mammals)

out multiple infection, are an example. In such cases, group selection might favor benign pathogens that do not kill the host too quickly by selfishly reproducing too fast. Some biologists think that the **myxoma virus** that infects rabbits has evolved from being highly lethal to being relatively benign, particularly in Australia, where the virus was deliberately used to kill off rabbits. (See Chapter 13 for more detail about myxoma virus.) Indeed, when the virus was first introduced it killed off rabbits at a great rate. Now the virus is not so deadly. On the other hand, the rabbit has evolved greater resistance to the virus.

There are other cases in which group selection may have been important. Some rodents, such as house mice, live in well-defined groups that rarely interbreed. There is some evidence that these small breeding populations of mice undergo group selection against selfish alleles, particularly alleles that disrupt the genetic system.

But these two cases do not suggest that group selection can explain the vast range of species that exhibit benign social behavior. Most altruistic species probably undergo little group selection. So group selection is not a general explanation for biological altruism, even it favors the evolution of altruism in some species. ❖

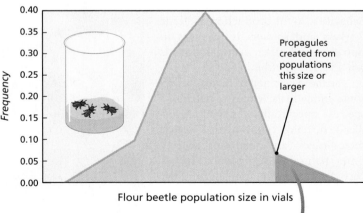

Propagules created from populations this size or larger

Flour beetle population size in vials

Each selected population is used to start enough new propagules to regenerate the complete set of populations for the next round of selection.

FIGURE 20.3A Experimental Group Selection on Population Size The inference from this experiment was that group-selected populations had a slower spread of selfish behavior.

KIN SELECTION

20.4 Selection can act on families

As already mentioned, the most spectacular examples of biological altruism come from the social insects. Soldier ants and soldier termites give their lives defending their colonies, even though they will never be able to reproduce. Indeed, their bodies are often greatly modified even from those of their fellow workers. Their mandibles are often much larger. Some ant soldiers have modified abdomens for spraying toxic fluids at other insects—a natural form of chemical warfare. And then there is the wide range of worker types, which also do not reproduce, yet go about building hives and mounds, gathering food, tending the young, and so on. In a bizarre way, these social insects resemble the social economies of human communities. But they take it farther, in that their "royalty" of queens and **kings** are the primary reproducers. (Figures 20.4A to 20.4C illustrate the lives of the social insects.) Not even human societies are so extreme.

These facts were well known to Darwin and others in the nineteenth century. Darwin devoted some thought to how such organisms could have evolved. One question was how a single species could have such different morphological forms and behavioral patterns among its members. This is not a profound difficulty. It is common enough to have males and females that differ substantially in morphology and behavior, and not all gender determination is genetic. Some of it is environmental. In reptiles, gender is often determined by the temperature at which the turtle or lizard matured. Within the single life spans of moths and butterflies, different life stages have very different morphologies and behaviors. So the production of sterile workers and soldiers by changes in nutrition and other manipulations is not a violation of biology.

The more pertinent difficulty is why some individuals within social insect species—specifically workers and soldiers—do not reproduce. Because these insects are organisms bearing genes, natural selection would be expected to favor the copying of their genes into the next generation. The natural conclusion is that sterile honeybees should be selected to reproduce—unless, perhaps, group selection has favored restraint.

But there is another possibility: **selection among families,** or "kin selec-

FIGURE 20.4A Life Inside an Ant Colony

tion." This is a common principle in plant and animal breeding, as Darwin knew. Plant varieties may be selected because of the flowers or fruit of sterile members of a family, even though the propagation of the variety takes place using the fertile members of the family. In the cultivation of fruit, for example, the seedless fruit may go to the market while the fruit having seed is used for cultivation, because people do not like to bite or swallow seeds. In meat production, breeders may choose bulls and cows according to the carcasses of their slaughtered offspring, even though the slaughtered will never be able to reproduce. In these cases, the entire family is being selected by the breeder.

In nature, when family groups function together ecologically, the same principle can still apply. Insect families that produce more fertile offspring than other families may do so by having some members of the family take on nonreproductive roles. We will consider the case of social insects in more detail shortly. But for now, the basic concept of kin selection is our provisional explanation for biological altruism when organisms function together as families. **Kin selection** is based on the transmission of an organism's genes into the next generation by means other than the production of immediate descendants. There is kin selection when one sister helps another sister raise the second sister's offspring. Kin selection also occurs when grandparents raise their grandchildren, or help their children do so. Kin selection also occurs when one first cousin dies to save the life of another first cousin. Note, however, that kin selection does not provide an explanation of biological altruism between unrelated individuals.

One way to think of this is that individual selection is about the survival and reproduction of an individual, and kin selection concerns *all other* means by which individuals can help their relatives so that copies of the helper's genes find their way into the next generation. Group selection is a process that occurs at another level from that of the family. Natural selection includes all forms of selection, including group selection. However, these are just terminologies. The differences between the levels of selection arise no matter what they are called.   ❖

FIGURE 20.4B Life Inside a Beehive

FIGURE 20.4C Life Inside a Termite Colony

20.5 Altruism toward relatives is favored when the cost is less than the benefit times relatedness

All the essential features of kin selection can be understood in relation to one story. This will enable you to understand the evolution and ecology of biological altruism in the family setting. Suppose that a chimpanzee finds a supply of a bananas that no other chimpanzee has found. He eats his fill of bananas, and leaves the rest for later. He is walking away with an uneaten banana and comes upon another chimp. Should he give the banana to the other chimp?

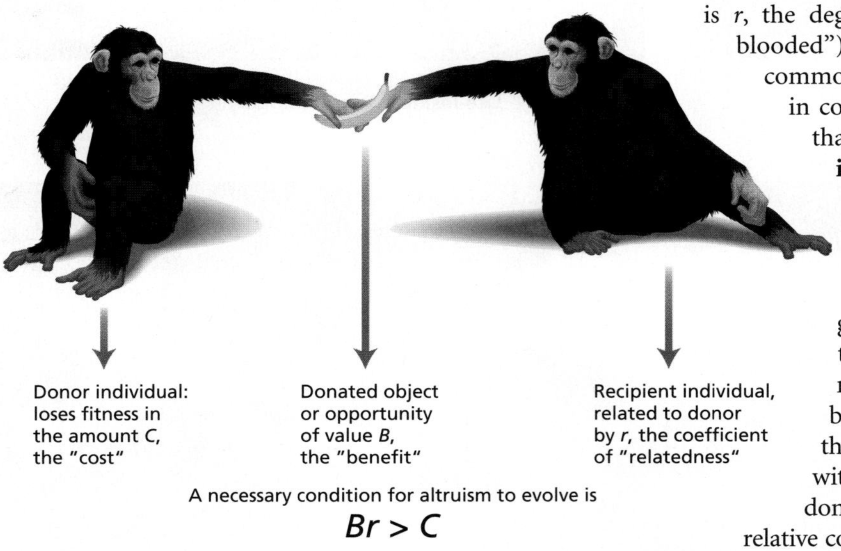

This decision is diagrammed in Figure 20.5A. The chimp with the banana, the potential donor, will lose some nutrition that might aid the donor's fitness if it does give up the banana. Call that cost C. The fitness of the hungry chimp will benefit from the banana by an amount B. The key parameter is r, the degree of **consanguinity** (which means "same-blooded"), or the number of genes the chimps have in common. For **identical twins**, r is 1; all their genes are in common. For unrelated individuals, r is 0. Note that r is sometimes called the **coefficient of genetic relatedness**.

If the benefit is greater than the cost, so $B > C$, and the two chimps are twins, then the chimp with the banana should give it up. He would be helping his own genes get into the next generation more by donating the banana, probably because he doesn't need much more banana, while his hungry twin might be starving. But if the other chimp is unrelated, then the logic of kin selection says that the chimp with the banana should not give it up. In short, the donation of a banana is entirely dependent on the relative costs and benefits to donor and recipient, respectively, weighted by how much they are in fact kin—r, the degree of consanguinity. In other words, according to the theorist W. D. Hamilton, altruism can evolve when $Br > C$. (It doesn't always evolve, but at least it becomes possible.)

Unrelated individuals and identical twins are the simplest cases, as well as the most extreme. In between are the usual types of kin: siblings, nieces, nephews, grandparents, and so on. The r values for these relationships are shown in the box, "Some Coefficients of Genetic Relatedness." In these relationships, the degree of consanguinity constrains the evolution of biological altruism between any two individuals. The value of r between the children of diploid parents provides a convenient benchmark. Each parent gives the child half its genes, leaving aside sex chromosomes, so $r = \frac{1}{2}$. A child has half the genetic makeup of the parent, from the standpoint of consanguinity. Then it is also logical that r between grandparent and grandchild should be $\frac{1}{4}$. The grandparent is two reproductive events removed from the grandchild. In this way, we expect r values to be larger for individuals who are closely related, and smaller for "distant relatives." Note also that inbreeding will systematically raise genetic relatedness, fostering the evolution of altruism.

Bear in mind that the benefit and cost parameters (B and C) are important too. In organisms where it is unlikely that an organism will have much to give to, or do for, its relatives, then there may be few opportunities for biological altruism to evolve. Because biological altruism is not that common, this may often be the situation. ❖

Donor individual: loses fitness in the amount C, the "cost"

Donated object or opportunity of value B, the "benefit"

Recipient individual, related to donor by r, the coefficient of "relatedness"

A necessary condition for altruism to evolve is

$$Br > C$$

FIGURE 20.5A The Essential Features of Altruism One chimp can help another chimp by donating a banana. Under what conditions will natural selection favor such an act of altruism?

Some Coefficients of Genetic Relatedness

If there is inbreeding, the individuals of a social group may have very high degrees of genetic relatedness.

But if we assume no inbreeding, and thus random mating, we have the following values for r in human populations:

Identical twins	1
Genetic clones	1
Full siblings	1/2
Parent-child	1/2
Half-siblings	1/4
Uncle-nephew	1/4
Aunt-niece	1/4
Uncle-niece	1/4
Aunt-nephew	1/4
Grandparent-grandchild	1/4
First cousins	1/8
Husband-wife	0
Unrelated individuals	0

With the basic rules of kin selection under our belts, especially the requirement that Br be greater than C, let us return to the high level of biological altruism among social insects. It is an extremely powerful example of the importance of kin selection in the evolution of social behavior. Social insects have evolved most often in the Order Hymenoptera, which in-

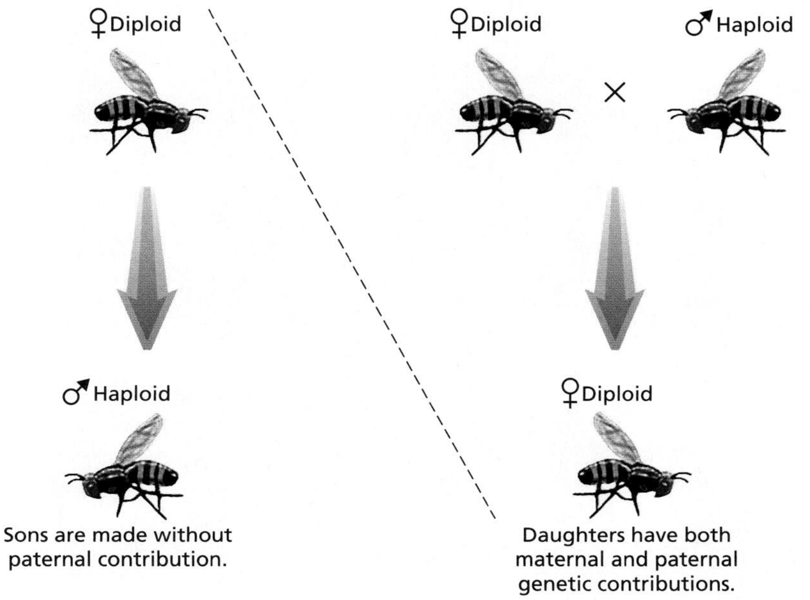

FIGURE 20.6A Haplodiploid Sex Determination Unfertilized eggs grow up to be sons, while fertilized eggs grow up to be daughters.

cludes all the ants, the bees, and the wasps. Not all hymenopterans are social, but many are. **Hymenoptera** have an unusual form of **sex determination** called **haplodiploid sex determination**. Unfertilized eggs, which are haploid, become male, as shown in Figure 20.6A. Fertilized eggs become female, and are diploid.

This unusual system of sex determination transforms the genetic relationships among the members of hymenopteran families. As shown in the box, "Genetic Relatedness in Hymenoptera," females are typically more related to each other than males are to other members of the family. The father-son relationship no longer exists. Notably, full sisters have at least a $3/4$ level of genetic relationship, because the half of their genome that comes from their haploid father is always identical and the half that comes from the diploid mother is identical half the time. (If there is inbreeding, this consanguinity value may be even higher than $3/4$.) Remarkably, this value is higher than that between mothers and their offspring. In effect, hymenopteran females can spread their genes more by "producing" sisters than by producing children.

What actually happens in nature is a demonstration of the power of a genetic system to shape social relationships. The societies that hymenoptera evolve are female dominated.

Their characteristic pattern is one of sisters helping each other to reproduce. In the most elaborate cases, such as honeybees and ants, there is a queen that does all the reproducing. The queen is tended by her numerous sterile daughters, so that more daughters can be produced—sisters to the sterile workers. In some social hymenoptera, the sterile daughters are specialized for combat, as soldiers, to defend the colony from potential invaders. Males are pushed to the margins of this animal society, supplying little work beyond fertilizing the queen.

The hymenopteran system of sex determination alters normal genetic relationships, transforming the organization of these insect societies. Note in particular that the sister-sister value for r is higher than that between all relatives in species with normal sex determination, leaving aside identical twins. There is a dramatic correspondence between genetic relationship and degree of biological altruism in hymenopteran females and males, which suggests that kin selection is indeed fundamental to the explanation of social behavior in such societies. ❖

Genetic Relatedness in Hymenoptera

Genetic relatedness involving males:

Father-son	0
Mother-son	1/2
Brothers	1/2
Brother-sister	1/4

Genetic relatedness between females:

Mother-daughter	1/2
Full-sib sisters	3/4
Half-sisters	1/4

The figure for half-sisters assumes that their fathers are unrelated. If the fathers are related to each other, then the coefficient of genetic relatedness of half-sisters will increase.

Males are less related to other hive members than females are, with sisters less related to brothers. This will tend to produce matriarchal societies, with peripheral males.

If the hymenopteran genetic system gives rise to a gender-biased social system, then a genetic system that is symmetrical between the genders should lead to the evolution of animal societies that are not biased with respect to gender. What do we find in nature?

The best-known examples of such societies are those of another social insect, termites, from the Isoptera. Termites feed on underground, woody, plant material, roots being an important item in their diet. This is an unusual ecological niche. Much of termite nutrition is dependent on breaking down cellulose, a polymer built by plants from molecules of glucose. The glucose molecules in cellulose are linked together by a chemical bond that most animals cannot break during digestion, so cellulose makes woody material indigestible for most species, including humans. Cellulose is roughage for us for that reason. But some microbes can digest cellulose. Termites maintain populations of these microbes in their guts and are nourished by the digestive action of these microbes. This dependence gives termites the problem of maintaining these microbe populations, which they solve by eating each other's vomit or feces. Not a pretty thought, but the microbes of the termite gut are absolutely fundamental to its way of life.

This dependence on ingesting each other's microbes greatly influences the social calculus of termite life. A termite that has lost its microbes stands to benefit to a high degree merely from the excreta of a termite that still has some of the microbes. This means that termite families can aid each other to a remarkable extent by staying together. This increases the value of B in the inequality $Br > C$. In addition, because they live underground and need to remain together, inbreeding is higher in termites than in other organisms, elevating r. Finally, because vomit and feces are not exactly hard to produce, the cost factor, C, is likely to be small. All these factors together explain why termite societies show so much biological altruism.

An important feature of termite societies is that they are symmetrical between sexes, as diagrammed in Figure 20.7A. There are both male and female sterile

workers and soldiers. Instead of queens with inconsequential **drone** "boy-friends," termites have kings and queens that do all the reproducing. Termites do not have significant sex-based asymmetry in their genetic system. Males are not haploid. Consanguinity does not depend on sex. Sisters are as related to their brothers as they are to their sisters. This leads to an expectation of symmetry between the sexes, according to kin selection theory. Indeed, the termite social system fits the expectations of evolutionary theory in its symmetry between sexes.

An intriguing parallel to termite social evolution is afforded by the naked mole rat of Africa, *Heterocephalus glaber*, shown in Figure 20.7B. This mammalian species also lives a subterranean life in family groups, like termites. There is a primary reproducing couple, the king and queen, along with nonreproductive males and females. This gender-role symmetry corresponds to the symmetry of the mammalian genetic system, where both males and females are diploid. Like the termites, naked mole rats depend on a microbe in their guts to digest woody and other plant material, making them exceptionally dependent on each other for reinfection. Again, inbreeding levels in naked mole rats will be elevated by their confined lifestyle. Termites and naked mole rats provide a remarkable example of parallel evolution, in which the ecology of a species determines its social evolution. ❖

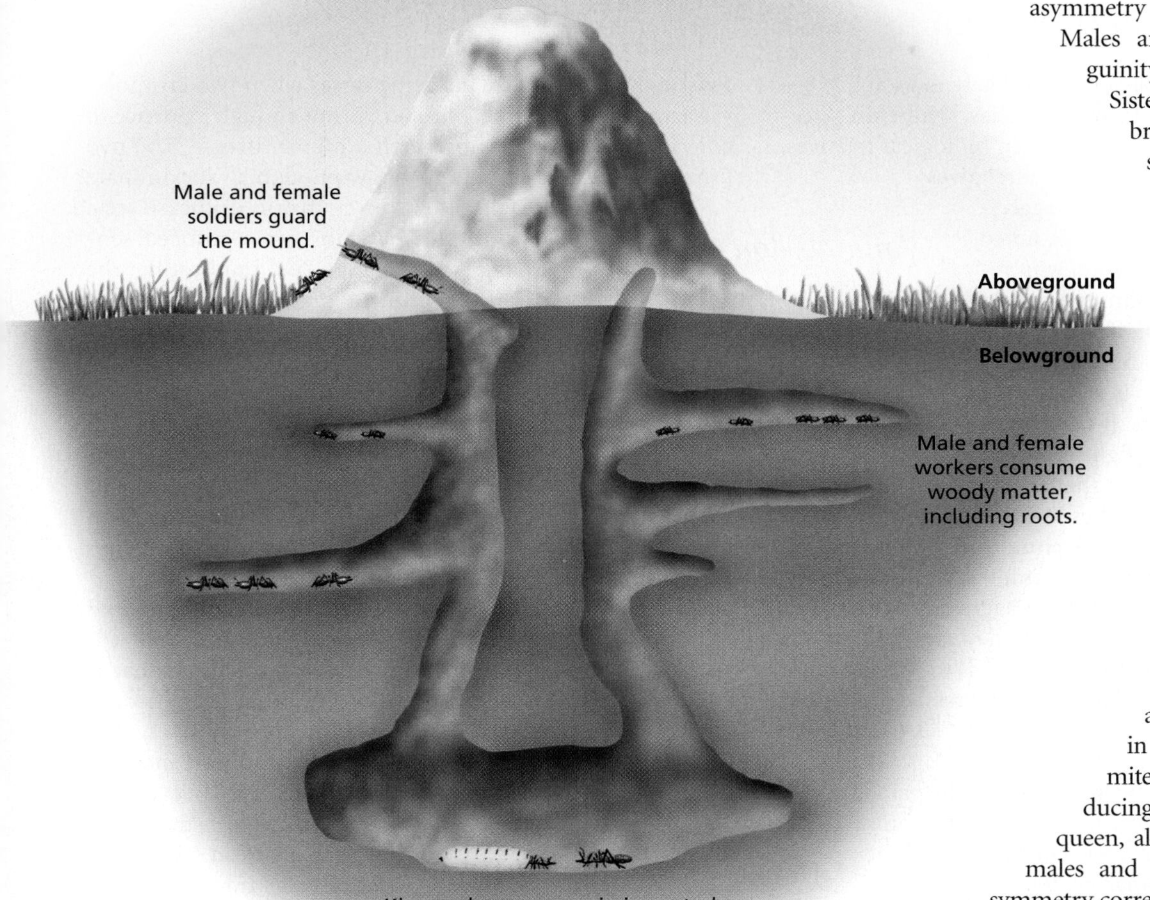

Male and female soldiers guard the mound.

Aboveground

Belowground

Male and female workers consume woody matter, including roots.

King and queen remain in central bower, away from physical danger.

FIGURE 20.7A The Ecology of Termite Societies

FIGURE 20.7B The naked mole rat is a mammal with an ecology like that of a termite.

EVOLUTIONARY GAMES

20.8 Animals balance aggression and peaceful behavior as if social interaction were a game

As we have seen, group selection is unlikely to establish cooperative social behavior in most species, while kin selection fosters altruism primarily between biological relatives. What about interactions between nonrelatives?

The basic paradox of restrained aggression remains to be resolved. As we noted earlier, there is no obvious reason why animals should not exhibit unlimited aggression toward nonrelatives, if there is a conflict over who should have some resource. Yet violent combat is rare among animals of the same species. It does occur in some situations, such as fights between lion males for sexual access to the females of a pride, as rendered in *The Lion King*. But these situations are rare. Much more common are fights that are **ritualized**—where antlers, fangs, and claws are not used with complete ruthlessness. Instead, animals seem to hold back their aggression. Often they are literally making more noise than they are fighting—as seen in animals ranging from chirping birds to roaring stags. Why do they do this? Attacking a nonaggressive opponent should lead to victory, as shown in Figure 20.8A, and the opportunity to feed or to mate.

The key step that led to the resolution of the paradox of restrained aggression came from the application of **game theory**. Game theory was originally a branch of mathematics applied to human economic and criminal behavior. Game theorists dealt with problems of cooperation among intelligent human strategists.

A classic problem of traditional game theory is called the

Animals don't bargain with district attorneys, and they don't play schoolyard games . . .

Prisoner's Dilemma. This game arises when two criminals are arrested for a crime, but there is not enough evidence to convict either one of more than a minor offense. The two criminals are quite likely to know enough to incriminate each other for a major crime. If one crook plea-bargains for a reduced sentence in exchange for implicating the other criminal, then the implicating crook ends up ahead, *provided* the other prisoner does not do the same. If each crook implicates the other, then the court can send both to prison for the full term; this is because plea bargains normally protect the indicted from self-incrimination, but they do not protect them from evidence that comes from other sources. This situation has the matrix of outcomes shown in Figure 20.8B.

In conventional game theory, the two prisoners should incriminate each other when they cannot coordinate their actions, because that is the only decision that each prisoner can make to improve his or her outcome. If prisoner 1 incriminates prisoner 2, but 2 does not incriminate 1, then prisoner 1 is freed. And vice versa. The traditional way this game is structured is to have the player who incriminates a nonincriminating partner (the Defector, or Rat) receive the highest payoff, greater than the payoff received when both stay silent. The worst-off prisoner is the one who stays silent (the Sucker) when the other one incriminates. But if the prisoners are incriminating each other, they are both going down for a long sentence anyway. Therefore, the best move is always to incriminate.

This theory seems paradoxical, but it arises from the nature of games. The results of your action depend on the choices of your opponent. In football, for example, if the defense assumes that the offense will pass, it will "defend the pass," which is a less effective strategy against a running play. And vice versa. The success of the offense will depend on choosing to do the opposite of what the defense expects. Still another example is the Rock-Scissors-Paper game, in which any choice may lose, win, or tie, depending on the opponent's choice. (For those of you who haven't played this game, Rock beats Scissors, Scissors beats Paper, and Paper beats Rock. You have to make your choice without knowing your opponent's choice.) Game outcomes depend on the choices of all players.

Animals don't bargain with district attorneys, and they don't play schoolyard games, so how does game theory apply to animal conflict? There is a key element in common—the consequences of aggression and cooperation depend on the actions of other animals, just as they do in humans. This realization led J. Maynard Smith and others to develop evolutionary game theory, the key to understanding the evolution of animal conflict between unrelated individuals. ❖

Aggressive behavior seems like it should be favored. In any contest between a nonaggressive animal and an aggressive one, the nonaggressive one should lose.

Yet even when there is no evidence for group selection or kin selection, animal behavior is normally nonviolent.

FIGURE 20.8A The Basic Paradox of Animal Social Behavior

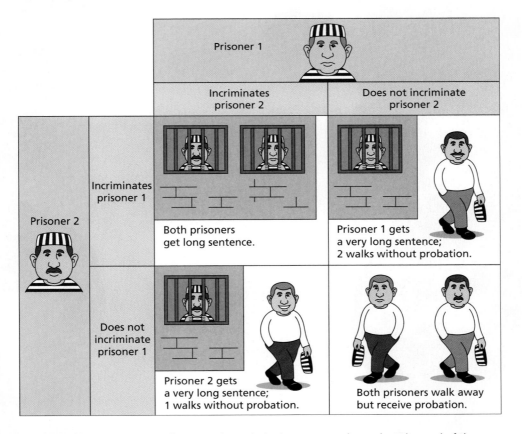

FIGURE 20.8B The Prisoner's Dilemma When criminals are arrested together, the goal of the police is to get them to incriminate each other. When the crooks do so, they both end up in prison. When they don't incriminate each other, they can both get probation for a lesser charge.

20.9 Fitness depends on strategies that specify an animal's behavior in its conflicts

Animals do not behave like humans. Individual animals rarely, if ever, adopt idiosyncratic social strategies previously unseen in their species. Humans do. Therefore, the kind of behavioral flexibility assumed in normal game theory is not appropriately assumed in evolutionary game theory. Instead, animals behave in a stereotypical manner, within a very limited range of options, or **strategy selections**.

This idea of a defined strategy can be applied to animal conflicts in a straightforward way. Suppose that two animals want the same food item, mate, or territory. These animals could be birds, mammals, or even insects. At each moment, they have three alternative behaviors: Attack, Retreat, or Display. An **Attack** may allow the attacker to drive off its opponent. Or the attacker could be injured by the opponent. In that case, it will probably **Retreat**. Or the animal could just Retreat, abandoning the conflict and letting the other animal take the prize. To

An evolutionary game strategy describes what an animal will do in its social behavior whenever there is a conflict.

There are two extreme strategies:

Hawk—Attack whenever there is conflict, continuing the attack until victory is achieved or injury (including fatal injury) forces a retreat.

Dove—Display (twitter, flap wings, open mouth, engage in electoral politics) until you or your adversary give up because other pursuits have become more interesting (looking for water, getting out of the sun, etc.). Also, if Attacked, immediately Retreat before injuries occur.

Display means to make enough noise, or other fuss, that the other animal knows that the prize is still disputed, but there is no direct aggression. Honking a car horn is a human display. Birdsong is a Display, as the flapping of insect or bird wings may be.

In a conflict situation, two extreme strategies are easily identified: Hawk and Dove. (Bear in mind that these terms do not refer to anatomy but to patterns of behavior.) Hawks always attack until they either get the prize or are injured and forced to retreat. The **Hawk strategy** embodies the nineteenth-century idea of the type of morality that people supposed Darwin's theory of evolution would foster—unrestrained ferocity. Doves, by contrast, only Display; they never attack. If another individual tries to attack them, the **Dove strategy** is to immediately retreat. Contests between Doves are decided when one or the other gives up out of boredom or fatigue. The Hawk and Dove strategies are summarized in the box, "Evolutionary Game Strategies."

What are the consequences of conflicts involving Hawks and Doves? Three cases are shown in Figure 20.9A. Let the fitness value of winning a conflict be V. Let the cost of being injured be W, which will normally be greater than V. Let the cost of a protracted period of display between two Doves be T (for time), where T is much less than V or W.

With these parameters, the *average* benefit that Hawks derive from a conflict with each other is $(V - W)/2 < 0$, as part (i) of Figure 20.9A shows. That is, Hawks beat up on each other. The average benefit that two Doves derive from a conflict with each other is $V/2 - T > 0$, as part (ii) of Figure 20.9A shows. That is, Doves resolve their conflicts at a lower cost than Hawks do. But contests between Hawks and Doves are resolved entirely in favor of Hawks, which always receive the prize, V. Doves, however, get out of trouble quickly, receiving 0. Part (iii) of Figure 20.9A shows this situation.

All this information defines animal conflicts in relation to alternative evolutionary strategies. Next, let's consider how these strategies actually fare in the arena of social evolution. ❖

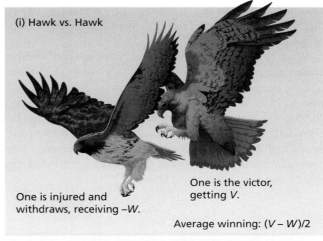

(i) Hawk vs. Hawk

One is injured and withdraws, receiving −W.

One is the victor, getting V.

Average winning: (V − W)/2

(ii) Dove vs. Dove

Both lose time T in displaying.

Average winning: V/2 − T

(iii) Hawk vs. Dove

Hawk receives V always.

Dove just runs away; it gets 0.

FIGURE 20.9A Conflicts Between Hawks and Doves There are three basic conflict possibilities: (i) Hawk vs. Hawk, (ii) Dove vs. Dove, and (iii) Hawk vs. Dove.

So what are the conditions for the evolution of peaceful social behavior? If the costs of injury are small, then violent aggression is selectively favored precisely because it is not actually that violent. Losers in conflicts do not lose heavily. This situation is reminiscent of the saying that the fighting was protracted because the stakes were so small, often applied to squabbling children and academics. In a sense, this implies that social evolution tolerates violence if it is ineffectual, and so not particularly deadly.

What happens when the costs of injury are considerable, greater than the benefits of victory in a conflict? Consider first the success of a Hawk invading a population that otherwise consists only of Doves, as shown in part (i) of Figure 20.10A. The only type of opponent that the Hawk faces is Dove, so the Hawk wins all its contests. It will be evolutionarily successful, establishing itself in the population and reproducing frequently. This suggests that Hawks should always be able to invade populations of Doves.

Consider a Dove invading a Hawk population, as shown in part (ii) of Figure 20.10A. This Dove will be retreating in every conflict situation, because it repeatedly faces aggression from the Hawks that it encounters. It receives nothing whenever there is a conflict. But Hawks, as pointed out earlier, receive less than nothing from most of their contests. Most of the contests involving Hawks are with other Hawks, when the Dove is first invading. Hawk contests with each other, Hawk vs. Hawk, net them an average payoff of $(V - W)/2$, which is less than zero. This arises because one of them will be injured, and we are considering the case with costly injury. As a result, Dove can always invade populations made up of Hawks.

Neither Hawk nor Dove can prevail when the cost of injury is greater than the benefit of victory. What will happen under these evolutionary conditions? The best way to understand the evolutionary outcome is in relation to the **evolutionarily stable strategy**, or **ESS**. The ESS is defined as an **unbeatable strategy** if almost all members of the population

(i) Hawk can invade an all-Dove population.

Hawk wins $V > V/2 - T$, Dove's payoff.

(ii) Dove can invade an all-Hawk population.

Dove wins $0 > (V - W)/2$ Hawk's payoff.

FIGURE 20.10A When Injury Is Very Harmful, Neither Hawk nor Dove Will Be an ESS

adopt it. The meaning of *unbeatable* is explained in the box, "Evolutionarily Stable Strategies." If a population has evolved an ESS, then it cannot be successfully invaded by any alternative strategy that is appropriate to the species. (For example, a strategy of using a high level of intelligence might win the day for a competing insect, but evolution does not allow large brains in insects. They're too small.) ESSs are patterns of social behavior that should dominate social evolution. Hawk and Dove, though, are not ESSs, which raises the question, what strategies should we expect to find in natural populations that have social conflict? We answer this question in the following modules. ❖

Evolutionarily Stable Strategies

An evolutionarily stable strategy is also called an ESS.

A strategy S is an ESS **if**, when S is nearly 100 percent of the population,

    S against S does better than all other strategies do against S.

or

    If there is another biologically appropriate strategy, say M, that does just as well against S as S does,

then

    the S strategy does better against M than M does against M.

Consider the evolution of a "peaceful" Hawk with a sense of "justice." Suppose this Hawk does not have the reflexive aggression of a normal Hawk, but instead tries to act like a Dove. When a Dove displays at it, it displays back, settling the conflict as if it were no different from a Dove. Then it would receive $V/2 - T$, on average, because it would win only half the time, but always pays the cost of display. But if this peaceful hawk is attacked by a normal Hawk, it gets mad and fights back. Indeed, it fights just as hard as a regular Hawk. Then it would receive $(V - W)/2$ from such conflicts—less than zero. This "peaceful Hawk" can be called a '**Retaliator**,' because that is its behavioral pattern. Its behavior is summarized in Figure 20.11A and the box, "Why Retaliator Is an ESS against Hawk Strategies."

Are Retaliators evolutionarily stable strategies, or ESSs? Can Hawks invade them? Hawks do as well against Retaliators as Retaliators do against Hawks. But in a mixed population of Hawks and Retaliators, the Retaliator-Retaliator contests are settled much more peacefully. Under these conditions, over all the contests that Hawks and Retaliators have, the Retaliators do better. They have a system of resolving conflicts that does not involve violent combat. It is better evolutionarily to

behave peacefully, when others are peaceful and the cost of injury is great. But when they come after you, you have to fight back. At least that is what the theory seems to say.

What about social behavior in nature? Do animals use Retaliator strategies? In a few cases, we have some direct evidence that they do. For example, the well-studied rhesus monkey, *Macaca mulatta*, uses ritualized combat to settle its conflicts, somewhat like wrestling. The loser accepts bites from the victor's incisors to conclude the combat. Incisor bites are fairly harmless, like "love bites." Such interactions are like those between Dove and Dove. But if the victor of this mock combat bites the loser with its canines, which are much more dangerous teeth, the "loser" then responds with an aggressive counterattack. Apparently, the canine bites are perceived as significant aggression, which brings out a Retaliator response.

There is circumstantial evidence for Retaliator strategies in many animal species. The best evidence of this kind is the many forms of weaponry that animals have but do not normally use in their conflicts with other members of their species—that is, fangs, claws, antlers horns, and so on. If all the species equipped for violence behaved only as Doves, they would have no need for these structures, except for defense against predators or capturing prey. How often such weapons have evolved solely for predator–prey interactions is unclear. Yet many animal species are well equipped to destroy their fellows, but rarely do so. This finding suggests that Darwinian evolution has led to less aggression than might have been intuitively expected in most animal species. ❖

We can get an ESS if

Payoff (Retaliator playing Retaliator)
> Payoff (Hawk playing Retaliator)

which is

$$V/2 - T > 0 > (V - W)/2,$$

assuming that the costs of display are small and the costs of injury in combat are great.

The biological meaning of this ESS is that the Retaliators have a system of resolving disputes that does not cost them too much, while the Hawks pay a steep price for their violence.

How can peaceful behavior be an ESS?

Consider the Retaliator strategy: Display unless Attacked, and then Attack back like Hawk.

Two Retaliators settle disputes like Doves, receiving $V/2 - T > 0$.

Against a Hawk, both Retaliator and Hawk receive $(V - W)/2 < 0$.

A Hawk attacking a Retaliator forces an aggressive response from the Retaliator.

FIGURE 20.11A Retaliator Is an ESS Against Hawk

One of the most surprising things about animal behavior is that many species seem to have some sense of ownership, or **territoriality**, even of marriage. Birds will avidly display against other birds, even humans, who wander into particular parts of their habitat. They behave as if they have a deed to the territory, as if it is their property. Similarly, males and females of many species behave very aggressively toward conspecifics that attempt to have sex with the individual they have been having sex with, even when they may not have had sex with their "mate" for some time. Even more amazing, many of the interlopers themselves act as if they have done something wrong and back off in the face of the protest of the "owner" or "spouse."

A simple evolutionary game strategy can be used to explain this type of territorial behavior: Bourgeois. The **Bourgeois strategy**, summarized in Figure 20.12A, is to defend a territory or other item if it is discovered first, but give it up if there is already an owner. Two Bourgeois animals will normally know which of them is owner and which is interloper, and the interloper will not fight or squabble with the owner. This allows them to settle their disputes more efficiently than Doves do, and at little risk of injury. Should a Hawk attempt an **invasion** of a territory, the Bourgeois fights back, much like a Retaliator. But if Hawk gets there first, Bourgeois concedes the territory to them right away. Therefore Bourgeois fights less than Retaliator and spends less time displaying than Dove.

Bourgeois is an ESS, because invading Hawks pay a penalty due to their violent fights with Bourgeois defending their "property." Doves settle their contests less efficiently than the Bourgeois do. Bourgeois strategy makes excellent sense theoretically, and seems to fit the behavior of many animal species. (See the box, "Why Bourgeois Is an ESS against Hawk Strategies.")

But is Bourgeois actually the strategy that territorial animals use? This hypothesis has been tested using an elegant experimental strategy—duping two animals into thinking that each is the owner. Male Hamadryas baboons, *Papio hamadryas,* keep specific females as mates, and other baboons normally respect such relationships. However, if two males are kept with the same females on alternate days, and never see each other with their shared mates, they will fight violently when they encounter each other in the presence of "their" females. Experiments with other species, including insects, have also found this pattern. If two animals from a bourgeois species both think they are owners, they will fight each other vigorously.

Animal behavior strangely conforms to many of the dictates of human morality—retaliation against aggressors, ownership of territory, even "marriage." Yet this behavior is

predictable on the basis of evolutionary game theory. There is no overwhelming tendency to violence "red in tooth and claw" in evolution, even though it sometimes occurs within animal societies, just as it does among humans. Twentieth-century research on the conjunction of evolution, ecology, and behavior has laid to rest nineteenth-century anxieties about Darwinism's implications for animal behavior. The remaining question is the implications of Darwinism for *human* behavior, the subject of Chapter 21. ❖

Bourgeois Strategy: If first to find territory or object, Attack if Attacked, like Hawk. If second to find territory, cede to owner and Retreat.

First Bourgeois there acquires object or territory.

Second Bourgeois to arrive does not get anything, but retreats without harm.

If a Bourgeois owner is attacked, the owner responds like a hawk.

FIGURE 20.12A Bourgeois Is an ESS Against Hawk

Why Bourgeois Is an ESS against Hawk Strategies

We can get an ESS if

> Payoff (Bourgeois playing Bourgeois)
> \> Payoff (Hawk playing Bourgeois)

which is

$$V/2 > V/2 + (V - W)/4,$$

Bourgeois ceding the territory to Hawk half the time and fighting for it half the time, because Hawk will find the territory first half the time. As before, we take $(V - W)/2$ less than zero, assuming that the costs of injury are great.

The biological meaning of this ESS is that the Bourgeois have a system of resolving disputes that does not cost them too much, while the Hawks pay a steep price for their violence.

SUMMARY

1. An outstanding problem in evolution and ecology is biological altruism: one animal aiding another, supplying food, perhaps, or helping another animal fend off an attack from a predator. Some of the most dramatic forms of altruism are found among the "social insects." Altruism is a problem for evolutionary biology because we expect that natural selection will oppose behavioral tendencies that might lead to death or sterilization.

2. In group selection, entire populations are selected in the same way that individuals are selected in individual selection, by the extinction and propagation of populations. The connection between group selection and social behavior is that "bad" behavior might foster individual fitness yet undermine the survival of the group as a whole. When individual selection and group selection are opposed, it is theoretically possible, but difficult, for group selection to prevail. One factor that affects group selection is the pattern of population propagation. The more propagation is itself a "group" process, the more likely it is that group selection will succeed in increasing altruistic behavior. Although group selection can be made to work in the laboratory, there are few cases in nature for which group selection is the most likely explanation of social evolution. These cases include pathogens that infect hosts as single propagules and mammals that live in small, well-defined groups.

3. How can we explain sterile worker insects, like honeybee and ant workers? Insect families that produce more fertile offspring may do so by having some members of the family take on nonreproductive roles. Kin selection is based on the transmission of an organism's genes into the next generation by means other than the production of immediate descendants. Individual selection is about the survival and reproduction of an individual, and kin selection concerns *all other* means by which individuals can help their relatives so that copies of the helper's genes find their way into the next generation. Kin selection favors aid to relatives when the cost of such aid is less than the benefit times the degree of consanguinity. We expect *r* values (the degree of consanguinity or genetic relatedness) to be larger for individuals that are closely related and smaller for "distant relatives."

4. Haplodiploid social insects have high consanguinity between sisters, with full sisters having a $^3/_4$ coefficient of genetic relatedness, which often leads to the evolution of a complex social system dominated by females. In the most elaborate cases, such as honeybees and ants, a queen does all the reproducing. Males are pushed to the margins of this animal society, supplying little work beyond fertilizing the queen. The correspondence between genetic relationship and degree of biological altruism in hymenopteran females compared with males is one of the most important scientific findings in the field of social evolution. In the burrowing societies of termites and naked mole rats, individuals aid each other by supplying each other with endosymbionts for digesting woody material. Such societies are symmetrical between genders. There are male and female sterile workers. Instead of queens with inconsequential drone "boyfriends," termites have kings and queens that do all the reproducing.

5. Animals seem to "hold back" their aggression against other members of the same species. Often they literally make more noise than they fight; this behavior is seen in animals ranging from chirping birds to roaring stags. The key to resolving the paradox of restrained aggression comes from evolutionary game theory. Evolutionary games are based on the concept of behavioral strategies that specify an animal's behavior in its conflicts. This idea of a defined strategy can be applied to animal conflicts in a straightforward way. When two animals want the same food item, mate, or piece of territory, two extreme strategies are easily identified—Hawk and Dove. Hawks always attack until they either get the prize or are injured and forced to retreat. Doves only Display. Violent Hawk behavior may not evolutionarily dominate if the costs of such violence are greater than the benefits of victory; but if the costs of injury are small, aggression will be selectively favored.

6. If the costs of injury are large, natural selection may favor the Retaliator strategy, which settles conflicts peacefully unless the Retaliator is attacked. Over all the contests that Hawks and Retaliators have, the Retaliators do better. It is better evolutionarily to behave peacefully, when others are peaceful. But when they attack, you have to fight back. There is at least circumstantial evidence for Retaliator strategies in many animal species. Bourgeois is another successful strategy for settling conflict without aggression. An animal using this strategy defends a territory or other item if it has been discovered first, but gives up the item if there is already an owner. Two Bourgeois animals will settle their disputes more efficiently than Doves will, and at little risk of injury. Invading Hawks pay a penalty because of their violent fights with Bourgeois animals defending their "property." Many animal species appear to use the Bourgeois strategy.

REVIEW QUESTIONS

1. What occurs when group selection and individual selection favor the same characteristics?

2. Are there any fly species that could be called 'social insects'?

3. Name a species in which you think group selection is occurring, and explain why you chose that species.

4. What is the difference between kin selection and group selection?

5. Analyze the evolutionary prospects for Anarchist, a strategy in which animals concede ownership to the next animal to find a territory.

6. Is kin selection important in the organization of human societies?

7. Do you think that group selection has ever been important in human evolution?

8. Are birds more or less likely to undergo group selection?

9. Why are male lions so violent when they compete for "ownership" of a pride?

10. Does the evolutionary analysis of social behavior suggest that animal behavior will be more or less violent than you first supposed?

KEY TERMS

altruism
Attack
Bourgeois strategy
coefficient of genetic relatedness
consanguinity
Display
Dove strategy
drones
Evolutionarily Stable Strategy (ESS)
evolutionary game theory
game theory
group selection
Hamilton, W. D.
haplodiploid sex determination
Hawk strategy
Hymenoptera
identical twins
invasion
king
kin selection (selection among families)
Maynard Smith, J.
migrant pool
myxoma virus
Prisoner's Dilemma
propagule
queen
Retaliator strategy
Retreat
ritualized conflict
selection among families (kin selection)
selfish behavior
sex determination
social insect
soldier
strategy selection
territoriality
unbeatable strategy
Wade, M. J.
worker

FURTHER READINGS

Hamilton, W. D. 1996. *Narrow Roads of Gene Land: The Collected Papers of W. D. Hamilton.* Vol. 1. *Evolution of Social Behavior.* New York: Freeman.

Krebs, J. R., and N. B. Davies. 1981. *An Introduction to Behavioral Ecology* and subsequent editions. Cambridge, UK: Blackwell.

Maynard Smith, J. 1982. *Evolution and the Theory of Games.* Cambridge, UK: Cambridge University Press.

Trivers, R. L. 1985. *Social Evolution.* Menlo Park, CA: Benjamin Cummings.

Wade, M. J. 1977. "An Experimental Study of Group Selection." *Evolution* 31:134–153.

Wade, M. J. 1978. "A Critical Review of the Models of Group Selection." *Quarterly Review of Biology* 53:101–114.

Wilson, E. O. 1971. *The Evolution of Insect Societies.* Cambridge, MA: Harvard University Press.

*An **Australopithecus africanus** skull*

Human Evolution and Human Behavior

A universal human desire is for knowledge of our ancestry, to learn about ourselves from learning about our ancestors. The study of human evolution promises to supply such knowledge in abundance. But have scientists actually revealed the "mysteries of mysteries" concerning our ancestry?

The answer is yes and no. Yes, we have discovered a great deal of information about human evolution, as a series of historical events. Despite long odds, intrepid fossil hunters have found numerous fossil bones that might have come from our ancestors, or their cousins (a picture of such fossilized bone is on this page). We will summarize this history of human evolution, but no—we do not have a complete accounting for our ancestry. Dark areas remain.

Yes, we know a great deal about human population genetics and human molecular genetics. Some of this recently collected information is staggering in its implications. But no, we are still very much in the dark as to why humans evolved. We know some of the story, but we don't know the motives behind the plot. The *causes* of human evolution remain open to debate. We will explore this controversy.

Finally, there is the evolutionary interpretation of present-day human behavior. This question has been the source of sputtering controversy, with biologists of various ideologies calling each other names. But beyond the rhetoric, there are issues of great importance. ❖

THE HOMINID PHYLOGENY

Humans evolved from Old World apes, which split from the rest of the primates about 20 million years ago

Humans are **primates**. The evolutionary tree of the primates is shown in Figure 21.1A. The primate evolutionary tree has two main branches: prosimian and anthropoid. **Prosimians** include tarsiers, lorises, lemurs, and pottos. These species have a variety of features in common with non-primate mammals, especially the insectivores. Prosimians cannot manipulate objects as well as other primates can, and they have relatively small brains. For these reasons, they are often thought to be similar to the earliest primates of 60 million years ago. However, it is also possible that contemporaneous species of prosimian have undergone specific selective processes to adapt them to their way of life.

Of more interest for human evolution is the **anthropoid** group, made up of monkeys, apes, and humans. There are three living anthropoid branches: New World monkeys, Old World monkeys, and the **hominoids** (apes and humans), this last group also being Old World in its distribution. The split between New World and Old World monkeys provides a wonderful example of biogeography. New World monkeys have prehensile tails, which function as a kind of fifth arm—very handy for animals that live in trees. Old World monkeys lack prehensile tails. This is almost certainly due to long-standing separate evolution in the two groups.

The hominoid branch of primate evolution is shown in more detail in Figure 21.1B (this figure corresponds to the rightmost branch in Figure 21.1A). Within the hominoids, the apes are represented by four genera: *Hylobates* (gibbons), *Pongo* (orangutans), *Gorilla* (gorillas), and *Pan* (chimpanzees). The apes are a very interesting group. None have tails, and they all have powerful arms and shoulders. The gibbons and orangutans are the most arboreal (tree living) in their behavioral ecology. Gibbons are not very good at walking, but they are extremely good at swinging from tree to tree. Gibbons exhibit a dramatic form of sexual dimorphism: Adult males and females have completely different colors in some species. There are 12 species of gibbon, also called the hylobatids. The other three ape genera have only one or two species. Gorillas and chimps share a very rare adaptation, **knuckle walking**. They use the knuckles of one hand to shuffle along the ground, leaving the other hand free to carry something. Humans are the upright hominids; we walk on two legs only. All hominoids are apes. The **great apes** are the orangutans, gorillas, chimpanzees, and humans, together grouped as members of the taxonomic family Hominidae, the hominids. Of course, some people do not like to be called great apes. ❖

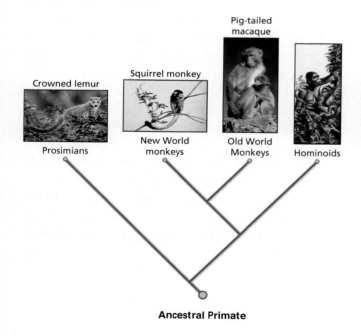

FIGURE 21.1A The Primate Evolutionary Tree

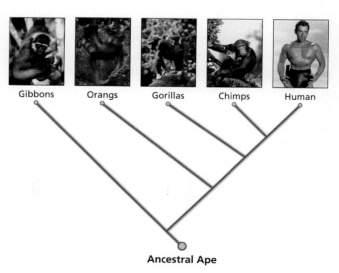

FIGURE 21.1B Evolution of the Apes

If you weren't human, you would be hard pressed to see the fundamental difference between gorillas, chimps, and humans. All share large brains, very dexterous hands, and excellent vision. All three are fairly large by the standards of most terrestrial animals—which are, indeed, insects. Perhaps the most obvious difference is that humans have relatively hairless bodies.

Molecular evidence suggests that upright hominids first evolved only 4–8 million years ago. However, fossil evidence indicates that our hominid ancestors were a distinct group as far back as 6–8 million years ago, suggesting that the earlier date for molecular divergence is likely the correct one.

There has been some controversy as to the order of divergence among the three hominid groups. The alternative possible phylogenies are shown in Figure 21.2A. One of the confusing features of this evolutionary radiation is that both chimps and gorillas are knuckle walkers. If knuckle walking evolved only once, then the gorilla and chimp species must have branched off from the pathway of hominid evolution before they evolved knuckle walking, by the principle of parsimony, described in Chapter 2. However, many morphological characters are shared by chimps and

> *There has been some controversy as to the order of divergence among the three hominid groups.*

humans, but not gorillas. These include skeletal structure as well as sexual maturation and anatomy.

Molecular data have provided the most consistent results. One body of data is shown in Figure 21.2B. Molecular studies concur in finding humans and chimps more closely related to each other than either of them are to gorillas. This suggests that knuckle walking either evolved separately in gorillas and chimps or that it was lost in the ancestral hominids, having evolved in the ancestors of all three groups. In any case, the molecular results support the following evolutionary scenario. Great apes diverged from the other primates around 20 million years ago. Orangutans split from the other great apes about 15 million years ago. Gorillas split from the ancestors of hominids and chimps about 10 million years ago. Chimps and hominids diverged from each other about 8 million years ago.

While there is still some controversy, the consensus among evolutionary biologists is that human ancestors diverged most recently from a common ancestor shared with chimpanzee species, though we are quite closely related to the other great apes. The greater controversies involve human evolution itself, to which we now turn. ❖❖

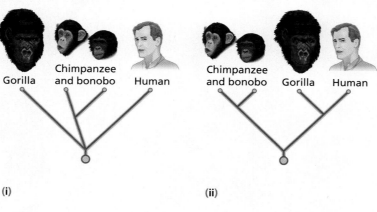

(i)   (ii)

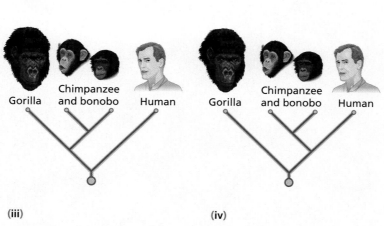

(iii)   (iv)

FIGURE 21.2A Four Possible Human-Ape Phylogenies

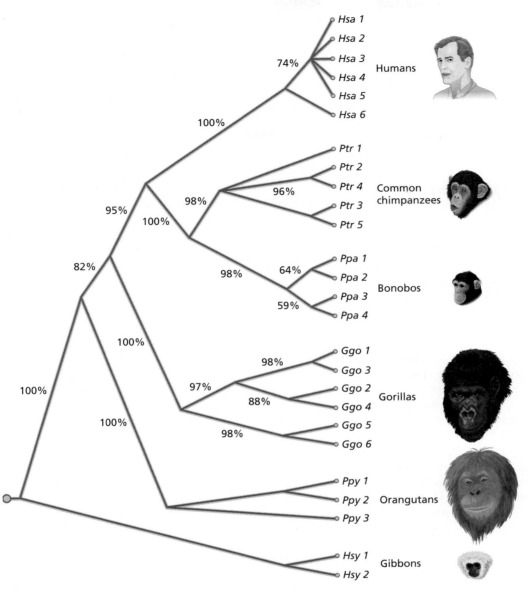

FIGURE 21.2B **Human-Ape Phylogeny Based on Mitochondrial Cytochrome Oxidase II Sequence** The three-character labels identify different alleles. The percentages give the statistical certainty of each branch of the evolutionary tree.

There were at least two major upright hominid lineages, which may have included multiple species each

It is difficult to learn hominid phylogeny, because it seems to change every other week with the announcement of a new hominid species or genus, based on a fossil found in or near Ethiopia. But this difficulty is more apparent than real. Hominid paleontologists as a group are almost certainly discovering less than they usually claim. Three major themes can be reliably derived from hominid paleontology.

1. The recent evolution of the genus *Homo* is being documented in greater detail, including important contributions from molecular biology. Some of this work is featured in this chapter. It reveals the phylogeny of human evolution in considerable, satisfying detail.

2. The hominid fossil record is being pushed farther and farther back in time, approaching the point where humans share a common ancestor with the chimpanzees. We have not yet reached that point in time, but we are getting closer; recent fossil finds are dated at more than 4 million years ago. Once we have 7- or 8-million-year-old fossils, they should be intermediate between chimpanzee and upright hominid species, roughly speaking. This creeping-backward pattern is shown in Figure 21.3A.

3. For several million years, multiple upright hominid species coexisted in the continent of Africa, possibly with more species in Europe and Asia. This finding came as some surprise in hominid paleontology, because many paleontologists had assumed that human evolution followed a "ladder" pattern, with larger-brained species superseding smaller-brained species, only one species being dominant at any one time. We now know without any doubt that the ladder model for hominid evolution is not correct.

The truth is that after the evolution of efficient bipedal locomotion, but some time before the evolution of modern humans, there were usually at least two distinct groups of upright hominids. This split began early in hominid evolution. How long it continued after that point is unclear. In any case, beginning about 3 million years ago, hominids were split into the **gracile** and **robust** lineages. (These lineages are sometimes put together in the ***Australopithecus*** genus, but sometimes they are placed in separate genera.)

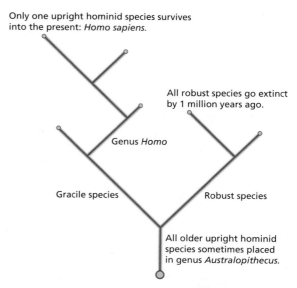

Common ancestor to all upright hominid species, 6–8 million years ago.

(i) A crude phylogeny

These three species are thought to have coexisted 1.9 million years ago. From left to right are *Homo rudolfensis*, *Homo ergaster*, and *Paranthropus boisei*.

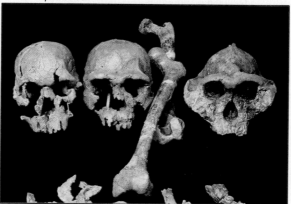

Three hominid species from East Turkana, Kenya

(ii) Coexistence of multiple upright hominid species

FIGURE 21.3B Human Evolution Simplified

FIGURE 21.3A Back to the Missing Link? The rough outlines of human evolution relative to that of the two chimpanzee species.

The gracile lineage appears to have contained species with relatively omnivorous diets, with smaller teeth and smaller jaws. The robust lineage had species with large molars and large jaws; they consumed a great deal of plant matter. Some paleontologists identify multiple coexisting gracile species, and other paleontologists find multiple robust species. There is little doubt that both types of lineage coexisted for a million years or more. But after that, the robust lineage(s) died out. The genus *Homo* is generally thought to have been derived from the gracile lineage,—or a gracile lineage, if there was more than one. This general scenario is depicted in Figure 21.3B. Many details of our ancestry have generated controversy, but the existence of separate gracile and robust lineages is not disputed. ❖

21.4 Human evolution featured expansion of the braincase, reduction in the jaws, and changes to the rest of the skeleton

How have upright hominids changed over the past several million years? The highlights of hominid morphological evolution are easy to understand. Three main types of morphological change took place, affecting the brain, feeding, and locomotion.

The obvious change during the last few million years of human evolution is a dramatic expansion in the **braincase**. Several million years ago, the braincase volume of upright hominids was around 400–500 cubic centimeters (cc). In modern humans, braincase volume ranges from 1000 cc to 2000 cc. Various hominid braincases are shown in Figure 21.4A.

At first sight, this change seems like a spectacular evolutionary increase in brain size. But the estimated body sizes of early hominids are much smaller than those of modern humans. Early hominid females, like the famous Lucy from several million years ago—a fairly complete *Australopithecus afarensis* skeleton—might grow to 40–48 inches tall. Brain size generally scales with body size in the mammals. The huge brains of whales do not necessarily indicate greater intelligence, because larger brains are required to animate larger bodies.

This argument almost implies that humans are not intelligent at all. But there are two additional factors to bear in mind. The first is that all the australopithecine species, whatever their size or diet, had brain sizes in the same range as that of gorillas and chimps. A reasonable interpretation is that the australopithecines were about as smart as these ape species. The second point is that over the last few million years, the size of the braincases of our ancestors increased faster than body size. This increase is shown in Figure 21.4B. If hominid brain size increased only because hominid body size increased, then the *Homo* data should have followed the same pattern as the data for other hominids. But *Homo* data are off this scale. Our ancestors' brains increased in size substantially more than body size would have required. We did get smarter.

The second major change that took place in upright hominids during the last 4 million years is a considerable reduction in the size of the upper and lower jaws. Basically, we

evolved flatter faces that look less apelike. This evolutionary change may have occurred because of dietary changes, or it may have occurred because of morphological constraints imposed by the expanding braincase. Conceivably, sexual selection might have been involved. We don't know.

Most hominid paleontology focuses on the skull. But in the background of hominid evolution is considerable change in the skeleton below the skull. Present thinking is that all known upright hominids were efficient bipeds and probably held simple implements in their hands. But that does not mean that all hominids were equally adept at walking or using tools. Between 4 and 8 million years ago, a period mostly unknown due to lack of fossils, the earliest upright hominids began to evolve bipedal locomotion and manual dexterity. The last 4 million years of hominid evolution may have seen further improvements in these central hominid adaptations. Upright walking, not a large brain, is the quintessential adaptation of our ancestors. ❖❖

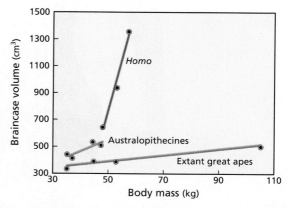

FIGURE 21.4B Brain Size versus Body Size in Hominids Note that the slope becomes much steeper in the genus *Homo*.

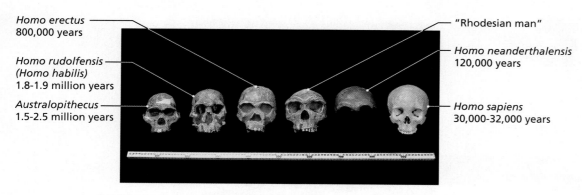

Homo erectus
800,000 years

Homo rudolfensis
(*Homo habilis*)
1.8–1.9 million years

Australopithecus
1.5–2.5 million years

"Rhodesian man"

Homo neanderthalensis
120,000 years

Homo sapiens
30,000–32,000 years

FIGURE 21.4A Six Hominid Braincase Reconstructions, Showing Their Age

Molecular genetic tools are crucial to unraveling the patterns of human evolution

The fossil record has helped us to understand some aspects of human evolution. There are, however, limits to what we can infer from the fossil record. Fossils, after all, are the phenotypic expression of only part of the genome. Genes shape many characteristics that are unseen or not preserved in fossils. In addition, the fossil record may reflect the influence of the environment in ways that have sometimes been hard to determine. For example, the size and shape of some bones can be affected by the environment. These facts have had major consequences for the study of human evolution.

For instance, when the first **Neanderthal** fossil was discovered in Europe in 1857, not all scientists were willing to acknowledge that these bones were from a hominid quite different from modern humans. Figure 21.5A shows the contrast between a Neanderthal skull and that of a member of our species. The bones were inspected by Professor **August Mayer**. He was convinced that the remains came from a Cossack soldier who died after deserting his troops. The curvatures of the leg bones and pelvis were attributed to riding horses. According to Mayer, these minor deformities may have been accentuated by a bad case of rickets, a vitamin D deficiency that can affect bone growth. Professor Mayer also noticed a badly healed fracture of the left arm. This injury, he implied, caused constant pain and the resultant frowning ultimately created the skull's thick browridge.

Nevertheless, scientists eventually became convinced that Neanderthals were ancient hominids. There were too many anatomical differences between Neanderthals and all modern

Genes shape many characteristics that are unseen or not preserved in fossils.

humans (we will give one example shortly). Eventually, the definitive data supporting this view came from the extraction of small samples of DNA from the Neanderthal bones. For the first time, DNA from an extinct hominid was compared to DNA from contemporary human populations. We will review those results.

The recognition that Neanderthals were ancient hominids did not end the controversy surrounding them. It was still not clear if modern humans were direct descendants of Neanderthals, or whether Neanderthals were contemporaries of modern man and simply went extinct.

Neanderthals notwithstanding, it is usually not possible to obtain genetic samples from our ancestors to chart the pattern of evolution. However, several clever studies with modern human populations have been useful for making inferences about this problem, which we will also review.

One of the more controversial topics in human genetics has been the study of race. Well before we knew the principles of genetics, humankind was acutely aware of differences in the physical appearances of members of different populations. These physical differences were used by racist groups to formulate systems of classification for humans. These classifications were used in turn to persecute members of other so-called races. This history has made many uneasy with the concept of race. There is no doubt that some human populations have become differentiated for genetic characteristics that affect physical features as well as biochemical and physiological features. These genetic differences arose because of isolation and random genetic drift, sometimes even natural selection (e.g., sickle-cell anemia).

But it is more important to understand that there is extensive overlap in the genetic constitution of human populations. Even at loci that show differentiation between populations, we often see the same alleles in different populations—though their frequencies differ. As we shall see shortly, interesting patterns also come out of human genetic variation and can help us understand important aspects of genetic and cultural evolution. ❖

FIGURE 21.5A The first Cro-Magnon skull (right) found in the Dordogne region of western France in 1868, next to a Neanderthal skull (left).

There is little doubt that most of us could pick out a Neanderthal male in a crowd. He would have a long trunk and powerful arms and shoulders. His head would be large relative to the rest of his body. He would have big eye sockets and a massive brow. Fossils of Neanderthals have been found in a variety of locations throughout Europe and the Middle East. Some of these fossils have been estimated to be only 34,000 years old, although Neanderthals were probably in Europe for a total of 100,000 years. One theory is that modern humans are direct descendants of Neanderthals, or at least exchanged genes freely with their populations. Recent evidence has cast doubt on this idea and supports the idea that Neanderthals were a separate subspecies, or even a separate species, that was eventually displaced by modern humans.

The differences between modern humans and Neanderthals are pervasive. Consider the bones of the semicircular **ear canals**. In Figure 21.6A, we see that a structure called the posterior canal (red) is situated well below a second structure called the lateral canal (blue) relative to its position in modern humans and chimpanzees. The Neanderthal ear structure is quite different from that of hominids like *Australopithecus* and *Homo erectus*. In France, samples of Neanderthals showing this type of ear canal have been found that are only 34,000 years old. It is unlikely that Neanderthals were exchanging genes with our immediate ancestors at any significant rate yet could still be so different from modern humans 34,000 years ago.

These conclusions have been strengthened by the recent analysis of **mitochondrial DNA** from Neanderthal fossils. The ability to get DNA from these samples is made possible by the relative stability of DNA as a molecule, the fact that single cells often contain many mitochondria and thus many copies of the same genome, and finally the polymerase chain reaction (PCR) that permits scientists to amplify DNA from very small samples. Recently (Figure 21.6B), mitochondrial samples from one Neanderthal and 663 humans have been compared. These data support the idea that Neanderthals separated from the human lineage well before modern populations of humans differentiated. Although these results are based on only a single gene region, they are consistent with the idea that Neanderthals were a distinct evolutionary unit, possibly even a species.

So what happened to Neanderthals? They probably coexisted in Europe with modern humans for 100,000 years. The numbers of both Neanderthals and modern humans could not have been very great during this period. However, during the period of coexistence, there were periods of climate change when glaciers spread south and almost certainly forced any hominids to

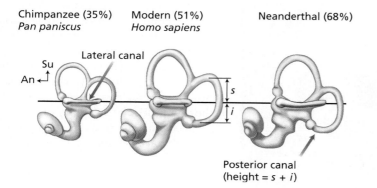

Chimpanzee (35%)
Pan paniscus

Modern (51%)
Homo sapiens

Neanderthal (68%)

Lateral canal

Su
An

s
i

Posterior canal
(height = $s + i$)

FIGURE 21.6A Neanderthal ears differed morphologically from the ears of modern humans. The Neanderthal posterior ear canal is lower than that of chimpanzees or humans.

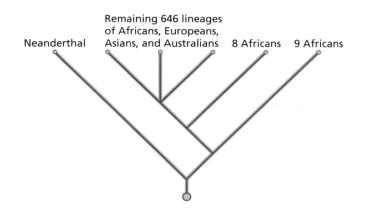

FIGURE 21.6B Genetic evidence shows that Neanderthals were different from modern humans. Mitochondrial DNA sequences from Neanderthals, humans, and chimpanzees show that Neanderthals are on a different evolutionary branch from all human populations. Human and Neanderthal mitochondrial genes shared a common ancestor between 300,000 and 800,000 years ago.

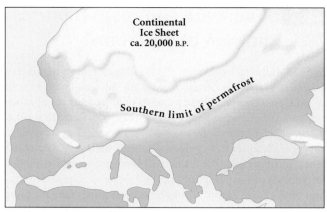

FIGURE 21.6C The spread of ice and permafrost during the last glacier advance in Europe, 20,000 years ago. Such dramatic climate changes may have earlier restricted the territories available to humans and Neanderthals.

move to the most southern regions of Europe (Figure 21.6C). These climatic changes may have made it more likely for modern human populations and Neanderthals to interact, either competitively for resources or as combatants. These interactions could then have led to the decline and ultimate extinction of the Neanderthals. ❖

There are two main theories concerning the ancestry of modern human populations

In 1868, railroad workers in the Dordogne region of western France uncovered a burial ground for ancient humans now called **Cro-Magnon**. These humans had the features of modern humans but were about 30,000 years old. Since that time, many other hominid fossils have been found and modern dating techniques have helped establish their age. As a result it has become clear that numerous hominid species coexisted for periods of time (Figure 21.7A). Of course, our interpretation of the fossil record would have been much easier if there was no overlap in the appearance of these fossils and there was a gradual progression from apelike hominids to modern humans. But life is not always simple. To explain these features of the fossil record, two major hypotheses about human evolution have been proposed.

One theory, called the **multiregional theory**, suggests that hominid evolution began in Africa; but perhaps 1 million years ago hominids radiated out of Africa and into other regions of the world (Figure 21.7B). Evolution continued giving rise to changing features of hominid fossils found in Europe, for instance, but there was appreciable genetic exchange among hominid populations then extant. This genetic exchange ensured that the features of modern humans would be fairly uniform around the world.

A second theory, called **out-of-Africa**, suggests there were several hominid lineages that had more or less independent histories from the one that gave rise to modern humans. The ancestors of modern humans left Africa more recently, perhaps as recently as 100,000 years ago. At that time they encountered other hominids, like Neanderthals, and these other lineages went extinct (see Figure 21.7B).

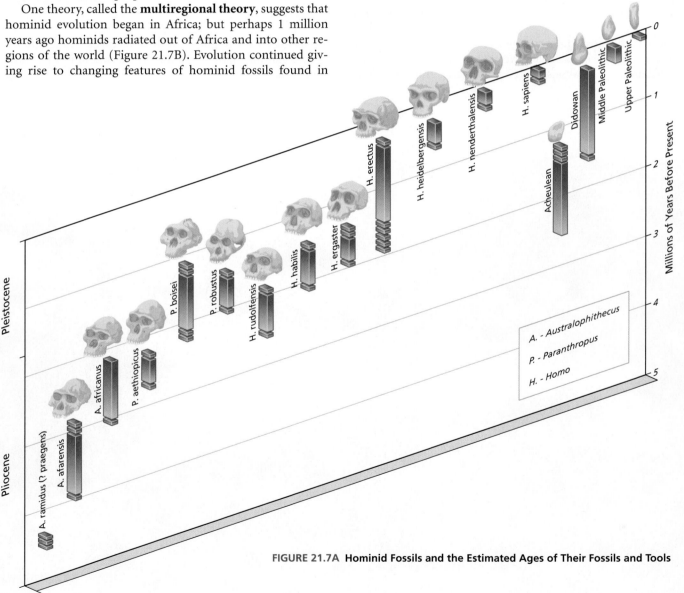

FIGURE 21.7A Hominid Fossils and the Estimated Ages of Their Fossils and Tools

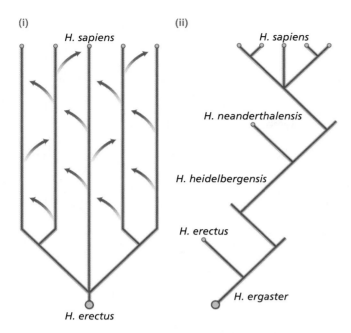

(i)

H. sapiens

H. erectus

(ii)

H. sapiens

H. neanderthalensis

H. heidelbergensis

H. erectus

H. ergaster

FIGURE 21.7B The Multiregional Theory of Human Evolution (i) and the Out-of-Africa Theory (ii)

These hypotheses lead to some general predictions about current levels of genetic variation in human populations. If we can estimate the divergence time between African and non-African populations, we would expect that number to be about 1 million years for the multiregional model, but 200,000 years or less for the out-of-Africa model. Genetic diversity should be greatest in African populations under the out-of-Africa model because that theory assume that all other populations were derived from them. Under the multiregional model, the long period of evolution with gene exchange would be more likely to give equal levels of genetic diversity between regions. A corollary of this idea is that non-African populations should have subsets of the alleles present in Africa under the out-of-Africa model. Again this follows from the derived nature of the other populations.

The strength of these predictions is not tremendous, because the two theories of human evolution depend on events that we don't know with any precision. For instance, under the multiregional theory, because we are certain that hominid evolution began in Africa, it might still be the case that the African population of *H. sapiens* will have greater genetic diversity. In any event, genetic data can be examined to test these ideas, as we will see in the next module. ❖

21.8 The latest data strongly support recent African proliferation of modern humans

A recent study of genetic variation in modern human populations looked at genetic variation in a large number of African populations and populations in Asia, Europe, South America, and Australia. They studied a gene on chromosome 12 that falls into a class of genes known as **short tandem repeats (STR)**. The genes have small sequences of DNA that are repeated a variable number of times, giving rise to alleles that are different sizes. The different alleles found are distinguished with integers indicating the number of repeated units (Figure 21.8A). There is also a second gene within 9,800 base pairs of the **STR** gene that has two alleles. The more common allele at the second locus is called $Alu(+)$. The less common allele is called $Alu(-)$ and represents a 256-base-pair deletion of the transposable element Alu. The $Alu(-)$ allele is not found in the great apes, and is presumed to be a rare event that probably occurred 4–6 million years ago, after our lineage diverged from the apes.

If we focus on just the variation at the STR locus, we find two key observations that support the out-of-Africa hypothesis. The first is that the levels of genetic variation are greatest in the sub-Saharan African populations. The second observation is that the types of alleles found in the non-African populations tend to be small subsets of the alleles that are found in sub-Saharan African populations. These observations are consistent with the idea that the human populations that colonized the non-African continents were small founding groups that ultimately came from the African continent, and therefore contained only a portion of the total genetic variability present in Africa.

FIGURE 21.8A Genetic Variation in Modern Human Populations Sarah Tishkoff and colleagues studied genetic variation for short tandem repeat (STR) alleles (numbered 4–15) found at a particular locus on human chromosome 12. Part (i) shows frequencies of alleles that are found on chromosomes with the $Alu(+)$ allele, a very close linked locus. Part (ii) shows similar information for alleles on chromosomes with the alternative Alu allele, $Alu(-)$. Two features of these data support the out-of-Africa theory: (1) There is more variation in the sub-Saharan African populations than in the others. (2) The variation found in the other populations appears to be subsets of genetic variation found in the sub-Sarahan African populations.

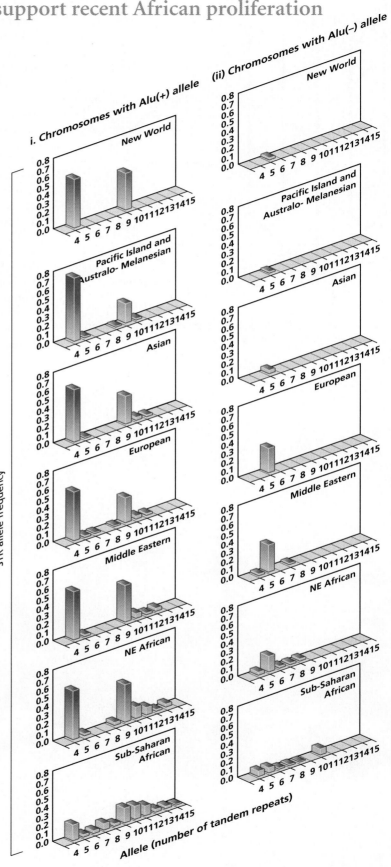

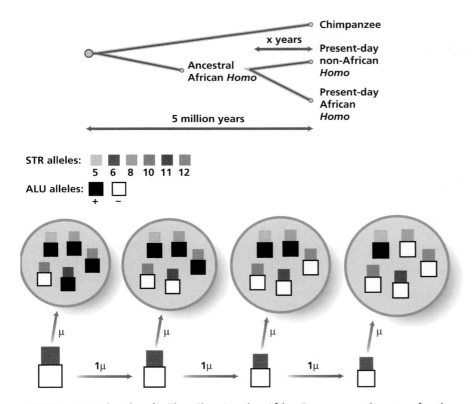

STR alleles: 5 6 8 10 11 12

ALU alleles: + −

FIGURE 21.8B Estimating the Time Since Leaving Africa Every generation some fraction, μ, of the gametes with the 6/− genotype (the gametes with the red box and the hollow box together) will mutate to one of the other STR alleles. This rate can be estimated from the accumulation of other gamete types in the present-day African populations under the assumption of 5 million years elapsing since the origin of the first 6/− gamete. Once the mutation rate has been estimated it can be used to estimate the time since non-African populations left Africa, x. This time estimate will depend on the frequency of non-6 STR alleles found in combination with $Alu(-)$.

An approximate date for the departure of the hominid populations that left Africa has been estimated (Figure 21.8B). The first $Alu(-)$ mutant is assumed to have occurred on a chromosome with the STR at allele 6. After the $Alu(-)$ mutation event, perhaps 5 million years ago, unequal crossing over generated new STR alleles, like 5/− and 11/− (see Chapter 5). The $Alu(-)$ allele is unlikely to mutate back to the $Alu(+)$ allele, and recombination between the two loci will contribute little to the new gametes observed. The frequency of $Alu(-)$ alleles that are found with STR alleles other than 6 in the sub-Saharan populations can be used to estimate the STR locus mutation rate, assuming the first $Alu(-)$ allele appeared 5 million years ago (Figure 21.8B).

We next assume that the migrants from Africa to other regions of the world always brought the $Alu(-)$ allele on a chromosome with the 6 allele at the STR locus. Then the frequency of $Alu(-)$ alleles associated with non-6 alleles in these non-African regions allows us to estimate how long the non-African populations have been isolated from Africa. This procedure produces an estimate of time since leaving Africa of 102,000 years. These estimates are subject to substantial statistical error, and the true value may be as high as 700,000 years ago. Even so, this higher time estimate is also consistent with a recent movement of *Homo* populations from Africa to the other continents. ❖

21.9 Modern human populations seem to be a patchwork of local differentiation, with little racial differentiation

It should come as no surprise that the rules of evolution and population genetics apply to humans in the same way they do to other organisms. Thus drift, mutation, natural selection, and inbreeding have historically played an important role in determining human population genetic structure, just as they do in other species. Many of the genetic differences among people are completely invisible to us. Thus, with respect to the *Alu* polymorphisms discussed in the previous pages, most of us have no idea whether we or our neighbors are *Alu*(+) or *Alu*(−) for that STR region. However, we probably can describe the phenotypic effects of the genes that affect our hair and skin color.

Genetic differences between human populations have arisen as a consequence of many factors, but isolation and lack of gene exchange with other populations are probably the most important factors. Using genetic data from many loci , we can develop genetic trees for the human population (Figure 21.9A). These tend to group populations that are either geographically close or have historically had gene flow between them.

The boundaries between human populations are not absolute. After all, we are all members of the same species, and the genetic data used to develop the trees in Figure 21.9A show there are always loci where different populations share some alleles. Thus in human population genetics there is little interest in developing categories or races into which people can be placed and more interest in studying patterns of genetic variation. For instance, there is a close association between the genetic trees of populations and the relationships between language groups (Figure 21.9A). The most reasonable explanation for this association is that when human populations become isolated, this isolation can lead to both genetic differentiation and language evolution.

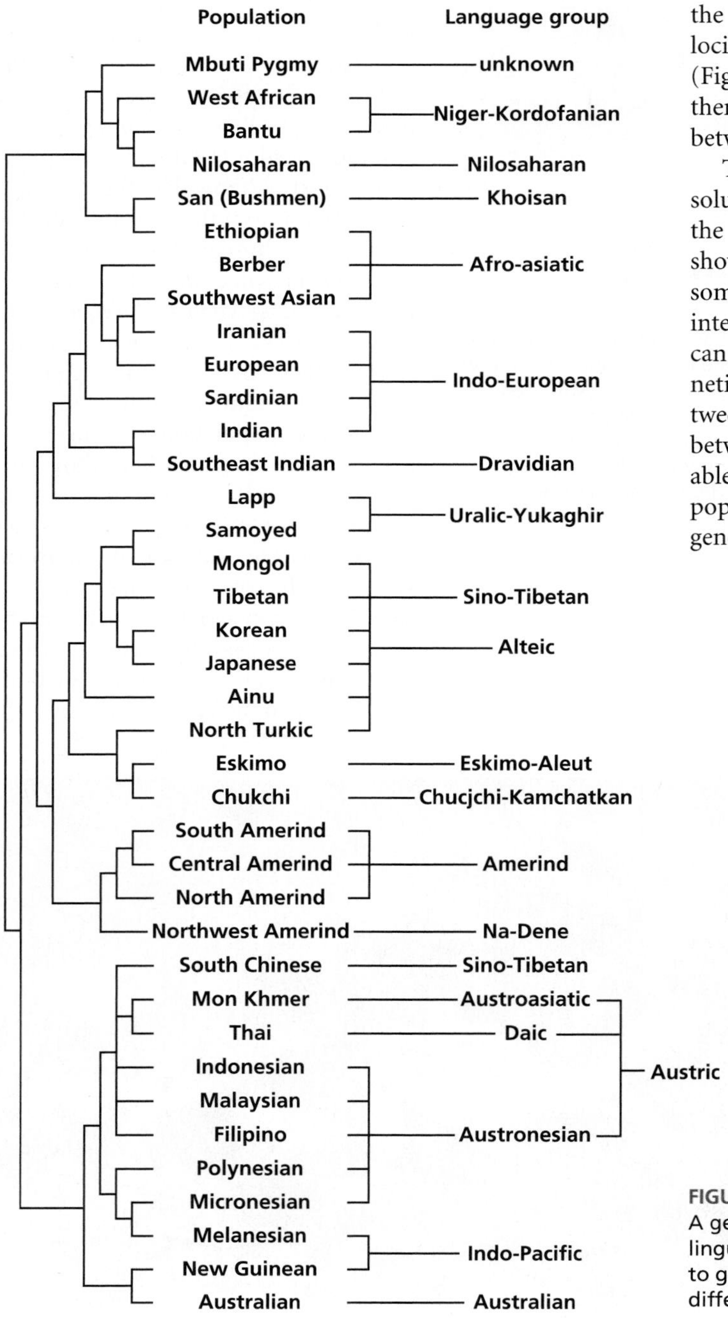

FIGURE 21.9A The Concordance between Genes and Language
A genetic tree based on 120 loci corresponds fairly well to linguistic diversity. This suggests that the same factors that lead to genetic differentiation can also lead to linguistic differentiation.

Occasionally these patterns break down. For instance, when populations or nations conquer territory, the resident languages may be virtually replaced by the language of the conquering population, except for isolated redoubts. This explains relic languages in marginal areas, such as the Basque language in Europe or the Dravidian languages in India. In other instances, the genes of a population may be replaced by exchange with a group that does not reflect the origins of the language of interest. This type of gene replacement can happen quite quickly. For instance, U.S. blacks in northern states obtained 25–30 percent of their genes from European sources. This gene flow has occurred over just 300 years since Africans were brought to the United States as slaves.

We can also find genetic patterns over geographic regions that parallel the movement of people out of Africa (Figure 21.9B). In another study of human population genetics, 120 polymorphic genes were used to characterize the genetic variation in the worldwide human population. As Figure 21.9B shows, populations in Africa are similar to each other, as we would expect. The non-African populations most similar to those from Africa are the populations just to the north and east of Africa—the European and Middle Eastern. This pattern continues across Asia and into North America. The populations least similar to those of Africa are found in Australian aboriginal populations. The out-of-Africa model is supported well by these results. ❖

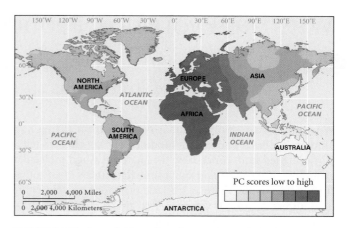

FIGURE 21.9B Genetic Map Showing Regions of Similarity For this map, a genetic distance measure was used to distill genetic information from 120 genes into a single number (the PC score) that represents the genetic differentiation of indigenous populations. The samples from the indigenous populations of Africa all have a similar value of this number. As we move eastward into Asia, the value of this number changes gradually, indicating greater homogeneity. These values are consistent with movement of human populations out of Africa.

WHY DID EVOLUTION PRODUCE HUMANS?

21.10 The puzzle of our evolution has generated both intellectual aversion and gratuitous speculation

Why did humans evolve? This is different from asking *how* humans evolved. The earlier modules of this chapter tell a fairly detailed story concerning the anatomical and molecular evolution of humans and our immediate ancestors. As that story makes clear, there is no lack of evidence concerning how humans evolved. Compared to the evolution of most forms of life, the human evolutionary record, both fossil and molecular, is quite good.

FIGURE 21.10A Alfred Russel Wallace (1823–1913)

But the "why question" remains a source of trouble. Even **Alfred Russell Wallace** (Figure 21.10A), the codiscoverer of evolution by natural selection, felt that human evolution required the intervention of spiritual forces, above and beyond mere natural selection.

Problems like Wallace's spiritualist views have tended to make evolutionary biologists somewhat averse to treating the question of why evolution produced humans. As a result, anthropologists, zoologists, and popular writers have tended to fill the void left by this aversion, and their speculations have been the most frequent statements addressing selection on humans.

One popular explanation of human evolution is based on the assumption that our evolution is a kind of straight line from chimpanzees to the **hunter-gatherer** societies of modern Africa to the suburbs of Chicago . Sometimes the favored explanation is that we became smart because of selection to be hunters. The problem with this sort of explanation is that a lot of other species are carnivores without being particularly intelligent. A variant of this hypothesis is that we had to become smart in order to hunt using tools. A problem with this theory is that ants use fairly complex technologies to find, devour, store, or otherwise exploit other species. Yet ants are not intelligent. Still other writers think that bad weather was important in making humans evolve, as if there aren't millions of other species on the planet that suffer through bad weather.

The problem with these stories is that they have a storybook quality, like Kipling's ***Just-So Stories,*** in which plausible explanations of peculiar things are given to children who have no way to check these explanations, test them, or even doubt them. Human evolution took place over the last few million years, and we were not there to observe it. This lets uninhibited writers, with few fears of contradiction, create stories about the African **savanna** 2 million years ago. But these stories have little scientific value because they are based on specific evolutionary scenarios that we cannot test directly. This problem arises because we can't test *any* very specific scenario for the evolution of any species millions of years ago. We don't have time machines, contrary to the stories of **H. G. Wells**.

Our inability to describe exactly why we evolved over the last 2 million years does not preclude useful scientific analysis of this question. Physicists commonly treat such difficult problems as the origin and early development of the universe, even though they are unlikely ever to go back in time to find out exactly what happened. Instead, they develop more general theories for the history of the universe, asking such basic questions as whether the universe has been expanding, whether there was a "big bang" at its origin, and so on. By reducing the specificity of their theories, physicists have been able to propose, examine, and test theories for the development of the universe.

In the same way, although evolutionary biologists will never know exactly what our ancestors were doing 2 million years ago, we can address very broad questions about human evolution. Questions of this kind approach a level at which human evolution can be discussed scientifically. That is, theories about human evolution that are general enough to be tested *without* a time machine are worth considering. The problem is, what type of theories are these? What do robust theories of human evolution look like?

At least two general questions about human evolution can be formulated to the point where we have some possibility of considering them scientifically. The first is whether or not the evolution of distinctive human attributes, such as a large brain, was driven by selection, or instead occurred primarily because of genetic drift. This is a central question for any discussion of human evolution. If selection did not drive our evolution, then we do not need to invent selection scenarios to solve the problem. We will address this question next.

The second general question about human evolution's distinctive features is as follows: If it *was* driven by selection, what kind of selection forced our evolution? And the *kind of selection* cannot be as specific as bad weather, a particular predator, or a new type of plant food. Kinds of selection worth comparing must instead be broad in nature, with significant consequences for data that we might some day obtain. These hypotheses are available, and we have at least some empirical information with which to test them, as we will see.   ❖

Evolutionary change does not require natural selection. As we saw in Chapter 3, evolution can change gene frequencies by processes like genetic drift alone. Therefore, the many evolutionary changes in humans relative to the other great apes do not prove, by themselves, that selection drove human evolution.

But other aspects of human biology suggest that intense selection shaped the evolution of our ancestors. What must be understood is that when considerable fitness costs are associated with the evolution of a feature, then genetic drift is unlikely to be a valid explanation of that feature's evolution. Under these conditions, stabilizing selection will prevent directional change. Therefore, the role of selection in human evolution depends on whether there is evidence of purifying or stabilizing selection against change in the characters involved.

And there certainly is. One of the best-known examples of stabilizing selection is on the birth weight of the human newborn. (See Module 4.11.) Babies with very large or very small weights are selected against. One of the biggest determinants of weight in the human newborn is the size of its brain. And brain size is by far the biggest problem in human birth. Once the head is through the birth canal, the rest is comparatively easy. Both child and mother are subject to relatively high levels of mortality in human childbirth. And the newborn human is a highly dependent, immobile, vulnerable mammal. Furthermore, it remains that way for more than a year, during which it is not even able to walk. One interpretation of the human pattern of child development is that the human infant is really a fetus for its first

One of the best-known examples of stabilizing selection is on the birth weight of the human newborn.

12–18 months. The problem is probably the difficulty of giving birth to a 1-year-old human. Human pelvises are simply not big enough.

Overall, it seems as if the processes of human gestation, birth, and early infancy have been radically modified to accommodate the development of the large human brain. Even the female pelvis has been evolutionarily remodeled to accommodate the human infant, rendering it less efficient for running. The many costs associated with the expanded brain of the human infant are diagrammed in Figure 21.11A.

Yet other costs are associated with the human brain. At rest, the brain generates a large fraction of the metabolism of the human body. Programming the human brain, or "learning," has considerable costs in missed opportunities for feeding or reproduction, as every bored high school student knows. Yet there is little point in having a powerful brain if it is not educated. Complex human languages are in large measure tools for instruction of the young, and massive amounts of neural processing underlie language.

Note that there are alternative lifestyles among other species. Even fairly sophisticated animals, like the other great apes, get by on very simple utterances and social signals, and their knowledge of biology and engineering is rarely impressive. The bonobos (*Pan paniscus*), for example, make do with a simple lifestyle, centered around an omnivorous diet and a very active sex life. Our failure to pursue such a way of life must have compensating selective advantages. In other words, increased brain size must have conferred abilities that provided advantages that outweighed its costs. ❖

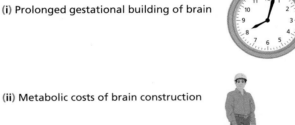

(i) Prolonged gestational building of brain

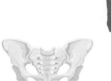

(ii) Metabolic costs of brain construction

(iii) Difficulty with birth

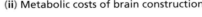

(iv) Cost of programming brain

(v) Prolonged period of dependency

FIGURE 21.11A The Costs of Being Smart

The hypothesis that we were selected only to use technology is undermined by the material simplicity of some cultures relative to their social complexity

So what are the advantages of a larger brain? The traditional proposal is the **tool-use theory** of human evolution—that natural selection favored human intelligence, and thus a large brain, because humans used that intelligence to develop useful technology, tool use first among these technologies. This is an old theory. One of its first proponents was **Friedrich Engels**, Karl Marx's collaborator, who wrote about it in the nineteenth century. This concept has been epitomized by the phrase, "man the tool-user." This phrase implies that we evolved to use tools, and that tool use is at the center of human nature.

In support of this tool-use theory is the vast technology associated with modern human civilizations. No other animal comes remotely close to this technological level. It takes significant intelligence to create, sustain, and use modern technologies. Yet this point of reference reveals a weakness in the argument. For almost all of human evolution, including the entire period during which the human brain evolved its present size—which was the last 2 million years, *excluding* the last 200,000 years—we had nothing like modern technology. We didn't even have extensive agriculture. The fact is that we evolved our intelligence before we developed modern technologies. Hominid life was quite simple two million years ago, when we started to evolve larger brains (Figure 21.12A).

A further complication is that some recent human societies have had minimal levels of technology. Many hunter-gatherer societies that have been studied in the last two centuries use extremely simple tools. Furthermore, historically recent human populations show wide differences in technology, yet there are no clear differences in intelligence. If technology were the driving force of selection that remodeled the human development pattern, the female pelvis, and human language,

FIGURE 21.12A Artist's conception of *Paranthropus robustus* in South Africa 1.7 million years ago.

(i) Very low levels of tool use in some humans

For example, Tasmanian aborigines used few tools, lacking even fire.

(ii) Homogeneity among human populations for average intelligence, despite wide variation in opportunities for tool use with environment

Barren, simple Arctic habitat

Complex rain forest environment

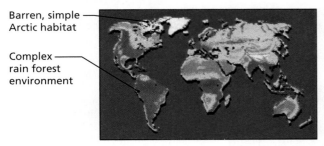

FIGURE 21.12B **Problems with the Tool-Use Theory of Human Evolution**

why wouldn't large differences in the use of technology over tens of thousands of years be associated with systematic, qualitative differences in intelligence? Figure 21.12B summarizes the main weaknesses in the tool-use theory.

Adding to these anomalies is the universality of complex human social behavior. Human societies generally invent elaborate social systems that give family and social status. Humans create complex rituals for coming of age, marriage, childbirth, and death. Humans spend vast amounts of time on gossip and rumor that is technologically unproductive. The individuals who spend their time making tools are usually a minority in human societies. Large numbers of people do not watch television shows about tractors. They watch television shows about love affairs, families, and heroes. Is this the behavioral pattern to be expected of a species that is supposed to have evolved to make and use tools?

Probably not. Instead, anthropologists have been interested in very different models for the evolutionary enlargement of the human brain—models based on social behavior, as we will see in the next module. ❖

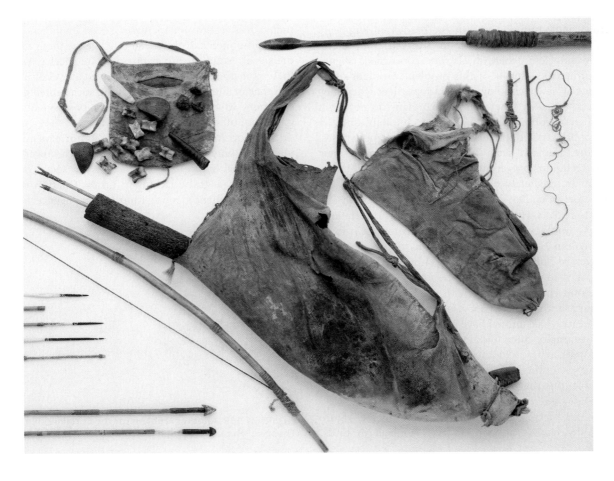

Why Did Evolution Produce Humans? **621**

21.13 The hypothesis that we were selected only for social calculation is undermined by our facility with complex material technologies

The main alternative theory for human evolution is the **social intelligence theory:** that we evolved "Machiavellian" intelligence in order to obtain relative social advantage. The idea is that being a little bit smarter gave some hominids an advantage in social competitions for mates or access to resources. This is a **mental arms race** because it is an individual's relative standing that matters. In some versions of this theory, it is supposed that such selection for social intelligence is virtually universal and relentless. Theories of this kind, at first sight, would seem to have all the features required to explain the evolution of human intelligence.

In conformity with this theory, humans spend a great deal of time and ingenuity outwitting each other, rather than doing real work. All human societies are notable for their maneuvering and back-stabbing. And humans are very interested in such behavior. Today, lawyers, politicians, and criminals are more popular subjects of fiction and drama, compared to scientists, engineers, and mathematicians. Social intelligence theories of human evolution have been popular among academics for the last 25 years. However, it might be noted that Charles Darwin proposed a version of this theory in his book, *The Descent of Man*, more than a century ago.

But this theory faces major problems too. The first problem is that, while humans are not the hardwired engineers that the technology selection model suggests, they aren't completely hopeless where tools are concerned. Figure 21.13A summarizes aspects of this problem. Humans of all cultures invent new technologies and adopt them when they have been invented in other cultures. Humans do not seem widely deficient in the ability to handle technology, unlike a species that has evolved intelligence only for social purposes. Furthermore, humans seem to delight in inventing new forms of technology, even technologies whose utilitarian value is not immediately obvious. Scientists and engineers seem to be on an endless quest to invent, from electron microscopes to rockets. Humans have had an enduring relationship with technology. Stone tools have been found in fossil strata more than 2 million years old. It is hard to believe that the evolution of human intelligence did not involve selection for the capacity to use tools efficiently.

The second major problem with the social intelligence theory is that this theory is too strong. Why shouldn't every social animal have higher intelligence ? Birds, for example, often live in large social groups called colonies. These colonies are characterized by squabbles over nesting space, food, and sex. Birds have social interactions that achieve considerable complexity, with behavior patterns that resemble such human practices as ownership, marriage, adultery, and so on (see Figure 21.13B). Why haven't large brain sizes evolved in birds,

(i) Humans have the capacity to develop elaborate technologies as the last 200 years have shown.

(ii) Once a technology is developed in one human culture, there appear to be few intellectual barriers to its adoption by other human cultures.

(iii) Humans even develop novel technologies of little immediate practical value, as illustrated by the electron microscope, shown at the right; we almost seem to have a technological drive.

FIGURE 21.13A Problems with the Social Intelligence Theory of Human Evolution

thanks to selection for increased social intelligence? Indeed, some proponents of the social intelligence theory apply it to the social evolution of various primate species. Yet none of these non-human primates has evolved a level of intelligence like that of humans. If selection for social intelligence is so common, why do humans uniquely possess complex language and other features of intelligence?

Selection for social intelligence cannot explain why humans, and only humans, have evolved such powerful intelligence. As with the tool-use theory, the social intelligence theory appears to be lacking as an explanation of the unprecedented phenomenon of human intelligence. ❖

There are many other highly social species, such as birds that live in colonies. Wouldn't they benefit from social intelligence too?

FIGURE 21.13B Another Problem for the Social Intelligence Theory of Human Evolution

Human evolution probably involved a combination of selection pressures favoring both technology and social behavior

Human intelligence is probably one of those characters that has been subjected to a combination of selection pressures. We exhibit high levels of both social and technological intelligence. It makes sense to suppose that the evolution of humans involved selection for both of these types of intelligence.

This idea still leaves the question of why humans should have been subjected to a uniquely powerful combination of these two selection mechanisms. One possible answer is that our ancestors, at some point in their evolution as upright tool-users, may have achieved hunting technologies that constituted deadly force for other hominids. Under these conditions, the normal type of strategy selection may have been superseded.

Consider the consequences for evolution of a stone axe blade tied to a wooden handle. Possession of such a weapon together with an understanding of how to use it would give its owner the status of a warrior. Invention of better weapons and weapon-using tactics would give a fitness advantage. But it would also destroy the evolutionary game rules that determine the best behavioral strategy in most animal species. For example, lions and eagles and spiders have evolved relatively simple behavior because their weapons are built in. In facing other lions, a lion is facing a predictable enemy. Standard reactions, tactics, and strategies are fine. But an armed hominid facing other armed hominids does not have a biologically standardized enemy. In this way, novel technology destroyed the social stability of hominid life (see Figures 21.14A and 21.14B).

Without evolutionary stability of social behavior, one possible characteristic for selection to favor is the ability to calculate immediately the best combination of social tactics and technology. In other words, hominid technology may have created an environment that selected for both social and technological intelligence.

An interesting feature of this type of selection is that brain mechanisms that foster both social and technological intelligence can give rise to a "free" arms race. If having a more generalized type of intelligence is roughly "paid for" at the level of food gathering, or in some other ecological context, then the use of that same intelligence in social competition is free. But since social competition is a selective process of relative advantage, natural selection will push for higher and higher levels of intelligence generalized enough to be of use in both social and technological arenas.

This unusual situation has no doubt arisen only rarely in the history of life on Earth, if indeed it does apply to human evolution. But such uniqueness is required to explain the historically unprecedented evolution of the human mind. That is, a theory of human evolution cannot be so general that it explains the evolution of higher intelligence in numerous species, because that never happened. An additional attractive feature of this theory is that it is based on a biological adaptation, learned and proficient tool-use, that is uniquely associated with hominids in the fossil record. That is, there are good reasons for supposing that this type of combined selection mechanism would have arisen only among upright hominids of the last few million years. ❖

Suppose more intelligence leads to the ability to make stone axes.

Axes are used to kill and butcher animals for food, giving environmental benefits.

Axes are used to wound or kill human enemies as part of the social contest for reproductive advantages.

FIGURE 21.14A Mental arms races are expensive unless there are spin-offs that pay for the costs of the arms race.

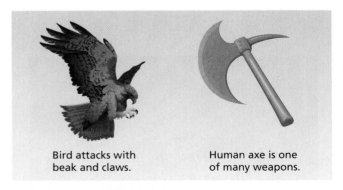

Bird attacks with beak and claws.

Human axe is one of many weapons.

FIGURE 21.14B How Did Free Mental Arms Races Start? Consider the evolutionary game situation of an animal that does not use tools. The parameters of its evolutionary games are fixed by its body. But humans that use tools are not limited to the weaponry built into their bodies; their intelligence will be a major limiting factor in their success in conflicts because it limits their tools and their tool use.

HUMAN BEHAVIOR FROM AN EVOLUTIONARY PERSPECTIVE

There is a long tradition of Darwinian analysis of human behavior, despite controversy about it

The importance of the Darwinian theory of evolution for our understanding of human behavior has been a perennial concern. However, it has received particularly intense attention since 1975. That year, the entomologist **E. O. Wilson** published a book, *Sociobiology,* that attempted to unite behavioral research on all animals, from insects to humans (E. O. Wilson is shown in Figure 21.15A). He explicitly connected selection on social behavior in insects with parallel selection on human behavior. A great many biologists and social scientists found Wilson's ideas interesting. We have summarized this type of thinking in Chapter 20, albeit without the application to humans.

But another large group of biologists, social scientists, and philosophers found Wilson's ideas repugnant. Not only did they object to his use of Darwinian thinking in reasoning about human social behavior, but they also found his scientific analyses of other animals inadequate. They were especially concerned about what they saw as a "reductionist" attempt to explain all behavior in terms of genes and natural selection. They regarded this view of behavior as grossly distorted. This difference of opinion generated one of the more rancorous Darwinian controversies.

One aftermath of this controversy was the avoidance of the term **sociobiology**. It was nonetheless reborn as a new academic movement, **evolutionary psychology**, which hardly differs from Wilson's original proposal. Evolutionary psychology has been very influential in the general culture, from articles in popular magazines to evolutionary "self-help" books. It is important to understand how much scientific support this discipline has received. This is our primary concern here.

As much as evolutionary psychology has attracted attention and criticism, there is nothing new about it. Its first practitioner was Charles Darwin, who made remarks about the evolution of human behavior in two books, *The Descent of Man* and *The Expression of the Emotions in Man and Animals* (see Figure 21.15B). Darwin's main interest was in showing the plausibility of human evolution from other mammals, especially from a common ancestor shared with the apes of Africa. Unlike Wallace, his co-inventor of the principle of natural selection, Darwin was unwilling to make a special case of human evolution, human behavior included.

Ever since Darwin, there have been many speculations about the causes of human evolution and the results of that evolution. Ethologist **Desmond Morris** wrote *The Naked Ape,* presenting humans as animals, which caused a sensation in the late 1960s (see Figures 21.15C and 231.15D). Even a playwright, **Robert Ardrey**, weighed in with speculative theories about the causes of human evolution and our resulting behavior. Like some others, Ardrey blamed the evolution of hunting behavior in the genus *Homo* for our continued aggression in the twentieth century.

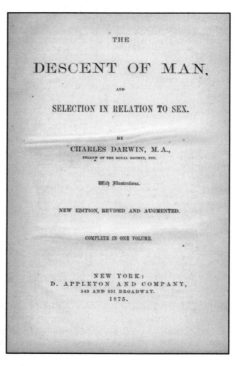

FIGURE 21.15B Title Page of *The Descent of Man* by Charles Darwin

FIGURE 21.15A E. O. Wilson

FIGURE 21.15C Desmond Morris

Before the sociobiology controversy, the controversies about the evolution of human behavior were not highly focused. Although quite a few scientists, especially evolutionary biologists, deplored the rampant speculation of some popular works, these works were generally considered harmless.

Since the sociobiology controversy, the discussion of the evolution of human behavior has become more intense. Ironically, even though Engels speculated about human evolution and Marx was a fan of Darwin, the left wing has disparaged attempts to analyze the evolution of human behavior. They have not presented much in the way of an alternative Darwinian analysis. Their main interest seems to be to stop the use of Darwinian foundations as an alternative to Marxist analysis of human behavior. As a final twist, the religious right wing doesn't like the Darwinian analysis of human evolution any more than the left wing does.

There is nothing new about Darwinian ideas coming under attack for ideological or religious reasons. It should be noted that ideologues have often used Darwinian thinking to support their causes. Darwinism was used an ideological support for nineteenth-century opposition to welfare. Even worse, in the twentieth century it was used a buttress for fascism and the extermination of minorities, homosexuals, and defective babies. There is nothing inherently benign about the extrapolation of science. It can be actively evil in its consequences.

Where does that leave the modern evaluation of evolutionary psychology? Should its Darwinian credentials be accepted, and political concerns dismissed? If present-day left-wing and right-wing groups are trying to censor discussions of the Darwinian basis of human behavior, should they be allowed to do so?

A few principles might provide useful guidelines.

1. We cannot assume that any particular group is infallible. Evolutionary psychologists who ally themselves with Charles Darwin are not necessarily correct. Even though both the left wing and right wing think they have ideological infallibility, they have no greater claim to the truth. Any of these groups could be right, and any of them could be wrong.

2. Because no one is infallible, the discussion of human evolution should be based on logic and evidence. Ideas that evolutionary biologists have assumed for decades might be wrong. Political consequences that might have been important in 1930 might no longer matter. But other societal consequences, never before significant, might now be devastating in their impact.

3. Because any discussion of human evolution is likely to be provisional at best, it seems useful to approach the issues tentatively.

All together, these three principles might be summarized as, "Try to be reasonable, and reasonably cautious, when considering the significance of evolution for human behavior." ❖

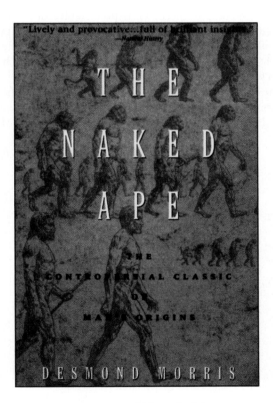

FIGURE 21.15D Cover of *The Naked Ape*

21.16 Some human behavior can be analyzed by the same methods used to study animal behavior

At the root of the debate about sociobiology and evolutionary psychology lies the fact that some human behavior can be analyzed using methods that are very similar to those used in the evolutionary analysis of animal behavior. There are two aspects to this situation.

First, evolutionary biology and the related field of quantitative genetics offer some of the best scientific tools for analyzing behavior in any species. In particular, using the tools of quantitative genetics, it is possible to transcend simple dichotomies like "nature vs. nurture." Though some erroneous publications have misused quantitative genetics to make polemical hay with issues like race and IQ, quantitative genetic methods themselves are the best tools for cleaning up such messes.

Figure 21.16A offers a brief overview of the causal complexity underlying human behavior. Here are some of the most important points. Few characters are completely free of genetic influence. And virtually no behavioral characters are completely free of environmental influences—including environmental influences like learning in species that learn, such as humans. Therefore, almost all behavior results from a combination of genes and environment. Sometimes this combination is "additive"; that is, a particular environmental change might always have the same kind of effect. But sometimes the combination of genes and environment will be nonadditive, and not particularly predictable. All of this was introduced in Chapter 3. Many genes are likely to shape any particular behavior, and many environments. Despite this complexity, behavior is not totally unpredictable. Much of it seems well organized.

The second aspect of human behavior to think about is that some specific patterns of human behavior are parallel to the patterns of animal behavior. These parallels are of two kinds. The first kind of parallel is the extent to which modern human behavior emulates adaptive behavior in animals. For example, animals often forage (look for food) very efficiently, maximizing their intake of calories and miscellaneous nutrients. Humans also seem to maximize their intake of calories per unit time; eating fast food is one example. But an animal species may have fed on a particular food, such as a hummingbird feeding on the nectar of a flower, for millions of years. It is reasonable to suppose that the feeding behavior of hummingbirds evolves by natural selection acting on genetic variation. Humans have had fast-food restaurants since the late 1940s, thanks particularly to the brothers McDonald. Our fast-food dining could not possibly have evolved by natural selection.

A second parallel between human and animal behavior might have involved evolution by natural selection. In a few contexts, the behavior of humans and animals has similar and devastating consequences for fitness. One example of this is incest. Incestuously produced human offspring are known to have reduced fitness, on average, and the production of even one human child is very costly . Many, but not all, animals systematically avoid incest. Often it is difficult to get siblings to mate with each other in animal colonies. As discussed in the box, "Incest," there is significant evidence that humans also avoid incest, and they do so in ways that suggest genetic determination of behavioral patterns . On the other hand, there are many known cases of human incest, sometimes with social approval. Both the ancient Egyptians (e.g., Ramses the Great) and Cleopatra's Ptolemy Dynasty routinely practiced incest, probably with adverse medical effects. So if humans avoid incest, they do not do so invariably (see box). ❖

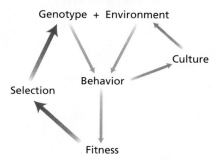

FIGURE 21.16A Sociobiology treats the evolution of all animal species as a process of natural selection in which genotype and environment determine behavior, which in turn determines both culture and the outcome of selection, which then mold the environment and the genotype, respectively.

Incest

The most revealing examples of human **incest avoidance** arise in pseudo-families. Israeli **kibbutzim** live together communally, with children raised together in day-care settings. The parents of these communally raised children encourage the children to date and marry each other, within the kibbutz; but they rarely do so.

In some Chinese families, the future brides of their sons are adopted into the family as young children, long before the wedding. Once the wedding occurs, many of these marriages have problems with sexual compatibility. The couples often divorce.

In both of these cases, the behavioral pattern can be explained on the basis of a genetically built-in incest avoidance, which interferes with socially condoned mating despite the absence of a real genetic problem. Apparently , young children imprint on peers raised with them as siblings, and therefore reject them as mates. This example is widely regarded as the single best illustration of sociobiological theory.

The apparently Darwinian avoidance of incest by humans suggests that human behavior might be straightforwardly organized along evolutionary genetic lines, like the behavior of social insects. But incest is one of the very few contexts in which this might be true. Most human behavior is organized in ways that seem difficult to explain genetically. For example, consider the ability of humans to navigate the modern world, with its many evolutionary novelties. Possums and deer face carnage crossing roads. Adult humans do not. We can even figure out ATM machines and Internet access.

An alternative approach to the flexibility of adult human behavior is to suppose that humans rationally, consciously organize their behavior to Darwinian ends. In this context, humans can be seen as immediate bearers of the Darwinian imperative to survive and reproduce . We would thus be expected to organize our behavior to maximize our Darwinian fitness with few genetic constraints. Notably, this "gene-free" behavior fits well with the theory that humans evolved large brains to be able to calculate the consequences of our behavior in both social and technological realms. That is, the immediate calculation of the consequences of our behavior for our Darwinian fitness fits both a plausible model of our evolution and a plausible interpretation of our present-day behavior. This basic model is shown in Figure 21.17A. One way to think of this theory is that it supposes that, unlike the brains of virtually all other animals, the human brain has replaced the genome as the controller for our behavior.

There is a huge problem facing this theory, however. Aside from a few committed sociobiologists, most of us do not consciously make decisions based on their impact on our Darwinian fitness. Indeed, many of us are quite uncertain as to our motives or desires. Therefore, it seems unlikely that human behavior is consciously and rationally organized toward the maximization of Darwinian fitness in our present environment, if not all environments.

But there is an alternative other than returning to gene-oriented sociobiology: To suppose that our thoughts, emotions, and behaviors are shaped to Darwinian ends as a result of the combined operation of both unconscious and conscious parts of the mind. Specifically, if some unconscious segment of our brain were to calculate consequences for fitness, it could then shape the motivations of the conscious segment toward enhancing fitness. This idea, sketched in Figure 21.17B, is admittedly more cumbersome than any of the theories considered so far. But it does allow the possibility of flexible, open-ended human behavior. At the same time, it explains our lack of the subjective experience of Darwinian thinking. These ideas need to be developed further before they can be considered proper scientific theories. But they indicate the diverse possibilities for the application of evolutionary thinking to human behavior. ❖

A Rational Processing Model for Adaptive Human Behavior

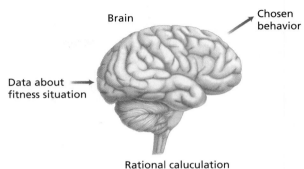

FIGURE 21.17A It might be imagined that the human brain directly, and as a matter of conscious experience, calculates the fitness consequences of each important action and chooses behavior accordingly.

Unconscious Darwinian Calculation

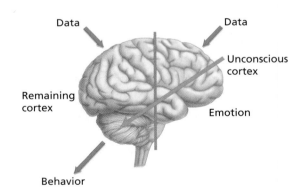

FIGURE 21.17B This model supposes that part of the cortex calculates the relative merits of alternative behaviors regarding their impact on fitness. That part of the cortex then signals the "conscious" part of the brain using emotion and biased perception to motivate behavior.

SUMMARY

1. Humans are primates. We belong to the apes of the old world, none of whom have tails. Our closest ape relatives are the gorillas and the chimpanzees, with the chimpanzees probably closer to us than the gorillas. Our ancestors were upright hominids that probably first evolved 7–8 million years ago. At first our evolution was dominated by selection for efficient upright walking and simple forms of tool use. Multiple upright hominid lineages have coexisted in evolutionary time. In the last 2 million years, at least one upright hominid lineage underwent a massive increase in brain size.

2. The population genetics of our species has been complex. We have coexisted with at least one other very similar species in the last 100,000 years: *Homo neanderthalensis. Homo sapiens,* our own species, probably spread out from Africa in the last few hundred thousand years. This recent diffusion from Africa means that there are no radical differences between human populations, only patchwork differentiation.

3. The why of human evolution remains unknown. We probably evolved for reasons more complex than hunting or bad weath-

er. One popular theory is that we evolved for tool use, but our elaborate social behavior does not fit that model. The alternative is that we evolved for social cunning, but this model has the difficulty that it does not uniquely fit human evolution. Many animals seemingly ought to evolve high levels of social calculation, but do not. There are reasons for supposing that instead our evolution may have involved selection for both technological and social acumen.

4. The theory of evolution is a plausible contender for preeminence in the theoretical explanation of human behavior, at first sight. But human behavior does not fit evolutionary genetic models well, especially with respect to our ability to invent novel behavior that is apparently adaptive. Therefore we need theories of human behavior that are not like those developed for other animals. One idea is that we can directly calculate the fitness consequences of our behavior and develop behavior accordingly. That is, the human brain may have replaced the genome as the ultimate arbiter of our behavior.

REVIEW QUESTIONS

1. How can we be sure that humans are apes and not monkeys?

2. Since the upright hominids split off from the common ancestor of upright hominids and knuckle-walking chimpanzees, what is the minimum number of periods when more than one upright hominid species was alive?

3. Were Neanderthals members of our species?

4. Are human "races" biological groups of long standing, or are they evolutionarily recent?

5. Why do human languages show similar geography to human genetic differentiation?

6. Give at least one piece of evidence suggesting that the evolution of the large human brain involved strong directional selection.

7. Why is it reasonable to say that the social intelligence theory is too strong to explain human evolution?

8. Why should organisms avoid incest?

9. Why is it unlikely that most of our behavior is genetically encoded in the same way that insect behavior is genetically encoded?

10. Do you think that you possess free will?

KEY TERMS

Alu	great ape	mental arms race	short tandem repeat (STR)
anthropoid	hominid	mitochondrial DNA	social intelligence
Ardrey, Robert	hominoid	Morris, Desmond	sociobiology
Australopithecus	*Homo*	multiregional theory	tool use
braincase	hunter-gatherer	Neanderthal	Wallace, Alfred Russel
Cro-Magnon man	incest avoidance	out-of-Africa theory	Wells, H. G.
ear canal	*Just-So Stories*	primate	Wilson, E. O.
Engels, Friedrich	kibbutzim	prosimian	
evolutionary psychology	knuckle walking	robust lineage	
gracile lineage	Mayer, August	savanna	

FURTHER READINGS

Barkow, J. H., L. Cosmides, and J. Tooby, eds. 1992. *The Adapted Mind: Evolutionary Psychology and the Generation of Culture*. New York: Oxford University Press.

Byrne, R., and A. Whiten, eds. 1988. *Machiavellian Intelligence, Social Expertise and the Evolution of Intellect in Monkeys, Apes, and Humans*. Oxford, UK: Clarendon Press.

Cavalli-Sforza, L. L. 1997. "Genes, Peoples, and Languages." *Proceedings of the National Academy of Sciences, USA*, 94:7719–24.

Cavalli-Sforza, L. L., P. Menozzi, and A. Piazza. 1994. *The History and Geography of Human Genes*. Princeton, NJ: Princeton University Press.

Hublin, J., F. Spoor, M. Braun, F. Zonneveld, and S. Condemi. 1996. "A Late Neanderthal Associated with Upper Palaeolithic Artefacts." *Nature* 381:224–26.

Kitcher, P. 1985. *Vaulting Ambition, Sociobiology and the Quest for Human Nature*. Cambridge, MA: MIT Press.

Krings, M., H. Geisert, R. W. Schmitz, H. Krainitzki, and S. Paabo. 1999. "DNA sequence of the Mitochondrial Hypervariable Region II from the Neandertal Type Specimen." *Proceedings of the National Academy of Sciences, USA*, **96**:5581–85.

Rose, M. R. 1998. *Darwin's Spectre: Evolutionary Biology in the Modern World*. Princeton, NJ: Princeton University Press.

Tattersall, I. 1999. *The Last Neanderthal*. New York: Nevramount.

Tishkoff, S. A., E. Dietzsch, W. Speed, A. J. Pakstis, J. R. Kidd, K. Cheung, B. Bonne-Tamir, A. S. Santachiara-Benerecetti, T. Jenkins, and K. K. Kidd. 1996. "Global Patterns of Linkage Disequilibrium at the CD4 Locus and Modern Human Origins." *Science* **271**:1380–87.

Wilson, E. O. 1975. *Sociobiology: The New Synthesis*. Cambridge, MA: Belknap Press.

———. 1978. *On Human Nature*. Cambridge, MA: Harvard University Press.

Wolpoff, M. H. 1997. *Paleoanthropology*. New York: McGraw-Hill.

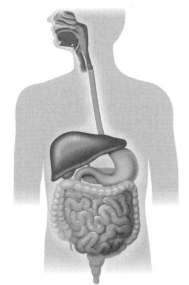

Darwinian biology can improve medicine.

Darwinian Medicine

Two hundred years ago, the practice of medicine was dangerous to the health of the patient. Physicians had very few tools to treat contagious disease. The septic conditions of surgery ensured that those who could not afford new instruments and clean bedding had a high likelihood of dying. Physicians also deliberately bled their patients, based on archaic and ill-founded notions about human physiology.

Some time after 1900, things changed. Going to the doctor actually increased your chances of survival. A revolution in medicine was under way, a revolution that has had cumulative benefits for human survival.

This revolution was brought about by the application of modern science, especially modern biology, to the problems of medicine. Microbiology, genetics, immunology, neurobiology, and organic chemistry are just some of the basic sciences that have helped to build an effective medicine. Much of modern biology has contributed to the development of a newly benign and efficacious medicine.

But some biological fields have contributed little to contemporary medicine, specifically the fields that derive from Darwinian thought. Evolutionary biology and ecology are generally excluded from the practice and teaching of medicine, with only a handful of exceptions. When medical school faculty and practicing physicians do use Darwinian ideas, they are more often based on outmoded evolutionary or ecological ideas from the nineteenth century, or simply illogical.

But some modern-day evolutionary biologists and ecologists, such as **George C. Williams, Randolph Nesse,** and **Paul W. Ewald,** have rejected this medical tradition. Instead, they take it as their business to introduce modern Darwinian analysis into medicine. Whether dealing with human deformity, contagious disease, aging, or brain disorders, modern Darwinians are making a strong case that their fields should contribute to the improvement of contemporary medicine. ❖

HUMAN IMPERFECTION

22.1 Some of our medical problems arise from our evolutionary history

Many medical problems arise from basic features of the human body. Death by drowning, for example, happens because we cannot efficiently extract oxygen from water. Such deaths do not require a pathogen, extreme age, or psychiatric delusion. They happen because of built-in limits to the human form, imperfections. There are two kinds of human imperfection: universal and idiosyncratic. The first kind of imperfection comes from evolutionary history. We evolved as land-dwelling animals with skeletons on the insides of our bodies. We cannot fly using our own limbs. We cannot make food from the sunlight that hits our skin. We are limited, or imperfect, in ways that are shared by all members of our species.

Idiosyncratic human imperfection is due to variation within our species. Some individuals are born with metabolic defects. Hemophiliacs, for example, have poor blood-clotting function. Minor injuries can cause them to bleed to death. (We will consider this type of imperfection in the next module.)

In many ways, physicians are practicing evolutionary biologists and ecologists, although they usually do not know it. A lot of the problems that they deal with arise from our evolutionary history and our ecology.

One problem with our evolutionary history is that we evolved over hundreds of millions of years from an ancestral aquatic chordate. Although we have some fossils of possible ancestors from about 500 million years ago, there is a modern-day species that is apparently much like them: *Amphioxus*. *Amphioxus* has very simple anatomy. Water flows in through the mouth, supplying food particles that are collected by the gills. The **pharynx** functions as the front end of the digestive tract. There are no specialized organs for respiration (see Figure 22.1A).

But humans cannot just push materials through the pharynx, willy-nilly. We deal with multiple kinds of materials: solid, liquid, and air. And we have very different destinations for these substances: air to the lungs, solids and liquids to the stomach.

We have adaptations for processing these substances differently. We chew solids and then swallow. While we swallow, a membrane called the **epiglottis** closes off the entry to our lungs, the **trachea**. We have nostrils that enable us to take in air even when we are chewing our food and are polite enough to keep our mouths shut (all this is shown in Figure 22.1B).

But it often happens that some food gets lodged in the trachea. Then we cough, to clear the obstruction. Sometimes

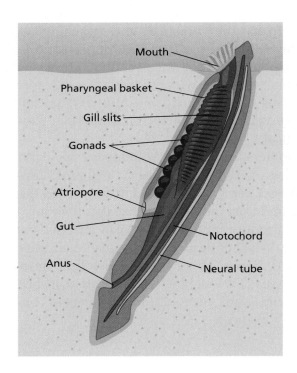

FIGURE 22.1A Anatomy of *Amphioxus*

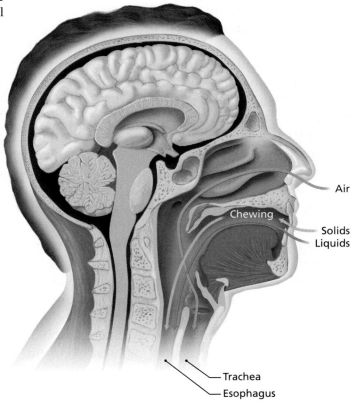

FIGURE 22.1B Cross Section of a Human Head, Showing Our Pharyngeal Anatomy

coughing does not work. If there is complete obstruction of the trachea, we **choke** to death. This is death by evolutionary history. We could have had respiration separated from food ingestion. Insects, for example, breathe using passageways distributed over the body, with many openings. They cannot choke. They can have their breathing blocked when their body surface is fouled. They can have their guts obstructed by small pieces of rock and other hard materials. But these two problems have nothing to do with each other. In mammals, with our evolutionary ancestry as a small aquatic animal, too many passages meet at the pharynx: nasal passages, mouth, esophagous, and trachea. So we choke. It should be noted that horses can drink and breathe at the same time. Horses are apparently ahead of us, evolutionarily.

Our pharyngeal structure is just one example of a universal human imperfection. Another is the spine. Most mammals are quadrupeds. The body is hung on the spine, with additional support coming from the pectoral and pelvic "girdles." The spine is only rarely stacked vertically, even in other primates. But humans are upright bipeds, forcing the spine to play a major structural role in keeping us vertical. Vertebrae are little suited to this anatomical task, so humans have chronic problems with posture, vertebral degeneration, and back pain. If you are too young to have experienced these medical problems, you are almost certain to do so as you get older.

A significant part of the future of medicine will consist of remedies for problems that have been bequeathed to us by our evolutionary history. ❖

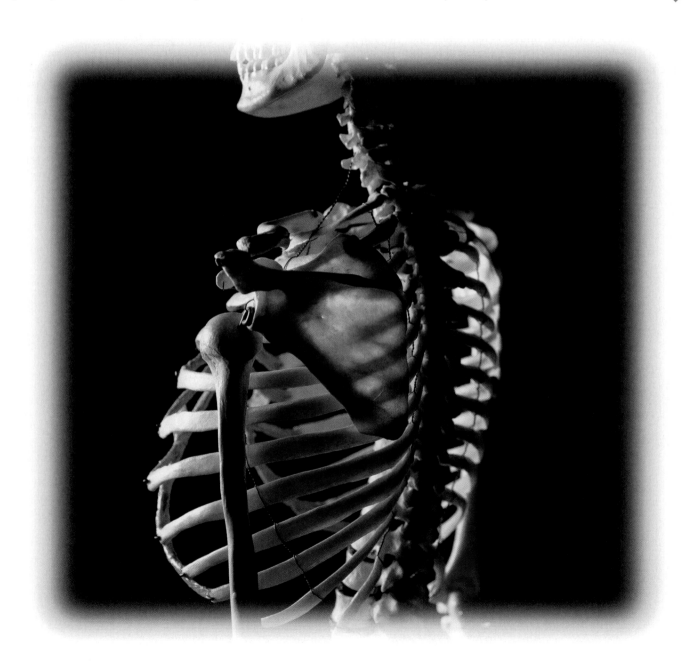

Genetic diseases are extreme forms of human imperfection generated by rare genotypes

The foremost examples of idiosyncratic human imperfection are the genetic diseases. Some individuals get genetic diseases. Most do not. **Genetic diseases** are disorders that occur in virtually all individuals that bear a particular genotype at a single locus. These diseases were already introduced in Chapter 4, because they are usually targets of purifying selection.

Although these diseases are not common, they absorb a substantial amount of medical resources and attention. Patients with disorders like cystic fibrosis and Tay-Sachs disease spend a large fraction of their lives receiving medical treatment. With cystic fibrosis, this medical care has greatly extended patients' lives.

In the case of **phenylketonuria (PKU)**, immediate medical care of the afflicted can largely treat the underlying condition. Specifically, dietary changes that eliminate **phenylalanine** from the patient's food prevent toxic buildup of phenylalanine. Even though PKU is a genetic disease with severe consequences when untreated, it has a simple, effective treatment. This particular medical disorder brings out a very general point about medicine: Medical problems have their roots in biochemistry, anatomy, and other features of the human organism. The cause of a disease, whether genetic or environmental, does not absolutely prevent or allow effective treatment.

Whatever their prospects for treatment, the underlying origins of genetic diseases are interesting for medical practice and evolutionary understanding. There are two main theories explaining these origins.

The first theory is that genetic diseases are produced by recurrent mutations of no benefit. These mutations frequently may be recessive, in which case the genotype that produces the disease is homozygous for a pathological mutant gene. But some disorders are caused by dominant alleles. Huntington's disease has this pattern of inheritance. For highly deleterious dominant alleles, such as those that cause childhood progeria—a disorder that is always fatal before reproductive maturity—only newly occurring mutations cause disease. Purifying selection eliminates these dominant alleles before the start of the next generation. Figure 22.2A shows the genetic and selection mechanisms that shape the incidence of genetic diseases caused by deleterious alleles.

The second main theory is that some genetic diseases have deleterious effects in homozygous combinations of alleles,

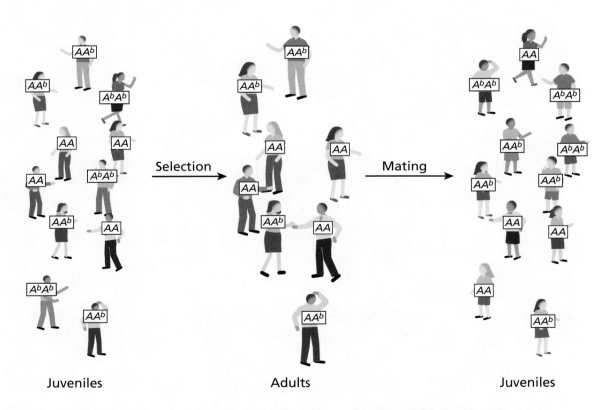

Juveniles   Selection   Adults   Mating   Juveniles

FIGURE 22.2A Genetic Diseases Caused by Recurrent Mutation Mutation adds individuals with alleles that cause genetic disorders, while selection removes individuals with these genetic disorders. Because mutations recur, selection never eliminates the genetic disease from the population.

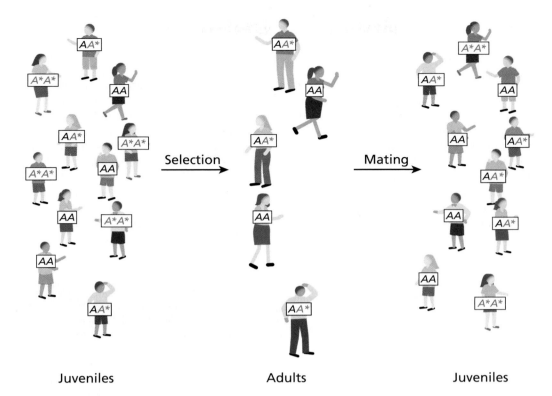

Juveniles   Adults   Juveniles

FIGURE 22.2B **Genetic Diseases Caused by Selection** The disease is caused by homozygosity for an allele (A*) that is beneficial in heterozygotes. Selection eliminates many of the homozygotes, but the heterozygotes generate more homozygotes with the genetic disease by segregation.

but are beneficial in heterozygous form. The type specimen for such disorders is sickle-cell anemia, a biological phenomenon that is an example of heterozygote advantage, as described in Chapter 4. In regions with active malaria outbreaks, the allele that causes red blood cell sickling ap-

pears to reduce the rate of malarial infection. The heterozygote is the genotype with greatest fitness, because it is resistant to malaria, but its level of sickling is minor. Figure 22.2B shows how this kind of genetic disease is evolutionarily maintained in human populations. ❖

Evolution of Medical Disorders

It is counterintuitive for genetic diseases to exist at all. Natural selection is intuitively expected to prevent bad things with genetic causes. The problem is that while natural selection acts on individuals and their reproduction, its historical, or eventual, effect on single individuals is not always benign. Only its average effect on the individuals in a population is eventually benign. Natural selection, as shown in Chapter 4, tends to increase the average individual fitness over the whole population. It may achieve this end by moving gene frequencies to values at which some individuals suffer medical harm.

Evolution is not a process guaranteeing benefits to all individuals. Natural selection does not take care of us. Individuals who have severe sickle-cell anemia are produced by natural selection in areas with malaria.

In addition, mutation will always occur. We have no prospect of forever eliminating mutations. Although many mutations have little effect on the human phenotype, other mutations have catastrophic effects. These mutations will produce genetic diseases.

The specter of genetic diseases raises the equally dramatic possibility of using molecular technology to engineer the genes that are passed to offspring. **Germline engineering** can be defined as altering the genetic information passed from one generation to the next. In its simplest, most powerful, form, germline engineering might take all individuals who carry a genetic disease allele and eliminate that allele. How this might be done will depend on molecular technology. The technological details are not of concern here. Let us suppose that we have the technology to eliminate gametes bearing particular alleles. Figure 22.3A shows how germline engineering might work.

But we have a more important problem to solve. Is this a good idea? If we could engineer the germline of the human species, should we? Many religions and political ideologies specifically proscribe such acts. The Roman Catholic Church is profoundly opposed to tampering with our hereditary nature. European Greens have a political ideology that similarly is opposed to such interventions. But then there are religions and political movements, like the eugenicists of the 1930s and the present-day Extropians, who very much favor such interventions. Therefore, it is not adequate to evaluate germline engineering by simply asserting that it's wrong. For some people, it is profoundly right.

Evolutionary biology and ecology supply useful information for this debate, but not because they can tell us what is right or wrong. There is no concept of the "natural" that can be imported into medical practice from Darwinian biology to tell us what is right or wrong. Indeed, the very idea of "nature" is almost the opposite of the implications of Darwinian biology, which reveals vast amounts of genetic variation within species. There is no specific human nature to conserve.

1. Isolate cells from subject.

2. Infect germline or somatic cells with the X retroviral vector. Retroviral DNA is integrated into the chromosomes of human cells.

Patient with Genetic Disease

3. Grow retrovirus-infected cells outside the body.

4. Infuse engineered cells that are expressing the X transgene back into the patient.

FIGURE 22.3A How Genetic Engineering Works

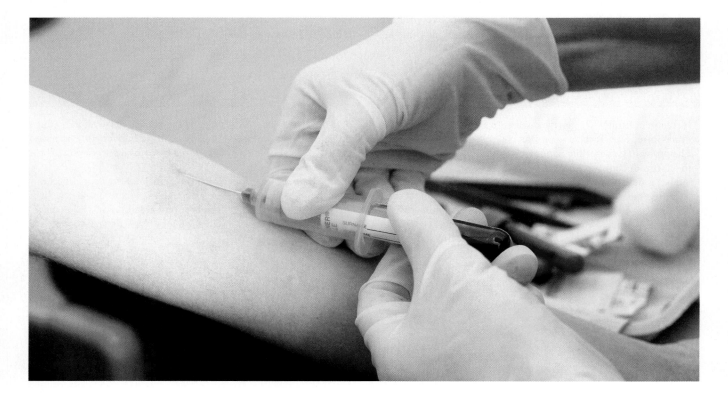

Darwinian biology can contribute two important matters of fact, which need to be considered in any germline engineering project. The first of these is that changing the germline constrains our descendants to our present level of knowledge. Suppose that germline engineers find a genetic substitution that reduces immunological problems like allergies and autoimmune diseases. This genetic change is unlikely to completely solve these problems, so we would hope to have better molecular genetic solutions for immune diseases in the future. The problem is that past rounds of germline engineering will have to be undone in future germline engineering. Even if this can be done at all, which is open to doubt, it adds a step that might go wrong.

The second important fact about germline engineering is that it is an engineering solution, based on molecular genetic knowledge. Natural evolution in real ecological settings, on the other hand, is an open-ended process incorporating alleles with beneficial effects that depend on the particular environment, whether we understand these beneficial effects or not. Germline engineering is likely to manipulate a small number of DNA sequences at a time, while natural evolution may act on hundreds of loci simultaneously. This contrast is illustrated in Figure 22.3B. In this sense, natural evolution has an inherent superiority in the volume of genetic information that it can sift.

There may be a few cases in which none of these technical problems will limit the success of germline engineering. One such case would be the removal of overwhelmingly deleterious alleles. But the evolutionary and ecological issues indicate that germline engineering may be subject to major problems and limitations, leaving aside any ethical or religious argument against its use. ❖

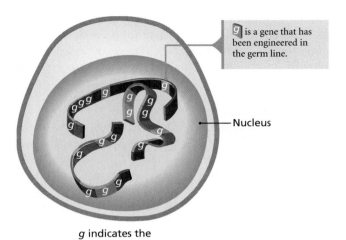

g is a gene that has been engineered in the germ line.

Nucleus

g indicates the genes affecting a particular character.

FIGURE 22.3B Many genes are likely to be unaffected by a single manipulation of the germline, and many of the unaffected genes will also control the medical problem.

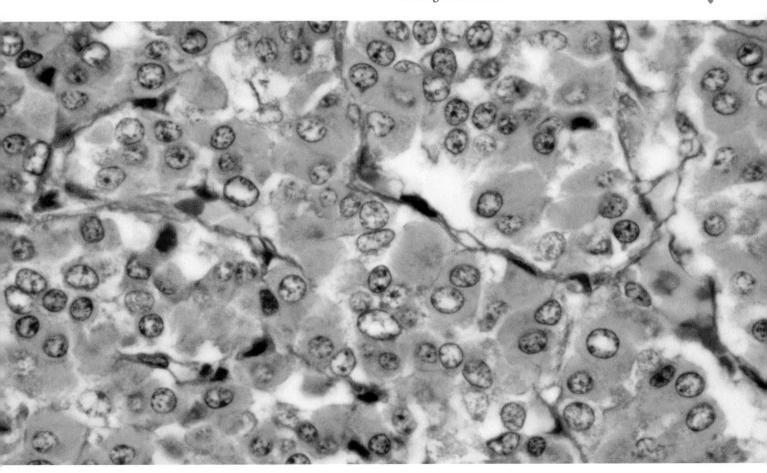

The term *genetic engineering* usually does not refer to germline engineering. Instead, it refers to molecular reengineering of somatic cells. For example, the artificial incorporation of a normal copy of a gene that is missing in the patient is used to produce a protein that is vital to metabolism *in that patient*.

Considerable progress has been made with **somatic engineering** to treat **adenosine deaminase (ADA) deficiency**, shown in Figure 22.4A. Lack of the ADA enzyme causes some incidental abnormalities, but its most important effect is to undermine immune response. Patients with this disorder are less resistant to viral, bacterial, and other infections, which greatly shortens their lives. Somatic gene therapy for this disorder has been successful in some cases. One method is to take cells from the patient's immune system and culture them under glass in the laboratory. The cultured cells are infected with a virus bearing the normal *ADA* gene. Once the cultured cells have been infected with this virus and have successfully incorporated the *ADA* gene into the cellular genome, the cul-

tured cells are reintroduced to the patient's body, as shown in Figure 22.4A. This procedure has been done clinically with success, and the treated patients have retained immune function since the somatic introduction of the *ADA* gene.

Although this type of therapy changes the DNA of patient cells, it is not evolutionarily different from providing insulin by injection to control hereditary diabetes. In either case, a molecular intervention saves a life; but it does not alter the gametes of the patient, thus leaving their children unchanged genetically. But this does not mean that such somatic therapy has no effect on evolution.

The profound point for evolutionary biology is that patients with genetic diseases get to have children at all. Because they do get to live long enough to have children, any genetic deficiency that predisposes them to their disease will be transmitted by them to the next generation, if they reproduce. Patently, such medical intervention will change the genetic future of the human species. For some individuals, this is a hot ethical issue. They are opposed to such effects on the

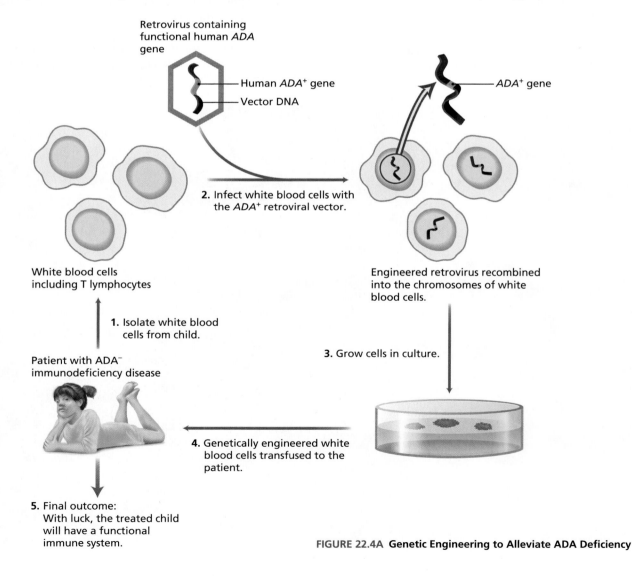

Retrovirus containing functional human *ADA* gene

Human *ADA*⁺ gene

Vector DNA

ADA⁺ gene

2. Infect white blood cells with the *ADA*⁺ retroviral vector.

White blood cells including T lymphocytes

Engineered retrovirus recombined into the chromosomes of white blood cells.

1. Isolate white blood cells from child.

Patient with *ADA*⁻ immunodeficiency disease

3. Grow cells in culture.

4. Genetically engineered white blood cells transfused to the patient.

5. Final outcome: With luck, the treated child will have a functional immune system.

FIGURE 22.4A Genetic Engineering to Alleviate ADA Deficiency

genetic future of the species. The Nazis, for example, were opposed to saving the lives of people *they* regarded as hereditarily unfit—such as mentally retarded children, homosexuals, gypsies, and Jews. Being systematic about their concern, the Nazis killed members of these groups. Other groups, such as the Roman Catholic Church, are opposed to tampering with our genetic future, regardless of the imperfections that are transmitted to the next generation.

To put this controversy into better focus, recall that a vast range of medical problems are influenced by genetic vulnerability—from nearsightedness to cardiovascular disease to cancer. All of these problems have some genetic component. Medical interventions from optometry to open-heart surgery help people survive; people who might otherwise never have had children instead have children who inherit a greater genetic susceptibility to these diseases. Virtually every human alive today carries genes that might incline their children to one medical disorder or another. Purifying the human species by eliminating the medically unfit, or merely denying medical care, is not feasible. Even if we wanted to create a genetically purified human species—and many of us do not want to—the project is illusory. Almost all of us would have to be killed or sterilized. This would then leave the genetically ideal survivors with a small population size, resulting in inbreeding. Over time, deleterious mutations would arise and accumulate in the tiny population of "supermen," making further rounds of eugenic cleansing necessary. There is no escaping genetic imperfection in humans. ❖

CONTAGIOUS DISEASE

22.5 | Human diseases are shaped by long-term evolution and global ecology

The treatment of contagious disease was one of the great successes of twentieth-century medicine, especially thanks to antibiotics and **vaccines**. It is easy to think of such figures as Pasteur, Fleming, and Florey as heroes because of their discoveries concerning the biology and prevention of contagious disease. Their work has saved millions of lives. But contagious disease is also singularly important as a Darwinian arena, where evolution, ecology, and medicine meet in the determination of the survival or death of medical patients.

Although it is natural to focus on acute contagious diseases, recent medical research has shown that many chronic diseases are caused by unsuspected pathogens. Just one example of this is that most stomach ulcers are now known to be produced by infection with the bacterium *Helicobacter pylori*—contrary to prior medical dogma, which did not imagine that a contagious disease was involved in ulcers. **Paul Ewald** has proposed that numerous chronic diseases, such as Alzheimer's disease, are products of our ecology rather than our genetics. Time will tell.

In many cases, the relationship between parasites and their hosts is one that is ripe for coevolution, as described in Chapter 14. The most important reason to expect coevolution is the dependence of the parasite on the host for the completion of its life cycle, and the possible death of the host from the parasitic infection. Together, these life-or-death possibilities will make natural selection intense.

Humans have coevolved with a number of diseases. One of the best examples of such coevolution is the **smallpox** virus. In seventeenth-century Europe and elsewhere well into the twentieth century, smallpox was a common cause of death.

In addition, smallpox caused scarring and, in some cases, blindness, providing the victim survived. But the smallpox virus had no alternative host to humans. This high degree of specialization made it possible to eradicate the disease through the use of vaccines. The only remaining threat of a smallpox outbreak comes from germ warfare stockpiles. Figure 22.5A shows a smallpox victim.

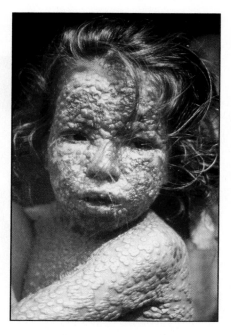

FIGURE 22.5A A Patient Suffering from Smallpox

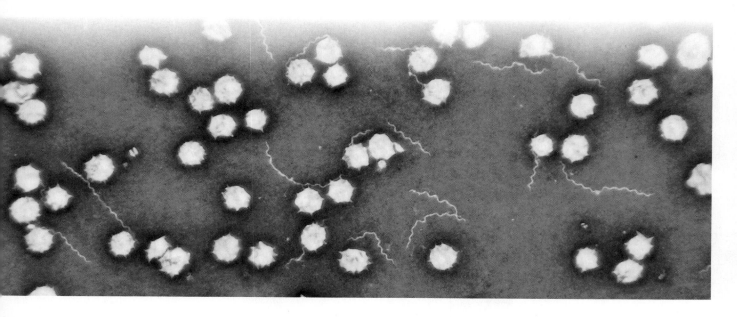

The prospects for medicine over the next few decades are not likely to involve further human evolution. A few decades is only one human generation, and relatively little genetic change will take place within one generation—unless more than 90 percent of the human population is killed off in that generation. This scenario would require thermonuclear war, a large body impact, or some other astronomical mishap. For the purpose of developing medicine on a Darwinian basis, it is a reasonable assumption that humans will not be evolving significantly in the immediate future.

On the other hand, many of our pathogens undergo hundreds or thousands of generations per calendar year. This gives them tremendous potential to evolve. At the evolutionary level, our diseases are moving targets—while we virtually sit still. This is an evolutionary asymmetry full of frightening possibilities, some of which, as we shall see, have been realized.

In addition to being evolutionarily inert, for most medical purposes, humans are distributed globally. With modern transportation, diseases that originate in one part of the world can quickly spread. Rural China, for example, is a major source of influenza epidemics that take only months to cause extensive school absenteeism in Wisconsin, thousands of miles away. Our global accessibility makes humans an excellent host organism for pathogens. The **severe acute respiratory syndrome (SARS)** outbreak is a perfect illustration of this pattern. Figure 22.5B shows the geographical course of SARS infection.

Coupled with easy access geographically, humans are numerous. With a population in the billions, we present many potential targets of infection. We are also a large organism, compared to most animals. We supply a lot of tissue for infection, making us a lush target for new pathogens. ❖

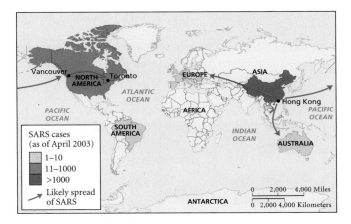

FIGURE 22.5B The Global Spread of SARS

A main theme of Darwinian medicine applied to contagious disease is the complexity of disease symptoms. For example, we get a cold and our noses run, our temperatures rise, and our throats hurt. A naïve way of looking at these cold symptoms is to interpret them as the effects of the cold viruses replicating in our bodies. And this is probably true of many disease symptoms: They are effects of pathogen replication.

But not all disease symptoms fit this pattern. Some symptoms are *defensive*, produced by the host to ward off the pathogen, or at least reduce its impact on the host. Fever is the classic example of a disease symptom that is a defense. Ironically, modern medicine does the best that it can to reduce fever. Yet the evidence is clear that the body elevates its temperature itself. Desert iguanas that regulate their body temperature by the amount of sun exposure seek more sunlight when they are infected. And such defenses work. If the iguanas are denied access to sunlight when they are infected, their disease becomes more severe. Many experiments with mammals give similar results: Reduction in fever by drugs tends to increase the severity of infections. Aspirin and acetaminophen have both been implicated in laboratory experiments. They reduce fever, and in so doing they often prolong infections, worsen other disease symptoms, and even increase death rates.

Does this mean that no one should ever bring their fever down? No—very high fevers can cause brain damage and even bring on death. Very high fevers should be treated as medical emergencies. But the pharmaceutical reduction of moderate fever may prolong minor illnesses.

Together with fever, the body has a wide range of disease defenses. Of particular importance are the defenses that actively expel pathogens: tearing, sneezing, coughing, vomiting, diarrhea, and copious urination (Figure 22.6A). Again, these include defensive symptoms that have been aggressively reduced by modern medicine, even though such medical treatment may have prolonged infections. This effect has

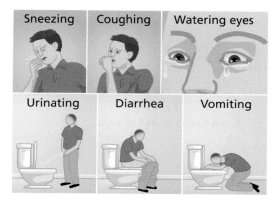

FIGURE 22.6A How the Body Defends Itself against Pathogens

been demonstrated quite well for diarrhea. Treatment with drugs that suppress diarrhea can prolong intestinal infections by bacteria.

But diarrhea brings up an additional category of symptom: *manipulative* symptoms. Some pathogens use the body's defense mechanisms to foster their infection of new hosts. **Cholera** is caused by the bacterium *Vibrio cholerae*, which infects the human body by ingestion (Figure 22.6B). Once it gets to the intestines, *Vibrio* faces competition from our normal bacterial colonists, such as *Escherichia coli*. To solve this problem, the pathogen induces acute diarrhea, flushing the *E. coli* out of our intestines. Once this has been accomplished, *Vibrio* has an intestinal environment better suited to its replication. Continued diarrhea also helps spread the pathogen to new hosts, particularly by infecting water supplies. In this case, early suppression of diarrhea might reduce the intensity of infection.

These examples show that there are no simple rules in the treatment of disease. A treatment that helps control one disease symptom might make another disease symptom worse. The most reliable generalization is that diseases involve complex interactions between pathogens and host. Treatment of contagious disease therefore requires some sophistication. Part of that sophistication must include knowledge of the evolutionary history and ecological circumstances that define a particular disease.   ❖

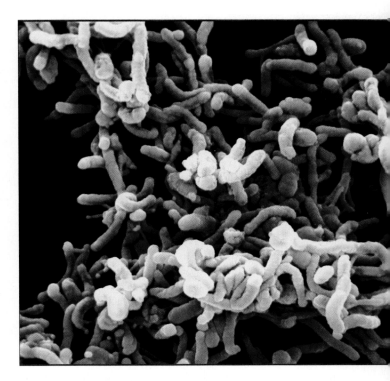

FIGURE 22.6B Cholera Bacteria

22.7 The evolution of pathogen virulence depends on the ecology of infection

Contagious disease is one of the most intimate couplings of evolution and ecology, and of course one of the most important. Perhaps no single phenomenon reveals this better than the relationship between the pattern of disease transmission and the lethality of infection.

Human diseases vary widely in their lethality, and that variation is of enormous interest to us. The **HIV** and **Ebola** viruses kill most of those whom they infect—HIV in years, Ebola in days. Cold rhinoviruses and many gut bacteria rarely kill humans. What is responsible for this disparity in the consequences of infection, the **virulence** of a pathogen?

One of the major factors is the ecology of disease transmission. Of particular importance is the likelihood that a particular pathogen will soon find a new host. And that in turn is shaped by a seemingly innocuous factor—whether the pathogen is transmitted directly from host-to-host contact. The pattern of lethality among different types of pathogen is shown in Figure 22.7A, taken from Paul Ewald's book, *Evolution of Infectious Disease*. Pathogens that are transmitted only from host to host are much less likely to kill their hosts relative to pathogens that have some intermediate phase that does not involve human infection, in their life cycle. Such intermediate phases usually involve organisms called **disease vectors**, which the pathogen can infect and which bear the pathogen from host to host.

Famous examples of such a vector-based pathogen are the species of the genus *Plasmodium* that cause **malaria**. (There is more information about malaria throughout this book; consult

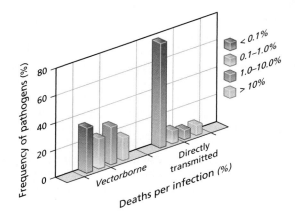

FIGURE 22.7A Mortality rates from untreated infections due to arthropod vectors vs. mortality rates with direct transfer from person to person.

the Index.) Their life cycles are summarized in Figure 22.7B. Malaria newly infects millions of people each year. It is the number one killer among vector-borne human diseases, in total number of deaths. An interesting aspect of malarial transmission is that it does not require its human host to be well enough to go out and mingle with others. The mosquitoes that are its primary transmitters come to the afflicted, suck their blood, and move on to infect individuals free of infection by biting them. Unlike medical doctors, the disease vector still makes house calls.

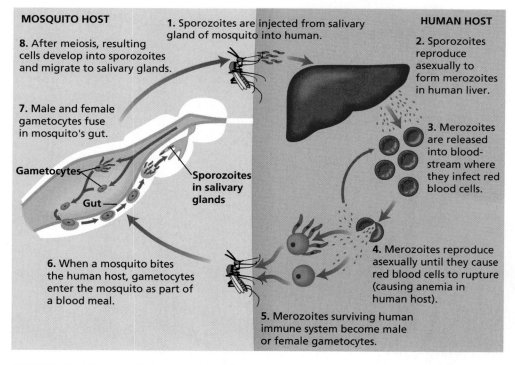

FIGURE 22.7B Life Cycle of the Malaria Parasite, *Plasmodium*

Compare this pattern with the common cold, typically caused by the benign rhinoviruses. People with colds spend very few days in isolation, returning to workplaces and schools in short order. In such settings, their sneezes, coughs, and nasal drippings falling onto pencils, paper, and furniture handily transmit the infection to new hosts. If those afflicted with colds were instead so debilitated that they could not return to public life readily, the infection might not make it to a new host due to a lack of opportunity to infect a new host.

These patterns reveal that pathogens that have more access to new hosts may evolve lethality. For human purposes, this suggests that when human behavior changes, pathogens may change their lethality by evolution. The upsurge in promiscuity among Western populations since World War II has provided a natural opportunity for more lethal venereal pathogens to infect new hosts quickly. The spread of HIV through the West since 1980 is perhaps a predictable consequence of more promiscuous human sexual behavior. ❖

Some aspects of present-day medicine can be understood using Darwinian ideas. But other aspects of medicine require active management using Darwinian principles. Perhaps no single area of medical practice requires evolutionary and ecological thinking more than the use of antibiotics.

Antibiotics were one of the most important medical discoveries of the twentieth century. **Antibiotics** are compounds that kill bacteria, often with very few side effects for the patients who use them. First developed in the period around World War II, antibiotics played an important role in that war, saving many lives that would have otherwise been lost and allowing the speedy return of soldiers to battle. For a time, it seemed as if antibiotics and **antivirals** would usher in a new era of medicine, in which contagious disease would be largely vanquished. This era of boundless optimism did not last long. By the 1960s, reports of bacterial strains resistant to antibiotics became common. Gonorrhea cultured from prostitutes commonly showed resistance to antibiotic treatment, requiring extended medication with one or more antibiotics.

The cause of **antibiotic resistance** in bacteria was not hard to determine: The bacteria had evolved genetically. A particularly common mechanism for the evolution of antibiotic resistance was the acquisition of plasmids bearing genes that thwarted the action of antibiotic medicines. These plasmids spread contagiously not only within bacterial species, but between bacterial species.

Some of the new bacterial strains have proven remarkably difficult to eliminate using antibiotics. Particularly in hospitals, in which new hosts are readily available, local outbreaks of bacterial infection have killed dozens of patients. Among those most susceptible to dying of antibiotic-resistant bacterial infections are the elderly and the newborn, two groups that are likely to be in hospitals. Doctors who examine a series of hospital patients in quick succession probably further foster the spread of potentially deadly infections among vulnerable groups.

Several features of modern medicine foster the evolution of bacterial resistance to antibiotics (Figure 22.8A).

1. The U.S. Food and Drug Administration makes it difficult to get approval for new antibiotics. Yet evolutionary first principles suggest that the use of a limited repertoire of antibiotics will make it easier for bacteria to evolve resistance.
2. Most physicians try just one antibiotic at a time, with a short period of medication. This gives bacteria the opportunity to adapt to one antibiotic at a time. The use of multiple antibiotics at one time is likely to allow fewer bacteria the opportunity to survive antibiotic treatment.
3. Collecting patients together in hospitals, and to a lesser extent clinics, maximizes the opportunities for exchange of resistant bacteria between patients. This practice also fosters the exchange of antibiotic-resistant plasmids between unrelated bacterial strains.

4. Most important, antibiotics are often mis-prescribed for viral and other diseases, especially at the insistence of patients who do not understand the biology involved.

With the widespread consumption of antibiotics, the human population is virtually an incubator for the evolution of bacteria that cannot be eliminated using known antibiotics.

As if the foregoing points were not sufficient cause for concern, two more problems highlight the likelihood that many bacteria will soon evolve resistance to antibiotics. First, though Western countries confine the use of antibiotics to patients under a doctor's care, many developing countries

1. Lack of new antibiotics
Turning off the pipeline of antibiotic development by stringent government regulation ensures that bacteria face a limited repertoire of antibiotics.

2. Using one antibiotic at a time
If a bacterium is resistant to an antibiotic, the only way to overcome its resistance may be to attack it with multiple antibiotics at the same time. But standard medical procedures dictate one medication at a time.

3. The hospital environment
Collecting sick and compromised patients together in one building gives bacteria excellent opportunities for infecting new hosts. Effective quarantine of infected patients is crucial, but hard to achieve.

4. Incorrect prescription
Sick patients commonly badger their physicians for antibiotics when they have viral infections, or no infection at all. Their use of antibiotics imposes selection on their bacteria for resisting the mis-prescribed antibiotic, with bad consequences when a bacterial infection does occur.

FIGURE 22.8A How We Sabotage Antibiotics

have virtually unregulated use of antibiotics, without a doctor's prescription. Because antibiotics are expensive by the standards of these countries, many do not take a full course of antibiotics. This makes such casual antibiotic use almost ideal for the evolution of antibiotic resistance in bacteria. Secondly, and almost unbelievably, Western farmers use feed that contains antibiotics. They do so because agricultural animals grow faster when they are routinely fed antibiotics. However, such use is again not calibrated to ensure the complete elimination of bacteria, allowing bacteria to undergo selection for antibiotic resistance.

These features of our medical ecology combine to undermine one of the most important weapons available to the medical profession. Antibiotics are still generally useful in the treatment of bacterial infection. How much longer this will continue depends on the reform of pharmaceutical development, medical practice around the world, and even agriculture. ❖

HIV illustrates the importance of rapid virus evolution in medicine

The human immunodeficiency virus (HIV) is probably the deadliest pathogen to have attacked the human species. HIV infection seems to eventually kill 95–99 percent of its victims, regardless of treatment. It has already killed somewhere between 10 and 20 million people, with a total of about 40 million infections so far. Given the lack of an effective vaccine for the foreseeable future, as well as no cure, these numbers are bound to grow. In sub-Saharan Africa, where the incidence of infection exceeds 10 percent of the total population, HIV is bringing about a demographic revolution—and a humanitarian disaster. In the worst-case scenario, HIV could eventually wipe out all members of our species, excepting only a tiny fraction of highly resistant individuals. This would leave our species with only some thousands of survivors—an impact comparable to that of thermonuclear war. How did our species acquire such a scourge? How does HIV work? And what can be done about it?

Other primates have immunodeficiency viruses like HIV, although they do not suffer lethal consequences from infection with these viruses. Humans acquired HIV strains from other primates; and other primates may have been infected with human HIV. Among the possible mechanisms of transmission between primate species are human consumption of "bushmeat," the flesh of other primates, and bites from flying insects like mosquitoes.

HIV is an unusual virus in that it infects the white blood cells that are used to fight off disease. HIV is a deadly disease because it eventually destroys this entire class of cells, undermining the immune system. Once their immune system no longer works, HIV patients have acquired immuno deficiency syndrome (or AIDS). **AIDS** patients cannot defend themselves against relatively benign infections, like tuberculosis, and thus they eventually die.

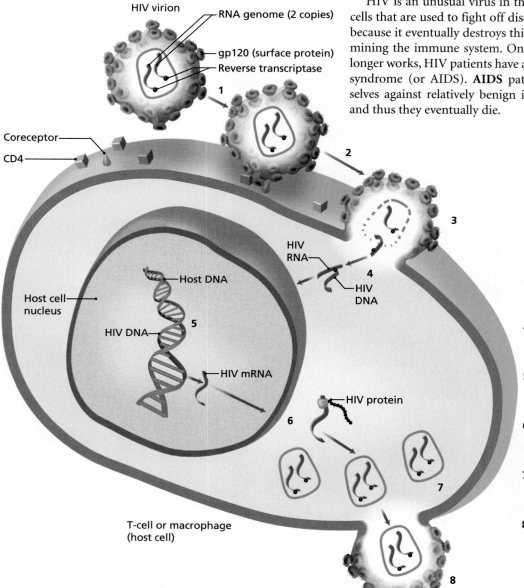

1. HIV's extracellular, or virion stage

2. HIV's gp120 protein binds to CD4 and coreceptor on host cell.

3. HIV's RNA genome and reverse transcriptase enter host cell.

4. Reverse transcriptase synthesizes HIV DNA from HIV's RNA template.

5. HIV DNA integrates into host genome and is transcribed to HIV mRNA.

6. HIV mRNA is translated to HIV protein by host cell's ribosomes.

7. New generation of virions assembles inside host cell.

8. New virions bud from host cell's membrane.

FIGURE 22.9A Life Cycle of HIV

Medications such as AZT can be used to control HIV infections. The problem is that the virus eventually evolves resistance to such medication and resumes proliferation. The use of multiple medications against HIV has shown promise in forestalling progression to AIDS, but it is likely that this medication strategy will eventually cease to work as well, because of HIV evolution. Why is HIV so resistant to antiviral drugs? Figure 22.9A shows the life cycle of this virus. HIV is an **RNA virus**: It uses RNA to encode its genes. When it enters a host cell, HIV makes a DNA copy of its RNA, using reverse transcriptase. This DNA is then incorporated into the host cell's genome, where it is transcribed to make many copies of the virus RNA. This complicated life cycle gives RNA viruses very high mutation rates. The virus also produces many descendants. The combination of a high mutation rate and abundant offspring gives HIV enormous evolutionary potential. It is probably one of the fastest-evolving things on the planet. It is possible that HIV will evolve gene sequence changes that will evade virtually any antiviral compound that we throw at it. This is a supremely dangerous pathogen: We can't stop it, and it kills off cells vital to our survival.

But there are two additional levels to the ecological and evolutionary story of HIV. The ecology of HIV infection varies. In some geographical areas, like sub-Saharan Africa, it has spread widely. In some particular groups found in other geographical areas, it has spread widely: intravenous (IV) drug abusers, homosexual men, and prostitutes. Among these groups HIV appears to be particularly virulent, quickly spreading and quickly killing its hosts. But among other groups, like heterosexuals in Western countries who do not use IV drugs or sell their bodies for sex, the virus has spread slowly. It appears to be less virulent among such groups. This fits evolutionary theory. The groups with virulent HIV appear to have more opportunities for virulence to evolve, because transmission is more common. Paul Ewald has predicted that HIV should evolve toward more benign infection in the groups that do not transmit it as often. If so, the widespread abandonment of promiscuity and IV drug abuse could "domesticate" HIV. This would forestall our virtual extinction.

But what if people, being people, do not inhibit themselves? What is the prognosis for the HIV epidemic? There is some evidence that a tiny number of people are naturally resistant to HIV infection, despite repeated exposure. They resist progression to AIDS. In the worst-case scenario, the rest of us will die and these individuals will be the relics of the species. ❖

AGING

Evolution and genetics offer new hope for the medical treatment of the elderly

In the West, the twentieth century saw the practice of medicine advance from success to success across a broad front. Newborns survive more often. A number of contagious diseases have mostly been fended off by a combination of vaccines, antibiotics, and hygiene. Accident victims now receive outstanding emergency care. Heroic surgery rescues victims of heart disease and cancer from certain death. Overall life expectancies have increased in countries with modern medicine.

When the elderly get a particular disease, such as cancer, medicine has more tools to treat the disease. But the elderly have poor survival rates after medical treatment. And nothing is done about the fact that they have been greatly debilitated by aging: from sagging skin to wasting muscles, from diminished memories to chronic pain.

Unlike virtually every other medical specialty in the twentieth century, even psychiatry, **geriatrics** has not made much progress. One might blame medical practitioners, except that in this case geriatrics has not been slow to build on the successes of its relevant biological discipline, **gerontology**. Over most of the last century, such successes have been almost wholly absent. Geriatrics has not worked, because gerontology did not make progress with the biology of aging. At least until recently.

As outlined in Chapter 7 on life histories, we now have a fairly successful theory for aging: Aging evolves because of a decline in the force of natural selection during adulthood. This theory is nicely developed mathematically, and it works in laboratory experiments, so far principally with insects.

Since 1950, gerontologists instead pursued theories of aging based on molecular and cell biology, theories based solely on physiological mechanisms of aging. But they had little success. Each of their theories, from somatic mutation to error catastrophe to limited cell replication, has failed as a way to explain aging in general. The intrusion of evolutionary biology into the field has not been universally welcomed, but it has had some influence. More and more of the work in gerontology is guided by evolutionary considerations.

The area of gerontology that has been most influenced by evolutionary ideas is the genetics of aging. Since 1980, with the demonstration that genetic change could be used deliberately to postpone aging in fruit flies, more and more geneticists have found ways to create organisms that live longer genetically.

The first substantial successes came with the genetics of aging in the nematode worm, *Caenorhabditis elegans* (Figure 22.10A). Starting with **age-1**, nematode geneticists have found a multiplicity of genes that give increased life span when mutated. Many of these genes are part of the pathway that regulates an immature stage called the **dauer**. The dauer worm delays reproductive maturation. A controversy has recently arisen concerning the metabolism of genetically longer-lived worms. In some studies, all mutant alleles that increase life span also have reduced metabolic rates. In *Drosophila* research, it was shown a long time ago that reduced metabolic rate in a small cold-blooded animal can give increased lifespan. Reduced metabolic rate appears to "stretch" life. On the other hand, the longer-lived flies (see Chapter 7) have no reduction in metabolic rate, and their total metabolic work per lifetime is greatly increased.

However the details work out, evolutionary and genetic research on aging is now supplying the field of gerontology with the scientific foundations that it used to lack. This should have two effects:

1. It is now possible to go through the field of gerontology with evolutionary theory as a guide, winnowing out the failed science and, at the same time, pursuing the attractive scientific leads with some hope of making progress. Gerontology can now be as successful as such biomedical fields as immunology, oncology, and the like, at least in principle.

2. It is now reasonable to consider reformulating geriatrics. In the past, geriatricians have been fairly circumspect. Holding out hope of changing the human aging process to this point has been the hallmark of the charlatan, because geriatrics had found no tools with which to ameliorate the aging process itself. Now this possibility is at least worth discussing, because altering the aging process genetically and dietarily has become routine within a newly successful gerontology. ❖

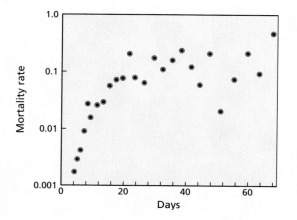

FIGURE 22.10A Aging in *C. elegans*

The starting point for Darwinian geriatrics must be the fact that humans are very long-lived organisms. This is partly because vertebrates are generally long-lived, compared with insects, nematodes, and other small invertebrate animals. There is nothing special about a vertebrate living for years, as opposed to the weeks and months that most animals live even when kept under excellent conditions. But humans can live for more than a few years. We can live for more than a century.

Part of the explanation for our longevity is related to size. Larger vertebrates tend to live longer. But that is not a complete explanation. We are also longer-lived for a large vertebrate. Some illustrative longevity data are shown in Table 22.11A.

What could explain greater human longevity? Old-style gerontologists were fond of explaining superior human longevity as a result of superior "homeostasis" or "canalization"—confusing and ambiguous concepts that derive from physiology.

But none of these theories ever worked, so we can safely dismiss them.

More data might be revealing. Table 22.11B shows some **maximum longevity** patterns associated with different ecologies. Certain patterns are striking. Organisms that have shells live much longer than similar organisms that lack

TABLE 22.11A Longevities of Some Large Mammals

Species	Common Name	Maximum Longevity
Balaena mysticetus	Bowhead whale	211 years
Homo sapiens	Human	122 years
Elephas indicus	Elephant	70 years
Rhinoceros unicornis	Rhinoceros	50 years
Hylobates lar	Gibbon	32 years
Papio papio	Baboon	27 years

TABLE 22.11B More Comparisons of Longevity among Species

A. Flying versus Nonflying Vertebrates

Taxonomic Group	Common Name	Maximum Longevity
Order Chiroptera	Bats	Up to 24 years, many over 10 years
Class Aves	Flying birds	Up to 68 years, many over 30 years
Class Aves	Non-flying birds	27–28 years
Order Rodenta	Rats, mice, and so forth	3–8 years

B. Reptiles

Taxonomic Group	Common Name	Maximum Longevity
Alligator sinensis	Alligator	52 years
	Snakes	Mostly more than 20 years
Genus *Testudo*	Tortoise	Several species more than 100 years

C. Invertebrates

Taxonomic Group	Common Name	Maximum Longevity
Coelenterata	Coelenterates	90 years (died by accident)
Schistosoma	Schistosomes	25–28 years
Lumbricus terrestris	Earthworm	5–6 years
Homarus	Lobster	50 years
Blaps gigas	Beetle	>10 years
Megalonaias gigantea	Bivalve	53–54 years
Octopus vulgaris	Octopus	3–4 years

them. Turtles, for example, often have maximum life spans of more than 70 years. Other reptiles have life spans averaging less than 20 years. Bivalves, like oysters, clams, and mussels, have life spans averaging more than 10 years, while other mollusks have life spans averaging less than 5 years. What is it about having shells that matters for aging?

A pattern that is perhaps even more obvious is the contrast between flying vertebrates and other vertebrates. Birds that have roughly the same size as rodents live far longer. Small rodents live just a few years, less than 10 maximum. Birds of the same size can live 40 to 60 years. Even within the birds, the flightless ostrich and emu have shorter life spans than all but a few flying bird species. Within the mammals, the flying bats have average life spans of 15–20 years, longer than rodents of comparable size.

The straightforward way to explain these patterns is the effect of organismal biology and ecology on the force of natural selection. Species that fly or have thick shells will tend to be preyed upon less often compared to species without these adaptations. So their natural ecology will usually be one in which they tend to live and reproduce to later ages. That difference in their ecology will in turn lead to a prolongation in the force of natural selection, following a pattern similar to that imposed on the fruit flies in the experiments on the evolution of aging. And, as those experiments showed, such a prolongation of the force of natural selection will tend to result in the evolution of postponed aging. This is how evolutionary biologists explain why flying and armored species tend to live longer.

But what of humans? Humans have a general-purpose mortality reducer—the human brain. With the human brain we are better able to avoid predators and accidents, thanks in part to prudence and in part to our tools. This must have had an effect analogous to that of flight or shells: Our survival and reproduction at later ages were enhanced, strengthening the force of natural selection at later ages. Strengthened natural selection then led to the evolution of a human body that could better withstand the ravages of time. We live longer because we are smarter. ❖

Postponement of human aging will be achieved by combining evolutionary and other biotechnologies

Despite the slow progression of human aging, the desire to postpone, slow, or reverse the process has been perennial. So long as medicine was generally impotent, the repeated failure of attempts to postpone human aging was not a notable blemish on the practice of medicine. Now that medicine has become generally effective, its failure to control the process of aging is prominent.

There have been two main responses to this situation. The medical establishment has generally eschewed the goal of postponing or slowing human aging, often with the implication that it is an impossible project. Recent successes in the postponement of aging in laboratory animals, however, suggest that this defeatism with respect to the postponement of human aging is outmoded.

The other response to the lack of medical interventions for aging has been to decry the medical establishment as conspirators intent on preventing the treatment of aging, perhaps because so much of medical practice derives from diseases that are associated with aging. What this criticism fails to address is the sheer difficulty of controlling aging. Many genes are involved, and finding even a few of them in simple organisms, like nematodes and fruit flies, has been difficult.

A rational compromise between these two views might be the following perspective from Darwinian medicine. Aging is a complex set of pathologies that are caused by a pervasive failure of adaptation at later ages. This pervasive failure in turn is due to a falling force of natural selection. Although this force of natural selection can be increased at later ages in simple animals, this is not a practical, or ethical, program for changing human aging. Therefore, any medical or biotechnological approach to postponing human aging will be difficult.

Difficult, but probably not impossible. The key is that many of the genes found in fruit flies and nematodes are represented in modified form in humans. If we can learn enough about manipulating the genes that control aging in simple lab animals, then we have the possibility of using such knowledge to develop antiaging medications and procedures that will work in humans.

This strategy cannot be based on the assumption that antiaging tricks that work in, say, fruit flies will always work in humans. But evolutionary homology connects many of the genes in fruit flies to genes in humans. Such homology has

been exploited experimentally in studies of development in fruit flies and humans. A major contributor to our scientific understanding of human development has been the study of the genetics and molecular biology of development in fruit flies, as described in the accompanying box. There is every reason to hope that a similar carryover can be achieved in research on aging, some of the time.

Figure 22.12A sketches a variety of research pathways that might lead to the medical postponement of aging in humans. Some of these pathways start with fruit flies, some with nematodes, and some with humans. The postponement of human aging requires that we use the full range of biological disciplines, from evolutionary biology to genetics to molecular biology. Because postponed aging has already been achieved in fruit flies and nematodes, we know that it is not impossible in human. But it will not be easy. ❖

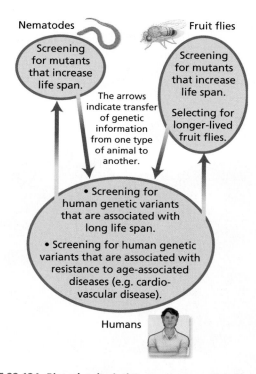

FIGURE 22.12A Biotechnological Postponement of Aging

From *Drosophila* to Mammal—Developmental Biology

Among the successes of recent cell biology has been the use of discoveries in fruit-fly research to understand mammalian development. Among the bizarre types of mutation in fruit flies are the homeotic mutations, which turn fly antennae into legs or duplicate the entire thorax, among other amazing effects. These effects occur because homeotic genes orchestrate the positional information underlying fruit-fly development.

But there's more. Mammals have genes with similar DNA sequences to those of fly homeotic genes, as do many other animals.

It turns out that the homologous mammalian genes play a similar homeotic role, controlling the use of positional information in mammalian development. The fruit-fly work set the stage for a major step forward in our understanding of mammalian development. A similar boost for geriatrics might come from studies of the genetics, evolution, and physiology of aging in fruit flies and nematodes.

It is brains that produce behavior. Therefore, aberrations in that behavior may reflect an underlying **brain disorder**. As with the symptoms of contagious disease, not all of the symptoms that concern psychiatrists are manifestations of brain disorder. Behavior that radically reduces fitness is likely to reflect an underlying brain disorder. But behavior that does not reduce fitness, however odd, may not be due to a brain disorder. Such behavior might even enhance fitness, in some cases.

Psychiatry tends to focus on behavior that is deviant from social norms. It is common for families to bring their deviant members to the attention of health-care professionals. In many cases the deviant behavior is a symptom of a significant brain disorder. But there are also cases of deviant behavior that do not lead to an impairment of fitness, and these cases are not medical problems from the standpoint of Darwinian medicine.

Phobias and other **anxiety** disorders are extremely common in human populations. People are afraid of snakes, heights, confinement in small spaces, crowds, and so on. Often they are troubled by their fears and anxieties, and seek out medical doctors to give them prescriptions for tranquilizers. All too often they then become addicted to these tranquilizers. Many people also use alcohol and tobacco to reduce their anxieties, and so become alcoholics and victims of lung cancer.

Yet anxiety and fear are basic brain functions that serve to keep us out of trouble. Anxiety and fear mobilize the autonomic nervous system in ways that foster "fight or flight": increased heart rate, rapid breathing, increased blood flow to the muscles, and so on. When severe, these symptoms are considered "panic attacks" by medical doctors. But what if fleeing is an appropriate reaction to danger, whether the danger comes from a snake or a vengeful enemy?

Anxiety and fear are mobilizing emotions that serve important functions. Individuals who suffer extremely high levels of these emotions may have a significant disorder. But in many cases these disorders may involve normal Darwinian brain functions that people find distressing. They could be analogs of diarrhea and fever, which can save lives but are not pleasant.

One way to view anxiety disorders is as part of a spectrum. Some individuals undoubtedly suffer too little anxiety. This causes them behavior problems that may result in their early death or injury. Other individuals are virtually paralyzed by fear, which prevents them from going outside to shop, socialize, or date. People at both of these extremes may suffer from some reduction in Darwinian fitness, as a result.

Most of the human population suffers from anxiety, but not so much as to reduce Darwinian fitness. Almost everyone would be fearful or anxious if kidnapped. Likewise, they would be fearful if a mugger pointed a gun at them. The fear would mobilize them to flee or fight, based on their body's physiology, even if they do not do either. In our ancestral environment, contact with poisonous snakes or large predators probably had the same effect. Fear and anxiety are adaptations to a dangerous world, not psychopathologies. Only at their extremes of predominance or absence are they problems of psychiatry, from a Darwinian perspective. And it is notable that the complete absence of anxiety and fear might be the greatest problem of all. This analysis is shown graphically in Figure 22.13A. ❖

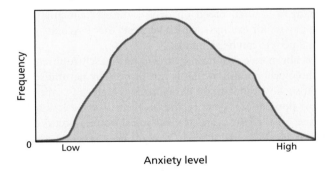

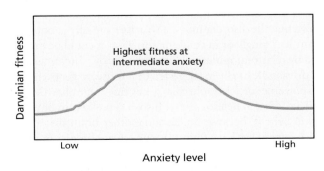

FIGURE 22.13A A Hypothetical Spectrum of Anxiety and Its Possible Consequences for Darwinian Fitness

Schizophrenia is characterized by delusions, hallucinations, and other failures to differentiate reality from other subjective experiences. Everyone experiences **hallucinations** and **delusions** during dreams. But we know that they are dreams, and we discount their relevance to our everyday lives. Schizophrenics have difficulty testing the reality of their hallucinations and delusions. The diagnostic medical criteria for schizophrenia are given in this module.

Along with this disconnection from reality, schizophrenics may experience paranoia, intense fear or anxiety, and depression. Their IQs tend to be reduced. Unmedicated schizophrenics have great difficulty keeping a job, sustaining a marriage, or retaining friends.

It is difficult to build a reasonable case for natural selection specifically favoring the maintenance of schizophrenia. Schizophrenics have reduced survival rates and reduced mating frequencies. They are often kept in psychiatric institutions, or they live on the streets as homeless vagrants. Medicated schizophrenics have remission of 30–70 percent of their symptoms and may function at a job or in a family. But sufferers often do not comply with medical treatment.

There is one historically important theory based on the idea of schizophrenia as a functional condition. This is the "schizophrenic shaman" theory of anthropology. It proposes that schizophrenics may have functioned as tribal shamans, using their hallucinations and delusions to guide the tribe spiritually. It is conceivable that schizophrenics could have played this role. But that leaves unaddressed the question of how these schizophrenics would have found a mate and produced children. Perhaps their shaman status helped them to obtain a mate? A further problem with this theory is that shamanism is often based on drug-induced hallucinations, using peyote for example, which suggests that psychiatrically normal people can be shamans.

The alternative evolutionary theory is that schizophrenia is a pathological condition that is not favored by natural selection. If so, how can it attain frequencies as high as 1 percent in human populations? Part of the problem may be lesions to sensitive areas of the brain. If mechanical trauma occurs during fetal development or birth, it might later generate schizophrenia. Or infections, chronic and acute, could cause such lesions. Ewald has proposed multiple pathogens as causes of schizophrenia, from *Toxoplasma gondii* to Borna disease virus. There is also evidence that schizophrenia is inherited. This suggests that the disorder might also arise from rare genotypes at many loci, singly or in combination. If so, it could be maintained by recurrent mutation, like cystic fibrosis and other genetic diseases. It has been suggested that recessive genes might be responsible for schizophrenia, genes that might have some beneficial effect in heterozygotes. If so, it is unlikely to be a behavioral benefit, because unlike some other brain disorders, no such benefit has been found among schizophrenics or their near relatives. (Such cases are discussed later in the chapter.) It is remotely possible that a gene for schizophrenia might have some beneficial effect on a totally unrelated medical problem, like heart disease, perhaps in heterozygotes. There is a little evidence for greater reproductive success among the near relatives of schizophrenics. For now, the most reasonable assumption is that schizophrenia is a deleterious disorder maintained in populations by developmental misfortune, disease lesions, recurrent deleterious mutations, or a combination of all these agents.

Diagnostic Criteria for Schizophrenia

A. Characteristic Symptoms: Two or more of the following
  1. Delusions
  2. Hallucinations
  3. Disorganized speech
  4. Catatonic behavior
  5. Flat affect; lack of speech

B. Social or Occupational Dysfunction
  Reduced functioning in one or more major areas:
  1. Work
  2. Interpersonal relations
  3. Self-Care
  4. Academics (when appropriate)

C. Duration
  Continuous signs of the disturbance must persist for at least 6 months

D. Exclusions
  1. Affect disorders have been ruled out as a cause of aberrant behavior.
  2. Substance abuse has been ruled out as a cause of aberrant behavior.
  3. Another medical condition, such as a brain tumor, has been ruled out.

E. History of a Major Developmental Disorder
  If autism or a similar disorder has been previously diagnosed, diagnosis of schizophrenia requires prominent delusions or hallucinations for at least a month.

From Diagnostic and Statistical Manual of Mental Disorders (DSM-IV), *4th ed. (Washington, DC: American Psychiatric Association, 1994), 285–86. Not to be used for medical diagnosis.*

Explanation of some of the terminology:

  Delusions—incorrect and persistent conclusions about the material or social environment

  Hallucinations—sensory experiences that do not correspond to externally observable events

  Catatonia—lack of responsiveness, lack of awareness of surroundings

  Flat affect—lack of joy or distress in circumstances where they are appropriate

  Self-care—getting adequate hygiene, nutrition, sleep

  Affect disorders—see below

  Autism—a pervasive failure of normal social and intellectual development in children not due to gross brain abnormality

22.15 It is unclear whether all affect disorders are actively sustained by natural selection

Most people get depressed. When they lose a job, a family member, or a significant other, many people go through a period of feeling little enthusiasm, disrupted sleep, changed eating habits, and difficulty concentrating. If this condition lasts for a few days, it is entirely normal. If it lasts for six months, then it is clinical depression. At least 10 percent of a human population will experience a clinical depression once in their lives. In some countries, this percentage is even higher. The diagnostic criteria for **depression** are given in this module.

A few people get high without drugs or alcohol. They experience great surges of energy during which they cannot sleep, cannot stop talking, are obsessed with sex, and spend money unwisely. This is **mania**. Many people experience a mild version of this pattern when they fall in love, graduate from college, or suddenly receive a large quantity of money. But for most of us, this "natural high" lasts no more than a few days. In manic depressives, the high may last for weeks, even months. Only 1 percent of a population, or less, experiences sustained mania. When mania ends, it is often followed by clinical depression, though there are rare manics who are never depressed. The diagnostic criteria for mania are also given in this module.

Clinical depression and sustained mania are **affect** disorders, the term *affect* referring to sustained positive or negative emotion. Affect is virtually a human universal. Affect disorders are very common, especially depression without a preceding period of mania.

Individuals who never experience mania, but do suffer from depression, are grouped separately from those who experience mania. Depression is analogous in many ways to anxiety disorders: Both are unpleasant experiences that interfere with functioning when they are extreme and sustained. But most people experience negative affect and anxiety. Doctors try to medicate either condition when their patients complain of suffering. But depression after the death of a child seems about as functional as phobia toward snakes or heights. If parents felt little sense of loss when a child dies, then they would be less likely to avoid circumstances that could kill or harm their other children. Depression, like anxiety, is an adaptive brain function that may reach an extreme in some patients. Being happy is not a Darwinian imperative. The Darwinian imperative is to propagate your genes. Depression can be thought of as an instance of signaling within the brain that gears behavior toward fitness enhancement. It may be hypertrophied in some individuals, in the same way that some of us are very tall or very short. But it is not necessarily a brain disorder.

Manic depression is quite different. Manic depressives have a 20–25 percent increase in their death rates, mostly from suicide, but also from substance abuse and other dangerous behaviors. Furthermore, manic depression is strongly inherited, with a heritability that exceeds 50 percent. If one identical twin is manic depressive, the other twin is also manic depressive about 70 percent of the time. Why are about 1 percent of humans subject to this disorder? It could be that manic depression is another brain disorder, like schizophrenia, maintained by mutation and brain damage. But there is evidence for this condition, unlike schizophrenia, of some beneficial effects. Both manic depressives and their close relatives show a tendency toward verbal creativity. Among great English poets, more than 50 percent probably had manic depression. There is no such association with disorders involving schizophrenia or anxiety. In some obscure way, manic depression is associated with an enhanced brain function. This makes it plausible that natural selection might maintain genes that foster manic depression. But in no way does it establish this conclusion.

Diagnosis of a Major Depressive Episode

A. Five (or more) of the following symptoms have been present during the same 2-week period and represent a change from previous functioning; at least one of the symptoms is either (1) depressed mood or (2) loss of interest or pleasure.

Note: Do not include symptoms that are clearly due to a general medical condition, delusions, or hallucinations.

1. Depressed mood most of the day, nearly every day, as determined from either subjective reports or observation of behavior. May also manifest as irritability.
2. Markedly diminished interest in most normal activities.
3. Significant weight loss without dieting, or appetite change.
4. Increased or decreased quantity of sleep.
5. Either marked sluggishness or hyperactivity.
6. Fatigue every day.
7. Excessive feelings of worthlessness or guilt.
8. Diminished ability to concentrate; Indecisiveness.
9. Recurrent thoughts of death, thoughts of suicide, planning suicide, or attempting suicide.

B. These symptoms cannot be accounted for by another psychiatric syndrome, drug abuse, hormonal problems, or recent bereavement.

C. These symptoms cause significant distress in social, occupational, or other important activities.

From Diagnostic and Statistical Manual of Mental Disorders (DSM-IV), *4th ed. (Washington, DC: American Psychiatric Association, 1994), 327. Not to be used for medical diagnosis.*

Diagnosis of Mania

A. Persistently elevated, expansive, or irritable mood, either lasting for more than one week or requiring hospitalization.

B. Three or more of the following symptoms have been significantly present:

1. Inflated self-esteem
2. Decreased need for sleep
3. Highly talkative
4. Racing thoughts
5. Distractibility
6. Increase in sexual or social behavior; physical agitation
7. Excessive spending or investment; sexual indiscretions

C. Symptoms do not meet criteria for other affect episodes.

D. Marked impairment of work, social, or family interactions; threat to self

E. Exclusions: Symptoms not due to

1. Medication
2. Substance abuse
3. General medical condition (e.g., hyperthyroidism)

From Diagnostic and Statistical Manual of Mental Disorders (DSM-IV), *4th ed. (Washington, DC: American Psychiatric Association, 1994), 332. Not to be used for medical diagnosis.* ❖

22.16 The sociopath combines subjective well-being with pathological Darwinian outcomes

Personality disorders give rise to behavior that is not usually "crazy," but law enforcement and families have considerable difficulty dealing with individuals who have personality disorders. Vague popular notions of personality disorders are common. **Narcissists**, **histrionics**, and **sociopaths** are just three personality disorders that are widely misunderstood. Many women think that their ex-husband or their attorney is a narcissist. Many men think that their ex-wife or girlfriend is a histrionic. And everybody thinks, at least sometimes, that unscrupulous politicians and bad drivers are sociopaths.

There may be some truth to all of these suspicions. There are more than ten clinically recognized personality disorders, each with a long list of diagnostic criteria and clinical features. Most of the personality disorders blend into each other, and patients with multiple disorders are commonly diagnosed by psychiatrists. However, one personality disorder stands out from the rest: antisocial personality disorder, which is often called sociopathy. This disorder is the most reliably diagnosed, among the personality disorders, and it has the most profound consequences. The diagnostic criteria for this disorder are given in this module.

Sociopaths generally lack loyalty, love, grief, and so on. Sociopaths commit a lot of crimes, perhaps as many as half of all crimes against property, despite making up only about 1 percent of most populations. This should give them a lot of experience at evading capture. Instead, they are remarkably easy to arrest. There is nothing unusual about a sociopath being captured within a few yards of a police station.

One of the most interesting features of sociopaths is their freedom from many types of psychological aberration or distress. Their intelligence is not markedly reduced. They do not exhibit clinical depression or mania. In many ways, they are relaxed. Yet their lives are completely disordered. They do not sustain marriages or careers. Often they are highly promiscuous; most unusually, sociopathic women may have a predilection for group sex and other dangerous sexual practices. Clinical case histories record a pattern of extensive social deviance, ranging from murder to inappropriate practical jokes.

It has been proposed repeatedly in the scientific literature that sociopaths are cheaters from the standpoint of evolutionary game theory; and this cheating, it has been argued, is evolutionarily beneficial. In other words, it has been argued that sociopaths do not suffer from a Darwinian brain disorder—unlike, say, schizophrenics. Sociopaths just take whatever they need and ignore the consequences.

This model might make sense of other personality disorders, perhaps narcissism or histrionics, which can be successfully manipulative. But the problem with applying it to sociopaths is that sociopaths do not follow through on long-term plans of any kind. It's not that they get into trouble because they are extraordinarily selfish in their choices. They are sometimes very generous. They just do not seem to be able to follow any sort of strategy.

An alternative Darwinian interpretation of this disorder is that it arises when there is a breakdown in signaling from one part of the brain to another, signaling that is normally experienced in the form of both positive and negative affect, anxiety and contentment. (This idea is outlined at the end of Chapter 21.) That is, sociopaths may suffer from a class of brain disorders in which the direction of behavior toward Darwinian ends fails. In a sense, then, sociopaths could be Darwinian idiots.

One piece of evidence in favor of this model is the phenomenon of pseudo-psychopaths. When people sitting in the front passenger seats of cars do not use seat belts, their crashes become experiments in brain function. It is not uncommon for people who go through car windshields to receive extensive damage to the **frontal lobes** of their brain, above and behind the eyes. Some of these patients exhibit a "couch potato" syndrome. They tend to lose motivation, overeat, and watch television programs endlessly. Other patients seem to become instant sociopaths: socially aggressive, self-centered, and uninhibited (see the box). Lesions to the frontal lobes thus can create most of the elements of the sociopathic personality. This suggests that sociopaths are not examples of a particular strategy choice. Instead, they lack at least some of a Darwinian capacity of coordinating choices to the end of increasing fitness. Some part(s) of their brains have lost key functions. Sociopathy is a Darwinian brain disorder, on this interpretation.

General Diagnostic Criteria for Personality Disorders

A. Consistent deviation from social expectations in two (or more) of the following respects:

1. Perception of self and other people
2. Appropriateness of affect
3. Social interactions
4. Impulsiveness

B. The deviant pattern is pervasive across a broad range of situations.

C. The deviant pattern causes distress or impairment of functioning in social, occupational, and other respects.

D. The pattern has been stable since adolescence or early adulthood.

E. The pattern cannot be accounted for in terms of some other psychiatric disorder.

F. The pattern cannot be explained by drug use, medication, trauma, or any other medical condition.

Diagnostic Criteria for Antisocial Personality Disorder

A. Pervasive disregard of the rights of others since adolescence, as indicated by three (or more) of the following symptoms:

   1. Repeated criminal acts

   2. Recurrent lying, assumption of false identities, "conning"

   3. Failure to plan ahead

   4. Repeated physical fights or assaults

   5. Reckless disregard for the safety of others

   6. Repeated dismissal at work, failure to repay loans

   7. Lack of remorse

B. The diagnosed individual is at least 18 years old.

C. Similar behavior before the age of 15 years.

D. Antisocial behavior is not confined to psychotic episodes of schizophrenia or mania.

From Diagnostic and Statistical Manual of Mental Disorders (DSM-IV), *4th ed. (Washington, DC: American Psychiatric Association, 1994), 633, 649. Not to be used for medical diagnosis.* ❖

The Mysterious Case of Phineas Gage

Phineas Gage was 25 years old in Vermont's summer of 1848. He was in charge of a railroad construction gang, and he was highly regarded by his employers. While setting up an explosive charge Phineas was distracted and inadvertently set off an explosion at close range. An iron rod blasted into his face, shooting through the front part of his brain and out through the top of his skull. Amazingly, Gage regained consciousness soon after the accident. Due to the close care of a physician, Gage survived the infections that resulted from his injury and was pronounced cured in two months. His physical recovery was excellent.

But Gage's personality was transformed. He seemed to have lost his social inhibitions, becoming newly irreverent, capricious, and profane in speech. He no longer carried through on his plans. His employers wouldn't take him back. The only steady work he could find was as a "freak" in a traveling circus, where his wounds were exhibited to the public. Gage's life was forever disordered after his accident.

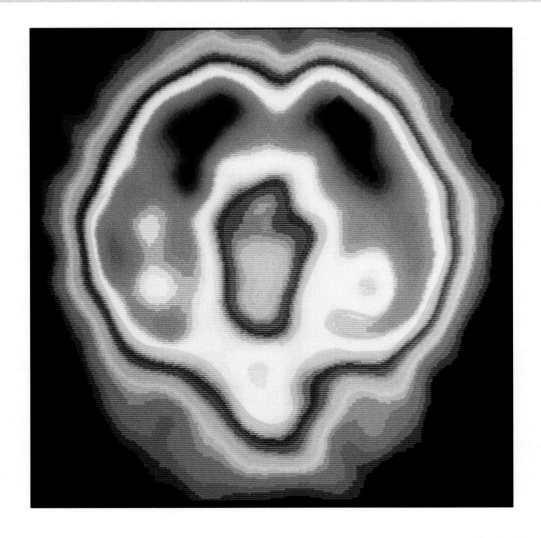

SUMMARY

1. Medicine has undergone considerable reform in the twentieth century, thanks to a substantial input of information from biological research. However, in the past this transfer of information from biology to medicine has largely neglected the Darwinian part of biology. In recent years, Darwinian biologists have attempted to rectify this oversight by contributing to the resolution of problems in medicine, ranging from contagious disease to aging to psychiatry.

2. One of the best-developed applications of Darwinian thinking to medicine has been the study of genetic diseases. One of these applications is the interpretation of the origin of genetic diseases. Darwinian biologists can also shed light on the prospects and value of engineering germlines and somatic cells.

3. The medicine of contagious disease has received some novel criticisms from Darwinian biologists. Of particular note is the differentiation of adaptive responses from pathological symptoms. Adaptive responses, like vomiting, may be distressing; but they may also save the patient's life if they are allowed to proceed. Physicians may overmedicate symptoms, thereby fostering the death of their patients.

4. Human aging is a problem with which conventional medicine has made little progress, at least in the past. Evolutionary analysis suggests that aging is a quintessential product of evolutionary mechanisms, not a side effect of cell biology. This insight suggests strategies for eventually treating human aging.

5. Psychiatry has never had a successful theoretical framework. Darwinian models of the human mind suggest one way to construct such a theoretical framework. As with the Darwinian analysis of contagious disease, it is important to distinguish between normal human brain functions that may cause distress and brain disorders that reduce Darwinian fitness. With such distinctions in mind, it is possible to make some sense of such phenomena as schizophrenia, phobia, depression, mania, and sociopathy.

REVIEW QUESTIONS

1. Why does evolutionary history produce universal human imperfections?

2. Pick one human genetic disease and analyze it from a Darwinian standpoint.

3. How plausible is it that HIV could eventually wipe out most of the human species?

4. If HIV did wipe out most humans, what would be the evolutionary prospects for the survivors?

5. Why should you take all the antibiotic that your physician prescribes?

6. If you had AIDS, would rather stay home or go to a hospital?

7. Are you, or someone you know, taking any measures to slow your aging? If you or they are, do you think that this measure will be effective?

8. Why do some people live more than 100 years, while others die from "natural causes" before reaching 50?

9. Do you have a relative with a major behavioral problem? If you do, analyze their condition from a Darwinian perspective.

10. Would you want to have a sociopath as a close friend?

KEY TERMS

ADA deficiency	delusion	hallucination	phenylalanine
affect	depression	histrionic	RNA virus
age-1	disease vector	HIV	SARS
AIDS	Ebola	malaria	schizophrenia
antibiotic	epiglottis	mania	smallpox
antibiotic resistance	Ewald, Paul W.	maximum longevity	sociopath
antiviral	frontal lobes	narcissist	somatic engineering
anxiety	Gage, Phineas	Nesse, Randolph M.	trachea
brain disorder	genetic disease	phobia	vaccine
choking	geriatrics	personality disorder	virulence
cholera	germline engineering	pharynx	Williams, George C.
dauer	gerontology	phenylketonuria (PKU)	

FURTHER READINGS

American Psychiatric Association. 1994. *Diagnostic and Statistical Manual of Mental Disorders, DSM*, 4th ed. Washington, DC: American Psychiatric Association.

Comfort, A. 1979. *The Biology of Senescence*, 3rd ed. Edinburgh: Churchill Livingstone.

Damasio, A. R. 1994. *Descartes's Error: Emotion, Reason, and the Human Brain*. New York: Avon.

Ewald, P. W. 1994. *Evolution of Infectious Disease*. New York: Oxford University Press.

Nesse, R. M., and G. C. Williams. 1994. *Why We Get Sick: The New Science of Darwinian Medicine*. New York: Random House.

Rose, M. R. 1991. *Evolutionary Biology of Aging*. New York: Oxford University Press.

Stearns, S. C., ed. 1999. *Evolution in Health and Disease*. Oxford, UK: Oxford University Press.

Stock, G., and J. Campbell, eds. 2000. *Engineering the Human Germline: An Exploration of the Science and Ethics of Altering the Genes We Pass to Our Children*. New York: Oxford University Press.

APPENDIX A

Random Variables and the Rules of Probability

Random Variables

Statistics is a process of making inferences about random phenomena. For instance, when we roll a die we can never know for certain what number will come up, but we can assign probabilities of observing any particular result. Random events may be a process, like flipping a coin or rolling a die. These phenomena fall into discrete categories. Random phenomena, like the weight of a randomly drawn person, vary continuously. Statistics methods can accommodate either type of random phenomena.

Our understanding of random processes starts by taking a random sample of observations from a well-defined population. When we take a sample, it is more likely that we will get samples from the middle of the distribution as shown in Figure A.1, since members of this population are more likely to fall into this middle range.

One important property of a sample is its central tendency or mean. Here we will symbolize the mean of a sample as $\bar{x}$. Suppose we have a sample of k observations. To compute the sample mean, we would simply add up all k observations and divide by the sample size, or k[1]. We show a numerical example of the sample mean in Table A.1. In this example the number of eggs laid by 21 different female fruit flies were used to compute a sample mean.

[1]If we represent the value of each observation as $x_1, x_2, \ldots, x_k$, then the *sample mean* is defined as

$$\bar{x} = \frac{1}{k}\sum_{i=1}^{i=k} x_i$$

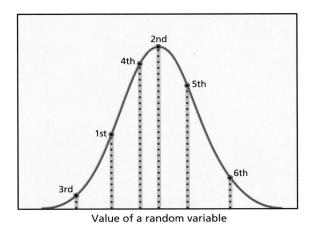

Value of a random variable

FIGURE A.1 A sample of six random variables showing the order in which the samples were taken. The values of each random variable would be read from the x-axis where the lines from each variable intersect the x-axis. The bell-shaped curve shows the relative probability of finding random variables at each of the different values.

It would be unusual for the sample mean to be exactly equal to the true mean. For instance, if we flip a fair coin 100 times, the number of heads is a random variable. The true mean for this random variable is 50. However, the chance of seeing exactly 50 heads out of 100 tosses in any single sample is only 0.08,—most samples will have slightly more or slightly less than 50 heads. It stands to reason that the larger the sample, the closer the sample mean will be to the true mean. Statistical theory can help us determine how far away from the sample mean the true mean might be.

To accomplish this last task, we first need some estimate of the variability of our random variables. By *variability* we mean how far, on average, the random variables are likely to be from the true mean. The standard estimate of variability is the sample variance. To estimate the variance, we compute the average of the squared differences of each observation and the mean, for example, $(x - \bar{x})^2$. In Table A.1 we have also calculated the sample variance for the fruit-fly egg data.[2]

Rules of Probability

An important concept in statistics is independence. Two random events, A and B, are statistically independent if the probability of event A does not depend on or is not altered by the probability of event B. Many important problems in genetics concern the calculation of the probability of two independent events simultaneously occurring. For instance, we may want to know the chance of inheriting an A allele from your mother and an a allele from your father. The chance of two independent events simultaneously occurring is just the product of each occurring separately. In Figure A.2, the probabilities of events A and B are proportional to their areas. The area of overlap represents the chance of both happening together and in the case of independent events would be equal to the product of the two probabilities.

In some problems we may be interested in the chance of either of two or more mutually exclusive events happening. For instance, we might want to know what the chance is that an offspring is either genotype AA or Aa. Obviously, a single offspring cannot be both of these genotypes at the same time, so the events are mutually exclusive. In this case the probability is equal to the sum of each of the events.

[2]Formally, the sample variance is estimated from a sample of k observations as

$$s^2 = \frac{1}{k-1}\sum_{i=1}^{i=k}(x_i - \bar{x})^2$$

The sum is divided by $k - 1$ rather than k to eliminate bias that arises when k is small.

TABLE A.1 Sample Calculations of the Mean and Variance for the Number of Eggs Laid per Week by 21 Fruit Flies		
Sample Number	Eggs per Female (x)	$(x - \bar{x})^2$
1	161	8890
2	116	19,401
3	90	27,319
4	206	2429
5	145	12,163
6	172	6937
7	152	10,668
8	187	4663
9	194	3756
10	193	3880
11	179	5820
12	212	1874
13	478	49,602
14	414	25,190
15	452	38,697
16	260	22
17	397	20,083
18	183	5225
19	346	8229
20	365	12,037
21	459	41,500
Sum	5361	Sum 308,382
Average = (sum/21) =	255	s^2 = (sum/20) = 15,419

Two independent events, *A* and *B*

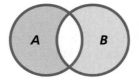

The white area shows the intersection of *A* and *B*, where both events will simultaneously occur.

Two mutually exclusive events, *A* and *B*

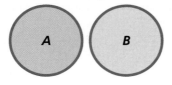

The chance of either event occurring will be equal to the sum of *A* + *B*.

FIGURE A.2 Each colored circle represents the probability of a different event. For events to simultaneously occur, they must show some overlap. The area of overlap is proportional to the chance of simultaneous occurrence. Events that do not overlap are mutually exclusive.

STATISTICAL DISTRIBUTIONS AND CORRELATION

Statistical Distributions

A statistical distribution is a mathematical description of the chance that a random variable assumes particular values. For a normal distribution (Figure B.1), the shape of the function is described by the mean and variance. The normal distribution is the familiar bell-shaped distribution that describes the distribution of many random variables with a continuous distribution. Because the normal distribution is symmetric, its mean value occurs at the peak of the bell shape. In Figure B.1 part (i), the two normal distributions have the same variance but different means; thus their peaks occur at different positions on the x-axis. We see that this simply causes one distribution to be shifted to the right or left of the second distribution. Figure B.1 part (ii) shows two distributions with the same mean and different variances. When the variance is small, random variables tend to be much closer to the mean; it is unlikely to find random variables with values much less than or much greater than the mean.

For continuously distributed random variables, the probability that the random variable falls into a certain range is equal to area under the curve over that range. For a normal distribution like those in Figure B.1, the area under the curve from $-\infty$ to $+\infty$ is equal to 1, because all values of the random variable must fall between this range. For a discrete random variable, we actually compute the probability of the possible values of the random variable.

A particularly useful discrete distribution is the binomial distribution. For random events having two possible outcomes that are typically referred to as success or failure, the binomial distribution tells us how likely it is to get x success-es out of N trials. As an example, the number of heads out of 10 coin flips would have a binomial distribution. If we sampled 100 mice from a field site, the number of males in the sample would also have a binomial distribution. To describe the binomial distribution, we need to know the total number of trials and the probability of a success (Figure B.2).

Confidence Intervals

Confidence intervals are the statistical tools used for computing uncertainty in estimates. Confidence intervals can be calculated for sample means, variances, or many other estimated statistics. The confidence interval will have a lower and upper bound that bracket the statistic of interest. Each interval has an associated confidence, which is often 95 percent but may be some other value.[1] This interval has the interpretation that we are 95 percent certain that the true value of the

---

[1] We need to use a t distribution to construct confidence intervals on a mean. The t distribution has a mean of zero and a variance that depends on a parameter called degrees of freedom. For a sample of k observations with a sample mean $\bar{x}$ and variance s^2, we estimate the standard error (se) as

$$\sqrt{s^2/k}.$$

As the sample size increases, the standard error gets smaller. A 95 percent confidence interval about the mean is ($\bar{x} - se \times t_{0.975,\, k-1}$, $\bar{x} + se \times t_{0.975,\, k-1}$), where $t_{0.975,\, k-1}$ is a t random variable with $k - 1$ degrees of freedom that is greater than 97.5 percent of all t values. The values of the t statistic can be obtained from tables in most standard statistical textbooks or from software like Excel or R.

(i) Same variance, different means

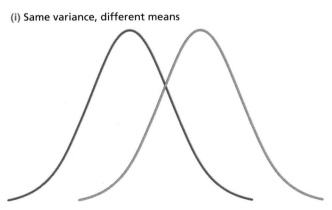

(ii) Same mean, different variances

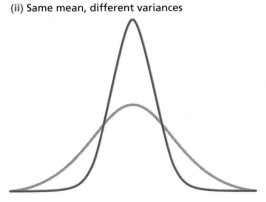

FIGURE B.1 Distributions with different means and variances.
(i) The two curves show normal distributions with the same variance but different means. (ii) These two curves have the same mean, but the narrower curve has a smaller variance than the wide curve.

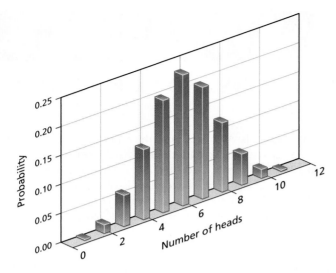

FIGURE B.2 The probability of obtaining the number of heads shown on the x-axis after flipping a fair coin 12 times. These probabilities are computed from a binomial distribution with 12 trials and a probability of success equal to 0.5.

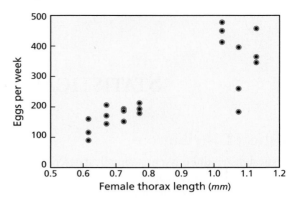

FIGURE B.3 The relationship between two random variables in individual female fruit flies: thorax length (size) and number of eggs laid during the first week of life.

statistic is greater than the lower limit but less than the upper limit. When making graphs of experimental results, confidence intervals may be displayed around estimated means. This is a useful application of confidence intervals since it conveys a level of uncertainty that has a direct probabilistic interpretation. You must always be careful, however, when interpreting error bars because many people display standard errors rather than confidence intervals. You cannot attach a simple probability statement to a standard error. Standard errors are smaller than confidence intervals, so many people use them because they believe it makes their data look better.

Correlation and Independence

Sometimes two random variables can be measured in a single sample. If we call the first random variable x and the second y, then a sample consists of a series of paired values. For example, we can sample people at random, measuring the height and weight of each person. A natural question is whether the two random variables are related to each other. In the case of height and weight, it would seem reasonable that in general taller people will weigh more. In Figure B.3 we show measurements of female size in fruit flies and the numbers of eggs laid by those females.

Although the relationship is not perfect, it is clear that the larger females tend to lay more eggs. One way to study the relationship between two random variables, like size and fecundity, is with the correlation coefficient. The correlation

coefficient[2] is a number between -1 and $+1$. Positive correlation coefficients imply that two random variables tend to change in the same direction. The results in Figure B.3 are one such example of a positive correlation. A negative correlation means that as one variable increases, the other decreases. For the data plotted in Figure B.3 the variance of female size is 0.0414 mm^2, the variance in fecundity is 15,400 eggs^2, and the covariance is 20.9 mm $\times$ eggs. The correlation coefficient is thus 0.83. This value supports the visual impact of Figure B.3 that there is a strong positive relationship between female size and fecundity.

An important concept in statistics is independence. Two random events, A and B, are statistically independent if the probability of event A does not depend on or is not altered by the probability of event B. Typically, while taking samples from a population, we need each of the samples to be independent of each other. If this is not true, then the sample will not be expected to have the same properties as the population. Because independent events cannot depend on each other, the correlation between two independent events has to be zero. However, it is not always the case that two random events with a zero correlation are independent.

---

[2]We can estimate the correlation coefficient from a sample of k observations—$(x_1, y_1), (x_2, y_2), \ldots (x_k, y_k)$—by estimating the variance of x (s_x^2), the variance of y (s_y^2) and the covariance of x and y. The covariance is defined as,

$$\text{cov}(xy) = \frac{1}{k-1} \sum_{i=1}^{i=k} (x_i - \bar{x})(y_i - \bar{y})$$

The correlation coefficient is then computed from the covariance as,

$$\rho_{xy} = \frac{\text{cov}(xy)}{\sqrt{s_x^2 s_y^2}}$$

LINEAR REGRESSION AND THE ANALYSIS OF VARIANCE

Linear Regression

In the absence of any other information, we might use the mean value of a random variable as our "best" guess at an unobserved value. However, for some random phenomena we can do better than this because the random variable of interest is related to a quantity we can measure.

In Figure C.1, the log of the observed metabolic rates appear to increase in a linear fashion with the log of the size of the organism. The solid line in Figure C.1 is the best-fitting line. Although this line does not go through every observed point, it comes very close to most of them. Clearly, this line does a much better job of predicting metabolic rate from the known size of the organism than simply using the mean metabolic rate of all these different species.

The general equation for a straight line is $y = mx + b$. The statistical question we are faced with is: How can we use a set of x and y observations to estimate the parameters, m and b, of a straight line? The answer is related to quantities we have already talked about. Thus, the slope of the line is estimated as the product of correlation coefficient and the square root of the variance of x divided by the square root of the variance of y. The y-intercept (b) is estimated by the mean of y minus the product of the slope and the mean of x.[1] From these definitions, it is obvious that the slope of the regression line will have the same sign as the correlation coefficient. These estimates are called *least squares*, because they minimize the squared difference between each observed value of y and its predicted value from the linear equation.

Regression can be used for several different purposes. One traditional use of regression is for prediction. For instance, from the data on female size versus fecundity, we have no observations for females between 0.77 and 1.03 mm. However, from our regression model we can make predictions of how many eggs females in this size range would be expected to produce in one week (Figure C.2). We would predict that a female with a thorax length of exactly 1 mm would lay 327 eggs in one week.

Regression can also be used to help us understand biological processes. For instance, in this book we discuss how the

growth of a population can be described by the logistic function that depends only on the current population size. There are other mathematical relationships between population density and population growth rate that might better describe the biological processes of density-dependent population growth. By using regression techniques to estimate model parameters, we can determine if there are substantial and important differences in the ability of these models to describe population growth.

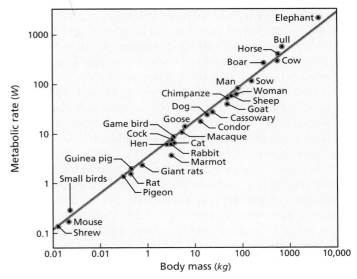

FIGURE C.1 The relationship between log metabolic rate vs. log body mass in a variety of different animals.

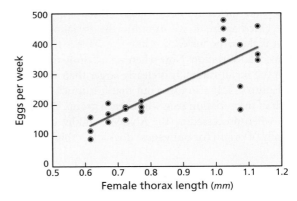

FIGURE C.2 The data in Figure B.2 has been replotted with the least-squares fit to a straight line (red). The least-squares estimates for the red lines are $b = -178$, $m = 505$.

[1]Suppose we have a set of k-observations, $(x_1, y_1), (x_2, y_2),.., (x_k, y_k)$. Then the least squares estimates are,

$$\hat{m} = \frac{\sum_{i=1}^{i=k}(x_i - \bar{x})(y_i - \bar{y})}{\sum_{i=1}^{i=k}(x_i - \bar{x})^2}, \text{ and } \hat{b} = \bar{y} - \hat{m}\bar{x}, \text{ where the "hat"}$$

indicates an estimate of the parameter.

Analysis of Variance

The analysis of variance, or ANOVA as it is often called, is a particularly powerful way to analyze experimental data. It is worthwhile to note that Sir Ronald Fisher, who also made numerous contributions to the theory of evolutionary biology, developed this type of statistical analysis. For example, one typical type of problem for which ANOVA is well suited is testing the effects of different fertilizers on plant growth. Random samples of plants would be treated with different fertilizers, and possibly with no fertilizer, but otherwise treated the same. The growth of these plants over a fixed time period would be measured. Then researchers would use ANOVA to determine if there are significant differences between the plants that received fertilizers and plants that received water alone. If there is a significant effect of fertilizers on plant growth, the researchers could further determine if some fertilizers are better than others.

In a genetic context, as biologists we might be interested in determining if a particular phenotype is affected by genetic variation, or whether it is affected only by the environment. If both genes and the environment affect the trait, then we might like to estimate the relative magnitude of each of the factors. Suppose we were to examine a phenotype that is affected only by the environment. If we sampled progeny from five families that had been raised in identical environments, we might obtain results similar to those at the top of Figure C.3. Within each group, the variance in phenotype is roughly the same. The red circles show the mean of each group, and we can see that the red circles show less variation than the black circles do. In fact, simple statistical theory tells us that when all the samples come from the same population, then the variation of the mean values will be about one-fifth the variation of the individual values, since the mean is based on five observations. ANOVA gives formal ways of examining this relationship. If we performed an ANOVA on the set of data at the top of Figure C.3, we would have concluded that there are no significant differences between the five groups, thus genetic background (family origin) does not affect this phenotype.

At the bottom of Figure C.3 we see quite a different set of results. Here the five groups (families) in fact come from populations with different mean values (these are the genetic effects). Once again, if we examine the variation within a group, it is the same for each of the five. This is the common environmental variation. Now when we examine the variation between the mean values, it is clearly larger than it was at the top of Figure C.3. It also turns out that it is much larger than one-fifth of the variation seen within each group. These observations, when processed by the ANOVA, will let us conclude that family of origin (or our genes) does affect this phenotype.

Furthermore, the ANOVA will give us a numerical estimate of relative effects of genes compared to the environment. These types of estimates are crucial for making predictions about the course of selection on quantitative genetic traits.

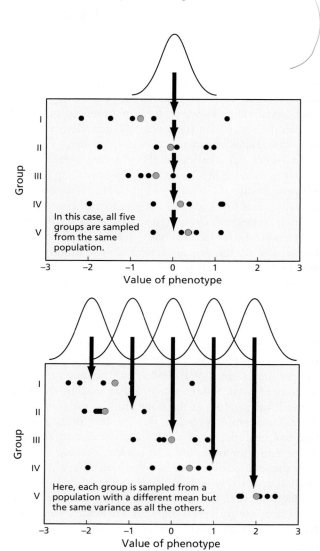

FIGURE C.3 The Types of Samples Used in the Analysis of Variance
In the top part of the figure, samples are derived from a single population. The samples are divided into five groups, each with five observations (black circles). A red circle represents the mean of each group. In the lower figure, the five groups each come from five different populations with different means. If we examine the variation that exists within each of the groups, it is roughly the same in the top five groups and in the bottom five groups. However, if we examine the variation between mean values, it is clear that this variation in greater in the bottom five means compared to the top five means.

GLOSSARY

acclimation   Physiological response to a new environment, normally one that improves fitness.

achondroplasia   A genetic disorder of cartilage development that impairs the growth of long bones; a common cause of dwarfism.

acid rain   Low pH (high acidity) rain that causes extensive ecological disruption.

active dispersal   Dispersal by locomotor adaptations, from walking to running to flying.

ADA   Adenosine deaminase, an enzyme that is not produced in a form of congenital immune system impairment called ADA deficiency.

adaptation   An attribute that increases Darwinian fitness; or a product of natural selection that increases Darwinian fitness; or the process of natural selection by which Darwinian fitness is increased.

adaptive radiation   The rapid generation of new species from a common ancestor species, due to natural selection.

addition rule   When either of two causal events can produce the same effect, the probability of observing the effect is obtained by adding the probabilities of the two causal events.

adultery   Sex with an animal that is not a partner by an animal that is part of an ostensibly monogamous pair; common in birds.

affect   Sustained pattern of emotion; e.g., depression, mania.

age-1   A mutant allele in the nematode *Caenorhabditis elegans* that increases lifespan; involved in the nematode dauer pathway.

age class   A group of individuals that share membership in a range of ages; e.g., humans between the ages of 6 and 10 years of age.

age structure   A population's age-class composition.

aging   The sustained decline in age-specific survival probability or fecundity that takes place after the start of reproduction in some species, regardless of general improvements in feeding, rest, etc.

AIDS   Acquired Immune Deficiency Syndrome, caused by HIV.

allele   A particular variant form of a gene; e.g., the allele for blue eyes in humans.

allogenic succession   A gradual change in the species composition of a community that takes place in response to changing physical or chemical conditions of the environment.

allometry   Quantitative analysis of data from multiple species taking into account body size and other structural features; or quantitative patterns between gross biological variables, such as height and weight.

allopatry   When populations of one species occupy distinct geographical regions.

allopolyploid   When genome size is increased by combinations of complete sets of chromosomes from more than one species.

altruism   Behavior that increases the Darwinian fitness of a target recipient while decreasing the fitness of the active donor.

Alu   A retrotransposable element that spreads through mammalian genomes by making RNA copies that are converted to DNA and then recombined back into the genome.

ammonotelism   Excretion, by animals, of nitrogen waste products in the form of ammonia.

anastomosis   In botany, when the branches of a tree fuse away from the trunk; in evolutionary biology, when independent lineages merge to form a new lineage.

anisogamy   When the gametes that undergo syngamy to form a zygote are different in size and possibly other attributes.

antibiotic   A compound that kills bacteria or prevents their reproduction.

antibiotic resistance   When bacteria evolve adaptations that reduce the effectiveness of antibiotics.

antiviral   A compound that kills viruses or prevents their reproduction.

anxiety   Sustained worry or distress; may or may not have a definable focus or area of concern.

aposematic coloration   Bright coloration in species that are toxic when eaten or have venom; also known as "warning colors."

apparent competition   When one species can exert a negative effect on the growth or reproduction of a second species even though they do not directly compete with each other.

arithmetic mean   The arithmetic value obtained by adding up a series of numbers and then dividing by the total count of these numbers.

artificial selection   When experimenters determine the organisms that reproduce based on specific characters, usually measured quantitatively, preventing natural selection as much as possible.

asexual reproduction   Reproduction without genetic exchange with another organism; may involve recombination among the genes of one genome; may involve seeds, buds, branching, or splitting into two symmetrical parts.

assimilation efficiency   The fraction of food energy that is consumed by an animal that is actually assimilated.

assortative mating   When organisms are more likely to mate because of phenotypic similarity to each other.

A-T syndrome   Also known as "ataxia-telangiectasia"; a recessive genetic disease that causes multiple degenerative pathologies, including increased carcinogenesis.

autogenic succession   The successional process that results from natural processes like glaciers or fires creating barren habitat.

autopolyploid   When genome size is increased by combinations of complete sets of chromosomes from only one ancestral species.

avirulence allele   Parasite alleles that are used by the host to recognize and attack the parasite.

balancing selection   When natural selection acts to maintain genetic polymorphism; includes cases with heterozygote superiority and frequency-dependent selection.

basal metabolic rate   The metabolic rate of an endotherm at rest.

Batesian mimics   Species that have evolved the appearance of a second species that is toxic for predators but are themselves not toxic.

Beagle   The Royal Navy ship on which Charles Darwin circumnavigated the world, collecting numerous specimens and experiencing a wide diversity of habitats.

biodiversity   A term indicating the numbers of different species and their relative abundance within a community.

biogeochemical cycles   The recycling of essential nutrients between living organisms and the non-living components of earth.

biogeography   The study of species distributions and diversity over a wide geographical range.

biological classification   The grouping of organisms into taxa based on their similarities of structure or function, or their ancestry.

biological control   The control of a pest species of plant or animals with other species that are natural predators or parasites.

biological species concept   The delimitation of species as groups of populations that do not successfully interbreed with other populations in nature.

biomes   Earth's major communities often classified by their major types of vegetation.

blood group   People who share a blood protein genotype.

bottom-up regulation   When biomass or number of species in a community is controlled by the amount of primary production in the community.

boundary layer   A thin layer of air above the stomates of plants.

Bourgeois   An evolutionary game strategy that plays Hawk when it has acquired a contested item first, but Dove otherwise.

braincase   The part of the vertebrate skull that encases the brain.

brood parasitism   Birds which deposit their eggs in the nest of a host species and have the host species care for and raise the parasitic young birds.

Burgess Shale   A geological formation that contained numerous fossils from the Cambrian Epoch, located in the Canadian Rocky Mountains.

Cambrian   A geological period that lasted from 543 million years ago to 506 million years ago.

cannibalism   When members of a species eat each other.

carrying capacity   The equilibrium number of individuals that can be supported by the environment.

castration   The complete destruction or removal of the primary reproductive structures of an organism.

Catastrophism   The explanation of major geological or biological changes in terms of large-scale abrupt events, such as Noah's Flood.

cell fusion   When the membranes of adjacent cells break down, resulting in shared cytoplasm.

Cenozoic   The most recent geological era, covering the last 65 million years.

chaos   Population dynamics that appear erratic and unpredictable but are due to the properties of density regulation.

chaparral   The communities found in the Mediterranean climates of California and Chile, characterized by drought resistant, evergreen plants.

character displacement   When the presence of an ecological competitor causes the evolutionary divergence of a population relative to that competitor.

child bride   The arranged marriage of a juvenile female to a member of another family which adopts her.

choking   The obstruction of a passageway for breathing.

cholera   A disease involving a bacterial (*Vibrio cholerae*) gastrointestinal infection and copious diarrhea.

chromosome   A long string of DNA, sometimes packaged with proteins, usually encoding genes.

climate   The long term average weather in a particular locality.

climax community   A community with a relatively stable species composition, often seen as the end product of succession.

clonal selection   Selection acting on populations that have asexual reproduction without any type of genetic recombination.

clone   A group of genetically identical organisms.

closed circulatory system   A system for circulating fluids in animals that does not open up into body cavities but insead relies on the transport of nutrients and oxygen across the membranes of small capillaries.

codon   A triplet of nucleotides that codes for an amino acid making up a protein.

coefficient of genetic relatedness   The proportion of alleles that are shared by two organisms because of inheritance from a common ancestor.

coevolution   Reciprocal evolutionary change in interacting species.

cohort   A group of individuals that began life at approximately the same time.

Cole's Paradox   Only a small increase in initial fecundity is required to counterbalance the advantages of repeated reproduction.

commensalism   An interaction between two species that benefits one species but has no apparent negative or positive effect on the second species.

common ancestor   Pertaining to organisms: an ancestor of two or more contemporaneous individuals. Pertaining to species: a species from which two or more contemporaneous lineages descend.

community   A collection of biological species sharing an ecosystem.

comparative method   The inference of patterns of evolutionary descent or adaptation from the comparison of contemporaneous species.

compensation point   The intensity of light at which the production of energy by photosynthesis just equals its consumption by respiration.

competition coefficient   A number which measures the effect of one species on the population growth of a second species.

competitive exclusion principle   No two species can coexist if they use the environment in precisely the same way.

conduction   The flow of heat energy that occurs when two bodies at different temperatures come into direct contact.

conjugation   The exchange of plasmids between bacterial cells by means of a pilus.

consanguinity   When two or more organisms share a common ancestor.

contest competition   When animals settle disputes by behavioral interaction, sometimes violent, sometimes not.

convection   The transfer of heat energy between two bodies at different temperatures via a third moving fluid body of either gas or liquid.

convergent evolution   The evolution of similar morphology or function as a result of selection in similar environments.

Coriolis effect   A deviation in the flow of currents due to the Earth's rotation, these deviations are to the right in the Northern Hemisphere and towards the left in the Southern Hemisphere.

cost of reproduction   A reduction in survival resulting from reproductive activities or structures.

cost of transport   The amount of energy used per meter traveled per kilogram of animal biomass.

countercurrent exchange   Two bodies of gas or liquid with different concentrations of solutes or at different temperatures that flow by each other in opposite directions to facilitate the exchange of solutes or heat.

Creationism   The doctrine that all species were created individually by an all-powerful Creator, rather than evolving.

Cretaceous   The last geological period of the Mesozoic era, and the last period in which dinosaurs were the common large vertebrates.

crude birth rate   The total number of births in any year divided by the population size.

crude death rate   The total number of deaths in any year divided by the population size.

crude rate of population increase   The total number births minus deaths in any year divided by the population size.

cryptic coloration   Coloration which makes it difficult to visually detect an organism in its natural habitat.

C-value   The amount of DNA present in a complete (usually haploid) set of chromosomes.

cystic fibrosis   The most common recessive genetic disease, involving the abundant production of mucous which in turn causes recurrent lung infections and other health problems.

cytochrome oxidase   One of the key enzymes in the intermediary energetic metabolism of the cell.

Darwinism   The scientific theory that the diversity of life on Earth was produced by descent from common ancestral species, directed by natural selection.

dauer   A developmental stage characterized by postponed reproductive maturation and enhanced survival under stressful conditions.

definitive host   The host in which a parasite reproduces.

degradative succession   The process of serial replacement of colonizing organisms in a new environment such as a dead organism or a piece of excrement.

delusion   A persistent, objectively false, opinion or belief.

demographic stochasticity   Variation that occurs in reproduction of small populations due to variation in offspring production of individual females.

demography   The study of population vital statistics such as age-specific birth and death rates.

denitrification   The conversion of nitrate to gaseous nitrogen by bacteria like Pseudomonas.

depression   A sustained period of lower energy, reduced enthusiasm, disrupted sleep, or reduced capacity to feel pleasure.

desertification   The process of converting productive land to desert.

diploid   When all chromosomes are present in exactly two copies.

direct transmission   When parasites are transmitted directly from the host's parent to their offspring.

directed link   A connection between two species or nodes in a food web that shows the flow of energy from one node to the second.

directional selection   When selection consistently favors an extreme phenotype.

discrete generations   When the parents of each generation never mate with individuals of any other generation.

disease vector   An organism that acts as a reservoir and transmission agent for a parasite that causes disease in another species.

display   A type of contest behavior that does not cause injury, but communicates continuing interest in the outcome of the contest.

disruptive selection When selection consistently favors two or more contrasting phenotypes. For example, selection for both low and high body weights.

divergence rate The number of genetic substitutions that have occurred during the divergence of two species, per unit time.

divergence time The time since two species shared a common ancestor.

DNA repair Replacement of nucleotides missing from a DNA molecule.

domain A functionally or structurally distinct part of a protein.

dominance When a heterozygote's phenotype is not precisely intermediate between the phenotypes of its respective homozygotes; or, the quantitative deviation of a heterozygous phenotype from the mid-point between the homozygous phenotypes.

Dove A contest strategy that Displays unless Attacked, in which case the strategy specifies Retreat.

drones Adult males in Hymenopteran eusocial insect societies.

dwarfism A genetic condition in which the growth of long bones is reduced.

dynamical stability hypothesis The stability of a community decreases with increasing links in the community food webs, but should increase in benign environments where populations do not fluctuate greatly in size.

ebola A viral disease characterized by rapid onset of hemorrhaging; often fatal.

Eckman flow A westward offshore flow of water.

ecological efficiency The overall efficiency of energy transfer from one trophic level to the next. It is calculated by dividing the energy content of the higher trophic level by the energy of the lower level.

ecological niche All of the habitat and resource requirements of a species.

ecosystem A community and its physical environment.

ecosystem size hypothesis The chain length of food webs increases with increasing volume of the ecosystem.

ectotherm Animals that maintain their body temperature with external sources of energy.

electrophoresis The separation of large molecules by moving them through charged gels.

endosymbiosis The acquisition of new cellular elements by evolutionarily engulfing other types of cell.

endotherm Animals that maintain their body temperature with internal sources of energy.

epilimnion The top layer of water in lakes and ponds.

epistasis When phenotypes that are determined by alleles at more than one locus are not predictable as additive combinations of the genotypes at individual loci; or, the quantitative deviation of phenotypes from such additive predictions.

equilibrium population size The population size where births exactly equal deaths.

escalated conflict Aggression within species that becomes progressively more severe, often resulting in the injury or death of one or more of the participants in the conflict.

eukaryote An organism with a nucleus distinct from the cytoplasm and other characteristic organelles, such as mitochondria or chloroplasts.

evapotranspiration The movement of water into the air via the evaporation of water on land or from plants and animals.

evolution A synonym for the Darwinian theory of life; or, change in the genetic composition of a population from one generation to the next.

Evolutionarily Stable Strategy (ESS) A strategy that cannot be beaten, on average, when almost all members of population adopt it.

evolutionary game theory The analysis of strategy selection.

evolutionary psychology The piecemeal analysis of human behavior in terms of specific selection mechanisms.

evolutionary tree The pattern of descent among a group of related species; or, a branching diagram showing the same.

exon A part of a gene's DNA sequence that codes for the amino or ribonucleic acids of the gene product.

exotic species An organism that has recently been placed in an environment in which it did not evolve.

exploitation efficiency The fraction of energy in one trophic level that is captured and consumed by members of the next higher trophic level.

exponential population growth The growth of a population that follows from the assumption of a constant per capita net rate of reproduction.

external fertilization When syngamy arises from the fusion of gametes released into the surrounding medium, instead of occurring inside the body of the female or the male.

extinction The dying off of an entire species, without direct descendants.

factorial aerobic scope The ratio of the maximum metabolic rate to the basal metabolic rate.

fecundity A female's output of eggs within a specified period of time.

fertility A female's output of viable offspring within a specific period of time.

fission The symmetrical division of a cell or a multicellular organism.

fitness The average reproduction of an individual or genotype, calibrated over a complete life cycle.

fixation When an allele rises to such a high frequency that all other alleles are eliminated from a population.

food chains A series of directed links in a food web, starting with the species or group of species that does not feed on any other species and ending with the species that is not fed on by any other species.

food web cycle When two nodes of a food web feed on each other.

food web loop When two or more species in a food web make a closed circuit.

force of natural selection The intensity of natural selection acting on an age-specific change in survival probability or fecundity.

fossil record The aggregate information derived by scientists from all known fossil specimens.

fragmentation The production of viable offspring by breaking off pieces of the parental body; or, the creation of isolated habitats from the loss of connecting habitat.

frontal lobes The lobes of the cerebrum located to the front of the human brain.

frugivory When animals derive the bulk of their nutrition from fruits.

fundamental niche The term used for the ecological niche by Hutchinson.

gamete A cell which can form a zygote by syngamy.

gene expression The extent to which a gene's DNA sequence is transcribed.

gene-for-gene system A relationship between a set of genes in a parasite that can be used by the host to recognize an infection and a set of genes in the host that can be used to fight the parasite.

generation time The average length of time between the reproduction of the parents of each generation.

genetic disease A medical disorder caused by a genetic change at a single locus.

genetic drift Fluctuation in the frequency of an allele as a result of accidents of genetic segregation, reproduction, and survival.

genetic engineering The deliberate modification of the genome of an organism's cells, whether somatic or gametic, using molecular technology.

genital herpes Dermal lesions caused by a herpes virus that is transmitted by genital contact.

genital warts Dermal growth caused by a papilloma virus that is transmitted by genital contact.

genome The complete set of DNAs transmitted during reproduction.

genotype The allelic content of a locus, chromosomal region, chromosome, or entire genome.

geriatrics The medical specialty concerned with the treatment of the elderly.

germline engineering Genetic engineering of the cells that will be used to make gametes.

gerontology The scientific study of aging in all organisms.

gills Structures used for gaseous exchange in aqueous media, chiefly water.

gonad An organ that produces gametes.

gonorrhea A sexually-transmitted disease caused by the bacterium *Neisseria gonorrhoeae*.

gracile lineage Hominids with relatively small teeth and jaws, probably ancestral to modern humans.

Gradualism The doctrine that changes in the physical world proceed primarily gradually, without large-scale catastrophes.

greenhouse effect   The heating of the Earth that occurs when high levels of CO_2 and other gases trap heat within the atmosphere.

group selection   Natural selection that arises from differences in the rates of extinction and colonization between groups that have different genetic compositions.

gyres   Directed flows of ocean water.

Hadley cell   An air flow pattern that is responsible for high rainfall at the equator and deserts 20°–30° north and south of the equator and characterized by air rising from the equator and sinking back to Earth just north and south of the equator.

hallucination   A sensory experience that is not a result of a physical stimulus.

halophile   An organism that thrives in an environment with a high level of salt.

handling time   The time needed for a predator to consume its prey.

haplo-diploid   A genetic system in which one sex is diploid and the other sex is haploid.

haploid   A full set of chromosomes without any duplicate chromosomes.

Hardy-Weinberg equilibrium   When the frequency of homozygotes is equal to the square of the frequency of homozygous allele and the frequency of heterozygotes is equal to twice the product of the frequencies of the alleles that make up the heterozygote; applies to each locus individually.

harvesting   The human removal of organisms from a natural population.

Hawk   A strategy that always chooses to Attack during conflicts.

herbivores   Animals that primarily derive their nutrition from the consumption of plant matter.

heritability   The ratio of the genetic variance to the total phenotypic variance, when inheritance is strictly additive; with more complex patterns of inheritance, heritability is no longer given by this ratio.

hermaphrodite   An organism with both male and female functional sex organs.

heterozygosity   When a diploid locus has two different alleles; or, the fraction of loci that are heterozygous.

heterozygote superiority   When the Darwinian fitness of a heterozygote is higher than that of homozygotes for its constitutent alleles.

histone   A protein that is used to package DNA into chromosomes in eukaryotes.

historical imperfections   Features of organisms that reduce their fitness in their present environments, but were favored in their evolutionary history.

histrionic   A personality disorder characterized by attention-seeking behavior and dress, together with an obsession with physical appearance and sexual allure.

HIV   Human Immunodeficiency Virus, one of a group of viruses that infects T lymphocytes, usually fatal.

home range   A characteristic region animals move within in search of food, mates, or shelter.

homology   When species have similar characteristics due to descent from a common ancestral species.

homoplasy   When parallel patterns of selection produce structural or physiological similarities between unrelated species.

homozygosity   When a diploid locus has two copies of the same allele.

hopeful monster   A mutant form that cannot mate with normal members of the species from which it is derived.

horizontal transmission   The transmission of DNA from one species to another, common in bacteria as a result of conjugative plasmids, transformation, and other forms of proto-sex.

host   A species that serves as the nutritional resource and often the site of reproduction for a parasite.

host race   Members of a species that feed on, or live in, a particular type of host organism.

hunter-gatherer   The human way of life before the advent of agriculture.

Huntington's disease   A dominant genetic disease characterized by progressive loss of central nervous system function.

hybrid breakdown   Reduced fitness of the offspring produced when the hybrids of two species mate with each other.

hybrid dysgenesis   Reduced fitness in the hybrids of old and new laboratory stocks of *Drosophila*, due to the proliferation of the P transposable element.

hybrid vigor   Increased fitness in offspring produced from crosses of differentiated stocks, breeds, varieties, or selected lines.

hybrid zone   An area where two species successfully mate and produce offspring.

hybridization   When two species successfully mate and produce offspring.

hyperosmotic   The condition where the cellular osmotic concentration is greater than the surrounding environment.

hypolimnion   Lower water levels in lakes and ponds.

hyposmotic   The condition where the cellular osmotic concentration is less than that of the surrounding environment.

identical twins   Individuals having genomic DNA sequences that are almost identical, except for recent mutations.

immortality, biological   The condition of roughly constant, positive, survival rates relative to age under good conditions; may not apply to all life-cycle stages.

immunoglobulin   An antibody protein.

inbred line   A lineage of experimental organisms that has been maintained with high levels of inbreeding for some time.

inbreeding   Mating of close relatives, which may be transitory or sustained.

inbreeding coefficient, F   The probability that the two alleles at an organism's loci are identical by descent.

inbreeding depression   Reduced fitness in an inbred organism; or, reductions in other functional characters in an inbred organism.

incest avoidance   When individuals who are related avoid mating; or, when individuals who are raised together avoid mating.

Incubator   An organism that supplies parental care to its offspring.

indefinite dormancy   When dormancy ends in response to an external cue, rather than after a fixed period of time.

indirect transmission   The transmission of an organism's alleles by related individuals other than offspring.

industrial melanism   The evolution of dark coloration in butterflies and moths in industrialized areas with heavy pollution.

intermediate host   Species that parasites use in their life-cycle but are not the sites of reproduction.

internal fertilization   When gametes are fertilized inside an organism's body.

interspecific competition   Competition between different species.

intraspecific competition   Competition between different members of the same species.

intrinsic rate of increase   The parameter r of the logistic equation that determines the maximum rate of population growth at low densities.

intron   DNA sequences that interrupt the coding sequences of genes; normally excised during transcription.

invasion   The rapid colonization of habitat by a new species; or, increasing frequency of novel behavioral strategy due to high fitness when at low frequencies.

iridium   A metallic element that is common in asteroids, but not in the Earth's crust.

isogamous   When a species lacks obviously differentiated gametes.

iteroparous   When life-histories involve repeated bouts of reproduction during adult life.

Just-So stories   Explaining the features of organisms in terms of untested, or untestable, selection hypotheses.

keystone predator   A predator whose effects on prey species reduces their competitive interactions and makes possible their coexistence.

kibbutzim   Israeli communes, typically having multiple generations living together.

kin selection   Selection mechanisms that depend on interactions between members of biological families.

king   A male who fathers most of the members of a colony of social insects.

Kingdom   A high-level taxonomic group made up of multiple phyla; e.g., Animalia.

knuckle-walking   Terrestrial locomotion that uses two feet and the knuckles of one hand.

K-T boundary   The point in time when the Cretaceous period gave way to the Tertiary; or, the layer of sediment in the Earth's crust that was produced at that time.

Lamarckism The doctrine that evolution occurs in independent lineages, without branching and without natural selection, the direction of evolution arising from some kind of endogenous adaptive process.

large-body impact When large asteroids and comets strike the Earth.

life cycle The sequence of birth, development, reproduction, and death that defines the biology of a species; or, synonym of life history.

life history The quantitative features of the survival and reproduction of a species.

life table A tabular summary of the quantitative life history of a species.

link A connection between two nodes of a food web indicating a feeding relationship between the two.

linkage disequilibrium When the frequencies of gametes are not given by the product of the frequencies of the alleles at individual loci.

linkage equilibrium When the frequencies of gametes are given by the product of the frequencies of the alleles at individual loci.

logistic population growth A model of population growth that assumes per capita growth rates decline as a linear function of population size.

Lotka-Volterra model Mathematical model of the population dynamics of interacting species developed independently by Alfred Lotka and Vito Volterra.

lottery model When natural selection favors specific genotypes in response to selection that varies widely in direction over time and among habitats.

malaria A disease that arises from the infection of erythrocytes (RBCs) by trypanosomes of the genus *Plasmodium*.

Malthusian parameter A synonym for the intrinsic rate of increase; the rate of growth of a population when it achieves its stable age distribution.

mania A psychiatric condition characterized by hyperactivity, pressure of speech, diminished need for sleep, indiscretion, grandiosity, and sometimes psychosis and paranoia.

mass extinction The extinction of a large fraction of the Earth's biological species at the boundaries of geological epochs.

materialism The scientific doctrine that objectively observable phenomena are to be explained in terms of material processes, rather than divine intervention.

maximum longevity The greatest longevity reliably recorded for a member of a particular species.

maximum parsimony The convention that the preferred evolutionary tree is the one that requires the fewest evolutionary transitions between character states.

maximum sustained yield The population density that yields the largest number of excess births and thus greatest potential for human harvesting.

median An indication of the average value of a random variable that has the property that 50% of all random variables are less than the median and 50% are greater than the median.

meiosis The process by which haploid gametes are produced by diploid eurkaryotes.

mental arms race Competition that depends on the innovation of appropriate evolutionary game strategies given variable game contingencies.

meroploid The presence of a variable number of copies of the genome in bacterial cells.

Mesozoic The middle era of the Phanerozoic eon, from about 251 million years ago to 65 million years ago.

messenger RNA The RNA transcribed from the DNA of a gene.

metapopulation Multiple populations linked through migration.

microclimate The climate conditions that are found on a small scale, perhaps only a few meters.

migrant pool When the process of interdemic selection features mixing of migrants in a common pool from which the founders of new populations are recruited.

migration When individuals disperse from one location to another, either seasonally or adventitiously.

mitochondrial DNA The bacteria-like genome of the eukaryotic mitochondrion.

molecular clock The hypothesis that nucleotide substitutions occur in genes with temporal regularity, like the marking of time by a clock.

monogamy Continued mating in a male-female pair, often with shared rearing of offspring; or, the preceding condition combined with complete absence of sexual contact with other members of the species (relatively unknown).

Mullerian mimics Species that have a similar form of protection against predators and have evolved similar warning coloration that advertises this protection to predators.

Muller's ratchet The progressive loss of individuals with low numbers of deleterious mutations due to genetic drift in finite populations.

multiplication rule When two or more independent events are required to produce a particular outcome, the probability of that outcome is calculated by multiplying the probabilities of these individual events together.

multiregional theory The hypothesis that hominid evolution began in Africa but ancestral species moved out to many different regions of the world about 1,000,000 years ago. Genetic exchange among these different populations ensured more or less similar evolution of these hominids to modern humans.

mutualism An interaction between two or more species that benefits each species.

mycorrhizae Structures in the roots of plants that involve symbiotic fungi that provide nitrogen to the plant in exchange for carbon.

narcissist A personality disorder characterized by a lack of empathy for others, together with grandiosity, exploitative behavior, and a need for admiration.

natural selection The differential net reproduction of genetically distinct entities, whether mobile genetic elements, organisms, demes, or entire species.

net reproductive rate The total reproductive output of an individual, discounted according to survival probabilities during juvenile and adult stages.

net production efficiency The fraction of energy assimilated by one trophic level that is made into new biomass.

nitrification The conversion of ammonia to nitrate by two different bacteria, *Nitrosomonas* and *Nitrobacter*.

node A species or groups of species in a food web with the same feeding relationships.

nutrition mutualism A mutualism that involves the exchange of nutrients between the participating species.

open circulatory system Fluid transport system that includes vessels that open up into body cavities.

origin of sex The evolutionary scenario by which sex is supposed to have originated.

outbred When populations, breeds, or varieties are not systematically inbred; or, the population is not inbred and the population size is large.

Out-of-Africa theory The hypothesis that several hominid lineages existed more or less independently, the lineage that gave rise to modern humans moved out of Africa about 100,000 years ago and is the only hominid lineage to survive today.

overpopulation When the numbers of individuals exceeds the numbers that can supported by the environment.

paleontology The study of fossils.

Paleozoic The first era of the Phanerozoic eon, from 543 million years ago to 251 million years ago.

parasite A species that completes most or all of its life cycle within another organism called a host, producing negative effects on the host although sometimes not the death of the host.

parental care The enhancement of the survival, development, or reproduction of an organism by its parent.

parthenogenesis The production of offspring without sex.

passive dispersal When organisms, their gametes, or their seeds are carried by currents of air or water or attached to moving animals.

pathogen-host arms race The dynamic process of evolution of resistance to a parasites by a host, followed by evolution in the parasite to overcome the host's new resistance.

pelagic Living on the surface of the oceans.

Permian The last geological period of the Paleozoic, from 290 million years ago to 251 million years ago.

personality disorder A psychiatric condition characterized by deviant behavior without obvious psychosis or aberrant affect, stable in manifestation and distressing to the families or associates of the afflicted, and sometimes to themselves.

Phanerozoic The most recent geological eon, from 543 million years ago to the present.

pharynx The upper part of the oral cavity, behind the nasal passage and the mouth.

phenotype A material attribute of an organism; or, the entire set of material attributes of an organism.

phenotypic selection Selection that depends on the phenotype of the organism.

phenylalanine A common amino acid.

phenylketonuria A recessive genetic disease in which the patient is unable to convert phenylalanine to tyrosine.

phobia A persistent fear of a particular stimulus or situation which causes marked distress.

phyletic gradualism The doctrine that evolutionary change proceeds gradually over long periods of geological time, with or without speciation.

phytoplankton Mostly unicellular plants that live in the top layer of open ocean waters.

polygamy Continued mating of an individual of one sex with several members of the opposite sex; or the above condition with an absence of sexual contact with other individuals.

polymorphism The presence of multiple distinct phenotypes in a population; or, the presence of multiple alleles at a genetic locus.

polyploidization An evolutionary event in which the normal number of haploid sets of chromosomes is increased.

polyteny Somatic cells in which chromosomes have a large number of parallel strands of genomic DNA.

population A group of sexual organisms that interbreed frequently; or, a group of closely related asexual organisms that share a particular habitat.

postzygotic reproductive isolation When gene flow is precluded between two species because their hybrids are inviable, infertile, or unable to produce fertile descendants.

prey A species that serves as food for another species.

prezygotic reproductive isolation When gene flow is precluded between two species because they (1) mate at different times or places, (2) are not sexually attracted to each other, or (3) are unable to undergo syngamy if sex occurs.

primary producers The plants within a community that capture energy from sunlight to make biomass that then serves as the source of energy either directly or indirectly for the remaining members of the community.

primary succession Succession that takes place in areas that have had all vegetation removed due to processes like glaciers or volcanoes.

Prisoner's Dilemma A game in which mutual cooperation is rewarded but the players do not know if their partner cooperates or not.

private alleles Alleles that are found in only one subpopulation of a species.

productivity hypothesis The mean chain length of a food web is proportional to the amount of energy in the primary producer level.

promiscuity The absence of persistent mating with one particular partner.

propagule A group of individuals from a common original population who found a new population.

protection mutualism A mutualism in which one of the participating species attacks or removes predators or competitors of another species.

pseudogene A DNA sequence that is similar to a transcribed gene but is not itself transcribed.

pulmonary circuit Blood flow that starts at the heart, goes through the lungs and then returns to the heart.

punctuated equilibrium The hypothesis that most evolutionary change occurs during speciation; or, the hypothesis that speciation proceeds by large-scale abrupt genetic revolutions with features unlike those of Darwinian evolution by natural selection.

purifying selection When natural selection eliminates individuals who deviate markedly from the average phenotype of a species.

Q_{10} The rate by which chemical reactions accelerate due to a 10°C increase in temperature.

quantitative character A biological character that can be characterized by a number, such as body weight.

queen In social insects, the female who produces all the eggs.

r- and K-selection Natural selection under conditions of low and high crowding, respectively.

rain shadow deserts Deserts that form on the leeward sides of mountains.

ram ventilation Fish that create water flow past their gills by swimming with their mouths open.

realized niche The reduction in the fundamental niche that results from the impact of other species.

recessive When an allele does not have as much influence on the phenotype as the allele(s) with which it is paired in heterozygotes.

recombination The re-assortment of alleles among gametes; may require physical breaking and re-joining of chromosomes when the alleles are located on the same chromosome; also occurs among alleles on separate chromosomes during meiosis.

reductive division The production of cells with lower ploidy, typical of meiosis.

resemblance between relatives When relatives resemble each other more than they resemble other members of a population; can be quantitatively estimated from the heritability in some cases.

reserve A habitat that has been set aside specifically for the protection of particular plants and animals.

resistance allele An allele in a host that confers resistance to a specific parasitic genotype.

respiratory quotient The amount of carbon dioxide produced by respiration divided by the amount of oxygen consumed.

Retaliator A contest strategy that Displays unless Attacked, in which case it Attacks in return.

retrogene A gene produced as a result of retrotransposition.

retrotransposon A transposable element that makes new copies by reverse transcription of an mRNA produced from a DNA copy located in a genome.

reverse transcriptase An enzyme that catalyzes the production of a DNA sequence complementary to the sequence of an RNA molecule.

risk analysis A statistical and mathematical assessment of the chances of population extinction.

ritualized conflict Animal contest characterized by displays that are unlikely to damage the contest participants.

RNA virus A virus that has a genome made of RNA.

robust lineage Early hominids that had large jaws and teeth suited to eating large amounts of plant material.

SARS Sever Acute Respiratory Syndrome: a potentially fatal respiratory disease characterized by fever, coughing, and difficulty breathing.

savanna A semiarid habitat dominated by long grasses and shrubs with few trees.

schizophrenia A psychiatric disorder characterized by persistent paranoia, delusions, hallucinations and general intellectual decline.

scramble competition The competition that results when all individuals have more or less equal access to a limiting resource.

seasonal dormancy Dormancy that occurs during specific seasons of the year.

secondary succession Succession that occurs in disturbed areas that already possess a mature community.

seed bank Viable seeds buried in soil, awaiting an environmental trigger for germination.

selection differential The phenotypic difference between a selected group and the population from which it was obtained.

selection response The phenotypic difference between the offspring of a selected group and the population from which the selected group was obtained.

self-fertilization, selfing When an organism mates with itself.

selfish behavior Behavior that does not include biological altruism.

semelparous When the life history has just one bout of reproduction.

senescence In general, a synonym for aging; in botany only, the deterioration and loss of deciduous leaves or flowers.

serial homology Similarities among specialized appendages that have evolved in species that have ancestors in which these appendages were similar to each other.

sex determination The determination of sex by genetic or environmental mechanisms.

sex ratio The ratio between males and females in a population, expressed as a percentage or a ratio.

sex switching When individual organisms can switch from one sex to another.

sexual reproduction Reproduction in which gametes are produced by meiosis and these gametes then undergo syngamy to form zygotes.

sexual selection When natural selection acts on attributes that determine mating success.

sib competition When natural selection acts on siblings that are in intense competition within a well-defined habitat.

sickle-cell anemia A genetic disease characterized by sickled erythrocytes (RBCs).

sink web A food chain in which all feeding relationships direct their energy to a common top predator.

smallpox A potentially fatal viral disease exclusive to humans characterized by numerous dermal blisters that give rise to small pock marks.

Social Darwinism A political doctrine that encourages the exploitation or destruction of the poor, the unintelligent, or the unfit, opposing welfare legislation and government regulation of the marketplace for jobs or products.

social insect An insect species in which kin selection reduces rates of individual reproduction, favoring altruism toward primary reproducers by less fertile workers.

social intelligence The use of the primate brain to better exploit conflict and cooperation with members of the same species in order to increase Darwinian fitness.

sociobiology The study of behavior in animals and humans based on the same general evolutionary principles and methods.

sociopath A personality disorder characterized by a lack of inhibition, remorse, reliability, conscience, and law-abiding behavior.

soldier A sterile social insect "caste" characterized by aggressive self-sacrificing defense of the insect colony.

somatic engineering The genetic engineering of somatic, not germline, cells.

source web A food chain in which all nodes derive their energy ultimately from a single food source.

speciation The formation of a new species by the evolution of reproductive isolation between an ancestral species and a population that evolutionarily derives from it.

species radiation The geologically-abrupt formation of numerous new species from one or a few common ancestor species.

specific metabolic rate The metabolic rate per kilogram of organism.

sperm Small, usually highly motile, gametes that produce zygotes when they fertilize egg cells.

stability The propensity of a population to return to an equilibrium when it is perturbed from that equilibrium.

stabilizing selection Selection against phenotypic extremes.

stable age-distribution A property of age-structured populations where the proportions of individuals in each age class remain constant over time.

stenohaline Organisms that can only live in a narrow range of environmental salinities.

stomata Small openings on the leaves of plants that allow gas exchange.

strategy selection When the Darwinian fitness of an animal depends on the combined actions of both itself and its conspecifics when there is conflict between them.

substitution, nonsynonymous Evolutionary substitutions in the DNA sequences of a gene that change the amino acid sequence of the protein encoded by that gene.

substitution, synonymous Evolutionary substitutions in the DNA sequences of a gene that do not change the amino acid sequence of the protein encoded by that gene.

succession The transitions in community structure that take place over time generally following some sort of disturbance.

symbiosis A long term intimate association between two species.

sympatry When two populations occupy the same geographical area, though not necessarily the same specific habitats within that area.

syngamy The fusion of two gametes to form a zygote.

syphilis A human venereal disease caused by the bacterium *Spirochaeta pallida*, characterized by eventual infertility, systemic infection, and sometimes death.

systemic circuit The vessels that transport blood from the body's tissues back to the heart and then under high pressure from the heart back to the tissues.

tandem array Sequential alignment of repeated DNA sequences in a chromosome.

Taoism An ancient Chinese cosmological system with religious and proto-scientific elements.

Tay-Sachs disease An incurable recessive genetic disease caused by a deficiency in fat metabolism resulting in the progressive poisoning of the brain; death usually occurs in childhood.

territoriality When animals defend pieces of habitat; see Bourgeois.

Tertiary The first part of the Cenozoic era, from 65 million years ago to 1.8 million years ago.

testes Male gonads.

thermocline A gradient of temperatures that develops in water bodies due to incomplete mixing of the water body.

thermophile An organism that successfully develops and reproduces at high temperatures.

tolerance The relatively small decline in fitness of plants due to a specific level of herbivory.

top-down regulation When biomass or number of species in a community is controlled by predation.

trachea A passageway in a respiratory structure, such as the passageway to the lungs in vertebrates.

tranposable element A DNA sequence that can make copies of itself that move to new locations in a host genome.

transcription The production of RNA sequences from a template of genomic DNA with the corresponding sequence of nucleic acids.

transduction The transmission of host nucleic acids by a viral capsule.

transformation The transmission of free-floating DNA molecules from one bacterium to another without transduction, conjugation, or cell fusion; generally, the insertion of exotic DNA molecules into a host genome.

transgenic construct An organism that has had its genome altered by the insertion of DNA from another organism.

translation The use of an RNA molecule to assemble a polypeptide based on the codon sequence of the RNA.

transportation mutualism A mutualism in which one mutualist species transports the seeds or individuals of another mutualist species.

transposase An enzyme that allows autonomous transposition by a transposable element.

trophic cascade The prediction that the reduction of the numbers of individuals at one trophic level will have a positive effect on the biomass of members in the next lowest trophic level but a negative effect on the numbers two trophic levels lower, and so on.

trophic level All organisms that occupy a similar place in the food chain.

unbeatable strategy A synonym for an Evolutionarily Stable Strategy, or ESS.

undirected link A connection between nodes of a food chain that indicates a feeding relationship exists but does not show the direction of energy flow.

unequal crossing over, unequal recombination When the elements of a tandem array are recombined out of register, producing gametes with unequal numbers of array elements.

Uniformitarian The doctrine that the processes now occurring in the universe also took place in the past, and vice versa.

univoltine When insects have one generation per calendar year.

upwelling The movement of deep cold water to the surface.

ureotelism Excretion of nitrogen primarily in the form of urea.

uricotelism   Excretion of nitrogen primarily in the form of uric acid.

UVB   High-energy ultraviolet radiation that is largely prevented from reaching the earth's surface by atmospheric ozone.

vaccine   A preparation used to confer immunity to a disease by inoculation.

vapor pressure   The force exerted by a gas in equilibrium with its liquid phase.

variance   The average squared deviations from the mean of a random variable.

variance, environmental   That part of the phenotypic variance which cannot be attributed to genetic causes.

variance, genetic   That part of the phenotypic variance which can be attributed to genetic causes, usually inferred from the resemblance of relatives.

variance, additive genetic   That part of the genetic variance which breeds true; that is, which causes predictable resemblance between relatives.

variance, phenotypic   The variance between organisms for a quantitative character.

variegation   Variation in the appearance, especially coloration, of a tissue arising from developmental processes, sometimes including somatic mutation.

vegetative reproduction   Reproduction without any form of genetic recombination.

venereal disease   Diseases transmitted by copulatory contact.

vertical transmission   Transmission of genetic information or, sometimes, disease from parent to offspring.

viability   The probability of survival from zygote to adulthood.

vicariance   The isolation of populations as a result of a geographical event, such as the geological creation of a mountain range or the diversion of a river.

virion   An intact virus particle consisting of a protein coat surrounding nucleic acid.

virulence   The ability of a pathogen to produce a disease; or, the likelihood of host death from infection by a particular pathogen.

Wahlund effect   A reduction in the frequency of heterozygotes when a population is subdivided.

waiting time   The time a predator must wait between encounters with prey.

worker   A sterile social insect "caste" that forages for food and cares for other members of the colony, especially the queen and the larvae.

X chromosome   In insects and mammals, a sex chromosome, present in two copies in females and one copy in males.

Y chromosome   In insects and mammals, a sex chromosome that is normally found only in males.

yeast   A unicellular fungal species, in general; more specifically, the yeast species *Saccharmoyces cerevisiae*.

zooplankton   Small animals and animal larvae that live in the top layers of the open ocean.

zygote   The cell stage at the start of organismal development, usually a diploid cell produced by syngamy of haploid egg and sperm.

BIBLIOGRAPHY

Agrawal, A.A., and R. Karban. 1997. "Domatia Mediate Plant-Arthropod Mutualism." *Nature* 387: 562–563.

Agrawal, A.A., F. Vala, and M.W. Sabelis. 2002. "Induction of Preference and Performance after Acclimation to Novel Hosts in a Phytophagous Spider Mite: Adaptive Plasticity?" *The American Naturalist* 159: 553–565.

Alberts, B., A. Johnson, J. Lewis, M. Raff, K. Roberts, and P. Walter. 2002. *Molecular Biology of the Cell*, 4th ed. New York: Garland Publishing.

Alvarez, W. 1997. *T. Rex and the Crater of Doom.* Princeton, NJ: Princeton University Press.

American Psychiatric Association. 1994. *Diagnostic and Statistical Manual of Mental Disorders, DSM,* 4th ed. Washington, DC: American Psychiatric Association.

Ayala, Francisco J., and John A. Kiger. 1980. *Modern Genetics.* Menlo Park, Calif.: Benjamin Cummings.

Bakker, K. 1961. "An Analysis of Factors Which Determine Success in Competition for Food among Larvae of *Drosophila melanogaster.*" *Archives Néerlandaises de Zoologie* 14: 200–281.

Barkow, J.H., L. Cosmides, and J. Tooby, eds. 1992. *The Adapted Mind: Evolutionary Psychology and the Generation of Culture.* New York: Oxford University Press.

Bates Smith, T. 1993. "Disruptive Selection and the Genetic Basis of Bill Size: Polymorphism in the African Finch *Pyrestes.*" *Nature* 353: 618–620.

Bell, Graham. 1982. *The Masterpiece of Nature; The Evolution and Genetics of Sexuality.* Berkeley, CA: University of California Press.

Bell, Graham. 1988. *Sex and Death in Protozoa; The History of an Obsession.* Cambridge, UK: Cambridge University Press,

Bernstein, Carol, and Harris Bernstein. 1991. *Aging, Sex, and DNA Repair.* San Diego: Academic Press.

Bijlsma, R., J. Bundgaard, and A.C. Boerema. 2000. "Does Inbreeding Affect the Extinction Risk of Small Populations? Predictions from *Drosophila.*" *Journal of Evolutionary Biology* 13: 502- 514.

Bongaarts, J. and R. A. Bulatao (eds). 2000. *Beyond Six Billion: Forecasting the World's Population.* Washington, DC: National Academy Press.

Borash, D. J., V. A. Pierce, A. G. Gibbs, and L. D. Mueller. 2000. "Evolution of Ammonia and Urea Tolerance in *Drosophila melanogaster*: Resistance and Cross-tolerance." *Journal of Insect Physiology* 46:763–769.

Botkin, D. B. and E. A. Keller. 2000. *Environmental Science*, 3rd ed. New York: John Wiley.

Briggs, D.E.G., D.H. Erwin, and F.J. Collier. 1994. *The Fossils of the Burgess Shale.* Washington, DC: Smithsonian Institution Press.

Bronstein, J. 2001. "The Exploitation of Mutualisms." *Ecology Letters* 4: 277–287.

Brooks, D.R. and D.A. McLennan. 1993. *Parascript.* Washington, DC: Smithsonian Institution Press.

Browne, Janet. 1995. *Charles Darwin, Voyaging.* New York: Knopf.

Browne, Janet. 2002. *Charles Darwin: The Power of Place.* New York: Knopf.

Bull, J. J., I.J. Molineux, and W.R. Rice. 1991. "Selection of Benevolence in a Host-Parasite System." *Evolution* 45: 875–882.

Burdon, J.J., R.H. Groves, P.E. Kaye and S.S. Speer. 1984. "Competition in Mixtures of Susceptible and Resistant Genotypes of *Chondrilla juncea* Differentially Infected with Rust." *Oecologia* 64: 199–203.

Byrne, R., and A. Whiten, eds. 1988. *Machiavellian Intelligence, Social Expertise and the Evolution of Intellect in Monkeys, Apes, and Humans.* Oxford, UK: Clarendon Press.

Campbell, D.R., N.M. Waser, and M.V. Price. 1996. "Mechanisms of Hummingbird-Mediated Selection for Flower Width in *Ipomopsis aggregata.*" *Ecology* 77: 1463–1472.

Caughley, G. and A. Gunn. 1996. *Conservation Biology in Theory and Practice.* Cambridge, MA: Blackwell Science.

Cavalli-Sforza, L.L. 1997. "Genes, Peoples, and Languages." *Proceedings of the National Academy of Sciences, USA,* 94: 7719–7724.

Cavalli-Sforza, L.L., and W.F. Bodmer. 1971. *The Genetics of Human Populations.* San Francisco: W.H. Freeman.

Cavalli-Sforza, L.L., P. Menozzi, and A. Piazza. 1994. *The History and Geography of Human Genes.* Princeton, NJ: Princeton University Press.

Charlesworth, Brian. 1980. *Evolution in Age-Structured Populations.* London: Cambridge University Press.

Charnov, E.L. 1982. *The Theory of Sexual Allocation.* Princeton, NJ: Princeton University Press.

Chippindale, A.K., T.J.F. Chu, and M.R. Rose. 1996. "Complex Trade-offs and the Evolution of Starvation Resistance in *Drosophila melanogaster.*" *Evolution* 50: 753–766.

Clark, Ronald W. 1984. *The Survival of Charles Darwin, A Biography of a Man and an Idea.* New York: Random House.

Cody, M. 1966. "A General Theory of Clutch Size." *Evolution* 20: 174–184.

Cohen, J.E., 1978. *Food Webs and Niche Space.* Princeton, NJ: Princeton University Press.

Cole, Lamont C. 1954. "The Population Consequences of Life History Phenomena." *Quarterly Review of Biology* 29: 103–137.

Comfort, A. 1979. *The Biology of Senescence,* 3rd ed. Edinburgh: Churchill Livingstone.

Connell, J.H. 1961. "The Influence of Interspecific Competition and the Other Factors on the Distribution of the Barnacle *Chthamalus stellatus.*" *Ecology* 42: 710–723.

Costantino, R.F., R.A. Desharnais, J.M. Cushing, and B. Dennis. 1997. "Chaotic Dynamics in an Insect Population." *Science* 275: 389–391.

Cox, G.W. 1997. *Conservation Biology.* Dubuque, IA: Wm. C. Brown.

Crnokrak, P. and D.A. Roff. 2002. "Trade-offs to Flight Capability in *Gryllus firmus:* The Influence of Whole-Organism Respiration Rate on Fitness." *Journal of Evolutionary Biology* 15: 388–398.

Cronin, H. 1992. *The Ant and the Peacock, Altruism and Sexual Selection from Darwin to Today.* Cambridge, UK: Cambridge University Press.

Currie, C.R., J.A. Scott, R.C. Summerbell, and D. Malloch. 1999. "Fungus-Growing Ants Use Antibiotic-Producing Bacteria to Control Garden Parasites." *Nature* 398: 701–704.

Damasio, A.R. 1994. *Descartes's Error; Emotion, Reason, and the Human Brain.* New York: Avon.

Darwin, Charles R. 1859. *On the Origin of Species by Means of Natural Selection, or The Preservation of Favoured Races in the Struggle for Life.* London: John Murray.

Darwin, Charles. 1896. *The Variation of Animals and Plants under Domestication*, facsimile edition. New York Appleton.

Davies N.B. 1977. "Prey Selection and Social Behaviour in Wagtails (*Aves: Motacillidae*)." *Journal of Animal Ecology* 46:37–57.

Dayton, P.K. 1971. "Competition, Disturbance, and Community Organization: The Provision and Subsequent Utilization of Space in a Rocky Intertidal Community." *Ecological Monographs* 41: 351–389.

Denno, R.F., M.A. Peterson, C. Gratton, J. Cheng, G.A. Langellotto, A.F. Huberty, and D.L. Finke. 2000. "Feeding-Induced Changes in Plant Quality Mediate Interspecific Competition between Sap-Feeding Herbivores." *Ecology* 81: 1814–1827.

Desmond, Adrian, and James Moore. 1991. *Darwin: The Life of a Tormented Evolutionist*. New York: Warner.

Dobzhansky, T.H. 1937. *Genetics and the Origin of Species*. New York: Columbia University Press.

Ehrlich, P.R. 1968. *The Population Bomb*. New York: Ballantine.

Ehrlich, P.R. and D.D. Murphy. 1987. "Conservation Lessons from Long-Term Studies of Checkerspot Butterflies." *Conservation Biology* 1:122–131.

Ehrlich, P.R. and P.H. Raven. 1964. "Butterflies and Plants: A Study in Coevolution." *Evolution* 18: 586–608.

Eldridge, J.L. and D.H. Johnson. 1988. "Size Differences in Migrant Sandpiper Flocks: Ghosts in Ephermal Guilds." *Oecologia* 77: 433–444.

Endler, J.A. 1991. "Interactions between Predators and Prey." In *Behavioural Ecology*, edited by J.R. Krebs and N.B. Davies, 3rd ed. Oxford, UK: Blackwell.

Ewald, P.W. 1994. *Evolution of Infectious Disease*. New York: Oxford University Press.

Falconer, D.S., and T.F.C. Mackay. 1996. *Introduction to Quantitative Genetics*, 4th ed. Harlow, Essex, England: Longman.

Finch, Caleb E. 1990. *Longevity, Senescence, and the Genome.* Chicago: University of Chicago Press.

Ford, E.B. 1971. *Ecological Genetics*, 3rd ed. London: Chapman and Hall.

Freeman, S., and J.C. Herron. 2001. *Evolutionary Analysis*, 2nd ed. Upper Saddle River, NJ: Prentice-Hall.

Gadgil, M. and O. T. Solbrig. 1972. "The Concept of r- and K-selection: Evidence from Wild Flowers and Some Theoretical Considerations." *American Naturalist* 106: 14–31.

Gadgil, M. and W. Bossert. 1970. "Life Historical Consequences of Natural Selection." *American Naturalist* 104: 1–24.

Gallagher, R. 1969. *Diseases That Plague Modern Man*. Dobbs Ferry, NY: Oceana Publications,

Gause, G.F. 1934. *The Struggle for Existence*. Baltimore: Williams and Wilkins. Reprint, New York: Dover, 1971.

Ghiselin, Michael. 1974. *The Economy of Nature and the Evolution of Sex*. Berkeley, CA: University of California Press.

Gillespie, John H. 1991. *The Causes of Molecular Evolution*. New York: Oxford University Press.

Halvorson, Harlyn O., and Alberto Monroy, eds. 1985. *The Origin and Evolution of Sex*. New York: Alan R. Liss.

Hamilton, W.D. 1996. *Narrow Roads of Gene Land: The Collected Papers of W.D. Hamilton*. Vol. 1. *Evolution of Social Behavior*. New York: W.H. Freeman.

Hamilton, W.D. 2002. *Narrow Roads of Gene Land*. Vol. 2. *Evolution of Sex*. Oxford, UK: Oxford University Press.

Hanski, I. 1999. *Metapopulation Ecology*. Oxford, UK: Oxford University Press.

Hartl, Daniel L. 2000. *A Primer of Population Genetics*, 3rd ed. Sunderland, MA: Sinauer.

Hartl, Daniel L., and Elizabeth W. Jones. 1999. *Essential Genetics*, 2nd ed. Sudbury, MA: Jones & Bartlett.

Harvey, P.H., and M.D. Pagel. 1991. *The Comparative Method in Evolutionary Biology*. Oxford, UK: Oxford University Press.

Hastings, A. 1997. *Population Biology*. New York: Springer-Verlag.

Hedrick, Philip W. 2000. *Genetics of Populations*, 2nd ed. Boston: Jones & Bartlett.

Herre, E.A. 1993. "Population Structure and the Evolution of Virulence in Nematode Parasites of Fig Wasps." *Science* 259: 1442–1445.

Himmelfarb, Gertrude. 1959. *Darwin and the Darwinian Revolution*. New York: Norton.

Hochwender, C.G., R.J. Marquis, and K.A. Stowe. 2000. "The Potential for and Constraints on the Evolution of Compensatory Ability in *Asclepias syriaca*." *Oecologia* 122: 361–370.

Holt, R.D. and J.H. Lawton. 1994. "The Ecological Consequences of Shared Natural Enemies." *Annual Review of Ecology and Systematics* 25: 495–520.

Holyoak, M. and S.P. Lawler. 1996. "Persistence of an Extinction-Prone Predator-Prey Interaction through Metapopulation Dynamics." *Ecology* 77: 1867–1879.

Hublin, J., F. Spoor, M. Braun, F. Zonneveld, and S. Condemi. 1996. "A Late Neanderthal Associated with Upper Palaeolithic Artefacts." *Nature* 381:224–226.

Hudson, P.J., A.P. Dobson, and D. Newborn. 1998. "Prevention of Population Cycles by Parasite Removal." *Science* 282: 2256–2258.

Jarosz, A.M. and J.J. Burdon. 1991. "Host-Pathogen Interaction in Natural Populations of *Linum marginale* and *Melampsora lini*: II. Local and Regional Variation in Patterns of Resistance and Racial Structure." *Evolution* 45: 1618–1627.

Jenkins, B., R.L. Kitching, and S.L. Pimm. 1992. "Productivity, Disturbance and Food Web Structure at a Local Spatial Scale in Experimental Container Habitats." *Oikos* 65: 249–255.

Judson, O. 2003. *Dr. Tatiana's Sex Advice to All Creation*. New York: Owl Books.

Karban, R., and A.A. Agrawak. 2002. "Herbivore Offense." *Annual Reveiw of Ecology and Systematics* 33: 641–664.

Karl, T.R., N. Nichols, and J. Gregory. 1997. "The Coming Climate." *Scientific American* 276: 78–83.

Kimura, Motoo. 1983. *The Neutral Theory of Molecular Evolution*. London: Cambridge University Press.

Kitcher, P. 1985. *Vaulting Ambition, Sociobiology and the Quest for Human Nature*. Cambridge, MA: MIT Press.

Krebs, J.R. and M.I. Avery. 1985. "Central Place Foraging in the European Bee-eater, *Merops apiaster*." *Journal of Animal Ecology* 54: 459–472.

Krebs, J.R., and N.B. Davies. 1981. *An Introduction to Behavioral Ecology* and subsequent editions. Cambridge, UK: Blackwell.

Krebs, J.R., J.T. Erichsen, M.I. Webber, and E.L. Charnov. 1977. "Optimal Prey Selection in the Great Tit (*Parus major*)." *Animal Behavior* 25: 30–38.

Krings, M., H. Geisert, R.W. Schmitz, H. Krainitzki, and S. Paabo. 1999. "DNA Sequence of the Mitochondrial Hypervariable Region II from the Neandertal Type Specimen." *Proceedings of the National Academy of Sciences, USA*, 96: 5581–5585.

Larson, A., and J.B. Losos. 1996. "Phylogenetic Systematics of Adaptation." In *Adaptation*, edited by M.R. Rose & G.V. Lauder, 187–220. San Diego: Academic Press.

Lawler, S.P. and P.J. Morin. 1993. "Food Web Architecture and Population Dynamics in Laboratory Microcosms of Protists." *The American Naturalist* 141: 675–686.

Lewontin, Richard C. 1974. *The Genetic Basis of Evolutionary Change.* New York: Columbia University Press.

Li, Wen-Hsiung and Daniel Graur. 2000. *Fundamentals of Molecular Evolution,* 2nd ed. Sunderland, MA: Sinauer.

Luckinbill, L.S. 1973. "Coexistence in Laboratory Populations of *Paramecium aurelia* and *Didinium nasutum.*" *Ecology* 54: 1320–1327.

Lutz, W., W. Sanderson, and S. Scherbov. 2001. "The End of World Population Growth." *Nature* 412: 543–545.

MacArthur, RH. 1958. "Population Ecology of Some Warblers of Northeastern Coniferous Forests." *Ecology* 39: 599–619.

Malthus, Thomas Robert. 1798. *An Essay on the Principle of Population.* Reprint, New York: Norton, 1976.

May, R.M. 1978. "Host-Parasitoid Systems in Patchy Environments: A Phenomenological Model." *Journal of Animal Ecology* 47: 833–843.

Maynard Smith, J. 1958. "The Effect of Temperature and of Egg-laying on the Longevity of *Drosophila subobscura.*" *Journal of Experimental Biology* 35: 832–842.

Maynard Smith, J. 1978. *The Evolution of Sex.* Cambridge, UK: Cambridge University Press.

Maynard Smith, J. 1982. *Evolution and the Theory of Games.* Cambridge, UK: Cambridge University Press.

Mayr, E. 1942. *Systematics and the Origin of Species.* New York: Columbia University Press.

Mayr, E. 1963. *Animal Species and Evolution.* Cambridge, MA: Harvard University Press.

McNaughton, S. J. 1975. "r- and K-selection in *Typha.*" *American Naturalist* 109: 251–261.

Meffe, G.K. and C.R. Carroll. 1994. *Principles of Conservation Biology.* Sunderland, MA: Sinauer.

Michod, R.E. 1995. *Eros and Evolution, A Natural Philosophy of Sex.* Reading, MA: Helix Box.

Michod, R.E., and B.R. Levin, eds. 1988. *The Evolution of Sex.* Sunderland, MA: Sinauer.

Moorcroft, P.R., S.D. Albon, J.M. Pemberton, I.R. Stevenson, and T.H. Clutton-Brock. 1996. "Density-Dependent Selection in a Fluctuating Ungulate Population." *Proceedings of the Royal Society of London B* 263: 31–38.

Moran, N.A. and J.J. Wernegreen. 2000. "Lifestyle Evolution in Symbiotic Bacteria: Insights from Genomics." *Trends in Ecology and Evolution* 15: 321–326.

Morin, P.J. 1999. *Community Ecology.* Malden, MA: Blackwell Science.

Mueller, L.D., 1988. "Evolution of Competitive Ability in *Drosophila* Due to Density-Dependent Natural Selection." *Proceedings of the National Academy of Science USA* 85: 4383–4386.

Mueller, L.D. and A. Joshi. 2000. *Stability in Model Populations.* Princeton, NJ: Princeton University Press.

Mueller, L.D., P.Z. Guo and F.J. Ayala. 1991. "Density-Dependent Natural Selection and Trade-Offs in Life History Traits." *Science* 253: 433–435.

Naeem, S., L.J. Thompson, S.P. Lawler, J.H. Lawton, and R.M. Woodfin. 1995. "Empirical Evidence that Declining Species Diversity May Alter the Performance of Terrestrial Ecosystems." *Philosophical Transactions of the Royal Society London* B 347: 249–262.

Nesse, R.M., and G.C. Williams. 1994. *Why We Get Sick, The New Science of Darwinian Medicine.* New York: Random House.

Nicholson, A., and V. Bailey. 1935. "The Balance of Animal Populations. Part 1." *Proceedings of the Zoological Society of London* 3: 551–598.

Niklas, K. J. 1994. *Plant Allometry.* Chicago: University of Chicago Press.

Novacek, M.J. 1996. "Paleontological Data and the Study of Adaptation." In *Adaptation,* edited by M.R. Rose & G.V. Lauder, 311–59. San Diego: Academic Press.

Otte, D., and J. Endler. 1989. *Speciation and its Consequences.* Sunderland, MA: Sinauer.

Paine, R.T. 1966. "Food Web Complexity and Species Diversity." *American Naturalist* 100: 65–75.

Paoletti, M.G., and D. Pimentel. 1996. "Genetic Engineering in Agriculture and the Environment." *Bioscience* 46: 665–673.

Post, D.M., M.L. Pace, and N.G. Hairston Jr. 2000. "Ecosystem Size Determines Food-Chain Length in Lakes." *Nature* 405: 1047–1049.

Power, M.E. 1990. "Effects of Fish in River Food Webs." *Science* 250: 811–814.

Raup, David M. 1991. *Extinction, Bad Genes or Bad Luck?* New York: W. W. Norton.

Rice, W.R., and E.E. Hostert. 1993. "Laboratory Studies on Speciation—What Have We Learned in 40 Years?" *Evolution* 47: 1637–1653.

Ridley, M. 1996. *Evolution,* 3rd ed. Cambridge, MA: Blackwell.

Robertson, O.H. 1961. "Prolongation of the Life of Kokanee Salmon (*Oncorhynchus nerka kennerlyi*) by Castration before Beginning of Gonad Development." *Proceedings of the National Academy of Sciences USA* 47: 609–621.

Roff, Derek A. 1992. *The Evolution of Life Histories: Theory and Analysis.* New York: Chapman and Hall.

Romer, A.S. 1970. *The Vertebrate Body,* 4th ed. Philadelphia: W.B. Saunders.

Rose, M.R. 1991. *Evolutionary Biology of Aging.* New York: Oxford University Press.

Rose, M.R. 1998. *Darwin's Spectre: Evolutionary Biology in the Modern World.* Princeton, NJ: Princeton University Press.

Rose, M.R., and G.V. Lauder, eds. 1996. *Adaptation.* San Diego, CA: Academic Press.

Roughgarden, J. 1971. "Density-Dependent Natural Selection." *Ecology* 52: 453–468.

Roughgarden, J. 1998. *Primer of Ecological Theory.* chapter 2. Upper Saddle River, NJ: Prentice Hall.

Roughgarden, J., S. Gaines, and H. Possingham. 1988. "Recruitment Dynamics in Complex Life Cycles." *Science* 241: 1460–1466.

Rowland, F.S. 1990. "Stratospheric Ozone Depletion by Chloroflurocarbons." *AMBIO* 19: 281–292.

Schluter, D. 2000. *The Ecology of Adaptive Radiation.* Oxford, UK: Oxford University Press.

Schluter, D., and J.D. McPhail. 1992. "Ecological Character Displacement and Speciation in Sticklebacks." *The American Naturalist* 140: 85–108.

Schmidt-Nielsen, K. 1990. *Animal Physiology.* Cambridge, UK: Cambridge University Press.

Schoener, T.W. 1989. "The Ecological Niches." In *Ecological Concepts,* edited by J.M. Cherrett, 79–113. Oxford, UK: Blackwell Scientific.

Selander, Robert K., Andrew G. Clark, and Thomas S. Whittam, eds. 1991. *Evolution at the Molecular Level.* Sunderland, MA: Sinauer.

Sinervo, Barry and Alexandra Basolo. 1996. "Testing Adaptation Using Phenotypic Manipulation." In *Adaptation* edited by M.R. Rose & G.V. Lauder, 149–185. San Diego: Academic Press.

Slatkin, M. 1981. "Estimating Levels of Gene Flow in Natural Populations." *Genetics* 99: 323–335

Smith, K.G.V. 1986. *A Manual of Forensic Entomology*. London: The Trustees of the British Museum of Natural History.

Sokolowski, M.B, H.S. Pereira, K. Hughes. 1997. "Evolution of Foraging Behavior in *Drosophila* by Density-Dependent Selection." *Proceedings of the National Academy of Science USA* 94: 7373–7377.

Sousa, W.P. 1979. "Disturbance in Marine Intertidal Boulder Fields: The Nonequilibrium Maintenance of Species Diversity." *Ecology* 60: 1225–1239.

Stearns, S.C., ed. 1999. *Evolution in Health and Disease*. Oxford, UK: Oxford University Press.

Stearns, Stephen C. 1992. *The Evolution of Life-Histories*. New York: Oxford University Press.

Stock, G., and J. Campbell, eds. 2000. *Engineering the Human Germline, An Exploration of the Science and Ethics of Altering the Genes We Pass to Our Children*. New York: Oxford University Press.

Stowe, K.A. 1998. "Experimental Evolution of Resistance in *Brassica rapa*: Correlated Response of Tolerance in Lines Selected for Glucosinolate Content." *Evolution* 52: 703–712.

Stowe, K.A., R.J. Marquis, C.G. Hochwender, and E.L. Simms. 2000. "The Evolutionary Ecology of Tolerance to Consumer Damage." *Annual Reveiw of Ecology and Systematics* 31: 565–595.

Swingland, I.R., and P.J. Greenwood. 1983. *The Ecology of Animal Movement*. Oxford, UK: Oxford University Press.

Tattersall, I. 1999. *The Last Neanderthal*. New York: Nevramount.

Thompson, J.N. 1982. *Interaction and Coevolution*. New York: John Wiley &Sons.

Thompson, J.N. 1994. *The Coevolutionary Process*. Chicago: University of Chicago Press.

Tilman, D. 1988. *Dynamics and Structure of Plant Communities*. Princeton, NJ: Princeton University Press.

Tishkoff, S.A., E. Dietzsch, W. Speed, A.J. Pakstis, J.R. Kidd, K. Cheung, B. Bonne-Tamir, A.S. Santachiara-Benerecetti, T. Jenkins, and K.K. Kidd. 1996. "Global Patterns of Linkage Disequilibrium at the CD4 Locus and Modern Human Origins." *Science* 271: 1380–1387.

Trivers, R.L. 1985. *Social Evolution*. Menlo Park, CA: Benjamin Cummings.

Trunbore, S. E., O.A. Chadwick, and R. Amundson. 1996. "Rapid Exchange between Soil Carbon and Atmospheric Carbon Dioxide Driven by Temperature Change." *Science* 272: 393–396.

Turelli, M. and A.A. Hoffmann. 1991. "Rapid Spread of an Inherited Incompatibility Factor in California *Drosophila*." *Nature* 353: 440–442.

Wade, M.J. 1977. "An Experimental Study of Group Selection." *Evolution* 31: 134–153.

Wade, M.J. 1978. "A Critical Review of the Models of Group Selection." *Quarterly Review of Biology* 53: 101–114.

Wassersug, Richard J., and Michael R. Rose. 1984. "A Reader's Guide and Retrospective to the 1982 Darwin Centennial." *Quarterly Review of Biology* 59: 417.

Weiner, Jonathan. 1994. *The Beak of the Finch, A Story of Evolution in Our Time*. New York: A.A. Knopf.

White, M.J.D. 1978. *Modes of Speciation*. San Francisco: W.H. Freeman.

Whitehouse, H.L.K. 1969. *Towards an Understanding of the Mechanism of Heredity*. 2nd ed. London: Edward Arnold.

Wickler, Wolfgang. 1968. *Mimicry*. New York: McGraw-Hill.

Wilbur, H. M. 1980. "Complex Life Cycles." *Annual Review Ecology and Systematics* 11: 67–93.

Williams, George C. 1975. *Sex and Evolution*. Princeton, NJ: Princeton University Press.

Willmer, P., G. Stone, and I. Johnston. 2000. *Environmental Physiology of Animals*. Oxford, UK: Blackwell Science.

Wilson, E.O. 1971. *The Evolution of Insect Societies*. Cambridge, MA: Harvard University Press.

Wilson, E.O. 1975. *Sociobiology: The New Synthesis*. Cambridge, MA: Belknap Press.

Wilson, E.O. 1978. *On Human Nature*. Cambridge, MA: Harvard University Press,

Wolpoff, M.H. 1997. *Paleoanthropology*. New York: McGraw-Hill.

Worthern, W. B. 1989. "Predator-Mediated Coexistence in Laboratory Communities of Mycophagous *Drosophila* (Diptera: Drosophilidae)." *Ecological Entomology* 14: 117–126.

Zamir, D. 2001. "Improving Plant Breeding with Exotic Genetic Libraries. *Nature Reviews Genetics* 2: 983–989.

Zera, A.J,. and R.F. Denno. 1997. "Physiology and Ecology of Dispersal Polymorphism in Insects." *Annual Review of Entomology*, 42: 207–231.

Frontmatter

p. i top Frank Greenaway©Dorling Kindersley **p.i bottom** Roger Phillips © Dorling Kindersley **p. iv** Michael Rose/Larry Mueller **p. x** Corbis/ Bettmann **p. xi** Peter Anderson©Dorling Kindersley **p. xiv** Tom McHugh/Photo Researchers, Inc. **p. xvi** Michael Fogden/Animals Animals/Earth Scenes **p. xviii** Karl Shone©Dorling Kindersley **p. xxi** Cyril Laubscher©Dorling Kindersley

Part 1

01 Alfred Pasieka/Photo Researchers, Inc. **02** Corbis/Bettmann

Chapter 1

CO.01 Mary Evans Picture Library Ltd. **p. 4** Dave King©Dorling Kindersley **1.1A** Punch Limited **1.1B** The Bridgeman Art Library International Ltd. **p. 5** Dave King©Dorling Kindersley **p. 6** Rob Reichenfeld©Dorling Kindersley **1.2A.i** Michael Nicholson/Corbis-NY **1.2A.ii** The Granger Collection, New York **1.2A.iii** Bettman Archives/ CORBIS/Corbis/Bettmann **1.2B.i** ARCHIV/Photo Researchers, Inc. **1.2B.ii** Bettmann Archive/Corbis/Bettmann **1.2B.iii** Corbis/Bettmann **1.3A** The Granger Collection, New York **p. 8** Rob Reichenfeld©Dorling Kindersley **p. 9** Robert Weiss/Silver Burdett Ginn **p. 11** Cyril Laubscher©Dorling Kindersley **p. 13** John Serafin **1.6A** The Granger Collection, New York **p. 14 left** John Davis©Dorling Kindersley **p. 14 right** Dave King©Dorling Kindersley **p. 15** ©Dorling Kindersley **p. 16** Irv Beckman©Dorling Kindersley **1.9A** The Academy of Natural Sciences of Philadelphia, Ewell Sale Stewart Library **p. 23** Peter Anderson©Dorling Kindersley **p. 24** Geoff Brightling©Dorling Kindersley **1.14A** The Burndy Library, Dibner Institute for the History of Science and Technology, Cambridge, Massachusetts **p. 25** Kim Taylor & Jane Burton©Dorling Kindersley **p. 26** M.I. Walker©Dorling Kindersley **p. 27** Frank Greenaway©Dorling Kindersley **p. 29** Steve Gorton©Dorling Kindersley **1.17A** Corbis/Bettmann **p. 30** Peter Dennis©Dorling Kindersley **p. 33** Harry Taylor©Dorling Kindersley **1.20A** George D. Lepp/Photo Researchers, Inc. **p. 34** Peter Chadwick©Dorling Kindersley **p. 35** Neil Fletcher©Dorling Kindersley **p. 36** Neil Fletcher©Dorling Kindersley **p. 39** Dave King©Dorling Kindersley **p. 40** Geoff Brightling©Dorling Kindersley **1.23A** Dave King©Dorling Kindersley

Chapter 2

CO.02 Corbis/Bettmann **2.3A** Courtesy of the Library of Congress **p. 49** Colin Keates ©Dorling Kindersley **p. 51** Jerry Young©Dorling Kindersley **p. 53** Rob Reichenfeld©Dorling Kindersley **2.7A.i** Keith Porter/Photo Researchers, Inc. **2.7A.ii** John Durham/Photo Researchers, Inc. **2.7A.iii** Dr. Dennis Kunkel/Visuals Unlimited **p. 54** M.I. Walker©Dorling Kindersley **p. 55** Matthew Ward©Dorling Kindersley **2.8A top** Karl O. Stetter, University of Regensburg, Germany **2.8A bottom** Eurelios/Phototake NYC **2.8B** National Institute for Biological Standards and Control (U.K.)/Science Photo Library/Photo Researchers, Inc. **2.9B** M.I. Walker©Dorling Kindersley **p. 58** Silver Burdett Ginn **2.11A** ©D.W. Miller **p. 61** Andy Crawford©Dorling Kindersley **2.12B top** Wolfgang Kaehler/CORBIS-NY **2.12B bottom** Buddy Mays/CORBIS-NY **p. 63** Shaen Adey©Dorling Kindersley **2.13A** American Museum of Natural History **p. 65** Frank Greenaway©Dorling Kindersley **2.14A.01** Cal Vornberger/Peter Arnold, Inc. **2.14A.02** Luiz C. Marigo/Peter Arnold, Inc. **2.14A.03** Frank Greenaway©Dorling Kindersley **2.14A.04** Richard Alan Wood/Animals Animals/Earth Scenes **p. 67** Kim Taylor & Jane Burton©Dorling Kindersley **p. 68** Frank Greenaway©Dorling Kindersley **p. 69** Kim Taylo ©Dorling Kindersley **2.17B** Paul Harris/Nature Portfolio

Part 2

01 & 03 Spots on the Spot **02** ©Dorling Kindersley

Chapter 3

CO.03 Dan McCoy/Rainbow/AGE Fotostock America, Inc. **3.1A** Archiv/ Photo Researchers, Inc. **3.1B.01** Photo Researchers, Inc. **3.1B.02** The Granger Collection, New York **p. 80** Pearson Learning Photo Studio **p. 81** Derek Hall©Dorling Kindersley **p. 83** Dr. Julian Thorpe©Dorling Kindersley **p. 85** Frank Greenaway©Dorling Kindersley **p. 86** Steven Wooster©Dorling Kindersley **3.7B** ©Dorling Kindersley **p. 91** Spike Walker (Microworld Services)©Dorling Kindersley **3.8A.i** Science Photo Library/Photo Researchers, Inc. **3.8A.ii** The Pearson Papers, Library Services, University College London **3.8A.iii** A. Barrington Brown/Photo Researchers, Inc. **p. 92** Paul Bricknell©Dorling Kindersley **p. 93** John Serafin **p. 95** Mike Peters/Silver Burdett Ginn **p. 96** ©Dorling Kindersley **p. 97** Roger Phillips©Dorling Kindersley **p. 98** Jules Selmes ©Dorling Kindersley **p. 99** Steve Shott©Dorling Kindersley **p.100** Matthew Ward©Dorling Kindersley **p. 103** ©Dorling Kindersley **p.105** Andy Crawford & Gary Ombler ©Dorling Kindersley **p. 107** EMG Education Management Group **p. 108** Andy Crawford©Dorling Kindersley **3.18A.01** Gary Randall/Getty Images, Inc.-Taxi **3.18A.02 & 3.18A.03** Dave King©Dorling Kindersley **3.19A** Renee Stockdale/ Animals Animals/Earth Scenes **p. 116** Kim Taylor & Jane Burton©Dorling Kindersley **3.21A** ©Historical Picture Archive/CORBIS **3.21B** Simon Harris/Robert Harding World Imagery **p. 119** Alan Keohan©Dorling Kindersley **p. 120**©Dorling Kindersley

Chapter 4

CO.04 John Pontier/Animals Animals/Earth Scenes **p. 126** Demetrio Carrasco©Dorling Kindersley **4.1A** The Granger Collection, New York **p. 129** Colin Keates©Dorling Kindersley, Courtesy of the Natural History Museum, London **p. 131** Paul Bricknell©Dorling Kindersley **p. 132** EMG Education Management Group **p. 134** Trish Gant©Dorling Kindersley **p. 135** Frank Greenaway©Dorling Kindersley **p. 136** ©Jerry Young/Dorling Kindersley **p. 140** Jerry Young©Dorling Kindersley **p. 142** Frank Greenaway©Dorling Kindersley **p. 144** Peter Anderson©Dorling Kindersley **4.13A** Kunsthistorisches Museum Wien **4.13B.01 & 4.13B.02** Corbis Bettman **4.13B.03** CORBIS-NY **4.13B.04** Corbis/Sygma **p. 148** Eddie Lawrence©Dorling Kindersley **p. 150** ©Jerry Young/Dorling Kindersley **p. 153** ©Judith Miller/Dorling Kindersley/Bucks County Antiques Center **p. 154** Spike Walker (Microworld Services) ©Dorling Kindersley **p. 156** ©Dorling Kindersley, Courtesy of Oxford Scientific Films **p. 157** Frank Greenaway©Dorling Kindersley **4.22A** Photo courtesy of Pfizer, Inc., New York **4.22B** Dr. L. Caro/Science Photo Library/Photo Researchers, Inc. **p. 159** ©Dorling Kindersley **4.23B** Breck P. Kent **4.24C.01** Mickey Gibson/Animals Animals/Earth Scenes **4.24C.02** Ed Reschke/Peter Arnold, Inc. **4.24C.03** Ralph Lee Hopkins/Wilderland Images **4.24C.04** Patti Murray/ Animals Animals/Earth Scenes **p. 162** Ariel Skelley/Corbis/Bettmann **4.25A** VU/Stanley Flegler/Visuals Unlimited **4.25C** Denis Finnin & Jackie Beckett©The American Museum of Natural History **p. 163** Frank Greenaway©Dorling Kindersley

Chapter 5

CO.05 A. Barrington Brown/Photo Researchers, Inc. **5.1A** Doug Scott/ **AGE** Fotostock America, Inc. **5.1D** ©Sinclair Stammers/Science Photo Library/Photo Researchers, Inc. **p. 169** Kim Taylor & Jane Burton©Dorling Kindersley **5.3A** Richard Kolar/Animals Animals/Earth

Scenes   **5.3B** Hermann Eisenbeiss/Photo Researchers, Inc.   **5.4C.01** AP Wide World Photos   **5.4C.02** Laurie Ann Achenbach, Southern Illinois University at Carbondale   **5.5C.01** Tony Gervis/Robert Harding World Imagery   **5.5C.02** Pete Saloutos/Corbis/Stock Market   **p. 174** Cyril Laubscher©Dorling Kindersley   **5.7A** Olga Shalygin/AP Wide World Photos   **5.7C** David Gifford/Photo Researchers, Inc.   **p. 179** Andy Crawford©Dorling Kindersley   **p. 180** Barnabas Kindersley©Dorling Kindersley   **p. 183** Jane Burton©Dorling Kindersley   **5.12A.01** Tom McHugh/Photo Researchers, Inc.   **5.12A.02** Rod Planck/Photo Researchers, Inc.

Chapter 6

CO.06 Douglas Henderron   **p. 188** Kim Taylor & Jane Burton©Dorling Kindersley   **p. 189** Mike Linley©Dorling Kindersley   **6.2A.01** Bob Langrish©Dorling Kindersley   **6.2A.02** Carl & Ann Purcell/CORBIS-NY   **6.2.A.03** Jerry Young©Dorling Kindersley   **p. 191** Paul Kenward©Dorling Kindersley   **6.3A** Kevin Schafer/Photo Researchers, Inc.   **6.3B** Teresa and Gerald Audesirk   **6.4C** Reprinted from Publication: *Experimental Cell Research*, Vol 281, No 1. pp 70, J. Shay et al, 1 figure. Copyright 2002 with permission from Elsevier Science Ltd.   **p. 197** Paul Kenward@Dorling Kindersley   **6.6B** Peter Gardner@Dorling Kindersley   **6.6C** Doug Sokell/Visuals Unlimited   **6.7Ba** Guy Bush, Michigan State University, and Jeff Feder, University of Notre Dame   **p. 201** Jerry Clive Boursnell@Dorling Kindersley   **p. 203** Dave King@Dorling Kindersley   **p. 204** Peter Anderson@Dorling Kindersley   **p. 205** Frank Greenaway@Dorling Kindersley   **p. 206** John Miller@Dorling Kindersley   **6.10B.i.01–03** Jerry Young@Dorling Kindersley   **6.10B.i.04** Max Gibbs@Dorling Kindersley   **6.11A.01, 05, & 07** Smithsonian Institution/Office of Imaging, Printing, and Photographic Services   **6.11A.02,03, 04, & 06** Chip Clark/Courtesy of the National Museum of Natural History   **6.11B** Chase Studio/Photo Researchers, Inc.   **p. 209** Peter Wilson@Dorling Kindersley   **p. 210** Nigel Hicks@Dorling Kindersley   **p. 213** Dave King@Dorling Kindersley   **p. 216** Frank Greenaway@Dorling Kindersley   **p. 217** Michael Pitts@Dorling Kindersley   **p. 219** Kim Taylor@Dorling Kindersley   **p. 221** Frank Greenaway@Dorling Kindersley   **p. 224** NASA©Dorling Kindersley   **6.19A.i** Allesandro Montanari/Geological Observatory of Coldigioco, Fronte di Apiro, Italy   **6.19A.ii & iv** Glen A. Izett/U.S. Geological Survey, Denver

Part 3

p. 228 Geoff Renner/Robert Harding World Imagery   **p. 229 top** ©Jerry Young@Dorling Kindersley   **p. 229 bottom** Frank Greenaway@Dorling Kindersley

Chapter 7

CO.07 Laurie Campbell/Getty Images Inc.-Stone Allstock   **7.1A.01** Jeremy Burgess/Science Photo Library/Photo Researchers, Inc.   **70.1A.02** Fred Bavendam/Peter Arnold, Inc.   **7.1B** Michael P. Gadomski/Photo Researchers, Inc.   **7.1C** Don Farrall/Getty Images, Inc.-Photodisc   **7.1D** B.G. Thomson/Photo Researchers, Inc.   **7.1E** John Eastcott, Yva Momatiuk/**Animals** Animals/Earth Scenes   **7.1F** Michel Viard/Peter Arnold, Inc.   **p. 236** @Dorling Kindersley   **p. 239** Roger Phillips@Dorling Kindersley   **p.243** Joe Cornish@Dorling Kindersley   **p. 244** Max Alexander@Dorling Kindersley   **p. 245** Cyril Laubscher@Dorling Kindersley   **7.10A** Michael Heron/Pearson Education/PH College   **7.10C** Heidelberg College Collection   **p. 247 top** Steve Gorton@Dorling Kindersley   **p. 247 bottom** Linda Whitwam@Dorling Kindersley   **7.11C** Ken Brate/Photo Researchers, Inc.   **p. 249** Steve Shott@Dorling Kindersley   **7.13B** ©Science Pictures Limited/CORBIS

Chapter 8

CO.08 Alistair Duncan@Dorling Kindersley   **p. 258** Dave King@Dorling Kindersley   **8.4A** Galen Rowell/Corbis/Bettmann   **p. 261** Alan Keohane@Dorling Kindersley   **p. 263** Alistair Duncan@Dorling Kindersley   **8.6A.01** CNRI/Science Photo Library/Photo Researchers, Inc.   **8.6A.02** Super Stock, Inc.   **8.6C.01** Jochen Tack/Das Fotoarchiv./Peter Arnold, Inc.   **8.6C.02** David T. Webb, Botany, University of Hawaii   **8.7B** Breck P. Kent/Animals Animals/Earth Scenes   **8.7D** Lynton

Gardiner@Dorling Kindersley, The American Museum of Natural History   **p. 269** @Dorling Kindersley

Chapter 9

CO.09 Dagli Orti/Picture Desk, Inc./Kobal Collection   **9.1B** Professor Therese Ann Markow   **p. 275** Ken Findlay@Dorling Kindersley   **9.5D** John D. Cunningham/Visuals Unlimited   **9.7B** Tom McHugh/Photo Researchers, Inc.   **p. 285** ©Jerry Young/Dorling Kindersley   **p. 286** Magnus Rew@Dorling Kindersley   **p. 289** Frank Greenaway@Dorling Kindersley   **p. 291** John Garrett@Dorling Kindersley   **p. 293** Karl Stone@Dorling Kindersley   **p. 294** Kim Taylor@Dorling Kindersley, Courtesy of the Natural History Museum, London   **9.13B** Thomas J. Walker

Chapter 10

CO.10 Laurence D. Mueller   **p. 298** David R. Frazier Photolibrary, Inc.   **10.1A** Corbis/Bettman   **p. 301 top** ©Barrie Watts/Dorling Kindersley   **p. 301 middle** Steve Gorton@Dorling Kindersley   **p. 302** UN/DPI Photo   **10.3C** Arnold Bocklin (1827–1901) Gefirnisste Tempera auf Tannenholz 149.5x104.5 cm. Kunstmuseum Basel Depostum der Gottfried Keller-Stiftung 1902. Inv. Nr. 114.   **p. 303** Martin Harvey/Peter Arnold, Inc.   **10.4A** Jim Zipp/Photo Researchers, Inc.   **10.5A** American Philosophical Society   **p. 307** Robert Brenner/PhotoEdit   **p. 308** Christopher & Sally Gable©Dorling Kindersley   **p. 310** Dave King©Dorling Kindersley   **p. 311** Frank Greenaway©Dorling Kindersley   **10.9A** Steve Gorton©Dorling Kindersley   **10.9B** Kim Taylor & Jane Burton©Dorling Kindersley   **10.9C** Princeton University Library   **10.10C** ©Dorling Kindersley   **10.12B** Laurence D. Mueller   **p. 317** ©Dorling Kindersley   **p. 318** Cecile Treal & Jean-Michel Ruiz©Dorling Kindersley   **10.14A** Phillippe Plailly/Eurelios Photographic Press Agency   **p. 321** Richard Leeney©Dorling Kindersley   **p. 323** James Sawders, The Workshop/Pearson Education/PH College   **10.16B.A** Brad Mogen/Visuals Unlimited   **10.16B.B** Jack Bostrack/Visuals Unlimited   **10.16B.C** Alfred Pasieka/Photo Researchers, Inc.   **10.16C** Ted Stefanski/Alamy Images   **p. 325** Peter Anderson©Dorling Kindersley

Chapter 11

CO.11 John Callahan/Getty Images Inc.-Stone Allstock   **11.1A** Runk/Schoenberger/Grant Heilman Photography, Inc.   **11.1B** Anthony J. Zera, University of Nebraska   **p. 331** Kim Taylor & Jane Burton©Dorling Kindersley   **p. 334** Alistair Duncan©Dorling Kindersley   **11.5A** Diane Campbell/University of California/Irvine   **11.6A.01** Stuart Westmorland/Getty Images.-Stone Allstock   **11.6A.02** David Cavagnaro/Peter Arnold, Inc.   **11.6B** Stouffer Prod./Animals Animals/Earth Scenes   **p. 338** Michael Heron/Pearson Education/PH College   **p. 339** Diana Miller©Dorling Kindersley   **11.9B** Roger Ressmeyer/Corbis/Bettmann   **p. 343** Peter Anderson©Dorling Kindersley   **p. 346** Kim Taylor & Jane Burton©Dorling Kindersley

Part 4

01 Michael Fogden/Animals Animals/Earth Scenes   **02** Stefan Meyer/Animals Animals/Earth Scenes   **03** Spots on the Spot   **p. 349** Jane Burton©Dorling Kindersley

Chapter 12

CO.12 Robert Brenner/PhotoEdit   **12.1A.01** G.R. "Dick" Roberts/The Natural Sciences Image Library (NSIL)   **12.1A.02** Hal Beral/Visuals Unlimited   **12.1A.03** Lior Rubin/Peter Arnold, Inc.   **p. 352** Jane Burton©Dorling Kindersley   **p. 353** Peter Anderson©Dorling Kindersley   **p. 354** Jerry Young©Dorling Kindersley   **p. 355** Frank Greenaway©Dorling Kindersley   **p. 356** ©Alan Watson/Dorling Kindersley   **p. 358** Peter Anderson©Dorling Kindersley   **12.4C** Northern Prairie Wildlife Research Center, U.S.G.S.   **12.4D** ©Dorling Kindersley   **p. 360** Frank Greenaway©Dorling Kindersley   **12.5B** Australian Academy of Science   **12.6A** M.I. Walker/Photo Researchers, Inc.   **p. 363** Alan Keohane©Dorling Kindersley   **p. 364** Irv Beckman©Dorling Kindersley   **p. 366** Richard Weiss/Silver Burdett Ginn   **p. 367** Andy Crawford©Dorling Kindersley   **p. 368** Irv Beckman©Dorling Kindersley   **p. 369** Chris Stowers©Dorling Kindersley   **12.10A.1** Museum of Vertebrate Zoology   **12.10A.2** Oxford

University Library Services **12.10A.3** Manuscripts and Archives, Yale University Library **p. 371** Dorota & Mariusz Jarymowicz©Dorling Kindersley **p. 373** ©Angus Beare/Dorling Kindersley **p. 374** Kim Taylor©Dorling Kindersley **p. 377** Alan Briere©Dorling Kindersley

Chapter 13

CO.13 Klaus Just/Peter Arnold, Inc. **pp. 382–83** ©Dorling Kindersley **13.1B** Fred Bruemmer/Peter Arnold, Inc. **pp. 384–85** Jane Burton©Dorling Kindersley **pp. 386–87** Geoff Dann©Dorling Kindersley **13.4A** Mark Newman/Photo Researchers, Inc. **13.4B** Friedrich Stark, Das Fotoarchiv./Peter Arnold, Inc. **13.4C** Nuridsany et Perrenou/Photo Researchers, Inc. **13.6B** Roger Wilmhurst/Bruce Coleman Inc. **p. 393** Steve Gorton©Dorling Kindersley **13.8A** Wayne & Karen Brown/Index Stock Imagery, Inc. **13.8B** Jose Azel/Aurora & Quanta Productions, Inc. **13.8C** John Neubauer/PhotoEdit **13.8D** David Fleetham/Gety Images, inc.-Taxi **13.8E** Karl Shone©Dorling Kindersley, Courtesy of the Natural History Museum, London **13.9A** Gary Buss/Getty Images, Inc.-Taxi **13.9B** Wayne Lynch/DRK Photos **13.9C** Fabio Colombini Medeuros/Grant Heilman Photography, Inc. **13.10A** The Zoological Society of London **13.10B** Juergen & Christine Sohns/Animals Animals/Earth Scenes **p. 399** Frank Greenaway©Dorling Kindersley **p. 400** ©Dorling Kindersley, Courtesy of Oxford Scientific Films **13.12D** Fred Habegger/Grant Heilman Photography, Inc.

Chapter 14

CO.14 20th Century Fox/Picture Desk, Inc./Kobal Collection **14.1A** John Gerlach/Tom Stack & Associates, Inc. **p. 407** Jane Miller©Dorling Kindersley **p. 409** Frank Greenaway©Dorling Kindersley **p. 411** Jane Miller©Dorling Kindersley **p. 415** Frank Greenaway©Dorling Kindersley **14.6A** Carl Roessler/Animals Animals/Earth Scenes **14.6B** Michel Viard/Peter Arnold, Inc. **14.6C** Ken Wagner/Visuals Unlimited **14.7A.01** Science Photo Library/Photo Researchers, Inc. **14.7A.02** Dr. Jeremy Burgess/Photo Researchers, Inc. **14.7A.03** M.F. Brown/Visuals Unlimited **14.7A.04** David T. Webb, Botany, University of Hawaii **14.7A.05** Pearson Education/PH College **14.7A.06** Ed Reschke/Peter Arnold, Inc. **14.7B.01** Christian Ziegler/University of Wisconsin. Reprinted by permission from *Nature*, 398, 701–704 copyright 1999. Macmillan Publishers Ltd. **14.7B.02** Mark W. Moffett/Minden Pictures **p. 419** ©Dorling Kindersley **14.8A.01** ©Ted Benton/Dorling Kindersley **14.8A.02** Robert A. Tyrrell Photography **14.8A.03** Manfred Danegger/Peter Arnold, Inc. **14.9A.01** Michael Fogden/Animals Animals/Earth Scenes **14.9A.02** P. Sharpe/OSF/Animals Animals/Earth Scenes **14.9B** AG Stock USA **p. 423** Irv Beckman©Dorling Kindersley **14.10B.i** ©Dorling Kindersley **14.10B.ii** Michael Fogden/DRK Photo **p. 425** Kim Taylor©Dorling Kindersley **p. 428** Deni Brown©Dorling Kindersley **p. 433** Peter Anderson©Dorling Kindersley

Chapter 15

CO.15 Foto Marburg/Art Resource, NY **15.1A** James P. Jackson/Photo Researchers, Inc. **15.1C** Cedar Creek Natural History Area, University of Minnesota **p. 439** Ken Findlay©Dorling Kindersley **15.2C.01** Stefan Mokrzecki/Photolibrary.com **15.2C.02** Rob Francis/Robert Harding World Imagery **15.2C.03** Francis Lepine/Animals Animals/Earth Scenes **15.2C.04** DRK Photo **15.2C.05** Rob Blakers/Photolibrary.com **15.2C.06** Roel Loopers/Photolibrary.com **15.2C.07** ©Japack Company/CORBIS **15.2C.08** Ted Mead/Photolibrary.com **p. 443** Ken Findlay©Dorling Kindersley **15.4A** David Weintraub/Photo Researchers, Inc. **15.4B** John S. Shelton **15.4C** Krafft-Explorer/Photo Researchers, Inc. **15.4D** Jacques Jangoux/Getty Images Inc.-Stone Allstock **15.5C** Douglas Faulkner/Corbis/Bettmann **15.7A.01** J.H. Robinson/Animals Animals/Earth Scenes **15.7A.02** Clive Druett Papilio/CORBIS-NY **15.7A.05** M.T. Fraxier/PSU/Photo Researchers, Inc. **15.7A.06** Nature's Images/Photo Researdhers, Inc. **15.7B** Jon Jefferson/DeathsAcre.com **p. 451** John Serafin **15.9A.01** J. Lotter Gurling/Tom Stack & Associates, Inc. **15.9A.02** Andreas von Einsiedel©Dorling Kindersley **15.9A.03** Dave King©Dorling Kindersley **15.9A.04 & 05** Matthew Ward©Dorling Kindersley **15.9A.06** Frans

Lanting/Photo Researchers, Inc. **15.9A.07 & 08** Maurice Nimmo/Photo Researchers, Inc. **15.9A.09** Alfred Pasieka/Photo Researchers, Inc. **p. 454** Dewey/EMG Education Management Group **15.10A.01** Phillip Dowell©Dorling Kindersley **15.10A.02** David Wrobel/Visuals Unlimited **15.10A.03** Dennis & Kristen Richardson **15.10A.04** Herve Chelle/Peter Arnold, Inc. **p. 457** EMG Education Management Group **p. 459** Alistair Duncan©Dorling Kindersley **15.13A** Luiz C. Marigo/Peter Arnold, Inc. **p. 460** Dr. Julian Thorpe©Dorling Kindersley **15.14A** Rob Wiltshire/University of Tasmania **15.14B** William H. Mullins/Photo Researchers, Inc. **15.15A** Qinetiq.Ltd/Peter Arnold, Inc. **p. 465** EMG Education Management Group **p. 469** ©Dorling Kindersley **p. 470** ©Dorling Kindersley **p. 471** Max Alexander©Dorling Kindersley

Chapter 16

CO.16 Geoff Higgins/Photolibrary.com **p. 480** Jon Spaull©Dorling Kindersley **p. 481** Susanna Price©Dorling Kindersley **p. 483** Angus Beare©Dorling Kindersley **p. 484** ©Dorling Kindersley **p. 485** Pearson Education/PH College **p. 486** Peter Anderson©Dorling Kindersley **16.6B.01** Corbis RF **16.6B.02** Corbis/Sygma **p. 487** C. Andrew Henley©Dorling Kindersley **p. 488** Dave King©Dorling Kindersley **p. 489** Alan Keohane©Dorling Kindersley **p. 491** Brian Cosgrove©Dorling Kindersley **p. 492** Kim Taylor & Jane Burton©Dorling Kindersley **16.9B.01** Tom McHugh/Photo Researchers, Inc. **16.9B.02** Dwight R. Kuhn Photography **16.9B.03** Jim Zipp/Photo Researchers, Inc. **16.9B.04** Mark Webster/Photolibrary.com **16.10A.01** U. S. Geological Survey, Denver **16.10A.02** Sunset Avenue Productions/Digital Vision Ltd. **16.10A.03** Peter Harrison/Photolibrary.com **16.11A.01** Peter Anderson©Dorling Kindersley **16.11A.02** Craig K. Lorenz/Photo Researchers, Inc. **16.11B** Carr Clifton/Minden Pictures **16.11C** Photo Researchers, Inc. **16.11D** Roger Aitkenhead/Animals Animals/Earth Scenes **16.11E** Michael & Patricia Fogden/Corbis/Bettmann **16.12A.01** Brian ParkerTom Stack & Associates, Inc. **16.12A.02** Walter Bibikow/Getty Images, Inc.-Taxi **p. 499** Geoff Dann©Dorling Kindersley **16.13A.01** Kathleen Campbell/Getty Images, Inc.-Liaison **16.13A.02** Getty Images-Photodisc **16.13B** Don & Pat Valenti **16.13C** AP Wide World Photos **p. 503** Alistair Duncan©Dorling Kindersley **p. 504** John Serafin **16.15A.01** Kelvin Aitken/Peter Arnold, Inc. **16.15A.02** Shaen Adey; Gallo Images/CORBIS-NY

Chapter 17

CO.17 Kevin Schaefer/Peter Arnold, Inc. **17.1A** Ted Benton©Dorling Kindersley **17.2A** C.C. Lockwood/Animals Animals/Earth Scenes **p. 511** Cyril Laubscher©Dorling Kindersley **17.3B.01** Leroy Simon/Visuals Unlimited **17.3B.02** P.R. Ehrlich/Ann McMillan **p. 513** Dave King©Dorling Kindersley **17.4A** OSF/Deeble & Stone/Animals Animals/Earth Scenes **p. 515** Frank Greenaway©Dorling Kindersley **p. 517** Nigel Hicks©Dorling Kindersley **p. 519** Peter Chen©Dorling Kindersley **17.7A** Bates Littlehales/Animals Animals/Earth Scenes **17.7B** Dianne Blell/Peter Arnold, Inc. **17.7C.01** George Bernard/SPL/Photo Researchers, Inc. **17.7C.02** Chase Studio/Photo Researchers, Inc. **17.8A.01 & 03** ©Dorling Kindersley **17.8A.02** Wisconsin Historical Society **p. 523** Frank Greenaway©Dorling Kindersley **p. 524** Geoff Brightling©Dorling Kindersley **17.9A.01** Michael Sewell/Peter Arnold, Inc. **17.9A.02** Domonic Johnson/Nature PL **17.9A.03** Martin Harvey/Peter Arnold, Inc. **17.10A.01** Australian Tourist Commission **17.10A.02** P. Jackson/Photo Researchers, Inc. **17.10A.03** Yva Momatiuk & John Eastcott/Photo Researchers, Inc.

Part 5

01 Photofest/Jagarts **02** Photo B.D.V./CORBIS-NY **03** CinemaPhoto/CORBIS-NY

Chapter 18

CO.18 Digital Vision Ltd. **18.1A** John Walsh/Photo Researchers, Inc. **18.1B** Wardene Wiesser/Bruce Coleman Inc. **18.1C** Robert C. Vrijenhoek, Monterey Bay Aquarium Research Institute **p. 533** Deni Brown©Dorling Kindersley **p. 535** Eddie Lawrence©Dorling Kindersley **p. 539** Ken Findlay©Dorling Kindersley **p. 541** Karl Shone©Dorling

Kindersley **p. 544** Matthew Ward©Dorling Kindersley **p. 547** ©Dorling Kindersley **p. 549** Kim Taylor & Jane Burton©Dorling Kindersley **p. 550** Derek Hall©Dorling Kindersley **p. 551** Karl Shone©Dorling Kindersley **p. 553** Denis Finnin and Jackie Beckett © The American Museum of Natural History

Chapter 19

CO.19 Marco Cristofori/Das Fotoarchiv./Peter Arnold, Inc. **19.1A** Dennis Kunkel/Phototake NYC **19.1B** Carolina Biological/Visuals Unlimited **19.3A** Michael & Patricia Fogden/CORBIS-NY **19.4A** Karen Huntt/CORBIS-NY **19.5B** K.G. Preston-Mafham/Animals Animals/Earth Scenes **p. 564** Kim Taylor©Dorling Kindersley **19.6A** Bruno P. Zehnder/Peter Arnold, Inc. **19.6B** Jose B. Ruiz/Nature Picture Library **19.7A.01** Linda Dunk/Getty Images, Inc.-Taxi **19.7A.02** Georgette Douwma/Getty Images, Inc.-Taxi **19.7A.03** W. Gregory Brown/Animals Animals/Earth Scenes **p. 567** Frank Greenaway©Dorling Kindersley **p. 568** Trish Gant©Dorling Kindersley **19.8B** Kurt Wittman/Omni-Photo Communications, Inc. **19.9A** Frans Lanting/Minden Pictures **19.9B.01 & 03** Carl R. Sams II/Peter Arnold, Inc. **19.9B.02** Frans Lanting/Minden Pictures **p. 572** Frank Greenaway©Dorling Kindersley **19.10B** Paul A. ZahlPhoto Researchers, Inc.

Chapter 20

CO.20 Ken Preston-Mafham/Animals Animals/Earth Scenes **20.1A** Steve P. Hopkin/Getty Images, Inc.-Taxi **p. 577** Jane Miller©Dorling Kindersley **p. 580-81** Jane Burton©Dorling Kindersley **20.4A** ©Dorling Kindersley **20.4B** Scott Camazine/Photo Researchers, Inc. **20.4C** Larry Miller/Photo Researchers, Inc. **p. 583** Frank Greenaway©Dorling Kindersley **p. 586** Alan Keohane©Dorling Kindersley **20.7B** Raymond A. Mendez/Animals Animals/Earth Scenes **p. 588** Dave King©Dorling Kindersley **p. 590** Dave King©Dorling Kindersley **p. 593** Jane Burton©Dorling Kindersley **p. 594** ©Jerry Young/©Dorling Kindersley **p. 597** Max Alexander©Dorling Kindersley

Chapter 21

CO.21 Harry Taylor©Dorling Kindersley **21.1A.01** Frans Lanting/Minden Pictures **21.1A.02** Jerry Young©Dorling Kindersley

21.1A.03 Roland Seitre/Peter Arnold, Inc. **21.1A.04** Christian Jejou/Photo Researchers, Inc. **21.1B.01** Tom Brakefield/CORBIS-NY **21.1B.02** W. Perry Conway/CORBIS-NY **21.1B.03** Gallo Images/CORBIS-NY **21.1B.04** Tom Brakefield/CORBIS-NY **21.1B.05** Sunset Boulevard/Corbis/Sygma **p. 603** Mike Dunning©Dorling Kindersley **p. 605** Geoff Brightling©Dorling Kindersley **21.3B** John Reader/Science Photo Library/Photo Researchers, Inc. **p. 607** John Davis©Dorling Kindersley **21.4A** The Natural History Museum, London **21.5A** John Reader/Science Photo Library/Photo Researchers, Inc. **p. 610** Harry Taylor©Dorling Kindersley **p. 611** Peter Wilson©Dorling Kindersley **p. 613** Roger de la Harpe©Dorling Kindersley **p. 615** John Serafin **p. 617** Jon Spaull©Dorling Kindersley **21.10A** LOC/Science Source/Photo Researchers, Inc. **21.12A** Neg./Transparency no. 4936(7). (Photo by D. Finnin/C. Chesek. Courtesy Dept. of Library Services, American Museum of Natural History) **p. 621** Dave King©Dorling Kindersley **21.13A.01** ©Dorling Kindersley, Courtesy of the National Motor Museum, Beaulieu **21.13A.02** EyeWire Collection/Getty Images-Photodisc **21.13A.03** Massachusetts Institute of Technology News Office **21.15A** Rick Friedman/CORBIS-NY **21.15B** Courtesy of the Library of Congress **21.15C** Frédèric Huijbregts/CORBIS-NY **21.15D** Desmond Morris/Random House, Inc.

Chapter 22

p. 633 John Davis©Dorling Kindersley **p. 636** Eddie Lawrence©Dorling Kindersley **p. 637** M.I. Walker©Dorling Kindersley **p. 639** Andy Crawford©Dorling Kindersley **22.5A** CDC/PHIL/CORBIS-NY **p. 640** M.I. Walker©Dorling Kindersley **p. 641** ©Dorling Kindersley **p. 642** Christopher & Sally Gable©Dorling Kindersley **p. 643** ©Dorling Kindersley **p. 645** Andy Crawford©Dorling Kindersley **p. 647** Susanna Price©Dorling Kindersley **p. 649** EMG Education Management Group **p. 651** Peter Wilson©Dorling Kindersley **p. 653** ©Jerry Young/Dorling Kindersley **p. 655** Jerry Young©Dorling Kindersley **p. 657** Laurence Pordes©Dorling Kindersley, Courtesy of the British Library **p. 659** Eddie Lawrence©Dorling Kindersley **p. 661** ©Dorling Kindersley

INDEX